AF595451

Structural Health Monitoring and Non-destructive Testing for Large-Scale Structures—Volume II

Structural Health Monitoring and Non-destructive Testing for Large-Scale Structures—Volume II

Editors

Phong B. Dao
Lei Qiu
Liang Yu
Tadeusz Uhl
Minh-Quy Le

Basel • Beijing • Wuhan • Barcelona • Belgrade • Novi Sad • Cluj • Manchester

Editors

Phong B. Dao
AGH University of Krakow
Krakow, Poland

Lei Qiu
Nanjing University of Aeronautics and Astronautics
Nanjing, China

Liang Yu
Northwestern Polytechnical University
Xi'an, China

Tadeusz Uhl
AGH University of Krakow
Krakow, Poland

Minh-Quy Le
Hanoi University of Science & Technology
Hanoi, Vietnam

Editorial Office
MDPI
St. Alban-Anlage 66
4052 Basel, Switzerland

This is a reprint of articles from the Topic published online in the open access journals *Aerospace* (ISSN 2226-4310), *Energies* (ISSN 1996-1073), *Materials* (ISSN 1996-1944), *Sensors* (ISSN 1424-8220), and *Applied Sciences* (ISSN 2076-3417) (available at: https://www.mdpi.com/topics/shmndtlss).

For citation purposes, cite each article independently as indicated on the article page online and as indicated below:

Lastname, A.A.; Lastname, B.B. Article Title. *Journal Name* **Year**, *Volume Number*, Page Range.

Volume II
ISBN 978-3-0365-9935-9 (Hbk)
ISBN 978-3-0365-9936-6 (PDF)
doi.org/10.3390/books978-3-0365-9936-6

Set
ISBN 978-3-0365-9931-1 (Hbk)
ISBN 978-3-0365-9932-8 (PDF)

Contents

About the Editors

Phong B. Dao

Phong B. Dao received an Engineering degree in Cybernetics in 2001, an M.Sc. degree in Instrumentation and Control in 2004, both from Hanoi University of Science and Technology in Vietnam, and a Ph.D. degree in Control Engineering in 2011 from the University of Twente, the Netherlands. In May 2020, Dr. Dao received a D.Sc. degree (Habilitation) in Mechanical Engineering from the AGH University of Science and Technology, Poland. He is currently an Associate Professor at the Department of Robotics and Mechatronics of the AGH University of Krakow. His research interests include structural health monitoring (SHM), non-destructive testing (NDT), wind turbine condition monitoring and fault diagnosis, statistical time series methods for SHM and NDT, advanced signal processing, and intelligent control. He has been the recipient of several awards for scientific achievements and has published over 50 articles, mostly as the first, single, or corresponding author, in major international peer-reviewed journals and conferences. He is the co-inventor of a European Patent on the homoscedastic nonlinear cointegration methodology for SHM. He is the author of the monograph entitled "Cointegration-Based Approach for Structural Health Monitoring: Theory and Applications" published in 2018. He delivered five invited lectures on topics about structural damage detection and wind turbine condition monitoring. He serves as an active reviewer for many international Top-10 journals such as IEEE Transactions on Industrial Electronics, Expert Systems with Applications, Applied Energy, and Mechanical Systems and Signal Processing.

Lei Qiu

Lei Qiu is a doctoral supervisor at the State Key Lab of Mechanics and Control for Aerospace Structures, College of Aerospace Engineering, Nanjing University of Aeronautics and Astronautics, and the Director of the Institute of Smart Materials and Structures. His interests include aircraft smart structures and structural health monitoring (SHM) technology, and he has made innovative advancements in SHM principles and methods of complex structures, reliable diagnostic mechanisms for structural damage under the influence of service, aircraft smart skin based on flexible electronics, and equipment development for aerospace applications. Prof. Qiu has presided over more than 40 projects and is currently undertaking projects such as the general program of the National Natural Science Foundation of China, the Outstanding Youth Foundation of Jiangsu Province of China, key project subjects of Foundation Strengthening of Military Commission of Science and Technology (173) and the pre-research program of the Army Armament Department. He has received nine important honors and technological awards, such as the China Youth Science and Technology Award, two first prizes in Science and Technology from Jiangsu Province, the Jiangsu Province Youth May Fourth Medal, the Youth Science and Technology Award of Chinese Society of Aeronautics and Astronautics, and the best paper award of the 8th European Workshop on SHM, etc.

Liang Yu

Liang Yu received a bachelor's degree in communication engineering from Hubei University of Technology, Wuhan, China, in 2007, a double M.S. degree in communication and information systems from Shanghai University, Shanghai, China, and Waseda University, Fukuoka, Japan, in 2011, and a Ph.D. degree in acoustics from INSA de Lyon, Lyon, France, in 2015. He is currently a Full Professor with Northwestern Polytechnical University, Xi'an, China. His research interests include fault diagnosis of mechanical equipment, intelligent signal processing methods and advanced acoustic measurement, and experimental aeroacoustics.

Tadeusz Uhl

Tadeusz Uhl graduated from the AGH University of Science and Technology, obtaining his doctorate in 1983 and the title of professor in 1997. He worked in research centers and universities abroad in the Netherlands, France, Belgium, Japan and the USA. Since 2020, he has been the Director of the AGH UST Space Technology Center. His research interests include mechatronics, structural health monitoring, hydrogen vehicles, robotics, and space technologies. He is the author of over 1,000 publications, with over 400 in JCR journals. He has a citation index of h = 32 (according to SCOPUS). He is the promoter of 49 defended doctoral theses and over 150 master's theses. He is the Coordinator of many European projects in topics related to structural diagnostics, transport and energy with partners such as Alstom, Siemens, Airbus, Dassault, EDF, FIAT, RENAULT, LMS, Bombardier, and ABB. He carries out extensive cooperation with universities and research centers around the world. He has 51 national and foreign patents. He was a member of HLG in the field of Key Enabling Technology at the European Commission in 2014 – 2017. In 2020, Stanford University placed him on the prestigious list of "World's Top 2% Scientists". He effectively implements the latest technologies in the economy, and he is the founder of 28 start-ups, several of which have grown to the size of a large company.

Minh-Quy Le

Minh-Quy Le received his BS degree from Hanoi University of Science and Technology, Vietnam, in 1995, and his MS degree in Mechanical Engineering from University of Poitiers, France, in 1997. He obtained his Ph.D. in Mechanical Engineering in 2004 from Kyungpook National University, Republic of Korea. In 2007–2008, he worked as a Postdoctoral Fellow at the Technical University of Dresden, Germany. Currently, he is a Full Professor at the School of Mechanical Engineering, Hanoi University of Science and Technology, Hanoi, Vietnam. His research interests include computational mechanics, solid mechanics, nanomechanics, fracture mechanics, finite element method, molecular dynamics simulation, and material testing methods.

Preface

Structural health monitoring (SHM) and non-destructive testing (NDT) have gained significant importance for civil, mechanical, aerospace, and offshore structures. Nowadays, we can find SHM and NDT applications being used on various structures with very different requirements. The SHM-NDT field involves a wide range of transdisciplinary areas, including smart materials, embedded sensors and actuators, damage diagnosis and prognosis, signal and image processing, wireless sensor networks, data interpretation, machine learning, data fusion, energy harvesting, etc.

Since the 1970s, there has been a large and increasing volume of research on SHM and NDT; a great deal of this effort has focused on developing cost-effective, automatic, and reliable damage detection technologies. However, few industrial and commercial applications can be found in the literature. The practical implementation of strategies for the detection of structural damage to real structures outside of laboratory conditions is always one of the most demanding tasks for engineers. One reason for the rare transfer of research into industrial practice is that most of the methods that have been developed have been tested on simple beam and plate structures in the laboratory, while many practical problems only manifest themselves in complex structures. Another reason is the influence of environmental and operational variations (EOVs) on damage-sensitive features. Thus, for the successful development of SHM and NDT for large structures, techniques should be enhanced to have the capability of dealing with the influence of EOVs. In addition, signal/data processing plays an important role in the implementation of SHM and NDT technologies. The processing and interpretation of the massive amount of data generated through the long-term monitoring of large and complex structures (e.g., bridges, buildings, ships, aircrafts, wind turbines, pipes, etc.) has become an emerging challenge that needs to be addressed by the community.

This Topic brings together the most established as well as newly emerging SHM and NDT approaches that can be used for the detection and evaluation of defects and damage development in large-scale or full-scale structures. After a strict peer-review process, 44 papers were published, which represent the most recent progress in SHM and NDT methods/techniques for aerospace, civil, mechanical, and offshore infrastructures.

Phong B. Dao, Lei Qiu, Liang Yu, Tadeusz Uhl, and Minh-Quy Le
Editors

Article

Angular Displacement Control for Timoshenko Beam by Optimized Traveling Wave Method

Huawei Ji [1,†], Chuanping Zhou [1,2,*,†], Jiawei Fan [1,*], Huajie Dai [3], Wei Jiang [3], Youping Gong [1], Chuzhen Xu [1], Ban Wang [1,4] and Weihua Zhou [4]

[1] School of Mechanical Engineering, Hangzhou Dianzi University, Hangzhou 310018, China; jhw76@hdu.edu.cn (H.J.); gyp@hdu.edu.cn (Y.G.); zhangjingxiangcrcc@foxmail.com (C.X.); bigban@zju.edu.cn (B.W.)
[2] School of Mechatronics Engineering, University of Electronic Science and Technology of China, Chengdu 611731, China
[3] Aerospace Systems Engineering Research Institute of Shanghai, Shanghai 201108, China; qinlongcrcc@foxmail.com (H.D.); chenmingcrcc@foxmail.com (W.J.)
[4] College of Electrical Engineering, Zhejiang University, Hangzhou 310027, China; dididi@zju.edu.cn
* Correspondence: zhoucp@hdu.edu.cn (C.Z.); 19205311@hdu.edu.cn (J.F.); Tel.: +86-1342-9168-099 (C.Z.)
† These authors contributed equally to this work.

Abstract: The vibration of flexible structures in spacecraft, such as large space deployable reflectors, solar panels and large antenna structure, has a great impact on the normal operation of spacecraft. Accurate vibration control is necessary, and the control of angular displacement is a difficulty of accurate control. In the traditional control method, the mode space control has a good effect on suppressing low-order modes, but there is control overflow. The effect of traveling wave control on low-order modes is worse than the former, but it has the characteristics of broadband control. It can better control high-order modes and reduce control overflow. In view of the advantages and disadvantages of the two control methods, based on Timoshenko beam theory, this paper uses vector mode function to analyze the modal of spacecraft cantilever beam structure, establishes the system dynamic equation, and puts forward an optimized traveling wave control method. As a numerical example, three strategies of independent mode space control, traditional traveling wave control and optimized traveling wave control are used to control the active vibration of beam angle. By comparing the numerical results of the three methods, it can be seen that the optimal control method proposed in this paper not only effectively suppresses the vibration, but also improves the robustness of the system, reflecting good control performance. An innovation of this paper is that the Timoshenko beam model is adopted, which considers the influence of transverse shear deformation and moment of inertia on displacement and improves the accuracy of calculation, which is important for spacecraft accessory structures with high requirements for angle control. Another innovation is that the optimized traveling wave control method is exquisite in mathematical processing and has good results in global and local vibration control, which is not available in other methods.

Keywords: Timoshenko beam; modal space control; optimized traveling wave method; angular displacement control

Citation: Ji, H.; Zhou, C.; Fan, J.; Dai, H.; Jiang, W.; Gong, Y.; Xu, C.; Wang, B.; Zhou, W. Angular Displacement Control for Timoshenko Beam by Optimized Traveling Wave Method. *Aerospace* **2022**, *9*, 259. https://doi.org/10.3390/aerospace9050259

Academic Editors: Phong B. Dao, Lei Qiu and Liang Yu

Received: 14 March 2022
Accepted: 7 May 2022
Published: 11 May 2022

1. Introduction

Flexible cantilever structures are often used in spacecraft structures, such as large antenna structures, large space deployable reflectors and solar panels. In the space environment, the damping is small, and the damping of the flexible structure itself is also small. When it is excited or disturbed by work, such as attitude adjustment and docking, orbit change and solar wind, if the vibration is not controlled, the vibration will last for a long time, which will affect the normal work of the structure and even reduce the service life of the structures [1–6].

At present, Euler–Bernoulli beam theory is mainly used in vibration control of beam structure, while Timoshenko beam theory is seldom used for the control difficulties in mathematics. The Euler–Bernoulli classical beam theory has some limitations in structural dynamic analysis, especially in some composite structures. Because the influence of transverse shear deformation and moment of inertia is considered, the analytical results of Timoshenko beam theory are closer to engineering practice [7,8]. Therefore, the Timoshenko beam theory begins to be applied in structural dynamics analysis. Based on Timoshenko beam theory and elastic wave theory, Carvalho and Zindeluk [9] studied the active control of bending waves in infinite Timoshenko beams. Halkyard and Mace [10] studied the bending vibration feedback adaptive control of beam structures by using wave control method. EL-Khatib et al. [11] used a tuned vibration absorber to study the bending wave suppression in a beam. Hu et al. [12] concerned with the control of flexural waves in a beam using a tuned vibration absorber. Su and Ma [13,14] compared three analysis methods: Laplace transform, ray and normal mode to study the dynamic transient response of cantilever Timoshenko beam under impact force. Cardoso [15] blended the Euler–Bernoulli beam theory with idealized transverse shear flows to study a new beam element for aircraft structural analysis. Xing and Liu [16] studied the dynamic modeling and adaptive boundary control of a three-dimensional Timoshenko beam to realize vibration suppression. Ishaquddin et al. [17] studied the flexure and torsion locking phenomena in out-of-plane deformation of the Timoshenko curved beam element. Endo et al. [18] studied a contact-force control problem for a flexible Timoshenko arm. Mei [19] studied a hybrid approach to active control of bending vibrations in beams based on the advanced Timoshenko theory. Pham et al. [20] developed a novel dynamic model of a variable-length Timoshenko beam attached to a translating base. The stability of the closed-loop system under the proposed boundary control law was analyzed. Eshag et al. [21] studied a global sliding mode boundary control method for the Timoshenko beam reduction of the vibration induced by rotation boundary disturbance, uncertainties and distributed disturbance. By using exponential reaching law, the chattering phenomena is avoided. Fleischmann and Könözsy [22] developed an explicit finite difference numerical method to approximately solve the bending equation of non-uniform fourth-order Euler–Bernoulli beams. The equation has velocity-dependent damping and second-order moments of area, and mass and elastic modulus distribution varying with beam spacing. The method is grid convergent and numerically stable. According to literature studies [9–24], it can be seen that the finite element method has become an important theoretical analysis method because it can deal with complex objects and has many general programs. However, the finite element method is more suitable for low-frequency vibration. For medium- and high-frequency vibration, many elements must be divided to accurately describe the vibration characteristics of the system, which will lead to large numerical errors and a sharp increase in the amount of calculation, and the overflow problem will inevitably be encountered when considering vibration control. In addition, enough grid elements must be divided to describe the dynamic characteristics of the structure, the grid of the element is continuously encrypted, and the accumulated error will increase. High requirements for computer hardware configuration are also needed. The modal space control method is to analyze the vibration of the structure from the perspective of the modal. The essence of modal analysis is coordinate transformation. It transforms the coupled motion equations in physical coordinates into modal coordinates for decoupling; thus, as to solve the modal parameters of the system. The above traditional numerical methods can deal with the response of structural vibration well in the low frequency band. However, there are limitations in the medium- and high-frequency bands. With the increase in analysis frequency, it will face many problems, such as the problem of dense and overlapping structural modes.

In this paper, based on Timoshenko beam theory, the vibration control of the cantilever beam is studied by using the optimal traveling wave control method. First, based on Timoshenko beam theory, the modal analysis of the cantilever beam structure commonly used in spacecraft is carried out by using vector vibration mode function, and the system dynamic equation is established. Then, based on the Timoshenko beam theory and the overall motion of the structure, the modal space control method is used to design the global vibration control of the structure. The traveling wave control is used to control the vibration energy flow in the specified area of the structure. According to the local characteristics of the structure, the wave control method is used to absorb the energy carried by the propagating wave in the structural vibration, and the reflection coefficient and transmission coefficient of the wave in the Timoshenko beam when the control torque is applied are obtained. Finally, combined with the advantages of modal space control method, we optimize the traveling wave control method and design an optimized waveform controller to control the rotation angle of the beam. The control results show that the optimized traveling wave control method is better than the traditional traveling wave control method or mode space control method, it and improves the robustness of the system.

2. Structural Dynamic Characteristics of Timoshenko Beam

In engineering practice, it is assumed that the cross-section size of the beam is small compared with its length, thus neglecting the influence of shear deformation and beam section rotation, which is suitable for most members. However, when the ratio of height to span is high or when the composite members are widely used in structures, the effects of shear deformation and cross-section rotation should be taken into account, which have great influence on the higher frequency and mode shape. At this time, Timoshenko beam theory needs to be used. The cantilever Timoshenko beam structure studied in this paper is shown in Figure 1.

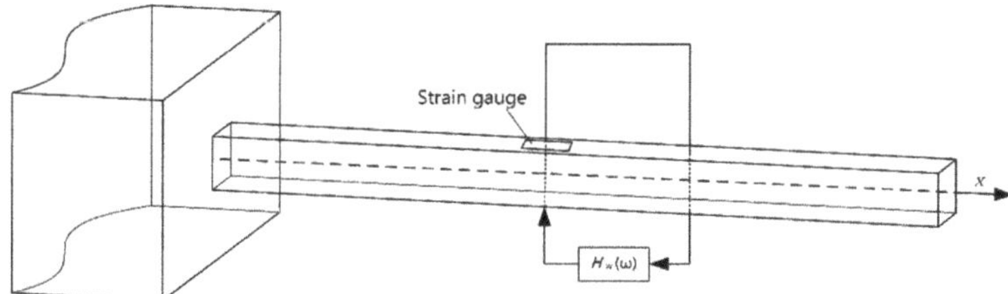

Figure 1. Schematic diagram of wave control in the cantilever Timoshenko beam.

It is assumed that the rotation angle of the cross section produced by the bending moment M is ψ. If there is no shear deformation, the cross section is perpendicular to the elastic axis, and Ψ equals the slope of the elastic axis. When shear deformation is taken into account, the rotation angle generated by shear force is γ. The actual cross section angle is $\alpha = \partial w / \partial x$; then, we can obtain:

$$\alpha = \psi - \gamma \tag{1}$$

The angle ψ and γ meet the following relationship:

$$\begin{cases} \frac{\partial \psi}{\partial x} = \frac{M}{EI} \\ \gamma = -\frac{Q}{\kappa G A} \end{cases} \tag{2}$$

where, $E,\ I,\ \kappa,\ G,\ A$ respectively represent the elastic modulus, the moment of inertia of the beam section, the shear conversion coefficient, the shear modulus and the section area, and $I = \frac{Ah^2}{12}$, $G = \frac{E}{1+\nu}$, $\kappa = \frac{\pi^2}{12}$, in which h, ν are respectively the section height and the Poisson's ratio of the material.

Respectively, ρ, q, $\overline{m}$ denoted the mass density, transverse load and bending moment of the beam. Considering the equilibrium of the beam element, the following equilibrium equations can be obtained.

$$\begin{cases} -Qdx + \frac{\partial M}{\partial x}dx - \rho I\frac{\partial^2 \psi}{\partial t^2}dx + \overline{m}dx = 0 \\ -\frac{\partial Q}{\partial x}dx - \rho A\frac{\partial^2 w}{\partial t^2}dx + qdx = 0 \end{cases} \tag{3}$$

Substituting Equation (2) into Equation (3), we can obtain:

$$\begin{cases} D\frac{\partial^2 \psi}{\partial x^2} + C\frac{\partial w}{\partial x} - C\psi + \overline{m} = \rho I\frac{\partial^2 w}{\partial t^2} \\ C\frac{\partial^2 w}{\partial x^2} - C\frac{\partial \psi}{\partial x} + q = \rho A\frac{\partial^2 w}{\partial t^2} \end{cases} \tag{4}$$

The bending moment and shear force in a beam are expressed as:

$$\begin{cases} M = D\frac{\partial \psi}{\partial x} = D\left(\frac{\rho}{GA}\frac{\partial^2 w}{\partial t^2} - \frac{\partial^2 w}{\partial x^2}\right) \\ Q = -C\left(\frac{\partial w}{\partial x} - \psi\right) \end{cases} \tag{5}$$

where $D = EI$ is the bending stiffness of beams, $C = \kappa GA$, G is the shear modulus.

Equation (4) can be written in matrix form as follows:

$$\overline{\mathbf{M}}\frac{\partial^2 \mathbf{\Phi}}{\partial t^2} - \overline{\mathbf{K}}\frac{\partial^2 \mathbf{\Phi}}{\partial x^2} + C\overline{\mathbf{E}}\mathbf{\Phi} = \mathbf{F}(x,t) \tag{6}$$

where $\mathbf{\Phi} = \begin{bmatrix} w & \psi \end{bmatrix}$ is the generalized displacement, w, ψ are the deflection of beam and angle of rotation, respectively; $\overline{\mathbf{M}}$ is the generalized mass; t is the time; $\mathbf{F}(x,t) = \begin{bmatrix} q & \overline{m} \end{bmatrix}$; $\mathbf{F}(x,t)$ is the external disturbance dynamic vector, and the expression of $\overline{\mathbf{M}}$, $\overline{\mathbf{K}}$, $\overline{\mathbf{E}}$ is

$$\overline{\mathbf{M}} = \begin{bmatrix} \rho A & 0 \\ 0 & \rho I \end{bmatrix}, \overline{\mathbf{K}} = \begin{bmatrix} C & 0 \\ 0 & D \end{bmatrix}, \overline{\mathbf{E}} = \begin{bmatrix} 0 & \partial/\partial x \\ -\partial/\partial x & 1 \end{bmatrix}$$

In order to analyze the frequency characteristics of the beam and consider the free vibration state of the beam, the following equations can be obtained by decoupling the displacement w and the rotation angle ψ in Equation (4):

$$D\frac{\partial^4 w}{\partial x^4} + \rho A\frac{\partial^2 w}{\partial t^2} - \rho I\frac{\partial^4 w}{\partial x^2 \partial t^2} - \frac{\rho AD}{C}\frac{\partial^4 w}{\partial x^2 \partial t^2} + \frac{\rho^2 AI}{C}\frac{\partial^4 w}{\partial t^4} = 0 \tag{7}$$

Equation (7) above shows two degrees of freedom of the Timoshenko beam: the free vibration characteristics w of deflection and rotation angle ψ are exactly the same, that is, the modes of corresponding order, both of which have the same frequency ω and wave number k. In Equation (7), the first and second items denote the basic conditions of vibration, the third and fourth items denote the effect of the cross section moment of inertia and shear deformation, respectively, and the last item denotes the coupling effect of moment of inertia and shear deformation.

3. The Simulation Results and Discussions in the Deorbiting Process

For flexible structures, the low frequency modes are dense; thus, it is difficult to design the controller directly by physical coordinates. In the modal space, the vibration control of the structural system can be transformed into a small number of modal coordinate vibration controls. The response of the structure can be expressed as the sum of infinite modes:

$$\mathbf{\Phi}(x,t) = \sum_{i=1}^{N} \boldsymbol{\varphi}_{\mathbf{i}}(x)\mathbf{q}_{\mathbf{i}}(t) \tag{8}$$

where $\varphi_i(x)$ is the ith-order mode function, $q_i(t)$ is the ith-order modal coordinates, and N is number of modal truncation.

By deriving Equation (4) from time variables and space variables, the following equation can be obtained at the left end of Equation (4) as:

$$\begin{cases} \overline{\mathbf{M}}\frac{\partial^2\mathbf{\Phi}}{\partial t^2} = \overline{\mathbf{M}}\sum\limits_{i=1}^{N} \boldsymbol{\varphi}_\mathbf{i}(x)\frac{\partial^2\mathbf{q}_\mathbf{i}(t)}{\partial t^2} \\ \overline{K}\frac{\partial^2\mathbf{\Phi}}{\partial x^2} - C\overline{\mathbf{E}}\mathbf{\Phi} = \sum\limits_{i=1}^{N} [\overline{\mathbf{K}}\frac{\partial^2\boldsymbol{\varphi}_\mathbf{i}(\mathbf{x})}{\partial x^2} - C\overline{\mathbf{E}}\boldsymbol{\varphi}_\mathbf{i}(\mathbf{x})]\mathbf{q}_\mathbf{i}(\mathbf{t}) \end{cases} \tag{9}$$

Substituting Equation (9) into Equation (6), we can obtain:

$$\overline{\mathbf{M}}\sum_{i=1}^{N} \boldsymbol{\varphi}_\mathbf{i}(x)\frac{\partial^2\mathbf{q}_\mathbf{i}(t)}{\partial t^2} - \sum_{i=1}^{N} [\overline{\mathbf{K}}\frac{\partial^2\boldsymbol{\varphi}_\mathbf{i}(x)}{\partial x^2} - C\overline{\mathbf{E}}\boldsymbol{\varphi}_\mathbf{i}(x)]\mathbf{q}_\mathbf{i}(t) = \mathbf{F}(x,t) \tag{10}$$

where $\boldsymbol{\varphi}_\mathbf{i}(x)$ is the modal vibration function and should satisfy the homogeneous differential equation as follows:

$$[\overline{\mathbf{K}}\frac{\partial^2\boldsymbol{\varphi}_i(x)}{\partial x^2} - C\overline{\mathbf{E}}\boldsymbol{\varphi}_i(x)] + \omega_i^2\overline{\mathbf{M}}\boldsymbol{\varphi}_i(x) = 0 \tag{11}$$

Then, the frequency characteristics of the system are analyzed. Assuming $\mathbf{\Phi} = \mathrm{Re}[\mathrm{e}^{\mathrm{i}(kx-\omega t)}\begin{bmatrix} 1 & 1 \end{bmatrix}^T]\overleftrightarrow{A}B$, and substituting into (7) to obtain:

$$k^4 - \left(\frac{\rho I}{D}\omega^2 + \frac{\rho A}{C}\omega^2\right)k^2 + \frac{\rho A}{D}\omega^2\left(\frac{\rho I}{C}\omega^2 - 1\right) = 0 \tag{12}$$

where $k_0 = \left(\rho A\omega^2/D\right)^{1/4}$ is the elastic traveling wave number of classical Euler beam and k is the wave number of the modal vibration function $\boldsymbol{\varphi}(x)$.

When the vibration frequency of the beam is $\omega \geq \omega_c$, the elastic wave number is $k_i^2 \geq 0 (i = 1,2)$, which indicates that there are two pairs of propagating waves in the structural beam, and the phase velocity is $c = \omega/k$. When the vibration frequency of the beam is $\omega < \omega_c$, the elastic wave numbers are $k_1^2 > 0$ and $k_2^2 < 0$, which indicate that there are a pair of propagating waves and a pair of attenuation waves in the structural beam. Where $\omega_c = \sqrt{C/\rho I}$ denotes the transition critical frequency of the extended state and the local state. Because the structure vibration can be regarded as the superposition of multiple reflection of the elastic wave mode, the propagation wave can form the whole vibration of the finite region of the structure, while the attenuation wave forms the localized vibration in the finite area of the structure. Thus, in the beam, a main vibration mode function is determined when a set of wave numbers k_1 and k_2 corresponding to the natural frequencies are determined.

$$\boldsymbol{\varphi}_\mathbf{i}(x) = \begin{bmatrix} w_i \\ \psi_i \end{bmatrix} = \begin{bmatrix} a_1 \\ b_1 \end{bmatrix}\cos(k_1x) + \begin{bmatrix} a_2 \\ b_2 \end{bmatrix}\sin(k_1x) + \begin{bmatrix} a_3 \\ b_3 \end{bmatrix}\cos(k_2x) + \begin{bmatrix} a_4 \\ b_4 \end{bmatrix}\sin(k_2x) \tag{13}$$

To study the cantilever beam, use the modal coordinate equation of the beam.

$$\ddot{\mathbf{q}}_\mathbf{i}(t) + \omega_i^2\mathbf{q}_\mathbf{i}(t) = \mathbf{f}_\mathbf{i}(t),\ i = 1,2,\cdots N \tag{14}$$

where $\mathbf{q}(t) = [q_1(t), q_2(t), \cdots, q_N(t)]^T$, $\mathbf{f}_\mathbf{i}(t) = \frac{1}{m_i}\int_0^l \boldsymbol{\varphi}_\mathbf{i}{}^\mathbf{T}\mathbf{F}(x,t)dx$ is the external disturbance force.

The essence of independent modal control is to reconfigure the poles of the vibration system by state feedback; thus, the state space description of the system is established first. Assuming the modal control force is $\mathbf{u}_\mathbf{i}(t) = \frac{1}{m_i}\int_0^l \boldsymbol{\varphi}_\mathbf{i}{}^\mathbf{T}\mathbf{U}(x,t)dx$, for the vibration system:

$$\ddot{\mathbf{q}}_\mathbf{i}(t) + \omega_i^2\mathbf{q}_\mathbf{i}(t) = \mathbf{f}_\mathbf{i}(t) + \mathbf{u}_\mathbf{i}(t),\ i = 1,2,\cdots N \tag{15}$$

Considering the control of the pth-order mode, and introducing the state vector $\mathbf{X}(t) = [\mathbf{X_C}(t) \vdots \mathbf{X_R}(t)]$, where $\mathbf{X_C} = [\mathbf{q}^{\mathbf{T}}(t), \dot{\mathbf{q}}^{\mathbf{T}}(t)]^T_{2p\times 1}$, the controlled partial state equation of the system is expressed as follows:

$$\dot{\mathbf{X}}_{\mathbf{C}}(t) = \mathbf{A_C X_C}(t) + \mathbf{B_C u}(t) + \mathbf{D_C f_C}(t) \tag{16}$$

where

$$\mathbf{A_C} = \begin{bmatrix} 0 & \mathbf{I} \\ -\boldsymbol{\omega}^2 & 0 \end{bmatrix}_{2p\times 2p}, \ \mathbf{B_C} = \mathbf{D_C} = \begin{bmatrix} 0 \\ \mathbf{I} \end{bmatrix}_{2p\times p}, \ \boldsymbol{\omega}^2 = \mathrm{diag}[\omega_i^2]_{p\times p}$$

In Equation (16), the 0 and I are a zero matrix and a unit matrix, respectively; $u(t)$, $f(t)$ are the modal control force and the modal disturbance force, respectively. When the controllability matrix $\mathrm{rank}([\mathbf{B}\ \mathbf{AB}]) = 2p$, based on the linear system theory, we know that the state of the system is completely controllable, and the pole can be arranged arbitrarily. Introducing the state feedback, the control force $\mathbf{u}(t)$ is given as follows:

$$\mathbf{u}(t) = -G\mathbf{X_C}(t) = -\begin{bmatrix} g & h \end{bmatrix} \begin{bmatrix} q \\ \dot{q} \end{bmatrix} \tag{17}$$

The closed-loop dynamic equation of the system (15) is as follows:

$$\dot{\mathbf{X}}_{\mathbf{C}}(t) = (\mathbf{A_C} - \mathbf{B_C}G)\mathbf{X_C}(t) + \mathbf{D_C f_C}(t) \tag{18}$$

According to Equation (18), the characteristic structure of the original vibration system can be changed arbitrarily by adjusting the feedback gain G.

Considering point force control input $\mathbf{U}(x,t) = \sum\limits_{i=1}^{p} \tilde{\mathbf{u}}_{\mathbf{i}}(t)\delta(x - x_i)$, point force disturbance input, $\mathbf{F}(x,t) = \sum\limits_{i=1}^{n} \tilde{\mathbf{f}}_i(t)\delta(x - x_i)$ in physical coordinates, where p, n are respectively the number of controlled modes and the number of disturbance forces, and the jth-order mode control force $\mathbf{u}_j(t)$ and disturbance force $\mathbf{f}_j(t)$ are expressed as follows:

$$\begin{cases} \mathbf{u_j}(t) = \frac{1}{m_j}\sum\limits_{i=1}^{p}\int\limits_0^1 \boldsymbol{\varphi}_{\mathbf{j}}(x)\tilde{\mathbf{u}}_{\mathbf{i}}(t)\delta(x-x_i) = \frac{1}{m_j}\sum\limits_{i=1}^{p} \boldsymbol{\varphi}_{\mathbf{j}}(x_i)\tilde{\mathbf{u}}_{\mathbf{i}}(t) \\ \mathbf{f_j}(t) = \frac{1}{m_j}\sum\limits_{i=1}^{n}\int\limits_0^1 \boldsymbol{\varphi}_{\mathbf{j}}(x)\tilde{\mathbf{f}}_{\mathbf{i}}(t)\delta(x-x_i) = \frac{1}{m_j}\sum\limits_{i=1}^{n} \boldsymbol{\varphi}_{\mathbf{j}}(x_i)\tilde{\mathbf{f}}_{\mathbf{i}}(t) \end{cases} \tag{19}$$

where the control input force vector is $\tilde{\mathbf{u}} = \left[\tilde{\mathbf{u}}_{\mathbf{x}1}(t)\ \tilde{\mathbf{u}}_{\mathbf{x}2}(t) \ldots, \tilde{\mathbf{u}}_{\mathbf{xp}}(t)\right]^T$, and the disturbance input force vector is $\tilde{f} = \begin{bmatrix} \tilde{\mathbf{f}}_{x1}(t) & \tilde{\mathbf{f}}_{x2}(t) & \cdots & \tilde{\mathbf{f}}_{xn}(t) \end{bmatrix}^T$. Thus, the transformation relations of the force vectors between the modal coordinates and the physical coordinates are written as follows

$$\mathbf{u} = \boldsymbol{\Psi}_c\tilde{\mathbf{u}}, \boldsymbol{\Psi}_c = \begin{bmatrix} \boldsymbol{\varphi}_1(x_1) & \boldsymbol{\varphi}_1(x_2) & \cdots & \boldsymbol{\varphi}_1(x_p) \\ \boldsymbol{\varphi}_2(x_1) & & \cdots & \boldsymbol{\varphi}_2(x_p) \\ \vdots & & \vdots & \vdots \\ \boldsymbol{\varphi}_p(x_1) & & \cdots & \boldsymbol{\varphi}_p(x_p) \end{bmatrix}_{p\times p}, \mathbf{f_d} = \boldsymbol{\Psi}_d\tilde{\mathbf{f}}, \boldsymbol{\Psi}_d = \begin{bmatrix} \boldsymbol{\varphi}_1(x_1) & \boldsymbol{\varphi}_1(x_2) & \cdots & \boldsymbol{\varphi}_1(x_n) \\ \boldsymbol{\varphi}_2(x_1) & & \cdots & \boldsymbol{\varphi}_2(x_n) \\ \vdots & & \vdots & \vdots \\ \boldsymbol{\varphi}_p(x_1) & & \cdots & \boldsymbol{\varphi}_p(x_n) \end{bmatrix}_{p\times n} \tag{20}$$

There is no corresponding relation between the position coordinates x_i of the above two equations, that is, the position of the control force and the disturbance force is not necessarily the same; and the selected position coordinates should at least ensure $\boldsymbol{\Psi}_c, \boldsymbol{\Psi}_d$ invertible.

Substituting Equation (20) with Equation (15), the state equation of the controlled part can be obtained as follows:

$$\dot{\mathbf{X}}_C(t)=\widetilde{\mathbf{A}}_C\mathbf{X}_C(t)+\widetilde{\mathbf{B}}_C\widetilde{\mathbf{f}}(t) \tag{21}$$

where

$$\widetilde{\mathbf{A}}_C = \begin{bmatrix} 0 & \mathbf{I} \\ -(\omega^2+\mathbf{g}) & -\mathbf{h} \end{bmatrix}_{2p\times 2p}, \widetilde{\mathbf{B}}_C = \begin{bmatrix} 0 \\ \mathbf{\Psi}_{dC} \end{bmatrix}_{2p\times n}$$

where $\mathbf{g}, \mathbf{h}$ are the generalized displacement gain of the modal control and the generalized velocity gain matrix, respectively. It is obvious that the modal structure of the system is reconstructed by state feedback, and the modal damping and stiffness of the system are improved effectively.

The state equation of the part of the uncontrolled mode should be:

$$\dot{\mathbf{X}}_R(t)=\widetilde{\mathbf{A}}_R\mathbf{X}_R(t)+\widetilde{\mathbf{B}}_R\widetilde{\mathbf{f}}(t) \tag{22}$$

where

$$\widetilde{\mathbf{A}}_R = \begin{bmatrix} 0 & \mathbf{I} \\ -\omega^2 & 0 \end{bmatrix}_{2(N-p)\times 2(N-p)}, \widetilde{\mathbf{B}}_R = \begin{bmatrix} 0 \\ \mathbf{\Psi}_{\mathbf{dR}} \end{bmatrix}_{2(N-p)\times n}$$

The deflection $w(x_0, t)$ at the position coordinate x_0 is selected as the output of the system, and the following results can be obtained from Equation (8);

$$\mathbf{w}(x_0,t) = \mathbf{w}_C(x_0,t) + \mathbf{w}_R(x_0,t) = \sum_{i=1}^{p} \varphi_i(x_0)\mathbf{q}_i(t) + \sum_{i=p+1}^{N} \varphi_i(x_0)\mathbf{q}_i(t) \tag{23}$$

The final output of the system is:

$$\mathbf{w}(x_0,t) = \mathbf{CX} \tag{24}$$

where

$$\begin{cases} \mathbf{C} = \begin{bmatrix} \mathbf{\Psi}_{0C}^T \vdots\, 0 \vdots\, \mathbf{\Psi}_{0R}^T \vdots\, 0 \end{bmatrix}_{1\times 2N} \\ \mathbf{\Psi}_{0C} = [\varphi_1(x_0)\ \varphi_2(x_0) \cdots \varphi_r(x_0)]_{1\times p}^T \\ \mathbf{\Psi}_{0R} = [\varphi_r(x_0)\ \varphi_{r+1}(x_0) \cdots \varphi_N(x_0)]_{1\times(N-p)}^T \end{cases}$$

Equations (21), (23) and (26) constitute a complete state space description of partially controlled vibration systems. It is clear that the controlled and uncontrolled modes are independent of each other.

4. Rotating Angle Traveling Wave Control of Timoshenko Beam

Vibration in structure is essentially the propagation of elastic wave in structure. An elastic wave can be regarded as the representation of vibration propagation in a continuous medium. When an elastic wave propagates in a bounded structure and various waveforms are superimposed and stable, its external performance is often regarded as a vibration. Structural vibration is a special form of wave expression, which can be regarded as the result of the superposition of traveling waves satisfying certain conditions. Therefore, the vibration of the structure can be regarded as the propagation, reflection and transmission of the elastic disturbance in the structure, and the whole structure shows a wave characteristic. Therefore, the dynamic response and vibration control of a missile structure can be studied from the perspective of a traveling wave. The vibration in the structure can be described by the superposition of wave modes. The active control method of a traveling wave is to control the energy propagation in the structure, that is, to reduce the transmission from one part of the structure to the other, or to absorb the energy carried by the traveling wave. In wave control, the control law is the transfer relation between the input and output of the controller. It can be divided into time domain design method and frequency domain design method. The control force can be a point force or a point moment. The frequency domain design method is used here. In this paper, a wave controller is designed to control the rotation angle of the beam, and the wave control torque is applied when the rotation angle is controlled.

There are two methods to describe vibration: modal mode description and wave description. In essence, the former belongs to standing wave expression, while the latter is traveling wave expression. Considering the free vibration of the Timoshenko beam, the time factor $\exp(-i\omega t)$ is omitted:

$$\Phi = A_1{}^{-}\exp(-ik_1x)+A_2{}^{+}\exp(ik_1x)+A_3{}^{-}\exp(-k_2x)+A_4{}^{+}\exp(k_2x) \tag{25}$$

In the above equation, + and − represent the positive and negative propagation waves, respectively. The modes with spatial factors $\exp(-ik_1x)$ and $\exp(ik_1x)$ represent the propagating waves carrying energy, while the modes with spatial factors $\exp(-ik_2x)$ and $\exp(ik_2x)$ represent the local vibrational modes that do not carry energy. The purpose of wave control is to dissipate and absorb the energy in the propagating wave. When there are no discontinuous points in the infinite beam structure, the above elastic waves will propagate to infinity, and the local vibration modes will decay quickly.

Considering the existence of excitation force in beam structure (shear discontinuity point), according to the traveling wave theory, the reflection and transmission of the elastic wave will occur at the discontinuous incidence. Assuming that a sequence of forward propagating waves is incident at $x = 0$, resulting in reflected, transmitted and near-field waves, the displacement of the beam at $x \le 0$ and $x \ge 0$ is as follows

$$\begin{cases} \boldsymbol{\varphi}_-(x) = \begin{bmatrix} w_- \\ \psi_- \end{bmatrix} = \begin{bmatrix} a^+ \\ b^+ \end{bmatrix}\exp(ik_1x) + \begin{bmatrix} a^- \\ b^- \end{bmatrix}\exp(-ik_1x) + \begin{bmatrix} a_N^- \\ b_N^- \end{bmatrix}\exp(k_2x) \\ \boldsymbol{\varphi}_+(x) = \begin{bmatrix} w_+ \\ \psi_+ \end{bmatrix} = \begin{bmatrix} c^+ \\ d^+ \end{bmatrix}\exp(ik_1x) + \begin{bmatrix} c_N^+ \\ d_N^+ \end{bmatrix}\exp(-k_2x) \end{cases} \tag{26}$$

In the feedback wave control, the sensor and actuator are located in a certain area of the structure to control the propagation of elastic wave. In this case, the control force is a discontinuous point. In the frequency domain feedback wave control, the applied wave control moment is taken as

$$\overline{\mathbf{T}}(\omega) = -\mathbf{H}(\omega)\boldsymbol{\Upsilon}(\omega) \tag{27}$$

According to the conditions of generalized displacement continuity and generalized force equilibrium, the function of beam satisfies at $x = 0$

$$\begin{cases} w_-(0) = w_+(0) \\ \psi_-(0) = \psi_+(0) \\ \mathbf{M}_+(0) - \mathbf{M}_-(0) = \mathbf{H}\psi_+ \\ Q_+(0) = Q_-(0) \end{cases} \tag{28}$$

In Equation (28), **H** is the transfer function of the wave controller, and the symbols − and + represent the corresponding mechanical quantities of the beam at $x \le 0$ and $x \ge 0$, respectively.

According to the expressions of bending and shear forces, Equation (28) can be written as

$$\begin{cases} \begin{bmatrix} 1 \\ ik_1 \end{bmatrix} b^+ + \begin{bmatrix} 1 & 1 \\ -ik_1 & k_2 \end{bmatrix}\mathbf{B}^- = \begin{bmatrix} 1 & 1 \\ ik_1 + \frac{H}{D} & \frac{H}{D} - k_2 \end{bmatrix}\mathbf{D}^+ \\ \begin{bmatrix} k_1^2 \\ ik_1^3 \end{bmatrix} b^+ + \begin{bmatrix} k_1^2 & -k_2^2 \\ -ik_1^3 & -k_2^3 \end{bmatrix}\mathbf{B}^- = \begin{bmatrix} k_1^2 & -k_2^2 \\ \frac{H\rho J\omega^2}{D^2} + ik_1^3 & \frac{H\rho J\omega^2}{D^2} + k_2^3 \end{bmatrix}\mathbf{D}^+ \end{cases} \tag{29}$$

where

$$\mathbf{B}^- = \begin{bmatrix} b^- \\ b_N^- \end{bmatrix} = \begin{bmatrix} r_1 \\ r_2 \end{bmatrix} b^+,\ \mathbf{D}^+ = \begin{bmatrix} d^+ \\ d_N^+ \end{bmatrix} = \begin{bmatrix} t_1 \\ t_2 \end{bmatrix} b^+$$

Denoted as $\overline{H} = \frac{H}{Dk_1}$, $\frac{\rho J\omega^2}{D} = \frac{k_0^4 h^2}{12}$, the following equation can be obtained:

$$\begin{cases} 1 + r_1 + r_2 = t_1 + t_2, \mathrm{i} - \mathrm{i}r_1 + \frac{k_2}{k_1} r_2 = \left(\mathrm{i} + \overline{H}\right) t_1 - \left(\frac{k_2}{k_1} - \overline{H}\right) t_2 \\ 1 + r_1 + r_2 = t_1 + t_2, \mathrm{i} - \mathrm{i}r_1 + \frac{k_2}{k_1} r_2 = \left(\mathrm{i} + \overline{H}\right) t_1 - \left(\frac{k_2}{k_1} - \overline{H}\right) t_2 \end{cases} \tag{30}$$

It is the same as the method of the wave controller designed to control the lateral displacement, denoted as $\overline{H}(\omega) = (1 + \mathrm{i})\omega g$. The energy reflected and transmitted by unit incident energy is $E(g) = |r_1|^2 + |t_1|^2$. The maximum energy dissipation in discontinuous structure is taken as the performance index of optimal control, and the optimal control law $\overline{H}(\omega)$ is determined.

$$\begin{aligned} E(g) = \ & (288k_1^2k_2^2 + 288k_2^4 + 24gh^2k_0^4k_1k_2\omega - 288gk_1^3k_2\omega + g^2h^4k_0^8\omega^2 - 24g^2h^2k_0^4k_1^2\omega^2 \\ & +144g^2k_1^4\omega^2 + 24g^2h^2k_0^4k_2^2\omega^2 - 144g^2k_1^2k_2^2\omega^2 + 144g^2k_2^4\omega^2)/(288k_1^2k_2^2 \\ & +288k_2^4 + 24gh^2k_0^4k_1k_2\omega - 288gk_1^3k_2\omega + 24gh^2k_0^4k_2^2\omega + 288gk_2^4\omega + g^2h^4k_0^8\omega^2 \\ & -24g^2h^2k_0^4k_1^2\omega^2 + 144g^2k_1^4\omega^2 + 24g^2h^2k_0^4k_2^2\omega^2 - 144g^2k_1^2k_2^2\omega^2 + 144g^2k_2^4\omega^2) \end{aligned} \tag{31}$$

When $E(g)$ reaches the minimum, the feedback gain is

$$g = \frac{a}{b} \tag{32}$$

where

$$\begin{cases} a = 12\sqrt{2}\sqrt{h^2k_0^4k_1^2k_2^2 + h^2k_0^4k_2^4 + 12k_1^2k_2^4 + 12k_2^6},\ b = \sqrt{g_1 + g_2} \\ g_1 = h^6k_0^{12}\omega^2 - 24h^4k_0^8k_1^2\omega^2 + 144h^2k_0^4k_1^4\omega^2 + 36h^4k_0^8k_2^2\omega^2 - 432h^2k_0^4k_1^2k_2^2\omega^2 \\ g_2 = 1728k_1^4k_2^2\omega^2 + 432h^2k_0^4k_2^4\omega^2 - 1728k_1^2k_2^4\omega^2 + 1728k_2^6\omega^2 \end{cases}$$

The corresponding reflection and transmission coefficients of incident waves are expressed as follows, respectively:

$$\begin{cases} r_1 = & -(gh^2k_0^4k_2\omega + 12gk_2^3\omega)\mathrm{i}/\{(k_1 - k_2\mathrm{i})\,[(-12 + 12i)\,k_1k_2 \\ & -(12 + 12i)k_2^2 - gh^2k_0^4\omega + 12gk_1^2\omega + 12gk_1k_2\omega i - 12gk_2^2\omega]\} \\ t_1 = & [(-12 + 12i)\,k_1^2k_2 - (12 - 12i)k_2^3 - gh^2k_0^4k_1\omega + 12gk_1^3\omega]/\{(k_1 - k_2i) \\ & [(-12 + 12i)\,k_1k_2 - (12 + 12i)k_2^2 - gh^2k_0^4\omega + 12gk_1^2\omega + 12gk_1k_2\omega i - 12gk_2^2\omega]\} \end{cases} \tag{33}$$

If the wave control moment is applied at the point $x = x_w$, the tuning PD control becomes:

$$m_w(x,t) = -[c_1 w(x,t) + c_2 \dot{w}(x,t)]\delta(x - x_w) \tag{34}$$

At this time, wave control Equation (34) is applied to the original vibration system, and the matrix form of the motion equation of the system is written as follows:

$$\overline{\mathbf{M}}_w\ddot{\mathbf{q}} + \overline{\mathbf{C}}_w\dot{\mathbf{q}} + \overline{\mathbf{K}}_w\mathbf{q} = \overline{\mathbf{f}} \tag{35}$$

where $\overline{\mathbf{f}} = \begin{bmatrix} \mathbf{f}_1 & \mathbf{f}_2 & \cdots & \mathbf{f}_\mathbf{n} \end{bmatrix}^T$, $\mathbf{f_i(t)} = \frac{1}{m_i}\int_0^l \boldsymbol{\varphi}_\mathbf{i}{}^\mathbf{T}\mathbf{F}(x,t)dx$, $\overline{\mathbf{M}}_w$, $\overline{\mathbf{C}}_w$, $\overline{\mathbf{K}}_w$ are mass matrix, damping matrix and stiffness matrix, respectively, and the expressions of them are as follows:

$$\overline{\mathbf{M}}_w = \mathbf{I}, \overline{\mathbf{C}}_w = c_2\overline{\boldsymbol{\Psi}}_w, \overline{\mathbf{K}}_w = \boldsymbol{\omega}^2 + c_1\overline{\boldsymbol{\Psi}}_w \tag{36}$$

where

$$\overline{\boldsymbol{\Psi}}_w = \begin{bmatrix} \boldsymbol{\varphi}_1(x_w)\boldsymbol{\varphi}_1(x_w) & \boldsymbol{\varphi}_1(x_w)\boldsymbol{\varphi}_2(x_w) & \cdots & \boldsymbol{\varphi}_1(x_w)\boldsymbol{\varphi}_N(x_w) \\ \boldsymbol{\varphi}_2(x_w)\boldsymbol{\varphi}_1(x_w) & \cdots & & \boldsymbol{\varphi}_2(x_w)\boldsymbol{\varphi}_N(x_w) \\ \vdots & \vdots & & \vdots \\ \boldsymbol{\varphi}_N(x_w)\boldsymbol{\varphi}_1(x_w) & \cdots & & \boldsymbol{\varphi}_N(x_w)\boldsymbol{\varphi}_N(x_w) \end{bmatrix}_{N\times N}$$

It is important to note that c_1, c_2 will also be a matrix, not a number, in subsequent programming calculations. For ease of use in optimized control, we can also write $\overline{\Psi}_w$ in block form:

$$\overline{\Psi}_w = \begin{bmatrix} \overline{\Psi}_{CC} & \overline{\Psi}_{CR} \\ \overline{\Psi}_{RC} & \overline{\Psi}_{RR} \end{bmatrix} \tag{37}$$

When there is no wave control force, the mass array $\boldsymbol{M}$ is $N \times N$ identity matrix, $\boldsymbol{C}$ is the $N \times N$ zero matrix, $\boldsymbol{K}$ is diagonal matrix with the square of natural frequency. However, the wave control moment is coupled with the vibration modes of the uncontrolled original system, and the mode of the wave control system changes as a result.

Introducing the state vector $\mathbf{X}(t) = [\mathbf{q}^T(t) : \dot{\mathbf{q}}^T(t)]^T$, Equation (39) is written in the form of a state space:

$$\dot{\mathbf{X}}(t) = \mathbf{A}\mathbf{X}(t) + \mathbf{B}\overline{\mathbf{f}} \tag{38}$$

where the coefficient matrix is:

$$\mathbf{A} = \begin{bmatrix} 0 & \mathbf{I} \\ -\overline{\mathbf{M}}_w{}^{-1}\overline{\mathbf{K}}_w & -\overline{\mathbf{M}}_w{}^{-1}\overline{\mathbf{C}}_w \end{bmatrix}, \mathbf{B} = \begin{bmatrix} 0 \\ \overline{\mathbf{M}}_w{}^{-1}\overline{\Psi}_d \end{bmatrix}, \overline{\Psi}_d = \begin{bmatrix} \varphi_1(x_1) & \varphi_1(x_2) & \cdots & \varphi_1(x_m) \\ \varphi_2(x_1) & \cdots & & \varphi_2(x_m) \\ \vdots & \vdots & & \vdots \\ \varphi_N(x_1) & \cdots & & \varphi_N(x_m) \end{bmatrix}_{N\times m}$$

5. Traveling Wave/Modal Optimized Control of Timoshenko Beam Rotation Angle

The optimized control method of the independent mode and traveling wave is used in this paper. First, the former pth-order mode is controlled by independent modal space control, and the partially controlled vibration system is obtained. The modal dynamic equation of the system is as follows:

$$\overline{\mathbf{M}}_m\ddot{\mathbf{q}} + \overline{\mathbf{C}}_m\dot{\mathbf{q}} + \overline{\mathbf{K}}_m\mathbf{q} = \overline{\mathbf{f}} \tag{39}$$

where

$$\overline{\mathbf{M}}_m = \mathbf{I},\ \overline{\mathbf{C}}_m = \begin{bmatrix} \mathbf{h} & 0 \\ 0 & 0 \end{bmatrix}, \overline{\mathbf{K}}_m = \begin{bmatrix} \omega_c^2 + \mathbf{g} & 0 \\ 0 & \omega_R^2 \end{bmatrix}$$

Applying wave control force to the system, combined with (35) and (36), the modal dynamics equation of the system is as follows:

$$\overline{\mathbf{M}}_h\ddot{\mathbf{q}} + \overline{\mathbf{C}}_h\dot{\mathbf{q}} + \overline{\mathbf{K}}_h\mathbf{q} = \overline{\mathbf{f}} \tag{40}$$

where

$$\overline{\mathbf{M}}_h = \mathbf{I},\ \overline{\mathbf{C}}_\mathrm{h} = \begin{bmatrix} (\mathbf{h} + c_2\overline{\Psi}_{CC}) & c_2\overline{\Psi}_{CR} \\ c_2\overline{\Psi}_{RC} & c_2\overline{\Psi}_{RR} \end{bmatrix}, \overline{\mathbf{K}}_h = \begin{bmatrix} \omega_C^2 + \mathbf{g} + c_1\overline{\Psi}_{CC} & c_1\overline{\Psi}_{CR} \\ c_1\overline{\Psi}_{RC} & \omega_R^2 + c_1\overline{\Psi}_{RR} \end{bmatrix}$$

Introducing the state vector $\mathbf{X}(t)=[\mathbf{q}^T(t):\dot{\mathbf{q}}^T(t)]^T$, and the equation of the state is:

$$\dot{\mathbf{X}}(t) = \widetilde{\mathbf{A}}\mathbf{X}(t) + \widetilde{\mathbf{B}}\overline{\mathbf{f}} \tag{41}$$

where

$$\mathbf{A} = \begin{bmatrix} 0 & \mathbf{I} \\ -\mathbf{M}_h{}^{-1}\mathbf{K}_h & -\mathbf{M}_h{}^{-1}\mathbf{C}_h \end{bmatrix}, \mathbf{B} = \begin{bmatrix} 0 \\ \mathbf{M}_h{}^{-1}\Psi_d \end{bmatrix}, \Psi_d = \begin{bmatrix} \varphi_1(x_1) & \varphi_1(x_2) & \cdots & \varphi_1(x_m) \\ \varphi_2(x_1) & \cdots & & \varphi_2(x_m) \\ \vdots & \vdots & & \vdots \\ \varphi_N(x_1) & \cdots & & \varphi_N(x_m) \end{bmatrix}_{N\times m}$$

6. Numerical Example and Analysis and Discussion

Using the independent modal space method and optimized control method, the dynamic response of the structural beam subjected to control force is investigated. Taking the characteristic length as the length of the beam l, the following dimensionless quantities: Poisson ratio $\nu = 0.30$; high to length ratio $h/l = 0.1$; damping factor before control $\zeta = 1.0 \times 10^{-4}$; ${\omega_i}^2/{\omega_0}^2 = (k_{0i}L)^4/(k_{01}L)^4$; k_{0i} is the elastic traveling wave number corresponding to the ith-order natural frequency. Table 1 shows the first 10 dimensionless natural frequencies of the Timoshenko beam.

Table 1. The first 10 dimensionless frequency of the Timoshenko beam.

Order of Mode	**1**	**2**	**3**	**4**	**5**	**6**	**7**	**8**	**9**	**10**
natural frequency	1.00	6.02	15.99	29.35	45.34	63.34	82.98	104.02	126.30	149.59

Figures 2–7 show the frequency response of the structure before and after the control with three different control methods under different parameters. In Figures 2–7, the horizontal coordinates are dimensionless frequency values, while the vertical coordinates are the common logarithmic values of the frequency response. In Figures 2–4, the position of disturbance action is $x_d = 0.15l$. Figure 2 shows the frequency response before and after modal control; the positions of mode control force applied are $x_{m1} = 0.215l$, $x_{m2} = 0.525l$, $x_{m3} = 0.775l$, $x_{m4} = 1.0l$; the first four modes are controlled, and the measuring position of the dynamic response is $x_s = 0.265l$, where the occurrence point of the maximum value of the total deflection of the first 10 modes is. Figure 3 shows the frequency response of the structure before and after the control when the position of the wave control force is $x_w = 0.40l$. Figure 4 describes the frequency response of the structure before and after the traveling wave/modal-optimized control.

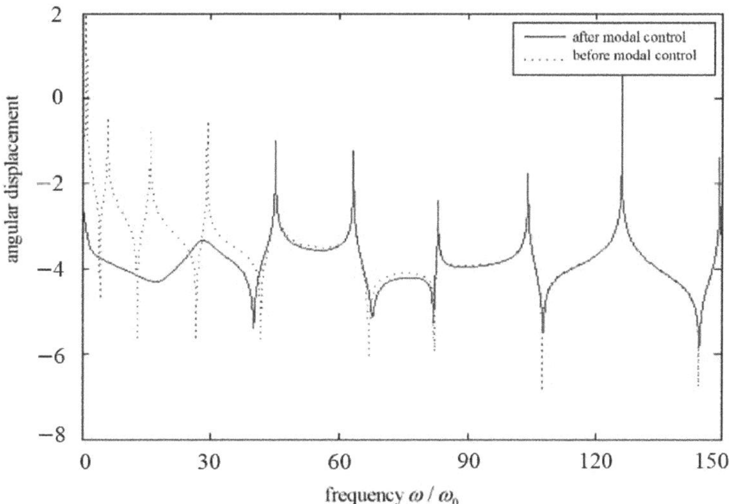

Figure 2. Comparison of frequency response before and after modal control ($x_d = 0.15l, x_s = 0.265l$).

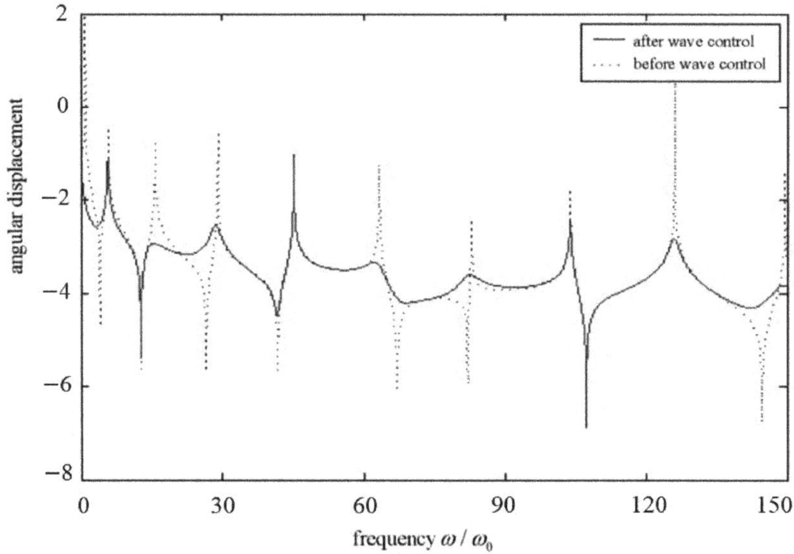

Figure 3. Comparison of frequency response before and after wave control ($x_d = 0.15l, x_w = 0.40l$).

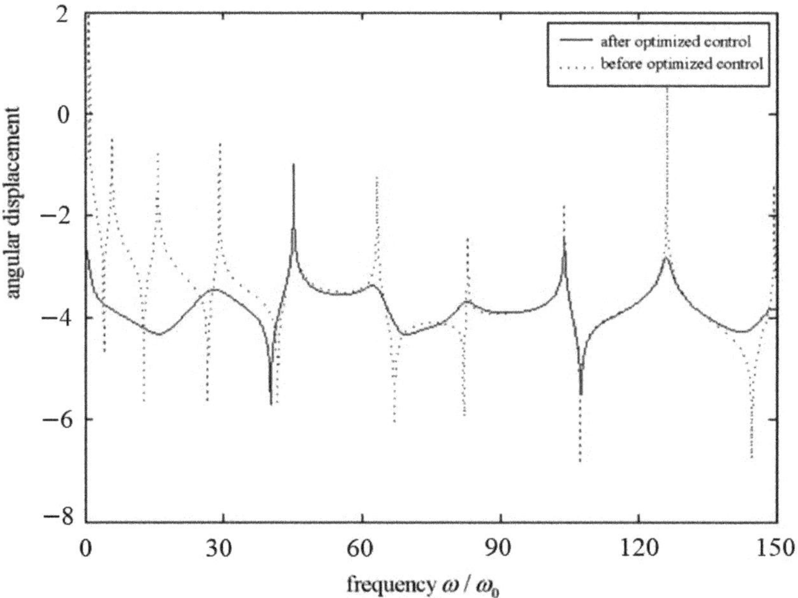

Figure 4. Comparison of frequency response before and after optimized control ($x_d = 0.15l$).

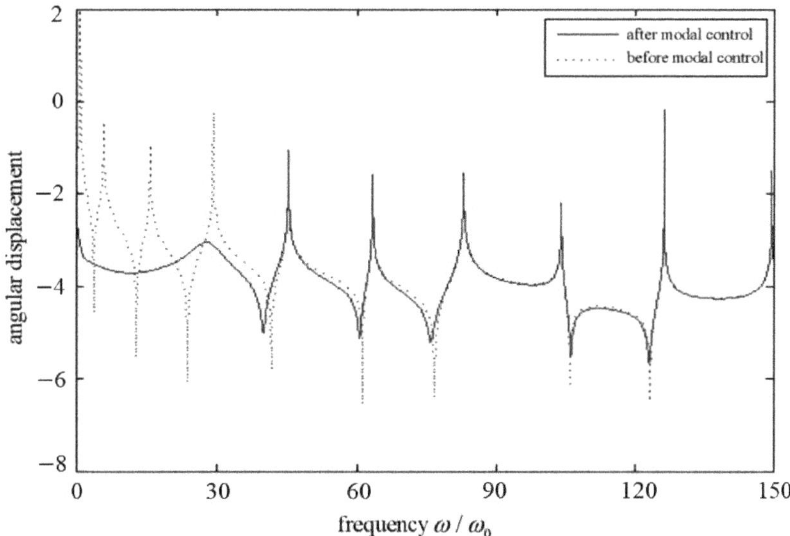

Figure 5. Comparison of frequency response before and after modal control ($x_d = 0.10l, x_s = 0.295l$).

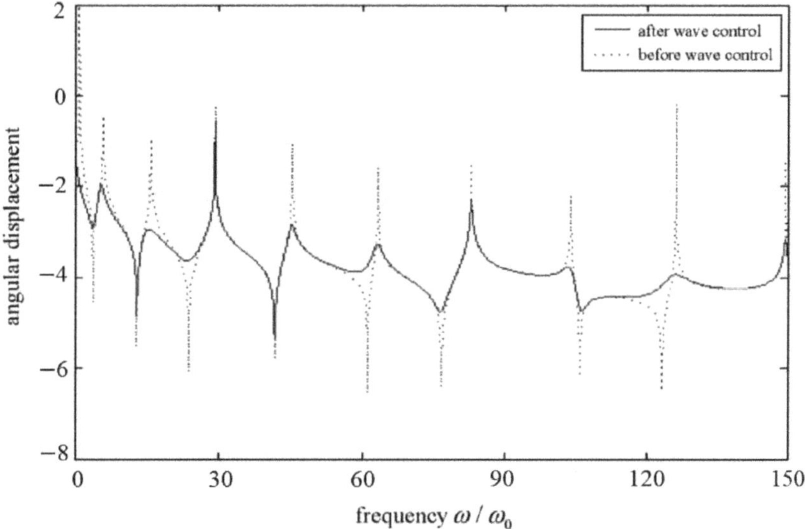

Figure 6. Comparison of frequency response before and after wave control ($x_d = 0.10l, x_w = 0.30l$).

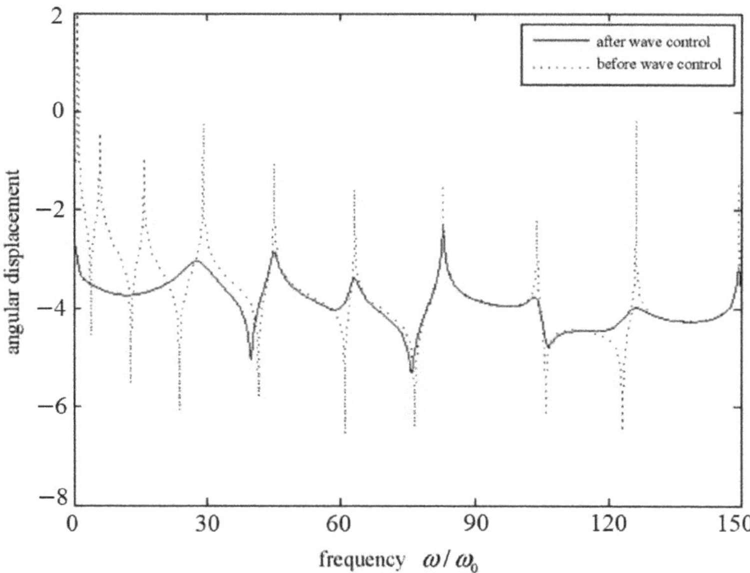

Figure 7. Comparison of frequency response before and after optimized control ($x_d = 0.10l$).

In Figures 5–7, the position of disturbance action is $x_d = 0.10l$. Figure 5 shows the frequency response before and after modal control. The positions of the modal control force applied are $x_{m1} = 0.215l$, $x_{m2} = 0.525l$, $x_{m3} = 0.775l$, $x_{m4} = 1.0l$, the first four modes are controlled, and the measuring position of dynamic response is $x_s = 0.295l$, where the occurrence point of the maximum value of the total deflection of the first 10 modes is. Figure 6 shows the frequency response of the structure before and after the control when the position of the wave control force is $x_w = 0.30l$, and Figure 7 describes the frequency response of the structure before and after the travelling wave/modal-optimized control.

When applying the wave control force, the node of the first 10 mode vibrations should be avoided as much as possible. The frequency responses of the single-use wave control, independent mode space control and the optimized traveling wave control, including the controlled and uncontrolled frequency response, are compared. In the approximation of the optimized control method, the controller is tuned to the optimum at the fifth mode. In the case of the independent modal space control, the first four-order damping factor of the structure is increased to 0.1 by using the pole assignment method. Through the analysis, we can obtain the discussion as follows:

Figures 2 and 5 show that modal control has good control effect on low-order modes of structure, and it has good independence and has little effect on other modes; thus, the rapid change of the uncontrolled modal response still exists.

Figures 3 and 6 show the frequency response of the structure before and after the use of the wave controller. The response of the structure before the control is sharp, but the wave control can be regarded as adding damping to the structure, absorbing the energy in the structure, and the sharp response is weakened. The figure also shows that the modal characteristics of the whole structure have changed after the wave control; that is, the natural vibration frequency has changed. Finally, we can see that the position of the wave controller is different for the control effect of each mode frequency.

Figures 4 and 7 show the frequency response of the structure before and after using wave control and independent mode space control. It reflects the excellent control effect of the optimized control. Because the optimized control combines the characteristics of the former two methods, that is, the wave control can control the response at higher frequencies better, and the independent modal space control can effectively control the first four modes.

It makes up for the poor control effect of wave control at low frequency. Therefore, the performance of using one of the above controls alone is improved.

7. Conclusions

(1) In this paper, the active vibration control of the Timoshenko beam structure is investigated in a more comprehensive way. First, considering the influence of section rotation and shear deformation, the vibration equation of the Timoshenko beam is established, and the modal and wave analysis are employed. The finite dimensional system dynamic equation is obtained by modal expansion. Then, we adopt three different control methods: traditional mode space control method, traditional traveling wave control method and optimized traveling wave control method. As a numerical example, three strategies of independent mode space control, traditional traveling wave control and optimized traveling wave control are used to control the active vibration of the beam angle. By comparing the numerical results of the three methods, it can be seen that: the traditional mode space control method has good effect on the low-frequency control, but it is easy to overflow because of omitting the high-order modes.

(2) Traveling wave control can effectively control the high-order modes of the system, while the effect of low-order mode control is far less than that of modal control. In addition, the application position of wave control torque has important influence on the control result.

(3) Using the optimized traveling wave control method and Timoshenko beam theory, the dispersion relationship of waves in structural vibration is determined, and the wave numbers of propagating waves and attenuation waves in the Timoshenko beam are given. The problem of solving the natural frequency and mode of the cantilever Timoshenko beam is effectively solved. The optimized traveling wave control method involves using the traveling wave control outside the modal nodes of each order, coupled with the global mode space control method. The optimized traveling wave control method not only effectively suppresses the main low-order modes, but also suppresses the high-order modes by wave control, effectively reduces the control overflow and enhances the robustness of the system.

In this paper, the Timoshenko beam model, which considers the influence of transverse shear deformation and moment of inertia on displacement, improves the accuracy of calculation. It is important for spacecraft accessory structures with high requirements for angle control. In terms of vibration control results, not only the vibration is well suppressed, but also the robustness of the system is improved by using the optimal traveling wave control method proposed in this paper. The optimized traveling wave control method is exquisite in mathematical processing and has good results in global and local vibration control, which is not available in other methods. Therefore, the optimized control method shows better vibration active control performance than any traditional control method.

Author Contributions: Conceptualization, H.J. and C.Z.; methodology, H.J. and B.W.; software, J.F.; validation, H.J., C.Z. and H.D.; formal analysis, J.F.; investigation, W.J. and C.X.; resources, H.D.; writing—original draft preparation, J.F.; writing—review and editing, C.Z.; supervision, H.J and W.Z.; project administration, Y.G.; funding acquisition, W.Z. All authors have read and agreed to the published version of the manuscript.

Funding: This paper is supported by Sichuan Science and Technology Program: 2022YFG0274; Key Research and Development Program of Zhejiang Province: 2021C03013; National Natural Science Foundation of China: 51875146; Natural Science Foundation of Zhejiang Province: LY21E050005; Key Laboratory for Technology in Rural Water Management of Zhejiang Province: ZJWEU-RWM-20200303B.

Institutional Review Board Statement: Not applicable.

Informed Consent Statement: Not applicable.

Data Availability Statement: The study did not report any data.

Conflicts of Interest: The authors declare no conflict of interest.

References

1. Rong, J.L.; Xu, T.F.; Wang, X.; Li, J.; Yin, X.; Xin, P. Dynamic stability analysis of flexible spinning flight vehicles under follower thrust. *J. Astronaut.* **2015**, *36*, 18–24.
2. Beck, A.T.; da Silva, C.R.A., Jr. Timoshenko versus Euler beam theory: Pitfalls of a deterministic approach. *Struct. Saf.* **2011**, *33*, 19–25. [CrossRef]
3. Li, D.X. *Structural Dynamics of Flexible Spacecraft*; Science Press: Beijing, China, 2010. (In Chinese)
4. Ruge, P.; Birk, C. A comparison of infinite Timoshenko and Euler-Bernoulli beam models on Winkler foundation in the frequency- and time-domain. *J. Sound Vib.* **2007**, *304*, 932–947. [CrossRef]
5. Shafiei, N.; Kazemi, M.; Ghadiri, M. Comparison of modeling of the rotating tapered axially functionally graded Timoshenko and Euler-Bernoulli microbeams. *Phys. E Low-Dimens. Syst. Nanostruct.* **2016**, *83*, 74–87. [CrossRef]
6. Sferza, M.; Ninić, J.; Chronopoulos, D.; Glock, F.; Daoud, F. Multidisciplinary Optimisation of Aircraft Structures with Critical Non-Regular Areas: Current Practice and Challenges. *Aerospace* **2021**, *8*, 223. [CrossRef]
7. Mei, C.; Mace, B.R.; Jones, R.W. Hybrid wave/mode active vibration control. *J. Sound Vib.* **2001**, *247*, 765–784. [CrossRef]
8. Mei, C.; Mace, B.R. Reduction of control spillover in active vibration control of distributed structures using multioptimal schemes. *J. Sound Vib.* **2002**, *251*, 184–192. [CrossRef]
9. Carvalho, M.O.M.; Zindeluk, M. Active control of waves in a Timoshenko beam. *Int. J. Solids Struct.* **2001**, *38*, 1749–1764. [CrossRef]
10. Halkyard, C.R.; Mace, B.R. Feedforward adaptive control of flexural vibration in a beam using wave amplitudes. *J. Sound Vib.* **2002**, *254*, 117–141. [CrossRef]
11. EL-Khatib, H.M.; Mace, B.R.; Brennan, M.J. Suppression of Bending Waves in a Beam Using a Tuned Vibration Absorber. *J. Sound Vib.* **2005**, *288*, 1157–1175. [CrossRef]
12. Hu, C.; Chen, T.; Huang, W.H. Active Vibration Control of Timoshenko Beam Based on Hybrid Wave/Mode Method. *Acta Aeronaut. Astronaut. Sin.* **2007**, *28*, 301–308.
13. Su, Y.C.; Ma, C.C. Theoretical analysis of transient waves in a simply-supported Timoshenko beam by ray and normal mode methods. *Int. J. Solids Struct.* **2011**, *48*, 535–552. [CrossRef]
14. Su, Y.C.; Ma, C.C. Transient wave analysis of a cantilever Timoshenko beam subjected to impact loading by Laplace transform and normal mode methods. *Int. J. Solids Struct.* **2012**, *49*, 1158–1176. [CrossRef]
15. Cardoso, R.P.R. A new beam element which blends the Euler-Bernoulli beam theory with idealised transverse shear flows for aircraft structural analysis. *Thin-Walled Struct.* **2020**, *157*, 107–118. [CrossRef]
16. Xing, X.Y.; Liu, J.K. Modelling and neural adaptive vibration control for three-dimensional Timoshenko beam with output restrictions and external disturbances. *Int. J. Syst. Sci.* **2021**, *52*, 1850–1867. [CrossRef]
17. Ishaquddin, M.; Raveendranath, P.; Reddy, J.N. Flexure and torsion locking phenomena in out-of-plane deformation of Timoshenko curved beam element. *Finite Elem. Anal. Des.* **2012**, *51*, 22–30. [CrossRef]
18. Endo, T.; Sasaki, M.; Matsuno, F. Contact-Force Control of a Flexible Timoshenko Arm. *IEEE Trans. Autom. Control* **2013**, *12*, 875–882. [CrossRef]
19. Mei, C. Hybrid wave/mode active control of bending vibrations in beams based on the advanced Timoshenko theory. *J. Sound Vib.* **2009**, *322*, 29–38. [CrossRef]
20. Pham, P.T.; Kim, G.H.; Hong, K.S. Vibration control of a Timoshenko cantilever beam with varying length. *Int. J. Control Autom. Syst.* **2022**, *20*, 175–183. [CrossRef]
21. Eshag, M.A.; Lei, M.; Sun, Y.; Zhang, K. Robust global boundary vibration control of uncertain Timoshenko beam with exogenous disturbances. *IEEE Access* **2020**, *8*, 72047–72058. [CrossRef]
22. Fleischmann, D.; Könözsy, L. On a novel approximate solution to the inhomogeneous Euler—Bernoulli equation with an application to aeroelastics. *Aerospace* **2021**, *8*, 356. [CrossRef]
23. Chen, X.; Liu, J.; Gao, B.; Chen, X.F. Analysis and implementation of a multiple-source multiple-channel active vibration control of large structures based on finite element model in-loop simulation system. In Proceedings of the 24th International Congress on Sound & Vibration, London, UK, 23–27 July 2017.
24. Liu, C.C.; Li, F.M.; Tang, L.; Huang, W.H. Vibration control of the finite L-shaped beam structures based on the active and reactive power flow. *Sci. China Phys. Mech. Astron.* **2011**, *54*, 310–319. [CrossRef]

 sensors

MDPI

Article

Structural Responses Estimation of Cable-Stayed Bridge from Limited Number of Multi-Response Data

Namju Byun [1], Jeonghwa Lee [1], Joo-Young Won [2] and Young-Jong Kang [2,*]

[1] Future and Fusion Laboratory of Architectural, Civil and Environmental Engineering, Korea University, Seoul 02841, Korea; skawn0702@naver.com (N.B.); qevno@korea.ac.kr (J.L.)
[2] School of Civil, Environmental and Architectural Engineering, Korea University, Seoul 02841, Korea; jywon27@korea.ac.kr
* Correspondence: yjkang@korea.ac.kr; Tel.: +82-2-3290-3317

Abstract: A cable-stayed bridge is widely adopted to construct long-span bridges. The deformation of cable-stayed bridges is relatively larger than that of conventional bridges, such as beam and truss types. Therefore, studies regarding the monitoring systems for cable-stayed bridges have been conducted to evaluate the performance of bridges based on measurement data. However, most studies required sufficient measurement data for evaluation and just focused on the local response estimation. To overcome these limitations, Structural Responses Analysis using a Limited amount of Multi-Response data (SRALMR) was recently proposed and validated with the beam and truss model that has a simple structural behavior. In this research, the structural responses of a cable-stayed bridge were analyzed using SRALMR. The deformed shape and member internal forces were estimated using a limited amount of displacement, slope, and strain data. Target structural responses were determined by applying four load cases to the numerical model. In addition, pre-analysis for initial shape analysis was conducted to determine the initial equilibrium state, minimizing the deformation under dead loads. Finally, the performance of SRALMR for cable-stayed bridges was analyzed according to the combination and number of response data.

Keywords: cable-stayed bridge; multi-response data; deformed shape; member internal force; SRALMR

Citation: Byun, N.; Lee, J.; Won, J.-Y.; Kang, Y.-J. Structural Responses Estimation of Cable-Stayed Bridge from Limited Number of Multi-Response Data. *Sensors* **2022**, *22*, 3745. https://doi.org/10.3390/s22103745

Academic Editors: Phong B. Dao, Liang Yu and Lei Qiu

Received: 8 April 2022
Accepted: 12 May 2022
Published: 14 May 2022

1. Introduction

The construction of bridges with longer spans is a challenging task in civil engineering. A cable-stayed bridge is widely adopted to construct long-span bridges, in which the cable connects the bridge deck and the pylon to support the bridge deck. The forces generated in cables can be divided into horizontal and vertical components. The horizontal component of the cable force causes a significant compressive force on the girder, and the girder has a complex distribution of bending moment owing to the vertical component of the cable force. In addition, horizontal and vertical components of the cable force cause a bending moment and compressive force on the pylon, respectively. The brokenness of one cable can affect the entire equilibrium system of the structure. Therefore, ensuring the integrity of cable-stayed bridges is a significant factor in structural maintenance. Various responses and environmental conditions have been measured using a monitoring system to evaluate the integrity of cable-stayed bridges [1–3].

If the Structural Health Monitoring (SHM) that can evaluate the integrity of the structure using measured data such as displacement, strain, slope, acceleration, and temperature is successfully applied to on-site structure, the limitations of manpower and budget for maintenance can be overcome. The main objective of SHM is the detection of degradation of structure and to provide useful data for maintenance decision making. The research field for SHM can be divided into three fields according to the objective: damage detection [4,5], response pattern prediction [6], and unmeasured response estimation (URE) [7–9].

This paper is a study on the field of URE. In previous studies for URE, measured data (displacement, slope, strain, and acceleration) related to the structural stiffness has been generally utilized.

Displacement, which is a structural response, is typically utilized to indirectly evaluate the performance of a structure. The performance of the structure was indirectly evaluated by comparing the measured displacement with the deformation limit determined via numerical analysis and a loading test. Linear variable differential transformers (LADTs) and laser Doppler vibrometers (LDVs) are widely used to measure the displacement of structures. However, these conventional displacement sensors have limitations, such as the requirement of additional fixed reference points and high sensitivity to the surrounding climate. To overcome these limitations, studies have been conducted on displacement estimation methods using other response data.

Among the analyzed structural responses, acceleration, strain, and slope data have been widely used to estimate the displacement of structures. Using acceleration data, displacement can be estimated based on the relation that displacement is a double integration of acceleration. Park et al. [10] estimated the displacement through double integration of acceleration, assuming that the initial displacement and average velocity were zero. Lee et al. [11] proposed a time-domain finite impulse response filter (FIR filter) for estimating displacement without initial conditions in a way that minimizes the square error between acceleration, which is assumed as the central finite difference of displacement, and measured acceleration. In addition, a wireless displacement measurement system using the FIR filter proposed by [11] was developed and validated with experimental acceleration data [12]. These methods use acceleration for accurate estimations at low prices; however, only the displacement at positions where the acceleration sensors are installed can be estimated because the spatial parameter is not included in the acceleration response.

In contrast, the slope and strain response, which include spatial parameters, can be used to estimate the displacement at positions where sensors are not installed. The shape superposition method (SSM) is mostly adopted for estimation methods that use slope and strain data. SSM consists of a shape function and weight factor. The weight factor is derived by minimizing the least-square error between the measured response and product of shape function and weight factor. Hou et al. [13] estimated displacement using slope data based on the SSM. The power series was utilized as a shape function, and the coefficient was considered to represent the boundary condition. The displacement estimation using strain data was firstly conducted by Foss and Haugse [14] based on the modal mapping. Modal mapping is SSM using the mode shape as a shape function. For general use of modal mapping, a theoretical mode shape composed of sine functions was adopted as the shape function, and its effectiveness was validated with experimental data from various structures [15]. A study using mode shape derived by frequency analysis of a finite element method (FEM) model was also conducted to improve the accuracy of displacement estimation [16]. The results of the research showed that the FEM mode shape can estimate the displacement more accurately than the theoretical mode shape. In addition, modal mapping proposed by [14] was widely adopted as a basic method to estimate displacement [17–21].

Recently, studies on displacement estimation using strain and acceleration data fusion have been conducted to cope with the drawback of estimation by one type of data. Park et al. [22] developed an indirect displacement estimation using acceleration and strain (IDEAS) method by combining FIR filter [11] and modal mapping [14]. The estimation by the IDEAS method showed better accuracy than that by only strain data. Then, the extendibility of the IDEAS method was investigated for various types of beam structures [23]. The Kalman filter, actively studied for the accuracy enhancement in aerospace engineering, was also applied to the IDEAS [24]. IDEAS using FIR and Kalman filter has been widely adopted to indirectly estimate displacement based on the strain and acceleration data [25–27].

Although the previously proposed methods can indirectly well estimate displacement, they use one or two types of response data and focus on the estimation of displacement at specific points. To accurately evaluate the integrity of the structure, the response shape for the entire structure is required, not the displacement at specific points. In addition, in order to apply the estimation method to the monitoring system in which the sensor is already installed, various combinations of response data need to be available to the algorithm. As the Global Navigation Satellite System (GNSS) has been developed, which overcomes the limitations of conventional displacement sensors, the "Structural Responses Analysis using a Limited amount of Displacement data" (SRALD) was proposed [28,29]. The deformed shape was estimated using the displacement data, and the internal force was determined using the deformed shape using the stiffness method. In addition, SRALD was validated for the beam, truss, and cable-stayed bridge finite element method (FEM) models by comparing the estimated responses with values estimated by the spline interpolation method. Subsequently, SRALD was improved to the "Structural Responses Analysis using a Limited amount of Multi-Response data" (SRALMR), which uses the slope and strain response as additional data to reduce the required number of expensive GNSS [30]. In addition, SRALMR was applied to the beam and truss FEM models, and its effectiveness was verified by comparing it with SRALD. The SRALMR was established by extending the shape superposition theory basically used in previous estimation studies to use the three response data in various combinations. Compared to the previously developed estimation methods, the major strength of the SRALMR is that it can utilize any combination of displacement, slope, and strain data and estimate the response shape of the entire structure rather than the response of a specific point. However, the validation of SRALMR for structures that have complex structural behavior compared to beams and trusses is required for on-site applications.

Therefore, the validation of SRALMR for a cable-stayed bridge FEM model that has complex structural behavior owing to the cable is conducted in this study. The deformed shape was estimated using a limited amount of displacement, slope, and strain response data. Additionally, the girder axial force, girder moment, and cable force were determined using the deformed shape estimated by using the stiffness method. The pylon is also an important element for the entire equilibrium system, but it has a simple distribution of internal force, and its integrity can be relatively simply estimated using displacement data at the top of the pylon. For this reason, internal force estimation in this study is limited to the cable and girder elements. For the deformed shape and internal force, the estimation performance of SRALMR for the cable-stayed bridge was analyzed according to the combination and number of response data.

2. Estimation Algorithm

Choi et al. [28] developed an SRALD technique that can estimate the deformed shape and internal force from limited displacement data. Byun et al. [30] introduced SRALMR by improving SRALD to use slope and strain data for estimation. SRALMR is based on the shape-superposition method for the structural shape of each response. In a linear problem, each response of the structure can be expressed as the product of the shape function Φ and the weight factor α, as shown in Equations (1)–(3). Φ_u, Φ_θ, and Φ_ϵ, are the shape functions of the displacement, slope, and strain, respectively. The weight factors for displacement u, slope θ, and strain ϵ are the same because the slope and strain are the first and second derivatives of the displacement, respectively. This is the basic concept of SRALMR.

$$\{\mathrm{u}\} = [\Phi_u]\{\alpha\} \tag{1}$$

$$\{\theta\} = [\Phi_\theta]\{\alpha\} = [\Phi_u]'\{\alpha\} \tag{2}$$

$$\{\epsilon\} = [\Phi_\epsilon]\{\alpha\} = [\Phi_u]''\{\alpha\} \tag{3}$$

In this study, the structural shape function (SSF) is derived by applying a unit load to each node of the numerical model. The *i*th SSF $\Phi_{i,u}$, $\Phi_{i,\theta}$, and $\Phi_{i,\epsilon}$ for displacement, slope, and strain are represented as Equations (4) and (6), respectively. Finally, comprehensive SSF Φ_i in Equation (7) can be derived by combining all the SSF values for each response in one matrix.

$$\Phi_{i,u}{}^T = \left[u_{i1}\ u_{i2}\ u_{i3}\ u_{i4}\ \cdots\ u_{iN_{dof,u}}\right] \tag{4}$$

$$\Phi_{i,\theta}{}^T = \left[\theta_{i1}\ \theta_{i2}\ \theta_{i3}\ \theta_{i4}\ \cdots\ \theta_{iN_{dof,\theta}}\right] \tag{5}$$

$$\Phi_{i,\epsilon}{}^T = \left[\epsilon_{i1}\ \epsilon_{i2}\ \epsilon_{i3}\ \epsilon_{i4}\ \cdots\ \epsilon_{iN_{dof,\epsilon}}\right] \tag{6}$$

$$\Phi_i = \begin{bmatrix} \Phi_{i,u} \\ \Phi_{i,\theta} \\ \Phi_{i,\epsilon} \end{bmatrix} \left(N_{dof} \times 1\right) \tag{7}$$

Here, $N_{dof} = N_{dof,u} + N_{dof,\theta} + N_{dof,\epsilon}$ is the number of degrees of freedom (DOF), $N_{dof,u}$, $N_{dof,\theta}$, and $N_{dof,\epsilon}$ are the numbers of DOF for each response, respectively; and $u_{iN_{dof,u}}$, $\theta_{iN_{dof,\theta}}$, and $\epsilon_{iN_{dof,\epsilon}}$ are the ith displacement, slope, and strain value at each DOF, respectively.

As indicated above, the structural response shape can be represented as a superposition of the SSF. An arbitrary response shape $\widetilde{ARS}$ needs to be defined to establish an error function with measurement data. As shown in Equation (8), $\widetilde{ARS}$ is defined as the superposed SSF multiplied by the weight factor. N_{sf} is the number of shape functions used to estimate the structural response, and the $\widetilde{ARS}$ in matrix form is represented by Equation (9).

$$\widetilde{ARS} = \alpha_1\Phi_1 + \alpha_2\Phi_2 + \alpha_3\Phi_3 + \cdots + \alpha_{N_{sf}}\Phi_{N_{sf}} = \sum_{i=1}^{N_{sf}} \alpha_i\Phi_i \tag{8}$$

$$\widetilde{ARS} = \left[\Phi_1\ \Phi_2\ \Phi_3\ \cdots\ \Phi_{N_{sf}}\right] \begin{bmatrix} \alpha_1 \\ \alpha_2 \\ \alpha_3 \\ \vdots \\ \alpha_{N_{sf}} \end{bmatrix} \left(N_{dof} \times 1\right) \tag{9}$$

$\widetilde{ARS}$ is composed of response values at each DOF. The measurement data matrix MD is composed of the response values measured at certain points where the sensors are installed. ω, $\varnothing$, and ε are the measurement data values for displacement, slope, and strain, respectively. The MDs for each response are defined in Equations (10)–(12), and N_{md} is the number of measurements. A comprehensive MD to establish the error function with $\widetilde{ARS}$ is constructed by combining the MD for each response.

$$MD_\omega{}^T = \left[\omega_1\ \omega_2\ \omega_3\ \omega_4\ \cdots\ \omega_{N_{md,\omega}}\right] \tag{10}$$

$$MD_\varnothing{}^T = \left[\varnothing_1\ \varnothing_2\ \varnothing_3\ \varnothing_4\ \cdots\ \omega_{N_{md,\varnothing}}\right] \tag{11}$$

$$MD_\varepsilon{}^T = \left[\varepsilon_1\ \varepsilon_2\ \varepsilon_3\ \varepsilon_4\ \cdots\ \omega_{N_{md,\varepsilon}}\right] \tag{12}$$

$$MD = \begin{Bmatrix} MD_\omega \\ MD_\varnothing \\ MD_\varepsilon \end{Bmatrix} (N_{md} \times 1) \tag{13}$$

The weight factor required to estimate the deformed shape can be calculated using MD and $\widetilde{ARS}$. Owing to the difference in the matrix size between MD and $\widetilde{ARS}$, the matrix size of the $\widetilde{ARS}$ needs to be adjusted to establish the error function. The $\widetilde{ARS}$ is converted to an ARS matrix including only response values for DOF where the measurement data exist. Finally, the error function E is defined as the sum of the square errors between MD and ARS in Equation (14). Partial differentiation is performed with respect to the weight factor to minimize the error, which must equal zero. The calculation process is represented by Equations (15)–(17).

$$E = \sum_{j=1}^{N_{md}} (MD_j - ARS_j)^2 \tag{14}$$

$$\frac{\partial E}{\partial \alpha_k} = 2\sum_{j=1}^{N_{md}} \left[(MD_j - ARS_j) \left(\frac{\partial MD_j}{\partial \alpha_k} - \frac{\partial ARS_j}{\partial \alpha_k} \right) \right] = 0 \text{ where } k = 1, 2, \cdots, N_{sf} \tag{15}$$

$$\frac{\partial E}{\partial \alpha_k} = \sum_{j=1}^{N_{md}} \left[\left(\sum_{i=1}^{N_{sf}} \alpha_i \Phi_{ij} \right) \left(\Phi_{kj} \right) \right] = \sum_{j=1}^{N_{md}} \left[(MD_j) \left(\Phi_{kj} \right) \right] \text{ where } k = 1, 2, \cdots, N_{sf} \tag{16}$$

$$[\Phi]^T_{N_{md}\times N_{sf}} [\Phi]_{N_{md}\times N_{sf}} \{\alpha\}_{N_{sf}\times 1} = [\Phi]^T_{N_{md}\times N_{sf}} \{MD\}_{N_{md}\times 1} \tag{17}$$

The weight factor that minimizes the square error between ASD and MD can be calculated by Equation (17), using the inverse matrix of $[\Phi]^T[\Phi]$. However, if the N_{sf} is greater than the N_{md}, $[\Phi]^T[\Phi]$ becomes a rank-deficient matrix. In general, the amount of measurement data is limited owing to the cost. To solve this problem, the singular value decomposition (SVD) method is adopted to solve Equation (17). By substituting the weight factor calculated using the SVD method in Equation (1), the structural deformation shape can be estimated.

Based on the stiffness method, the internal force can be determined using the relative rotation and displacement of the nodes attached to the element. Therefore, the internal structural force can be estimated using the estimated deformed shape (EDS). In this study, the internal force of a cable-stayed bridge was estimated by applying EDS to the numerical model as the displacement load.

3. Validation Process

The most representative characteristic of a cable-stayed bridge, which differs from other types of bridges, is the cable connected to girders and pylons. The cable force supports the girder and can be decomposed into vertical and horizontal forces. The horizontal component of the cable force causes a compression force in the girder. However, the girder has a complex moment shape owing to the vertical component of the cable force. Therefore, cable-stayed bridges consist of geometric nonlinearity, such as the cable sag effect, beam–column effect, and large displacement, whereas material nonlinearity arises when bridge elements exceed their individual elastic limits. The nonlinear effects must be considered when analyzing the ultimate behavior of a cable-stayed bridge [31].

However, cable-stayed bridges under normal operating conditions can be analyzed using a linear model. Ren [32] investigated the ultimate behavior of cable-stayed bridges up to failure, considering the material and geometric nonlinearity. The investigation results demonstrated that the behavior of a cable-stayed bridge is affected by geometric nonlinearity when the live load is four times greater than that of the dead load, and it is affected by material nonlinearity when the live load is two times greater than that of the dead load. In general, the dead load of a bridge is larger than the live load. Namely, the nonlinearity of a cable-stayed bridge is sufficiently small to be ignored under normal operating conditions. Therefore, SRALMR based on the superposition method can be applied to cable-stayed bridges.

The validation process of SRALMR for cable-stayed bridges is illustrated in Figure 1. The analysis of cable-stayed bridges is divided into two parts: initial shape analysis considering geometric nonlinearity and linear analysis with live load. The initial shape analysis was performed to determine the optimal cable forces that ensured minimal deformation of the structure under dead load conditions; whereas the linear analysis with the live load was performed to represent the structural behavior generated by the live load. The results of the linear analysis with the live load are used to assume the real response of the structure because the sensors installed after construction are complete and can only measure the response data generated by the live load. In addition, the response data (displacement, strain, and slope) at a limited point from the real response of the structure are utilized as measurement data for deriving the error function. The deformed shape and internal force of the target model are determined by superposing the results of the initial shape analysis and linear analysis with a live load.

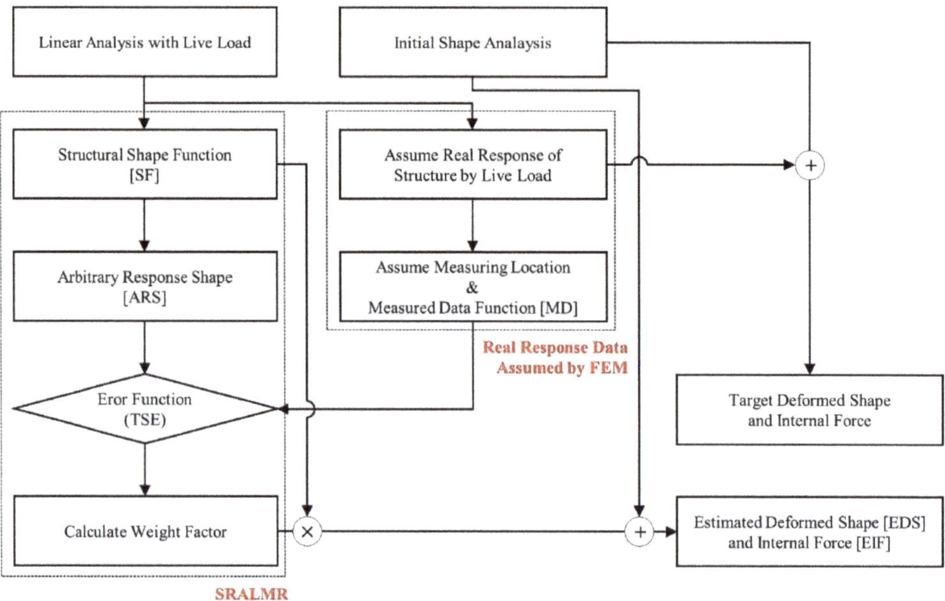

Figure 1. Validation analysis scheme of SRALMR for cable-stayed bridge.

SRALMR is applied to estimate the deformed shape and internal force generated by a live load because the sensor usually measures the response data generated by the live load. Therefore, the SSF and ARS required for estimation are derived from a linear analysis with a live load. The error function is then established using the ARS and MD. The deformed shape and internal force were estimated using the calculated weight factor to minimize the error function. Finally, the SRALMR was validated by comparing the results of the estimated and target models.

4. Numerical Model for Validation

4.1. Validation Model

The radiating-type cable-stayed bridge shown in Figure 2, which has three spans, two pylons, and forty cables, was used as the validation model in this study. The total length of the bridge was 920 m, and the lengths of the three spans were 220, 480, and 220 m, respectively. The height of each pylon was 165 m. The girder was composed of 185 nodes and 184 beam elements, and each pylon had 34 nodes and 33 beam elements. However, a truss element with no compression was used for the cable. Both ends of the girder were in the roller-supported condition, and both bottoms of the pylon were in a fixed condition. At the connections between the girder and pylon, a roller for the girder and free for pylon were applied for the boundary condition. Table 1 presents the materials and geometric properties of the main members. Abaqus 2022 is used for numerical analysis of the validation model. Girder and pylon are modeled by the beam element, and the cable is modeled by a truss element with no compression. In addition, all analyses for the initial shape, structural shape function, and target model are material and geometrically linear.

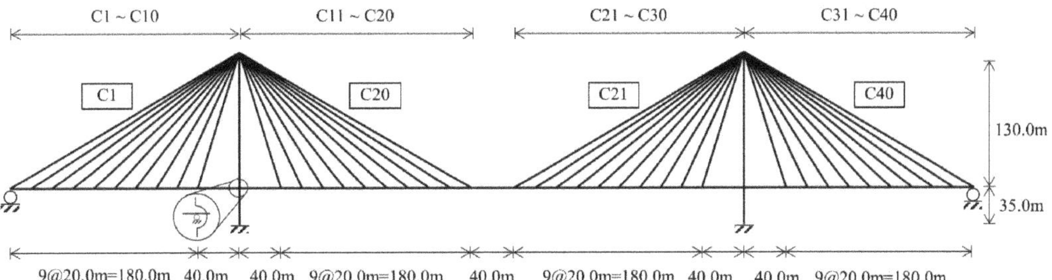

Figure 2. Validation model configuration.

Table 1. Material and geometric properties of main members.

	Girder	Pylon	Cable
Elastic modulus E (kN/m^2)	2.1×10^8	2.1×10^8	2.1×10^8
Sectional A (m^2)	0.749	0.374	0.02
2nd moment of inertia I (m^4)	1.446	3.143	–
Unit weight γ (kN/m^3)	218.27	76.90	76.90

4.2. Initial Shape Analysis

The deformation and internal force of a cable-stayed bridge are critically dependent on the cable force. Therefore, it is important to determine the cable force and initial equilibrium state to minimize the deformation under dead loads. Various methods of initial shape analysis, such as the trial-and-error approach, successive substitution method, and target configuration under dead load (TCUD), have been suggested [33–35]. In this study, the initial force method was used to determine the initial equilibrium state of the validation model. In the initial force method, the geometric nonlinear analysis is repeated considering the internal force of all members at the current iteration as an initial internal force at the next iteration. Figure 3 presents the displacement of the girder center and pylon top according to the number of iterations. The cable forces for the original model are zero, and the appropriate cable forces that satisfy the equilibrium state under a dead load are determined by iteration. As the iterations progressed, the displacement converged to zero. The displacements of iteration 6 were 3.86 cm for the girder center and 0.38 cm for the pylon top. This is negligibly small compared to the span length of 450 m. Therefore, the deformed shape and internal force at iteration 6 were used for the results of the initial shape analysis.

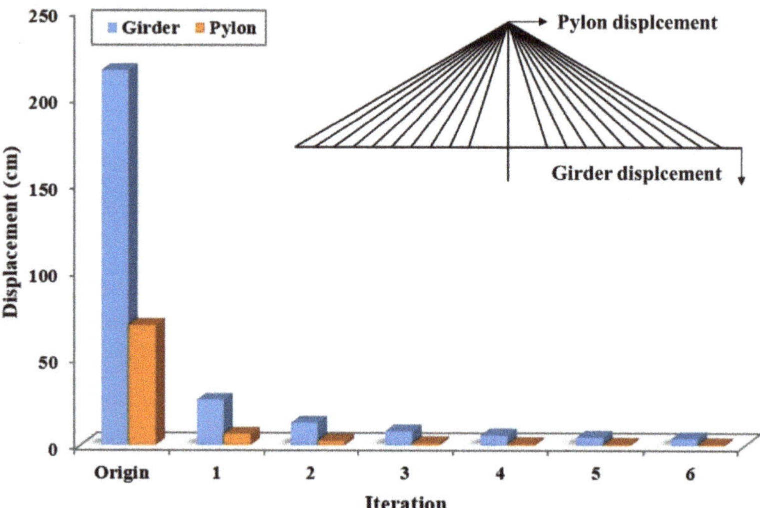

Figure 3. Pylon and girder displacement in each stage of initial shape analysis.

4.3. Target Model

In this study, the target models used for the validation of SRALMR were derived using the numerical model. Figure 4 presents the deformed shape of the target model according to the applied force, and the magnitudes of each force are presented in Table 2. The static load is applied for target models, and the magnitudes of each load are determined for sufficient variation of internal force compared to the initial internal force generated by self-weight. Only one concentrated load is applied to each span for TM1 and TM2. Different loads act on two points for TM3. TM4 is the most complex case, in which three different loads are applied. The displacement, slope, and strain data of each target model were used as the measurement data for the validation of SRALMR.

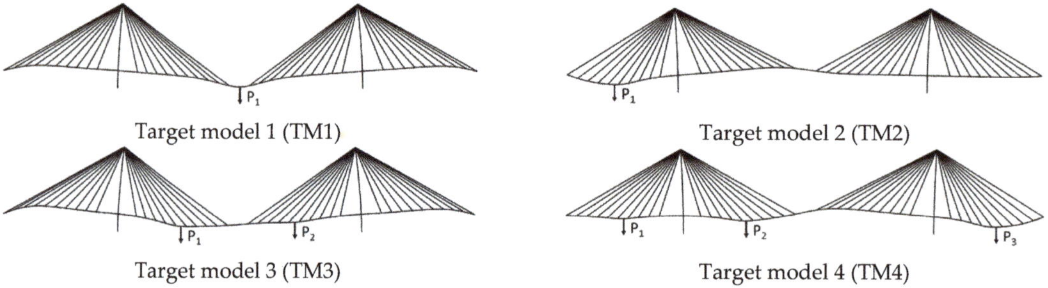

Figure 4. Deformed shape of target model.

Table 2. Details of applied force for target models.

	Force (kN)			
	TM1	**TM2**	**TM3**	**TM4**
P_1	5000	5000	5000	2000
P_2	-	-	3000	2500
P_3	-	-	-	3000

4.4. Structural Shape Function (SSF)

The SSF for the validation model was first derived to estimate the deformed shape and the internal force. The SSF can be derived by applying a unit load on each node of the girder, excluding the boundary conditions. The total number of SSFs was 181. As indicated above, each SSF is composed of displacement, slope, and strain response data. The horizontal and vertical directions were considered for the displacement data, whereas only the longitudinal direction was considered for the slope and strain data. In addition, the displacement and slope data were derived from the nodes, whereas the strain data were derived from the elements. The number of nodes and elements were 253 and 290, respectively. Therefore, the matrix size of the SSF established for the validation model is 1049×181.

4.5. Measurement Location

Each element of the response data of the target model at the limited points was used as measurement data for validation in this study. The location of the measurement data can significantly affect estimation accuracy. Therefore, an effective sensor placement method has been studied. Kammer [36] first introduced the effective independence (EI) method, which maximizes both the spatial independence and signal strength of the shape function by maximizing the determinant of the associated Fisher information matrix. Papadopoulos and Garcia [37] proposed a driving point residue (DPR) coefficient to overcome the drawback of the EI method, which can select a location with a low energy content. However, the EI and EI-DPR methods can only be applied when the number of shape functions is larger than the number of measurement data. Therefore, the EI-DPR-distance method proposed by Byun et al. [30] was adopted to estimate the deformed shape and internal force. Equations (18)–(20) express the EI-DPR-distance method. The distance coefficient d_i is the minimum value among the distances from the ith candidate location to each previously selected sensor location. The ith candidate location with the highest effective independence distribution E_{D_i} was selected as the sensor location, and the distance coefficient was recalculated by considering the previously selected sensor locations. This procedure was repeated until the selected number of locations for each response was equal to the planned number of sensors. The EI-DPR-distance method can overcome the limitations of the EI and EI-DPR methods and consider the effects of different response sensor locations.

$$[E] = [\Phi]\left([\Phi]^T[\Phi]\right)^{-1}[\Phi]^T \tag{18}$$

$$DPR_i = \sum_{j=1}^{N_{sf}} \frac{\Phi_{ij}^2}{|\Phi_j|_{max}} \tag{19}$$

$$E_{D_i} = diagonal([E])_i \times DPR_i \times d_i \tag{20}$$

If the sensor location is determined via the EI-DPR-distance method, the sensor layout case is numerous according to the number of each response sensor due to the distance coefficient. For this reason, the only example of measurement location using one displacement sensor and seven displacement sensors is represented in Figures 5 and 6, respectively. As shown in Figures 5a and 6a, the measurement location of the slope is different due to the measurement location of the displacement. In addition, Figure 5b,c shows the different measurement locations of the strain due to the measurement location of the slope. All combinations of response data are used for deformed shape validation and only three combinations of response data are used for internal force validation.

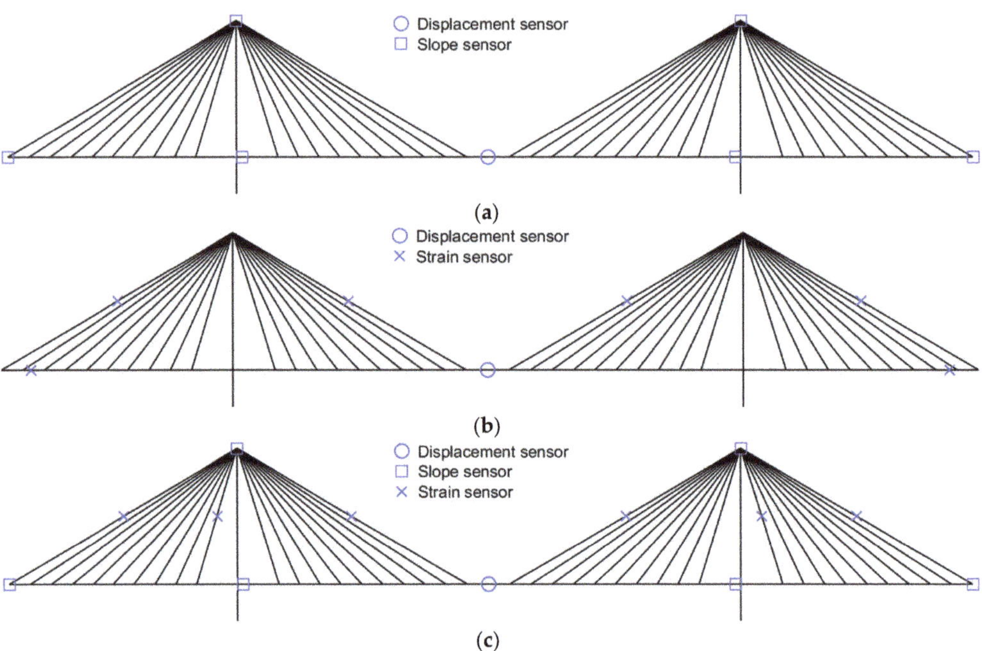

Figure 5. Example of measurement location using 1 displacement sensor. (**a**) $\omega = 1$, $ø = 6$, $\varepsilon = 0$; (**b**) $\omega = 1$, $ø = 0$, $\varepsilon = 6$; (**c**) $\omega = 1$, $ø = 6$, $\varepsilon = 6$.

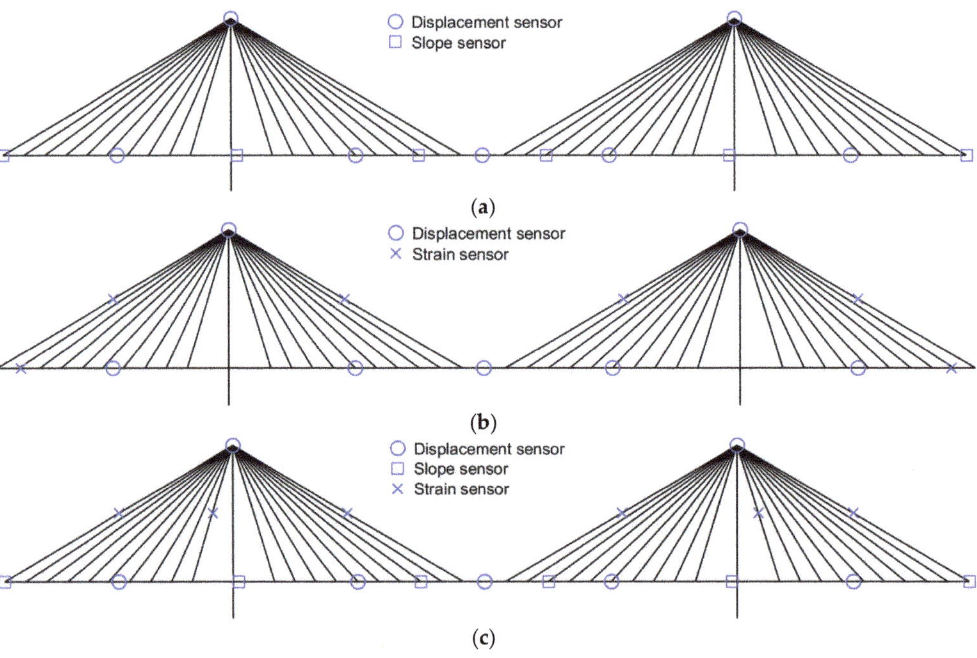

Figure 6. Example of measurement location using 7 displacement sensors. (**a**) $\omega = 7$, $ø = 6$, $\varepsilon = 0$; (**b**) $\omega = 7$, $ø = 0$, $\varepsilon = 6$; (**c**) $\omega = 7$, $ø = 6$, $\varepsilon = 6$.

5. Validation Results

In this study, the application of SRALMR was validated using a numerical model of a cable-stayed bridge. The deformed shape and internal forces (girder axial force, girder moment, and cable force) were estimated from limited multi-response data. The estimated response shape (ERS) was then compared with the target response shape (TRS). The accuracy of the estimation according to the number of measurement data was analyzed using the normalized mean absolute percent error (NMAPE) represented in Equation (21), where N_{dof} is the number of DOFs.

$$NMAPE(\%) = \frac{100}{N_{dof}} \sum_{i=1}^{N_{dof}} \frac{|TRS_i - ERS_i|}{|TRS|_{max}} \tag{21}$$

5.1. Deformed Shape

In this section, the deformed shapes for TM1-4 are estimated using the SSF and EI-DPR-distance methods to identify the internal force. The parameter for estimating the deformed shape was the number of measurement data points for each response. The number of measurement data varied from one to seven for the displacement and zero to six for the slope and strain. The number of slope and strain data points increases equally. Based on the EI-DPR-distance method, the position of the displacement sensors was first determined by considering the location of the boundary conditions, and the position of the slope sensors was determined according to the position of the displacement sensors. Finally, the strain sensors were placed considering the displacement and slope sensors.

Figure 7 presents the NMAPE of EDS according to the number of measurements. The NMAPE is calculated considering the vertical and horizontal deformed values of all the nodes. The maximum NMAPEs are 11.04% for TM1, 11.10% for TM2, 11.51% for TM3, and 13.51% for TM4. The difference in the maximum NMAPE for each TM is not large because the deformed shape of the pylon can be estimated using only one displacement data point on the girder center. The NMAPE of the EDS decreased as the number of measurement data points increased. The estimation results demonstrated that displacement data, including vertical and horizontal values, are the most effective estimation responses. However, slope and strain data, including only one-dimensional values, can also be used to enhance the estimation accuracy. Namely, slope and strain data can reduce the number of expensive GNSSs required to estimate the proper deformed shape. Figure 8 presents the position of the sensors for each response and the enhancement of the estimation accuracy by the slope and strain data. The target deformed shape (TDS), EDS with only one displacement data, and EDS with one displacement data, six slope data, and six strain data are compared in Figure 8.

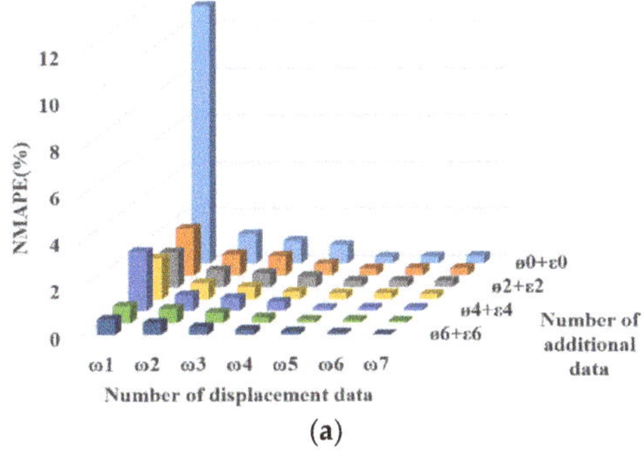

(a)

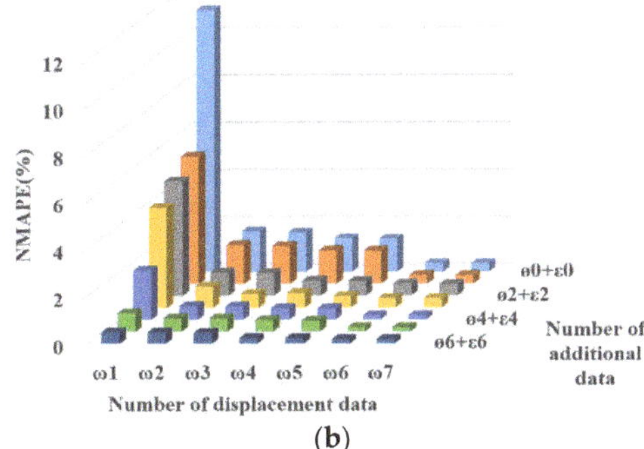

(b)

Figure 7. *Cont.*

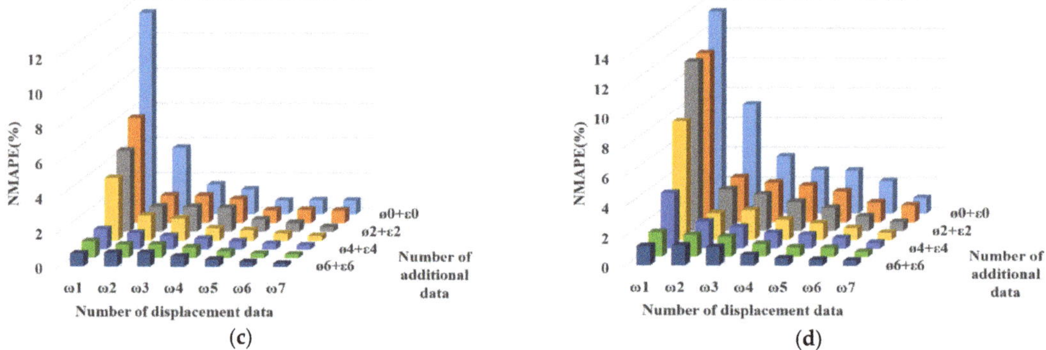

Figure 7. NMAPE of the EDS according to the number of measurement data. (**a**) TM1; (**b**) TM2; (**c**) TM3; (**d**) TM4.

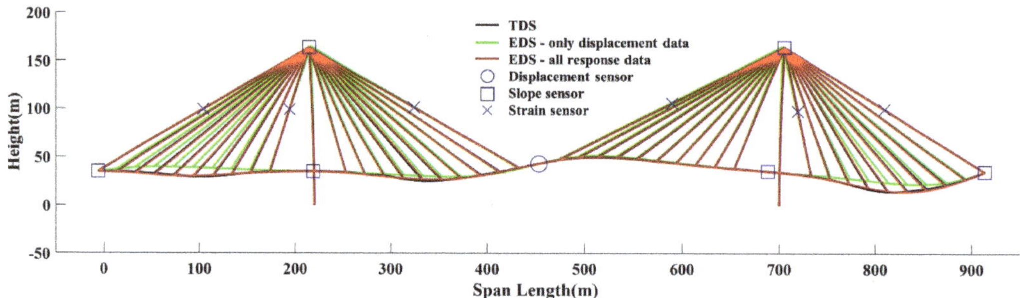

Figure 8. Example of EDS for TM4.

5.2. Internal Force

The displacement value at a certain point of the structure has generally been used to indirectly evaluate the performance of a structure by comparing it with a limited standard displacement. Therefore, most studies have focused on the estimation of displacement at a certain point, such as the girder center, using other response data. However, if internal forces such as axial force and moment can be estimated, the performance of the structure can be more precisely evaluated by comparing the internal force with the section strength. Based on the stiffness method, the internal force of the structure can be derived using displacement and rotation. The internal forces of each element were determined by using the relative displacement and rotation of the end nodes. The deformed shape was estimated in this study, including the relative displacement at all the nodes of the cable-stayed bridge. Then, the internal forces (girder axial force, girder moment, and cable force) were estimated by applying the deformed shape as a displacement force to the numerical model. The following three cases according to the number of each response data are considered for the estimation of the internal force: case 1 (ω 1, ø 0, ε 0), case 2 (ω 1, ø 6, ε 6), and case 3 (ω 7, ø 6, ε 6). Case 1 represents insufficient data and case 3 represents sufficient data. The results of case 2 were used to validate the effect of the slope and strain data. The displacement measurement location of case 1 is the center of the two-span girder. The measurement locations of case 2 and case 3 are shown in Figures 5c and 6c, respectively.

5.2.1. Girder Axial Force

The girder axial force that significantly occurs owing to the horizontal component of the cable force should be considered to evaluate the performance of cable-stayed bridges. Figure 9 presents examples of the initial, target, and estimated girder axial forces for TM4.

The initial line represents the axial force derived from the initial shape analysis, considering the dead load. The maximum axial force is generated at the junction between the girder and pylon, and the minimum axial force is generated at the center of the girder.

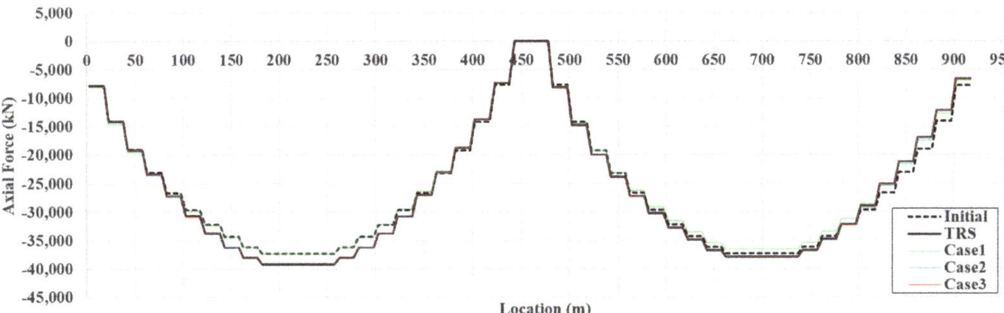

Figure 9. Example of estimated girder axial force for TM4.

Figure 10 presents the normalized absolute percentage error (NAPE) of the axial force along the girder span length for TM4. The estimation error of case 2 dramatically decreases compared to that of case 1, and the estimation error of case 3 is almost zero. These results indicate that slope and strain data can be used as additional data to improve the estimation accuracy, and the exact axial force can be estimated if sufficient multi-response measurement data are provided. The tendency for improvement in the estimation accuracy is represented equally in Table 3. For all cases and TMs, the girder axial force of the cable-stayed bridge can be properly estimated using SRALMR.

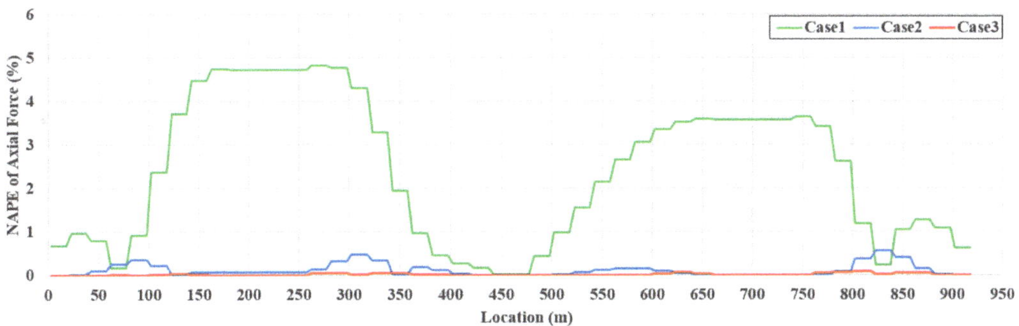

Figure 10. NAPE of estimated girder axial force for TM4.

Table 3. NMAPE of estimated girder axial force for each TM and case.

	TM1	TM2	TM3	TM4
Case 1	1.471	2.275	2.819	2.382
Case 2	0.332	0.074	0.157	0.115
Case 3	0.002	0.021	0.022	0.019

5.2.2. Girder Moment

A girder moment is an important factor that should be considered when evaluating the performance of bridges with long spans, such as cable-stayed bridges, where the distribution of girder moment is complex owing to the vertical component of the cable force. An example of the initial, target, and estimated girder moments for TM4 is shown in Figure 11. In the initial shape analysis, the positive moment was mainly large at the girder center, and the negative moment was mainly large at the connection between the girder and pylon. The final girder moment is then determined according to the location and magnitude of the live load.

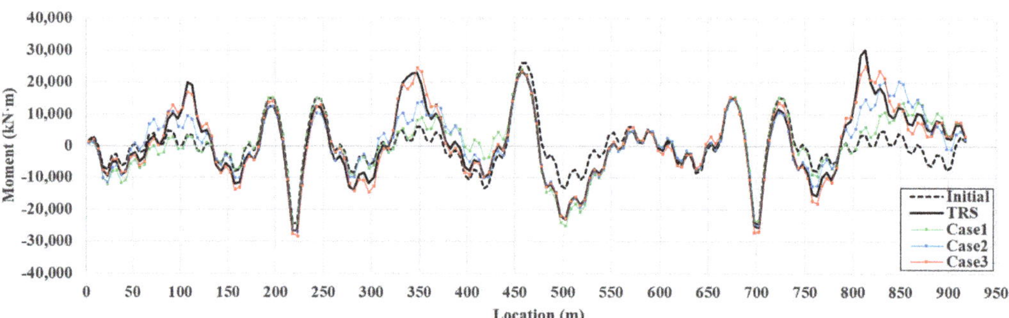

Figure 11. Example of estimated girder moment for TM4.

Figure 12 presents the NAPE of the moment along the girder span length for TM4. Based on the stiffness method, the girder moment was determined by the relative displacement and rotation of the end node at the element. Even a small error in the relative displacement and rotation can cause a large moment-estimation error. SRALMR estimates the deformed shape by minimizing the difference between the ARS and MD only at the locations where the sensors are installed. Additionally, the concentrated load applied to the target model caused a large relative displacement and rotation near the location where the load was applied. Therefore, the estimation error for the moment is generally larger than that for the axial force, and a large error occurs, particularly at the load locations. The maximum NAPEs of the girder moment for TM4 were 80.15, 52.45, and 18.98% for case 1, case 2, and case 3, respectively. The addition of measurement data significantly decreased the estimation accuracy.

The NMAPEs for all cases and TMs are presented in Table 4. The estimated moments for TM1 and TM2 considering one load were more accurate than those for TM3 and TM4 considering multiple loads because the moment distributions of TM1 and TM2 are simpler than those of TM3 and TM4, respectively. In addition, using additional slope and strain data, case 2 can estimate the girder moment more accurately than case 1 for all TMs, and the estimation error of case 3 using a sufficient number of response data is lower than 5%. Therefore, SRALMR can be applied to estimate the girder moment of a cable-stayed bridge if a sufficient amount of response data can be used.

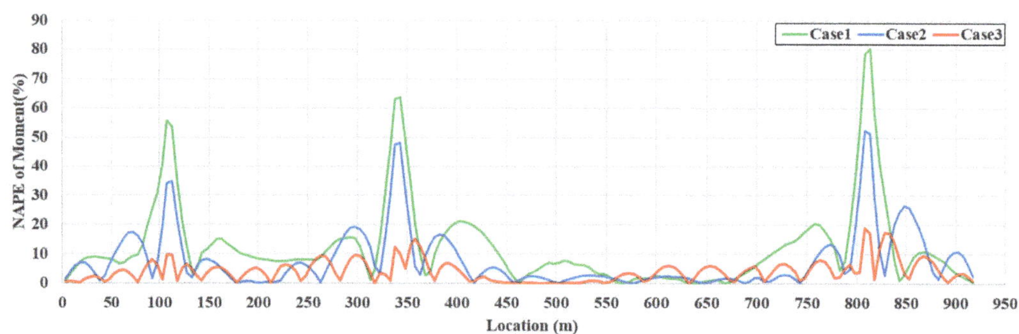

Figure 12. NAPE of estimated girder moment for TM4.

Table 4. NMAPE of estimated girder moment for each TM and case.

	TM1	TM2	TM3	TM4
Case1	6.024	7.127	12.891	12.410
Case2	2.240	3.261	6.647	8.035
Case3	0.345	2.238	3.458	4.164

5.2.3. Cable Axial Force

A cable that allows the bridge to have a long span is the main characteristic of a cable-stayed bridge. Owing to this characteristic, the cable suffers a significant tension force and should be considered when evaluating the performance of cable-stayed bridges. The cable force that suffers only from tension can be easily determined using the relative displacement at both ends of the cable. Therefore, the deformation of the girder and pylon should be estimated precisely to determine the proper cable axial force.

Figure 13 represents examples of the initial, target, and estimated cable forces for TM4. The target and estimated cable forces were determined by superposing the cable forces generated by the dead and live loads. Owing to the dead load, the cable forces near both ends of the girder and the girder center are relatively large. The NMAPE for each case for TM4 is shown in Figure 14. A large estimation error of approximately 10% occurred at a certain cable for case 1. However, the estimation error for case 2 dramatically decreases compared to that of case 1, and the estimation error of case 3 is almost zero. Similar to the girder axial force and moment, strain and slope data can be used as additional data to improve the estimation accuracy for the cable force. In addition, the exact cable force can be determined if sufficient multi-response data are provided. The tendency for improvement in the estimation accuracy is represented equally in Table 5.

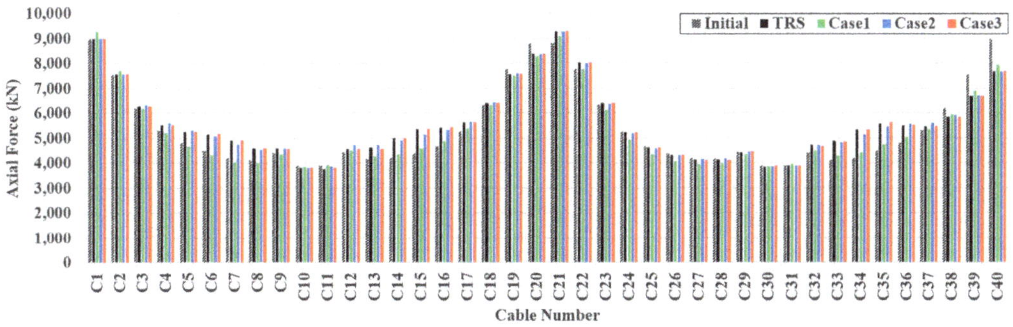

Figure 13. Example of estimated cable axial force for TM4.

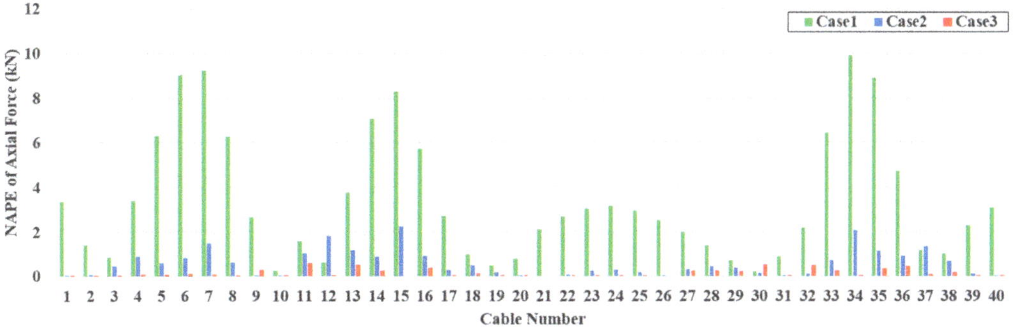

Figure 14. NAPE of estimated cable axial force for TM4.

Table 5. NMAPE of estimated cable axial force for each TM and case.

	TM1	TM2	TM3	TM4
Case1	2.905	3.320	3.495	3.406
Case2	0.773	0.335	0.615	0.580
Case3	0.008	0.143	0.132	0.146

6. Conclusions

In this study, the applicability of SRALMR was extended to the cable-stayed bridge that has complex structural behavior beyond the beam and truss previously validated in [30]. To apply SRALMR to the cable-stayed bridge, an analysis scheme including initial shape analysis was proposed, and performance was verified with four numerical target models. The deformed shape of the cable-stayed bridge was estimated according to the various combinations of limited displacement, slope, and strain data. Then, the girder axial force, girder moment, and cable force were determined for three sensor layout cases using the deformed shape as the displacement force. The numerical validation can be summarized as follows:

1. The deformed shape of the cable-stayed bridge can be well estimated by SRALMR using various combinations of displacement, slope, and strain data. In addition, estimation results show that slope and strain data can enhance the estimation accuracy and reduce the required number of displacement data.
2. From the deformed shape estimated by SRALMR, internal force (girder axial force, girder bending moment, cable axial force) can be properly determined according to the limited amount of response data. A greater amount of used response data enhances the accuracy of internal force estimation.

The results of this study evidence the applicability of SRALMR to the cable-stayed bridges that have complex behavior compared to beams and trusses. However, future studies for validation using experimental data in laboratory and on-site schemes are required to apply SRALMR to the real structure.

Author Contributions: Supervision, Y.-J.K.; Writing original draft, N.B.; Writing—review and editing, J.L. and J.-Y.W. All authors have read and agreed to the published version of the manuscript.

Funding: This work was supported by a National Research Foundation of Korea (NRF) grant funded by the Korea Government (MIST) (No. 2020R1A2C2014450).

Institutional Review Board Statement: Not applicable.

Informed Consent Statement: Not applicable.

Data Availability Statement: The data presented in this study are available on request from the corresponding author.

Conflicts of Interest: The authors declare no conflict of interest.

References

1. Jang, S.; Jo, H.; Cho, S.; Mechitov, K.; Rice, J.A.; Sim, S.; Jung, H.; Yun, C.; Spencer, B.F.J.; Agha, G. Structural Health Monitoring of a Cable-Stayed Bridge Using Smart Sensor Technology: Deployment and Evaluation. *Smart Struct. Syst.* **2010**, *6*, 439–459. [CrossRef]
2. Cho, S.; Jo, H.; Jang, S.; Park, J.; Jung, H.; Yun, C.; Spencer, B.F.J.; Seo, J. Structural Health Monitoring of a Cable-Stayed Bridge Using Wireless Smart Sensor Technology: Data Analyses. *Smart Struct. Syst.* **2010**, *6*, 461–480. [CrossRef]
3. Zhang, L.; Qiu, G.; Chen, Z. Structural Health Monitoring Methods of Cables in Cable-Stayed Bridge: A Review. *Measurement* **2021**, *168*, 108343. [CrossRef]
4. Entezami, A.; Sarmadi, H.; Behkamal, B.; Mariani, S. Big Data Analytics and Structural Health Monitoring: A Statistical Pattern Recognition-Based Approach. *Sensors* **2020**, *20*, 2328. [CrossRef]
5. Son, H.; Pham, V.-T.; Jang, Y.; Kim, S.-E. Damage Localization and Severity Assessment of a Cable-Stayed Bridge Using a Message Passing Neural Network. *Sensors* **2021**, *21*, 3118. [CrossRef]

6. Lee, Y.; Park, W.J.; Kang, Y.J.; Kim, S. Response pattern analysis-based structural health monitoring of cable-stayed bridges. *Struct. Control. Health Monit.* **2021**, *28*, e2822. [CrossRef]
7. Oh, B.K.; Glisic, B.; Kim, Y.; Park, H.S. Convolutional neural network-based wind-induced response estimation model for tall buildings. *Comput. Aided Civ. Infrastruct. Eng.* **2019**, *34*, 843–858. [CrossRef]
8. Wu, R.-T.; Jahanshahi, M.R. Deep Convolutional Neural Network for Structural Dynamic Response Estimation and System Identification. *J. Eng. Mech.* **2019**, *145*, 04018125. [CrossRef]
9. Pamuncak, A.P.; Salami, M.R.; Adha, A.; Budiono, B.; Laory, I. Estimation of structural response using convolutional neural network: Application to the Suramadu bridge. *Eng. Comput.* **2021**, *38*, 4047–4065. [CrossRef]
10. Park, K.-T.; Kim, S.-H.; Park, H.-S.; Lee, K.-W. The Determination of Bridge Displacement Using Measured Acceleration. *Eng. Struct.* **2005**, *27*, 371–378. [CrossRef]
11. Lee, H.S.; Hong, Y.H.; Park, H.W. Design of an FIR Filter for the Displacement Reconstruction Using Measured Acceleration in Low-Frequency Dominant Structures. *Int. J. Numer. Methods Eng.* **2010**, *82*, 403–434. [CrossRef]
12. Park, J.W.; Sim, S.H.; Jung, H.J.; Jr, B. Development of a Wireless Displacement Measurement System Using Acceleration Responses. *Sensors* **2013**, *13*, 8377–8392. [CrossRef] [PubMed]
13. Hou, X.; Yang, X.; Huang, Q. Using Inclinometers to Measure Bridge Deflection. *J. Bridge Eng.* **2005**, *10*, 564–569. [CrossRef]
14. Foss, G.C.; Haugse, E.D. Using Modal Test Results to Develop Strain to Displacement Transformation. In Proceedings of the 13th International Modal Analysis Conference, Nashville, TN, USA, 13–16 February 1995; pp. 112–118.
15. Shin, S.; Lee, S.-U.; Kim, Y.; Kim, N.-S. Estimation of Bridge Displacement Responses Using FBG Sensors and Theoretical Mode Shapes. *Struct. Eng. Mech.* **2012**, *42*, 229–245. [CrossRef]
16. Cho, S.; Yun, C.-B.; Sim, S.-H. Displacement Estimation of Bridge Structures Using Data Fusion of Acceleration and Strain Measurement Incorporating Finite Element Model. *Smart Struct. Syst.* **2015**, *15*, 645–663. [CrossRef]
17. Kang, L.-H.; Kim, D.-K.; Han, J.-H. Estimation of dynamic structural displacements using fiber Bragg grating strain sensors. *J. Sound Vib.* **2007**, *305*, 534–542. [CrossRef]
18. Rapp, S.; Kang, L.-H.; Han, J.-H.; Mueller, U.C.; Baier, H. Displacement field estimation for a two-dimensional structure using fiber Bragg grating sensors. *Smart Mater. Struct.* **2009**, *18*, 025006. [CrossRef]
19. Li, L.; Zhong, B.-S.; Li, W.-Q.; Sun, W.; Zhu, X.-J. Structural shape reconstruction of fiber Bragg grating flexible plate based on strain modes using finite element method. *J. Intell. Mater. Syst. Struct.* **2017**, *29*, 463–478. [CrossRef]
20. Deng, H.; Zhang, H.; Wang, J.; Zhang, J.; Ma, M.; Zhong, X. Modal learning displacement–strain transformation. *Rev. Sci. Instrum.* **2019**, *90*, 075113. [CrossRef]
21. Kliewer, K.; Glisic, B. A Comparison of Strain-Based Methods for the Evaluation of the Relative Displacement of Beam-Like Structures. *Front. Built Environ.* **2019**, *5*, 118. [CrossRef]
22. Park, J.-W.; Sim, S.-H.; Jung, H.-J. Displacement Estimation Using Multimetric Data Fusion. *IEEE/ASME Trans. Mechatron.* **2013**, *18*, 1675–1682. [CrossRef]
23. Cho, S.; Sim, S.; Park, O.; Lee, J. Extension of indirect displacement estimation method using acceleration and strain to various types of beam structures. *Smart Struct. Syst.* **2014**, *14*, 699–718. [CrossRef]
24. Cho, S.; Park, J.-W.; Palanisamy, R.P.; Sim, S.-H. Reference-Free Displacement Estimation of Bridges Using Kalman Filter-Based Multimetric Data Fusion. *J. Sens.* **2016**, *2016*, 1–9. [CrossRef]
25. Sarwar, M.Z.; Park, J.-W. Bridge Displacement Estimation Using a Co-Located Acceleration and Strain. *Sensors* **2020**, *20*, 1109. [CrossRef]
26. Won, J.; Park, J.-W.; Park, J.; Shin, J.; Park, M. Development of a Reference-Free Indirect Bridge Displacement Sensing System. *Sensors* **2021**, *21*, 5647. [CrossRef]
27. Zhang, Q.; Fu, X.; Sun, Z.; Ren, L. A Smart Multi-Rate Data Fusion Method for Displacement Reconstruction of Beam Structures. *Sensors* **2022**, *22*, 3167. [CrossRef]
28. Choi, J.; Lee, K.; Lee, J.; Kang, Y. Evaluation of quasi-Static Responses Using Displacement Data from a Limited Number of Points on a Structure. *Int. J. Steel Struct.* **2017**, *17*, 1211–1224. [CrossRef]
29. Choi, J.; Lee, K.; Kang, Y. Quasi-Static Responses Estimation of a Cable-Stayed Bridge from Displacement Data at a Limited Number of Points. *Int. J. Steel Struct.* **2017**, *17*, 789–800. [CrossRef]
30. Byun, N.; Lee, J.; Lee, K.; Kang, Y.-J. Estimation of Structural Deformed Configuration for Bridges Using Multi-Response Measurement Data. *Appl. Sci.* **2021**, *11*, 4000. [CrossRef]
31. Wang, P.H.; Yang, C.G. Parametric Studies on Cable-Stayed Bridges. *Comput. Struct.* **1996**, *60*, 243–260. [CrossRef]
32. Ren, W.X. Ultimate Behavior of Long-Span Cable-Stayed Bridges. *J. Bridge Eng.* **1999**, *4*, 30–37. [CrossRef]
33. Wang, P.H.; Tseng, T.C.; Yang, C.G. Initial Shape of Cable-Stayed Bridges. *Comput. Struct.* **1993**, *47*, 111–123. [CrossRef]
34. Kim, J.C. Determination of Initial Equilibrium State and Construction Geometry on Cable-Stayed Bridges. Ph.D. Dissertation, Seoul National University, Seoul, Korea, 1999.
35. Kim, K.S.; Lee, H.S. Analysis of Target Configurations Under Dead Loads for Cable-Supported Bridges. *Comput. Struct.* **2001**, *79*, 2681–2692. [CrossRef]
36. Kammer, D.C. Sensor Placement for On-Orbit Modal Identification and Correlation of Large Space Structures. *J. Guid. Control Dyn.* **1991**, *14*, 251–259. [CrossRef]
37. Papadopoulos, M.; Garcia, E. Sensor Placement Methodologies for Dynamic Testing. *AIAA J.* **1998**, *36*, 256–263. [CrossRef]

Article

Assessment of Cracking in Masonry Structures Based on the Breakage of Ordinary Silica-Core Silica-Clad Optical Fibers

Sergei Khotiaintsev [1,*] and Volodymyr Timofeyev [2]

[1] Faculty of Engineering, Universidad Nacional Autonoma de Mexico, Coyoacan, Mexico City 04510, Mexico
[2] Faculty of Electronics, National Technical University of Ukraine "Igor Sikorsky Kyiv Polytechnic Institute", Peremohy Street 37, 03056 Kyiv, Ukraine; v.timofeyev@kpi.ua
* Correspondence: sergeikh@unam.mx; Tel.: +52-55-5468-5757

Abstract: This paper presents a study on the suitability and accuracy of detecting structural cracks in brick masonry by exploiting the breakage of ordinary silica optical fibers bonded to its surface with an epoxy adhesive. The deformations and cracking of the masonry specimen, and the behavior of pilot optical signals transmitted through the fibers upon loading of the test specimen were observed. For the first time, reliable detection of structural cracks with a given minimum value was achieved, despite the random nature of the ultimate strength of the optical fibers. This was achieved using arrays of several optical fibers placed on the structural element. The detection of such cracks allows the degree of structural danger of buildings affected by earthquake or other destructive phenomena to be determined. The implementation of this technique is simple and cost effective. For this reason, it may have a broad application in permanent damage-detection systems in buildings in seismic zones. It may also find application in automatic systems for the detection of structural damage to the load-bearing elements of land vehicles, aircraft, and ships.

Keywords: structural health monitoring; seismic assessment; detection of structural cracks; masonry buildings; optical fiber; distributed optical fiber sensors

Citation: Khotiaintsev, S.; Timofeyev, V. Assessment of Cracking in Masonry Structures Based on the Breakage of Ordinary Silica-Core Silica-Clad Optical Fibers. *Appl. Sci.* **2022**, *12*, 6885. https://doi.org/10.3390/app12146885

Academic Editors: Phong B. Dao, Tadeusz Uhl, Liang Yu, Lei Qiu and Minh-Quy Le

Received: 13 April 2022
Accepted: 3 July 2022
Published: 7 July 2022

1. Introduction

1.1. Earthquakes: A Need for Rapid Assessment of Structural Damage

Earthquakes destroy buildings and cause the loss of human lives in various regions of the world. The crisis situations that follow have many complex aspects [1]. Among them, a quick assessment of damage to buildings and the scope of destruction in the affected region is crucial to making the right decisions. Thus, it is better to have rapidly available, "good enough" analysis, rather than "perfect" information and analysis that comes too late. Late analysis, no matter how good, is of little use in designing immediate life-saving interventions" [2].

Modern buildings are usually quite resistant to damaging factors, but some still suffer from strong earthquakes, the degradation of materials, land sinking, etc. [3,4]. Moreover, many old buildings are exposed to great risks from earthquakes and other damaging factors [5,6]. The rapid detection of partial, frequently non-visible damage to buildings is particularly important since the degree of risk that it poses is high. However, such a risk is not always visible to the occupants of the buildings or to rescue teams. Moreover, timely information allows decision-makers to take appropriate measures, and thus avoid more serious consequences. Permanent seismic structural health monitoring systems can significantly reduce the human and economic losses from earthquakes. Given the millions of public and residential buildings, and the common dwellings in seismic zones, installing permanent structural health-monitoring systems on such buildings is feasible, provided that such systems are simple and cost-effective [7].

This paper presents studies on the suitability and accuracy of detecting structural cracks in brick masonry by exploiting the breakage of ordinary silica optical fibers bonded to its surface. Brick masonry is common in buildings in many locations where baked clay brick is the main building material [8,9]. Furthermore, virtually the same technique can be applied to other masonry structures found worldwide [10,11].

1.2. Methods for Detecting Damage to Civil Engineering Structures

There is a variety of visual, mechanical, electrical, acoustical, computer vision, global dynamic behavior, and other methods that are used for damage detection in civil engineering structures [12–21]. Nevertheless, many are susceptible to the adverse effects of moisture, chemical corrosion, electromagnetic interference, and lightning discharges. These factors make it rather difficult to use these tools on large structures, especially outdoors.

Over the past few decades, optical fiber sensors have reached a high level of sophistication and performance, and they are increasingly used for structural health monitoring (SHM). The advantages of using optical fiber sensors in civil engineering are numerous. Firstly, the optical fibers (shown schematically in Figure 1) are made of dielectric materials, and, therefore, they do not conduct electricity and are immune to electromagnetic interference. Secondly, they are chemically inert, and therefore immune to adverse climatic factors.

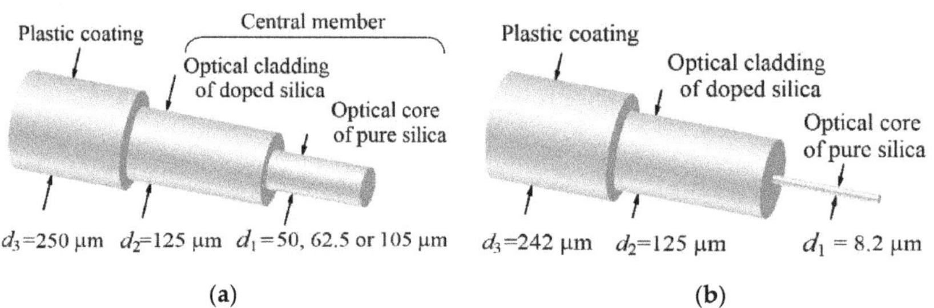

Figure 1. Schematic drawings of ordinary communication-grade silica-core silica-clad optical fibers: (**a**) multimode; (**b**) single mode.

The optical-fiber sensor systems that are most commonly used for SHM employ Optical fiber Bragg gratings (FBGs), optical frequency domain reflectometry (OFDR), Brillouin scattering optical time-domain analysis (BOTDA), and optical fiber Fabry–Perot resonators (FPRs). These optical-fiber sensor systems are highly sensitive to strain and have a distributed sensing capability (for details, see the reviews [22–28] and references therein).

Monitoring strain in structural elements is a useful way to confirm and ensure that bridges, buildings, etc., are indeed behaving as expected. Even in cases of seemingly proper structural behavior, these sensors provide useful information about the degree of coincidence between the predicted and observed behavior of the structure. Moreover, monitoring strain in existing buildings makes it possible to ensure their safety during nearby construction work or other hazardous activities. In addition, monitoring strain in buildings threatened by environmental or man-made factors helps specialists to take timely remedial measures. Some buildings from the past centuries are already equipped with distributed optical fiber sensor systems that provide useful information about the structural behavior of the buildings [29–33].

Moreover, distributed optical fiber sensors provide indirect information about structural damage. Anomalous strain concentrations in the vicinity of cracks indicate the cracking of structural elements [34–40]. This allows specialists to locate cracks and estimate their width using strain measurements and indirect numerical methods developed for this purpose. Such estimations have been carried out on reinforced concrete (RC) beams [34–38]. However, the maximum width of the cracks evaluated using this technique did not exceed 0.4 mm. In a previous study, structural cracks were detected in masonry, but the relationship between an anomalous strain and the respective crack width was not established [40].

Despite the impressive achievements in assessing the strain and temperature distribution in structures with the help of FBG, BOTDA, OFDR, and FPR-based sensors, the drawbacks of these sensors include the high cost of optical signal analyzers (also called interrogators) and the FBG, OFDR, and FPR gauges (the estimated cost of a high-resolution OTDR is USD 30,000–40,000; the cost of an FBG interrogator is USD 10,000–50,000; the cost of an OFDR interrogator is USD 60,000–100,000; and the cost of a high-performance BOTDA analyzer reaches USD 100,000, with the cost being highly dependent on the specific configuration and technical specifications. In addition, FBG and OFDR sensor systems require special optical fibers with embedded Bragg gratings, which have an estimated cost of USD 500–1000 per 100 m), as well as the need for specially trained personnel to work with these systems and interpret the results of the strain measurements.

Given the high cost and complexity of the said optical fiber sensor systems, FBG, BOTDA, OFDR, and FPR-based sensor systems are mainly used to monitor new original structures and precarious buildings to ensure that they behave properly and safely in real-world conditions [41,42]. For mass application in permanent damage-detection systems, which this work targets, only a much simpler and cheaper method of detecting structural damage is suitable.

1.3. *Cracking of Load-Bearing Structural Elements*

The cracking of the load-bearing structural elements of buildings is a significant indication of structural damage. The degree of danger of a crack is influenced by many factors—the type of building, its structural features, and other factors. Generally, the wider the crack, the greater the structural risk that it poses. There is no generally accepted classification of cracks in buildings. Various manuals and handbooks treat the danger posed by cracks differently [43–45]. In most documents, microcracks and cracks less than 1 mm wide are considered insignificant. According to the manual in [43], cracks from 1 mm to 3 mm wide in masonry present a medium structural risk, while in the case of wider cracks, the structural risk is high. The handbook in [44] states that cracks wider than 3 mm in concrete and masonry pose a serious structural risk and require a detailed structural evaluation. This is in contrast with the classification in [45], which considers that only a few 3-millimeter-wide cracks or a single crack over 5 mm wide in masonry buildings present "serviceability" issues. A stringent crack-width limit provides more safety. For this reason, this work aimed to detect cracks with a width of 1 mm or more, as they are capable of posing a structural hazard in masonry after earthquakes and causing other damaging effects.

1.4. *Optical Fiber Breakage as an Indication of Cracking of Structural Elements*

The appearance of structural cracks can be indicated by the breakage of optical fibers that are attached to the element or embedded in it and cross the crack. A fiber break causes a sharp decrease in the intensity of optical radiation passing through the fiber, which is easily detected by appropriate instruments. This method was proposed and applied to detect cracks in steel and aluminum components, as well as in composite materials, as early as the 1980s [46,47]. However, it was believed that this method produced poor results in detecting structural damage [48]. This opinion may be the result of the following:

- First, the optical coupling between the two parts of a broken fiber depends not only on their axial displacement but also on many random factors, including the angle between the crack and the fiber axis, the shear displacement of the two parts, and the roughness of their end faces. Therefore, signal loss cannot be uniquely related to the expansion of the crack.
- Second, optical fibers, as a brittle material, break at slightly different structural crack widths each time (this property is discussed in detail in the following sections).
- Third, the pioneering work did not demonstrate a practical cost-effective implementation of this method. These might be the three main reasons as to why it has never been used to detect structural damage in buildings.

1.5. Ultimate Strength of Optical Fibers

The ultimate strength is the maximum stress that a material can withstand before rupture. The ultimate tensile strength of optical fibers varies among supposedly identical specimens. As with any brittle material, the ultimate tensile strength of optical fibers of silica depends on the size and distribution of flaws, which are random [49,50]. Therefore, a breakage of an optical fiber cannot be attributed to a single strain or a single structural crack opening. Instead, a probabilistic rather than a deterministic definition describes the width of a structural crack when the crack is detected through a breakage of an optical fiber [49,51].

It should be noted that silica optical fibers are subjected to a tensile proof test at a specific stress of σ_{pt} = 0.69 GPa (100 kpsi) at production facilities. That is, the silica optical fibers available on the market are guaranteed to withstand a stress of σ_{pt} = 0.69 GPa. This stress corresponds to a tensile force of F = 8.6 N in the case of a silica-core silica-clad optical fiber with a cladding diameter of d_2 = 125 um. Furthermore, the ultimate tensile strain, ε_u, of thin silica optical fibers is much higher than that of brick masonry, with their values being about 0.035 and 0.0001, respectively [52,53]. That is, masonry cracks are under a much lower strain than the silica optical fibers bonded to the masonry. Therefore, the optical fiber bonded to masonry initially stretches under masonry deformation and the onset of small structural cracks. It then breaks at a certain crack width, δ_c. A particular crack width, δ_c, depends on the ultimate strain of the optical fiber, ε_u; the bond between the silica cladding and the plastic coating; the coating material and its stiffness; the bond between the coating and the binder; and the stiffness of the binder [54–58].

1.6. The Authors' Previous Research

The authors previously investigated the possibility of detecting cracks in RC beams using fiber breaks [59–61]. Several embedding techniques were tested and twelve types of optical fibers were studied both in terms of their survival rate and their abilities to detect cracks. Large-diameter optical fibers successfully detected the progressive cracking of the RC beams under increasing load up to the ultimate failure of the beams. However, the correlation between fiber breakage and crack width could not be established due to the impossibility of assessing the width of internal cracks in a concrete beam.

Moreover, the prospects of detecting structural cracks in brick masonry using the present method were investigated in [62]. In this work, optical fibers were bonded to separate baked clay bricks that were pressed tightly together and laid on a platform. Then, the bricks were pushed apart at a strain rate of 0.25/min until the optical fiber broke [63]. Optical fibers of silica core, silica cladding, various coating materials, and various adhesives were tested. The most rigid bonding was achieved in the case of acrylate-coated optical fibers bonded to bricks with an epoxy adhesive. Further progress was described in the conference publication [64], which reported the detection of cracks with a minimum width of 1 mm in masonry using optical fibers.

1.7. Objectives

The general objective of this work was to evaluate the suitability and accuracy of detecting structural cracks in brick masonry by exploiting the breakage of ordinary silica optical fibers bonded to its surface. Of particular interest were the behavior of optical signals upon the deformation and cracking of the test specimen and the relationship between fiber breakage and the crack width that caused the breakage.

The specific objectives included the choice of the most suitable optical fiber for this application, the fabrication of a brick masonry test specimen, the instrumentation of the specimen with an array of optical fibers, the design and implementation of the monitoring system, and the load testing of the sample. Moreover, the specific objectives included analyses of the results: deformations of the specimen under load, the behavior of the optical signals under the deformation and cracking of the test specimen, and a statistical and probabilistic evaluation of the results.

2. Materials and Methods

2.1. Choosing the Right Optical Fiber

There are a variety of silica-core silica-clad optical fibers of different core and cladding diameters available on the market. The silica core and cladding constitute a monolithic silica member. The tensile strength of a silica-core silica-clad optical fiber depends on the overall diameter of the silica member, d_2, as well as on the ultimate stress of the fiber, σ_u. In practice, the tensile strength of 20 m fiber samples averages about 4 GPa [65]. The average ultimate strength, σ_u, increases with a decrease in the length of fiber samples, since the probability of a flaw in a sample is inversely proportional to its length.

However, the type of coating material affects the strength of optical-fiber bonding to a structural element. Based on the above considerations and previous research [61], and with the aim of detecting structural cracks in the mm range [43,44], acrylic-coated optical fibers of the smallest available cladding diameter d_2 = 125 um were chosen for this study. Among the fibers in this group, the optical fibers with the largest available core diameter d_1 = 105 um were chosen and used (a large core diameter facilitates the efficacious coupling of inexpensive light sources—light-emitting diodes (LEDs) and optical fibers). The optical fibers chosen had an acrylate coating with a diameter of d_3 = 250 um, numerical aperture of NA = 0.22, and optical attenuation of α = 15 dB/km @ λ = 633 nm.

The suitability of these optical fibers for the present application was verified by a simple tensile test, similar to that in [62], on 20 samples of these optical fibers. The test consisted of bonding an optical fiber sample to two separate ceramic bricks and pushing the two bricks apart at a strain rate of 0.25/min until the breakage of the fiber [63]. This led to a steep decrease in the pilot optical signal transmitted through the fiber to almost zero. The test yielded a minimum detected crack of about 1 mm wide, a mean structural crack width of M = 2.34 mm, a variance of σ^2 = 0.79 mm^2, and a standard deviation of σ = 0.89 mm (see Figure 2). (The zero probability of failure at a structural crack width range of $0 < \delta_c$ 1 mm is due to the screening out of optical fibers that failed the proof tests of σ_{pt} = 0.7 GPa at the production facility.) These cracks fitted a range of 1 mm to 5 mm wide, which corresponds to a medium-to-high structural hazard for masonry [43,44]. Therefore, the above-named optical fiber was used in this work.

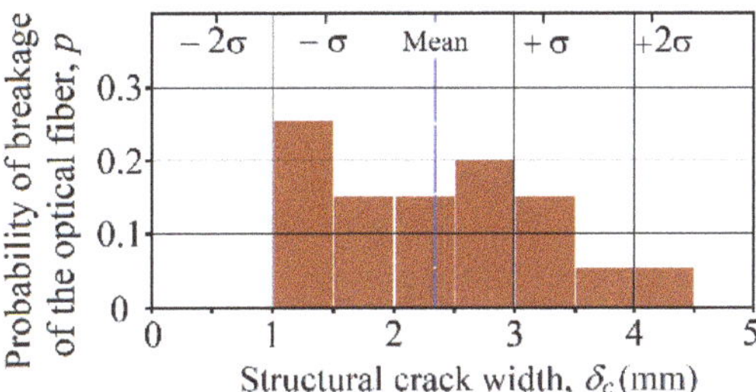

Figure 2. Probability of breakage of the optical fiber, p, as a function of structural crack width that causes the crack, δ_c, obtained by performing a simple tensile test and determined for sequential intervals of a width of 0.5 mm.

2.2. Test Specimen

A clay brick masonry specimen was designed, fabricated, instrumented with the optical fibers, and subjected to a vertical load of the magnitude encountered in earthquakes. It was a stack with dimensions of 520 (W) $\times$ 520 (D) $\times$ 510 (H) mm^3 of baked clay bricks of 60 $\times$ 120 $\times$ 240 mm^3 and a sand–cement mortar (proportion 2:1, type II Portland cement and fine pumice sand). All the aforesaid components are typical of the clay brick masonry elements of dwellings and other brick masonry structures in central Mexico. To achieve a uniform compression load distribution on the brick stack, two RC platens, each about 100 mm thick, were built on the top and bottom of the stack. There were five reinforcing steel bars of diameter $3/4''$ in each platen. The total height of the stack with two platens was 710 mm. The stack is shown in Figure 3a–c.

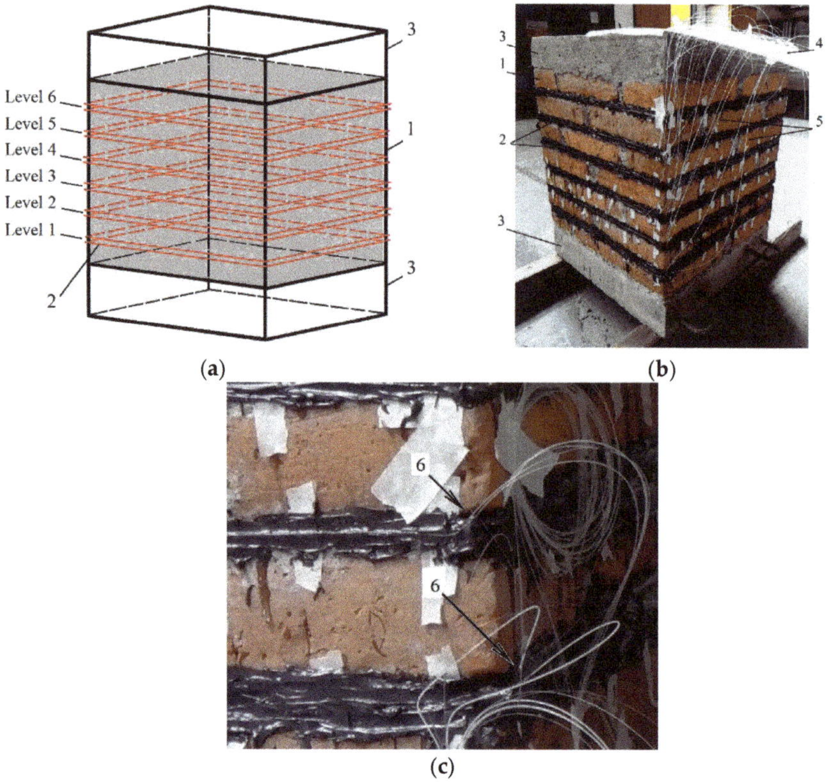

Figure 3. Brick masonry specimen instrumented with the optical fibers: (**a**) schematic; (**b**) photograph; (**c**) an enlarged view of the points of connection of optical fibers to the specimen (photos). (1) Brick masonry specimen; (2) segments of optical fibers bonded to masonry that serve as crack detectors (not visible under opaque epoxy adhesive); (3) RC platen; (4) tray with free ends of the optical fibers; (5) free ends of the optical fibers; (6) points of connection of the optical fibers to the masonry specimen.

The mortar of the brick specimen was cured for 28 days. Then the specimen was instrumented with the optical fibers, transported to the load machine, and installed in its test frame, as shown in Figure 4a. Moreover, the specimen was instrumented with eight electrical linear variable displacement transducers (LVDTs), in compliance with compression test methodology generally accepted in structural mechanics. The LVDTs were installed on the four faces of the brick specimen in the horizontal and vertical directions, as shown in Figure 4b,c. The transducers measured the horizontal and vertical deformations of the specimen in the central part of each lateral face at a length of 250 mm.

2.3. Optical Fiber Array

Segments of optical fibers, each about 5 m long, were prepared in advance. The ends of the optical fibers were cut at a right angle, and the quality of the cut and the integrity of the fibers were checked. The optical fibers were placed on a special tray that protected the fibers from possible damage, and they were delivered to the test site. The specimen was instrumented with twelve optical fibers. The optical fibers were bonded to the specimen over its perimeter at several (six) equidistant horizontal planes. A horizontal position for the optical fibers was chosen because cracks in masonry stacks subjected to a compression load arise predominantly in the vertical direction. Two optical fibers were bonded near each other in each of the six planes of the specimen at about a 70 mm distance, as shown schematically in Figure 3a, using a two-component industrial-grade epoxy adhesive.

Figure 4. Brick masonry specimen: (**a**) instrumented with the optical fibers and LVDTs and installed in the test frame of the load machine; (**b**) the specimen front face (A); (**c**) the specimen back face (C). (1) Universal load machine; (2) brick masonry specimen; (3) opto-electronic monitoring system; (4) horizontal LVDT; (5) vertical LVDT.

The optical transmission of all 12 fibers was verified after bonding and again after 24 h. The aim was to identify possible damage to the optical fibers that could occur during bonding to the brick specimen or could be caused by the volumetric shrinkage of the adhesive due to its polymerization. The check was carried out with a common LED lamp. Each optical fiber was illuminated at one end, and the light coming out of the other end was observed. The inspection showed that all the fibers transmitted light normally; that is, none of the fibers were damaged during the process of bonding and polymerization shrinkage.

2.4. Opto-Electronic Monitoring System

To monitor the integrity and breakage of the optical fibers of the crack detection array, a special optoelectronic system was used, the configuration of which is schematically shown in Figure 5.

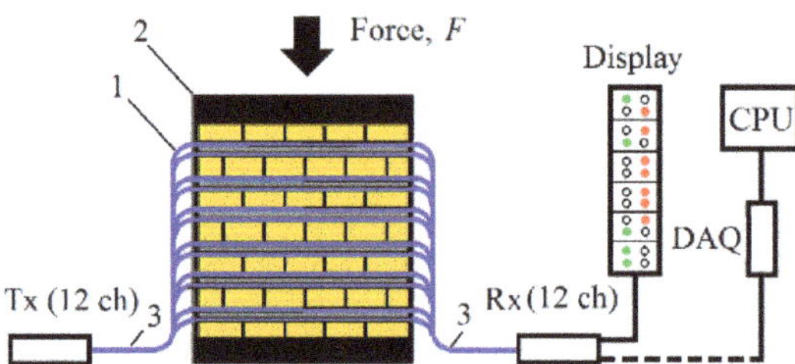

Figure 5. Schematic diagram of the opto-electronic monitoring system: (1) optical fibers; (2) test specimen; (3) array of 12 optical fibers for detection of structural cracks; Tx—optical transmitter, 12-channel; Rx—photoreceiver, 12-channel; DAQ—data acquisition module (multichannel analog-to-digital converter); CPU—computer central processing unit for storing the measured optical transmission in digital form; Display—an indicator of the structural damage in visual form.

The system monitored the intensity of the optical radiation transmitted through all optical fibers of the array by the multichannel optical transmitter, Tx. The breakage of the optical fiber led to a decrease in the signal intensity. The signals of all optical fibers were received by the multichannel optical receiver, Rx, and converted from analog to binary (two-level) form, that is, "1" or "0", by comparing the signal values against a given value. The zero level was interpreted as a structural crack, and the state of the corresponding color indicator on the display changed from green to red.

The optical radiation generated by the transmitter Rx was a continuous 1 kHz sine optical pilot signal. The light sources were LEDs operating at a wavelength of 633 nm (visible red light). Each LED was coupled to one of the optical fibers of the array. The outputs of the optical fibers were connected to the respective photoreceivers. Each measuring channel had a threshold amplitude device, the purpose of which was to automatically signal a fiber break. This device turned on a light indicator on a special display when the signal dropped below a certain level. The threshold was set at 50% of the nominal, as it was assumed that such a threshold would minimize the false alarms of the signaling system caused by possible fluctuations in light sources and all kinds of noise.

For this experiment only, the monitoring system was complemented with a 16-channel data acquisition device (DAQ) and a CPU that stored the data of all measuring channels in digital form. It should be noted that neither a DAQ nor a CPU is needed for a practical crack detection system. Instead, a practical system would require a means of transmitting alarms to the system users. (The cost of all elements and components of this monitoring system is about USD 300. The wholesale cost of the multimode optical fiber is about USD 50 per 1 km.).

2.5. Structural Load Test

The compression load test was carried out by means of a 200 kN universal load machine. The test was carried out in several stages of loading and unloading, until the specimen began to split off (at $F = 1334$ kN). The vertical and horizontal deformations of the specimen were measured by the electrical LVDTs installed in the central part of each lateral face of the specimen over a length of 250 mm.

The optical signals of all fibers were measured continuously by the above monitoring system, stored in the CPU, and post-processed after finalizing the test. The breakage of the optical fibers was easy to observe visually due to the red light emanating from the broken optical fibers. Photographs were taken of the locations where the fiber breakage occurred. The width of the structural cracks causing the fiber breakage was determined by measuring the width of the cracks in the photographs. With this method, the actual width of the cracks may be somewhat smaller than that measured, since the cracks may expand from the time the fiber breaks to the time the photo is taken. Nevertheless, the use of other, more accurate methods to measure the width of structural cracks at the moment of fiber breakage requires a much more complex and expensive infrastructure, and was therefore not possible in this experiment.

3. Results

3.1. Data Overview

The changes in the load and deformation of the specimen, measured by the electrical LVDTs in the central part of each face of the specimen over a length of 250 mm, are represented by graphs in Figure 6. Some differences in the horizontal and vertical deformations were observed on the four lateral faces of the specimen. These can be explained by the heterogeneity of the materials, the geometric asymmetry of the specimen, and eccentric loading. It is difficult to avoid these factors in these types of tests.

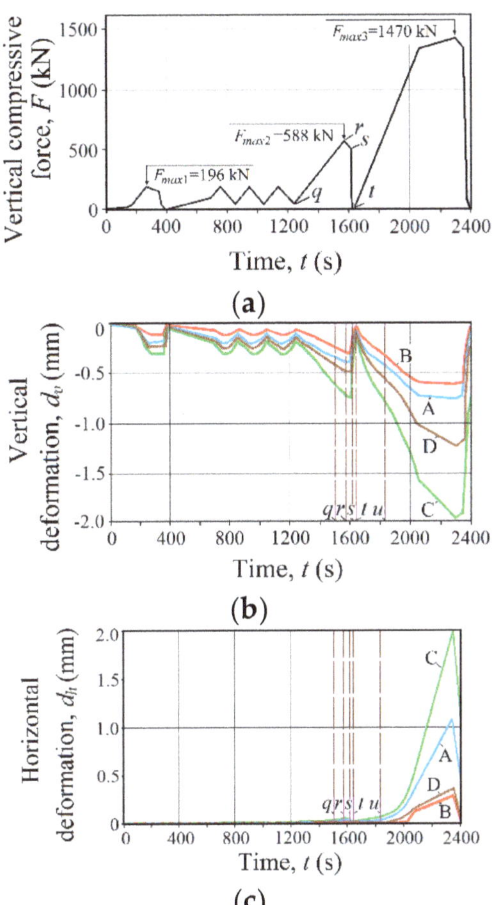

Figure 6. Specimen behavior under load: (**a**) variation in compressive force, F; (**b**) specimen vertical deformation, d_v; (**c**) specimen horizontal deformation, d_h, with time, t, as measured with electrical LVDTs at the center of each specimen lateral face—A, B, C, and D—during the test.

All vertical and horizontal deformations of the specimen returned to zero in the first three cycles of the load after the vertical compressive force F was removed. This indicates an elastic behavior of the specimen under a vertical compressive force of $F \leq 530$ kN. With $F > 530$ kN in the last load cycle, the vertical deformation increased non-linearly with force F. Such a behavior was especially noticeable on faces C and D of the specimen, which showed the change from an elastic to a plastic deformation regime.

It is worth noting that the electrical LVDT sensors did not detect the onset of cracks in the specimen, because they were installed in the center of the four faces of the specimen, while cracks formed on one of its corners. At the same time, the development of structural cracks led to the breakage of the optical fibers crossing the cracks, and accordingly, to a decrease in the output signals of the monitoring system. In some optical fibers, the breakage was observed visually due to the red light coming from the broken optical fibers (Figure 7).

(a)

(b)

Figure 7. Broken optical fibers in locations where they intersect structural cracks: (**a**) completely broken optical fibers; (**b**) optical fibers with broken silica cores and claddings that stay in the intact but stretched elastic coatings.

The plots in Figure 8 show the variation in the applied vertical compressive force on the stack, F, and the respective dimensionless amplitude of the output signal, A, of the twelve optical fibers with time, t. The graphs of the dimensionless amplitude of the signal, A, are displayed in groups of four for levels 1 and 2, 3 and 4, and 5 and 6 of the specimen.

The sharp decrease in the signal amplitude observed in these diagrams was caused by optical fiber breaks, complete or partial, due to the occurrence of structural cracks. There are several characteristic points on the plots in Figure 8, denoted with the lowercase italic letters q, r, s, t, and u. Some optical fibers (1a, 1b, 3a, 3b, 4a, and 4b) broke and consequently decreased their optical transmission at a load of about 400 kN in the fourth load cycle (interval q-r), and then they partially recovered it when the load returned to zero (interval r-s). Other optical fibers (2a, 2b, 5a, 5b, 6a, and 6b) sharply decreased their optical transmission at a larger load of about 620 kN in the fifth load cycle (point u).

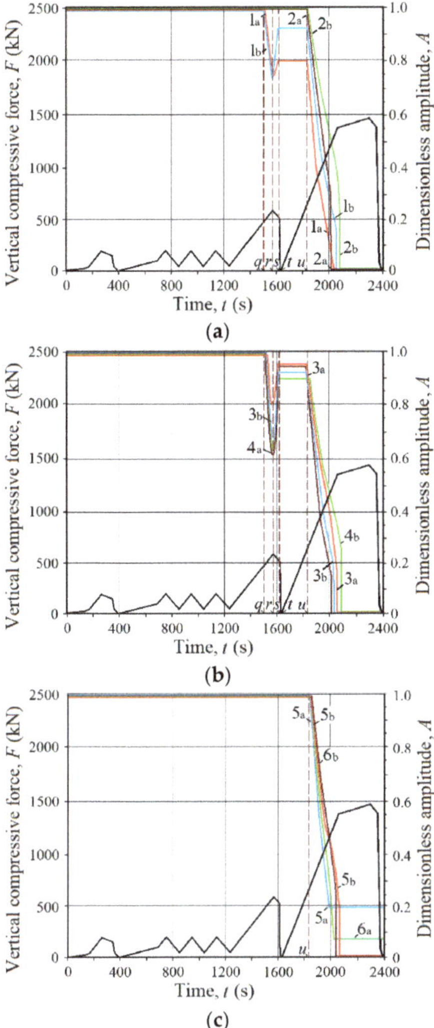

Figure 8. Plots of applied vertical compressive force F (—) and dimensionless amplitude of the output signal A (color lines) of the optical fibers mounted at different specimen levels vs. time t: (**a**) levels 1 and 2; (**b**) levels 3 and 4; (**c**) levels 5 and 6.

The observed differences in the behavior of some of the optical fibers can be attributed to the structural inhomogeneity of the specimen. This led to inhomogeneous cracking of the specimen under load, and the first structural cracks developed in the most strained parts of the specimen. The optical fibers that crossed the structural cracks were stretched, and their silica members were broken, while the elastic plastic coatings were stretched but not broken. The optical transmission decreased but was not completely interrupted (interval *q*-*r*). With the load dropped to zero, the structural cracks partially closed in the generally elastic specimen. The elastic plastic coatings of the optical fibers contracted and brought the end faces of the broken optical fibers back to a closer position. In this way, optical transmission was partially restored in the said optical fibers (point *t*). In the fifth load cycle, the structural cracks in the specimen opened again and continued to grow. At a load exceeding 620 kN, the optical transmission of all optical fibers dropped sharply (point

u). This was caused by the onset and widening of two large vertical structural cracks in the two adjacent lateral faces of the specimen. The width of these cracks varied along the sample height and ranged from 3 mm to 4.6 mm. These cracks caused the breakage of all optical fibers. However, the optical signal of some fibers (5a and 6a) did not vanish completely, as the plastic coatings of these fibers stretched but did not break. As can be seen in Figure 7, the coatings kept both parts of the fibers' broken silica members in a position that provided a finite optical coupling of the two parts and a non-zero amplitude of the optical signal.

3.2. Statistical and Probabilistic Evaluation

Statistical analyses yielded a mean structural crack width of M = 2.68 mm, a variance of σ^2 = 1.12 mm^2, and a standard deviation of σ = 1.06 mm (n = 12). The relationship between the probability of the optical fiber breakage, p, and the width of the structural crack in the test specimen that caused the fiber breakage, δ_c, determined for sequential intervals of a crack width of 0.5 mm is illustrated in Figure 9.

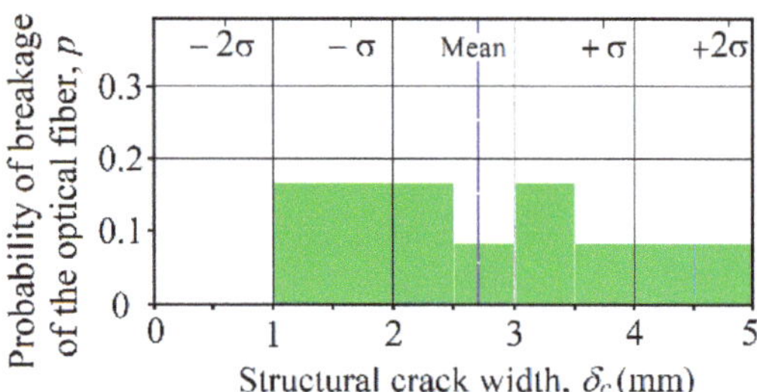

Figure 9. Probability of breakage of the optical fiber, p, as a function of the width of the structural crack that caused the fiber breakage, δ_c, measured in the test specimen of brick masonry and determined for sequential intervals of a width of 0.5 mm.

As can be seen in the diagram, all optical fibers broke at cracks 1–5 mm wide. Half of all of the optical fibers broke at cracks 1–2.5 mm wide, and three-quarters of all optical fibers broke at cracks 1–3.5 mm wide. Cracks of this width present a medium-to-severe structural hazard, which is targeted in this work.

Furthermore, the probability that at least one fiber of an array of n optical fibers breaks due to a crack with a width in the range $\Delta\delta_c$, P_1, is [66]

$$P_1 = 1 - (1 - p)^n \quad (1)$$

where p is the probability of the optical fiber breakage in the range $\Delta\delta_c$ (see Figure 9), and n is the number of optical fibers.

Table 1 shows the probability of breaking at least one fiber in an array of n optical fibers, P_1, calculated for some ranges of crack widths $\Delta\delta_c$. Calculations were made for the detection arrays of 1, 2, 4, 6, 8, 10, and 12 optical fibers.

The data in Table 1 show that the probability of the breakage of at least 1 fiber of an array of 12 optical fibers, due to a crack in different parts of the 1–5 mm range, is high ($0.888 < P_1 < 1.0$). A high probability of the breakage of at least one fiber exists even for rather small cracks 1–1.5 mm wide ($P_1 \cong 0.888$). Reducing the size of the detection array even by a factor of two (to 6 fibers) reduces the probability P_1, but only the probability of detecting the narrowest cracks degrades significantly (P_1 = 0.665@$\Delta\delta_c$ = 1–1.5 mm). The above data illustrate the high probability of detecting cracks with detection arrays of a moderate size, as well as the possibility of trade-off between the size of the detection array and the desired probability of detecting cracks of a certain width.

Table 1. Probability that at least one fiber of an array of n optical fibers breaks, P_1, calculated for several ranges of crack widths $\Delta\delta_c$ and detection arrays of 1, 2, 4, 6, 8, 10, and 12 optical fibers.

Number of Optical Fibers of the Detection Array	The Range of Crack Width in Which the Fiber Break Occurs, $\Delta\delta_c$ (mm)				
n	1–1.5 (p = 0.167)	1–2 (p = 0.333)	1–3 (p = 0.583)	1–4 (p = 0.833)	1–5 (p = 1.0)
	Probability that at least one fiber of an array of n optical fibers breaks, P_1				
1	0.167	0.333	0.583	0.833	1.0
2	0.306	0.556	0.826	0.972	1.0
4	0.518	0.802	0.970	>0.999	1.0
6	0.665	0.912	0.995	>0.999	1.0
8	0.767	0.961	>0.999	>0.999	1.0
10	0.838	0.983	>0.999	>0.999	1.0
12	0.888	0.992	>0.999	>0.999	1.0

4. Discussion

Silica-core silica-clad optical fibers can be installed in new and existing buildings and serve as artificial "nerves" that indicate structural failure, just as nerves indicate pain in the body, during the lifetime of the buildings. This technique is simple and inexpensive, for the following reasons:

- The use of ordinary optical fibers of silica without any built-in special sensing elements, such as gratings and tapers. This contributes to the low cost of the optical fiber detector array.
- The use of multi-mode optical fibers rather than single-mode optical fibers. Multi-mode optical fibers have a large core diameter, which allows the use of inexpensive and reliable LEDs instead of expensive and delicate semiconductor lasers needed to efficiently launch light into the small diameter core of single-mode optical fibers.
- The use of optical fibers with their original polymer coating. The removal of the polymer coating used in the manufacture of FBGs with subsequent recoating compromises the strength of the optical fiber. Keeping the original coating intact contributes to the long service life and reliability of optical fibers.
- The use of a sufficiently large number of optical fibers in the crack detection array. This makes it possible to reliably detect structural cracks of a certain width.
- The use of a binary conversion and interpretation of the optical signals.
- The use of time division multiplexing (TDM) to monitor the optical transmission of all fibers of the crack detection array. TDM is much simpler and cheaper than the optical spectral multiplexing used in distributed FBG and OFDR sensors.

Many of the characteristics of this method, as well as the singularities of its application in buildings of different types, require further collaborative study by experts in civil engineering, and measurement and information technologies. It is preferable to install optical fiber arrays at locations of maximum stress and in the direction of the maximum equivalent strain. This ensures the predominantly perpendicular crossing of the optical fibers and the structural cracks. For a given crack, signal loss with a perpendicular crossing is less than that in other cases. More specifically, the signal loss depends on the angle between the crack trace and the fiber axis. Therefore, the expected geometry of the intersection of a fiber with a crack should be taken into account when placing optical fibers on structures. Typical crack patterns in structures under different loadings are generally well known. Moreover, proper stress and strain fields can be obtained by modeling the stress–strain state in the most loaded areas of the structure using numerical methods, such as the finite element method and artificial neural networks [67,68].

There are many options for informing users of the occurrence of structural risk. The simplest form is to notify users about the appearance of a new crack on the monitored building using special displays and sirens at the facility, automatic calls, etc. More advanced and more efficient forms involve the connection of various damage-monitoring modules to data processing facilities by means of different kinds of communication links or networks [69].

5. Conclusions

In this work, cracks in a masonry specimen were detected by exploiting the breakage of ordinary silica optical fibers bonded to its surface. For the first time, a reliable detection of structural cracks with a width exceeding a given minimum value was obtained despite the random nature of the ultimate strength of the optical fibers. This was achieved using arrays of several optical fibers placed on the structural element. This paper established a relationship between the number of fibers in the detector array and the probability of detecting a structural crack of a given width. A fairly moderate number of optical fibers (about 6 to 12) was sufficient to detect cracks 1–3 mm wide with a probability of more than 0.99, while the probability of detecting cracks with a width of 1–5 mm tended toward unity.

Overall, this work confirmed the suitability of ordinary silica optical fibers for reliably assessing cracks in masonry that poses a structural risk. The simplicity of this technique and the low cost of ordinary optical fibers make it suitable for broad application in permanent damage-detection systems in buildings in seismic zones. This technique can also potentially be applicable in the detection of structural damage to the load-bearing elements of large land vehicles, aircraft, and ships.

Further work is needed to test this method in various scenarios, to better characterize it, and to obtain more data. In addition, our future objectives are to explore the possibility of detecting narrower and wider structural cracks than those detected in this work.

Author Contributions: Conceptualization, methodology, experimentation, and writing—original draft preparation, S.K.; interpretation of the results, data analyses, and validation, V.T. All authors have read and agreed to the published version of the manuscript.

Funding: This study was funded by the General Directorate for Academic Staff Affaires (DGAPA), Universidad Nacional Autonoma de Mexico (grant number IT102021).

Informed Consent Statement: Not applicable.

Acknowledgments: The authors acknowledge Eng. Abraham Roberto Sanchez-Ramirez and Eng. Laura Justina Olivares-Sanchez de Tagle, Institute of Engineering, Universidad Nacional Autonoma de Mexico, for providing structural elements of brick masonry and carrying on the compression tests of the said structural elements, respectively. The authors acknowledge Oleg Kolokoltsev, Institute of Applied Sciences and Technology (ICAT), Universidad Nacional Autonoma de Mexico, Mykola Bobyr, Institute of Mechanical Engineering, National Technical University of Ukraine "Igor Sikorsky Kyiv Polytechnic Institute," Kyiv, Ukraine, and Anastasia K. Lopez, Faculty of Science, University of British Columbia, Vancouver, BC, Canada, for helpful suggestions to the manuscript.

Conflicts of Interest: The authors declare no conflict of interest.

References

1. Jiang, Y.; Yuan, Y. Emergency logistics in a large-scale disaster context: Achievements and challenges. *Int. J. Environ. Res. Pub. Health* **2019**, *5*, 779. [CrossRef] [PubMed]
2. Patrick, J. Haiti earthquake response: Emerging evaluation lessons. *Eval. Insights* **2011**, *1*, 1–14. Available online: https://alnap.org/help-library/haiti-earthquake-response-emerging-evaluation-lessons (accessed on 12 April 2022).
3. Ruiz-Sandoval, M.E.; Rivera, E.; Fernandez-Sola, L.R. Mexico's earthquake: Description, effects, and reconstruction. A particular case. In Proceedings of the International Conference on Critical Thinking in Sustainable Rehabilitation and Risk Management of the Built Environment, Iași, Romania, 7–9 November 2019; CRIT-RE-BUILT 2019; Rotaru, A., Ed.; Springer: Berlin/Heidelberg, Germany, 2019; pp. 47–57. [CrossRef]

4. Tena-Colunga, A.; Hernandez-Ramirez, H.; Godinez-Dominguez, E.A.; Perez-Rocha, L.E.; Grande-Vega, A.; Urbina-Californias, L.A. Seismic behavior of buildings in Mexico City during the 2017 Puebla–Morelos earthquake. *Asian J. Civ. Eng.* **2021**, *4*, 649–675. [CrossRef]
5. Stepinac, M.; Lourenço, P.B.; Atalić, J.; Kišiček, T.; Uroš, M.; Baniček, M.; Šavor Novak, M. Damage classification of residential buildings in historical downtown after the ML5.5 earthquake in Zagreb, Croatia in 2020. *Int. J. Disast. Risk Reduct.* **2021**, *56*, 102140. [CrossRef]
6. Lamego, P.; Lourenço, P.B.; Sousa, M.L.; Marques, R. Seismic vulnerability and risk analysis of the old building stock at urban scale: Application to a neighborhood in Lisbon. *Bull. Earthq. Eng.* **2017**, *7*, 2901–2937. [CrossRef]
7. Giordano, P.F.; Iacovino, C.; Quqa, S.; Limongelli, M.P. The value of seismic structural health monitoring for post-earthquake building evacuation. *Bull. Earthq. Eng.* **2022**. [CrossRef]
8. Pinto, J.; Gülten, G.; Vieira, J.; Meltem, V.; Varum, H.; Bal, I.E.; Costa, A. Save the *Tabique* Construction. In *Structural Rehabilitation of Old Buildings*; Costa, A., Guedes, J., Varum, H., Eds.; Springer: Berlin/Heidelberg, Germany, 2014. [CrossRef]
9. Pinto, J.; Cunha, S.; Soares, N.; Soares, E.; Cunha, V.M.C.F.; Ferreira, D.; Sá, A.B. Earth-based render of tabique walls—An experimental work contribution. *Int. J. Archit. Herit.* **2016**, *2*, 185–197. [CrossRef]
10. Karic, A.; Atalić, J.; Kolbitsch, A. Seismic vulnerability of historic brick masonry buildings in Vienna. *Bull. Earthq. Eng.* **2022**, *20*, 4117–4145. [CrossRef]
11. Hendry, E.A.W. Masonry walls: Materials and construction. *Constr. Build. Mat.* **2001**, *8*, 323–330. [CrossRef]
12. Brownjohn, J.M.W. Structural health monitoring of civil infrastructure. *Phil. Trans. R. Soc. Lond. A Math. Phys. Eng. Sci.* **2007**, *1851*, 589–622. [CrossRef]
13. Glisic, B.; Inaudi, D.; Casanova, N. SHM process as perceived through 350 projects. *Proc. SPIE* **2010**, *7648*, 26. [CrossRef]
14. Goldfeld, Y.; Quadflieg, T.; Ben-Aarosh, S.; Gries, T. Micro and macro crack sensing in TRC beam under cyclic loading. *J. Mech. Mat. Struct.* **2017**, *5*, 579–601. [CrossRef]
15. Cigada, A.; Corradi Dell'Acqua, L.; Mörlin Visconti Castiglione, B.; Scaccabarozzi, M.; Vanali, M.; Zappa, E. Structural health monitoring of an historical building: The main spire of the duomo di Milano. *Int. J. Archit. Herit.* **2017**, *4*, 511–518. [CrossRef]
16. Ramos, L.; Marques, L.; Loureno, P.; DeRoeck, G.; Campos-Costa, A.; Roque, J. Monitoring historical masonry structures with operational modal analysis: Two case studies. *Mech. Syst. Signal Proc.* **2010**, *5*, 1291–1305. [CrossRef]
17. Ubertini, F.; Comanducci, G.; Cavalagli, N. Vibration-based structural health monitoring of a historic bell-tower using output-only measurements and multivariate statistical analysis. *Struct. Health Monit.* **2016**, *4*, 438–457. [CrossRef]
18. Godínez-Domínguez, E.A.; Tena-Colunga, A.; Perez-Rocha, L.E.; Archundia-Aranda, H.I.; Gomez-Bernal, A.; Ruiz-Torres, R.P.; Escamilla-Cruz, J.L. The September 7, 2017 Tehuantepec, Mexico, earthquake: Damage assessment in masonry structures for housing. *Int. J. Disast. Risk Reduct.* **2021**, *56*, 102123. [CrossRef]
19. Senaldi, I.; Magenes, G.; Ingham, J.M. Damage assessment of unreinforced stone masonry buildings after the 2010–2011 Canterbury earthquakes. *Int. J. Archit. Herit.* **2014**, *5*, 605–627. [CrossRef]
20. Palmisano, F. Rapid diagnosis of crack patterns of masonry buildings subjected to landslide-induced settlements by using the load path method. *Int. J. Archit. Herit.* **2014**, *4*, 438–456. [CrossRef]
21. Yosef, L.; Goldfeld, Y. Smart self-sensory carbon-based textile reinforced concrete structures for structural health monitoring. *Struct. Health Monit.* **2020**. [CrossRef]
22. Bao, X.; Chen, L. Recent progress in distributed fiber optic sensors. *Sensors* **2012**, *7*, 8601–8639. [CrossRef]
23. Ferdinand, P. The Evolution of Optical Fiber Sensors Technologies during the 35 Last Years and Their Applications in Structure Health Monitoring. In Proceedings of the EWSHM-7th European Workshop on Structural Health Monitoring, Nantes, France, 8–11 July 2014; pp. 914–929. Available online: https://hal.inria.fr/hal-01021251 (accessed on 10 April 2022).
24. Ye, X.W.; Su, Y.H.; Han, J.P. Structural health monitoring of civil infrastructure using optical fiber sensing technology: A comprehensive review. *Sci. World J.* **2014**, *2014*, 652329. [CrossRef] [PubMed]
25. Barrias, A.; Casas, J.R.; Villalba, S. A review of distributed optical fiber sensors for civil engineering. *Sensors* **2016**, *5*, 748. [CrossRef] [PubMed]
26. Wu, T.; Liu, G.; Fu, S.; Xing, F. Recent progress of fiber-optic sensors for the structural health monitoring of civil infrastructure. *Sensors* **2020**, *16*, 4517. [CrossRef] [PubMed]
27. Bado, M.F.; Casas, J.R. A review of recent distributed optical fiber sensors applications for civil engineering structural health monitoring. *Sensors* **2021**, *5*, 1818. [CrossRef]
28. Tang, F.; Zhou, G.; Li, H.N.; Verstrynge, E. A review on fiber optic sensors for rebar corrosion monitoring in RC structures. *Constr. Build. Mat.* **2021**, *313*, 125578. [CrossRef]
29. Kapogianni, E.; Kalogeras, I.; Psarropoulos, P.; Michalopoulou, D.; Eleftheriou, V.; Sakellariou, M. Suitability of optical fibre sensors and accelerographs for the multi-disciplinary monitoring of a historically complex site: The case of the Acropolis circuit wall and hill. *Geotech. Geol. Eng.* **2019**, *37*, 4405–4419. [CrossRef]
30. Bellagamba, I.; Caponero, M.; Mongelli, M. Using fiber-optic sensors and 3D photogrammetric reconstruction for crack pattern monitoring of masonry structures at the Aurelian Walls in Rome, Italy. *WIT Transact. Build. Env.* **2019**, *191*, 457–465. [CrossRef]
31. Cocking, S.; Thompson, D.; DeJong, M. Comparative evaluation of monitoring technologies for a historic skewed masonry arch railway bridge. In Proceedings of the International Conference on Arch Bridges, Porto, Portugal, 2–4 October 2019; Arêde, A., Costa, C., Eds.; Springer: Cham, Switzerland, 2019; Volume 11, pp. 439–446. [CrossRef]

32. Alexakis, H.; Lau, F.D.H.; DeJong, M.J. Fibre optic sensing of ageing railway infrastructure enhanced with statistical shape analysis. *J. Civ. Struct. Health Monit.* **2021**, *11*, 49–67. [CrossRef]
33. Acikgoz, S.; Fidler, P.R.A.; Pascariello, M.N.; Kechavarzi, C.; Bilotta, E.; DeJong, M.J.; Mair, R.J. A fibre-optic strain measurement system to monitor the impact of tunnelling on nearby heritage masonry buildings. *Int. J. Archit. Herit.* **2021**. [CrossRef]
34. Berrocal, C.G.; Fernandez, I.; Bado, M.F.; Casas, J.R.; Rempling, R. Assessment and visualization of performance indicators of reinforced concrete beams by distributed optical fibre sensing. *Struct. Health Monitor.* **2021**, *6*, 3309–3326. [CrossRef]
35. Berrocal, C.G.; Fernandez, I.; Rempling, R. Crack monitoring in reinforced concrete beams by distributed optical fiber sensors. *Struct. Infrastruct. Eng.* **2021**, *1*, 124–139. [CrossRef]
36. Wu, J.; Liu, H.; Yang, P.; Tang, B.; Wei, G. Quantitative strain measurement and crack opening estimate in concrete structures based on OFDR technology. *Opt. Fib. Technol.* **2020**, *60*, 102354. [CrossRef]
37. Brault, A.; Hoult, N. Monitoring reinforced concrete serviceability performance using fiber-optic sensors. *ACI Struct. J.* **2019**, *1*, 57–70. [CrossRef]
38. Rodriguez, G.; Casas, J.R.; Villaba, S. Cracking assessment in concrete structures by distributed optical fiber. *Smart Mat. Struct.* **2015**, *24*, 035005. [CrossRef]
39. Chen, R.; Zaghloul, M.A.S.; Yan, A.; Li, S.; Lu, G.; Ames, B.C.; Zolfaghari, N.; Bunger, A.P.; Li, M.J.; Chen, K.P. High resolution monitoring of strain fields in concrete during hydraulic fracturing processes. *Opt. Express* **2016**, *4*, 3894. [CrossRef]
40. Verstrynge, E.; De Wilder, K.; Drougkas, A.; Voet, E.; Van Balen, K.; Wevers, M. Crack monitoring in historical masonry with distributed strain and acoustic emission sensing techniques. *Const. Build. Mat.* **2018**, *162*, 898–907. [CrossRef]
41. Chan, T.H.T.; Yu, L.; Tam, H.Y.; Ni, Y.Q.; Liu, S.Y.; Chung, W.H.; Cheng, L.K. Fiber Bragg grating sensors for structural health monitoring of Tsing Ma bridge: Background and experimental observation. *Eng. Struct.* **2006**, *5*, 648–859. [CrossRef]
42. Liu, H.B.; Zhang, Q.; Zhang, B.H. Structural health monitoring of a newly built high-piled wharf in a harbor with fiber Bragg grating sensor technology: Design and deployment. *Smart Struct. Syst.* **2017**, *2*, 163–173. [CrossRef]
43. Rodriguez, M.; Castrillon, E. *Manual de Evaluacion Postsismica de la Seguridad Estructural de Edificaciones*; Informe No. 569; Instituto de Ingenieria, UNAM: Mexico City, Mexico, 1995; Available online: https://datosabiertos.unam.mx/IINGEN:RUSI:569 (accessed on 10 April 2022).
44. Federal Emergency Management Agency. *FEMA P-154 Rapid Visual Screening of Buildings for Potential Seismic Hazards: A Handbook*, 3rd ed.; Federal Emergency Management Agency: Washington, DC, USA, 2015. Available online: https://fema.gov/sites/default/files/2020-07/fema_earthquakes_rapid-visual-screening-of-buildings-for-potential-seismic-hazards-supporting-documentation-third-edition-fema-p-155.pdf (accessed on 12 April 2022).
45. Building Research Establishment. *Assessment of Damage in Low-Rise Buildings with Particular Reference to Progressive Foundation Movement*; Digest 251; Building Research Establishment: Watford, UK, 1990; ISBN 978-0-85125-461-6.
46. Hale, K.F.; Hockenhuli, B.S.; Christodoulou, G. The application of optical fibers as witness devices for the detection of plastic strain and cracking. *Strain* **1980**, *4*, 150–154. [CrossRef]
47. Crane, R.M.; Macander, A.B.; Gagorik, J. Fiber optics for a damage assessment system for fiber reinforced plastic composite structures. In *Review of Progress in Quantitative Nondestructive Evaluation. Library of Congress Cataloging in Publication Data, Vol 2A*; Thompson, D.O., Chimenti, D.E., Eds.; Springer: Boston, MA, USA, 1983; pp. 141–1430. [CrossRef]
48. Balageas, D.; Bourasseau, S.; Dupont, M.; Bocherens, E.; Dewynter-Marty, V.; Ferdinand, P. Comparison between non-destructive evaluation techniques and integrated fiber optic health monitoring systems for composite sandwich structures. *J. Intell. Mat. Syst. Struct.* **2000**, *6*, 426–437. [CrossRef]
49. Romaniuk, R. Tensile Strength of Tailored Optical Fibers. *Opto-Electron. Rev.* **2000**, *2*, 101–116. Available online: https://citeseerx.ist.psu.edu/viewdoc/download?doi=10.1.1.474.6168&rep=rep1&type=pdf (accessed on 10 April 2022).
50. Matusewicz, M.A.; Wójcik, J.; Mergo, P.; Walewski, A. Mechanical strength of photonic crystal fibres. *Photon. Lett. Pol.* **2010**, *1*, 19–21. [CrossRef]
51. Todinov, M.T. The cumulative stress hazard density as an alternative to the Weibull model. *Int. J. Solid. Struct.* **2010**, *24*, 3286–3296. [CrossRef]
52. Antunes, P.; Lima, H.; Monteiro, J.; Andre, P.S. Elastic constant measurement for standard and photosensitive single mode optical fibres. *Microw. Opt. Technol. Lett.* **2008**, *9*, 2467–2469. [CrossRef]
53. Backes, H.P. Tensile Strength of Masonry. In Proceedings of the 7th International Brick Masonry Conference, Melbourne, Australia, 17–20 February 1985; pp. 779–790. Available online: https://hms.civil.uminho.pt/ibmac/1985/779.pdf (accessed on 12 April 2022).
54. Shena, W.; Wanga, X.; Xua, L.; Zhao, Y. Strain transferring mechanism analysis of the substrate-bonded FBG sensor. *Optik* **2018**, *154*, 441–452. [CrossRef]
55. Wang, H.; Jiang, L.; Xiang, P. Improving the durability of optical fiber sensor based on strain transfer analysis. *Opt. Fib. Technol.* **2018**, *1*, 97–104. [CrossRef]
56. Wang, H.; Jiang, L.; Xiang, P. Priority design parameters of industrialized optical fiber sensors in civil engineering. *Opt. Laser Technol.* **2018**, *100*, 119–128. [CrossRef]
57. Wang, H.; Jiang, L.; Xiang, P. Strain transfer theory of industrialized optical fiber-based sensors in civil engineering: A review on measurement accuracy, design, and calibration. *Sens. Actuators A Phys.* **2019**, *285*, 414–426. [CrossRef]

58. Alj, I.; Quiertant, M.; Khadour, A.; Grando, Q.; Terrade, B.; Renaud, J.C.; Benzarti, K. Experimental and numerical investigation on the strain response of distributed optical fiber sensors bonded to concrete: Influence of the adhesive stiffness on crack monitoring performance. *Sensors* **2020**, *18*, 5144. [CrossRef]
59. Khotiaintsev, S.; Beltran-Hernandez, A.; Gonzalez-Tinoco, J.E.; Guzman-Olguin, H.; Aguilar-Ramos, G. Structural health monitoring of concrete elements with embedded arrays of optical fibers. In Proceedings of the Health Monitoring of Structural and Biological Systems 2013, San Diego, CA, USA, 10–14 March 2013; Volume 8695, pp. 267–272. [CrossRef]
60. Gonzalez-Tinoco, J.E.; Gomez-Rosas, E.R.; Guzman-Olguin, H.; Khotiaintsev, S.; Zuñiga-Bravo, M.A. Monitoring of transverse displacement of reinforced concrete beams under flexural loading with embedded arrays of optical fibers. In Proceedings of the Structural Health Monitoring and Inspection of Advanced Materials, Aerospace, and Civil Infrastructure 2015, San Diego, CA, USA, 8–12 March 2015; Volume 9437, pp. 155–161. [CrossRef]
61. Gonzalez-Tinoco, J.E.; Martinez-Gonzalez, D.M.; Miron-Carrasco, J.; Khotiaintsev, S.; Guzman-Olguin, H.J.; Lopez-Bautista, M.C.; Sanchez-Ramirez, A.R. Specialized optical fibre sensor array for structural damage detection. In Proceedings of the 2016 10th International Symposium on Communication Systems, Networks and Digital Signal Processing (CSNDSP), Prague, Czech Republic, 20–22 July 2016; IEEE: Piscataway, NJ, USA, 2016. [CrossRef]
62. Hurtado-De Mendoza-Lopez, A.; Khotiaintsev, S.; Guzman-Olguin, H.J.; Reyes, D.A.H.; Lopez-Mancera, J.A.; Zuniga-Bravo, M. Application of fiber optic light guides with quartz core and cladding and plastic coating as indicators of the emergence of cracks in brickwork. *J. Opt. Technol.* **2020**, *2*, 132–136. [CrossRef]
63. *TIA-455-28 FOTP-28*; Measuring Dynamic Strength and Fatigue Parameters of Optical Fibers by Tension. Telecommunications Industry Association: Arlington, VA, USA, 1999. Available online: https://standards.globalspec.com/std/734963/tia-455-28 (accessed on 11 April 2022).
64. Gonzalez-Tinoco, J.E.; Khotiaintsev, S.; Lopez-Bautista, M.C.; Timofeyev, V.I.; Vountesmery, Y.V. Arrays of optical fibers as detectors of cracks in brick structural elements. In Proceedings of the 2018 IEEE 38th International Conference on Electronics and Nanotechnology (ELNANO), Kyiv, Ukraine, 24–26 April 2018; pp. 717–720. [CrossRef]
65. Glaesemann, G.S.; Dainese, P.; Edwards, M.; Dhliwayo, J. *The Mechanical Reliability of Corning Optical Fiber in Small Bend Scenarios*; WP1282; Corning: Glendale, AZ, USA, 2007; Available online: https://www.corning.com/media/worldwide/coc/documents/Fiber/white-paper/wp1282.pdf (accessed on 10 April 2022).
66. Hines, W.W.; Montgomery, D.C.; Goldsman, D.M.; Borror, C.M. *Probability and Statistics in Engineering*, 4th ed.; John Wiley & Sons Inc.: Hoboken, NJ, USA, 2003; 655p, ISBN 0-471-24087-7.
67. Aguilar Espinosa, A.A.; Fellows, N.A.; Durodola, J.F.; Fellows, L.J. Development of numerical model for the determination of crack opening and closure loads, for long cracks. *Fatigue Fract. Eng. Mat. Struct.* **2017**, *4*, 571–585. [CrossRef]
68. Ramachandra, S.; Durodola, J.F.; Fellows, N.A.; Gerguri, S.; Thite, A. Experimental validation of an ANN model for random loading fatigue analysis. *Int. J. Fatigue* **2019**, *126*, 112–121. [CrossRef]
69. Hersent, O.; Boswarthick, D.; Elloumi., O. *The Internet of Things: Key Applications and Protocols*; John Wiley & Sons Inc.: Hoboken, NJ, USA, 2012; 376p, ISBN 978-1-11999-435-0.

Article

Carbon Nanofibers Grown in CaO for Self-Sensing in Mortar

Lívia Ribeiro de Souza [1,*], Matheus Pimentel [2], Gabriele Milone [1], Juliana Cristina Tristão [2] and Abir Al-Tabbaa [1]

[1] Department of Engineering, University of Cambridge, Cambridge CB2 1PZ, UK; gm683@cam.ac.uk (G.M.); aa22@cam.ac.uk (A.A.-T.)
[2] Instituto de Ciências Exatas e Tecnológicas, Campus de Florestal, Universidade Federal de Viçosa, Florestal 35690-000, Brazil; matheus.pimentel@ufv.br (M.P.); juliana@ufv.br (J.C.T.)
* Correspondence: lrds2@cam.ac.uk

Abstract: Intelligent cementitious materials integrated with carbon nanofibers (CNFs) have the potential to be used as sensors in structural health monitoring (SHM). The difficulty in dispersing CNFs in cement-based matrices, however, limits the sensitivity to deformation (gauge factor) and strength. Here, we synthesise CNF by chemical vapour deposition on the surface of calcium oxide (CaO) and, for the first time, investigate this amphiphilic carbon nanomaterial for self-sensing in mortar. SEM, TEM, TGA, Raman and VSM were used to characterise the produced CNF@CaO. In addition, the electrical resistivity of the mortar, containing different concentrations of CNF with and without CaO, was measured using the four-point probe method. Furthermore, the piezoresistive response of the composite was quantified by means of compressive loading. The synthesised CNF was 5–10 μm long with an average diameter of ~160 nm, containing magnetic nanoparticles inside. Thermal decomposition of the CNF@CaO compound indicated that 26% of the material was composed of CNF; after CaO removal, 84% of the material was composed of CNF. The electrical resistivity of the material drops sharply at concentrations of 2% by weight of CNF and this drop is even more pronounced for samples with 1.2% by weight of washed CaO. This indicates a better dispersion of the material when the CaO is removed. The sensitivity to deformation of the sample with 1.2% by weight of CNF@CaO was quantified as a gauge factor (GF) of 1552, while all other samples showed a GF below 100. Its FCR amplitude can vary inversely up to 8% by means of cyclic compressive loading. The method proposed in this study provides versatility for the fabrication of carbon nanofibers on a tailored substrate to promote self-sensing in cementitious materials.

Keywords: carbon nanofibers; self-sensing; CVD; piezoresistivity

Citation: de Souza, L.R.; Pimentel, M.; Milone, G.; Tristão, J.C.; Al-Tabbaa, A. Carbon Nanofibers Grown in CaO for Self-Sensing in Mortar. *Materials* **2022**, *15*, 4951. https://doi.org/10.3390/ma15144951

Academic Editors: Phong B. Dao, Tadeusz Uhl, Liang Yu, Lei Qiu and Minh-Quy Le

Received: 12 May 2022
Accepted: 12 July 2022
Published: 15 July 2022

1. Introduction

Self-sensing concrete refers to concrete materials and structures possessing intrinsic properties that sense various physical and chemical parameters. This property can be harnessed for traffic monitoring [1,2] and structural health monitoring (SHM), including monitoring of load [3], strain sensing [4,5], crack formation [6], freeze-thaw [3], electric magnetic shielding [7] and self-healing performance [8]. To achieve sensing, functional fillers are distributed in the cementitious matrix in order to promote electrically conductive properties [9]. The resultant self-sensing composite is easily prepared and offers good compatibility with concrete structures, making it a more attractive alternative to conventional sensing devices [5]. Among the different types of conductive fillers, steel and carbon materials are commonly used, as recently discussed in several reviews [9,10]. Although the macroscale of the former allows for easier production and use, as well as a lower cost, its main drawback is associated with its susceptibility to corrosion [11]. Carbon nanomaterials, on the other hand, offer high electrical conductivity and durability [9]. Graphene nanoplatelets [12,13], carbon nanotubes (CNTs) [14–16] and carbon nanofibers (CNFs) alone, as well as the synergic effect of the combination containing these materials [17–19], are examples of conductive fillers used to promote the electrical properties of cementitious systems. Beyond increasing conductivity, the inclusion of carbon-based nanomaterials in cementitious matrices has been gaining attention due to its contribution to accelerating hydration [20] and its effect on durability and mechanical properties [21].

Percolation theory is used to describe the electrical behaviour of a system containing a conductive filler. Above a critical concentration, also known as the percolation threshold, these functional fillers form a conductive network inside the matrix [7], leading to changes in the electrical properties as external forces change. Nanofibrous materials, such as CNFs/CNTs, have very high aspect ratios and specific surface areas, resulting in percolation at lower concentrations than for other functional fillers [22]. However, because of the high van der Waals forces between the materials, the high aspect ratios and large surface areas of CNTs/CNFs lead to agglomerations. When dispersed in the cementitious matrix, these functional fillers are likely to cluster rather than be uniformly dispersed [23,24]. As a result, poor dispersion increases the amount of material required to achieve adequate conductivity in the sample [8]. To address this issue, a combination of sonication and surfactants is typically used to increase the dispersion of the nanomaterials. However, sonication is a time-consuming and energy-intensive step that poses a significant challenge in large-scale applications. Experimental work has revealed that cement paste containing 2% CNF by weight of cement is sensitive to its own structural damage [4]. In mortar, xperiments and modelling of the dispersion of CNF in mortar indicate a threshold value between 1.6–2%vol [25,26]. However, previous research also demonstrates a significant decrease in resistivity for cement paste containing 0.1% CNF, as well as changes in resistivity under compressive loads of up to 5% [27]. When comparing CNT and CNF, a smaller fractional change in resistance has been observed when comparing 0.6 wt% of CNF and CNT under identical loading conditions [28]. Thus, the exact values associated with the percolation threshold are unknown, with significant factors associated with the size and length of the CNF, the cementitious matrix used (cement paste or mortar), the water-to-cement ratio and the superplasticizer content used in the mixing [9,29]. To facilitate substrate dispersion in the hydrophilic cement paste, we propose using CaO as a substrate for the growth of CNF, resulting in an amphiphilic composite.

Chemical vapour deposition (CVD) offers a versatile route to obtaining carbon nanofibers in the laboratory. In this process, a vaporised carbon source is catalytically reduced and deposits itself on a substrate as graphite and CNTs/CNFs [30,31]. Catalytic CVD has also been used for large-scale commercial production of multiwalled carbon nanotubes (MWCNT) [32,33]. In this case, the price for MWCNTs is in the range of 0.1–0.15 €/g, depending on the material quality and acquired quantities [34]. In addition to its low cost and scalability [35], an attractive feature of this method is the wide variety of carbon sources, catalyst particles and substrates that can be used for the production of CNTs/CNFs. Thus, the synthesis can be fine-tuned towards sustainable sources of carbon [36], maximising production of CNTs/CNFs by catalyst particle selection and substrate selection based on the final application. For example, in order to facilitate dispersion in the cementitious matrix, carbon nanofibers have recently been synthesised using Portland cement as a substrate [37]. In this case, iron (III) oxide (Fe_2O_3) naturally present in the cement acts as a catalyst for the growth of carbon nanotubes. However, the limited amount of catalyst leads to small amounts of carbon nanofibers being produced—typically 2.5 to 3.2% by weight [37,38]. Alternatively, CVD results using Portland as a substrate and containing conversion powder as a catalyst have shown CNT/CNF concentrations as high as 12% by weight [39,40], while using Ni as the catalyst resulted in the production of composites with ~25 wt% of CNT [2]. Due to the high natural iron content, fly ash has also been investigated, using ferrocene as an extra catalyst, yielding CNF ~33% by weight [41]. Furthermore, CNT synthesised on the surface of fly ash showed good dispersion in the mortar and excellent piezoresistive response [42]. Based on a similar principle, we suggest the use of CaO as a substrate for the production of CNF, since the material has an amphiphilic character—hydrophobic CNF grows on top of a hydrophilic substrate [43]. In addition, it offers an excellent bond to the cement and a route to easily remove the substrate by acidic attack, increasing the dispersion of the carbon nanofibers.

Here, we report the production of amphiphilic carbon nanofibers (CNF) and, for the first time, its application to increase the conductivity and promote the piezoresistivity of mortar. We produce CNF using chemical vapour deposition (CVD) of ethanol over a calcium oxide substrate doped with iron nanoparticles. The width and length of the CNF are characterised using SEM, as well as the deposition of the fibres on the surface of CaO. The encapsulation of iron nanoparticles inside the fibres was visualised using TEM. Thermogravimetric analysis of the produced fibres is used to investigate the thermal stability and concentration of the CNF over the CaO substrate. Once the CNF@CaO is dispersed in the matrix, SEM is also used to investigate the dispersion and interaction of the composite. To demonstrate the role of the CNF in decreasing the electrical resistivity of mortar, different concentrations of fibres are dispersed in the matrix and the conductivity is measured using the four-point probe method. To assess the piezoresistive response of the material, a cyclic compressive load is applied and the gauge factor quantified. The approach in this study demonstrates versatility for fabrication of carbon nanofibers with a tailored substrate to achieve better dispersion of the nanofibers in the cementitious matrix.

2. Materials and Methods

Magnetic carbon nanofibers supported in CaO (CNF@CaO) were produced via chemical vapour deposition using ethanol as a reagent. First, mixed calcium iron oxide was produced via wet impregnation by adding 0.8 g of calcium oxide to an aqueous solution containing 1.4 g of iron (III) nitrate to produce samples with Fe/CaO contents of 20 wt%. The resulting suspension was heated to 90 °C, under constant stirring, until complete evaporation of the water. Then, the material was calcined at 800 °C for 1 h. The resulting substrate is composed of $Ca_2Fe_2O_5$, as indicated in Equation (1). The presence of this phase was confirmed in previous work [43], where XRD and Mössbauer spectroscopy confirmed the presence of $Ca_2Fe_2O_5$. The presence of mixed iron oxide, instead of phases of stable oxides such as hematite, indicates that the iron was homogenously added to the substrate [43].

$$2Fe(NO_3)_3 \cdot 9H_2O + 2CaO + O_2 \rightarrow Ca_2Fe_2O_5 + 6NO_x + 18H_2O \quad (1)$$

After calcination, a nitrogen flow of 100 mL min^{-1} was purged through a gas-washing bottle containing ethanol at room temperature. The resulting nitrogen, saturated with ~6 vol% of ethanol, passed through a quartz tube containing 200 mg of mixed calcium iron oxide. The quartz tube was placed horizontally inside a furnace (Tubular furnace 1200, Sanchis, Porto Alegre, Brazil) and the temperature was raised at a heating rate of 5 °C min^{-1} up to 900 °C and maintained for 1 h. The production of carbon nanofibers is shown schematically in Figure 1. The iron nanoparticles dispersed in CaO acted as catalysts for the dissociation of the gaseous phases and templates for the nucleation and growth of the carbon nanofibers [44]. After CVD, Mössbauer analysis was performed in the as-synthetised samples to identify the iron phases after the reduction with ethanol. Phases of α-Fe and iron carbide were primarily identified (75%) [43]. Other phases were associated with $CaFe_2O_4$ and a solid solution of iron and carbon (γ-Fe(C)). XRD showed different phases of $CaCO_3$, $Ca(OH)_2$, CaO and Fe formed. The calcium oxide substrate could be easily removed after acid attack, but also acted as a hydrophilic moiety, conferring amphiphilic properties to the CNF@CaO composite. In this work, to remove the substrate, 2.6 g of CNF@CaO as grown was mixed with 50 mL of HCl 1 M. After mixing, the CNF was separated with a magnet until the solution was clear. Then the acid was removed and the CNF was washed with water until neutralisation.

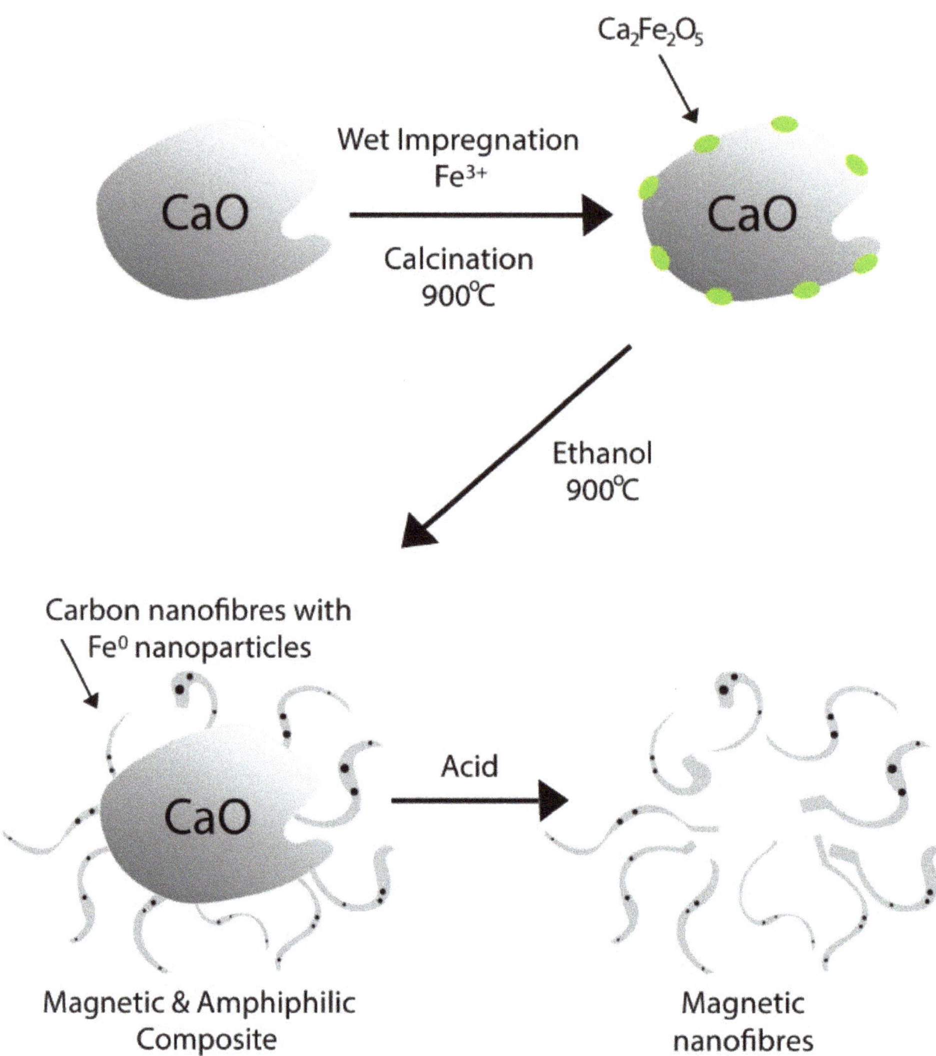

Figure 1. Schematics of production of mixed calcium iron oxide, carbon nanofibers supported in CaO and pure magnetic nanofibers.

Scanning electron microscopy (SEM) was carried out using a Evo LS15 (Carl Zeiss, Cambridge, UK). The samples were gold coated using a rotary pumped sputter coater (Agar Sputter Coater B7367A, Agar Scientific Ltd., Essex, UK) to improve the conductivity of the surface of the samples and to prevent overcharging. For the CNF deposited in the surface of CaO, the SEM operated at an accelerating voltage of 8 kV; for the mortar composite containing CNF, the voltage was 6 kV. It was difficult to distinguish ettringite from nanofibers in the overall surface sample at the maximum distance in the SEM. As a result, the air bubbles in the mortar were chosen for investigation because they contained small agglomerates of fibres. To measure the diameter and length of the CNF, as well as observe the encapsulation of iron nanoparticles and graphene layering of the carbon nanofibers, a transmission electron microscope was used (TEM, FEI Tecnai Osiris FEGTEM, Hillsboro, OR, USA). The CNF@CaO samples were washed with acid, followed by water neutralisation. The suspended material was then placed over a carbon film with a Cu grid for analysis.

Thermogravimetrical analysis (TGA) was used to assess the weight ratio of as-grown CNF@CaO using a PerkinElmer STA6000 (Shelton, CO, USA) between 100 and 800 °C at a rate of 5 °C min^{-1} under air atmosphere. After acid washing the material with HCl 1 M, followed by washing with water until the pH was ~5, another TGA was performed of the dried material in the same conditions as before. Approximately ~5 mg of material was analysed. To investigate the magnetic properties of the samples, the hysteresis of CNF@CaO and washed CNF were measured using a Lakeshore Cryotronics (Westerville, OH, USA) vibrating sample magnetometer as a function of the magnetic field up to 15 kOe at room temperature. Raman spectra were collected on a XploRA Plus (Horiba Jobin Yvon, Villeneuve d'Ascq, France) using a 638 nm laser.

Ordinary Portland Cement (CEM I 42.5) provided by Heidelberg-UK was used for the production of the mortar samples. Sieved and cleaned sand ranging between 0.18–2 mm was used as fine aggregate. The workability of the mixture was aided by adding 0.3 wt% by weight of cement (bwoc) with modified polycarboxylic ether as the superplasticiser (MasterGlenium 315C, BASF). All mortar mixes had a water–cement ratio (w/c) of 0.6 and a sand–cement ratio (s/c) of 3. In situ grown carbon nanofibers were added to the mortar matrix at 0.4, 1.2 and 2 wt% by weight of cement (bwoc) or 0.36, 1.08 and 1.8% by volume of cement, respectively, considering a density of CNF ~1.6 g/cm^3 and the cement density as 1.44 g/cm^3. The quantity of CNF@CaO to achieve the mentioned concentrations of carbon nanofibers was calculated with the weight ratio obtained from TGA. To assist in the dispersion of carbon nanofibers, the CNF@CaO composite was sonicated (Fisherbrand FB11203, Singer, Germany, 80 kHz frequency and 100% power) for 30 min in a solution of water and superplasticiser. The ready-dispersed CNT and superplasticizer suspension were mixed into a rotary mixing according to the following protocol: sand and cement were dry-mixed in a mixer pan for 3 min, and then half of the water containing the superplasticiser and the nanomaterials was added and mixed in for another 1 min, before adding the rest of the water and mixing for another 1 min. The mixture was then mixed at maximum velocity for 0.5 min, followed by 1 min at minimum speed before adding to the oiled mould. The process is schematically represented in Figure 2.

The mix proportions are shown in Table 1. Triplicates of all samples were cast into oiled moulds to produce samples of 20 × 20 × 80 mm. Four perforated steel sheets of dimension 40 × 17 × 0.55 mm, with a 3 mm hole diameter, were embedded in the sample (Figure 3A). All samples were placed in an electric vibrator (Controls Automatic Sieve Shaker D407, Cernusco, Italy) for 30 s for good compaction and to reduce air bubbles, then covered with plastic film for 24 h, curing at room temperature. The specimens were demoulded and cured in a moist container at 20 ± 1 °C and with a relative humidity ≥95% for 28 days.

Table 1. Mix proportions used for test mortars.

Specimen	w/c	s/c	CNF (% bwoc)	CNF@ CaO (% bwoc)	Superplasticiser (% bwoc)
Control	0.6	3	-	-	0.3
CNF 0.4	0.6	3	0.4	1.6	0.3
CNF 1.2	0.6	3	1.2	4.8	0.3
CNF 2	0.6	3	2.0	8.0	0.3
CNF 1.2(A)	0.6	3	1.2	-	0.3
CTRL-CaO	0.6	3	-	1.9	0.3

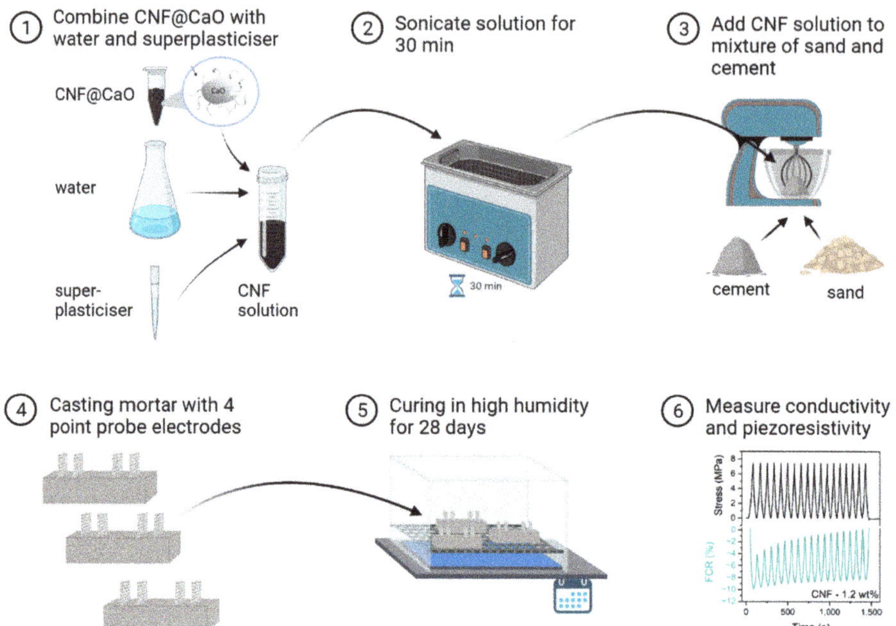

Figure 2. Workflow used for dispersion of CNF in mortar, followed by measurement of electrical properties.

The electrical resistance (R) was measured using the four-probe method with a digital multimeter (TTi 1604, Aim & Thurlby Thandar Instruments, Huntingdon, UK), as shown in Figure 3B. Direct current (DC) of 20 V was applied between the two outer electrodes and the electric potential was measured between the two inner electrodes (Figure 3B). An insulator was also added in between the compressive plates and the sample. The electrical resistivity (ρ) was calculated using the following equation:

$$\rho = R \cdot \frac{A}{L} \tag{2}$$

where ρ is electrical resistivity in ohm meters, L is the internal electrode distance in meters, A is the electrode area in square meters and R = V/I is the measured resistance determined by measuring the voltage drop across the specimen (V) in volts and the applied current (I) in amperes. To tackle the polarization effect, the values were collected after 15 min of constant tension applied to the outer two electrodes [45]. The application of compressive load to the mortar sample was accomplished with a hydraulic press (Instron 5567, Norwood, MA, USA—30 kN capacity) operating under a distance control of 0.8 mm min^{-1}, up to 7.5 MPa. Figure 3C shows a schematic of the PC-controlled acquisition system that was used to collect the data for load, strain gauge, voltage and current from the samples while a voltage of 20 V was applied.

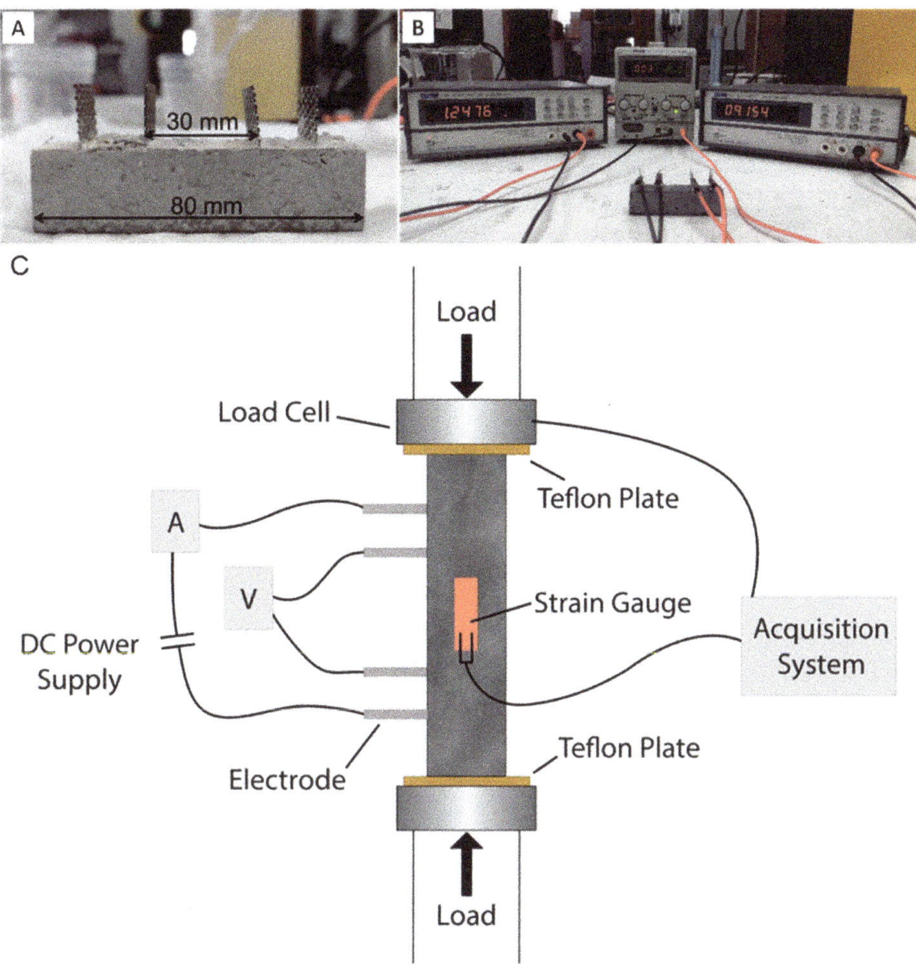

Figure 3. Set-up for measuring the conductivity of the samples. (**A**) Mortar sample with 4 electrodes; (**B**) multimeters for measurements; (**C**) experimental set-up used to measure the piezoresistive behaviour of the mortar samples containing CNF.

3. Results

3.1. Production and Characterisation of Carbon Nanofibers

The carbon nanofibers (CNF) containing magnetic nanoparticles were synthesised using chemical vapour deposition (CVD) with CaO as the substrate. Calcium oxide was selected as a substrate due to its compatibility with cementitious materials and its polar nature, which grants an amphiphilic property to the CNF@CaO composite. Through CVD, the substrate containing mixed calcium iron oxides as catalyst reacted with nitrogen saturated with ethanol at 900 °C. At this temperature, the reduction of $Ca_2Fe_2O_5$ by ethanol results in the formation of Fe^0, Fe_3C, $CaFe_2O_4$ amorphous carbon and CNF [43]. The production of carbon nanofibers is shown schematically in Figures 1 and 4A,B shows entangled CNF grown on the surface of CaO.

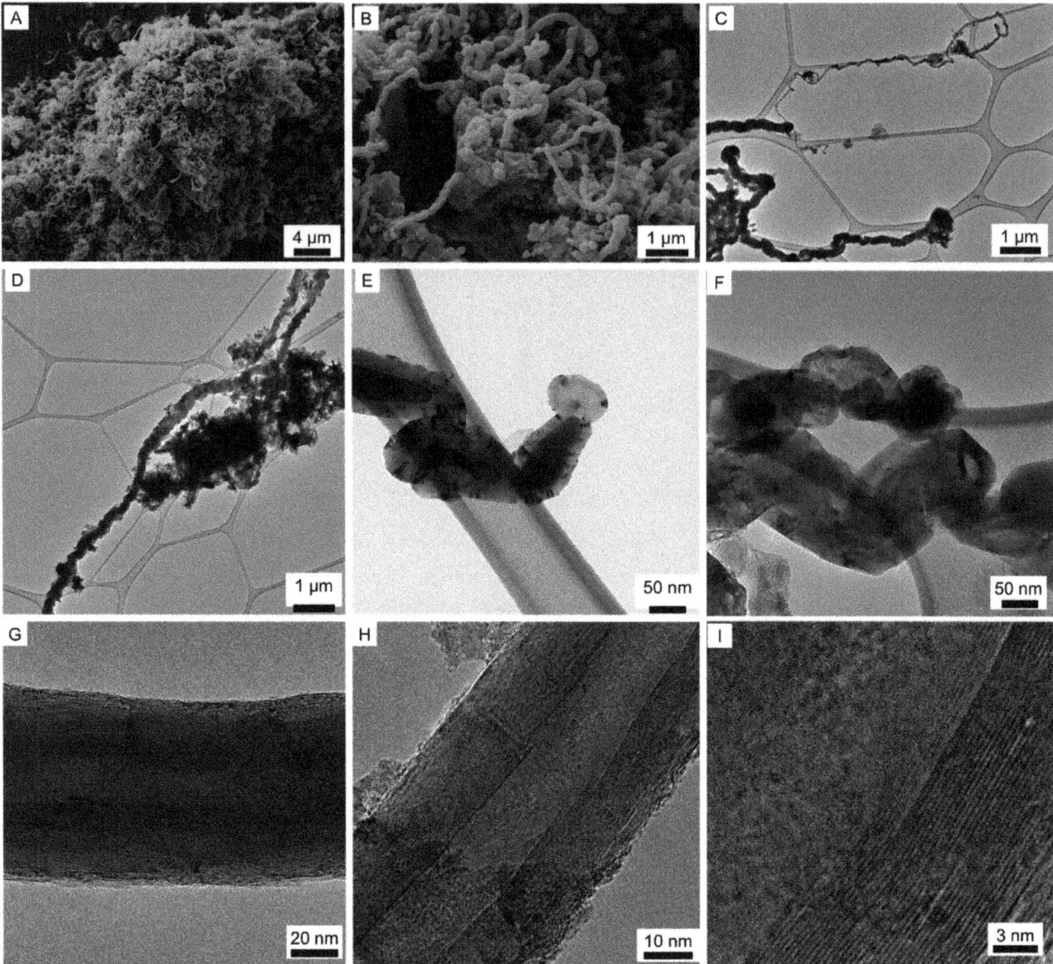

Figure 4. Carbon nanofibers supported in calcium oxide. (**A**,**B**) SEM images from the as-grown samples showing the growth of the CNF on the surface of calcium oxide particles. (**C–I**) TEM images after the acid washing for the removal of the calcium oxide support.

The calcium oxide substrate was then removed by acid washing, resulting in free CNF to be observed using TEM and shown in Figure 4C–I. Figure 4C shows the different widths of the CNF, with fine CNF around 40 μm and typical ones ~160 μm. The high length-to-width ratio of the CNF is demonstrated in Figure 4D, with fibrous structures ~11 μm long and 280 μm wide. In general, the length of the CNF ranged between 5–10 μm and the diameter was between 40–300 nm, with average values around 160 nm. The high length-to-diameter aspect ratio of the CNF is particularly suitable for sensing, as it contributes to more conduction paths [9]. Amorphous carbon was observed through TEM analysis, as shown in the central part of Figure 4D. After the CNF has grown, the catalyst particles are trapped inside the graphene layers, as shown in Figure 4E, and empty channels are inside the CNF (Figure 4F). The angle between the graphite basal planes and the tube axis is different from zero, resulting in the term carbon nanofibers of composites. In contrast, the term carbon nanotube (CNT) refers to graphene sheets rolled up in concentric cylinders with walls parallel to the axis [46]. This is demonstrated in Figure 4F–I, where the angle between the nanosheets and the tube axis is different from zero; Figure 4I is a zoom of the graphene layers in Figure 4H. Although the recorded images identify the presence of carbon nanofibers, multi-walled carbon nanotubes are likely to also occur. Controlling the size distribution of the catalyst particles is a standard way to tune the CNF diameter dispersion [44]. As a result, the nanoparticles trapped inside the nanofibers have an elongated shape and vary between 46–200 nm, slightly smaller than the diameter of the CNF. These trapped nanoparticles inside the nanofibers are responsible for the magnetic behaviour of the CNF, even after washing with acid.

Thermogravimetric analysis (TGA) was performed to quantify the metal content after full oxidation of carbon and to examine the thermal stability of the as-grown CNF over substrate. The oxidation temperature was determined by TGA curves (deflexion point—Figure 5A) and more precisely by the DTA curves (Figure 5B). The curves of the as-grown CNF show: (1) oxidation between 372–427 °C, attributed to amorphous carbon, where ~2% of the mass was lost; (2) a second weight loss, attributed to the oxidation of CNF, peaking at 602 °C, where 26% of the mass was lost; and (3) at around 688 °C, loss of 26% of the mass, corresponding to the transformation of $CaCO_3$ [43] into CaO. The residual material was ~45%, representing the CaO support and the iron-based catalyst. After acid washing the sample, the CaO was removed, together with most of the amorphous carbon during the neutralisation process. Consequently, the TGA (black curve) presented only one thermal event, corresponding to the oxidation of the CNF, with a maximum weight loss rate at 590 °C, where 84% of the material was lost. The residual mixture of catalyst and any residual support made up ~11%, comprised of catalyst metallic oxides. This is consistent with the previous characterisation of the washed CNF, in which the elemental analysis revealed a carbon concentration of ~80 wt% [43]. At ~580 °C, an anomalous mass gain is observed, where a sharp decrease in specimen mass is accompanied by a rapid increase in temperature by 6 °C, followed by a quick decrease. Similar behaviour has been observed on as-produced, unpurified and uncompacted nanotubes [47] and attributed to spontaneous combustion; i.e., the heat released in the exothermic reaction is enough to sustain rapid burning of the sample [48]. This behaviour seems to be mainly associated with single-walled carbon nanotubes, which could imply their presence in the sample, despite not being observed in the TEM.

The Raman spectra obtained for the pure CNF are shown in Figure 6C (bottom). In the first-order region, two bands can be observed in the red laser (638 cm^{-1}). The so-called D band is sited around 1325 cm^{-1}, and it is associated with disordered structures in carbon materials [49]. The peak around 1575 cm^{-1} is the G band associated with the high degree of symmetry and order of carbon materials [50]. Finally, a third, weak band is observed at 1614 cm^{-1}, associated with D′. The strong, dispersive band around 2643 cm^{-1} is designated as the G′ band (called 2D sometimes). For the sample CNF@CaO (Figure 6C—top), the bands are in a similar region, but much more intense. When comparing the I_D/I_G ratios calculated from the intensities in the D and G bands, a small rise is observed: the ratio for CNF@CaO is 0.57, whereas the ratio for washed CNF is 0.64. These findings suggest that washing the sample increases the ratio of disordered to organised carbon. This could be attributed to defects created in the graphitic structure during the acid washing for the removal of CaO.

To determine the magnetic properties of the CNF samples, a vibrating sample magnetometer was employed at room temperature. As indicated in Figure 6D, CNF and CNF@CaO have saturation magnetisation values of 7.3 and 11.8 emu/g, respectively. Similar values were reported for iron-doped diatomite, which exhibited a saturation magnetisation in the range of 10–15 emu/g [51]; other reports of CNT doped with a high iron content demonstrated a saturation magnetisation as high as 38 emu/g [52]. Coercivity values (H_c) for pure CNF and CNF@CaO are −414 and −277 Oe, respectively. This low coercivity of remanence is indicative of a superparamagnetic property (i.e., responsiveness to an applied field without retaining any magnetism after removal of the same). This property allows for simple separation of the washed carbon nanofibers suspended in solution.

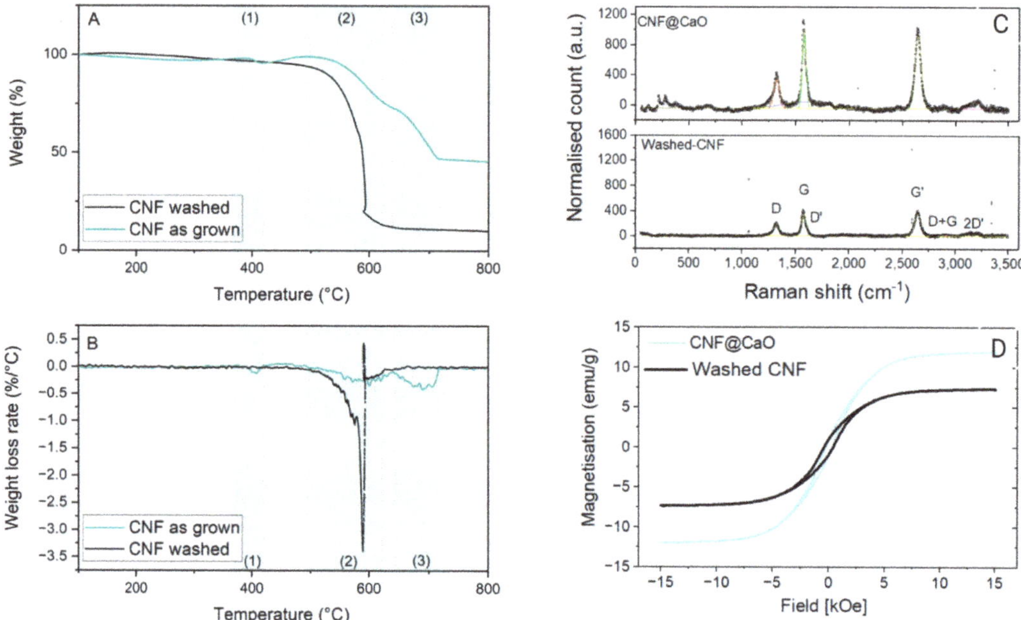

Figure 5. Characterisation of CNF@CaO as produced and CNF after acid washing. (**A**) Thermogravimetric analysis of as-grown CNF in the surface of CaO (teal) and acid-washed nanomaterial, purified for the removal of CaO (black); (**B**) Weight loss rate for the as-grown CNF (purple) and purified CNF (teal); (**C**) Raman spectra of CNF@CaO (top) and CNF—washed (bottom); (**D**) Magnetisation curve of CNF@CaO (teal) and CNF—washed (black).

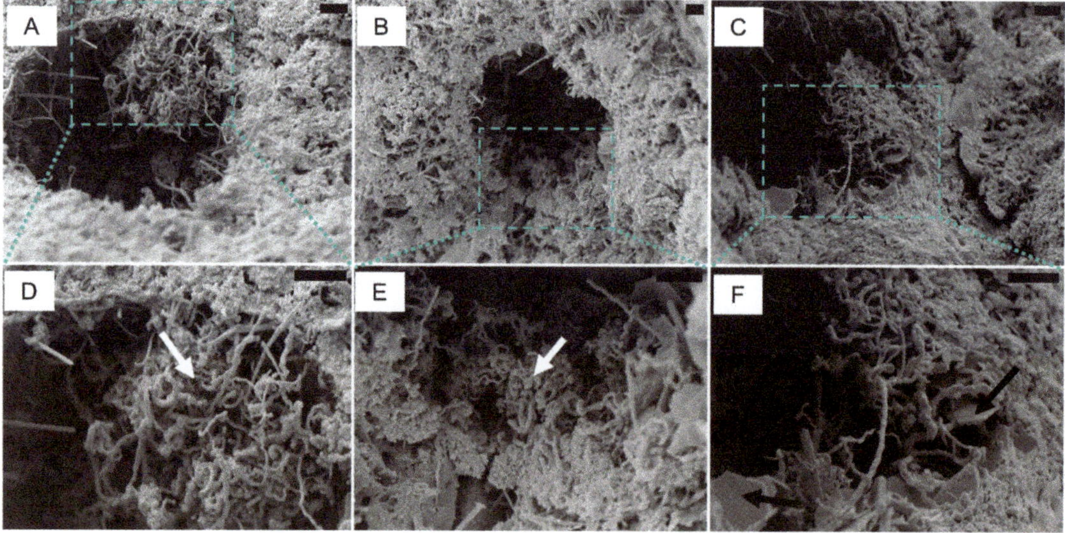

Figure 6. Scanning electron microscopy of carbon nanofibers supported in CaO, dispersed in mortar. Scale bar 2 µm. (**A–C**) mortar surface containing the nest of CNF; (**D–F**) details of the CNF next, white arrows indicate CNF and black arrow indicate portlandite.

3.2. Dispersion of Carbon Nanofibers in Mortar

Scanning electron microscopy was used to investigate the dispersion of the carbon nanofibers in mortar, using the sample with 2 wt% of CNF. The highest content of carbon nanofibers was used due to the challenge in locating the carbon nanofibers in the matrix—the carbon nanofibers are difficult to differentiate from hydration products, particularly ettringite. After the CNF was dispersed and the material cured in high humidity for 28 days, the sample was dried in air, followed by vacuum drying before SEM. The fibres were mainly observed as small conglomerates, ranging between 3 and 7 µm, inside the pores in the mortar (Figure 6A–C). No conglomerate larger than 7 µm was observed, indicating the nanofibers are distributed in small nest-like bundles throughout the matrix. Previous investigations of carbon nanofibers and nanotubes distributed in cementitious matrix have shown that a poor dispersion may lead to agglomerated CNT; alternatively, a better dispersion of the fibres leads to high-density products dispersed as clumps [53,54]. Figure 6F also shows portlandite entangled with the carbon nanofibers (Figure 6D,E), demonstrating the hydration of the CaO substrate.

3.3. Influence of CNF Concentration on Electrical Resistivity

The addition of carbon nanofibers to the mortar has the effect of increasing the electrical conductivity of the specimens. The electrical resistivity measured with the four-probe point method is presented in Figure 7 and differed with the hydration levels of the specimen, as well as the content of carbon nanofiber and presence of the CaO support. The electrical resistivity was much lower for samples immediately after 28 days of curing in a high-humidity (>95%) environment, as shown in Figure 7A. For the damp samples, the voltage between the internal electrodes at the central electrodes was ~8–10 V and the current ranged from 400 to 1400 µA. As a result, the electrical resistivity of the mortar samples with high relative humidity was between 80–230 Ω m, with a slight decrease in conductivity for samples with washed CNF and CaO. Similar values were found in water-saturated mortar samples containing CNF [3,8]. These lower values of electrical resistivity to the water-saturated specimen were attributed to the high conductivity of the pore solution. The electrical resistivity of mortar samples after drying (Figure 7B) was markedly increased—between 0.77–42 kΩ m. For the dried samples, the voltage measured in the internal electrodes was ~4–8 V whereas the current was significantly lower—between 1 and 110 µA. Furthermore, for the dried samples, the contribution of the content of CNF to the changes in conductivity was more pronounced when compared with the damp samples. In this case, the samples with control and with 0.4 wt% of CNF in mortar presented similar values of resistivity, ~2×10^4 Ω m. This indicates that, up to 0.4 wt% CNF, the mortar is still presenting DC electrical resistivity in the range of 10^4–10^7 Ω m, i.e., acting as an quasi-insulator [9]. However, when the content of CNF increases to 2 wt%, the resistivity drops to 5×10^3 Ω m, which is at least one order of magnitude less than the control. For this sample, an increased standard deviation was observed, possibly associated with the sample's non-uniform conductivity and bonding issues with the steel electrode. In addition, the values of electrical resistivity were 7.7×10^2 Ω m, two orders of magnitude lower, for the samples containing 1.2 wt% of CNF after the material was washed with acid, i.e., when the CaO was removed. Unfortunately, two specimens containing CNF(A) −1.2 wt% fractured between the matrix and the steel electrode; thus, there was no standard deviation associated with this sample. The decrease in resistivity highlights the strong effect of the CNF in the composite. These results are in agreement with other reports, where the required dosage of CNF to achieve a well-established current through tunnelling varies between 0.6 wt% [28] and 2.25 vol% [23]. The results also show the increased conductivity of the composites once the support is removed.

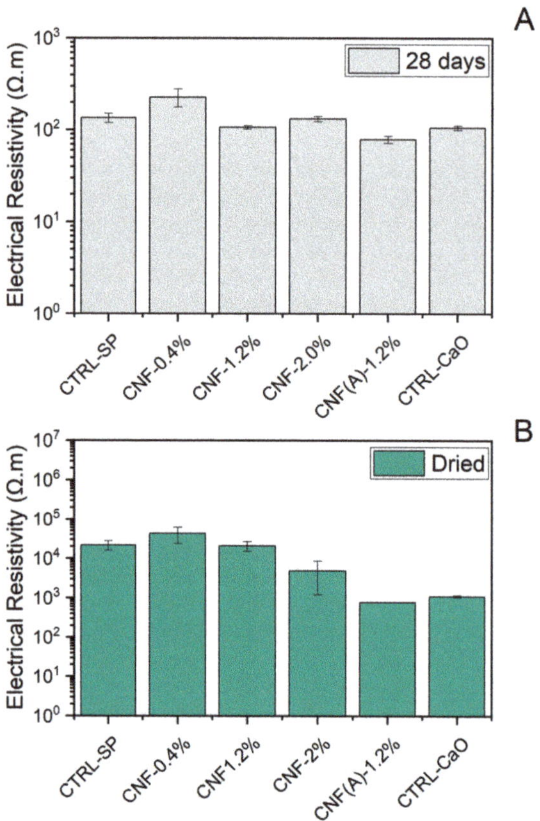

Figure 7. Effect of CNF concentration on the electrical resistivity of mortar specimens (**A**) immediately after 28 days in a high-humidity chamber and (**B**) after drying.

Interestingly, the control with only the substrate, i.e., the precursor used for the deposition of the material, was also investigated and the values of resistivity were ~1.1 kΩ m. The substrate was produced from the reaction between CaO and Fe^{3+} to produce mixed oxides with iron and calcium, $Ca_2Fe_2O_5$, as shown in Equation (1). It is the reduction of $Ca_2Fe_2O_5$ by ethanol that results in the formation of the carbon nanofibers, and also the reduced forms of iron, including nanoparticles of iron and iron carbide. In this case, ~1.8% by weight of cement (bwoc) of calcium oxide containing mixed oxides with iron and calcium was added to the mortar. Other authors have also investigated the use of nanoparticles of iron oxide to successfully increase the conductivity of cementitious matrix [55]. However, it is interesting to notice that the conductivity of the samples with iron oxides is markedly increased when compared with samples containing more iron and CNF. The reason behind the difference is not immediately apparent. Iron oxides are poorer conductors when compared with the reduced forms of iron [56]. However, it could potentially highlight how the pure powder dispersion in water is a lot more favourable than the material coated with CNF. Therefore, the dispersion is favoured and increases the conductivity. However, more studies are necessary.

3.4. Influence of CNF Concentration on the Piezoresistive Response under Compressive Loading

The piezoresistivity of the mortar samples under compressive load was investigated in the elastic region. This focuses on the potential application of the composite for load detection and structural soundness, i.e., for non-destructive systems. The four electrodes in the mortar samples were positioned perpendicular to the plane of compression (Figure 3a), with a load exerted at a rate of 0.8 mm min^{-1} up to 7.5 MPa. The linear response of the stress–strain curve aspect (Figure 8) confirms the elastic behaviour of the sample and the slope of the curve was 29 GPa for the control sample with SP (CTRL-SP), 42 GPa for the control sample with CaO (CTRL-CaO) and between 23.7 and 35.3 GPa for the samples with CNF. The addition of CNF/CNTs in cementitious composites has been proven effective in developing the mechanical properties of the materials [57], since fibres can bridge microcracks, fill pores and accelerate hydration. However, the obtained stiffness parameters are not accurate due to the presence of electrodes and the cuboid shape of the samples. Compression tests following standards were not performed in this study due to the limited amount of material.

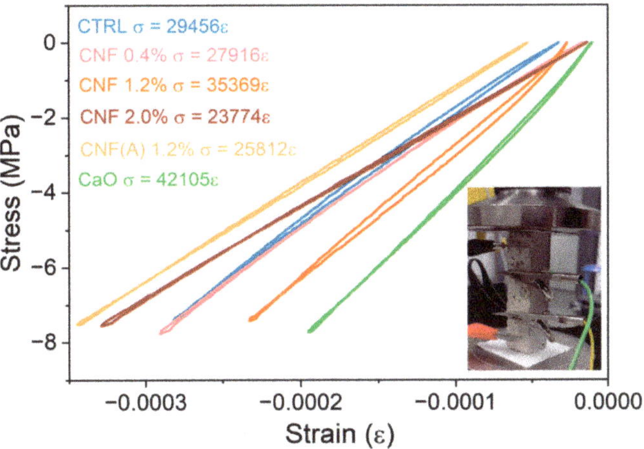

Figure 8. Stress–strain relationship of mortar specimen under compressive loading. Inside detail: experimental setup for the piezoresistivity testing.

Mortar samples containing CNF showed piezoresistive behaviour dependent on the concentration of CNF. To investigate the stability and repeatability of the piezoresistive response to compressive loading, 18 cycles with an amplitude of 7.5 MPa and a loading rate of 0.8 mm·min^{-1} were applied to the mortars. The fractional change in resistivity (FCR) was calculated by dividing the difference in resistivity at each point with the no-load resistivity (or baseline resistivity). Figure 9 shows that the electrical resistivity of all mortars is consistent with expected response under compression and decreases with the increase in the compressive load due to the shortening of the conduction path. Typically, samples take some time to stabilise, with the first few cycles still not returning to zero. The increase in the baseline electrical resistivity over time is caused by the polarisation effect, mainly due to the presence of water and dissolved ions in the water. This also indicates that residual water is still present in the sample, as dry samples are less prone to variations on the curve [27]. Furthermore, micro damages separating adjacent nanofibers may also lead to an increase in resistivity, consequently, an increase in the baseline [6]. The control mortar without CNF (Figure 9A), as well as the sample with only CaO and mixed iron and calcium oxides (Figure 9B), present a small amplitude for the FCR values on loading, with a maximum modular FCR (FCR_{max}) of 1%, indicating a negligible piezoresistive behaviour. The FCR of the samples with CNF reached FCR_{max} of 0.5, 8.4 and 1.3% in concentrations of 0.4, 1.2% and 2 wt% CNF, respectively. For the sample with washed CNF at 1.2 wt%, the FCR_{max} was 1.0%.

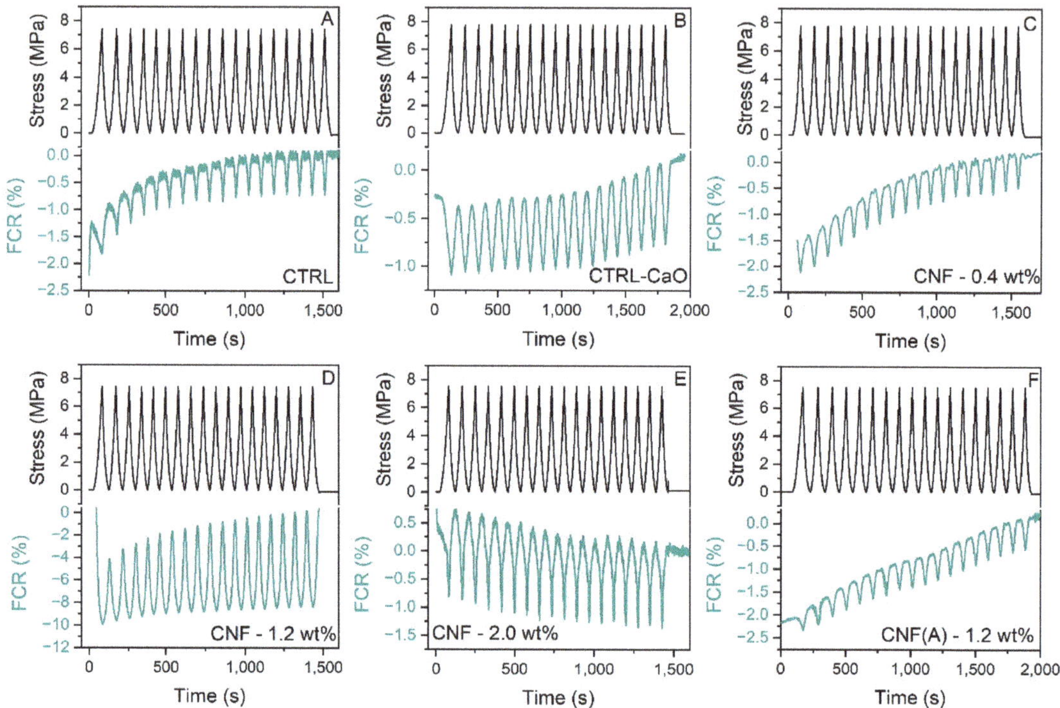

Figure 9. Piezoresistive behaviours of mortars under cyclic compressive loading. (**A**) Control mortar; (**B**) control mortar with CaO substrate; (**C**) mortar with 0.4 wt% of CNF; (**D**) mortar with 1.2 wt% CNF; (**E**) mortar with 2.0 wt%; (**F**) mortar with 1.2 wt% washed CNF.

The obtained values, as well as the pronounced change in resistivity for the sample with 1.2 wt% of CNF, were interpreted in relation to percolation theory [9]. For the samples with CNF 0.4 wt%, we conjecture that the concentration may not be sufficient to achieve a conductive network capable of sensing the variation in the state of strain. Thus, values of resistivity (Figure 7) and FCR under compressive loading were similar to the control. On the other hand, when the CNF@CaO concentration increases to 1.2 wt%, changes in contact resistance increase due to the formation/breakage of CNF junctions. In addition, under compressive loading, the distance between CNF decreases, facilitating a tunnelling effect. The unique nest-like morphology of the CNF (Figure 6D,E) distributed in the mortar may also facilitate the high FCR observed in this sample, providing many possible locations for triggering CNF contact and tunnelling [42]. When the concentration of CNF increases further, by the addition of 2 wt% of CNF and 1.2 wt% of washed CNF, the sample already has a more pronounced and stable conductive network. Thus, its electrical properties were not particularly affected by any external load; that is, the piezoresistive effect caused by the variation in proximity between CNFs is not so pronounced. As a consequence, the resistivity values (Figure 7) were lower than the control, but the FCR_{max} is similar to the control. This indicates that, to obtain a piezoresistive behaviour for the sample with washed CNF, reduced concentrations of material should be used. When compared to the literature, a maximum FCR of 9% was observed for carbon nanofibers at 2.25 vol% [23] and an FCR maximum of 2% for mortar samples with 2.5 vol% of CNF [58], thus indicating a lower concentration of the material here also shows a good result for the sample at 1.2 wt% (1.08 vol%). Alternatively, an FCR max of 5.5% was observed for the cement samples with 0.1 wt% of CNF, indicating a better piezoresistive response under loading [27]. This could be attributed to a conductive network being obtained in this sample under compressive loading at 0.1 wt% of CNF. It could be that the dispersion of the material is better, since the material does not form clumps. This could have been the case for washed samples if the CaO substrate had not been used.

To examine the relationship between strain and variations in electrical resistivity in the mortar, the strain versus FCR curves are presented in Figure 10. To quantify the strain sensitivity, the gauge factor (*GF*) was used, as it represents the relative change in electrical resistivity due to the mechanical resistance. The relationship between strain amplitude and resistance change can be described as follows:

$$GF = \frac{d\rho/\rho}{\varepsilon_1} \quad (3)$$

where ε_1 is compressive strain measured by a strain gauge and $d\rho/\rho$ is equal to the FCR. The *GF* is then obtained by applying Equation (3) via fitting with a linear regression to the FCR–strain curve, as shown in Figure 10.

For the mortar without CNF, as well as the mortar with 0.4, 2 wt% of CNF and 1.2 wt% of washed CNF, the GF is ~30, confirming that the specimens are not applicable for strain sensing under a compressive load. However, when 1.2 wt% of CNF is added, the GF markedly increases to 1552, demonstrating the high sensitivity of the material. Moreover, the sample with 1.2 wt% shows a more constant variation, without noise. This indicates the potential of mortar embedded with CNF for more stable piezoresistive behaviours. Recently, Ding et al. (2022) investigated the piezoresistivity of cement composites containing 0–5 wt% of CNT synthesised in the surface of cement, resulting in gauge factors of 22–748 [2]. Likewise, CNT@cement embedded in mortar at a concentration between 0.4–2 wt% resulted in gauge factors between 744 to 1170 [42]. For future work, a different design of experiments could be considered, including more points around 1.2 wt% CNF, to better understand the behaviour of the material.

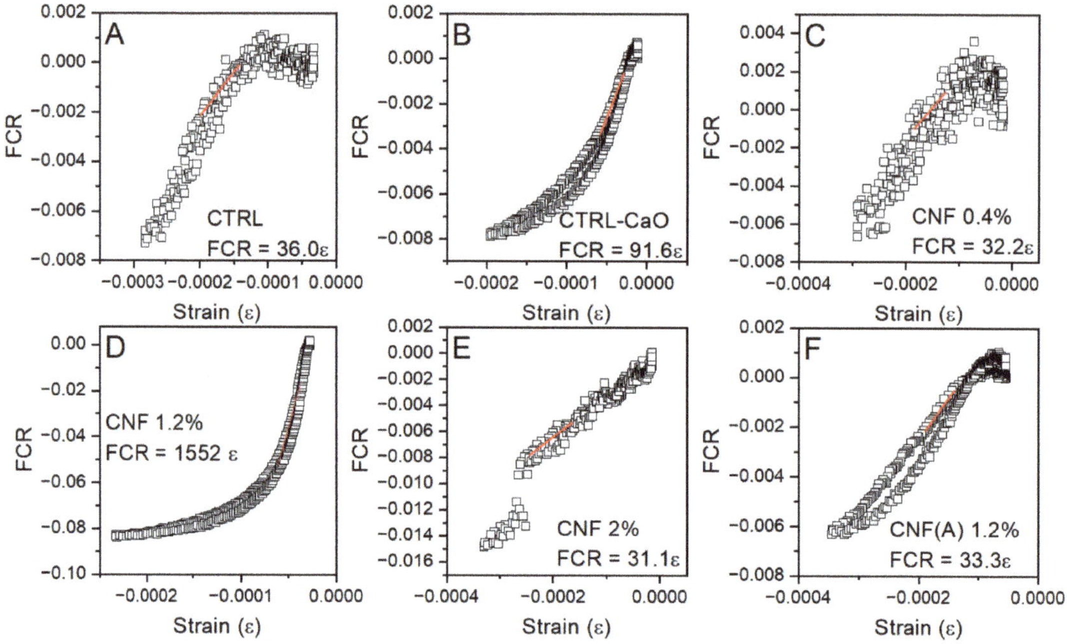

Figure 10. FCR versus strain for mortar specimens at different CNF concentrations under a monotonic uniaxial compressive loading with an amplitude of 7.5 MPa. (**A**) Control mortar; (**B**) control mortar with CaO substrate; (**C**) mortar with 0.4 wt% of CNF; (**D**) mortar with 1.2 wt% CNF; (**E**) mortar with 2.0 wt%; (**F**) mortar with 1.2 wt% washed CNF.

4. Conclusions

This study presents a versatile method for fabricating carbon nanofibers on substrates designed to facilitate the dispersion of the nanomaterial. Using iron as a catalyst, CNF was produced by chemical vapour deposition (CVD) on the surface of calcium oxide (CaO). This substrate was chosen due to its ease of removal, which results in free CNF, and its polarity, which lends an amphiphilic quality to the composite and facilitates its dispersion in mortar. Small agglomerates of loose fibres were observed with a SEM after the fibres were dispersed in the mortar, indicating a good distribution of the material. Increasing the concentration of CNF in the mortar resulted in a decrease in resistivity, with the lowest values occurring at around 2 wt%. Additionally, after removing CaO, the electrical resistivity decreased to 0.8 kΩ m for contents containing approximately 1.2 wt% CNF. Under compressive loading, the piezoresistive response of CNF was studied, and the composite containing 1.2 wt% of unwashed CNF exhibited an excellent variation in electrical resistivity. The gauge factor (GF) was used to quantify the sensitivity to deformation, and the sample containing 1.2% by weight of CNF@CaO had a gauge factor of 1552 while the others had gauge factors below 100. This sample's exceptional deformation sensitivity suggests that the contact points formed between small adjacent CNF@CaO clusters can be easily formed and broken, thereby increasing the nanocomposite's sensitivity. Potential applications for the enhanced electrical properties include evaluating the condition of civil engineering structures.

Author Contributions: Conceptualization, L.R.d.S., A.A.-T. and J.C.T.; Data curation, L.R.d.S., M.P. and J.C.T.; Investigation, L.R.d.S. and M.P.; Methodology, L.R.d.S.; Project administration, A.A.-T.; Resources, A.A.-T. and J.C.T.; Supervision, A.A.-T.; Validation, L.R.d.S.; Writing—original draft, L.R.d.S. and G.M.; Writing—review and editing, L.R.d.S., G.M., M.P. and J.C.T. All authors have read and agreed to the published version of the manuscript.

Funding: This work is supported by UKRI-EPSRC (Grant No. EP/P02081X/1, Resilient Materials 4 Life, RM4L) and by CAPES (grant no. 88887.653331/2021-00).

Institutional Review Board Statement: Not applicable.

Informed Consent Statement: Not applicable.

Data Availability Statement: The raw/processed data required to reproduce these findings cannot be shared at this time as the data also forms part of an ongoing study.

Acknowledgments: The authors kindly acknowledge the comments from Christos Vlachakis and Sripriya Rengaraju.

Conflicts of Interest: The authors declare that they have no known competing financial interests or personal relationships that could have appeared to influence the work reported in this paper.

References

1. Han, B.; Yu, X.; Kwon, E. A Self-Sensing Carbon Nanotube/Cement Composite for Traffic Monitoring. *Nanotechnology* **2009**, *20*, 445501. [CrossRef] [PubMed]
2. Ding, S.; Xiang, Y.; Ni, Y.Q.; Thakur, V.K.; Wang, X.; Han, B.; Ou, J. In-Situ Synthesizing Carbon Nanotubes on Cement to Develop Self-Sensing Cementitious Composites for Smart High-Speed Rail Infrastructures. *Nano Today* **2022**, *43*, 101438. [CrossRef]
3. Wang, H.; Gao, X.; Liu, J. Effects of Salt Freeze-Thaw Cycles and Cyclic Loading on the Piezoresistive Properties of Carbon Nanofibers Mortar. *Constr. Build. Mater.* **2018**, *177*, 192–201. [CrossRef]
4. Galao, O.; Baeza, F.J.; Zornoza, E.; Garcés, P. Strain and Damage Sensing Properties on Multifunctional Cement Composites with CNF Admixture. *Cem. Concr. Compos.* **2014**, *46*, 90–98. [CrossRef]
5. Lee, S.J.; Ahn, D.; You, I.; Yoo, D.Y.; Kang, Y.S. Wireless Cement-Based Sensor for Self-Monitoring of Railway Concrete Infrastructures. *Autom. Constr.* **2020**, *119*, 103323. [CrossRef]
6. Fu, X.; Chung, D.D.L. Self-Monitoring of Fatigue Damage in Carbon Fiber Reinforced Cement. *Cem. Concr. Res.* **1996**, *26*, 15–20. [CrossRef]
7. Singh, A.P.; Gupta, B.K.; Mishra, M.; Govind; Chandra, A.; Mathur, R.B.; Dhawan, S.K. Multiwalled Carbon Nanotube/Cement Composites with Exceptional Electromagnetic Interference Shielding Properties. *Carbon N. Y.* **2013**, *56*, 86–96. [CrossRef]
8. Siad, H.; Lachemi, M.; Sahmaran, M.; Mesbah, H.A.; Hossain, K.A. Advanced Engineered Cementitious Composites with Combined Self-Sensing and Self-Healing Functionalities. *Constr. Build. Mater.* **2018**, *176*, 313–322. [CrossRef]
9. Ding, S.; Dong, S.; Ashour, A.; Han, B. Development of Sensing Concrete: Principles, Properties and Its Applications. *J. Appl. Physics.* **2019**, *24*, 241101. [CrossRef]
10. Abedi, M.; Fangueiro, R.; Gomes Correia, A. A Review of Intrinsic Self-Sensing Cementitious Composites and Prospects for Their Application in Transport Infrastructures. *Constr. Build. Mater.* **2021**, *310*, 125139. [CrossRef]
11. Marcos-Meson, V.; Michel, A.; Solgaard, A.; Fischer, G.; Edvardsen, C.; Skovhus, T.L. Corrosion Resistance of Steel Fibre Reinforced Concrete–A Literature Review. *Cem. Concr. Res.* **2018**, *103*, 1–20. [CrossRef]
12. Papanikolaou, I.; Ribeiro de Souza, L.; Litina, C.; Al-Tabbaa, A. Investigation of the Dispersion of Multi-Layer Graphene Nanoplatelets in Cement Composites Using Different Superplasticiser Treatments. *Constr. Build. Mater.* **2021**, *293*, 123543. [CrossRef]
13. Dong, W.; Li, W.; Zhu, X.; Sheng, D.; Shah, S.P. Multifunctional Cementitious Composites with Integrated Self-Sensing and Hydrophobic Capacities toward Smart Structural Health Monitoring. *Cem. Concr. Compos.* **2021**, *118*, 103962. [CrossRef]
14. Han, B.; Yu, X.; Zhang, K.; Kwon, E.; Ou, J. Sensing Properties of CNT-Filled Cement-Based Stress Sensors. *J. Civ. Struct. Heal. Monit.* **2011**, *1*, 17–24. [CrossRef]
15. Song, C.; Choi, S. Moisture-Dependent Piezoresistive Responses of CNT-Embedded Cementitious Composites. *Compos. Struct.* **2017**, *170*, 103–110. [CrossRef]
16. Yoo, D.Y.; You, I.; Zi, G.; Lee, S.J. Effects of Carbon Nanomaterial Type and Amount on Self-Sensing Capacity of Cement Paste. *Measurement* **2019**, *134*, 750–761. [CrossRef]
17. Abedi, M.; Fangueiro, R.; Correia, A.G. Effects of Multiscale Carbon-Based Conductive Fillers on the Performances of a Self-Sensing Cementitious Geocomposite. *J. Build. Eng.* **2021**, *43*, 103171. [CrossRef]
18. Abedi, M.; Fangueiro, R.; Correia, A.G. Ultra-Sensitive Affordable Cementitious Composite with High Mechanical and Microstructural Performances by Hybrid CNT/GNP. *Materials* **2020**, *13*, 3484. [CrossRef]
19. Ding, S.; Ruan, Y.; Yu, X.; Han, B.; Ni, Y.Q. Self-Monitoring of Smart Concrete Column Incorporating CNT/NCB Composite Fillers Modified Cementitious Sensors. *Constr. Build. Mater.* **2019**, *201*, 127–137. [CrossRef]
20. Makar, J.M.; Chan, G.W. Growth of Cement Hydration Products on Single-Walled Carbon Nanotubes. *J. Am. Ceram. Soc.* **2009**, *92*, 1303–1310. [CrossRef]
21. Mendoza Reales, O.A.; Dias Toledo Filho, R. A Review on the Chemical, Mechanical and Microstructural Characterization of Carbon Nanotubes-Cement Based Composites. *Constr. Build. Mater.* **2017**, *154*, 697–710. [CrossRef]

22. Yoo, D.Y.; You, I.; Lee, S.J. Electrical Properties of Cement-Based Composites with Carbon Nanotubes, Graphene, and Graphite Nanofibers. *Sensors* **2017**, *17*, 1064. [CrossRef] [PubMed]
23. Wang, H.; Gao, X.; Wang, R. The Influence of Rheological Parameters of Cement Paste on the Dispersion of Carbon Nanofibers and Self-Sensing Performance. *Constr. Build. Mater.* **2017**, *134*, 673–683. [CrossRef]
24. Gao, D.; Sturm, M.; Mo, Y.L. Electrical Resistance of Carbon-Nanofiber Concrete. *Smart Mater. Struct.* **2009**, *18*, 095039. [CrossRef]
25. Liu, Y.; Wang, M.; Wang, W. Ohmic Heating Curing of Electrically Conductive Carbon Nanofiber/Cement-Based Composites to Avoid Frost Damage under Severely Low Temperature. *Compos. Part A Appl. Sci. Manuf.* **2018**, *115*, 236–246. [CrossRef]
26. Wang, H.; Gao, X.; Liu, J.; Ren, M.; Lu, A. Multi-Functional Properties of Carbon Nanofiber Reinforced Reactive Powder Concrete. *Constr. Build. Mater.* **2018**, *187*, 699–707. [CrossRef]
27. Konsta-Gdoutos, M.S.; Aza, C.A. Self Sensing Carbon Nanotube (CNT) and Nanofiber (CNF) Cementitious Composites for Real Time Damage Assessment in Smart Structures. *Cem. Concr. Compos.* **2014**, *53*, 162–169. [CrossRef]
28. Dalla, P.T.; Dassios, K.G.; Tragazikis, I.K.; Exarchos, D.A.; Matikas, T.E. Carbon Nanotubes and Nanofibers as Strain and Damage Sensors for Smart Cement. *Mater. Today Commun.* **2016**, *8*, 196–204. [CrossRef]
29. Dinesh, A.; Abirami, B.; Moulica, G. Carbon Nanofiber Embedded Cement Composites: Properties and Promises as Sensor—A Review. *Mater. Today Proc.* **2021**, *44*, 4166–4172. [CrossRef]
30. Yan, Y.; Miao, J.; Yang, Z.; Xiao, F.X.; Yang, H.B.; Liu, B.; Yang, Y. Carbon Nanotube Catalysts: Recent Advances in Synthesis, Characterization and Applications. *Chem. Soc. Rev.* **2015**, *44*, 3295–3346. [CrossRef]
31. Kumar, M.; Ando, Y. Chemical Vapor Deposition of Carbon Nanotubes: A Review on Growth Mechanism and Mass Production. *J. Nanosci. Nanotechnol.* **2010**, *10*, 3739–3758. [CrossRef] [PubMed]
32. Mathur, A.; Wadhwa, S.; Sinha, S. Say Hello to Carbon Nanotubes. In *Introduction to Carbon Nanomaterials*; Current and Future Developments in Nanomaterials and Carbon Nanotubes; Narang, J., Pundir, C., Eds.; Bentham Science Publishers: Sharjah, United Arab Emirates, 2018; Volume 1, pp. 1–79. [CrossRef]
33. Pirard, S.L.; Douven, S.; Pirard, J.-P. Large-Scale Industrial Manufacturing of Carbon Nanotubes in a Continuous Inclined Mobile-Bed Rotating Reactor via the Catalytic Chemical Vapor Deposition Process. *Front. Chem. Sci. Eng.* **2017**, *11*, 280–289. [CrossRef]
34. Fecht, H.-J.; Brühne, K.; Gluche, P. Carbon-Based Nanomaterials and Hybrids: Synthesis, Properties, and Commercial Applications. *Carbon-Based Nanomater. Hybrids* **2016**. [CrossRef]
35. De Volder, M.F.L.; Tawfick, S.H.; Baughman, R.H.; Hart, A.J. Carbon Nanotubes: Present and Future Commercial Applications. *Science* **2013**, *339*, 535–539. [CrossRef]
36. Maruyama, S.; Kojima, R.; Miyauchi, Y.; Chiashi, S.; Kohno, M. Low-Temperature Synthesis of High-Purity Single-Walled Carbon Nanotubes from Alcohol. *Chem. Phys. Lett.* **2002**, *360*, 229–234. [CrossRef]
37. Buasiri, T.; Habermehl-Cwirzen, K.; Krzeminski, L.; Cwirzen, A. Piezoresistive Load Sensing and Percolation Phenomena in Portland Cement Composite Modified with In-Situ Synthesized Carbon Nanofibers. *Nanomaterials* **2019**, *9*, 594. [CrossRef]
38. Ghaharpour, F.; Bahari, A.; Abbasi, M.; Ashkaran, A.A. Parametric Investigation of CNT Deposition on Cement by CVD Process. *Constr. Build. Mater.* **2016**, *113*, 523–535. [CrossRef]
39. Ludvig, P.; Calixto, J.M.; Ladeira, L.O.; Gaspar, I.C.P. Using Converter Dust to Produce Low Cost Cementitious Composites by in Situ Carbon Nanotube and Nanofiber Synthesis. *Materials* **2011**, *4*, 575–584. [CrossRef]
40. de Souza, T.C.; Pinto, G.; Cruz, V.S.; Moura, M.; Ladeira, L.O.; Calixto, J.M. Evaluation of the Rheological Behavior, Hydration Process, and Mechanical Strength of Portland Cement Pastes Produced with Carbon Nanotubes Synthesized Directly on Clinker. *Constr. Build. Mater.* **2020**, *248*, 118686. [CrossRef]
41. Dunens, O.M.; Mackenzie, K.J.; Harris, A.T. Synthesis of Multiwalled Carbon Nanotubes on Fly Ash Derived Catalysts. *Environ. Sci. Technol.* **2009**, *43*, 7889–7894. [CrossRef]
42. Zhan, M.; Pan, G.; Zhou, F.; Mi, R.; Shah, S.P. In Situ-Grown Carbon Nanotubes Enhanced Cement-Based Materials with Multifunctionality. *Cem. Concr. Compos.* **2020**, *108*, 103518. [CrossRef]
43. de Souza, L.R. *Síntese e Caracterização de Nanofibras Magnéticas de Carbono Suportadas Por Óxido Em Cálcio*; Universidade Federal de Minas Gerais: Belo Horizonte, Brazil, 2010.
44. Jourdain, V.; Bichara, C. Current Understanding of the Growth of Carbon Nanotubes in Catalytic Chemical Vapour Deposition. *Carbon* **2013**, *25*, 2–39. [CrossRef]
45. D'Alessandro, A.; Rallini, M.; Ubertini, F.; Materazzi, A.L.; Kenny, J.M. Investigations on Scalable Fabrication Procedures for Self-Sensing Carbon Nanotube Cement-Matrix Composites for SHM Applications. *Cem. Concr. Compos.* **2016**, *65*, 200–213. [CrossRef]
46. Meyyappan, M.; Delzeit, L.; Cassell, A.; Hash, D. Carbon Nanotube Growth by PECVD: A Review. *Plasma Sources Sci. Technol.* **2003**, *12*, 205–216. [CrossRef]
47. Arepalli, S.; Freiman, S.; Hooker, S.; Migler, K. *Measurement Issues in Single-Wall Carbon Nanotubes*; Special Publication (NIST SP); National Institute of Standards and Technology: Gaithersburg, MD, USA, 2008. Available online: https://tsapps.nist.gov/publication/get_pdf.cfm?pub_id=852726 (accessed on 11 May 2022).
48. Arepalli, S.; Nikolaev, P.; Gorelik, O.; Hadjiev, V.G.; Holmes, W.; Files, B.; Yowell, L. Protocol for the Characterization of Single-Wall Carbon Nanotube Material Quality. *Carbon N. Y.* **2004**, *42*, 1783–1791. [CrossRef]

49. Sato, K.; Saito, R.; Oyama, Y.; Jiang, J.; Cançado, L.G.; Pimenta, M.A.; Jorio, A.; Samsonidze, G.G.; Dresselhaus, G.; Dresselhaus, M.S. D-Band Raman Intensity of Graphitic Materials as a Function of Laser Energy and Crystallite Size. *Chem. Phys. Lett.* **2006**, *427*, 117–121. [CrossRef]
50. Sadezky, A.; Muckenhuber, H.; Grothe, H.; Niessner, R.; Pöschl, U. Raman Microspectroscopy of Soot and Related Carbonaceous Materials: Spectral Analysis and Structural Information. *Carbon N. Y.* **2005**, *43*, 1731–1742. [CrossRef]
51. Alijani, H.; Beyki, M.H.; Shariatinia, Z.; Bayat, M.; Shemirani, F. A New Approach for One Step Synthesis of Magnetic Carbon Nanotubes/Diatomite Earth Composite by Chemical Vapor Deposition Method: Application for Removal of Lead Ions. *Chem. Eng. J.* **2014**, *253*, 456–463. [CrossRef]
52. Jiao, Q.; Hao, L.; Shao, Q.; Zhao, Y. In Situ Synthesis of Iron-Filled Nitrogen-Doped Carbon Nanotubes and Their Magnetic Properties. *Carbon N. Y.* **2013**, *61*, 647–649. [CrossRef]
53. Kim, H.K.; Park, I.S.; Lee, H.K. Improved Piezoresistive Sensitivity and Stability of CNT/Cement Mortar Composites with Low Water-Binder Ratio. *Compos. Struct.* **2014**, *116*, 713–719. [CrossRef]
54. Kim, H.K.; Nam, I.W.; Lee, H.K. Enhanced Effect of Carbon Nanotube on Mechanical and Electrical Properties of Cement Composites by Incorporation of Silica Fume. *Compos. Struct.* **2014**, *107*, 60–69. [CrossRef]
55. Vipulanandan, C.; Mohammed, A. Smart Cement Modified with Iron Oxide Nanoparticles to Enhance the Piezoresistive Behavior and Compressive Strength for Oil Well Applications. *Smart Mater. Struct.* **2015**, *24*, 125020. [CrossRef]
56. Kharton, V.V.; Tsipis, E.V.; Kolotygin, V.A.; Avdeev, M.; Viskup, A.P.; Waerenborgh, J.C.; Frade, J.R. Mixed Conductivity and Stability of $CaFe_2O_{4-\delta}$. *J. Electrochem. Soc.* **2008**, *155*, P13. [CrossRef]
57. Gao, Y.; Zhu, X.; Corr, D.J.; Konsta-Gdoutos, M.S.; Shah, S.P. Characterization of the Interfacial Transition Zone of CNF-Reinforced Cementitious Composites. *Cem. Concr. Compos.* **2019**, *99*, 130–139. [CrossRef]
58. Wang, H.; Zhang, A.; Zhang, L.; Wang, Q.; Yang, X.; Gao, X.; Shi, F. Electrical and Piezoresistive Properties of Carbon Nanofiber Cement Mortar under Different Temperatures and Water Contents. *Constr. Build. Mater.* **2020**, *265*, 120740. [CrossRef]

Communication

Torsional Low-Strain Test for Nondestructive Integrity Examination of Existing High-Pile Foundation

Yunpeng Zhang [1,2], M. Hesham El Naggar [2], Wenbing Wu [1,*] and Zongqin Wang [1]

[1] Faculty of Engineering, China University of Geosciences, Wuhan 430074, China; zypsky@cug.edu.cn (Y.Z.); wzongqin@cug.edu.cn (Z.W.)

[2] Geotechnical Research Centre, Department of Civil and Environmental Engineering, Western University, London, ON N6A 5B9, Canada; helnaggar@eng.uwo.ca

* Correspondence: wuwb@cug.edu.cn

Abstract: Low-strain tests are widely utilized as a nondestructive approach to assess the integrity of newly piled foundations. So far, the examination of existing pile foundations is becoming an indispensable protocol for pile recycling or post-disaster safety assessment. However, the present low-strain test is not capable of testing existing pile foundations. In this paper, the torsional low-strain test (TLST) is proposed to overcome this drawback. Both the upward and downward waves are considered in the TLST wave propagation model established in this paper so that a firm theoretical basis is grounded for the test signal interpretations. A concise semi-analytical solution is derived and its rationality is verified by comparisons with the existing solutions for newly piled foundations and the finite element results. The main conclusions of this study can be drawn as follows: (1). by placing the sensors where the incident wave is applied, the number of reflected signals can be minimized; (2). the defects can be more evidently identified if the incident wave/sensors are input/installed close to the superstructure/pile head.

Keywords: nondestructive test; existing pile integrity; low-strain test; wave propagation

Citation: Zhang, Y.; El Naggar, M.H.; Wu, W.; Wang, Z. Torsional Low-Strain Test for Nondestructive Integrity Examination of Existing High-Pile Foundation. *Sensors* **2022**, 22, 5330. https://doi.org/10.3390/s22145330

Academic Editors: Phong B. Dao, Lei Qiu, Liang Yu, Tadeusz Uhl and Minh-Quy Le

Received: 14 June 2022
Accepted: 15 July 2022
Published: 16 July 2022

1. Introduction

Among many structure health monitoring approaches (static load test [1], image-based displacement measurement [2], low-strain test [3], and high-strain test [4]), the low-strain test is so far the most intuitive and economical way to assess the integrity of deep foundations, especially pile foundations [5–8]. This is because the test signal of the low-strain test is easily identifiable and it involves no disposable equipment or gauges. The traditional low-strain test utilizes longitudinal harmonic excitation as the incident wave so that an exposed cross-section of the foundation is needed to conduct the test [9,10]. Hence, the low-strain test is commonly used as the integrity inspection for newly piled foundations instead of existing ones. However, after decades of vigorous developments in infrastructure construction, the testing demands in major global construction markets have shifted from the newly piled foundations to the existing ones [11–14]. As a result, upgrading the low-strain test to satisfy the testing of existing foundations is especially urgent.

The fundamental theory of the low-strain test for pile foundations originates from the longitudinal vibration theory of the pile [15,16]. The combination of one-dimensional rod theory and the subgrade reaction model forms the mathematical prototype of the low-strain test [17–19]. High-frequency interferences often occur during the tests of large diameter piles, which are not revealed by the one-dimensional rod theory. The high-frequency interference can be addressed by simulating the soil and pile as three-dimensional continuum media [20,21]. However, due to the massive computation involved in rigorous 3D continuum models, digital signal filters (e.g., Savitzky–Golay) are preferred by engineers. Compared to the longitudinal vibration of piles, torsional vibrations receive less attention because they are not that common in nature. For most studies, torsional vibrations of piles are only regarded as additional problems caused by eccentric loadings [22]. However, since torsional vibration is less common in nature than longitudinal or horizontal vibration, it is an ideal subject for studying pile testing, as its strain wave signal may not be easily jammed or suppressed by other environmental loads. Moreover, because the velocity of the torsional wave is much smaller than that of the longitudinal wave, the torsional low-strain test has a smaller detection blind zone than the traditional low-strain test [23,24]. The torsional vibration theory is initially established on a similar basis to the longitudinal one: by simplifying the soil medium to infinitely thin layers, the rigorous 3D continuum theory for soil medium can be reduced to the plane strain model, based on which the straightforward closed-form solutions can be derived [25–29]. As the torsional vibration of pile foundations gained interest in the most recent decade, the finite element method (FEM) [30–32], finite integration technique [33], and boundary element method [34,35] all considerably fulfilled the knowledge of wave propagation across the soil-pile system during vibration.

In the literature mentioned above, the torsional incident wave is input at the pile head, under which circumstance there will only be an upward wave or a downward wave inside the intact pile body at the same time and neither will exist simultaneously [36–38]. However, when conducting the test for existing high-pile foundations, the incident wave can only be input at the shaft of the pile, because the pile head is fixed into the superstructure firmly. As a result, the upward and downward waves propagate inside the pile body simultaneously, dramatically increasing the complexity of strain wave signals. To account for this phenomenon, a rigorous torsional wave propagation model, taking both the upward and downward waves into account, is established in this paper to guide the signal interpretations of the TLST for existing high-pile foundations. Based on the proposed model, the optimal excitation and signal receiving layouts in the TLST for existing high-pile foundations are revealed.

2. Mathematical Model and Assumptions

The layout of the TLST for the existing high-pile foundation is depicted in Figure 1. Due to the head of the existing pile being firmly fixed in the superstructure, the torsional incident wave can only be input at the extending pile shaft so that both the upward and downward strain waves are generated. The pile is modeled as a one-dimensional rod in the proposed mathematical model and the surrounding soil is modeled as a three-dimensional viscoelastic medium. The interactions between the pile and the superstructure are simplified to springs and dashpots. Further, the fictitious soil-pile model [28] is introduced herein to authentically simulate the wave reflection at the interface of the pile bottom and the pile end soil. Other general assumptions adopted are listed as follows:

1. Throughout the TLSTs, the soil-pile system only undergoes small strain deformations so that the surrounding soil and the pile shaft are assumed to remain in perfect contact.
2. The incident wave utilized as the input of TLSTs in this paper is a half-sine harmonic impulse.
3. There are no normal and shear stresses at the ground surface and the amplitude of the strain wave diminishes to zero in radial infinity in the soil.
4. The displacement and forces at the interfaces of the fictitious soil pile and the real pile are continuous. By increasing the length and modulus of the fictitious soil pile, the foundation can shaft from the end-bearing piles to the floating piles.
5. The velocity response at the pile shaft is acquired to simulate the test results collected from the velocity or acceleration sensors installed at the pile shaft.

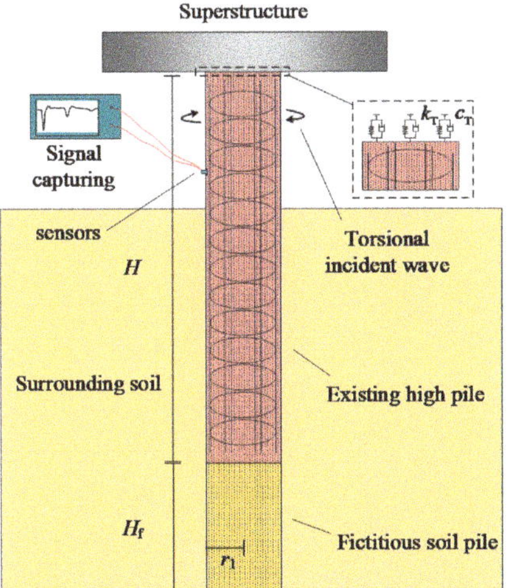

Figure 1. Schematics of torsional low-strain test for existing high-pile foundations.

3. Governing Equations and Boundary Conditions

3.1. Governing Equations

Based on the continuum theories, the equilibrium equations for the soil medium in a cylindrical coordinate system can be written as

$$\left(G_j^{\mathrm{s}} + \eta_j^{\mathrm{s}} \frac{\partial}{\partial t}\right) \nabla^2 u_j^{\mathrm{s}}(z, r, t) = \rho_j^{\mathrm{s}} \frac{\partial^2 u_j^{\mathrm{s}}(z, r, t)}{\partial t^2} \tag{1}$$

where G_j^{s}, η_j^{s}, u_j^{s}, and ρ_j^{s} denote the shear modulus, material damping, circumferential displacement, and density of the jth (vertically labeled) soil layer, respectively. $\nabla^2 = \frac{\partial^2}{\partial r^2} + \frac{\partial}{r \partial r} + \frac{\partial^2}{\partial z^2} - \frac{1}{r^2}$ is the Laplacian written in the cylindrical coordinates.

The three-dimensional rod theory can better reveal the wave propagation during the TLSITs. However, its adoption would significantly increase the mathematical complexity of the problem, resulting in terrible computational efficiency. Further, it was reported by Zhang et al. [24] that the wave signal captured at the pile edge is limitedly influenced by the three-dimensional effect during the TLSITs. Hence, the pile is modeled through the one-dimensional rod theory in pursuit of a more efficient closed-form solution. Commonly, the high-pile foundation can be divided into two parts: one embedded in the soil, the other one extending out of the soil. For the part that is embedded in the soil, the equilibrium equation can be written as

$$\left(G_j^{\mathrm{p}} I_j^{\mathrm{p}} + \eta_j^{\mathrm{p}} I_j^{\mathrm{p}} \frac{\partial}{\partial t}\right) \frac{\partial^2 \varphi_j^{\mathrm{p}}(z, t)}{\partial z^2} - 2\pi r_1^2 f_j^{\mathrm{s}}(z, t) = \rho_j^{\mathrm{p}} I_j^{\mathrm{p}} \frac{\partial^2 \varphi_j^{\mathrm{p}}(z, t)}{\partial t^2} \tag{2}$$

For the part extending out of the soil, the equilibrium equation can be written as

$$\left(G_m^{\mathrm{p}} I_m^{\mathrm{p}} + \eta_m^{\mathrm{p}} I_m^{\mathrm{p}} \frac{\partial}{\partial t}\right) \frac{\partial^2 \varphi_m^{\mathrm{p}}(z, t)}{\partial z^2} = \rho_m^{\mathrm{p}} I_m^{\mathrm{p}} \frac{\partial^2 \varphi_m^{\mathrm{p}}(z, t)}{\partial t^2} \tag{3}$$

where G_j^{p}, η_j^{p}, I_j^{p}, φ_j^{p}, ρ_j^{p}, f_j^{s}, and r_1 are the shear modulus, material damping, polar moment of inertia, twist angle, density, pile-side resistance, and the radius of the jth pile segment.

3.2. Boundary and Initial Conditions

The displacement and stress in the soil medium diminish at the radial infinite so that the following boundary conditions can be acquired:

$$u_j^{\mathrm{s}}(z,r,t)\Big|_{r\to\infty}=0 \tag{4}$$

$$\tau_j^{\mathrm{s}}(z,r,t)\Big|_{r\to\infty}=0 \tag{5}$$

The interactions between the soil layers are simulated by a distributed Kelvin–Voigt model, whose formulas can be presented as

$$\left[\left(G_j^{\mathrm{s}}+\eta_j^{\mathrm{s}}\frac{\partial}{\partial t}\right)\frac{\partial u_j^{\mathrm{s}}(z,r,t)}{\partial z}-\left(k_j+c_j\frac{\partial}{\partial t}\right)u_j^{\mathrm{s}}(z,r,t)\right]\Bigg|_{z=h_j}=0 \tag{6}$$

$$\left[\left(G_j^{\mathrm{s}}+\eta_j^{\mathrm{s}}\frac{\partial}{\partial t}\right)\frac{\partial u_j^{\mathrm{s}}(z,r,t)}{\partial z}+\left(k_{j-1}+c_{j-1}\frac{\partial}{\partial t}\right)u_j^{\mathrm{s}}(z,r,t)\right]\Bigg|_{z=h_{j-1}}=0 \tag{7}$$

The transient impulse is subjected to the side of the extending part. Considering that the stress distribution inside the pile shaft is continuous, these boundary conditions can be written as

$$\left[\left(G_m^{\mathrm{p}}I_m^{\mathrm{p}}+\eta_m^{\mathrm{p}}I_m^{\mathrm{p}}\frac{\partial}{\partial t}\right)\frac{\partial\varphi_m^{\mathrm{p}}(z,t)}{\partial z}\right]\Bigg|_{z=h_m}+T(t)=\left[\left(G_{m+1}^{\mathrm{p}}I_{m+1}^{\mathrm{p}}+\eta_{m+1}^{\mathrm{p}}I_{m+1}^{\mathrm{p}}\frac{\partial}{\partial t}\right)\frac{\partial\varphi_m^{\mathrm{p}}(z,t)}{\partial z}\right]\Bigg|_{z=h_m} \tag{8}$$

$$\varphi_j^{\mathrm{p}}\Big|_{z=h_j}=\varphi_{j+1}^{\mathrm{p}}\Big|_{z=h_j} \tag{9}$$

$$\tau_j^{\mathrm{p}}\Big|_{z=h_j}=\tau_{j+1}^{\mathrm{p}}\Big|_{z=h_j} \tag{10}$$

The pile end soil-pile end interaction is modeled by the fictitious soil pile. At the end of the fictitious soil pile, the displacement is supposed to be zero.

$$\varphi_1^{\mathrm{fp}}(z,t)\Big|_{z=L}=0 \tag{11}$$

The interaction between the pile and the upper structure is simplified to elastic springs and dashpots:

$$\left[\left(G_{m+1}^{\mathrm{p}}I_{m+1}^{\mathrm{p}}+\eta_{m+1}^{\mathrm{p}}I_{m+1}^{\mathrm{p}}\frac{\partial}{\partial t}\right)\frac{\partial\varphi_m^{\mathrm{p}}(z,t)}{\partial z}+\left(k_{\mathrm{T}}+c_{\mathrm{T}}\frac{\partial}{\partial t}\right)\varphi_m^{\mathrm{p}}(z,t)\right]\Bigg|_{z=0}=0 \tag{12}$$

where k_{T} and c_{T} denote the elastic and damping coefficients of the springs and dashpots, respectively.

When conducting the low-strain integrity test, both the soil and pile only go through tiny deformations, under which circumstance the motions of the soil and pile can be regarded as simultaneous.

$$\varphi_j^{\mathrm{p}}(z,t)\Big|_{r=r_1}\cdot r_1=u_j^{\mathrm{s}}(z,r,t)\Big|_{r=r_1} \tag{13}$$

At the initial moment, the system has no velocity nor acceleration, and the transient pile-side impulse is the only reason for the system vibration.

$$u_j^{\mathrm{s}}(z,r,t)\Big|_{t=0}=0 \tag{14}$$

$$\left.\frac{\partial u_j^{\mathrm{s}}(z,r,t)}{\partial t}\right|_{t=0} = 0 \tag{15}$$

$$\left.\varphi_j^{\mathrm{P}}(z,t)\right|_{t=0} = 0 \tag{16}$$

$$\left.\frac{\partial \varphi_j^{\mathrm{P}}(z,t)}{\partial t}\right|_{t=0} = 0 \tag{17}$$

4. Solution of Dynamic Equilibrium Equations

4.1. Solution of the Governing Equation in the Surrounding Soil

By performing Laplace Transform on both sides of Equation (1) and conducting the variable separation method, Equation (1) can degenerate to the following two differential equations:

$$r^2 R_j^{\mathrm{s}\prime\prime}(r,s) + rR_j^{\mathrm{s}\prime}(r,s) - \left[\kappa_j^2 r^2 + 1\right] R_j^{\mathrm{s}}(r,s) = 0 \tag{18}$$

$$Z_j^{\mathrm{s}\prime\prime}(z,s) + \beta_j^2 Z_j^{\mathrm{s}}(z,s) = 0 \tag{19}$$

where $\kappa_j^2 = \frac{\rho_j^{\mathrm{s}} s^2}{G_j^{\mathrm{s}} + \eta_j^{\mathrm{s}} s} + \beta_j^2$. Therefore, the general solution of Equation (1) can be acquired through the combination of general solutions of Equations (18) and (19) as

$$U_j^{\mathrm{s}}(z,r,s) = \left[E_j \mathrm{K}_1\left(\kappa_j r\right) + F_j \mathrm{I}_1\left(\kappa_j r\right)\right] \cdot \left[M_j \sin(\beta_j z) + N_j \cos(\beta_j z)\right] \tag{20}$$

where U_j^{s} is the Laplace Transform of u_j^{s}, while E_j, F_j, M_j and N_j are all undetermined coefficients. Meanwhile, $\mathrm{I}_1(\cdot)$ and $\mathrm{K}_1(\cdot)$ are modified Bessel Function of order one of the first and second kind, respectively.

Submitting Equation (20) into Equations (4) and (5), one obtains

$$U_j^{\mathrm{s}}(z,r,s) = \left[M_j \sin(\beta_j z) + N_j \cos(\beta_j z)\right] \cdot \mathrm{K}_1\left(\kappa_j r\right) \tag{21}$$

Further considering the interaction between soil layers, as listed in Equations (6) and (7), the following transcendental equations can be established:

$$G_j^{\mathrm{s}*}\beta_j^2\left[\tan(\beta_j h_j) - \tan(\beta_j h_{j-1})\right] + G_j^{\mathrm{s}**}\beta_j\left[\tan(\beta_j h_j)\tan(\beta_j h_{j-1}) + 1\right] + \left[\tan(\beta_j h_{j-1}) + \tan(\beta_j h_j)\right] = 0 \tag{22}$$

where $G_j^{\mathrm{s}*} = \frac{\left(G_j^{\mathrm{s}} + \eta_j^{\mathrm{s}} s\right)^2}{\left(k_j + c_j s\right)\left(k_{j-1} + c_{j-1} s\right)}$ and $G_j^{\mathrm{s}**} = \frac{\left(G_j^{\mathrm{s}} + \eta_j^{\mathrm{s}} s\right)\left(k_j + c_j s + k_{j-1} + c_{j-1} s\right)}{\left(k_j + c_j s\right)\left(k_{j-1} + c_{j-1} s\right)}$. With the introduction of local coordinates $[0, l_j]$, the transcendental equation can be simplified to

$$\tan(\beta_j l_j) - \frac{\left(G_j^{\mathrm{s}} + \eta_j^{\mathrm{s}} s\right)\left(k_j + c_j s + k_{j-1} + c_{j-1} s\right)\beta_j}{\left(G_j^{\mathrm{s}} + \eta_j^{\mathrm{s}} s\right)^2 \beta_j^2 - \left(k_j + c_j s\right)\left(k_{j-1} + c_{j-1} s\right)} = 0 \tag{23}$$

Through numerical iterations, the above transcendental equations can be solved with a series of numerical answers, which can be denoted as $\beta_{j1}, \beta_{j2}, \beta_{j3}, \ldots, \beta_{jn}$. Then, Equation (21) can be written as

$$U_j^{\mathrm{s}}(z,r,s) = \sum_{n=1}^{\infty} A_{jn} \sin(\beta_{jn} z + \varphi_{jn}) \cdot \mathrm{K}_1\left(\kappa_{jn} r\right) \tag{24}$$

where $A_{jn} = \sqrt{M_{jn}^2 + N_{jn}^2}$, $\varphi_{jn} = \arctan\left(\frac{N_{jn}}{M_{jn}}\right)$, $\frac{M_{jn}^s}{N_{jn}^s} = \frac{(k_j + c_j s)}{\left(G_j^s + \eta_j^s s\right)\beta_{jn}}$. The resistance force of soil acting on the pile side can be expressed as

$$f_j^s = \left(G_j^s + \eta_j^s s\right) \sum_{n=1}^{\infty} A_{jn}\kappa_{jn} \sin(\beta_{jn} z + \varphi_{jn}) \mathrm{K}_2\left(\kappa_{jn} r_1\right) \tag{25}$$

4.2. Solution of the Governing Equation of the Pile

Similarly, by performing Laplace Transform on both sides of Equations (2) and (3), one obtains

$$\left(G_j^p I_j^p + \eta_j^p I_j^p s\right) \frac{\partial^2 \phi_j^p(z,s)}{\partial z^2} - 2\pi r_1^2 f_j^s(z,t) = \rho_j^p I_j^p s^2 \phi_j^p(z,s) \tag{26}$$

$$\left(G_m^p I_m^p + \eta_m^p I_m^p s\right) \frac{\partial^2 \phi_m^p(z,s)}{\partial z^2} = \rho_m^p I_m^p s^2 \phi_m^p(z,s) \tag{27}$$

where ϕ_j^p is the Laplace Transform of φ_j^p with respect to t. It can be found that Equations (26) and (27) are non-homogeneous and homogeneous functions, respectively. The general solution of the corresponding homogeneous function of Equation (26) can be given as

$$\phi_j^p(z,s) = C_j^p \sin(\lambda_j z) + D_j^p \cos(\lambda_j z) \tag{28}$$

where $\lambda_j = \sqrt{-\frac{\rho_j^p s^2}{G_j^p + \eta_j^p s}}$. The specific solution of Equation (26) is found as

$$\sum_{n=1}^{\infty} A_{jn}\kappa_{jn} k_{jn}^s \sin(\beta_{jn} z + \varphi_{jn}) \cdot \mathrm{K}_2\left(\kappa_{jn} r_1\right) \tag{29}$$

where $k_{jn}^s = -\frac{2\pi r_1^2 \left(G_j^s + \eta_j^s s\right)}{\left(G_j^p I_j^p + \eta_j^p I_j^p s\right)\beta_{jn}^2 + \rho_j^p I_j^p s^2}$. The general solutions for the buried and extending pile segments can then be written as

Embedded pile segments:

$$\begin{aligned} \phi_j^p(z,s) &= C_j^p \sin(\lambda_j z) + D_j^p \cos(\lambda_j z) \\ &+ \sum_{n=1}^{\infty} A_{jn}\kappa_{jn} k_{jn}^s \sin(\beta_{jn} z + \varphi_{jn}) \cdot \mathrm{K}_2\left(\kappa_{jn} r_1\right) \end{aligned} \tag{30}$$

Extending pile segments:

$$\phi_m^p(z,s) = C_m^p \sin(\lambda_m z) + D_m^p \cos(\lambda_m z) \tag{31}$$

Based on the small strain assumption, the displacements at the soil-pile interface for embedded pile segments are continuous. Next, Equation (30) is substituted into (13), herein introducing the orthogonality of the following equations:

$$\int_0^{l_j} \sin(\beta_{jn} z + \varphi_{jn}) \sin(\beta_{in} z + \varphi_{in}) dz = \begin{cases} 0, & j \neq i \\ \frac{l_j}{2} - \frac{\sin(2\beta_{jn} l_j + 2\varphi_{jn}) - \sin(2\varphi_{jn})}{4\beta_{jn}}, & j = i \end{cases} \tag{32}$$

With the utilization of Equation (32), one obtains

$$A_{jn}\kappa_{jn} = C_j^p \delta_{jn1} + D_j^p \delta_{jn2} \tag{33}$$

In addition, the undetermined coefficients C_j^{p} and D_j^{p}, and other parameters, can be derived from the following relations:

$$\delta_{jn1} = \frac{r_1 \kappa_{jn} \chi_{jn1}}{\chi_{jn3}\left[\mathrm{K}_1\left(\kappa_{jn} r\right) - k_{jn}^{\mathrm{s}} r_1 \mathrm{K}_2\left(\kappa_{jn} r_1\right)\right]} \tag{34}$$

$$\delta_{jn2} = \frac{r_1 \kappa_{jn} \chi_{ji2}}{\chi_{jn3}\left[\mathrm{K}_1\left(\kappa_{jn} r\right) - k_{jn}^{\mathrm{s}} r_1 \mathrm{K}_2\left(\kappa_{jn} r_1\right)\right]} \tag{35}$$

$$\chi_{jn1} = \frac{1}{2} \cdot \left[\frac{\sin(\lambda_j l_j - \beta_{jn} l_j - \varphi_{jn}) + \sin(\varphi_{jn})}{(\lambda_j - \beta_{jn})} - \frac{\sin(\lambda_j l_j + \beta_{jn} l_j + \varphi_{jn}) - \sin(\varphi_{jn})}{\left(\lambda_j + \beta_{jn}\right)}\right] \tag{36}$$

$$\chi_{ji2} = \frac{1}{2} \cdot \left[\frac{\cos(\lambda_j l_j - \beta_{jn} l_j - \varphi_{jn}) - \cos(\varphi_{jn})}{(\lambda_j - \beta_{jn})} - \frac{\cos(\lambda_j l_j + \beta_{jn} l_j + \varphi_{jn}) - \cos(\varphi_{jn})}{\left(\lambda_j + \beta_{jn}\right)}\right] \tag{37}$$

$$\chi_{jn3} = \frac{l_j}{2} - \frac{\sin(2\beta_{jn} l_j + 2\varphi_{jn}) - \sin(2\varphi_{jn})}{4\beta_{jn}} \tag{38}$$

After the above derivation, the general solution for the embedded pile segments can then be given in a homogeneous equation form.

$$\begin{aligned} \phi_j^{\mathrm{p}}(z,s) = C_j^{\mathrm{p}}&\left[\sin(\lambda_j z) + \sum_{n=1}^{\infty} k_{jn}^{\mathrm{s}} \delta_{jn1} \sin(\beta_{jn} z + \varphi_{jn}) \cdot \mathrm{K}_2\left(\kappa_{jn} r_1\right)\right] \\ +D_j^{\mathrm{p}}&\left[\cos(\lambda_j z) + \sum_{n=1}^{\infty} k_{jn}^{\mathrm{s}} \delta_{jn2} \sin(\beta_{jn} z + \varphi_{jn}) \cdot \mathrm{K}_2\left(\kappa_{jn} r_1\right)\right] \end{aligned} \tag{39}$$

To acquire the undetermined coefficients (C_j^{p} and D_j^{p}), the continuous deformation boundary conditions at pile segment interfaces are utilized. Substituting Equations (30) and (31) into Equations (9) and (10), the iteration relations between different embedded pile segments can be obtained as

$$\begin{bmatrix} C_{j+1}^{\mathrm{p}} \\ D_{j+1}^{\mathrm{p}} \end{bmatrix} = \begin{bmatrix} \psi_{j+1,1}(l_{j+1}) & \psi_{j+1,2}(l_{j+1}) \\ \psi_{j+1,3}(l_{j+1}) & \psi_{j+1,4}(l_{j+1}) \end{bmatrix}^{-1} \times \begin{bmatrix} \psi_{j1}(0) & \psi_{j2}(0) \\ \psi_{j3}(0) & \psi_{j4}(0) \end{bmatrix} \begin{bmatrix} C_j^{\mathrm{p}} \\ D_j^{\mathrm{p}} \end{bmatrix} \tag{40}$$

$$\psi_{j1}(z) = \sin(\lambda_j z) + \sum_{n=1}^{\infty} \delta_{jn1} \zeta_{jn}^1(z) \tag{41}$$

$$\psi_{j2}(z) = \cos(\lambda_j z) + \sum_{n=1}^{\infty} \delta_{jn2} \zeta_{jn}^1(z) \tag{42}$$

$$\psi_{j3}(z) = G_j^{\mathrm{p}*}\left[\lambda_j \cos(\lambda_j z) + \sum_{n=1}^{\infty} \delta_{jn1} \zeta_{jn}^2(z)\right] \tag{43}$$

$$\psi_{j4}(z) = G_j^{\mathrm{p}*}\left[-\lambda_j \sin(\lambda_j z) + \sum_{n=1}^{\infty} \delta_{jn2} \zeta_{jn}^2(z)\right] \tag{44}$$

$$\zeta_{jn}^1(z) = k_{jn}^{\mathrm{s}} \sin(\beta_{jn} z + \varphi_{jn}) \cdot \mathrm{K}_2\left(\kappa_{jn} r_1\right) \tag{45}$$

$$\zeta_{jn}^2(z) = k_{jn}^{\mathrm{s}} \beta_{jn} \cos(\beta_{jn} z + \varphi_{jn}) \cdot \mathrm{K}_2\left(\kappa_{jn} r_1\right) \tag{46}$$

$$G_j^{\mathrm{p}*} = G_j^{\mathrm{p}} I_j^{\mathrm{p}} + \eta_j^{\mathrm{p}} I_j^{\mathrm{p}} s \tag{47}$$

Similarly, the coefficient transform relations between the embedded and the extending pile segments can be expressed as

$$\begin{bmatrix} C_m^{\mathrm{P}} \\ D_m^{\mathrm{P}} \end{bmatrix} = \begin{bmatrix} \sin(\lambda_m l_m) & \cos(\lambda_m l_m) \\ G_m^{\mathrm{P}*}\lambda_m \cos(\lambda_m l_m) & -G_m^{\mathrm{P}*}\lambda_m \sin(\lambda_m l_m) \end{bmatrix}^{-1} \times \begin{bmatrix} \psi_{m-1,1}(0) & \psi_{m-1,2}(0) \\ \psi_{m-1,3}(0) & \psi_{m-1,4}(0) \end{bmatrix} \begin{bmatrix} C_{m-1}^{\mathrm{P}} \\ D_{m-1}^{\mathrm{P}} \end{bmatrix} \tag{48}$$

The continuous stress conditions at the location of the pile side impulse can be written as

$$\begin{aligned} \begin{bmatrix} C_{m+1}^{\mathrm{P}} \\ D_{m+1}^{\mathrm{P}} \end{bmatrix} = & \begin{bmatrix} \sin(\lambda_{m+1} l_{m+1}) & \cos(\lambda_{m+1} l_{m+1}) \\ G_{m+1}^{\mathrm{P}*}\lambda_{m+1} \cos(\lambda_{m+1} l_{m+1}) & -G_{m+1}^{\mathrm{P}*}\lambda_{m+1} \sin(\lambda_{m+1} l_{m+1}) \end{bmatrix}^{-1} \times \begin{bmatrix} 0 & 1 \\ G_m^{\mathrm{P}*}\lambda_m & 0 \end{bmatrix} \begin{bmatrix} C_m^{\mathrm{P}} \\ D_m^{\mathrm{P}} \end{bmatrix} \\ & + \begin{bmatrix} \sin(\lambda_{m+1} l_{m+1}) & \cos(\lambda_{m+1} l_{m+1}) \\ G_{m+1}^{\mathrm{P}*}\lambda_{m+1} \cos(\lambda_{m+1} l_{m+1}) & -G_{m+1}^{\mathrm{P}*}\lambda_{m+1} \sin(\lambda_{m+1} l_{m+1}) \end{bmatrix}^{-1} \begin{bmatrix} 0 \\ T(\omega) \end{bmatrix} \end{aligned} \tag{49}$$

where $T(\omega) = \frac{T}{\pi^2 - T^2\omega^2}(1 + e^{-i\omega T})$ represents the half-sine harmonic impulse acted on the pile shaft in the frequency domain. Combing the boundary conditions at the pile top and end, one obtains

$$\frac{C_1^{\mathrm{P}}}{D_1^{\mathrm{P}}} = -\frac{\psi_{1,2}(l_j)}{\psi_{1,1}(l_j)} \tag{50}$$

$$\frac{C_{m+1}^{\mathrm{P}}}{D_{m+1}^{\mathrm{P}}} = -\frac{k_{\mathrm{T}} + c_{\mathrm{T}}s}{G_{m+1}^{\mathrm{P}*}\lambda_{m+1}} \tag{51}$$

The deformation and stress at the interfaces of different pile segments are continuous so that

$$\begin{bmatrix} C_{m+1}^{\mathrm{P}} \\ D_{m+1}^{\mathrm{P}} \end{bmatrix} = \begin{bmatrix} \chi_1 & \chi_2 \\ \chi_3 & \chi_4 \end{bmatrix} \begin{bmatrix} C_1^{\mathrm{P}} \\ D_1^{\mathrm{P}} \end{bmatrix} + \begin{bmatrix} \mu_1 \\ \mu_2 \end{bmatrix} \tag{52}$$

where the matrices $\begin{bmatrix} \chi_1 & \chi_2 \\ \chi_3 & \chi_4 \end{bmatrix}$ and $\begin{bmatrix} \mu_1 \\ \mu_2 \end{bmatrix}$ can be derived from

$$\begin{aligned} \begin{bmatrix} \chi_1 & \chi_2 \\ \chi_3 & \chi_4 \end{bmatrix} = & \begin{bmatrix} \sin(\lambda_{m+1} l_{m+1}) & \cos(\lambda_{m+1} l_{m+1}) \\ G_{m+1}^{\mathrm{P}*}\lambda_{m+1} \cos(\lambda_{m+1} l_{m+1}) & -G_{m+1}^{\mathrm{P}*}\lambda_{m+1} \sin(\lambda_{m+1} l_{m+1}) \end{bmatrix}^{-1} \\ & \times \begin{bmatrix} 0 & 1 \\ G_m^{\mathrm{P}*}\lambda_m & 0 \end{bmatrix} \begin{bmatrix} \sin(\lambda_m l_m) & \cos(\lambda_m l_m) \\ G_m^{\mathrm{P}*}\lambda_m \cos(\lambda_m l_m) & -G_m^{\mathrm{P}*}\lambda_m \sin(\lambda_m l_m) \end{bmatrix}^{-1} \\ & \times \begin{bmatrix} \psi_{m-1,1}(0) & \psi_{m-1,2}(0) \\ \psi_{m-1,3}(0) & \psi_{m-1,4}(0) \end{bmatrix} \cdots\cdots \begin{bmatrix} \psi_{3,1}(l_{j+1}) & \psi_{3,2}(l_{j+1}) \\ \psi_{3,3}(l_{j+1}) & \psi_{3,4}(l_{j+1}) \end{bmatrix}^{-1} \\ & \times \begin{bmatrix} \psi_{2,1}(0) & \psi_{2,2}(0) \\ \psi_{2,3}(0) & \psi_{2,4}(0) \end{bmatrix} \begin{bmatrix} \psi_{2,1}(l_{j+1}) & \psi_{2,2}(l_{j+1}) \\ \psi_{2,3}(l_{j+1}) & \psi_{2,4}(l_{j+1}) \end{bmatrix}^{-1} \times \begin{bmatrix} \psi_{1,1}(0) & \psi_{1,2}(0) \\ \psi_{1,3}(0) & \psi_{1,4}(0) \end{bmatrix} \end{aligned} \tag{53}$$

$$\begin{bmatrix} \mu_1 \\ \mu_2 \end{bmatrix} = \begin{bmatrix} \sin(\lambda_{m+1} l_{m+1}) & \cos(\lambda_{m+1} l_{m+1}) \\ G_{m+1}^{\mathrm{P}*}\lambda_{m+1} \cos(\lambda_{m+1} l_{m+1}) & -G_{m+1}^{\mathrm{P}*}\lambda_{m+1} \sin(\lambda_{m+1} l_{m+1}) \end{bmatrix}^{-1} \begin{bmatrix} 0 \\ T(\omega) \end{bmatrix} \tag{54}$$

Equation (52) can be further simplified to

$$C_{m+1}^{\mathrm{P}} = \chi_1 C_1^{\mathrm{P}} + \chi_2 D_1^{\mathrm{P}} + \mu_1 \tag{55}$$

$$D_{m+1}^{\mathrm{P}} = \chi_3 C_1^{\mathrm{P}} + \chi_4 D_1^{\mathrm{P}} + \mu_2 \tag{56}$$

in which,

$$C_1^{\mathrm{P}} = \frac{G_{m+1}^{\mathrm{P}*}\lambda_{m+1}\mu_1 + (k_{\mathrm{T}} + c_{\mathrm{T}}s)\mu_2}{\left[G_{m+1}^{\mathrm{P}*}\lambda_{m+1}\chi_2 + (k_{\mathrm{T}} + c_{\mathrm{T}}s)\chi_4\right]\frac{\psi_{1,1}(l_j)}{\psi_{1,2}(l_j)} - \left[G_{m+1}^{\mathrm{P}*}\lambda_{m+1}\chi_1 + (k_{\mathrm{T}} + c_{\mathrm{T}}s)\chi_3\right]} \tag{57}$$

$$D_1^{\mathrm{P}} = \frac{G_{m+1}^{\mathrm{P}*}\lambda_{m+1}\mu_1 + (k_{\mathrm{T}} + c_{\mathrm{T}}s)\mu_2}{\left[G_{m+1}^{\mathrm{P}*}\lambda_{m+1}\chi_1 + (k_{\mathrm{T}} + c_{\mathrm{T}}s)\chi_3\right]\frac{\psi_{1,2}(l_j)}{\psi_{1,1}(l_j)} - \left[G_{m+1}^{\mathrm{P}*}\lambda_{m+1}\chi_2 + (k_{\mathrm{T}} + c_{\mathrm{T}}s)\chi_4\right]} \tag{58}$$

$$C_{m+1}^{\mathrm{P}} = \left[\chi_2 - \chi_1 \frac{\psi_{1,2}(l_j)}{\psi_{1,1}(l_j)}\right] D_1^{\mathrm{P}} + \mu_1 \tag{59}$$

$$D_{m+1}^{\mathrm{P}} = \left[\chi_4 - \chi_3 \frac{\psi_{1,2}(l_j)}{\psi_{1,1}(l_j)}\right] D_1^{\mathrm{P}} + \mu_2 \tag{60}$$

Then, the undetermined coefficients of the near-ground pile segment can be acquired through the inverse transfer function as

$$\begin{aligned}\begin{bmatrix} C_m^{\mathrm{P}} \\ D_m^{\mathrm{P}} \end{bmatrix} = \begin{bmatrix} 0 & 1 \\ G_m^{\mathrm{P}*}\lambda_m & 0 \end{bmatrix}^{-1} \begin{bmatrix} \sin(\lambda_{m+1} l_{m+1}) & \cos(\lambda_{m+1} l_{m+1}) \\ G_{m+1}^{\mathrm{P}*}\lambda_{m+1}\cos(\lambda_{m+1} l_{m+1}) & -G_{m+1}^{\mathrm{P}*}\lambda_{m+1}\sin(\lambda_{m+1} l_{m+1}) \end{bmatrix} \\ \times \left[\begin{bmatrix} C_{m+1}^{\mathrm{P}} \\ D_{m+1}^{\mathrm{P}} \end{bmatrix} - \begin{bmatrix} 0 & 1 \\ G_m^{\mathrm{P}*}\lambda_m & 0 \end{bmatrix}^{-1} \begin{bmatrix} 0 \\ T(\omega) \end{bmatrix} \right]\end{aligned} \tag{61}$$

The twist angle and velocity response of the near-ground pile segment can be obtained as

$$\phi_m^{\mathrm{P}}(z,s) = C_m^{\mathrm{P}} \sin(\lambda_m z) + D_m^{\mathrm{P}} \cos(\lambda_m z) \tag{62}$$

$$V_m^{\mathrm{P}}(z,t) = \frac{1}{2\pi} \int_{-\infty}^{+\infty} \phi_m^{\mathrm{P}}(z,s) \cdot s \cdot e^{i\omega t} d\omega \tag{63}$$

5. Model Verification

To verify the correctness of the proposed model, the results calculated from the present solution are compared with those derived from the TLST theory aimed at the newly piled foundation and those computed from the finite element method (FEM). The soil-pile parameters utilized in this section are presented in Tables 1 and 2.

Table 1. Soil parameters utilized for model verification and parametric studies.

Density	Young's Modulus	Poisson's Ratio	Shear Modulus
1800 kg/m^3	12 MPa	0.3	4.6 MPa

Table 2. Default pile parameters utilized for model verification and parametric studies.

Density	Young's Modulus	Poisson's Ratio	Shear Modulus	Length	Radius
2500 kg/m^3	24 GPa	0.2	10 GPa	10 m	0.5 m

5.1. Comparisons with the TLIST Signals of Newly Piled Foundations

As mentioned, the classic TLST theory is established for newly piled foundations. Consequently, it is only capable of simulating the specific testing case in which the incident wave is input at the pile head. Unlike the newly piled foundations, the selections of incident wave input and signal receiving locations can be diverse for the testing of existing high-pile foundations. h_e and h_r are defined as the distances from the incident wave input location and the signal receiving location to the pile head. By adopting the soil and pile parameters in Tables 1 and 2, the present solution is compared with the classic TLST theory established in Ref. [28]. The length of the fictitious soil pile is set as zero to simulate an end-bearing condition. As shown in Figure 2, the reflection of the upward wave at the pile head would result in an inverse wave signal after the incident wave. Further, as the incident wave is input more away from the pile head, the time intervals between the incident wave and the reflection of the upward wave would increase. Once the incident wave is input close enough to the pile head, the incident wave and the reflection of the upward wave will combine into one signal. It is also noticed that the reflection at the pile end in the newly piled foundation signal would always match the second reflection of the downward wave in the existing foundation signal, as long as the sensors are installed in the same place as the input of the incident wave. This is because the distance traveled by the strain wave at this time is exactly equal to twice the length of the pile, as the reflection at the pile end in the newly piled foundation signal does.

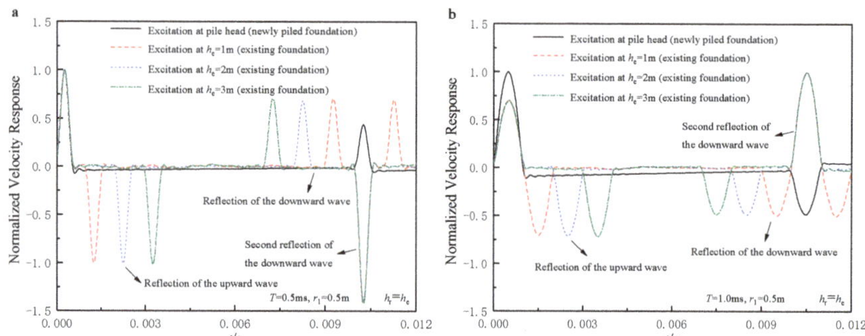

Figure 2. Comparisons of velocity response between existing and newly piled foundations: (**a**) viscoelastic boundary at the pile end (T = 0.5 ms); (**b**) fixed boundary at the pile end (T = 1.0 ms).

5.2. Comparisons with the FEM Results

To verify the accuracy of the proposed solution in simulating the simultaneously propagating upward and downward strain waves, the results calculated from the present model are compared with those computed from FEM. The finite element model is established and solved using Abaqus Explicit solver and C3D8R is utilized as the elements for both the soil and pile, the mesh of which is depicted in Figure 3.

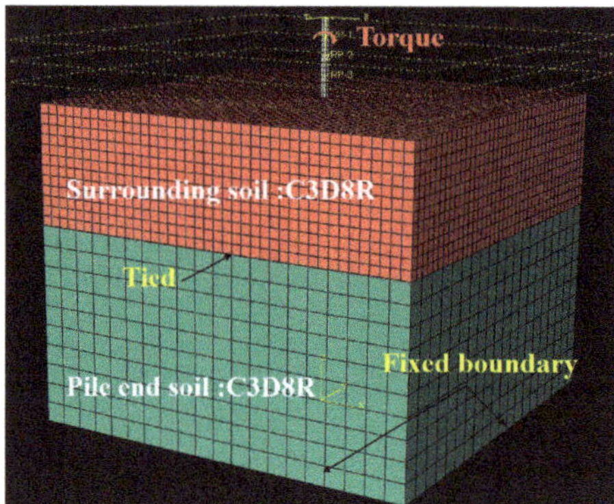

Figure 3. Mesh of the Finite element model.

As shown in Figure 4, the results derived from the present solution show good agreement with those calculated from the FEM, especially for the occurrence time of each reflection. However, there can be seen some deviations in the amplitudes of the reflected signal, mainly because of more significant strain wave energy dissipation in the 3D FEM model than in the present solution. In addition, by inputting the incident wave as close to the pile head as possible, the incident wave and the reflection of the upward wave at the pile head are more likely to be identified as one signal so that the difficulties of signal interpretation can be considerably reduced.

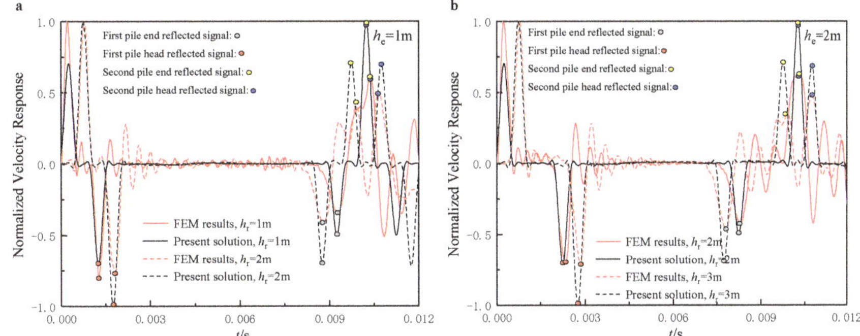

Figure 4. Comparisons of velocity response between present solution and FEM results: (**a**) $h_e = 1$ m; (**b**) $h_e = 2$ m.

6. Parametric Studies

6.1. Layouts of the Input and Signal Receiving Locations

Based on the above analysis, a preliminary conclusion is drawn: by inputting the incident wave as close to the pile head as possible, the difficulty in signal identification and interpretation can be reduced. This section investigates the influence of the layouts at the input and signal receiving locations on the velocity response, aiming to find the optimal layouts for the TLSTs. In order to simulate the test for floating piles, the length of the fictitious soil pile is set as 5 m.

As shown in Figure 5, once the sensors are placed where the incident wave is applied, the number of the reflected signals is minimized, making the signal spectrum clearer. In addition, the time intervals between the incident wave and the reflection of the upward wave collected by the sensors installed above the input position of the incident wave would not vary with the changes in the input position. In contrast, the time intervals collected by the sensors installed below the input position would increase when the input position moves away from the pile head. For cases where inputting the incident wave close to the pile head is difficult, installing the sensors close to the pile head can be an alternative. However, the optimal layouts of the TLST are inputting the incident wave close to the pile head and installing the sensors close to the pile head as well.

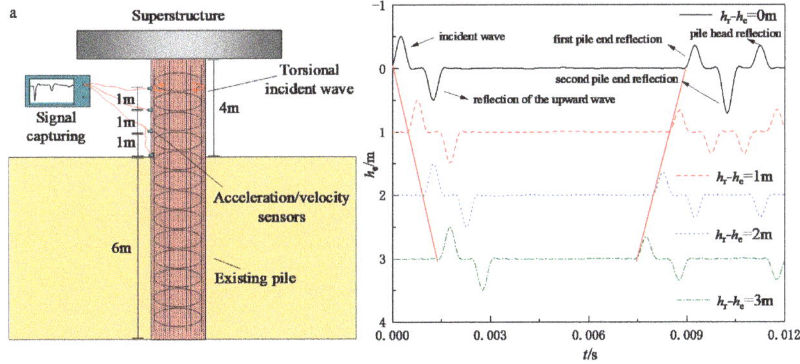

Figure 5. *Cont.*

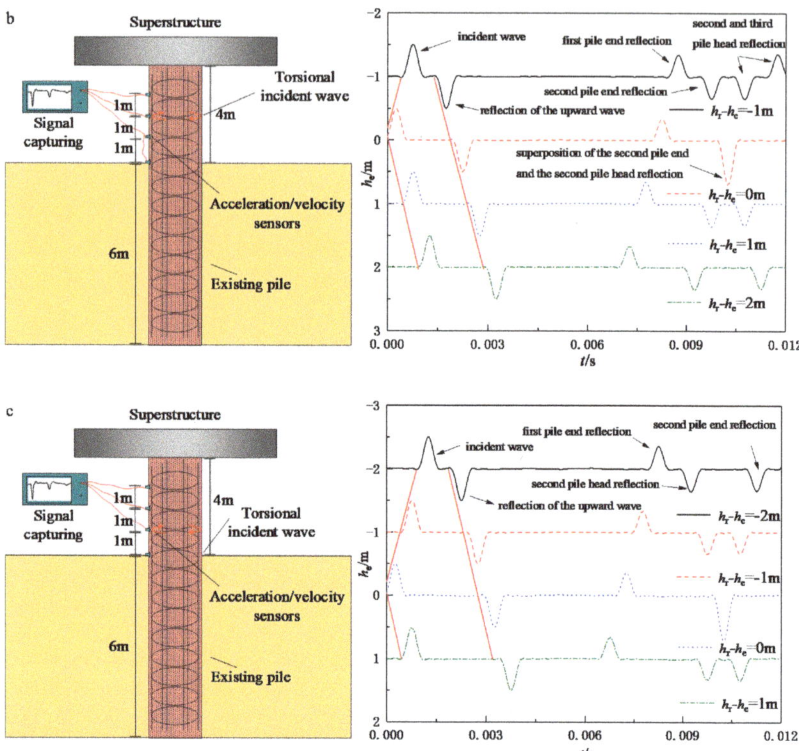

Figure 5. Influence of the layouts of the input and signal receiving locations on the TLSIT spectrums: (**a**) $h_e = 1$ m; (**b**) $h_e = 2$ m; (**c**) $h_e = 3$ m.

6.2. Identification of Defects from the TLST Spectrums

Defect identification is one of the major tasks for the integrity examination of the existing pile foundations. Further, the neckings and concrete segregations are the two most commonly found defects in practice. This section investigates the identification ability of these two defects utilizing the TLSTs.

The results shown in Figures 6 and 7 again justified the rationality of the optimal layouts of the TLSTs proposed in the above paragraphs. As shown in Figures 6a and 7a, both the necking and concrete segregation defects can be clearly identified, as long as the incident wave is input close to the pile head and the sensors are installed close to the pile head as well. However, once the incident wave is input far from the pile head, identifying the reflected signals at the defect turns out to be extremely difficult because the reflected signals can no longer be identified as one signal but as several separate signals reflecting all the time, making the signal spectrum a mess. In addition, the concrete segregation defects would alter the pile body's wave velocity so that each reflected signal's occurrence time would vary, while the reflected signals caused by necking defects would not.

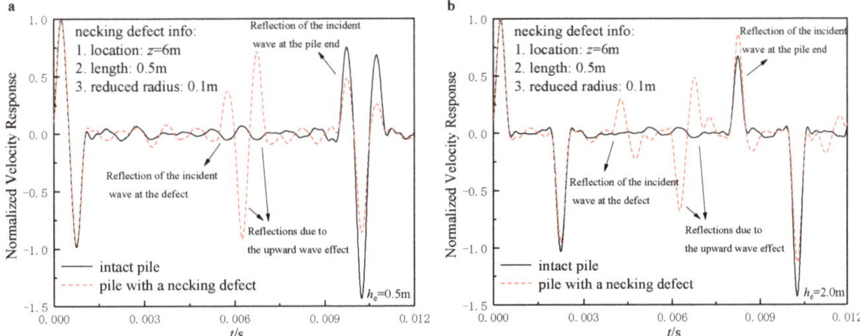

Figure 6. Identification of necking defects through TLSIT: (**a**) $h_e = 0.5$ m; (**b**) $h_e = 2.0$ m.

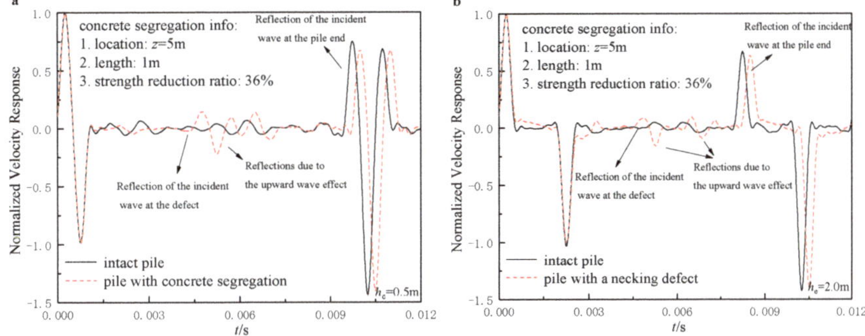

Figure 7. Identification of concrete segregation through TLSIT: (**a**) $h_e = 0.5$ m; (**b**) $h_e = 2.0$ m.

7. Conclusions

This paper establishes a rigorous mathematical model to simulate the strain wave propagation during the torsional low-strain test (TLST) for existing high-pile foundations. In the proposed model, the simultaneous propagation of the upward and downward strain waves inside the pile body is considered. The parametric analysis reveals the optimal layouts of the TLSTs for the existing high-pile foundation. The main conclusions can be drawn as follows:

1. By placing the sensors where the incident wave is applied, the number of reflected signals can be minimized to acquire a more precise signal spectrum.
2. The optimal layouts of the TLST are inputting the incident wave close to the pile head and installing the sensors close to the pile head as well. By doing this, the defects can be more easily identified from the signal spectrum.
3. The existence of concrete segregation defects would influence the occurrence time of each reflected signal, while the necking defects would not. Hence, this is a helpful tip for distinguishing the concrete segregation defects (decrease in strength of pile body material) from the necking defects.

Author Contributions: Conceptualization, Y.Z. and W.W.; methodology, Y.Z. and W.W.; validation, Y.Z. and M.H.E.N.; formal analysis, Y.Z. and M.H.E.N.; writing—original draft preparation, Y.Z. and Z.W.; writing—review and editing, M.H.E.N. and W.W.; supervision, M.H.E.N. and W.W.; funding acquisition, W.W. All authors have read and agreed to the published version of the manuscript.

Funding: This research is supported by the National Natural Science Foundation of China (Grant No. 52178371), the Outstanding Youth Project of Natural Science Foundation of Zhejiang Province (Grant No. LR21E080005), the Fundamental Research Founds for National University, China University of Geosciences (Wuhan) (Outstanding Ph.D. Innovation Fund), and the Engineering Research Center of Rock-Soil Drilling & Excavation and Protection, Ministry of Education (Grant No. 202203).

Institutional Review Board Statement: Not applicable.

Informed Consent Statement: Not applicable.

Data Availability Statement: The data presented in this study are available on request from the corresponding author.

Conflicts of Interest: The authors declare no conflict of interest.

References

1. Bersan, S.; Bergamo, O.; Palmieri, L.; Schenato, L.; Simonini, P. Distributed strain measurements in a CFA pile using high spatial resolution fibre optic sensors. *Eng. Struct.* **2018**, *160*, 554–565. [CrossRef]
2. Zhang, D.; Yu, Z.; Xu, Y.; Ding, L.; Ding, H.; Yu, Q.; Su, Z. GNSS Aided Long-Range 3D Displacement Sensing for High-Rise Structures with Two Non-Overlapping Cameras. *Remote Sens.* **2022**, *14*, 379. [CrossRef]
3. Zhang, Y.P.; Di, T.Y.; El Naggar, M.H.; Wu, W.B.; Liu, H.; Jiang, G.S. Modified Rayleigh-Love rod model for 3D dynamic analysis of large-diameter thin-walled pipe pile embedded in multilayered soils. *Comput. Geotech.* **2022**, *149*, 104853. [CrossRef]
4. Tu, Y.; El Naggar, M.H.; Wang, K.H.; Rizvi, S.M.F.; Qiu, X.C. Dynamic multi-point method for evaluating the pile compressive capacity. *Soil. Dyn. Earthq. Eng.* **2022**, *159*, 107317. [CrossRef]
5. Loseva, E.; Lozovsky, I.; Zhostkov, R. Identifying small defects in cast-in-place piles using low strain integrity testing. *Indian Geotech. J.* **2022**, *52*, 270–279. [CrossRef]
6. Chow, Y.K.; Phoon, K.K.; Chow, W.F.; Wong, K.Y. Low strain integrity testing of piles: Three-dimensional effects. *J. Geotech. Geoenviron.* **2003**, *129*, 1057–1062. [CrossRef]
7. Cui, D.-M.; Yan, W.; Wang, X.-Q.; Lu, L.-M. Towards Intelligent Interpretation of Low Strain Pile Integrity Testing Results Using Machine Learning Techniques. *Sensors* **2017**, *17*, 2443. [CrossRef]
8. Li, Q.; Li, X.; Wen, M.; Hu, L.; Duan, W.; Li, J. Dynamic Responses of a Pile with a Cap under the Freezing and Thawing Processes of a Saturated Porous Media Considering Slippage between Pile and Soil. *Appl. Sci.* **2022**, *12*, 4214. [CrossRef]
9. Meng, K.; Cui, C.Y.; Liang, Z.M.; Li, H.J.; Pei, H.F. A new approach for longitudinal vibration of a large-diameter floating pipe pile in viscoelastic soil considering the three dimensional wave effects. *Comput. Geotech.* **2020**, *128*, 103840. [CrossRef]
10. Cui, C.Y.; Meng, K.; Xu, C.S.; Liang, Z.M.; Li, H.J.; Pei, H.F. Analytical solution for longitudinal vibration of a floating pile in saturated porous media based on a fictitious saturated soil pile model. *Comput. Geotech.* **2021**, *131*, 103942. [CrossRef]
11. Vrouwenvelder, T.; Scholten, N. Assessment Criteria for Existing Structures. *Struct. Eng. Int.* **2010**, *20*, 62–65. [CrossRef]
12. Zhang, X.; Ni, Y.; Song, C.; Xu, D. Research on non-destructive testing technology for existing bridge pile foundations. *Struct. Monit. Maint.* **2020**, *7*, 43–58. [CrossRef]
13. Wu, J.T.; El Naggar, M.H.; Ge, J.; Wang, K.H.; Zhao, S. Multipoint traveling wave decomposition method and its application in extended pile shaft integrity test. *J. Geotech. Geoenviron. Eng.* **2021**, *147*, 04021128. [CrossRef]
14. Kou, H.-L.; Diao, W.-Z.; Liu, T.; Yang, D.-L.; Horpibulsuk, S. Field performance of open-ended prestressed high-strength concrete pipe piles jacked into clay. *Sensors* **2018**, *18*, 4216. [CrossRef] [PubMed]
15. Novak, M. Dynamic stiffness and damping of piles. *Can. Geotech. J.* **1974**, *11*, 574–598. [CrossRef]
16. Militano, G.; Rajapakse, R.K.N.D. Dynamic response of a pile in a multi-layered soil to transient torsional and axial loading. *Géotechnique* **1999**, *49*, 91–109. [CrossRef]
17. Mamoon, S.M.; Banerjee, P.K. Time-domain analysis of dynamically loaded single piles. *J. Eng. Mech.* **1992**, *118*, 140–160. [CrossRef]
18. El Naggar, M.H. Vertical and torsional soil reactions for radially inhomogeneous soil layer. *Struct. Eng. Mech.* **2000**, *10*, 299–312. [CrossRef]
19. Anoyatis, G.; Mylonakis, G. Dynamic Winkler modulus for axially loaded piles. *Géotechnique* **2012**, *62*, 521–536. [CrossRef]
20. Zheng, C.J.; Ding, X.M.; Kouretzis, G.P.; Liu, H.L.; Sun, Y. Three-dimensional propagation of waves in piles during low-strain integrity tests. *Géotechnique* **2018**, *68*, 358–363. [CrossRef]
21. Zhang, Y.P.; Liu, H.; Wu, W.B.; Wang, L.X.; Jiang, G.S. A 3D analytical model for distributed low strain test and parallel seismic test of pipe piles. *Ocean Eng.* **2021**, *225*, 108828. [CrossRef]
22. Nghiem, H.M. Variational approach for torsional dynamic response of a single pile in multi-layered soils. *Geomech. Geoengin.* **2022**. [CrossRef]
23. Liu, D.J.; Liu, Y.Z.; Wang, J.Y. Theoretical study on torsional wave applied in low strain dynamic testing of piles. *Chin. J. Geotech. Eng.* **2003**, *25*, 283–287. (In Chinese)
24. Zhang, Y.P.; Wang, Z.Q.; El Naggar, M.H.; Wu, W.B.; Wang, L.X.; Jiang, G.S. Three-dimensional wave propagation in a solid pile during torsional low strain integrity test. *Int. J. Numer. Anal. Methods. Geomech.* **2022**. [CrossRef]

25. Veletsos, A.S.; Doston, K.W. Vertical and torsional vibration of foundations in inhomogeneous media. *J. Geotech. Eng.* **1988**, *114*, 1002–1021. [CrossRef]
26. Budkowska, B.B.; Szymczak, C. Sensitivity analysis of piles undergoing torsion. *Comput. Struct.* **1993**, *48*, 827–834. [CrossRef]
27. Zheng, C.J.; Liu, H.L.; Ding, X.M.; Lv, Y. Torsional dynamic response of a large-diameter pipe pile in viscoelastic saturated soil. *Int. J. Numer. Anal. Methods Geomech.* **2014**, *38*, 1724–1743. [CrossRef]
28. Wu, W.B.; Liu, H.; El Naggar, M.H.; Mei, G.X.; Jiang, G.S. Torsional dynamic response of a pile embedded in layered soil based on the fictitious soil pile model. *Comput. Geotech.* **2016**, *80*, 190–198. [CrossRef]
29. Zhang, Y.P.; Yang, X.Y.; Wu, W.B.; El Naggar, M.H.; Jiang, G.S.; Liang, R.Z. Torsional complex impedance of pipe pile considering pile installation and soil plug effect. *Soil. Dyn. Earthq. Eng.* **2020**, *131*, 106010. [CrossRef]
30. Chow, Y.K. Torsional response of piles in non-homogeneous soil. *J. Geotech. Geoenviron. Eng.* **1985**, *111*, 942–947. [CrossRef]
31. Zidan, A.F.; Ramadan, O.M.O. Three-dimensional analysis of pile groups subject to torsion. *Geotech. Res.* **2020**, *7*, 103–116. [CrossRef]
32. Cheng, X.; Cheng, W.; Wang, P.; El Naggar, M.H.; Zhang, J.; Liu, Z. Response of offshore wind turbine tripod suction bucket foundation to seismic and environmental loading. *Ocean Eng.* **2022**, *257*, 111708. [CrossRef]
33. Lu, Z.T.; Wang, Z.L.; Liu, D.J.; Xiong, F.; Ma, H.C.; Tan, X.H. Propagation characteristics of flexural wave and the reflection from vertical cracks during pipe-pile integrity testing. *Int. J. Numer. Anal. Methods. Geomech.* **2022**, *46*, 1660–1684. [CrossRef]
34. Basack, S.; Sen, S. Numerical solution of single piles subjected to pure torsion. *J. Geotech. Geoenviron. Eng.* **2014**, *140*, 74–90. [CrossRef]
35. Basack, S.; Sen, S. Numerical solution of single pile subjected to simultaneous torsional and axial loads. *Int. J. Geomech.* **2014**, *14*, 06014006. [CrossRef]
36. Zhang, Y.P.; Jiang, G.S.; Wu, W.B.; El Naggar, M.H.; Liu, H.; Wen, M.J.; Wang, K.H. Analytical solution for distributed torsional low strain integrity test for pipe pile. *Int. J. Numer. Anal. Methods. Geomech.* **2022**, *46*, 47–67. [CrossRef]
37. Zheng, C.J.; Gan, S.S.; Luan, L.B.; Ding, X.M. Vertical dynamic response of a pile embedded in a poroelastic soil layer overlying rigid base. *Acta Geotech.* **2021**, *16*, 977–983. [CrossRef]
38. Qu, L.M.; Yang, C.W.; Ding, X.M. A continuum-based model on axial pile-head dynamic impedance in inhomogeneous soil. *Acta Geotech.* **2021**, *16*, 3339–3353. [CrossRef]

Review

Piezoelectric Impedance-Based Structural Health Monitoring of Wind Turbine Structures: Current Status and Future Perspectives

Thanh-Cao Le [1,2,3], Tran-Huu-Tin Luu [2,4], Huu-Phuong Nguyen [1,2], Trung-Hau Nguyen [2,5], Duc-Duy Ho [1,2,*] and Thanh-Canh Huynh [6,7,*]

[1] Faculty of Civil Engineering, Ho Chi Minh City University of Technology (HCMUT), 268 Ly Thuong Kiet, District 10, Ho Chi Minh City 700000, Vietnam; 1880698@hcmut.edu.vn or caolt@ntu.edu.vn (T.-C.L.); nhphuong.sdh20@hcmut.edu.vn (H.-P.N.)
[2] Vietnam National University Ho Chi Minh City (VNU-HCM), Linh Trung Ward, Thu Duc District, Ho Chi Minh City 700000, Vietnam; lthtin@vnuhcm.edu.vn (T.-H.-T.L.); haunguyen85@hcmut.edu.vn (T.-H.N.)
[3] Faculty of Civil Engineering, Nha Trang University, Nha Trang 650000, Vietnam
[4] Vietnam National University Ho Chi Minh City—Campus in Ben Tre, Ben Tre City 930000, Vietnam
[5] Faculty of Applied Science, Ho Chi Minh City University of Technology (HCMUT), 268 Ly Thuong Kiet, District 10, Ho Chi Minh City 700000, Vietnam
[6] Institute of Research and Development, Duy Tan University, Danang 550000, Vietnam
[7] Faculty of Civil Engineering, Duy Tan University, Danang 550000, Vietnam
* Correspondence: hoducduy@hcmut.edu.vn (D.-D.H.); huynhthanhcanh@duytan.edu.vn (T.-C.H.)

Abstract: As an innovative technology, the impedance-based technique has been extensively studied for the structural health monitoring (SHM) of various civil structures. The technique's advantages include cost-effectiveness, ease of implementation on a complex structure, robustness to early-stage failures, and real-time damage assessment capabilities. Nonetheless, very few studies have taken those advantages for monitoring the health status and the structural condition of wind turbine structures. Thus, this paper is motivated to give the reader a general outlook of how the impedance-based SHM technology has been implemented to secure the safety and serviceability of the wind turbine structures. Firstly, possible structural failures in wind turbine systems are reviewed. Next, physical principles, hardware systems, damage quantification, and environmental compensation algorithms are outlined for the impedance-based technique. Afterwards, the current status of the application of this advanced technology for health monitoring and damage identification of wind turbine structural components such as blades, tower joints, tower segments, substructure, and the foundation are discussed. In the end, the future perspectives that can contribute to developing efficient SHM systems in the green energy field are proposed.

Keywords: wind turbine; structural health monitoring; impedance-based technique; damage detection; piezoelectric material

Citation: Le, T.-C.; Luu, T.-H.-T.; Nguyen, H.-P.; Nguyen, T.-H.; Ho, D.-D.; Huynh, T.-C. Piezoelectric Impedance-Based Structural Health Monitoring of Wind Turbine Structures: Current Status and Future Perspectives. *Energies* **2022**, *15*, 5459. https://doi.org/10.3390/en15155459

Academic Editors: Frede Blaabjerg, Filipe Magalhães and Massimiliano Renzi

Received: 19 February 2022
Accepted: 15 July 2022
Published: 28 July 2022

1. Introduction

Wind energy has overgrown and is now becoming one of the most cost-effective means of providing electricity and decarbonizing the energy industry in many countries [1,2]. It is forecasted that a typical onshore wind turbine installed in 2035 will have a capacity of 3.25 MW with a 174-m rotor diameter and 130-m hub height (see Figure 1). The offshore wind turbine's capacity will be 17 MW with a 250-m rotor diameter and 130-m hub height by 2035 [2]. With the increase of turbine sizes to efficiently harvest more energy, the wind turbine structure becomes more complex and prone to damage [3]. A comprehensive wind turbine failure analysis reported that the blade failure is the most common damage accounting for 23%, followed by the fire-induced failure with 19%; the structural failure and the environmental damage accounted for 12% and 9%, respectively (see Figure 2) [4].

Those failures could be caused by multiple factors such as moisture absorption, fatigue, wind gusts, thermal stress, corrosion, fire, and lightning strikes [5,6].

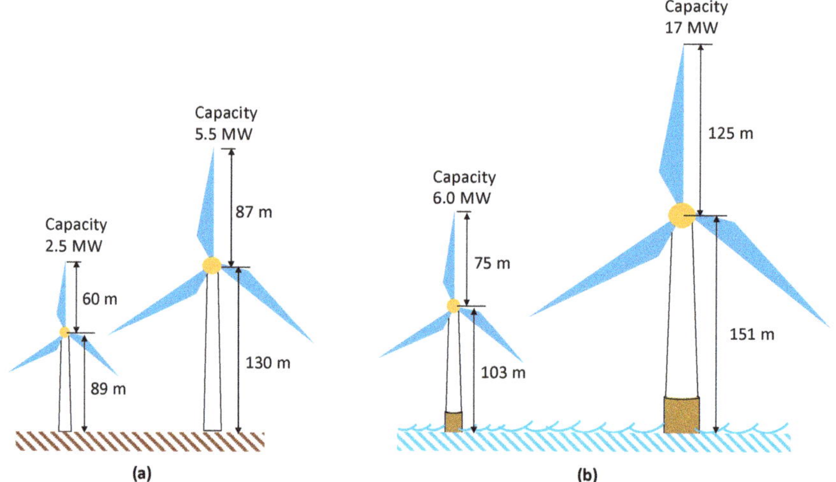

Figure 1. Expected turbine size in 2035 for onshore and offshore wind, compared to 2019 medians: (**a**) onshore wind turbines and (**b**) offshore wind turbines.

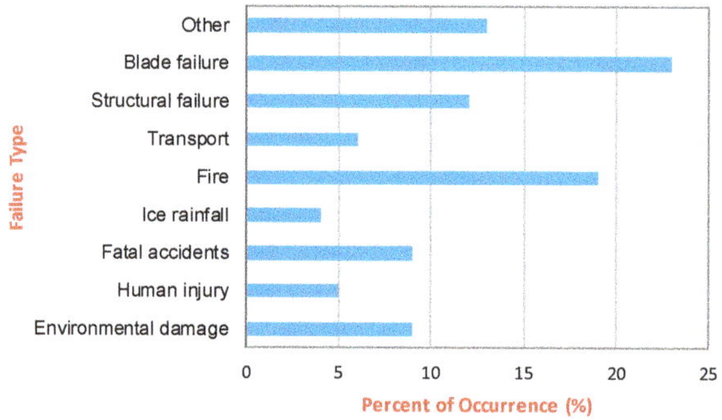

Figure 2. Failure types of wind turbine accidents.

The operation, repair, and maintenance expenditures make up a large percentage of wind energy projects' total life cycle costs. It is reported that those costs for a 500-MW offshore wind farm were about 26% over a service lifetime of 25 years [7]. Thus, developing innovative condition monitoring technologies that can enhance the safety and reliability of wind farms is a crucial priority for the wind energy sectors [8,9]. The implementation of structural failure detection and characterization procedure for engineering structures is known as structural health monitoring (SHM) [10]. Generally, there are five levels of SHM: detection of structural damage (Level 1), damage localization (Level 2), damage quantification (Level 3), damage classification (Level 4), and structural integrity assessment (Level 5). To the wind industry, a low-cost and reliable integrated SHM system could reduce the wind turbine life cycle costs and secure the efficiency of a wind farm [3]. The SHM data can be used to prevent unnecessary component replacement and unexpected catastrophic failures, minimize the inspection times, make the wind energy supply chain run smoothly and confidently, support further development of a wind turbine, and provide supervision at remote sides and remote diagnosis.

So far, there have been several SHM technologies that can be applied for the damage assessment in wind turbine structures, such as acoustic emission monitoring [11], imaging technique [12], ultrasonic method [13], impedance-based technique [14], modal properties-based approach [15,16], and strain monitoring [17,18]. Among those SHM technologies, the impedance-based method has been extensively studied and constantly shown its practicality in assessing the structural damage in civil, mechanical, and aerospace structures [19–22]. Moreover, the impedance-based technique is suitable for damage detection and severity quantification in the early stage, which is particularly important in SHM systems. In brief, the technique utilizes a non-instructive piezoelectric transducer, which is surface-mounted on a host structure, to sense the electromechanical impedance response in ultrasonic frequency bands. The impedance response represents the local dynamics of the host structure, so it will be altered when the structural damage occurs [23]. The use of short wavelengths in high-frequency bands enables the technique to easily detect minor damages close to the vicinity of the transducer. Unlike the local vibration method, the local impedance technique utilizes the local vibrations in a high frequency, which could be up to 1 MHz in some cases.

A comparison of SHM technologies considering their damage assessment capabilities is outlined in [24]. The piezoelectric impedance-based SHM technology has achieved Level 3 of SHM [24] and is approaching Level 4, thanks to advanced machine learning algorithms [19,21,25]. It is noted that a Level 5 damage assessment is still a concept for SHM technologies so far [24,26]. As summarized in [22,27], the significant advantages of the impedance-based technology include: (i) cost-effectiveness, since the technique uses inexpensive, lightweight, fast-response, self-diagnostic piezoelectric transducers (such as PZT) to sense the impedance response of the host structure; (ii) being able to detect various types of damage such as fatigue cracks, corrosion, or loosened bolts; (iii) easy implementation for complex structures regardless their materials; (iv) robustness to early-stage damages due to the use of high-frequency and short-wavelength excitation; (v) advances in low-cost, onboard-computing, multi-functional, wireless data acquisition systems have reduced the cost of large SHM systems [28]; (vi) the technique is potential for the integration with other SHM technologies to enhance the safety and the optimization of maintenance costs. Additionally, impedance-based SHM still has some limitations, including limited sensing range of the transducers, temperature/noise effects on impedance responses, and required impedance data before damage (i.e., the reference). Nonetheless, many advanced signal processing algorithms have been developed for compensating the noise and environmental effects [19,21,29] and reference-free damage assessments [30]. Those have significantly enhanced the performance of the impedance-based technique, making this technology an ideal candidate for developing an effective SHM system for wind turbines.

Over the past decades, the impedance-based technique has shown its capacity and effectiveness for detecting and quantifying structural damages in pipeline systems [31,32], reinforced concrete [33–35], steel joints [36–38], and aerospace engineering systems [39,40]. It also can be applied for corrosion monitoring [41,42], soil monitoring [43–45], and cable monitoring [46,47]. However, a very small number of studies have reported the applications of the impedance-based technique for the SHM of wind turbine structures, and most of them have focused on blade monitoring such as fatigue detection [48], added mass and stiffness change [49], crack monitoring [14], and ice monitoring [50]. Recently, the reference [51] showed the potential of the impedance response for damage detection in the grouted connection of offshore wind turbines. Another study reported the feasibility of the impedance-based technique for bolted joint monitoring in a wind turbine tower [52]. There are several challenges with the SHM of wind turbines, such as: (1) difficulties in the field inspection and maintenance due to the height of the structure, the operation of the blades, (2) the effect of environmental conditions, especially under offshore atmosphere, and (3) the remote site of wind farms [3,9]. Since it is measured in the high-frequency band, the impedance response is insensitive to changing boundary conditions and any operational vibrations [3]. Therefore, the impedance-based SHM technology is ideal for damage assessment in wind turbines under operation.

In 2008, Ciang et al. [3] published a comprehensive review reporting different techniques for SHM of wind turbines and discussed the unique advantages of the impedance-based technique. In 2015, Antoniadou et al. [53] published a mini-review discussing advanced signal processing approaches for SHM strategies in the wind energy sector. In 2016, Martinez-Luengo et al. [9] reviewed many SHM techniques for offshore wind turbines with a focus on statistical pattern recognition methods and highlighted the need for developing novel effective SHM methods to measure and analyze the meaningful responses of the wind turbines. The specific issues of structural monitoring of wind turbine blades have been well-presented in recent papers [54,55]. In 2022, Civera and Surace [56] presented a comprehensive review of nondestructive techniques performed during the last 20 years for SHM and condition monitoring of wind turbine systems. They also highlighted important works that have been receiving significant interest from the academic community.

An excellent review of the piezoelectric-based impedance-based SHM was presented by Na et al. [22]. The advances and challenges of this SHM technology were discussed by Huynh et al. [27]. Physics-based and data-driven methods for the impedance-based SHM were also discussed in the review paper by Fan et al. [57]. However, no recent and updated review papers have focused on the applications of the impedance-based method in the wind turbine monitoring field. Therefore, in this review paper, we aim to give the readers an overview of how this advanced technology has been applied for the damage identification and safety evaluation of wind turbine systems and future research areas that can turn the technology into practical SHM applications. At first, the potential damages to wind turbine structures are presented. Afterwards, the fundamentals of the impedance-based technique are outlined with two critical aspects: impedance sensing technology and damage identification algorithms. Then, the paper discusses the current status of the application of this SHM technology for wind turbine monitoring, focusing on critical structural components such as blades, towers, substructures, and foundations. From that, we highlight potential applications and outline new research topics for the future development of low-cost impedance-based SHM systems in the green energy field, including the concept of impedance-based smart blades, the development of wearable sensor devices, corrosion probes for towers, the concept of a smart sensor bar for scour monitoring, and the development of smart aggregates for concrete foundation monitoring.

2. Potential Damages in Wind Turbines

2.1. Blade Failure

A typical construction layout of a wind turbine blade is a thin-walled multicellular hollow airfoil-shaped cross-section structure with different materials, including fiber composites and sandwich composite systems, primers, UV gel coats, paint, bolted joints, and so on. As a complex structural element, wind turbine blades are easily damaged. The blade damage type depends on the surrounding environment and the land morphology of the installation area of the wind turbines [6]. The blade damages impose not only severe repairing costs but also the income lost due to the unwanted interruption of the wind turbine operation and the reduction in the aerodynamic performance of the blades. As reported in [6], the blade failures can be classified into four main types: damage from lightning, leading-edge erosion, structural fatigue damage, and damage from icing, as depicted in Figure 3.

i. Damage from lightning is regarded as the most frequent damage to the wind turbine blades [6]. Light strikes a blade could cause delamination, debonding in the upper and lower shells, and shell and tip detachments [58].
ii. Leading-edge erosion is mainly due to airborne particulates in the form of rain, hailstone, sea spray, dust and sand, and UV light and humidity/moisture. As reported in [59], leading-edge erosion can occur after only two years of wind turbine operation, and it is dependent on the site.
iii. Structural fatigue damage can occur when the wind turbine is subjected to repeated loads during its life cycle induced by wind. The turbine's cyclic starts and stops, with yaw error, yaw motion, and vibrational resonance-induced loads from the dynamics of the structure.

iv. Damage from icing is caused by the accretion of ice or snow on the blade structures exposed to the icing atmosphere. As reported in [6], there are two distinguished types of atmospheric icing, in-cloud icing (rime ice or glaze) and precipitation icing (freezing rain or drizzle, wet snow). Icing causes added mass and could induce unbalanced vibrations of the blade, resulting in considerable reductions in the economic efficiency of the wind energy project.

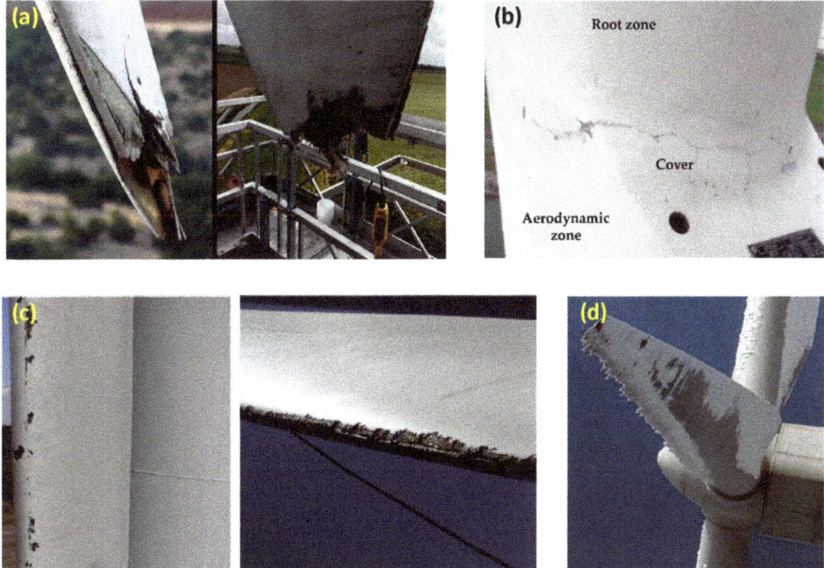

Figure 3. Potential damage sources of wind turbine blades: (**a**) damage from lightning, (**b**) failures due to fatigue, (**c**) leading-edge erosion, and (**d**) damage from icing (adapted from [6]).

2.2. Tower Failure

The structural failure averagely accounted for 12% of the total failure cases, as reported in [4] and shown in Figure 2. Fatigue cracks, joint failure, storm-induced failure, improper installation, and even lighting damage can contribute to the structural failure of wind turbine towers [3,52,60].

i. Fatigue cracks. Different materials can be used for constructing wind turbine towers. From the beginning, a wind turbine tower was made of steel truss-like structures with many connections that could be easily corroded. With increased turbine capacity, steel tubular tubes are preferred for the wind turbine tower design. Harte et al. [61] reported that a steel tubular tower with a height of over 85 m is extremely challenging to balance the dynamical excitation. A concrete tower can be an alternation, but it faces thermal constraints that cause cracking damages in the tower, reducing the stiffness of the whole system. On the other hand, wind turbine towers are often subjected to repeated loads during their life cycle, which can result in fatigue cracks that can expand and carry potential, leading to the tragic collapse of a whole tower.

ii. Joint failure. In the steel tubular tower, the segments are assembled by high-strength bolts, which are indeed hot spots in the tower. The self-looseness is one of the main reasons for the failure of a bolted joint, especially under transverse loads [62,63]. Regarding the wind tower joint, the bolt failures such as loose and broken could be found during harvesting wind energy [64]. Imbalanced loads across a clamping bolted connection could cause local plastic zones and create proper conditions for corrosion damages [60]. Figure 4a shows the scene of the tower's structural failure at the wind farm Abuela de Santa Ana in Spain in 2008 [65]. A field investigation found signs of corrosion in the damaged flange section of the joint, which eventually led to the failure of the whole tower, as shown in Figure 4b.

iii. Storm-induced failure. A turbine tower could be broken and toppled over during a storm, as reported in [66]. Accidents are ranked the third most common cause of failure [60]. Cheng and Xu [67] investigated the structural failure of wind turbines in China's Shanwei City caused by the super typhoon Usagi in 2013. The representative tower failure is presented in Figure 4c. The local buckling was observed at the shell wall thickness transition zone in the tower during the typhoon, as depicted in Figure 4d. Their study further reveals that the tubular tower collapsed at a hub wind speed lower than the design survival wind speed. A finite element model was constructed to predict the failure modes and failure locations in the tubular towers caused by the Usagi typhoon [68].

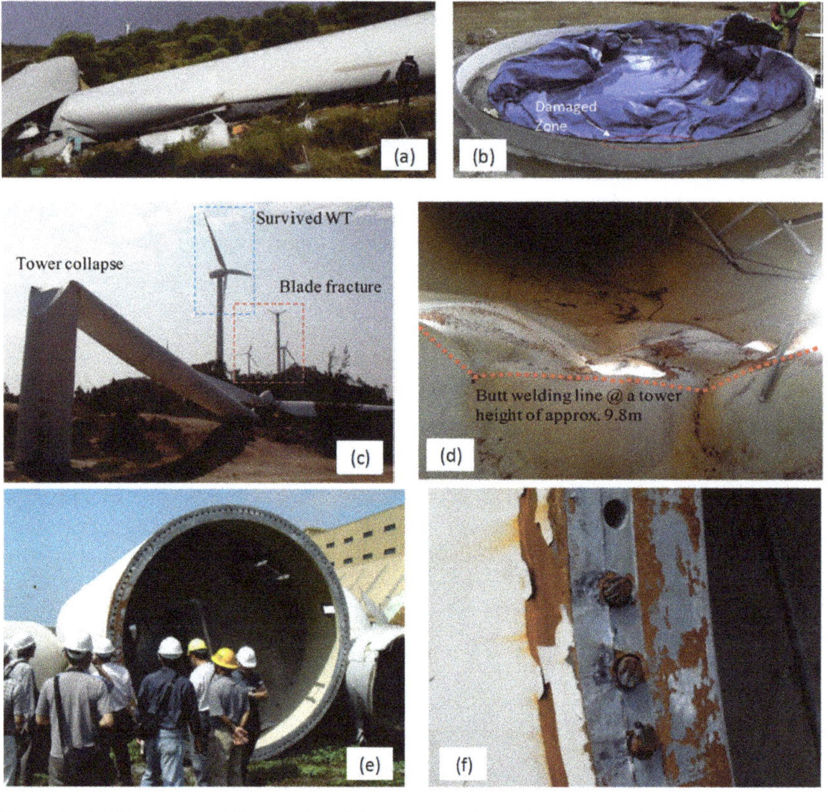

Figure 4. (**a**) The scene of the wind turbine tower collapsed at the wind farm Abuela de Santa Ana, Spain caused by joint failure. (**b**) The view of the damaged joint where the failure began. (**c**) The structural failure of a wind turbine tower in China's Shanwei city caused by super typhoon Usagi. (**d**) The view of local buckling of the tower at wall-thickness transition zone. (**e**) The view of the fractured tubular tower section of a wind turbine in Taiwan. (**f**) The scene of fractured bolts ((**a**,**b**) adapted from [65], (**c**,**d**) adapted from [67], and (**e**,**f**) adapted from [4]).

iv. Improper installation. In some realistic situations, faulty construction and poor quality control are responsible for the collapse of new turbine structures [3]. As reported in [64], faulty bolt installation related to an insufficient preload at the joint induced the failure of a wind turbine tower in Sweden. In 2008, the Jangmi typhoon struck Taiwan, resulting in strong winds and heavy rains that caused the collapse of a wind turbine [4]. The actual views of the fractured tubular tower section and fractured bolts are depicted in Figure 4e,f, respectively. Chou et al. [4] conducted a failure analysis of the collapsed wind turbine and found that the designed strength of the bolts and the used ones during the construction was different.

v. Lightning damage. Lightning can cause severe damage and even destroy the tower. Wind turbine towers are getting so high and attractive to lightning, primarily when they are frequently located in flat areas with nothing around. Unlike the blades, the tower's damage related to lightning is rarely observed. A survey of wind turbine tower collapse cases reported that lightning produced only two incidents of the tower collapse, with the occurrence of lightning-related damage being 2.8% [60].

2.3. Substructure and Foundation Failures

Structural damage can be anything in the substructure and the foundation of wind turbines. Different types of potential cracks and damages in concrete foundations have been reported for onshore wind turbines [69]. They occur in the foundations for many reasons, including using substandard concrete mixes, faults in the design, or multi-stage concreting in extreme weather conditions [70].

i. Concrete foundation damage. In Germany, in 2000, four turbines experienced sudden and total collapse due to concrete damage at the base. This led to the shutdown of forty-four similar turbines for pending investigation [66]. The steel tower is clamped to the foundation through pre-stalled anchor bolts, and there are many problems in the mortar grout between the steel flange and the foundation. The potential damages can be the vertical shrinkage cracks, the side excess material, the weak mortar grout caused by the separation of the motor, and the voids between the concrete and the tower segments [69]. The combination of the stresses caused by the serviceability load and the thermal stresses could result in cracks in a high concrete foundation pedestal. Cracks can propagate and lead water and dirt from the outside to the inside of the foundation pedestal. Further, the cracks could potentially occur in the transition zones between the insert ring of the tower and the concrete foundation, as reported in [69].

ii. Corrosion, scouring, grounded connection failure, and fatigue damage in substructure. The offshore environment is one of the harshest. Under the offshore atmosphere, many potential damages can occur in the substructure and foundation of offshore wind turbines, including corrosion [71,72], scours in the sea bed [73], failure of the grouted connections [74], and cracks in welding/bolted joints [75]. An example of corrosion that has occurred inside a substructure is illustrated in Figure 5a. Many factors influence the rate of corrosion in the offshore atmosphere, such as temperature, humidity, biological organisms, and airborne contaminants [3]. When a wind turbine is placed offshore, locally increased current and wave motions are induced around the structure, resulting in scouring, as depicted in Figure 5b [76]. Grouted connections are commonly used as the structural joint between the substructure and the foundation of an offshore wind turbine [74]. A typical failure of a grouted connection is shown in Figure 5c. Under the different ambient conditions and high dynamical loads caused by waves and wind, the grouted connections are potentially damaged, leading to a reduction in their mechanical stability [77,78], especially severe fatigue damages [74]. The combination of wave and wind loading could also result in fatigue damage in a tripod support structure. The previous analysis of fatigue damage assessment [75] shows that the hot spots (the most severe regions) in the tripod are at the joints between the central column and the bracing members and between the pile and the brace (see Figure 5d).

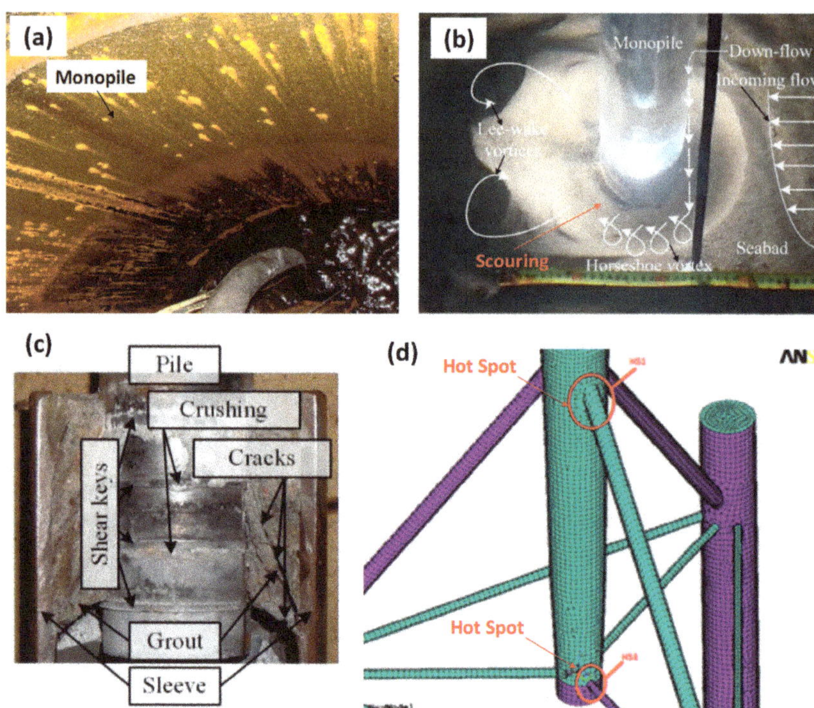

Figure 5. Structural damages in offshore wind turbine substructure and foundation: (**a**) corrosion in the substructure (obtained from [72]), (**b**) scouring in the sea bed (obtained from [73]), (**c**) failures of the grouted connection (obtained from [74]), and (**d**) potential fatigue damage in the joint of the offshore wind turbine tripod (obtained from [75]).

3. Fundamentals of Impedance-Based SHM

Figure 6 shows two main aspects that must be considered in developing an impedance-based SHM system for a wind turbine structure, including (i) the hardware for impedance sensing and (ii) the interpretation algorithm for damage detection. The hardware consists of a PZT patch surface bonded to a host structure to sense the electromechanical impedance in a frequency domain through an impedance analyzer. Afterwards, interpretation algorithms are implemented to process the impedance signals and assess the host structure's health status. Since the impedance data consists of the local dynamic information of the structural region near the PZT patches, any structural damages in the host structure could be identified by interpreting the impedance change. In the following sections, those two aspects are briefly described.

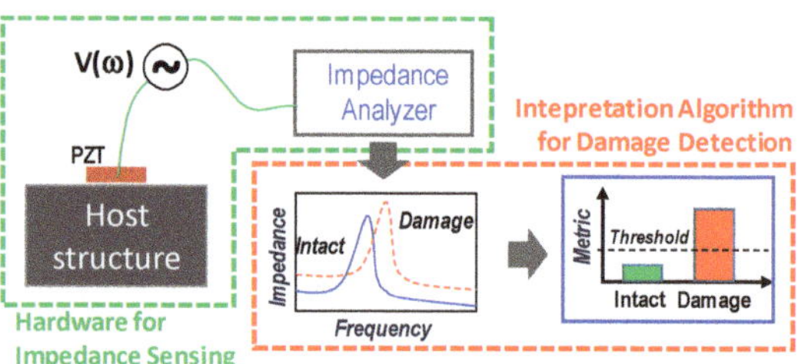

Figure 6. Scheme of the impedance-based SHM process.

3.1. Impedance Sensing Technology

3.1.1. PZT-Driven System

Piezoelectric materials are at the heart of the impedance-based technique. They are smart and low-cost materials that show the potential of converting mechanical energy into electrical energy and vice versa, so they can provide an excellent opportunity to create intelligent, efficient, and effective SHM systems. PZT is today considered one of the most economical piezoelectric elements with a considerable product market. A PZT patch costs only a few dollars, but it can be applied to high-frequency applications at hundreds of kHz and above.

By utilizing the unique piezoelectric effect, Liang et al. [79] developed the electromechanical impedance technique for dynamic analysis and damage detection in structural systems. A theoretical model of the PZT-driven system that can predict the impedance response was proposed in [79]. As shown in Figure 7, the PZT is electrically excited by a harmonic voltage. Under the piezoelectric effect, the PZT is mechanically expanded, and its deformation causes an exciting force at the contact between the PZT and the host structure. The ability of the structure to resist this exciting force is defined as the structural impedance, consisting of the dynamic properties as follows:

$$Z_s(\omega) = c + m\frac{\omega^2 - k/m}{\omega}i \tag{1}$$

where Z_s is the structural impedance of the host structure, the terms m, c, and k are the mass, the damping coefficient, and the stiffness of the host structure, respectively; the term ω is the exciting frequency of the harmonic voltage.

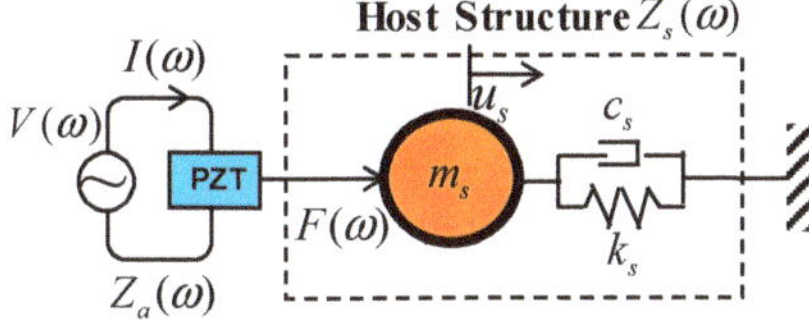

Figure 7. A theoretical impedance model of a PZT-driven system.

The PZT patch has itself the electrical impedance Z_a, and the overall electromechanical impedance is a combination of Z_a and Z_s, as expressed in [80]:

$$Z(\omega) = \frac{V}{I} = \left\{ i\omega\frac{b_a h_a}{t_a}\left[\hat{\varepsilon}_{33}^T - \frac{1}{Z_a(\omega)/Z_s(\omega)+1}d_{31}^2\hat{Y}_{11}^E\right]\right\}^{-1} \tag{2}$$

where $\hat{Y}_{11}^E = (1+i\eta)Y_{11}^E$ is the complex form of Young's modulus of the PZT at a zero electric field; $\hat{\varepsilon}_{33}^T = (1-i\delta)\varepsilon_{33}^T$ is the complex form of the dielectric constant at zero stress; d_{31} is the 1-directional piezoelectric coupling constant at zero stress; and b_a, h_a, and t_a are the width, the height, and thickness of the PZT, respectively. The terms η is the structural damping loss factor, and δ is the dielectric loss factor of the PZT.

As expressed in Equations (1) and (2), any changes in the structural properties (m, c, and k) as the result of structural damages would lead to the variation in the measured impedance response of the PZT-driven system. Therefore, the structural damage could be assessed by monitoring the change in the impedance response. It is noted that the frequency band of the impedance response should be carefully selected to realize the structural damage. As demonstrated in [81], the effective frequency band should cover the resonant frequencies of the host structure to enhance the opportunity of diagnosing small-size damages. The effective frequency range for a given host structure is obtained by trial-and-error methods or finite element modelling.

3.1.2. Impedance Analyzer

Wired Impedance Analyzers

The impedance or admittance (inverse of impedance) signatures are acquired over a high-frequency range (typically 30–400 kHz). Commercial high-performance impedance analyzers such as Agilent E4980A, LCR meter HIOKI 3532, Wayne Kerr impedance analyzer, and HP 4192A/4194A are commonly-used in impedance-based SHM practices. Generally, a user interface (UI) package is installed on a computer to control the impedance analyzer via an I/O interface. After setting the parameters in the UI, the PZT is interrogated in the impedance analyzer and excited by a harmonic voltage to generate the short wavelength and high-frequency waves into the host structure. The recorded impedance data is then transferred to a computer and saved for data interpretation. Although the commercial impedance analyzers can measure the impedance in a very high-frequency band, up to MHz, they are often bulky, expensive, and inconvenient for field applications.

Wireless Impedance Analyzers

Many researchers developed new, low-cost, and portable hardware to reduce the cost and enhance the portability of impedance measurement systems. In 2007, Mascarenas et al. [82] presented the first portable, low-energy consumption, and wireless impedance sensor node, as shown in Figure 8a. The node uses a cheap impedance chip AD5934 for recording the impedance from the PZT [83]. The chip is embedded with multifunctional circuits: function generator, current-to-voltage amplifier, antialiasing filter, analog-to-digital converter (ADC), and discrete Fourier transform (DFT) analyzer. A microcontroller ATmega128L is used for controlling and computing, and a radio frequency module XBee (2.4 GHz Zigbee) wirelessly transmits the recorded data. A microwave wireless energy transmission module powers the node to maximize the portability of the device.

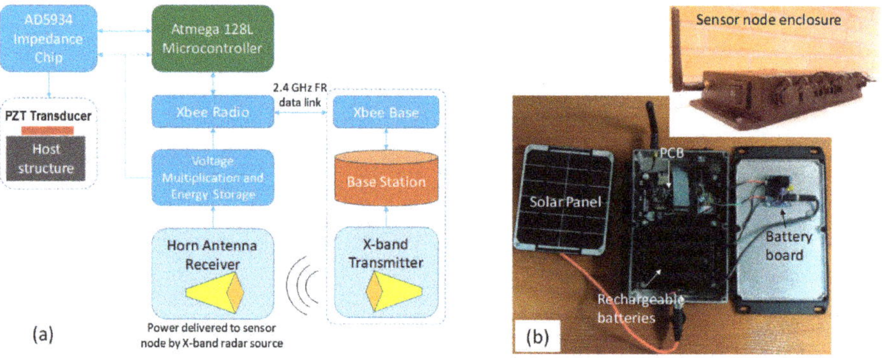

Figure 8. (**a**) Block diagram of the first wireless impedance sensor node [82]; (**b**) Prototype of a low-cost wireless impedance sensor node [28].

Based on the pioneering study by Mascarenas et al. [82], many researchers have improved the initial prototype to achieve a low-cost multifunctional wireless impedance sensor node. In 2009, Park et al. [84] embedded more functions into the sensor node, such as temperature recording, multi-channel measurement, and a SD memory card slot. A year later, Min et al. [85] embedded damage detection/sensor self-diagnosis algorithms and power management with energy harvesters. In 2012, Nguyen et al. [86] developed an Imote2-platformed impedance sensor node for wireless, autonomous, cost-efficient, and multi-channel monitoring. The Imote2 board could enable high operating speed, low power requirement, large storage memory, and onboard computing capability for onsite applications [87]. To provide stable software and reliable hardware suitable for full-scale and autonomous SHM, Perera et al. (2017) [28] proposed a flexible wireless impedance sensor node, which was also developed on the low-cost impedance chip AD5933. Figure 8b depicts a prototype of the sensor node in which a printed circuit board (PCB) is designed with an AD5933 chip, a multiplexer, an Atmega644PA microcontroller, a wireless XBee S2C 802.15.4 RF module, and a real-time clock DS3221. The battery module includes a solar panel, a battery board MCP73871 and rechargeable batteries for power supply and management.

The in situ applicability of the wireless impedance analyzers was verified for impedance-based SHM of cable-stayed bridges [88] and large girder bridges [85] and building roofs [85]. The solar energy harvesting and consumption, as well as the survivability of the Imote2-based wireless sensor nodes during field testing, were discussed in [88], revealing the stable operation of the designed SHM system. Despite many benefits that a wireless impedance sensor network can offer, the existing wireless sensor nodes can provide only a limited frequency band for impedance measurements, since they have relied on the impedance chip AD5933/AD5934 that can generate signals ≤ 100 kHz.

3.2. Damage Identification Method

3.2.1. Traditional Metrics

For damage detection, the impedance change is traditionally quantified using statistical damage metrics such as root mean square deviation (RSMD), covariance (COV), correlation coefficient deviation (CCD), and mean absolute percentage deviation (MAPD) [22]. The formulas of those standard metrics are expressed in Equations (3)–(6). Each metric exhibits different behaviors in quantifying the impedance change, and the selection of optimal metrics depends on the detection's purpose or target structure. The RMSD and MAPD metrics are found to be more appropriate for localizing and characterizing the damage growth. Meanwhile, the COV and CCD metrics are more suitable for diagnosing the damage size increment at a fixed location [22]. For monitoring concrete curing and strength gain, it is found that the MAPD metric is better than the RMSD and CCD metrics [89]. As reported in Le et al. [47], the CCD metric is an excellent indicator to detect the prestress force change in the piezoelectric-based smart strand. The RMSD index showed good performance for sensor fault diagnosis, such as sensor debonding and breakage problems [90]. Some researchers used other, less common metrics such as average square deviation (ASD), R_x/R_y, chessboard distance (CD), united mechanical impedance (UMI), ellipse damage index (EDI), and so on. Hu et al. [91] developed an alternative metric, so-called R_x/R_y, and showed that it was more suitable for damage identification in a concrete slab. Another researcher used the chessboard distance (CD) and showed that it was better than RMSD for detecting damage in a composite structure under different temperatures.

$$\mathrm{RMSD} = \left(\frac{\sum\limits_{k=1}^{N} \left[\mathrm{Re}(Z_k)_j - \mathrm{Re}(Z_k)_i) \right]^2}{\sum\limits_{k=1}^{N} [\mathrm{Re}(Z_k)_i)]^2} \right) \tag{3}$$

$$\mathrm{MAPD} = \frac{1}{N} \sum_{k=1}^{N} \left| \frac{\left[\mathrm{Re}(Z_k)_j - \mathrm{Re}(Z_k)_i \right]}{\mathrm{Re}(Z_k)_i} \right| \tag{4}$$

$$\mathrm{COV} = \frac{1}{N} \sum_{k=1}^{N} \left[\mathrm{Re}(Z_k)_j - \mathrm{Re}(\overline{Z})_j \right] \cdot \left[\mathrm{Re}(Z_k)_i - \mathrm{Re}(\overline{Z})_i \right] \tag{5}$$

$$\mathrm{CCD} = 1 - \frac{\frac{1}{N} \sum\limits_{k=1}^{N} \left[\mathrm{Re}(Z_k)_j - \mathrm{Re}(\overline{Z})_j \right] \cdot \left[\mathrm{Re}(Z_k)_i - \mathrm{Re}(\overline{Z})_i \right]}{\sigma_{Z_j} \sigma_{Z_i}} \tag{6}$$

in which $\mathrm{Re}(Z_k)_i$ denotes the real part of the reference impedance signature (i.e., intact state), and $\mathrm{Re}(Z_k)_j$ is the real part of the current impedance signature (i.e., unknown state),

N is the number of swept frequencies (i.e., the number of data points), and the terms $\overline{Z}$ and σ_Z signify the mean and the standard deviation of an impedance signature, respectively.

3.2.2. Advanced Damage Identification Algorithms

For damage detection under environmental conditions, alternative damage metrics should be employed. It is noted that the piezoelectric material and the host structure are temperature-dependent [92], and the traditional damage metrics are just statistical comparisons of the two impedance signatures. Thus, the temperature change will indeed induce alternations in the damage metrics, leading to a false damage detection unless the temperature effect is well-compensated. Fabricio et al. [93] observed that the peak frequencies of impedance signatures were reduced when the temperature increased. They also found that the frequency shift was more significant at higher-frequency bands. Several advanced signal processing algorithms have been developed to deal with the effect of temperature change on damage detection results.

Advanced Statistical Index

To remove the temperature effect, Park et al. [94] proposed one of the first temperature compensation algorithms. They verified the proposed algorithm for damage detection in gears, composite reinforced structures, and bolted joints under varying temperatures of 25–75 °C. The temperature compensation algorithm is developed based on the reconstruction of the damage metric through a correction scheme, as expressed by the following equation:

$$M = \sum_{i,j=1}^{n} \left[\mathrm{Re}(Y_{i,1}) - \mathrm{Re}(Y_{j,2})\right]^2 \text{ where } \mathrm{Re}(Y_{j,2}) = \mathrm{Re}(Y_{j,2})_{measured} + \delta^s \quad (7)$$

In Equation (7), the variable M is the sum of the real impedance change squared, $Y_{i,1}$ is the impedance of the pre-damaged structure at the ith frequency, and $Y_{j,2}$ is the impedance of the post-damaged structure at the jth frequency; δ^s is defined as the average difference between the reference impedance and the measured impedance, which can be determined by minimizing the value of damage metric via an iteration process. The experimental result shows that, by incorporating the developed compensation technique into health monitoring applications, the impedance-based technique can detect early-stage structural damage, even with severe temperature variation.

Assuming that the temperature change mainly causes the shift in the impedance pattern, Koo et al. [95] developed a temperature-compensated damage index based on the concept of effective frequency shift. The Koo's method looks for the maximum value of the CC metric (maxCC) by shifting the effective frequency band. The formula for the maxCC metric can be expressed as [95]:

$$\max_{\widetilde{\omega}} CC = \max_{\widetilde{\omega}} \left\{ \frac{1}{N} \sum_{i=1}^{N} (x(\omega_i) - \overline{x})(y_i(\omega_i - \widetilde{\omega}) - \overline{y}) \right\} / (\sigma_x \sigma_y) \quad (8)$$

In Equation (8), $\overline{x}$ and $\overline{y}$ are the mean of the impedance signature $x(\omega_i)$ and $y(\omega_i)$, respectively, σ_x and σ_y are the corresponding standard deviations, $\widetilde{\omega}$ is the effective frequency shift, and ω_i is the ith frequency. They proposed a strategy combining the maxCC metric with the outliner analysis for damage assessment. They applied the maxCC method to monitor the structural damage in a lab-sized steel truss bridge member. The results show that the proposed strategy can accurately detect a 2-mm cut of the test specimen with a 99.5% confidence level.

Machine Learning Algorithms

Over the past decade, machine learning algorithms have enhanced the automation capability of impedance-based damage detection, which is mainly required for real-time SHM under different environmental conditions. Sepehry et al. [96] presented a radial basis function network (RBFN) for compensating temperature effects on the impedance response of steel plates and bolted joints of a gas pipe. The RBFN learned the impedance signatures corresponding to different temperatures to predict the RMSD metric. Lim et al. [97] developed a kernel principal component analysis (PCA)-based data normalization technique to minimize false alarms caused by varying temperature and loading conditions. Moreover, the proposed technique was successfully verified on a composite aircraft wing under a temperature range of −30 °C to 50 °C (5 °C interval) and a loading range of 10–40 MN (5-MN intervals). The simulated damage was bolt-loosening with a half-turn severity, and the impedance data was recorded in a range of 60–70 kHz (20 Hz interval). The ability of the PCA-based technique for the problem of temperature filtering/compensation is also confirmed by other research groups [98].

Recently, Gianesihi et al. [99] developed a general poly-nominal regression-based methodology to remove the temperature effect during impedance-based SHM. The proposed method was successfully applied to two aluminum beams and one steel pipe with a considered temperature range from −40 °C to 80 °C and a studied frequency range from 10 to 90 kHz. Du et al. [100] developed a convolutional neural network (CNN) model to compensate the effects of temperature change on the impedance response for bolt loosening detection. The model is based on a modified Unet for temperature compensation and a lightweight subnetwork to identify bolt looseness. The proposed model can achieve good damage detection results under the varying temperature condition, even when the network is trained by limited data. The validation accuracy was 97.71% when the model was trained by only about 30 samples from each damage state. The developed model also showed good generality to untrained temperatures and bolt torques.

A promising feature of the machine learning-based models is that they can be retrained to learn new nonlinear temperature and frequency dependences of impedance signatures, so it is readily incorporated into real-time systems. However, one of the important issues is that most machine learning models need to be trained not only by the impedance signature before the failure but also by the data after the failure to ensure the accuracy of damage detection. However, acquiring the training data for damaged states from the structures being used without corrupting them is very challenging. To overcome this issue, Huynh et al. [29] presented a different temperature compensation strategy using a set of the RBFN models to predict the baseline impedance signature at any temperature. The proposed method requires only the impedance data of a healthy state to train the network. As shown in Figure 6, this strategy consists of two main phases: training RBFN for temperature compensation and detecting damage using temperature-compensated impedance signatures, as detailed in Figure 9. In the first phase, the impedance data of the healthy state is recorded under varying temperature conditions and used to build the training data. Then, a set of RBFNs corresponding to all scan frequencies are set up and trained on the recorded data. In the next phase, the impedance data of an unknown state at a temperature T is measured, and the impedance baseline at T is predicted using the trained RBFNs. RMSD and CCD indices are then computed and compared with an upper control limit (UCL) to assess the structural integrity of the monitored structure. The experimental verification on a post-tensioned reinforced concrete girder showed that the proposed method could detect as small as 1 ton prestress loss (roughly 7% damage) under varying temperatures ranging from 6.72 °C to 22.33 °C. Those advanced signal processing algorithms for compensating the operational and environmental effects have enhanced the accessibility of the impedance-based technique for in-field measurements.

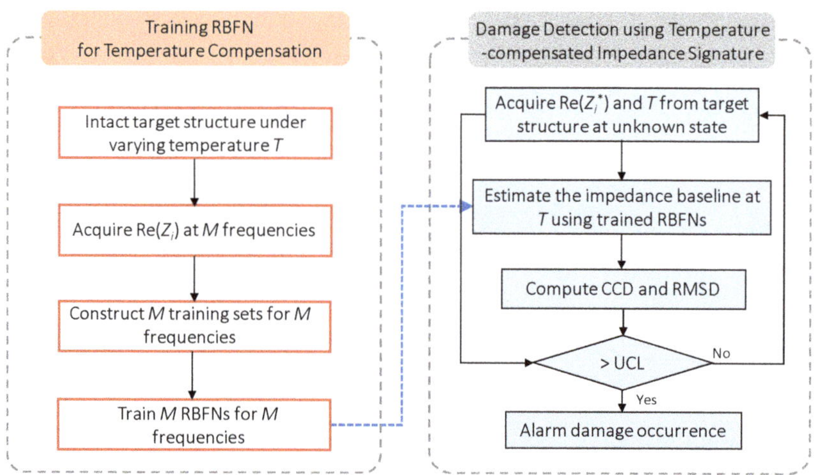

Figure 9. RBFN-based temperature compensation algorithm for impedance-based damage detection [29].

4. Current Status of Impedance-Based SHM of Wind Turbines

4.1. SHM of Wind Turbine Blades

An assessment of wind turbine failure cases showed that the blade failure rate is the highest at 23%; see Figure 2 [4]. Therefore, avoiding wind turbine blade failures will lead to significant cost reductions related to the maintenance of wind turbines. Several studies have implemented the impedance-based technique for the SHM of wind turbine blades. Pitchford et al. [49] investigated the experimental feasibility of implementing an impedance system integrated inside turbine blades as a field method to detect turbine blade failure. The experiment procedure can be briefly illustrated in Figure 10a. The test specimen is an actual wind turbine blade section (TX-100 blade) developed at Sandia National Laboratories [49]. Three PZT transducers (made of 20 × 20 × 0.27 PSI-5H4E material, Piezo Systems) are internally bonded to the structural components of the blade, such as skin (Skin PZT) and spar (spar PZT and carbon PZT). Three damage locations (1, 2, and 3) were considered. As shown in Figure 10a, damage location 1 is where the carbon fiber spar cap meets the balsa skin, and damage locations 2 and 3 are the adhesive between the spar flange and the spar cap on the curved side and the other side of the blade, respectively. An HP 4194A impedance analyzer is used to measure the impedance of the PZTs with a 10-Hz resolution. The RMSD index was used to detect and estimate the damage to the blade.

In [49], two different tests, including indirect damage testing and actual damage testing, were performed on the testing blade. In the indirect damage testing, as shown in Figure 10b, magnets of 25 g and a C-clamp were attached to the blade to simulate the added mass and the added stiffness, respectively. To study the sensing range, the mass was moved downward the blade away from the PZTs and stopped at 13, 25, and 40 cm, respectively. The actual damage testing was conducted at locations 1 and 3. As shown in Figure 10c, the notch damage was created at location 1 with a progressive damage severity (deep/wide: 8/3, 15/5, and 19/5 mm). Several notch damages in the form of 2.5-cm-long gaps in location 1 were also created outside the blade with distances of 3, 12, 22, and 31 cm, respectively. As shown in Figure 10d, holes of progressive sizes (4 cm in depth with 1.6, 3.2, 4.8, and 6.4 mm in diameter) were created in location 3. Additionally, holes were drilled in location 3 roughly 5 to 6 cm in depth and 6.5 cm in diameter with distances down the length of the blade of 0, 5, 13, 20, and 28 cm, respectively. The results showed that the impedance in the frequency band 10–60 kHz was suitable for damage detection. All sensors could sense the added mass, the added stiffness, and the notch damage in the blade. The sensors should be positioned on both sides of the blade to sense the damage. It is found that the detection range is about 10–30 cm, depending on the transducer and the type of damage. However, considering the size of the wind turbine blades, such a sensing range seems pretty narrow. Challenges remain to improve the sensing range of PZT patches and optimize their placement on actual wind turbine blades.

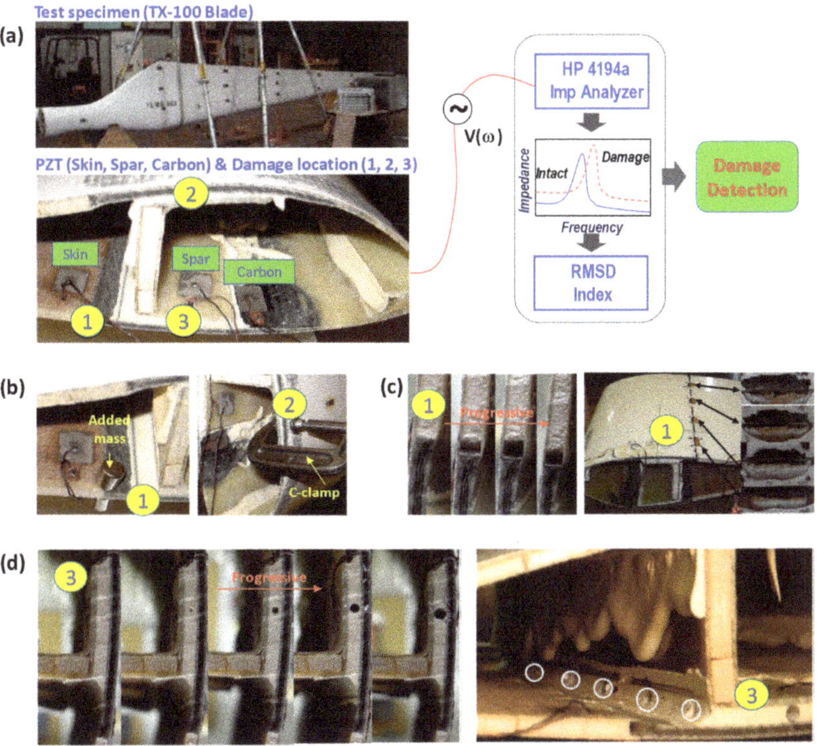

Figure 10. Detecting structural damages on the spar and the skin of a blade segment using the impedance-based technique [49]: (**a**) test specimen and procedure, (**b**) indirect damage testing, (**c**) and actual damage testing at location 1 and (**d**) at location 3.

The fatigue damage detection in a wind turbine blade was investigated by Huh et al. [48]. In their study, four piezoelectric transducers (PVDF film sensor) were attached to the fixing end of a 10-kW wind turbine blade. The impedance response in 1–200 MHz, the local strain values of the blade, and the maximum deflection were recorded under fatigue loading up to 508,249 cycles. The obtained data revealed the local damage or geometrical change during the experiment. The RMSD metric was used to quantify the impedance change that showed noticeable variations with different sensor locations and fatigue loads. By observing the RMSD change, the fatigue damage was successfully identified.

The crack detection in a commercially purchased wind turbine (WINDMAX) when it was loaded until failure was studied by Ruan et al. [14]. The PZT patches were attached to the top and bottom of the wind turbine blade near the fixing root, and the Agilent 4294A impedance analyzer was used to record the impedance signatures. The blade was undergone increasing loading cycles in which the blade tip was loaded by a displacement of 50, 100, 150, 225, and 250 mm, respectively, while returning to zero displacement before each cycle. The impedance change was also quantified by the RMSD metric. Visual failure of the blade was observed around 225-mm and 250-mm tip displacements and a 0.57-kN load. The recorded impedance data showed rapid changes when the blade nearly failed, leading to significant variations in the RMSD index. The RMSD changes from different sensors provided information about the health status of the blade and the damage location.

Rumsey et al. [101] examined the applicability of MFC (macro-fiber composite) actuators for monitoring the fatigue damage that occurred in a TX-100 blade, which was more massive than the blade tested in the reference [49]. Six MFC patches were installed on the low-pressure (downward-facing) side between 1 and 3 m from the root, where the potential fatigue failure can be expected. The impedance data in 5–60 kHz recorded from MFC patches before the fatigue test showed no resonant peaks and, therefore, no structural information. The recording was continued throughout the fatigue test, but the damage was not successfully detected. This result suggested that the MFC actuators failed to activate local dynamical modes of the tested blade. Thus, selecting sensing material in the impedance-based SHM is a critical issue to secure the success of a damage detection process. Challenges remain for developing novel piezoelectric materials for better sensing mechanisms with higher damage sensitivity and a more extended sensing range for SHM of large-scale wind turbine blades.

4.2. SHM of Wind Turbine Tower

4.2.1. Monitoring of Tower Joint

So far, there have been minimal investigations on the impedance-based SHM of wind turbine towers. Nguyen et al. [52] conducted a pioneered experiment on the feasibility of impedance-based SHM for bolt looseness monitoring in a wind turbine tower. The experimental setup is shown in Figure 11. The test structure was a 300-W wind turbine (1.845 m in height) fixed to the floor through a bolted joint. In their study, several PZTs were patched on the bolted flanges of a lab-scale wind turbine structure, and the joint damage was simulated by sequentially loosening the bolt nuts: B1 and B2, as shown in Figure 11. The PZTs were excited by a high-frequency voltage using a HIOKI 3532 LCR HiTester, and the impedance responses of the joint were then recorded. It is observed that the impedance response in the frequency band of 100–300 kHz, reflecting the local resonances of the joints, was sensitive to the bolt looseness. The looseness of the bolted joint was successfully identified by monitoring the change in the RMSD and CCD indices of the impedance signature. In [52], the impedance-based technique was combined with the vibration-based technique to enhance the damage detection in wind turbine towers. However, the experimented tower model is made of rectangular hollow cross-sections, not tubular cross-sections. Additionally, the effects of the operation of wind turbines and the temperature variation on the impedance response remain unsolved.

Some impedance-based SHM investigations have been conducted on structures with similar geometries to a wind turbine tower. Jiang et al. [102] attempted to identify a minor degree of looseness in flange bolts of a tubular joint. Figure 12 shows the experimental setup, including a NI impedance acquisition system (see Figure 12a) and an eight-bolt flange joint with a PZT patch (PZT-5A φ16 mm in diameter and 1 mm thick) (see Figure 12b) [102]. The impedance was recorded in the frequency band of 350–650 kHz (0.75-kHz intervals). The bolts were initially torqued by 40 Nm in the healthy state. Then, the #7 bolt was loosened with monitor torque reductions ranging from 39.5 Nm to 36 Nm with a 0.5-Nm interval. Pseudo-random noise (ranging from 5 to 30 dBW with 5-dBW intervals) was added to the excitation signal of the PZT during the impedance measurement. The significant impact of the noise on the conductance was observed in Figure 12c. The results show that traditional damage index RMSD, MAPD, CCD, and RMSCR (root mean square of the change ratio) were less effective in detecting minor bolts under the added noise. A new damage evaluation metric (the so-called T index) was developed to deal with the problem of distinguishing minor damage from noise [102]. The test on the eight-bolt flange showed that the indication ranges of the T index were distinguishable for the minor looseness and the noise (i.e., 0.07–0.3 and 0.3–1.2, respectively); see Figure 12d. By comparing the value of the T index with the set threshold, it was able to identify minor bolt looseness in the test specimen.

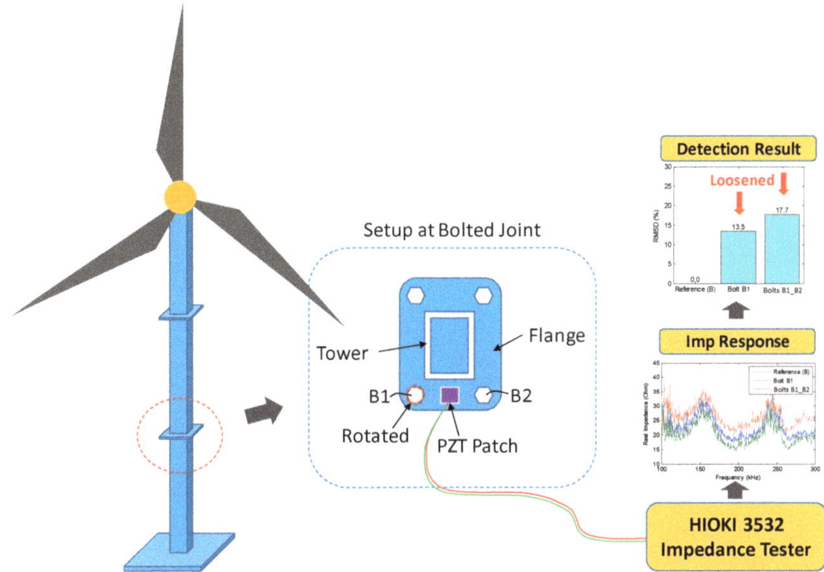

Figure 11. Impedance-based SHM of a lab-scaled wind turbine model [52].

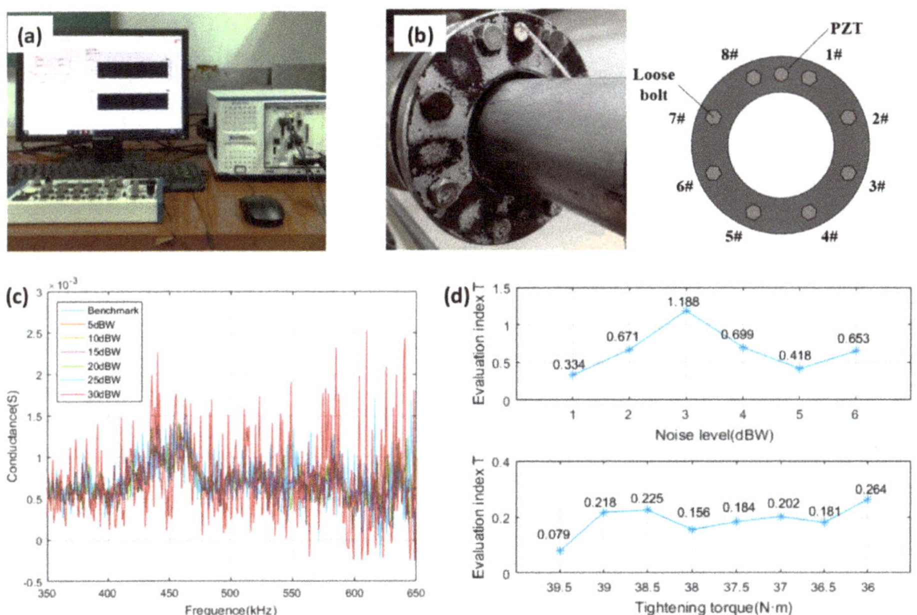

Figure 12. Impedance-based SHM of a bolted flange (**a**) impedance analyzer, (**b**) PZT mounted on an eight-bolt flange and layout diagram, (**c**) noise effect on the conductance, and (**d**) evaluation index under different working conditions (adapted from [102]).

To enhance the portability, Wang et al. [103] developed a wearable sensor device and successfully applied it to monitor the change in the bolt preload of a bolted flange. As shown in Figure 13a,b, the device comprises PZT transducers protected by a rubber layer and a metallic outer layer that plays as a protection layer. It is also designed with a clamp-through-screw mechanism, allowing the device to be easily attached and removed from the host flange. With an adequate clamping force, the PZTs are contacted with the flange body without using any permanent adhesive layers. Thus, the device can be worn on the tubular structure, thus enabling the applications for wind turbine towers. As shown in Figure 13c, the device was mounted on a 12-bolt flanged valve from which the bolt's torque was gradually changed from 30 to 180 lb/ft with 30-lb/ft intervals. The impedance was recorded in 40 Hz–110 MHz through an Agilent 4294A impedance analyzer. The normalized RMSD was computed as a bolt looseness index, showing a significant variation under torque changes, as shown in Figure 13d.

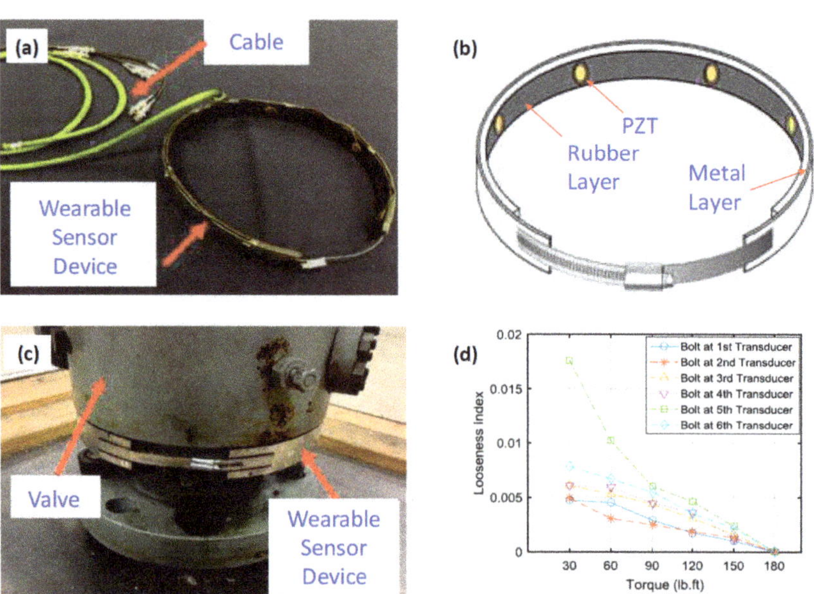

Figure 13. Impedance-based SHM of a valve using wearable technique: (**a**) wearable sensor device, (**b**) layout of the device, (**c**) device mounted on a valve, and (**d**) loosening index vs. bolt torque (adapted from [103]).

4.2.2. Monitoring of Tower Segment

The impedance-based technique can also detect the local cracks in the tower segment. So far, there has been no study investigating this problem. However, several research attempts have been made to assess the damages in tubular structures such as wind turbine towers by the impedance-based technique [31,32]. For example, Du et al. [31] studied the feasibility of detecting multi cracks in pipeline structures using the PZT array. Two artificial cracks were inflicted on the tested pipe, with the severity ranging from 0 mm to 9 mm. The RMSD index of the impedance was computed as a damage indicator. Their results showed that the impedance-based method could realize the quantitative analysis of the cracks. A theoretical impedance model was developed to provide an insight into the impedance response of a PZT pipe system [32]. Based on this model, a new damage-sensitive feature factor was derived from the mechanical impedance equation for damage identification of pipe structures.

Recently, Antunes et al. [104] mounted the PZT on the outer surface of the pipes (Pipe 1 and Pipe 2) to acquire the impedance response in the frequency range of 5–120 kHz, as seen in Figure 14a,b. Structural damage was simulated by adding the masses to the pipes. They also investigated the varying temperature conditions, ranging from −40 to +80 °C, by putting the pipes into a temperature chamber. Figure 14c shows the shifts in the impedance resonances in a representative frequency range (i.e., 95–105 kHz) along with the temperature change; both vertical and horizontal shifts were observed. A temperature compensation method was designed to improve the damage detection result. The compensation method is efficient and can be applied for the SHM of tubular towers under varying ambient conditions. Since most PZT sensors are flat and brittle, they are not convenient to mount on the curved surfaces of tubular towers unless their sizes are small enough. Challenges remain to design novel piezoelectric transducers, both flexible and durable, to record the impedance response from tubular structures such as wind turbine towers.

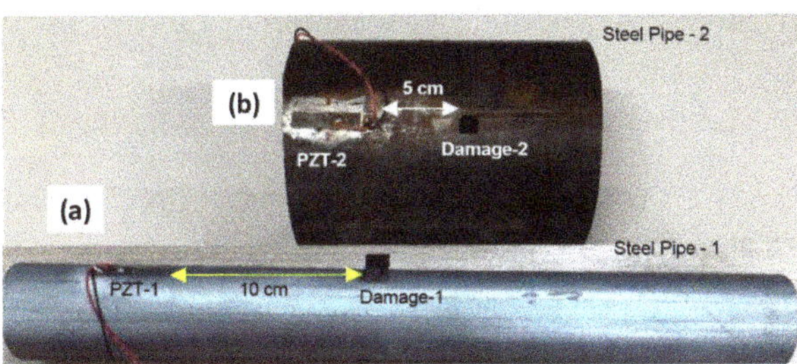

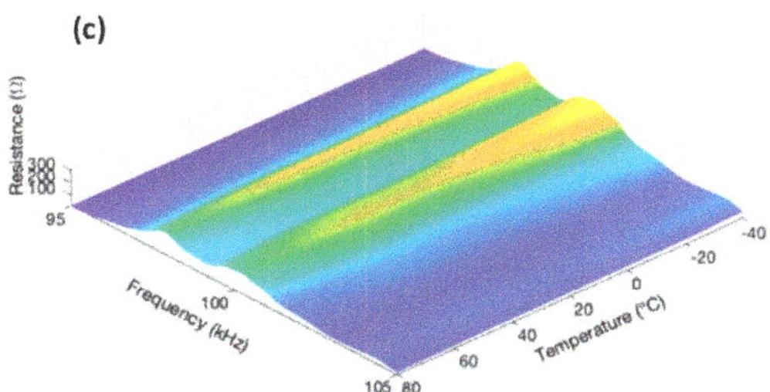

Figure 14. PZTs mounted on steel pipes: (**a**) pipe 1, (**b**) pipe 2, and (**c**) shifts in the impedance resonances in 95–105 kHz due to the temperature effect (adapted from [104]).

Recently, the concept of the wearable sensor device (presented in Section 4.2.1) was extended for pipes' corrosion and bearing wear monitoring [105]. The lab-scale experiment showed that the corrosion-induced changes in the thickness of the pipe led to a shift in the peak frequency. This experimental investigation has not only evidenced the feasibility of the wearable sensor device but also provided a promising solution for the problem of corrosion detection in the wind turbine tower.

4.3. SHM of Substructure and Foundation

Only a few studies reported the application of the impedance-based technique for SHM of the wind turbine substructure and foundation. Choi et al. [106] investigated the numerical feasibility of the impedance-based technique for welded joint monitoring of an offshore wind turbine substructure. As illustrated in Figure 15, a PZT transducer (type PZT-5A) was supposed to be attached to the central upper joint of a tripod, and its impedance response of 10–24 kHz under different levels of the crack in welding was simulated through the COMSOL program. They observed the apparent shifts in the signature of the impedance peaks that can be used for reliable damage identification. The numerical simulation also investigated the effect of the PZT's size on the impedance response. Three sizes were considered, including 3 × 3 cm, 4 × 4 cm, and 5 × 5 cm. They observed that the impedance curve was shifted downward with the increasing size of the PZT, while the impedance resonances were relatively stable. However, more numerical analyses and experimental verifications are still needed to verify the proposed idea and optimize the PZT size and placement.

Moll [51] presented a successful preliminary application of the impedance-based technique for monitoring the structural damage occurring in a lab-scale grouted connection. Figure 16 shows the experimental setup of a grouted connection for impedance-based SHM. Six circular piezoelectric transducers of φ10 mm were circumferentially mounted on the outer metal part of the connection. The impedance measurement system includes a computer, a multiplexer, an amplifier, and a frequency generator (HS3), as illustrated in Figure 16. A 4-mm hole with a 35-mm depth was created in the concrete part to simulate the concrete defect, and the corresponding impedance response was recorded at 25–300 kHz. The RSMD index was computed as the damage indicator. The recorded impedance response was altered as the grounded connection was damaged in the concrete part. Moreover, the damage was successfully detected by monitoring the change in the RMSD index. Nonetheless, the simulated damage was not realistic. Additionally, challenges remain in evaluating the durability of the transducer under the marine atmosphere and the effect of varying environmental and loading conditions on the damage-detectability.

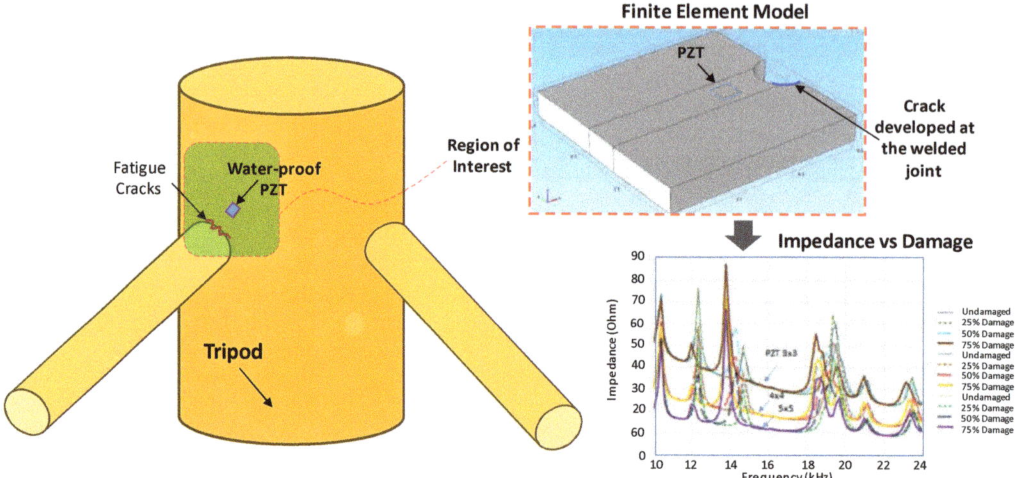

Figure 15. Impedance-based fatigue crack monitoring of the upper central joint of the offshore wind turbine tripod [106].

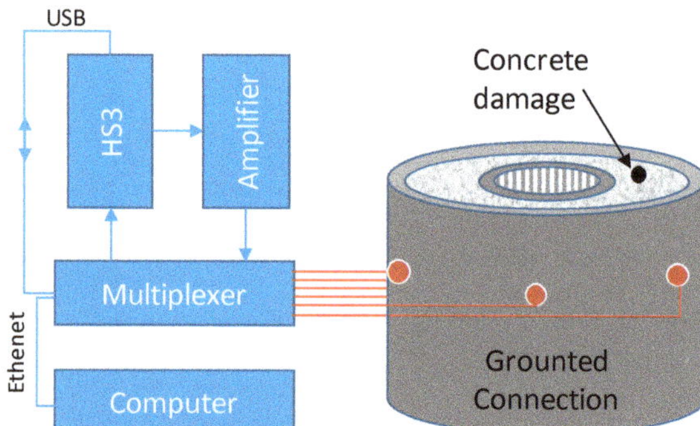

Figure 16. Experimental setup of impedance-based SHM of a grouted connection [51].

5. Future Perspectives of Impedance-Based SHM of Wind Turbines

The above literature review shows that the impedance-based technique has been insufficiently explored for wind turbines. Due to its potential for developing one of the most effective SHM systems, new applications of the impedance-based technique for the SHM of wind turbines will be indeed discovered. In Figure 17, we suggest five main research topics that can contribute to the translation of the impedance-based technique to practical applications in wind turbine monitoring. They include blade monitoring, tower monitoring, substructure monitoring, foundation monitoring, and scour monitoring. In the following sections, those suggested research topics are discussed.

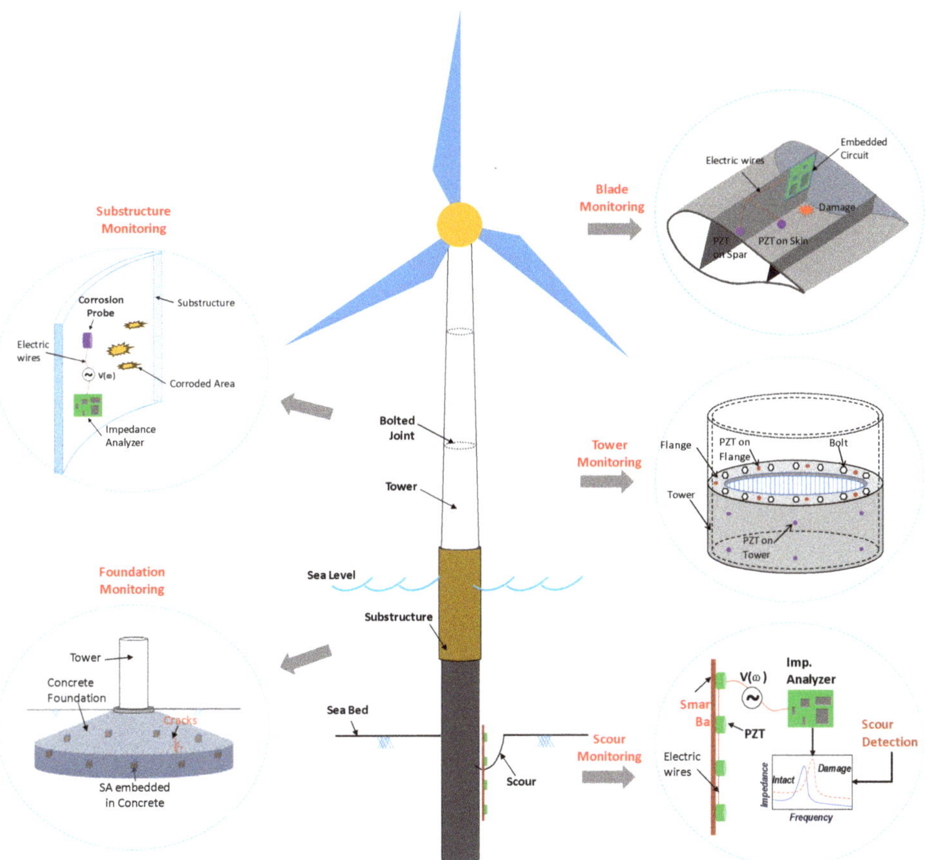

Figure 17. Impedance-based SHM of wind turbine structures.

5.1. Blade Monitoring

So far, the impedance-based technique has been mainly applied for fatigue damage detection in wind turbine blades. Therefore, more research is needed to explore this innovative technology for detecting other types of blade failures, such as damage from lightning, damage from atmospheric icing, and leading-edge erosion. Since the impedance characteristics corresponding to those damages could be different, the damage severity prediction model corresponding to each failure should be developed for quantitative SHM of wind turbine blades.

A previous experiment by Pitchford et al. [49] showed that an added mass to a wind turbine blade led to variations in the impedance signature. Since the added mass can be considered a freezing growth, this experiment holds promise for detecting ice on the blade [107]. However, more numerical and experimental investigations are needed to make this idea a reality. Leading-edge erosion is related to the loss of the material and the change in the geometrical parameters of the wind turbine blade. The impedance-based technique could feasibly detect the wind turbine blade's mass, according to the reference [49]. The lightning causes delamination, shell debonding, and tip detachment in the blade, and such damages could be potentially identified through the impedance-based technique based on some previous investigations [108,109].

A typical structural design has a critical bolted joint at the blade root, which should be adequately monitored. The impedance-based technique has been regarded as a promising tool for loosened bolt monitoring [37,110,111]. Therefore, there is a need to develop a bolt looseness monitoring system for the blade root of a wind turbine structure. As reported in [101], the MFC was insensitive to a thick blade's damage due to its weak piezoelectric excitation. Therefore, selecting the piezoelectric transducers corresponding to a target wind turbine blade could play an essential role in the success of an impedance-based damage detection process. This deserves a future intensive investigation.

The previous study proposed a smart blade concept for guided wave-based SHM [112]. Based on this concept, we introduce an impedance-based smart blade for the impedance-based SHM, as shown in Figure 18. The piezoelectric-based smart blade is embedded with self-powered PZT nerves and embedded signal conditioning and data acquisition circuits for wireless and autonomous sensing. The blade vibration-induced energy can be harvested to power the SHM system. Since the impedance-based technique has a limited sensing range for each PZT transducer, a smart blade should be developed with dense PZT nerves to provide the health status of any points on the blade.

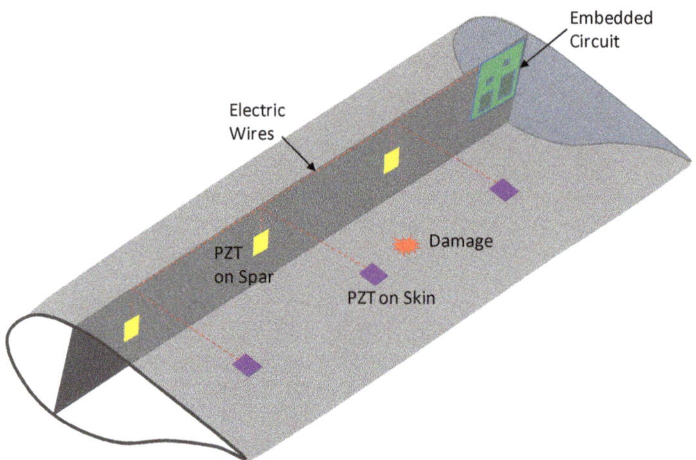

Figure 18. Proposed prototype of an impedance-based smart blade for impedance-based SHM.

Additionally, the survivability of PZT sensor systems during a lightning occurrence is also an important issue. The PZT sensors close to the area struck by lightning can be heated up to tens of thousands of degrees Celsius. During such high temperatures, the piezoelectric materials, bonding layers, and soldering components of the sensors can be damaged, threatening the regular operation of the monitoring system. Nonetheless, one of the promising features of piezoelectric impedance-based SHM technology is sensor self-diagnosis. Thus, a sensor (i.e., breakage or debonding) damaged can be timely identified using the inverse of impedance signature, which further improves the reliability of operating SHM system during a lightning occurrence [22,27]. The impact of lightning on the survivability of the PZT sensors is a good research topic that needs to be studied further in the future.

5.2. Tubular Tower Monitoring

Researchers have paid less attention to the tower failures than blade failures [65]. Although there have been some successful applications of the impedance-based technique to structures with similar geometry to a wind turbine tower, such as pipelines and bolted flanges [102,104], there is a long way to go for this technology to be transferred into real SHM applications. To guarantee the successful translation, we need a comprehensive study of the impedance characteristics, the sensing range, and the effective frequency band of the PZT mounted on both lab-scale and real-scale wind turbine towers. Realistic/near-realistic damages such as fatigue cracks, bolt looseness, lightning, and corrosion should be investigated to evaluate the practicality of the impedance-based technique. Additionally, there is a need to extensively investigate the influence of offshore environmental changes on the impedance responses and develop robust compensation techniques.

Figure 19 illustrates an example of PZT deployment on the tower for impedance-based SHM. Lessons learned from the previous accidents [67,68] show that the shell wall-thickness transition areas of the tower are hot spots where local buckling could have occurred. Thus, a PZT network should be designed and deployed in those hot spots to monitor the local buckling damage; see Figure 19. Additionally, the fracture could have happened at the bolted joint of the tubular tower segments [4]. It is therefore necessary to install PZT on the flange to monitor multiple bolts simultaneously. A multi-channel impedance board is developed to record the impedance from the number of PZT patches in real-time. Efficient deep learning models are developed to autonomously extract the damage-sensitive impedance features from the recorded data and identify the damage type, location, and extent of the quantitative SHM of the tower (e.g., convolutional neural networks [25,113] and graph convolutional network [114]). Besides the fracture and local buckling, the PZT network can also be used to monitor the tower's cracks and corrosion.

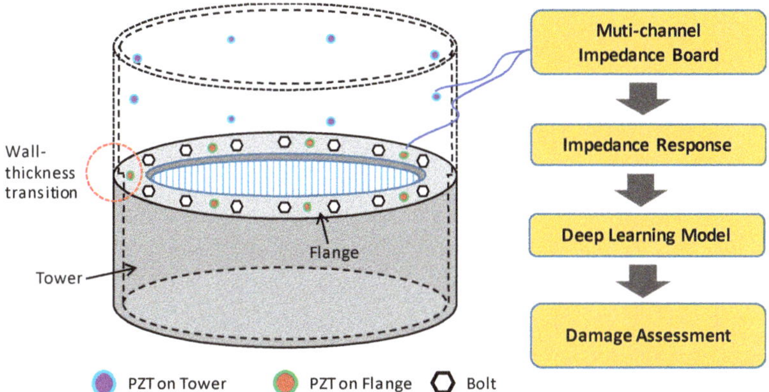

Figure 19. An example of PZT deployment on the tower for impedance-based SHM.

As reported in [115], the sensing range of an individual PZT patch is dependent on the material and the density of the monitored structure. Considering the limited sensing range of a PZT patch, it is necessary to develop an optimal PZT network that can be used to monitor the whole wind turbine tower effectively. For a large and high tower, it is necessary to determine the hot spots in the structure and optimize transducer placement to reduce the number of transducers and the cost of the SHM system. The research on PZT placement, considering the installation cost and the efficiency, also deserves intensive investigation.

To reduce the cost of SHM systems, along with enhancing the damage detectability, a research trend is to combine the local impedance-based technique with other techniques (i.e., guided wave method, thermal imaging technique, fiber optic method, laser vibrometer technique, and vision-based approach) to build a hybrid system. In [52], the authors developed a hybrid SHM strategy that combines the local impedance-based technique with the global vibration-based technique to enhance the damage detection result of a wind turbine tower. Both accelerometers and PZT patches were attached to the flanges of the tower, and their recorded responses are fused to detect potential damages. However, this study did not consider other potential damages, such as cracks and corrosion. Additionally, the effect of the rotor's vibration on the impedance response was not investigated.

Developing impedance-based low-cost portable/wearable devices for the SHM of wind turbine towers is suggested for future research. Some preliminary studies have developed piezoelectric wearable devices with cost-effectiveness, portability, a predictable frequency band, and real-time applicability [46,86,103]. For example, Huo et al. [110] embedded a PZT transducer into a ring to develop the concept of the smart washer for single-bolt monitoring. Wang et al. [103] developed a wearable sensor device for bolt group monitoring. The prototype of the wearable sensor device and its experimental verification for impedance-based SHM were previously described in Section 4.2.1. The device (comprised of many PZT transducers) can be worn on a tubular joint/segment of the tower through a clamping mechanism to collect the impedance responses, which are then used for further health monitoring and damage assessment. Despite the preliminary successful verification results [103,105,110], the application of wearable sensor devices for the impedance-based SHM of wind turbine towers has not been explored. For this purpose, a large-size wearable sensor device needs to be designed, and its durability and repeatability performance should be sufficiently investigated before real applications.

Recent developments in machine learning and artificial intelligence have opened alternative doors toward processing impedance signals. Early studies have explored the application of artificial neural networks (ANN) for impedance-based damage detection [116] and compensation of the effect of temperature on the impedance [29]. There is also some research developing principal component analysis (PCA) models for the impedance-based SHM [98]. Some authors explored combining a convolutional neural network (CNN) with the impedance-based technique for damage detection [113]. However, those models were not fully automated due to the requirement of preprocessing impedance signals. More research is needed to develop new machine learning-based algorithms for processing impedance data with fast operations, higher accuracy, automatic damage-sensitive feature extraction, and real-time performances.

5.3. Substructure and Foundation Monitoring

Over the past decades, the impedance-based technique was extensively studied and applied for the SHM of concrete structures. Many numerical and experimental investigations have been conducted to monitor concrete crack, curing, reinforced rebar corrosion, and crack repair [33,34,117–119]. For example, Liu et al. [120] applied the technique for assessing concrete's freezing–thawing and crack damage and proposed a mathematical method to evaluate the development of concrete damage. Their result showed that the RMSD index increased with the development of the freezing–thawing cycles and crack depth, and the most effective frequency band was found at 100–150 kHz. Zhang et al. [121] studied the relationship between the stage of the cement setting process and the resonant frequency shift of impedance peaks to monitor early-age hydration and the setting of cement paste. Ahmadi et al. [122] monitored the corrosion of rebars embedded in concrete blocks and the admittance responses under different levels of corrosion. In their study, the equivalent structural parameters extracted from the admittance response were found as effective indicators for detecting and quantifying the corrosion damage in rebars. However, the PZT patches are brittle and easily damaged. To overcome such issues, some authors embedded a PZT patch inside a concrete block to form a smart aggregate (SA) [33,123]. As a multifunctional device, a SA can be used as both actuator and sensor, making it ideal for impedance-based SHM of concrete parts in wind turbine systems.

Despite the impedance-based technique has been a mature technology and the SA technique has been extensively studied, their application for offshore/onshore foundation monitoring has not been explored so far. In Figure 20, we propose an idea to embed the SAs inside the foundation of the onshore wind turbine to monitor inner cracks. The impedance response of the distributed SAs is continuously monitored, and the impedance shift can be quantified to predict crack locations. The idea presented in Figure 20 needs sufficient numerical and experimental investigations to convert it into a reality. Additionally, intelligent algorithms based on machine learning to predict damage locations are also suggested for future studies.

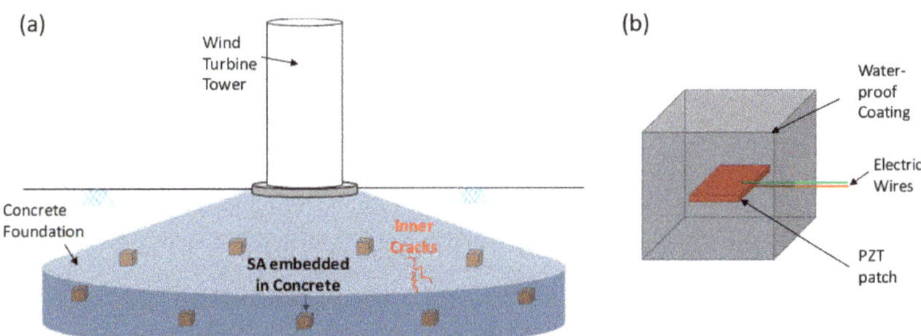

Figure 20. (**a**) Impedance-based SHM of an onshore wind turbine foundation using the SA technique; (**b**) an SA prototype.

To offshore wind turbines, scour is one of the main concerns. Due to its merits, the impedance-based technique could play a vital role in developing low-cost scouring systems for offshore foundations. In Figure 21, we propose a piezoelectric-based smart bar that can be used for monitoring foundation scouring. Several PZT patches are waterproof and bonded to a stainless-steel bar to form a smart bar, which is then inserted into the sea bed next to the foundation (such as a monopile in Figure 21). The operating principle is that the scour will cause the looseness/removal of the material around each PZT patch, leading to the variations of the impedance responses. The scouring depth could be identified by analyzing the damage metrics of the PZT patches.

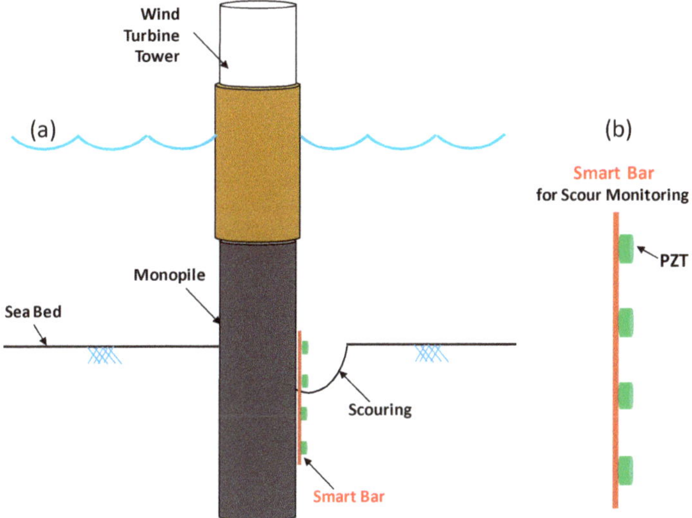

Figure 21. (**a**) Proposed impedance-based scour monitoring of a wind turbine foundation; (**b**) smart bar prototype for scour monitoring.

It is also needed to explore the application of the impedance-based technique for detecting corrosion in offshore wind turbines. The early study showed that the resonant frequency shift of the impedance peaks was proportional to the corroded area [124]. Zhu et al. [125] investigated the use of mechanical impedance to estimate the corrosion damage progression of a steel beam quantitatively. Li et al. [41] developed a PZT-based smart corrosion coupon and established the relationship between the corrosion-induced thickness loss and the impedance variation for comprehensively monitoring the corrosion amount. Future research efforts are required to develop piezoelectric-based corrosion transducers that can be applied for the SHM of offshore towers and substructures. Under the marine condition, protection technologies should also be developed for those piezoelectric devices to secure their functionality during the impedance monitoring process.

6. Conclusions

In the past decades, the impedance-based technique has been extensively studied for SHM of different civil engineering structures. Thanks to its merits, including cost-effectiveness, easy implementation for complex structures, robustness to early-stage damages, and online and autonomous damage assessment, the impedance-based technique has the potential to create one of the most effective SHM systems. Nonetheless, its SHM applications in the wind energy industry have been limited. Therefore, this paper was motivated to present the current status and propose new research topics for impedance-based SHM of wind turbines.

The contributions of this review paper lie in:

(1) Focusing on the applicability of the impedance-based technique for SHM of different structural components such as blades, towers, substructures, and foundations;
(2) Identifying research needs for improving the performance of the impedance-based SHM of wind turbines;
(3) Proposing innovative concepts for the future development of an effective impedance-based SHM system in the green energy field.

The proposed concepts include (i) the piezoelectric-based smart wind turbine blade with a self-powering and embedded circuit for real-time SHM purposes, (ii) the piezoelectric-based smart bar to monitor the scour damage occurring around an offshore wind turbine foundation, (iii) the PZT-based corrosion probe for monitoring the structural conditions of a tower and substructure, (iv) the wearable sensor device for the SHM of wind turbine towers and bolted joints, and (v) innovative SA technologies for quantitative damage estimations on the concrete foundations of wind turbines.

There are still many challenges to substantiate the application of the impedance-based technique for wind turbine tower monitoring and to ensure the smooth transition of this technology from research to practice. Besides the above-mentioned topics, several engineering issues remain for (i) enhancing the sensing range of PZT patches and optimizing their geometrical parameters considering the sizes of wind turbine structures; (ii) designing a method of protection for PZT patches from harsh environmental conditions, especially extreme offshore conditions, to secure their functionality and reliable damage detection; (iii) developing new piezoelectric materials with better sensing mechanisms, higher damage sensitivity, and a more extended sensing range and durability under a marine atmosphere; (iv) developing innovative algorithms based on artificial intelligence and machine learning for compensating the effects of marine conditions on the impedance response and minimizing false damage identifications; and (v) designing novel wireless impedance sensor nodes with higher measurable frequency bands, up to MHz, for practical applications.

We are in the early stage of the discovery journey towards the practical application of the impedance-based method in the wind energy industry. Many research efforts need to be carried out to verify the feasibility and evaluate the practicality of the impedance-based technique for lab-scale and real-scale wind turbine towers. However, the unique advantages of the impedance-based SHM technology pose it as a possible important tool for the SHM of wind turbines in the green energy field.

Author Contributions: Conceptualization, T.-C.H., D.-D.H. and T.-C.L.; methodology, T.-C.H., D.-D.H. and T.-C.L.; formal analysis, T.-C.H., T.-H.N., T.-H.-T.L. and H.-P.N.; writing—original draft preparation, T.-C.H. and T.-C.L.; writing—review and editing, T.-C.H., T.-H.-T.L. and D.-D.H. All authors have read and agreed to the published version of the manuscript.

Funding: This research received no external funding.

Acknowledgments: We acknowledge the support of time and facilities from Ho Chi Minh City University of Technology (HCMUT), VNU-HCM for this study.

Conflicts of Interest: The authors declare no conflict of interest.

References

1. Luderer, G.; Pietzcker, R.C.; Carrara, S.; de Boer, H.S.; Fujimori, S.; Johnson, N.; Mima, S.; Arent, D. Assessment of wind and solar power in global low-carbon energy scenarios: An introduction. *Energy Econ.* **2017**, *64*, 542–551. [CrossRef]
2. Wiser, R.; Rand, J.; Seel, J.; Beiter, P.; Baker, E.; Lantz, E.; Gilman, P. Expert elicitation survey predicts 37% to 49% declines in wind energy costs by 2050. *Nat. Energy* **2021**, *6*, 555–565. [CrossRef]
3. Ciang, C.C.; Lee, J.-R.; Bang, H.-J. Structural health monitoring for a wind turbine system: A review of damage detection methods. *Meas. Sci. Technol.* **2008**, *19*, 122001. [CrossRef]
4. Chou, J.-S.; Tu, W.-T. Failure analysis and risk management of a collapsed large wind turbine tower. *Eng. Fail. Anal.* **2011**, *18*, 295–313. [CrossRef]
5. Adedipe, O.; Brennan, F.; Kolios, A. Review of corrosion fatigue in offshore structures: Present status and challenges in the offshore wind sector. *Renew. Sustain. Energy Rev.* **2016**, *61*, 141–154. [CrossRef]
6. Katsaprakakis, D.A.; Papadakis, N.; Ntintakis, I. A Comprehensive Analysis of Wind Turbine Blade Damage. *Energies* **2021**, *14*, 5974. [CrossRef]
7. Shafiee, M.; Brennan, F.; Espinosa, I.A. A parametric whole life cost model for offshore wind farms. *Int. J. Life Cycle Assess.* **2016**, *21*, 961–975. [CrossRef]
8. Shafiee, M.; Zhou, Z.; Mei, L.; Dinmohammadi, F.; Karama, J.; Flynn, D. Unmanned Aerial Drones for Inspection of Offshore Wind Turbines: A Mission-Critical Failure Analysis. *Robotics* **2021**, *10*, 26. [CrossRef]
9. Martinez-Luengo, M.; Kolios, A.; Wang, L. Structural health monitoring of offshore wind turbines: A review through the Statistical Pattern Recognition Paradigm. *Renew. Sustain. Energy Rev.* **2016**, *64*, 91–105. [CrossRef]
10. Balageas, D. Introduction to Structural Health Monitoring. In *Structural Health Monitoring*; Wiley Publisher: New York, NY, USA, 2006; pp. 13–43.
11. Gómez Muñoz, C.Q.; García Márquez, F.P. A new fault location approach for acoustic emission techniques in wind turbines. *Energies* **2016**, *9*, 40. [CrossRef]
12. Rizk, P.; Younes, R.; Ilinca, A.; Khoder, J. Wind turbine blade defect detection using hyperspectral imaging. *Remote Sens. Appl. Soc. Environ.* **2021**, *22*, 100522. [CrossRef]
13. Wang, P.; Zhou, W.; Bao, Y.; Li, H. Ice monitoring of a full-scale wind turbine blade using ultrasonic guided waves under varying temperature conditions. *Struct. Control Health Monit.* **2018**, *25*, e2138. [CrossRef]
14. Ruan, J.; Ho, S.C.M.; Patil, D.; Song, G. Structural health monitoring of wind turbine blade using piezoceremic based active sensing and impedance sensing. In Proceedings of the 11th IEEE International Conference on Networking, Sensing and Control, Banff, AB, Canada, 7–9 April 2014; pp. 661–666.
15. Nguyen, T.-C.; Huynh, T.-C.; Kim, J.-T. Numerical evaluation for vibration-based damage detection in wind turbine tower structure. *Wind Struct.* **2015**, *21*, 657–675. [CrossRef]
16. Nguyen, C.-U.; Huynh, T.-C.; Kim, J.-T. Vibration-based damage detection in wind turbine towers using artificial neural networks. *Struct. Monit. Maint.* **2018**, *5*, 507.
17. Khadka, A.; Afshar, A.; Zadeh, M.; Baqersad, J. Strain monitoring of wind turbines using a semi-autonomous drone. *Wind Eng.* **2021**, *46*, 296–307. [CrossRef]
18. Wu, R.; Zhang, D.; Yu, Q.; Jiang, Y.; Arola, D. Health monitoring of wind turbine blades in operation using three-dimensional digital image correlation. *Mech. Syst. Signal Process.* **2019**, *130*, 470–483. [CrossRef]
19. de Castro, B.A.; Baptista, F.G.; Ciampa, F. Comparative analysis of signal processing techniques for impedance-based SHM applications in noisy environments. *Mech. Syst. Signal Process.* **2019**, *126*, 326–340. [CrossRef]
20. Singh, S.K.; Soman, R.; Wandowski, T.; Malinowski, P. A variable data fusion approach for electromechanical impedance-based damage detection. *Sensors* **2020**, *20*, 4204. [CrossRef]
21. de Rezende, S.W.F.; de Moura, J.d.R.V.; Neto, R.M.F.; Gallo, C.A.; Steffen, V. Convolutional neural network and impedance-based SHM applied to damage detection. *Eng. Res. Express* **2020**, *2*, 035031. [CrossRef]
22. Na, W.S.; Baek, J. A Review of the Piezoelectric Electromechanical Impedance Based Structural Health Monitoring Technique for Engineering Structures. *Sensors* **2018**, *18*, 1307. [CrossRef]
23. Huynh, T.-C.; Lee, K.-S.; Kim, J.-T. Local dynamic characteristics of PZT impedance interface on tendon anchorage under prestress force variation. *Smart Struct. Syst.* **2015**, *15*, 375–393. [CrossRef]
24. Kralovec, C.; Schagerl, M. Review of Structural Health Monitoring Methods Regarding a Multi-Sensor Approach for Damage Assessment of Metal and Composite Structures. *Sensors* **2020**, *20*, 826. [CrossRef] [PubMed]
25. Nguyen, T.-T.; Tuong Vy Phan, T.; Ho, D.-D.; Man Singh Pradhan, A.; Huynh, T.-C. Deep learning-based autonomous damage-sensitive feature extraction for impedance-based prestress monitoring. *Eng. Struct.* **2022**, *259*, 114172. [CrossRef]

26. Viechtbauer, C.; Schröder, K.-U.; Schagerl, M. Structural Health Control—A Comprehensive Concept for Observation and Assessment of Damages Applied on a Darrieus Wind Turbine. In Proceedings of the 9th International Workshop on Structural Health Monitoring, Stanford, CA, USA, 10–13 September 2013; Volume 1.
27. Huynh, T.-C.; Dang, N.-L.; Kim, J.-T. Advances and challenges in impedance-based structural health monitoring. *Struct. Monit. Maint* **2017**, *4*, 301–329.
28. Perera, R.; Pérez, A.; García-Diéguez, M.; Zapico-Valle, J.L. Active wireless system for structural health monitoring applications. *Sensors* **2017**, *17*, 2880. [CrossRef]
29. Huynh, T.C.; Kim, J.T. RBFN-based temperature compensation method for impedance monitoring in prestressed tendon anchorage. *Struct. Control Health Monit.* **2018**, *25*, e2173. [CrossRef]
30. Zheng, Y.; Liu, K.; Wu, Z.; Gao, D.; Gorgin, R.; Ma, S.; Lei, Z. Lamb waves and electro-mechanical impedance based damage detection using a mobile PZT transducer set. *Ultrasonics* **2019**, *92*, 13–20. [CrossRef]
31. Du, G.; Huo, L.; Kong, Q.; Song, G. Damage detection of pipeline multiple cracks using piezoceramic transducers. *J. Vibroengineering* **2016**, *18*, 2828–2838. [CrossRef]
32. Zuo, C.; Feng, X.; Wang, J.; Zhou, J. The Damage Detection for Pipeline Structures Based on Piezoceramic Transducers. In *Earth and Space 2021*; American Society of Civil Engineers: Reston, VA, USA, 2021; pp. 591–598.
33. Fan, S.; Zhao, S.; Kong, Q.; Song, G. An embeddable spherical smart aggregate for monitoring concrete hydration in very early age based on electromechanical impedance method. *J. Intell. Mater. Syst. Struct.* **2021**, *32*, 537–548. [CrossRef]
34. Zhao, S.; Fan, S.; Yang, J.; Kitipornchai, S. Numerical and experimental investigation of electro-mechanical impedance based concrete quantitative damage assessment. *Smart Mater. Struct.* **2020**, *29*, 055025. [CrossRef]
35. Sabet Divsholi, B.; Yang, Y. Combined Embedded and Surface-Bonded Piezoelectric Transducers for Monitoring of Concrete Structures. *NDT E Int.* **2014**, *65*, 28–34. [CrossRef]
36. Huynh, T.-C.; Dang, N.-L.; Kim, J.-T. Preload Monitoring in Bolted Connection Using Piezoelectric-Based Smart Interface. *Sensors* **2018**, *18*, 2766. [CrossRef]
37. Huynh, T.-C.; Ho, D.-D.; Dang, N.-L.; Kim, J.-T. Sensitivity of piezoelectric-based smart interfaces to structural damage in bolted connections. *Sensors* **2019**, *19*, 3670. [CrossRef] [PubMed]
38. Nguyen, T.-H.; Phan, T.T.V.; Le, T.-C.; Ho, D.-D.; Huynh, T.-C. Numerical Simulation of Single-Point Mount PZT-Interface for Admittance-Based Anchor Force Monitoring. *Buildings* **2021**, *11*, 550. [CrossRef]
39. Hoshyarmanesh, H.; Abbasi, A. Structural health monitoring of rotary aerospace structures based on electromechanical impedance of integrated piezoelectric transducers. *J. Intell. Mater. Syst. Struct.* **2018**, *29*, 1799–1817. [CrossRef]
40. Hoshyarmanesh, H.; Ghodsi, M.; Kim, M.; Cho, H.H.; Park, H.-H. Temperature Effects on Electromechanical Response of Deposited Piezoelectric Sensors Used in Structural Health Monitoring of Aerospace Structures. *Sensors* **2019**, *19*, 2805. [CrossRef]
41. Li, W.; Liu, T.; Zou, D.; Wang, J.; Yi, T.-H. PZT based smart corrosion coupon using electromechanical impedance. *Mech. Syst. Signal Process.* **2019**, *129*, 455–469. [CrossRef]
42. Thoriya, A.; Vora, T.; Nyanzi, P. Pipeline corrosion assessment using electromechanical impedance technique. *Mater. Today Proc.* **2021**, *56*, 2334–2341. [CrossRef]
43. Zhang, J.; Zhang, C.; Xiao, J.; Jiang, J. A pzt-based electromechanical impedance method for monitoring the soil freeze–thaw process. *Sensors* **2019**, *19*, 1107. [CrossRef]
44. Zhang, C.; Wang, X.; Yan, Q.; Vipulanandan, C.; Song, G. A novel method to monitor soft soil strength development in artificial ground freezing projects based on electromechanical impedance technique: Theoretical modeling and experimental validation. *J. Intell. Mater. Syst. Struct.* **2020**, *31*, 1477–1494. [CrossRef]
45. Wu, J.; Yang, G.; Wang, X.; Li, W. PZT-Based soil compactness measuring sheet using electromechanical impedance. *IEEE Sens. J.* **2020**, *20*, 10240–10250. [CrossRef]
46. Ryu, J.-Y.; Huynh, T.-C.; Kim, J.-T. Tension Force Estimation in Axially Loaded Members Using Wearable Piezoelectric Interface Technique. *Sensors* **2019**, *19*, 47. [CrossRef] [PubMed]
47. Le, T.-C.; Phan, T.T.V.; Nguyen, T.-H.; Ho, D.-D.; Huynh, T.-C. A Low-Cost Prestress Monitoring Method for Post-Tensioned RC Beam Using Piezoelectric-Based Smart Strand. *Buildings* **2021**, *11*, 431. [CrossRef]
48. Huh, Y.-H.; Kim, J.; Hong, S. Damage Monitoring for Wind Turbine Blade using Impedance Technique. *J. Korean Soc. Nondestruct. Test.* **2013**, *33*, 452–458. [CrossRef]
49. Pitchford, C.; Grisso, B.; Inman, D. Impedance-based structural health monitoring of wind turbine blades. *Proc. SPIE Int. Soc. Opt. Eng.* **2007**, *6532*, 653211. [CrossRef]
50. Madi, E.; Pope, K.; Huang, W.; Iqbal, T. A review of integrating ice detection and mitigation for wind turbine blades. *Renew. Sustain. Energy Rev.* **2019**, *103*, 269–281. [CrossRef]
51. Moll, J. Damage detection in grouted connections using electromechanical impedance spectroscopy. *Proc. Inst. Mech. Eng. Part C J. Mech. Eng. Sci.* **2018**, *233*, 947–950. [CrossRef]
52. Nguyen, T.-C.; Huynh, T.-C.; Yi, J.-H.; Kim, J.-T. Hybrid bolt-loosening detection in wind turbine tower structures by vibration and impedance responses. *Wind Struct.* **2017**, *24*, 385–403. [CrossRef]
53. Antoniadou, I.; Dervilis, N.; Papatheou, E.; Maguire, A.E.; Worden, K. Aspects of structural health and condition monitoring of offshore wind turbines. *Philos. Trans. R. Soc. A Math. Phys. Eng. Sci.* **2015**, *373*, 20140075. [CrossRef]

54. Kalkanis, K.; Kaminaris, S.; Psomopoulos, C.; Ioannidis, G.; Kanderakis, G. Structural Health Monitoring for the Advanced Maintenance of Wind Turbines: A review. *Int. J. Energy Environ.* **2018**, *12*, 69–79.
55. Metaxa, S.; Kalkanis, K.; Psomopoulos, C.S.; Kaminaris, S.D.; Ioannidis, G. A review of structural health monitoring methods for composite materials. *Procedia Struct. Integr.* **2019**, *22*, 369–375. [CrossRef]
56. Civera, M.; Surace, C. Non-Destructive Techniques for the Condition and Structural Health Monitoring of Wind Turbines: A Literature Review of the Last 20 Years. *Sensors* **2022**, *22*, 1627. [CrossRef] [PubMed]
57. Fan, X.; Li, J.; Hao, H. Review of piezoelectric impedance based structural health monitoring: Physics-based and data-driven methods. *Adv. Struct. Eng.* **2021**, *24*, 3609–3626. [CrossRef]
58. Garolera, A.C.; Madsen, S.F.; Nissim, M.; Myers, J.D.; Holboell, J. Lightning Damage to Wind Turbine Blades From Wind Farms in the U.S. *IEEE Trans. Power Deliv.* **2016**, *31*, 1043–1049. [CrossRef]
59. Hasager, C.B.; Vejen, F.; Skrzypiński, W.R.; Tilg, A.-M. Rain Erosion Load and Its Effect on Leading-Edge Lifetime and Potential of Erosion-Safe Mode at Wind Turbines in the North Sea and Baltic Sea. *Energies* **2021**, *14*, 1959. [CrossRef]
60. Ma, Y.; Martinez-Vazquez, P.; Baniotopoulos, C. Wind turbine tower collapse cases: A historical overview. *Proc. Inst. Civ. Eng. Struct. Build.* **2019**, *172*, 547–555. [CrossRef]
61. Harte, R.; Van Zijl, G.P.A.G. Structural stability of concrete wind turbines and solar chimney towers exposed to dynamic wind action. *J. Wind Eng. Ind. Aerodyn.* **2007**, *95*, 1079–1096. [CrossRef]
62. Huynh, T.-C. Vision-based autonomous bolt-looseness detection method for splice connections: Design, lab-scale evaluation, and field application. *Autom. Constr.* **2021**, *124*, 103591. [CrossRef]
63. Fort, V.; Bouzid, A.-H.; Gratton, M. Analytical modeling of self-loosening of bolted joints subjected to transverse loading. *J. Press. Vessel Technol.* **2019**, *141*, 031205. [CrossRef]
64. Backstrand, J.; Hurtig, A. *Final Report RO 2017: 01*; Statens Haverikommission: Stockholm, Sweden, 2017.
65. Alonso-Martinez, M.; Adam, J.M.; Alvarez-Rabanal, F.P.; del Coz Díaz, J.J. Wind turbine tower collapse due to flange failure: FEM and DOE analyses. *Eng. Fail. Anal.* **2019**, *104*, 932–949. [CrossRef]
66. Ashley, F.; Cipriano, R.J.; Breckenridge, S.; Briggs, G.A.; Gross, L.E.; Hinkson, J.; Lewis, P.A. Bethany Wind Turbine Study Committee Report. Available online: www.townofbethany.com (accessed on 15 June 2022).
67. Chen, X.; Xu, J. Structural failure analysis of wind turbines impacted by super typhoon Usagi. *Eng. Fail. Anal.* **2015**, *60*, 391–404. [CrossRef]
68. Chen, X.; Li, C.; Tang, J. Structural integrity of wind turbines impacted by tropical cyclones: A case study from China. *J. Phys. Conf. Ser.* **2016**, *753*, 042003. [CrossRef]
69. Hassanzadeh, M. Cracks in onshore wind power foundations: Causes and consequences. *Elforsk* **2012**, *11*, 56.
70. McAlorum, J.; Perry, M.; Fusiek, G.; Niewczas, P.; McKeeman, I.; Rubert, T. Deterioration of cracks in onshore wind turbine foundations. *Eng. Struct.* **2018**, *167*, 121–131. [CrossRef]
71. Technology, F. Online Monitoring of Sub-Structure. Available online: https://forcetechnology.com/-/media/force-technology-media/pdf-files/4501-to-5000/4635-online-monitoring-of-sub-structures.pdf (accessed on 15 June 2022).
72. Hilbert, L.R.; Black, A.; Andersen, F.; Mathiesen, T. Inspection and monitoring of corrosion inside monopile foundations for offshore wind turbines. *Eur. Corros. Congr.* **2011**, *3*, 2187–2201.
73. Yang, B.; Wei, K.; Yang, W.; Li, T.; Qin, B. A feasibility study of reducing scour around monopile foundation using a tidal current turbine. *Ocean Eng.* **2021**, *220*, 108396. [CrossRef]
74. Schaumann, P.; Raba, A.; Bechtel, A. Fatigue behaviour of grouted connections at different ambient conditions and loading scenarios. *Energy Procedia* **2017**, *137*, 196–203. [CrossRef]
75. Yeter, B.; Garbatov, Y.; Guedes Soares, C. Fatigue damage assessment of fixed offshore wind turbine tripod support structures. *Eng. Struct.* **2015**, *101*, 518–528. [CrossRef]
76. Petersen, T.U.; Sumer, B. Scour around Offshore Wind Turbine Foundations. Ph.D. Thesis, Technical University of Denmark, Lyngby, Denmark, 2014.
77. Dallyn, P.; El-Hamalawi, A.; Palmeri, A.; Knight, R. Experimental investigation on the development of wear in grouted connections for offshore wind turbine generators. *Eng. Struct.* **2016**, *113*, 89–102. [CrossRef]
78. Li, W.; Wang, D.; Han, L.-H. Behaviour of grout-filled double skin steel tubes under compression and bending: Experiments. *Thin-Walled Struct.* **2017**, *116*, 307–319. [CrossRef]
79. Liang, C.; Sun, F.P.; Rogers, C.A. An Impedance Method for Dynamic Analysis of Active Material Systems. *J. Vib. Acoust.* **1994**, *116*, 24061–24261. [CrossRef]
80. Sun, F.P.; Chaudhry, Z.; Liang, C.; Rogers, C.A. Truss Structure Integrity Identification Using PZT Sensor-Actuator. *J. Intell. Mater. Syst. Struct.* **1995**, *6*, 134–139. [CrossRef]
81. Nguyen, K.-D.; Kim, J.-T. Smart PZT-interface for wireless impedance-based prestress-loss monitoring in tendon-anchorage connection. *Smart Struct. Syst.* **2012**, *9*, 489–504. [CrossRef]
82. Mascarenas, D.L.; Todd, M.D.; Park, G.; Farrar, C.R. Development of an impedance-based wireless sensor node for structural health monitoring. *Smart Mater. Struct.* **2007**, *16*, 2137. [CrossRef]
83. Usach, M. How to Configure the AD5933/AD5934. *Analog. Devices* **2013**, 1–12.
84. Park, S.; Shin, H.-H.; Yun, C.-B. Wireless impedance sensor nodes for functions of structural damage identification and sensor self-diagnosis. *Smart Mater. Struct.* **2009**, *18*, 055001. [CrossRef]

85. Min, J.; Park, S.; Yun, C.-B.; Song, B. Development of a low-cost multifunctional wireless impedance sensor node. *Smart Struct. Syst.* **2010**, *6*, 689–709. [CrossRef]
86. Nguyen, K.-D.; Lee, S.-Y.; Lee, P.-Y.; Kim, J.-T. Wireless SHM for Bolted Connections via Multiple PZT-Interfaces and Imote2-Platformed Impedance Sensor Node. In Proceedings of the 6th International Workshop on Advanced Smart Materials and Smart Structures Technology–ANCRiSST2011, Dalian, China, 25–26 July 2011.
87. Nachman, L.; Huang, J.; Shahabdeen, J.; Adler, R.; Kling, R. Imote2: Serious computation at the edge. In Proceedings of the 2008 International Wireless Communications and Mobile Computing Conference, Chania Crete Island, Greece, 6–8 August 2008; pp. 1118–1123.
88. Ho, D.-D.; Lee, P.-Y.; Nguyen, K.-D.; Hong, D.-S.; Lee, S.-Y.; Kim, J.-T.; Shin, S.-W.; Yun, C.-B.; Shinozuka, M. Solar-powered multi-scale sensor node on Imote2 platform for hybrid SHM in cable-stayed bridge. *Smart Struct. Syst.* **2012**, *9*, 145–164. [CrossRef]
89. Tawie, R.; Lee, H.K. Monitoring the strength development in concrete by EMI sensing technique. *Constr. Build. Mater.* **2010**, *24*, 1746–1753. [CrossRef]
90. Huynh, T.-C.; Nguyen, T.-D.; Ho, D.-D.; Dang, N.-L.; Kim, J.-T. Sensor Fault Diagnosis for Impedance Monitoring Using a Piezoelectric-Based Smart Interface Technique. *Sensors* **2020**, *20*, 510. [CrossRef]
91. Hu, X.; Zhu, H.; Wang, D. A Study of Concrete Slab Damage Detection Based on the Electromechanical Impedance Method. *Sensors* **2014**, *14*, 19897–19909. [CrossRef] [PubMed]
92. Huynh, T.-C.; Kim, J.-T. Quantification of temperature effect on impedance monitoring via PZT interface for prestressed tendon anchorage. *Smart Mater. Struct.* **2017**, *26*, 125004. [CrossRef]
93. Baptista, F.G.; Budoya, D.E.; Almeida, V.A.D.d.; Ulson, J.A.C. An Experimental Study on the Effect of Temperature on Piezoelectric Sensors for Impedance-Based Structural Health Monitoring. *Sensors* **2014**, *14*, 1208–1227. [CrossRef] [PubMed]
94. Park, G.; Kabeya, K.; Cudney, H.H.; Inman, D.J. Impedance-Based Structural Health Monitoring for Temperature Varying Applications. *Jsme Int. J. Ser. A-Solid Mech. Mater. Eng.* **1999**, *42*, 249–258. [CrossRef]
95. Koo, K.-Y.; Park, S.; Lee, J.-J.; Yun, C.-B. Automated Impedance-based Structural Health Monitoring Incorporating Effective Frequency Shift for Compensating Temperature Effects. *J. Intell. Mater. Syst. Struct.* **2008**, *20*, 367–377. [CrossRef]
96. Sepehry, N.; Shamshirsaz, M.; Abdollahi, F. Temperature variation effect compensation in impedance-based structural health monitoring using neural networks. *J. Intell. Mater. Syst. Struct.* **2011**, *22*, 1975–1982. [CrossRef]
97. Lim, H.J.; Kim, M.-K.; Park, C.Y. Impedance based damage detection under varying temperature and loading conditions. *Ndt E Int.* **2011**, *44*, 740–750. [CrossRef]
98. Huynh, T.-C.; Dang, N.-L.; Kim, J.-T. PCA-based filtering of temperature effect on impedance monitoring in prestressed tendon anchorage. *Smart Struct. Syst* **2018**, *22*, 57–70.
99. Gianesini, B.M.; Cortez, N.E.; Antunes, R.A.; Vieira Filho, J. Method for removing temperature effect in impedance-based structural health monitoring systems using polynomial regression. *Struct. Health Monit.* **2021**, *20*, 202–218. [CrossRef]
100. Du, F.; Wu, S.; Xu, C.; Yang, Z.; Su, Z. Electromechanical impedance temperature compensation and bolt loosening monitoring based on modified Unet and multitask learning. *IEEE Sens. J.* **2021**. [CrossRef]
101. Rumsey, M.A.; Paquette, J. Structural Health Monitoring of Wind Turbine Blades. Available online: https://www.osti.gov/servlets/purl/1146392 (accessed on 15 June 2022).
102. Jiang, X.; Zhang, X.; Zhang, Y. Evaluation of Characterization Indexes and Minor Looseness Identification of Flange Bolt Under Noise Influence. *IEEE Access* **2020**, *8*, 157691–157702. [CrossRef]
103. Wang, C.; Wang, N.; Ho, S.-C.; Chen, X.; Pan, M.; Song, G. Design of a novel wearable sensor device for real-time bolted joints health monitoring. *IEEE Internet Things J.* **2018**, *5*, 5307–5316. [CrossRef]
104. Antunes, R.A.; Cortez, N.E.; Gianesini, B.M.; Vieira Filho, J. Modeling, Simulation, Experimentation, and Compensation of Temperature Effect in Impedance-Based SHM Systems Applied to Steel Pipes. *Sensors* **2019**, *19*, 2802. [CrossRef] [PubMed]
105. Wang, J.; Li, W.; Lan, C.; Wei, P.; Luo, W. Electromechanical impedance instrumented piezoelectric ring for pipe corrosion and bearing wear monitoring: A proof-of-concept study. *Sens. Actuators A Phys.* **2020**, *315*, 112276. [CrossRef]
106. Choi, S.-H.; Kim, T.-H.; Huynh, T.-C.; Kim, J.-T. Impedance-based Crack Monitoring in Tubular Joint of Wind-turbine Tower: Numerical Study. In Proceedings of the International Conference on Civil and Environmental Engineering—ICCEE, Taoyuan, Taiwan, 8–11 November 2015.
107. Zhang, X.; Zhou, W.; Li, H. Electromechanical impedance-based ice detection of stay cables with temperature compensation. *Struct. Control Health Monit.* **2019**, *26*, e2384. [CrossRef]
108. Djemana, M.; Hrairi, M. Impedance Based Detection of Delamination in Composite Structures. *IOP Conf. Ser. Mater. Sci. Eng.* **2017**, *184*, 012058. [CrossRef]
109. Wandowski, T.; Malinowski, P.; Ostachowicz, W. Delamination detection in CFRP panels using EMI method with temperature compensation. *Compos. Struct.* **2016**, *151*, 99–107. [CrossRef]
110. Huo, L.; Chen, D.; Liang, Y.; Li, H.; Feng, X.; Song, G. Impedance based bolt pre-load monitoring using piezoceramic smart washer. *Smart Mater. Struct.* **2017**, *26*, 057004. [CrossRef]
111. Huynh, T.-C. Structural parameter identification of a bolted connection embedded with a piezoelectric interface. *Vietnam J. Mech.* **2020**, *42*, 173–188. [CrossRef]
112. Schulz, M.; Sundaresan, M. Smart Sensor System for Structural Condition Monitoring of Wind Turbines. In *National Renewable Energy Laboratory*; Cole Boulevard: Golden, CO, USA, 2016; pp. 80401–83393.

113. De Oliveira, M.A.; Monteiro, A.V.; Vieira Filho, J. A New Structural Health Monitoring Strategy Based on PZT Sensors and Convolutional Neural Network. *Sensors* **2018**, *18*, 2955. [CrossRef]
114. Zhou, L.; Chen, S.-X.; Ni, Y.-Q.; Choy, A.W.-H. EMI-GCN: A hybrid model for real-time monitoring of multiple bolt looseness using electromechanical impedance and graph convolutional networks. *Smart Mater. Struct.* **2021**, *30*, 035032. [CrossRef]
115. Park, G.; Sohn, H.; Farrar, C.R.; Inman, D.J. Overview of piezoelectric impedance-based health monitoring and path forward. *Shock Vib. Dig.* **2003**, *35*, 451–464. [CrossRef]
116. Lopes, V., Jr.; Park, G.; Cudney, H.H.; Inman, D.J. Impedance-based structural health monitoring with artificial neural networks. *J. Intell. Mater. Syst. Struct.* **2000**, *11*, 206–214. [CrossRef]
117. Kim, H.; Liu, X.; Ahn, E.; Shin, M.; Shin, S.W.; Sim, S.-H. Performance assessment method for crack repair in concrete using PZT-based electromechanical impedance technique. *NDT E Int.* **2019**, *104*, 90–97. [CrossRef]
118. Lim, Y.Y.; Smith, S.T.; Padilla, R.V.; Soh, C.K. Monitoring of concrete curing using the electromechanical impedance technique: Review and path forward. *Struct. Health Monit.* **2021**, *20*, 604–636. [CrossRef]
119. Talakokula, V.; Bhalla, S.; Gupta, A. Corrosion assessment of reinforced concrete structures based on equivalent structural parameters using electro-mechanical impedance technique. *J. Intell. Mater. Syst. Struct.* **2014**, *25*, 484–500. [CrossRef]
120. Liu, P.; Wang, W.; Chen, Y.; Feng, X.; Miao, L. Concrete damage diagnosis using electromechanical impedance technique. *Constr. Build. Mater.* **2017**, *136*, 450–455. [CrossRef]
121. Zhang, C.; Panda, G.P.; Yan, Q.; Zhang, W.; Vipulanandan, C.; Song, G. Monitoring early-age hydration and setting of portland cement paste by piezoelectric transducers via electromechanical impedance method. *Constr. Build. Mater.* **2020**, *258*, 120348. [CrossRef]
122. Ahmadi, J.; Feirahi, M.H.; Farahmand-Tabar, S.; Fard, A.H.K. A novel approach for non-destructive EMI-based corrosion monitoring of concrete-embedded reinforcements using multi-orientation piezoelectric sensors. *Constr. Build. Mater.* **2021**, *273*, 121689. [CrossRef]
123. Pham, Q.-Q.; Dang, N.-L.; Ta, Q.-B.; Kim, J.-T. Optimal Localization of Smart Aggregate Sensor for Concrete Damage Monitoring in PSC Anchorage Zone. *Sensors* **2021**, *21*, 6337. [CrossRef]
124. Park, S.; Park, S.-K. Quantitative corrosion monitoring using wireless electromechanical impedance measurements. *Res. Nondestruct. Eval.* **2010**, *21*, 184–192. [CrossRef]
125. Zhu, H.; Luo, H.; Ai, D.; Wang, C. Mechanical impedance-based technique for steel structural corrosion damage detection. *Measurement* **2016**, *88*, 353–359. [CrossRef]

Article

Research on Mechanical Properties and Damage Evolution of Pultruded Sheet for Wind Turbine Blades

Ying He [1], Yuanbo Wang [1,*], Hao Zhou [2], Chang Li [1], Leian Zhang [1] and Yuhuan Zhang [1]

[1] Shandong University of Technology, Zibo 255049, China
[2] Sinoma Technology (Funing) Wind Power Blade Co., Ltd., Yancheng 224400, China
* Correspondence: sdlgdxtgyx@163.com

Abstract: In order to explore the mechanical properties, failure mode, and damage evolution process of pultruded sheets for wind turbine blades, a tensile testing machine for pultruded sheets for wind turbine blades was built, and the hydraulic system, mechanical structure, and control scheme of the testing machine were designed. The feasibility of the mechanical structure was verified by numerical simulation, and the control system was simulated by MATLAB software. Then, based on the built testing machine, the static tensile test of the pultruded sheet was carried out to study the mechanical properties and failure mode of the pultruded sheet. Finally, an infrared thermal imager was used to monitor the temperature change on the surface of the test piece, and the temperature change law and damage evolution process of the test piece during the whole process were studied. The results show that the design scheme of the testing machine was accurate and feasible. The maximum stress occurred in the beam after loading the support, the maximum stress was 280.18 MPa, and the maximum displacement was 0.665 mm, which did not exceed its structural stress-strain limit. At the same time, the control system met the test requirements and had a good follow-up control effect. The failure load of the pultruded sheet was 800 kN. The failure deformation form included three stages of elasticity, yield, and fracture, and the finite element analysis data were in good agreement with the test results. The failure modes were fiber breakage, delamination, and interfacial debonding. The surface temperature of the specimen first decreased linearly, and then continued to increase. The strain and temperature trend were consistent with time.

Keywords: wind power blade; pultruded sheet; composite material; static tensile test; infrared thermal imaging method; testing machine

Citation: He, Y.; Wang, Y.; Zhou, H.; Li, C.; Zhang, L.; Zhang, Y. Research on Mechanical Properties and Damage Evolution of Pultruded Sheet for Wind Turbine Blades. *Materials* **2022**, *15*, 5719. https://doi.org/10.3390/ma15165719

Academic Editors: Phong B. Dao, Lei Qiu, Liang Yu, Tadeusz Uhl and Minh-Quy Le

Received: 1 July 2022
Accepted: 2 August 2022
Published: 19 August 2022

1. Introduction

According to the outline of the 14th Five-Year Plan and the proposal for 2035, China will accelerate the development of new energy. Compared with traditional carbon-based energy, wind energy has huge advantages in terms of sustainable development. In order to reduce the cost of electricity, the size and power of wind turbines are increasing. The average power rating of wind turbines has increased from 0.1 MW in 1985 to an average of 10 MW today [1,2]. It is expected that the power of the wind turbine will reach 20 MW in the future, and the diameter of the impeller will be about 200 m [3]. The traditional manual lay-up technology can no longer meet the increasing blade size, and the plates produced by the pultrusion process with high forming quality, high production efficiency and low production cost can be used for key structural parts such as the main beam of wind turbine blades [4]. Due to technical risks and high costs, full-scale blade destructive testing still faces great challenges, so pultruded sheet testing has received more and more attention. Therefore, studying the failure mode and damage process of pultruded sheets representing blade design can be used to supplement full-scale blade testing and provide detailed mechanical performance parameters for the entire blade, which is of great engineering significance for reducing blade costs.

At present, a lot of work has been done on the research of pultruded glass fiber composites. A comprehensive presentation of pultruded fiber-reinforced polymers by Alessandro et al. [5] showed that pultruded fiber reinforcements had undergone extensive experimental and numerical studies to evaluate their performance as structural components. A critical review of the state-of-the-art in numerical simulation for predicting mechanical behavior at the limit state of failure was also presented, and a variety of numerical simulation methods, including finite element methods, were introduced. Al-Saadi et al. [6,7] conducted tensile, compression and shear tests on pultruded glass fiber round tubes and square tubes to explore the mechanical properties and failure modes of the tubes and compared the numerical simulation with the test results to verify the accuracy of the test. The study explored the different specifications of materials produced by the pultrusion process, but not pultruded sheets. Madenic et al. [8] analyzed the bending mechanical properties and damage process of pultruded glass fiber composites by a three-point bending mechanical test. The theoretical solution and finite element analysis were carried out, and the obtained load-displacement curve was compared with the experimental results. It was found that the numerical simulation and theoretical solution results were consistent with the experimental values, but the article only analyzed the bending mechanical properties. Since the glass fiber is an anisotropic composite material, only the bending direction analysis is far from enough. Silva et al. [9] studied the flexural properties of dry and wet composites of polyester-based glass fiber reinforced plastic flat specimens cut from I-beam pultruded profiles, and found that the flexural modulus and strength of the material specimens wetted to the saturation level decreased. Paciornik et al. [10] explored the microstructure of pultruded glass fiber sheets and used an image analysis system to characterize the fiber size, spatial distribution and filler fraction of the sheets. The bending mechanical behavior of the sheet was determined, and it was found that in the pultrusion process, even if all parameters were controlled, the pultruded part still produced uneven fiber distribution and voids. In this paper, the microstructure of pultruded sheet was explored, but its mechanical properties and macroscopic damage evolution were not explored. Harizi et al. [11] used infrared thermal imaging technology to monitor the surface temperature changes of glass fiber composites under static load and stage load. Through experiments, it was found that infrared thermal imaging technology could monitor the evolution process of material damage, and the thermal imaging map could explore the material damage generated under large load conditions for analysis. Aniskevich et al. [12] designed a novel equipment to provide a good connection with a conventional existing pultrusion machine, allowing the production of pultruded pipes that involved cork and polyurethane foam preforms in a continuous way. The equipment had been developed and tested to obtain high flexural strength hybrid pultruded products. The above research shows that the technology of infrared thermal imaging monitoring of the surface temperature change of the test piece is very mature, and can be directly used for the study of the damage evolution of pultruded sheets, but the research on pultruded glass fiber sheets for wind turbine blades has not been involved. Its failure mode and damage evolution need to be specifically explored.

In order to solve the above problems, with the help of the conclusions of many scholars, this paper mainly studied the tensile mechanical properties and failure modes of pultruded glass fiber sheets, and explored the damage evolution process through infrared thermal imaging. First, a wind turbine blade pultrusion fiberglass sheet tensile testing machine was built. The finite element analysis of its structure was carried out and the control system was simulated by MATLAB software. Secondly, the static tensile test of the pultruded sheet was carried out using the built testing machine and compared with the numerical calculation results to study the mechanical properties and failure mode of the pultruded sheet. Finally, the damage evolution process was explored by using an infrared thermal imager to monitor the surface temperature of the pultruded sheet, which provided a reference for the mechanical properties of the entire wind turbine blade.

2. Test Principle

2.1. Pultruded Sheet Theory

Pultruded glass fiber reinforced polymer sheet, referred to as pultruded sheet for short, is made of pultruded fiberglass and epoxy resin through a pultrusion process [13]. In wind turbine blades, several pultruded sheets are laid to form the main beam structure [14–17]. The main beam structure is mainly determined by the mechanical properties, geometry and boundary conditions of the glass fiber. It is known that glass fiber is an orthotropic material, and each unit in the formula is an international standard unit. The constitutive relation in the plane stress state is:

$$\{\sigma\} = Q\{\varepsilon\} \tag{1}$$

In the formula, σ is the stress, ε is the strain, and Q is the positive axis stiffness matrix.

After the positive axis stiffness matrix is obtained from the elastic constant of the glass fiber, its main direction is converted to the overall coordinate xyz, and the coordinate conversion matrix T and the corresponding relationship between stress and strain are obtained. Finally, we calculate the in-plane stiffness coefficient

$$A_{ij} = \sum_{k=1}^{n} (\overline{Q_{ij}}) f \ (i=1,2,3; j=1,2,3) \tag{2}$$

where $\overline{Q}$ is the transformation stiffness matrix of Q:

The in-plane flexibility S is as follows:

$$S = A^{-1} \tag{3}$$

$$S = \begin{bmatrix} 1/E_1 & -v_{21}/E_2 & 0 \\ v_{21}/E_2 & 1/E_2 & 0 \\ 0 & 0 & 1/G_{12} \end{bmatrix} \tag{4}$$

The orthotropic equilibrium equation and deformation coordination equation are as follows

$$\frac{\partial \sigma_x}{\partial x} + \frac{\partial \tau_{xy}}{\partial y} = 0 \tag{5}$$

$$\frac{\partial \tau_{xy}}{\partial x} + \frac{\partial \tau_y}{\partial y} = 0 \tag{6}$$

$$\frac{\partial^2 \varepsilon_x}{\partial y^2} + \frac{\partial^2 \varepsilon_{xy}}{\partial x^2} = \frac{\partial^2 \gamma_{xy}}{\partial x \partial y} \tag{7}$$

where τ_{xy} is the shear stress between the x and y axes and γ_{xy} is the shear strain.

The stress function $F = (x, y)$ is introduced, and the relationship between the stress functions and σ_x, σ_y, and τ_{xy} is as follows:

$$\begin{cases} \sigma_x = \dfrac{\partial^2 F}{\partial y^2} \\ \sigma_y = \dfrac{\partial^2 F}{\partial x^2} \\ \tau_{xy} = -\dfrac{\partial^2 F}{\partial x \partial y} \end{cases} \tag{8}$$

The plane equation of the orthotropic blade composite material is obtained from Equations (1)–(8)

$$S_{22}\frac{\partial^4 F}{\partial x^4} + (2S_{12} + S_{66})\frac{\partial^4 F}{\partial x^2 \partial y^2} + S_{11}\frac{\partial^4 F}{\partial y^2} = 0 \tag{9}$$

where $S_{11} = 1/E_1$, $S_{12} = -v_{21}/E_2$, $S_{22} = 1/E_2$, $S_{66} = 1/G_{12}$.

The elastic modulus in the x-direction after regularization is as follows

$$E_x = \frac{1}{tS_{11}} \tag{10}$$

where t is the total thickness of the plate.

According to the macroscopic treatment method, the fracture study of pultruded sheet can be regarded as a homogeneous anisotropic material. The fracture criterion of isotropic materials is extended to plate composite materials, and the stress field and displacement field near the origin are calculated by the elastic mechanics criterion

$$\begin{cases} \sigma_x = \dfrac{K}{\sqrt{2\pi r}} \cos\frac{\theta}{2}(1 - \sin\dfrac{\theta}{2}\sin\dfrac{3\theta}{2}) \\ \sigma_y = \dfrac{K}{\sqrt{2\pi r}} \cos\frac{\theta}{2}(1 + \sin\dfrac{\theta}{2}\sin\dfrac{3\theta}{2}) \\ \tau_{xy} = \dfrac{K}{\sqrt{2\pi r}} \sin\frac{\theta}{2} \cdot \cos\dfrac{\theta}{2} \cdot \cos\dfrac{3\theta}{2} \\ u_x = \dfrac{K}{8G}\sqrt{\frac{2r}{\pi}}\left[(2\kappa - 1)\cos\dfrac{\theta}{2} - \cos\dfrac{3\theta}{2}\right] \\ u_y = \dfrac{K}{8G}\sqrt{\frac{2r}{\pi}}\left[(2\kappa + 1)\cos\dfrac{\theta}{2} - \sin\dfrac{3\theta}{2}\right] \end{cases} \tag{11}$$

where κ is the modulus of elasticity, G is the shear modulus, and K is the stress intensity factor.

The energy release rate G is the energy released per unit area of material by crack propagation. According to elastic fracture mechanics, its relationship with the strength factor K is described as

$$\begin{cases} G_1 = \dfrac{1 - v^2}{E} K^2 \\ G_2 = \dfrac{1}{E} K^2 \end{cases} \tag{12}$$

where G_1 is the plane strain energy release rate and G_2 is the plane stress energy release rate.

2.2. Infrared Imaging Theory

Any object can emit infrared radiation. According to Stephen Boltzmann's law in thermodynamics, the total energy radiated per unit area of an object's surface in a unit time is proportional to the fourth power of the thermodynamic temperature of the object itself [18,19]. The formula is as follows:

$$j^* = \varepsilon\sigma T^4 \tag{13}$$

Among them, j^* is the radiance, T is the absolute temperature, ε is the emissivity of the black object, and σ is the Stefan constant.

Infrared thermal imaging detection technology is a non-destructive testing technology that converts invisible infrared radiation into visible images [20]. It is divided into active infrared thermal imaging and passive infrared thermal imaging [21,22]. Active infrared thermal imaging needs to excite the test piece by heating, so that the test piece forms a temperature difference [23]. Then, infrared imaging technology is used to test the specimen, and the defect is judged according to the difference of temperature [24,25]. In this test, it is assumed that a small change in temperature during the static tensile test of the testing machine will cause a large change in the radiation power, so there is no need to stimulate the curing of the test piece. The passive thermal imager is used for detection, which is presented in the form of a thermal image, and the detection is realized by comparison.

In addition, heat is transferred from a hotter part of an object to a cooler part, or from a hotter object to another cooler object in contact with it. This heat transfer process is called heat conduction [26]. The heat flow in the pultruded sheet can be determined according to Fourier's theorem [27]:

$$q(r,t) = -\lambda \nabla T(r,t) \tag{14}$$

In the formula, $q(r, t)$ is the heat flow per unit area per unit time in the direction of decreasing temperature gradient; λ is the thermal conductivity of wind turbine blades, W/(m·K); and $T(r, t)$ is the spatial and temporal distribution of the temperature in the wind turbine blade.

Equation (14) reveals the relationship between heat flow and temperature. The internal relationship of the temperature field in the spatiotemporal domain is usually described by the heat conduction differential Equation (15)

$$\nabla T(r,t) + \frac{q_v}{\lambda} = \frac{\rho c}{\lambda} \nabla T(r,t) \tag{15}$$

where q_v is the loading heat source term; ρ is the density of the wind turbine blade, kg/m^3; and c is the specific heat capacity of the wind turbine blade, J/(kg·°C).

3. Overall Test Plan and Simulation

3.1. Test Machine Structure Scheme

The structure of the pultruded sheet testing machine is composed of a hydraulic system and a loading mechanism, including loading brackets, fixtures, force sensors and oil cylinders, which can perform horizontal tension and compression tests. At the same time, in order to offset the radial force generated during loading, the testing machine is equipped with a ball head mechanism on the front beam and the rear beam. The pultruded sheet is connected to the fixture by bolts, and the bolts are subjected to axial load to fasten the specimen in the fixture. The schematic diagram of the mechanical structure is shown in Figure 1.

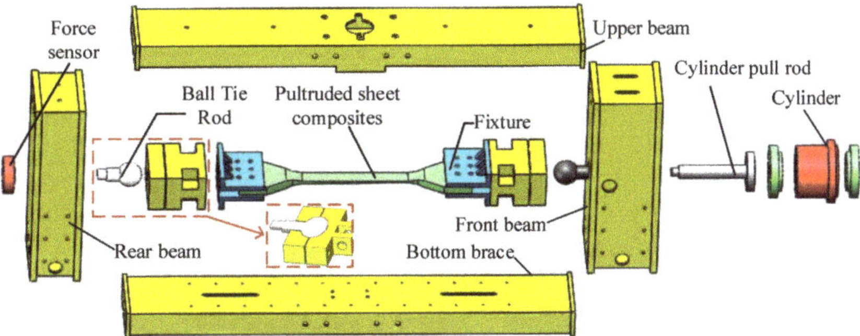

Figure 1. Test machine structure diagram.

When testing in the horizontal direction, the hydraulic cylinder installed on the front beam exerts an axial force on the component through the cylinder tie rod, which is then transmitted to the sensor installed on the rear beam. From this, it can be seen that the rear beam is the main component that bears the load. The design parameters are shown in Table 1. The force sensor of the front beam converts the electricity into electrical signals through the transmitter, and feeds them back to the host computer.

Table 1. Design parameters.

Name	Parameter
Maximum specimen length (m)	2
Loading frequency (Hz)	1–3
Maximum load (kN)	1300
Motor power (kW)	36
Force sensor (kN)	150
Force sensor diameter (mm)	100
Sample thread	M36

3.2. Structural Finite Element Analysis

The loading bracket of the testing machine was numerically simulated in Ansys workbench, and its finite element model was shown in Figure 2. The pre-processing steps for static analysis using the finite element method included assigning material properties, dividing meshes, selecting elements, applying constraints and loads, etc. [28–30]. The holes and grooves that generated stress concentration and did not affect the structural strength were simplified. The loading bracket material was Q345 structural steel, and the material parameters were shown in Table 2. The Solid185 element was used to construct three-dimensional solid structures. Due to its large deformation and large strain capacity [5], this element was used. Since the mesh division had a great influence on the results of the finite element analysis [31], the mesh of the sensor directly bearing the force on the rear beam was refined. The element type, total number of nodes, and mesh size after meshing are shown in Table 2. The front beam and the rear beam of the testing machine have four supports respectively, which are connected with ground bolts to prevent the testing machine from overturning, and the contact surface with the ground is set as a restraint surface. When an external load is applied, the pultruded sheet produces a load towards the inside of the contacting load bracket.

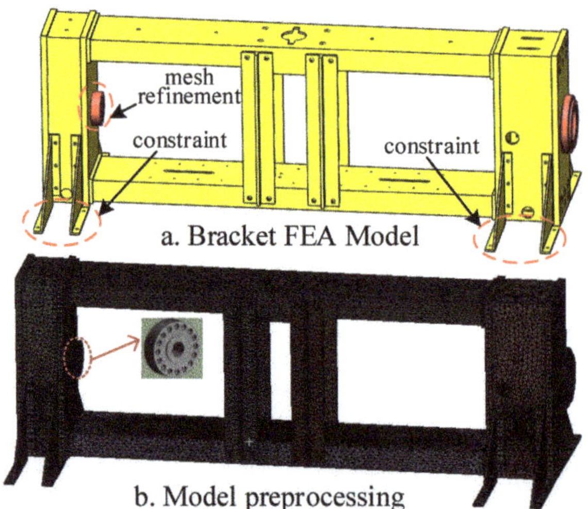

Figure 2. Finite element modeling and pre-processing: (**a**) bracket FEA model; (**b**) model preprocessing.

Grid independence verification was performed first. In Ansys workbench, by setting the target value, the software automatically calculated many times to determine the grid-independent solution. Inserted Convergence in the resulting Equivalent Stress. Since it was simpler to load the scaffold model, we inserted a delta of 0%. After each solution, the software automatically re-divided the mesh in the area that needs to be refined, calculated the corresponding equivalent stress, and compared it with the previous result until the difference between the results met the set requirements. Then, numerical simulation was carried out. After 4 calculations, the maximum stress of the structure was calculated to be 280.18 MPa, which was located at the contact point between the pressure sensor and the front beam. In addition, the stress at the bottom connection reached 249.05 MPa, as shown in Figure 3. The result was not the allowable stress of the metamaterial, which could meet the testing requirements. The safety of the structure was demonstrated.

Table 2. Simulation parameters.

Name	Parameter
Gross weight (kg)	3912.3
Elastic modulus (GPa)	209
Poisson's ratio	0.269
Yield strength (MPa)	345
Total number of nodes	270,136
Total number of units	142,078
Grid size (mm)	50
Front/rear beam grid size (mm)	15

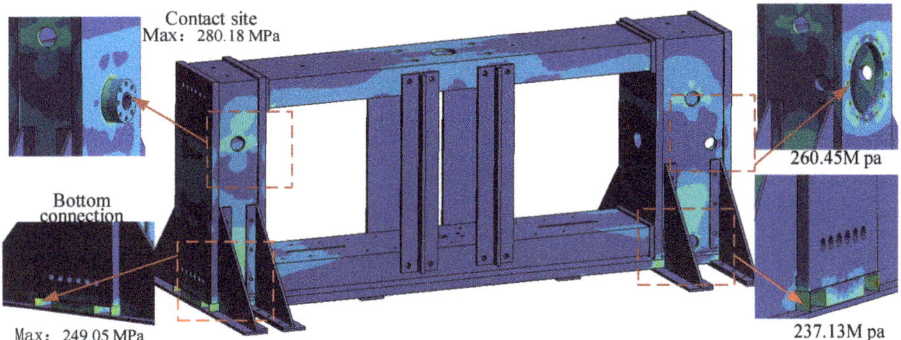

Figure 3. Loading bracket static stress analysis cloud diagram.

At the same time, the displacement results of the loading bracket and other components of the testing machine obtained by running the example are shown in Figure 4. The maximum deformation is 0.665 mm, and the combined deformations of the front-end fixture, rear-end fixture, and bolt of the specimen are 1.381 mm, 2.101 mm, and 0.4485 mm, respectively, all of which meet the design requirements.

3.3. Test Control Scheme and Simulation

The design scheme of the static tensile test of pultruded sheet is shown in Figure 5. The test consists of a tensile testing machine and an infrared thermal imaging test system. The control system of the testing machine is composed of the actuator, the lower computer, the upper computer, and the data acquisition system. The host computer is written in LabVIEW language, and the overall control of the platform is carried out through the PC terminal. The upper computer has the functions of programming the control algorithm, collecting data, storing and displaying the strain curve. The lower computer is composed of PLC and RMC in parallel, in which PLC obtains signals such as force sensor, oil temperature sensor and liquid level sensor. RMC control electro-hydraulic servo valve has the advantages of fast response speed, high control precision and good dynamic performance. The execution structure adopts the MOOG servo valve, which has the advantages of high control precision and fast dynamic response. It can perform high-frequency reversing action and drive sub-components to perform fatigue tests through the pulling and pressing action of the oil cylinder. At the same time, the infrared thermal imager observes the workpiece in real time and collects data, and transmits the surface temperature data and thermal image of the test piece to the PC.

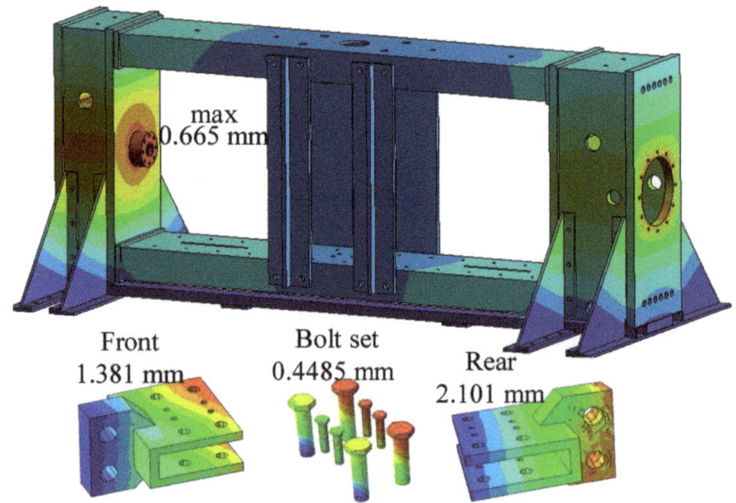

Figure 4. Displacement analysis cloud diagram of testing machine parts.

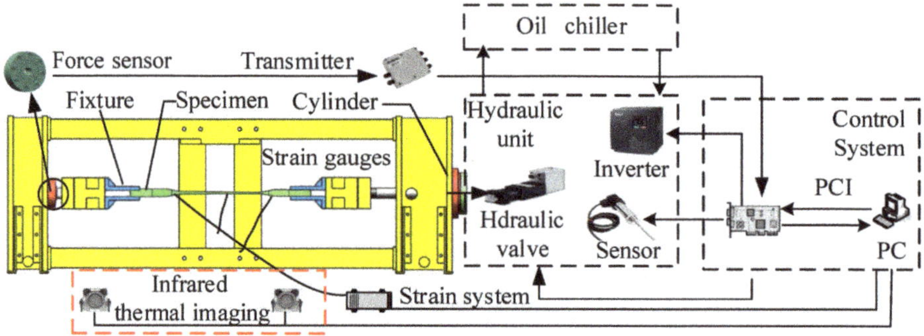

Figure 5. Experimental control scheme diagram.

During the test, the host computer sends a control signal to the controller and then controls the servo valve to make the oil circuit on and off and change direction, so as to load the plate with tension. The PLC collects the signal of the force sensor and feeds it back to the host computer. The signal is processed by the control algorithm in the host computer to realize the follow-up control of the actual load and the expected load. During the test, according to the IEC 61400-5 standard [32], it provides a data reference for the selection of the safety factor of blade material layout, as shown in the following Formula (16):

$$\gamma_{\text{m}} = \gamma_{\text{m0}} \cdot \gamma_{\text{m1}} \cdot \gamma_{\text{m2}} \cdot \gamma_{\text{m3}} \cdot \gamma_{\text{m4}} \cdot \gamma_{\text{m5}} \tag{16}$$

In the formula, γ_{m0} is the safety factor of the base material; γ_{m1} is the environmental deterioration factor (irreversible influence); γ_{m2} is the temperature effect factor (reversible influence); γ_{m3} is the manufacturing effect factor; γ_{m4} is the method calculation accuracy and verification factor; and γ_{m5} is the load characteristic factor.

Since the loading accuracy and loading frequency of the pultruded sheet are required to be high during the test process, the actual loading and the expected loading curve should ensure high synchronization, so that the slight difference between the two should be as close as possible, at 0. This means slight differences between the sinusoidal responses may also have a large impact on the test results. On the MATLAB/Simulink platform, the Sim Hydraulics module is used to build the simulation model of the control system of the testing machine, as shown in Figure 6. The figure shows the simulation of the hydraulic loading system control of the testing machine to achieve precise control of the axial load of the pultruded sheet, so as to ensure a good follow-up control effect between the actual axial load and the expected load of the pultruded sheet, and finally verify that the control system meets the test requirements. According to the test needs, the electro-hydraulic servo force control system is adopted, and the servo control is realized through the servo valve. The actuator of the system is a hydraulic cylinder, and one end of the hydraulic cylinder is connected with the fixture. The servo mechanism is a direct-acting servo valve, and the servo valve and the hydraulic cylinder form a valve-controlled cylinder structure. The hydraulic system is powered by a plunger pump, which is connected to the motor through a coupling. In order to reduce the flow pulsation caused by the plunger pump and ensure the smooth movement of the hydraulic cylinder, an accumulator is added to the oil circuit to reduce the corresponding vibration and noise. It should be noted that, since the components such as the cooler on the hydraulic station mainly play an auxiliary role in the hydraulic system, the simulation model is simplified to improve the efficiency of the control algorithm [33,34].

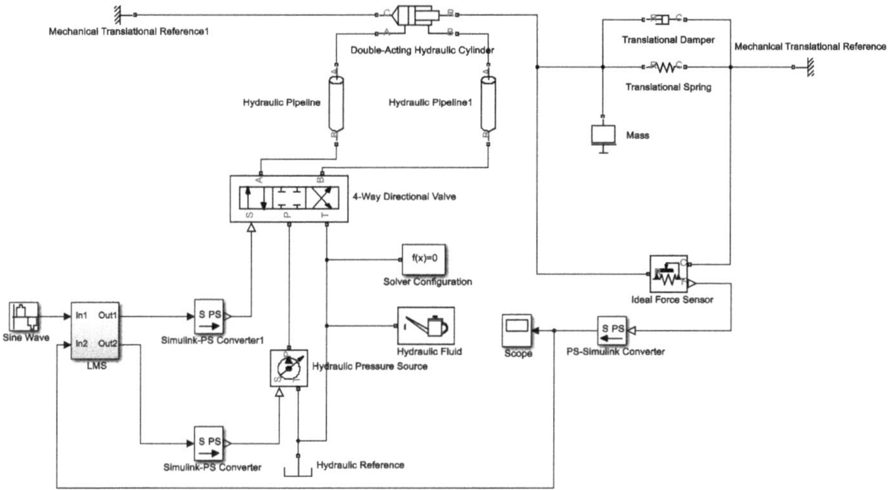

Figure 6. Simulation model of testing machine.

The simulation curve obtained based on the above model is shown in Figure 7, in which the actual response is the load of the specimen fed back by the force sensor, the expected response is the desired ideal sinusoidal waveform, and the error is the deviation between the actual response and the expected response. It can be seen from the figure that the actual response gradually follows the expected response after 1 s, and its followability can meet the design requirements of the algorithm, except for a small range of amplitude attenuation at the peak. Therefore, the control system meets the test requirements, and has a good follow-up effect.

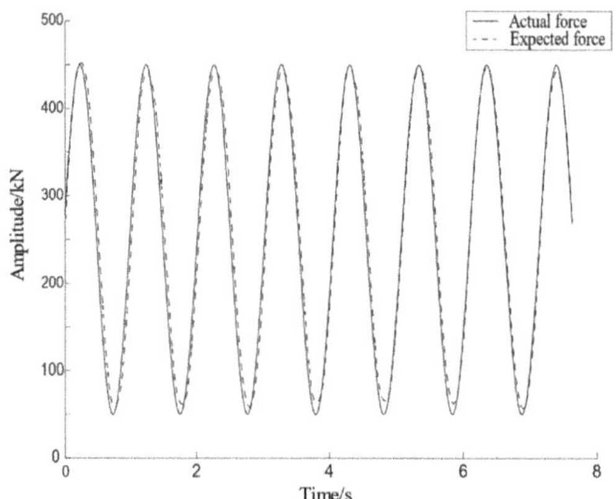

Figure 7. Simulation curve of testing machine.

4. Test Platform Construction and Test

4.1. Construction of Test Platform

Based on the design of the wind turbine blade pultrusion plate testing machine and the static stress analysis of the loading bracket, the construction of the testing machine was completed. The design parameters are shown in Table 3. The servo motor drives the hydraulic pump to convert the mechanical energy into hydraulic energy [35], and the hydraulic oil enters the hydraulic cylinder through the servo valve to work, as shown in Figure 8. The infrared thermal imager is UTi260B infrared thermal imager from Unitech. This thermal imager can monitor the temperature change on the surface of the specimen throughout the whole process. The acquisition frequency is 25 Hz, the pixel size is 320 × 240, and the thermal imaging sensitivity is less than 60 mK.

Table 3. Component test platform design parameters.

Name	Parameter
Specimen length (mm)	1760
Specimen thread	M36
Loading frequency (Hz)	1
Maximum load (kN)	1000
Servo valve	MOOG D661-G60KOAA4NSM2HA
Force sensor (kN)	150
Inverter	Invt CHF100A-022G-2
Motor speed (r/min)	1440
Motor power (kW)	36
Maximum working pressure (bar)	200
Pump maximum output pressure (bar)	31.5
Rotor brake feedback voltage (V)	24
Accumulator 1 volume (L)	2.8
Pump flow (L/min)	210
Air-cooled chiller	STSF-10
Rated throughput (T/h)	12

The static tensile test site of the pultruded sheet is shown in Figure 9. During the test, the ambient temperature was always 13.1°. The infrared thermal imager mainly monitored the surface temperature of the test piece. The infrared thermal imager was located on the outside of the test piece, and the lens was perpendicular to the test piece surface. During the experiment, a unified measurement standard was maintained, and the distance from the test piece was 30.00 cm. The strain measured after the ball head cancels the torque was the tensile and compressive strain. The fixture was fastened to the test piece by means of bolts. In order to obtain the strain of the pultruded sheet, the strain gauge was used in combination with the digital image system. The linear strain gauges A-4 to A-9 used for strain measurement were attached to the upper and lower surfaces of the expected failure area, and the rest of the strain gauges were at the transition position between the reinforced and unreinforced parts of the pultruded sheet, which could detect the detailed force data of the pultruded sheet.

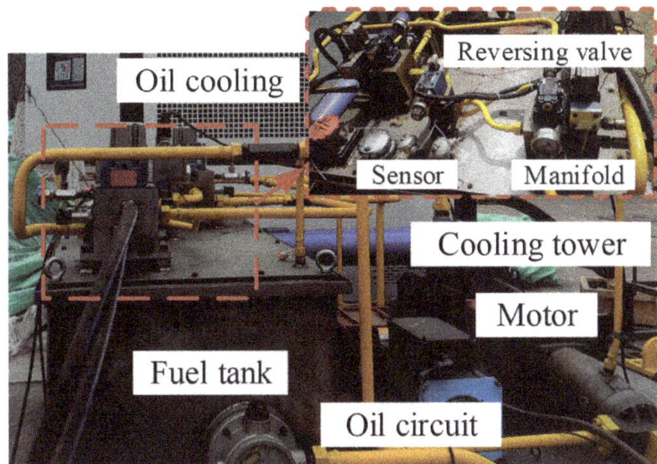

Figure 8. Testing machine hydraulic system.

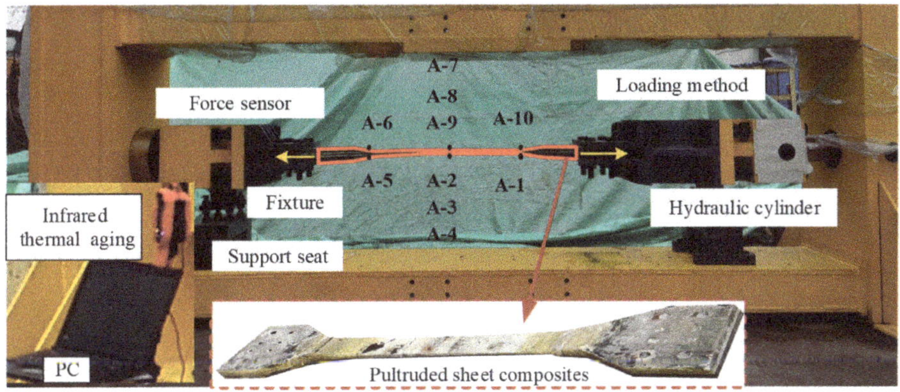

Figure 9. Static tensile test site of pultruded sheet.

4.2. Analysis of Results

After the test, the strain fed back by the force sensor and the digital image system and the temperature change of the surface of the pultruded sheet by the infrared thermal imager are shown in Figure 10. It was found that the strain of the specimen was consistent with the trend of the infrared temperature change during the static tensile test. The loading process went through elastic, plastic, and fracture stages. It can be seen from the curve that the change trend of the pultruded composite sheet in the static test is as follows:

(1) From the beginning of loading to 20 s, the strain and surface temperature of the pultruded sheet decreased slightly, and the temperature decreased by about 1 °C. The strain drop in this process is due to the linear elastic deformation of the material. In the elastic deformation stage of the material, the material is subjected to less

stress and will not yield, and the deformation can be quickly recovered. The reason for the temperature drop is the thermoelastic effect of the pultruded sheet without any damage to the surface. In the elastic deformation stage, the specimen is in the initial stage of tension, and the matrix stress is small, which is not enough to cause matrix damage.

(2) After the specimen passed through the elastic stage, it continued to be subjected to the applied load. The strain and temperature values started to rise after reaching the minimum value, and the curve was nonlinear, which was manifested as a sharp rise at first and then a slow rise. At this time, the plate changed from linear elasticity to plastic deformation. When the specimen was stretched for 100 s, the infrared thermal image of the specimen changed rapidly, and a heat source with a significantly higher temperature appeared in the delamination defect. In this stage, the surface temperature of the specimen increased because the irreversible plastic deformation of the specimen began at this stage, and the mechanical work was converted into heat dissipation, which increased the temperature at the defect. Heat dissipation continued to increase cumulatively, making the heat source range larger with increasing load. When the temperature of the specimen reached the lowest point, the strain of the specimen presented a turning point. The stress value at this time was the yield value of the composite material, and the yield value of the specimen could be quickly determined by infrared thermal imaging technology.

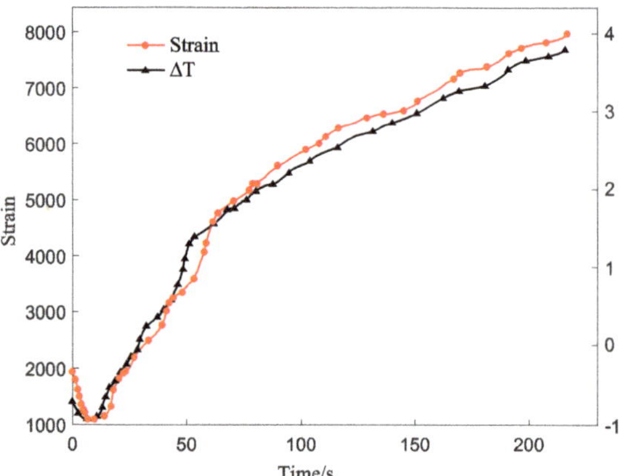

Figure 10. Specimen strain time domain diagram.

In the plastic stage, the strain continued to increase, and no obvious change was found on the surface of the pultruded sheet. The number of damages at this stage increased, and the sound of fiber breaking was heard during the process, but the damaged part could not be seen with the naked eye. According to the preparation characteristics of the composite material, some damage phenomena such as resin cracking, delamination, and interfacial debonding had already occurred in the specimen during this tensile process.

(3) After a long period of plastic deformation, the specimen continued to be subjected to the applied load. After the specimen was loaded for 210 s, the specimen entered the fracture stage, and there was still no change on the surface. When the load reached 800 kN, the specimen broke rapidly.

After analyzing the variation trend of temperature and strain with time, the infrared thermal image obtained during the experiment was processed by the special infrared thermal image processing software that came with the infrared thermal imager. We selected the infrared thermal images at 20 s, 75 s, and 210 s with obvious changes during the test, as shown in Figure 11. From the infrared thermal images of Figure 11a–c, the temperature change trend and heat source distribution of the pultruded sheet composite material during the entire experimental process are obtained. Figure 11a shows that during the elastic deformation of the specimen, the surface temperature of the specimen is 0.97 °C lower than the environment, and the reason for the temperature drop is the thermoelastic effect of the pultruded sheet. Figure 11b shows that the specimen is in the plastic deformation period; at this time, the surface temperature of the specimen rises, and the temperature at the delamination damage position is higher than that in other places. Figure 11c shows that the specimen is fractured, and the surface temperature of the specimen reaches the highest at this time. The damage of the pultruded sheet in the static stretching process was verified through the elastic, plastic, and fracture stages by infrared thermal image.

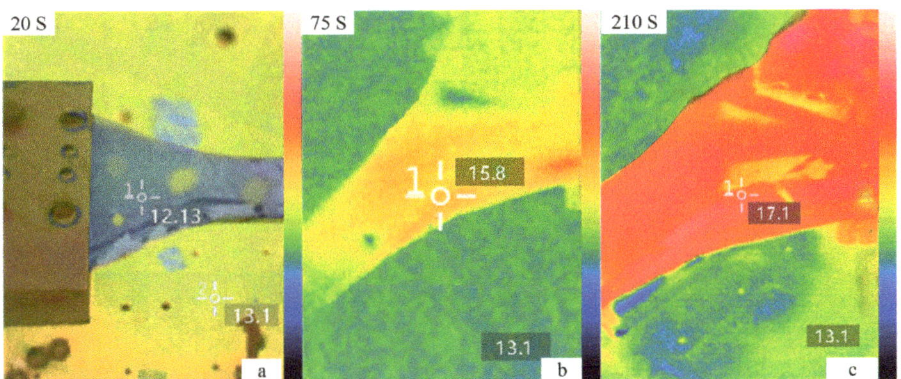

Figure 11. Infrared thermal images at different times. (**a**) Infrared thermal image at 20 s (**b**) Infrared thermal image at 75 s (**c**) Infrared thermal image at 210 s.

Figure 12a shows the time domain diagram of the force sensor obtained. Each 100 kN increase in load during the loading process is a stage. The figure shows that the force increases linearly with time. After reaching 800 kN, the pultruded sheet broke and the load on the force sensor decreased sharply. At this time, the pultruded sheet still bores a large load. Over time, the pultruded sheet failed completely. Figure 12b is a comparison of experimental and numerically calculated strain-displacement data. During the numerical simulation calculation, the pultruded sheet is considered to be a uniform material, and the material parameters do not change with space. However, in the pultrusion process, the uniformity of the pultruded sheet cannot be strictly guaranteed [10], so the numerical simulation can only obtain a benchmark result [31], which cannot be strictly consistent with the test. Therefore, it can be seen that the linearization characteristic of the finite element model is stronger than that of the experimental data.

Figure 13 reveals the internal failure mechanism of the pultruded sheet. After the failure of the specimen, the surface is intact and there is no obvious fracture, which proves that the pultruded sheet composite material has a great damage tolerance. The specimen has no obvious signs before fracture, mainly because the fiber in the composite material is the main bearing capacity, and the layup angle is ±45°, which will prevent the crack propagation at the interface between the fiber and the matrix. Figure 13 shows fiber breakage along the 45° direction with interfacial debonding and delamination. According to the damage characteristics and on-site failure of composite materials, it is inferred that the fracture failure modes of pultruded sheets are fiber fracture, delamination and interfacial debonding.

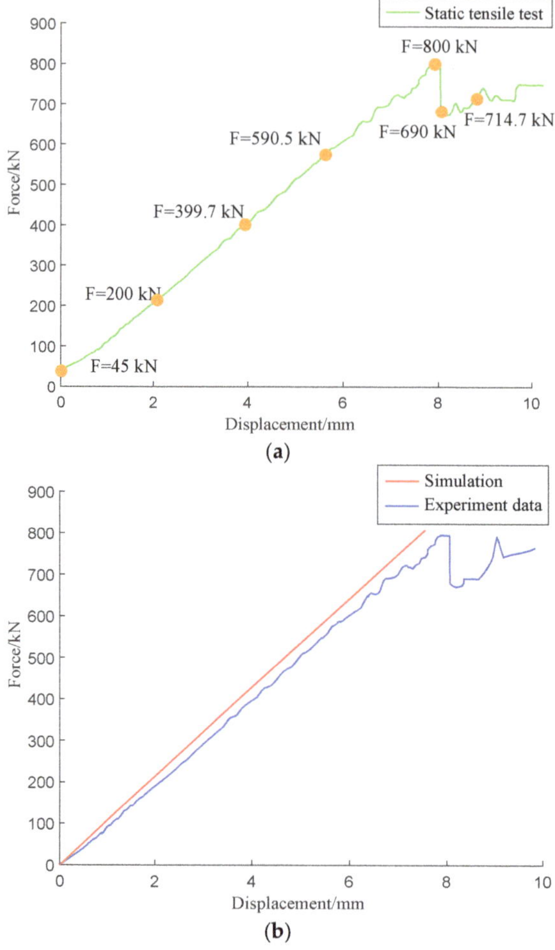

Figure 12. Force-displacement curve: (**a**) force sensor curve; (**b**) test-simulation curve.

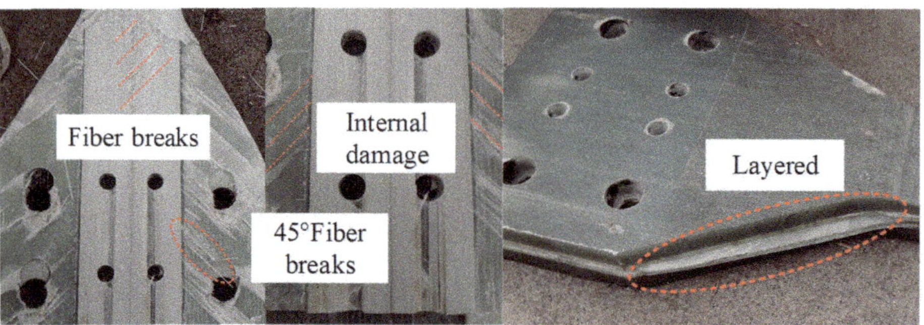

Figure 13. Failure area of pultruded sheet.

5. Conclusions

In this paper, taking the mechanical properties and damage evolution of pultruded sheet composites for wind turbine blades as the research object, a pultruded sheet static tensile testing machine was built. Based on this, the static tensile test of the pultruded sheet composite material was carried out to explore the mechanical properties of the pultruded sheet for wind turbine blades. At the same time, the temperature change of the pultruded sheet composite material was monitored by infrared thermal imaging technology to explore the damage evolution of the pultruded sheet. The main conclusions are as follows:

(1) Through the finite element analysis of the loading bracket of the testing machine, it was obtained that the beam after the loading bracket was the key part to bear the load. When the design load of the loading bracket was applied, the maximum stress of the loading bracket was 280.18 MPa, and the maximum deformation was 0.665 mm, which was lower than the yield limit of the Q345 material and met the requirements of the test machine. Using the MATLAB/Simulink platform to simulate the electro-hydraulic servo force control system, it was verified that the control system met the test requirements and had a good follow-up control effect.

(2) When the pultruded sheet composite was subjected to static load, the failure load was 800 kN, and it would go through elastic, plastic, and fracture stages during the loading process. In the elastic deformation stage, the material was under less stress and could quickly recover from deformation. In the plastic stage, the strain first rose sharply and then rose slowly, which was irreversible at this stage. Finally, in the fracture stage, the pultruded sheet composites had damages such as fiber fracture, delamination, and debonding under static load. During the test, there was internal damage but no obvious fracture, which proved that the specimen had a large damage tolerance. The full-scale test of wind turbine blades was supplemented by exploring the mechanical properties of pultruded sheets to provide data reference for the mechanical performance parameters of the entire blade.

(3) When the damage mode of the pultruded sheet was detected by infrared thermal imaging, the surface temperature of the specimen in the elastic stage decreased by about 1 °C. As the load increased, the specimen went through the plastic stage and the fracture stage, and the surface temperature gradually increased. The temperature variation trend was consistent with the strain, and its variation range was 5 °C.

(4) Infrared thermal imaging technology can analyze the damage evolution process of layered composites under load from the monitored infrared thermal images. When the temperature of the surface of the test piece is monitored by an infrared thermal imager, the yield limit of the test piece can be determined, which is beneficial to quickly determine the performance parameters of the test piece, and provides a simple method for engineering applications.

Author Contributions: Conceptual-ization, Y.H. and Y.W.; methodology, Y.H.; software, H.Z.; validation, H.Z., Y.Z. and L.Z.; formal analysis, Y.H.; investigation, C.L.; resources, C.L.; data curation, Y.H.; writing—original draft preparation, Y.H.; writing—review and editing, Y.H.; visualization, Y.W.; supervision, Y.W.; pro-ject administration, C.L.; funding acquisition, Y.W. All authors have read and agreed to the published version of the manuscript.

Funding: This research was funded by Natural Science Foundation of National Natural Science Foundation of China (grant number 52075305), Shandong Provincial Key Laboratory of Precision Manufacturing and Non-Traditional Machining, Zhoucun District School City Integration Development Project (grant number 2020ZCXCZH01) and A Project of Shandong Province Higher Educational Science and Technology Program (grant number 2019KJB013).

Institutional Review Board Statement: No applicable.

Informed Consent Statement: No applicable.

Data Availability Statement: No applicable.

Conflicts of Interest: The authors declare no conflict of interest.

References

1. Cao, Q.; Xiao, L.; Cheng, Z.; Liu, M. Dynamic responses of a 10 MW semi-submersible wind turbine at an intermediate water depth: A comprehensive numerical and experimental comparison. *Ocean Eng.* **2021**, *232*, 109138. [CrossRef]
2. Liu, D.; Liu, Y.; Sun, K. Policy impact of cancellation of wind and photovoltaic subsidy on power generation companies in China. *Renew. Energy* **2021**, *177*, 134–147. [CrossRef]
3. Lucena, J.D.A.Y. Recent advances and technology trends of wind turbines. *Recent Adv. Renew. Energy Technol.* **2021**, *1*, 177–210.
4. Zou, C.; Xiong, B.; Xue, H.; Zheng, D.; Ge, Z.; Wang, Y.; Jiang, L.; Pan, S.; Wu, S. The role of new energy in carbon neutral. *Pet. Explor. Dev.* **2021**, *48*, 480–491. [CrossRef]
5. Fascetti, A.; Feo, L.; Abbaszadeh, H. A critical review of numerical methods for the simulation of pultruded fiber-reinforced structural elements. *Compos. Struct.* **2021**, *273*, 114284. [CrossRef]
6. Al-Saadi, A.U.; Aravinthan, T.; Lokuge, W. Effects of fibre orientation and layup on the mechanical properties of the pultruded glass fibre reinforced polymer tubes. *Eng. Struct.* **2019**, *198*, 109448. [CrossRef]
7. Lee, H.G.; Park, J. Static test until structural collapse after fatigue testing of a full-scale wind turbine blade. *Compos. Struct.* **2016**, *136*, 251–257. [CrossRef]
8. Madenci, E.; Özkılıç, Y.O.; Gemi, L. Experimental and theoretical investigation on flexure performance of pultruded GFRP composite beams with damage analyses. *Compos. Struct.* **2020**, *242*, 112162. [CrossRef]
9. Silva, F.; Amorim, E.; Baptista, A.; Pinto, G.; Campilho, R.; Castro, M. Producing hybrid pultruded structural products based on preforms. *Compos. Part B Eng.* **2017**, *116*, 325–332. [CrossRef]
10. Paciornik, S.; Martinho, F.; de Mauricio, M.; D'Almeida, J. Analysis of the mechanical behavior and characterization of pultruded glass fiber–resin matrix composites. *Compos. Sci. Technol.* **2003**, *63*, 295–304. [CrossRef]
11. Harizi, W.; Chaki, S.; Bourse, G.; Ourak, M. Mechanical damage assessment of Glass Fiber-Reinforced Polymer composites using passive infrared thermography. *Compos. Part B Eng.* **2014**, *59*, 74–79. [CrossRef]
12. Aniskevich, K.; Arnautov, A.; Jansons, J. Mechanical properties of pultruded glass fiber-reinforced plastic after moistening. *Compos. Struct.* **2012**, *94*, 2914–2919. [CrossRef]
13. Leong, M.; Overgaard, L.C.; Thomsen, O.T.; Lund, E.; Daniel, I.M. Investigation of failure mechanisms in GFRP sandwich structures with face sheet wrinkle defects used for wind turbine blades. *Compos. Struct.* **2012**, *94*, 768–778. [CrossRef]
14. Lal, H.M.; Xian, G.; Thomas, S.; Zhang, L.; Zhang, Z.; Wang, H. Experimental Study on the Flexural Creep Behaviors of Pultruded Unidirectional Carbon/Glass Fiber-Reinforced Hybrid Bars. *Materials* **2020**, *13*, 976.
15. Zhang, S.; Caprani, C.C.; Heidarpour, A. Strain rate studies of pultruded glass fibre reinforced polymer material properties: A literature review. *Constr. Build. Mater.* **2018**, *171*, 984–1004. [CrossRef]
16. Correia, J.R.; Gomes, M.M.; Pires, J.M.; Branco, F.A. Mechanical behaviour of pultruded glass fibre reinforced polymer composites at elevated temperature: Experiments and model assessment. *Compos. Struct.* **2013**, *98*, 303–313. [CrossRef]
17. Menna, C.; Asprone, D.; Caprino, G.; Lopresto, V.; Prota, A. Numerical simulation of impact tests on GFRP composite laminates. *Int. J. Impact Eng.* **2011**, *38*, 677–685. [CrossRef]
18. Zhou, B.; Zhang, X.; Li, H. Study on Air Bubble Defect Evolution in Wind Turbine Blade by Infrared Imaging with Rheological Theory. *Appl. Sci.* **2019**, *9*, 4742. [CrossRef]
19. Cui, A.; Zhijun, W.U.; Jianxun, L.I. Research on method of infrared target implantation based on temperature field consistency. *Comput. Eng. Appl.* **2017**, *53*, 177–182.
20. Xin, H.; Mosallam, A.; Liu, Y.; Veljkovic, M.; He, J. Mechanical characterization of a unidirectional pultruded composite lamina using micromechanics and numerical homogenization. *Constr. Build. Mater.* **2019**, *216*, 101–118. [CrossRef]
21. Shigui, L.; Liu, X.; Pla, U.O. Development and research status of infrared thermal image detection technology. *Infrared Technol.* **2018**, *40*, 214–219.
22. Yang, N.; Yuan, M.; Wang, P.; Zhang, R.; Sun, J.; Mao, H. Tea diseases detection based on fast infrared thermal image processing technology. *J. Sci. Food Agric.* **2019**, *99*, 3459–3466. [CrossRef]
23. Zhu, W.J.; Lin, L.I.; Mei-Qing, L.I. Rapid detection of tomato mosaic disease in incubation period by infrared thermal imaging and near infrared spectroscopy. *Spectrosc. Spectr. Anal.* **2018**, *38*, 2757–2762.
24. Zhang, S.; Li, X.; Zhang, X.; Zhang, S. Infrared and visible image fusion based on saliency detection and two-scale transform decomposition. *Infrared Phys. Technol.* **2021**, *114*, 103626. [CrossRef]
25. Kim, J. Non-Destructive Characterization of Railway Materials and Components with Infrared Thermography Technique. *Materials* **2019**, *12*, 4077. [CrossRef]
26. Jfa, B.; Djb, A.; Ott, B. On the source of the thermoelastic response from orthotropic fibre reinforced composite laminates. *Compos. Part A Appl. Sci. Manuf.* **2021**, *149*, 106515.
27. Malheiros, F.C.; Figueiredo, A.A.A.; Ignacio, L.H.D.S.; Fernandes, H.C. Estimation of thermal properties using only one surface by means of infrared thermography. *Appl. Therm. Eng.* **2019**, *157*, 113696. [CrossRef]
28. Altanopoulos, T.I.; Raftoyiannis, I.G.; Polyzois, D. Finite element method for the static behavior of tapered poles made of glass fiber reinforced polymer. *Mech. Adv. Mater. Struct.* **2020**, *28*, 2141–2150. [CrossRef]
29. Xin, H.; Liu, Y.; Mosallam, A.; He, J.; Du, A. Evaluation on material behaviors of pultruded glass fiber reinforced polymer (GFRP) laminates. *Compos. Struct.* **2017**, *182*, 283–300. [CrossRef]

30. Nan, B.; Wu, Y.; Sun, H. Buckling behavior of pultruded carbon fiber reinforced polymer pipes under axially compressive load. *Harbin Gongcheng Daxue Xuebao/J. Harbin Eng. Univ.* **2015**, *36*, 779–783.
31. Jw, A.; Xh, A.; Cw, B. Failure analysis at trailing edge of a wind turbine blade through subcomponent test. *Eng. Fail. Anal.* **2021**, *130*, 105596.
32. Freudenreich, K.; Argyriadis, K. The Load Level of Modern Wind Turbines according to IEC 61400-1. *J. Phys. Conf. Ser.* **2007**, *75*, 012075. [CrossRef]
33. Mizumoto, I.; Fujii, S.; Tsunematsu, J. Adaptive Combustion Control System Design of Diesel Engine via ASPR Based Adaptive Output Feedback with a PFC. *J. Robot. Mechatron.* **2016**, *28*, 664–673. [CrossRef]
34. Terada, K.; Miura, H.; Okugawa, M.; Kobayashi, Y. Adaptive Speed Control of Wheeled Mobile Robot on Uncertain Road Condition. *J. Robot. Mechatron.* **2016**, *28*, 687–694. [CrossRef]
35. Liu, Y.; Ao, D.U.; Xin, H. Experiment on bolted joints of pultruded GFRP laminates. *Zhongguo Gonglu Xuebao/China J. Highw. Transport.* **2017**, *30*, 223–229.

MDPI

Article

Thermal Analysis and Prediction Methods for Temperature Distribution of Slab Track Using Meteorological Data

Qiangqiang Zhang [1], Gonglian Dai [1] and Yu Tang [2,*]

[1] School of Civil Engineering, Central South University, Changsha 410083, China
[2] School of Resources and Safety Engineering, Central South University, Changsha 410083, China
* Correspondence: tangyu12@csu.edu.cn

Abstract: The structural temperature distribution, especially temperature difference caused by solar radiation, has a great impact on the deformation and curvature of the concrete slab tracks of high-speed railways. Previous studies mainly focused on the temperature prediction of slab tracks, while how the temperature distribution is affected by environmental conditions has been rarely investigated. Based on the integral transformation method, this work presents an analytical method to determine and decompose the temperature distribution of the concrete slab track. A field temperature test of a half-scaled specimen of concrete slab track was conducted to validate the developed methodology. In the proposed method, we decompose the temperature distribution of the slab track into an initial temperature component and a boundary temperature component. Then, the boundary temperature components caused by solar radiation and atmospheric temperature are investigated, respectively. The results show that the solar radiation plays a significant role in the nonlinear temperature distribution, while the atmospheric temperature has little effect. By contrast, the temperature change in the slab surface resulting from the atmospheric temperature accounts on average for only 5% in the hot weather condition. The proposed method establishes a relation between the structural temperature and meteorological parameters (i.e., the solar radiation and atmospheric temperature). Consequently, the temperature distribution of the concrete slab track is predicted via the meteorological parameters.

Keywords: scale model test; slab track; thermal analysis; theoretical study; meteorological data

Citation: Zhang, Q.; Dai, G.; Tang, Y. Thermal Analysis and Prediction Methods for Temperature Distribution of Slab Track Using Meteorological Data. *Sensors* **2022**, *22*, 6345. https://doi.org/10.3390/s22176345

Academic Editors: Phong B. Dao, Tadeusz Uhl, Liang Yu, Lei Qiu and Minh-Quy Le

Received: 6 July 2022
Accepted: 19 August 2022
Published: 23 August 2022

1. Introduction

Slab tracks have been widely used in high-speed railways because of their advantages of high stability and smoothness. By 2020, nearly 29,000 km of slab tracks had been built in China, which accounts for more than 80% of all slab tracks in the world [1]. At present, there are three types of China Railway Track System (CRTS) slab tracks, namely, CRTS-I, CRTS-II, and CRTS-III. In particular, the CRTS-III slab track, which is technically improved from the German Bögl slab track, has been used in new railway lines. Among these, a CRTS-III slab track laid over a steel bridge was first built, as shown in Figure 1, which consists of a precast slab, self-compacting concrete, and a concrete base, and can be defined as a multilayered structure. Different from slab tracks on concrete bridges, the heat flow through the slab track to the steel bridge is easier and faster due to the limited height of the concrete members of the bridge-track system. Consequently, this may lead to a larger temperature difference. The novel construction of a slab track on a steel bridge requires more attention to be paid to the temperature evolution of concrete slab tracks under the environmental conditions.

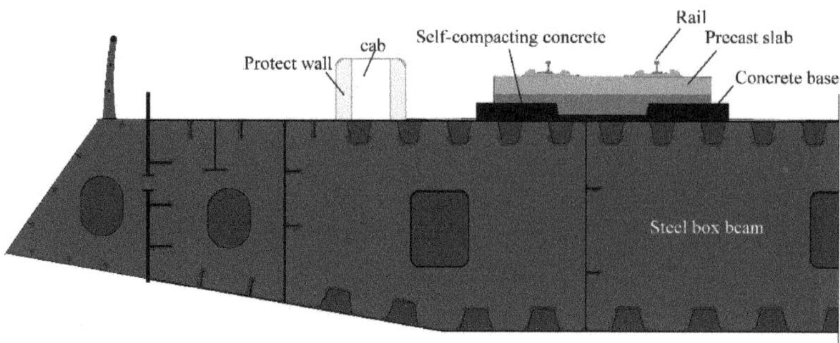

Figure 1. A concrete slab track on a steel box girder section.

Slab tracks are affected by environmental conditions, such as solar radiation, air temperature, and wind, and can hypothetically produce two periodic thermal actions. First, the seasonal changing temperature may cause the overall concrete slab to rise or drop, and its macro-performance consists of the expansion and contraction deformation. Second, the daily solar radiation and air temperature variations may lead to a temperature difference that causes bending deformation in the slab track [2–4]. Considering some temperature-induced damage problems in other slab tracks, the CRTS-III slab track and France's New Ballastless Track have considered the temperature effect on slab tracks [5]. In addition, some physical methods to reduce the slab temperature have also been tested [6–8]. These involve coating the slab surface with composite materials to reduce the radiation absorption or increase the reflected radiation.

Temperature is one of the most critical parameters related to the behavior and response of slab tracks. Temperature tests of slab tracks in the laboratory under specific conditions have attracted more attention. Yang et al. [2] analyzed the temperature distribution of the slab track based on a full-scale temperature test and found that a long-term daily mean air temperature and stronger solar radiation caused the whole slab temperature to rise under continuous hot weather conditions. Zhang et al. [9] developed a 1:4 scaled laboratory test for CRTS-II slab track, which was used to quantify the stresses in the various components of the track system resulting from sudden temperature variations. Zhong et al. [4] investigated the impact of the daily air temperature on the interface stress of a full-scale specimen of CRTS-II slab track in the construction stage. Zhou et al. [10] compared the effect of the constraint conditions on the temperature distribution of a slab track in two 1:4 scaled specimens and found that the fixed constraint condition decelerated the temperature transfer of the track slab to the cement asphalt mortar. Zhou et al. [11,12] carried out a temperature test of a 1:4 scaled specimen for a CRTS-II slab track on three simply-supported box girders with a heat device and then analyzed the distribution of the three-dimensional thermal fields in the slab. They also found that the strains in the track structure increased nonlinearly with the environmental temperature increase [13]. Although there were some satisfactory conclusions to be drawn from the temperature test of slab tracks, there is currently a lack of research on the different effects of solar radiation and air temperature on the temperature distribution of slab tracks on a steel bridge.

Slab tracks built in a natural environment undergo a complicated heat transfer, and it is theoretically necessary to investigate the temperature evolution of slab tracks subjected to environmental thermal loads. Ou et al. [14] investigated the temperature distribution of CRTS-II slab track during four seasons based on a simplified solution of heat conduction. It was found that air temperature was one of main factors affecting the temperature distribution inside the track structure, and the temperature gradient was biggest in the summer. Liu et al. [15] offered a simplified solution of the thermal field of concrete slab track to reveal the relation between temperature distributions and environmental conditions. The results showed that the combined action of the mean air temperature and solar radiation impacted the overall concrete slab temperature, and the air temperature amplitude greatly influenced the temperature gradient. Zeng et al. [16] provided an analytical solution to a semi-infinite thermal field model of concrete slab track, which effectively predicted the temperature gradient of the track structure in different cities in China. Riding et al. [17] calculated the thermal field of concrete structures using three recommended methods in specifications and proved that the error based on heat conduction equations was the smallest. In summary, previous researchers have sought a simplified method for solving the heat problem of slab tracks under external environmental conditions, which is convenient for demonstrating the temperature evolution of slab tracks in one day. However, they ignore the coupling actions of solar radiation and air temperature on the temperature distribution of slab tracks. Therefore, it is necessary to find an analytical method to distinguish the different effects of solar radiation from air temperature on the temperature evolution of slab tracks.

As the slab track system is similar with the multilayered concrete pavement, the analytical method used for concrete pavement temperature can be consulted in [18–23]. The Green's function method was used for the analytical solution of thermal field in the multilayered pavement [18]. A one-dimensional temperature model of the pavement under site conditions was solved by the Laplace transform method [19]. The heat problem with the measured surface temperature was solved without considering the environmental conditions [24]. Compared to the semi-infinite pavement system, the slab track is a three-dimensional thermal field due to the finite size of the structure. The different types and numbers of boundary conditions make the governing equations hard to be solved by analytical methods.

Focusing on the different effects of solar radiation and air temperature on concrete slab tracks, this paper proposes an analytical prediction method for the temperature distribution of concrete slab tracks via meteorological parameters. The structure of this paper is as follows. First, an experimental program of a half-scaled specimen of a concrete slab track was described for an outdoor temperature test. Then, a one-dimensional temperature model of the concrete slab was developed while considering the meteorological parameters and solved based on the integral transformation method. The calculation accuracy was approved using other methods. Combined with the solution, the temperature distribution in a generic concrete section is decomposed theoretically and the different effects of solar radiation and air temperature on the time–space temperature distribution in slab tracks is discussed.

2. Experimental Setup

2.1. Specimen Design

The test site is located on the top of a hillside in Changsha, China (112° E, 27° N), with an altitude of 150 m. The region has a subtropical climate with an annual average temperature of 28 °C. Before the construction, a suitable location was chosen in an open area without shade to meet the requirement of natural solar radiation thermal loads. The temperature test of the track structure is shown in Figure 2. The experimental beam segment is 1.2 m in width, 5.7 m in length, and 0.95 m in height. The thickness of the flange and web of the steel beam are 20 mm and 15 mm, respectively. The concrete slab was cast to the left side of the steel beam segment, while the other side was manufactured as a ballast track structure. Since the objective of this paper was to investigate the thermal field of a concrete slab track structure, only the monitoring results of the left side of the specimen were used, as shown in Figure 3.

In general, the temperature change in the concrete members along the longitudinal section is small [15,16,25], so concrete slab tracks can be reduced to a half-scaled concrete slab specimen, with a specific size of 2.4 m × 1.2 m × 0.3 m. The detailed dimension of the temperature test specimen is shown in Figure 3. The concrete slab is made of ordinary cement, water, crushed sand, and gravel with a mixing ratio of 210, 355, 703, and 887 $\text{Kg}\cdot\text{m}^{-3}$, respectively. The beam segment is made of Q235b steel with a grayish color coating. In addition, to prevent the slippage of the concrete slab resulting from the thermal action, vertical reinforcements with a diameter of 20 mm were welded on the top plate of the steel beam at a spacing of 1.2 m.

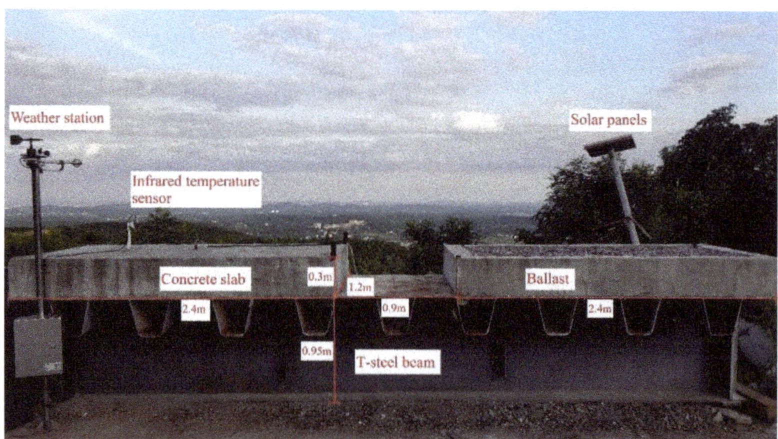

Figure 2. Temperature test of track structures on a T-steel beam segment.

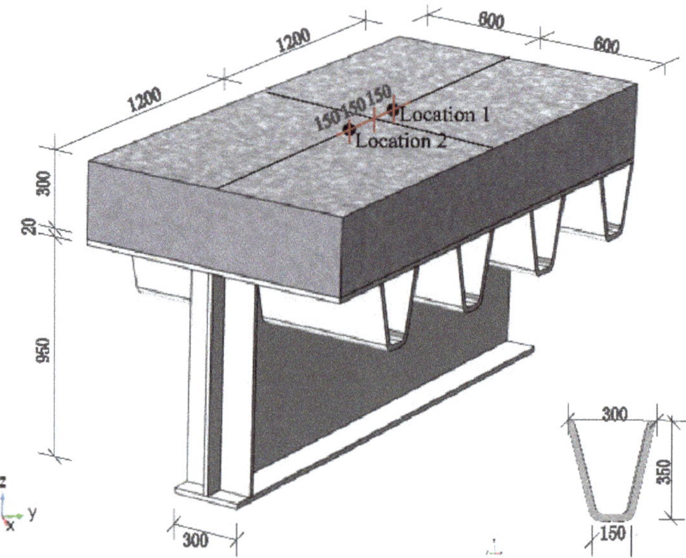

Figure 3. Geometric dimensions and measuring positions of the testing specimen (unit: mm).

2.2. Arrangement of Temperature Sensors

In this study, the test is to measure the vertical temperature distribution in the concrete slab. To prevent the influence of the transverse heat flow on the measuring point, the center location of the concrete slab was selected as the test section, namely, Location 1 (Figure 3). Five temperature sensors (H1–H5) were embedded in the concrete slab along the depth at Location 1. The distribution of all temperature sensors is clearly shown in Figure 4.

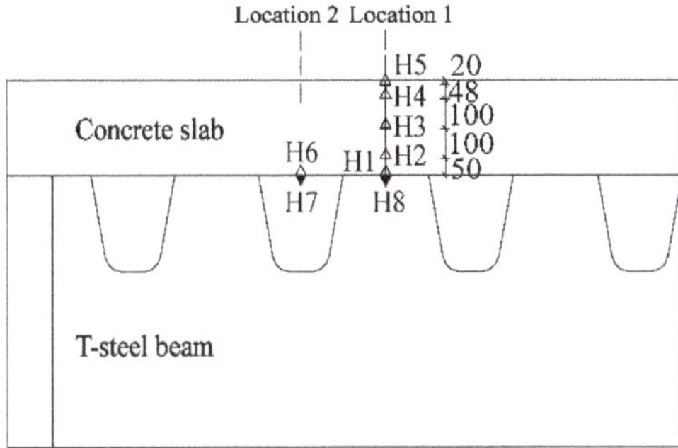

Figure 4. Distributions of temperature sensors (unit: mm).

The temperature of the contact interface was measured to obtain the bottom boundary condition. Two pairs of measuring points were arranged at different locations of the contact interface. The one pair of measuring points with the numbers H1 and H8 was in Location 1 and another point with the numbers H6 and H7 was in Location 2. Points H7 and H8 were pasted on the bottom surface of the top plate of the steel beam along Locations 1 and 2.

The PT 100 platinum thermal sensor of 30 mm in length, 8 mm in width, and 4 mm in thickness was used for the measurement. The working range of the sensor is −50~200 °C, and the testing precision is ±0.2 °C. To prevent the influence of reflected radiation on the exterior sensor (H7 and H8), an aluminum square shell was constructed to cover the sensor, as shown in Figure 5.

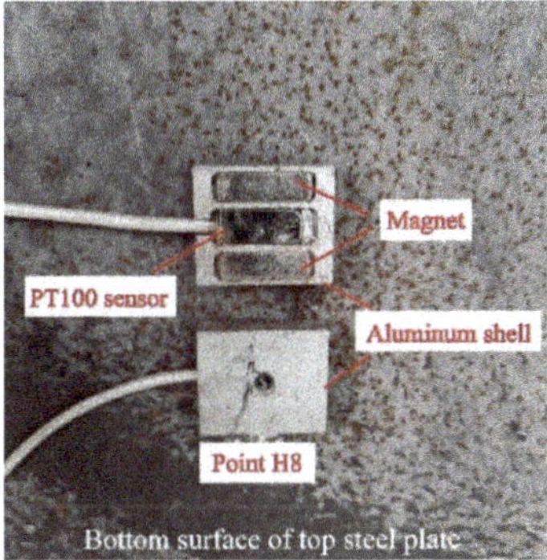

Figure 5. The exterior sensor with an aluminum square shell.

2.3. Meteorological Parameters

The monitoring meteorological parameter is essential for the prediction and analysis methods [26]. For accurate boundary conditions of the thermal field, a weather station was built at the experimental site, as shown in Figure 6. The weather station monitors five meteorological parameters. The solar radiometer is used to observe the total solar radiation intensity on the horizontal plane. The temperature probe in the instrument shelter records the air temperature. The anemometer and anemoscope are applied to measure wind direction and wind speed. The net solar radiometer is used to observe the instantaneous heat budget.

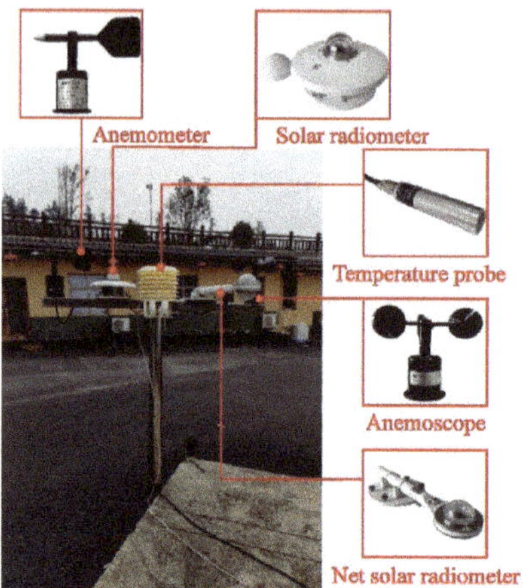

Figure 6. Environment monitoring system.

A solar-powered system was adopted to provide energy for the structural temperature and meteorological monitoring system. The data were collected every 30 min by an automatic acquisition instrument since 1 October 2019.

3. Analytical Prediction Method

3.1. Analytical Solution of a One-Dimensional Temperature Distribution

In practical engineering, the vertical temperature distribution of a slab track is the main consideration, especially the most unfavorable temperature distribution under extreme environmental conditions. The calculated result of the multi-dimensional temperature model was close to that of the one-dimensional model [25]. For mathematical simplicity, we only consider a one-dimensional heat conduction problem with double convection boundaries in the finite region, representing change through depth, as shown in Figure 7.

x=0 BC1: h_1、$f_1(t)$

T_0 $T(x,t)$

x=d BC2: h_2、$f_2(t)$

Figure 7. One-dimensional model of the temperature distribution.

In Figure 7, The surfaces BC1 and BC2 of the slab are heated by the convection of two ambient fluids: $f_1(t)$ and $f_2(t)$. It is necessary to develop a unique function for the boundary condition variables $f_1(t)$ and $f_2(t)$ when applying analytical methods. A third-order polynomial function had been used to achieve approximation [19]. For a special case of the boundary conditions of a period convection in this problem, the mathematical model of the ambient fluid temperature is assumed to be a cosine function:

$$f_i(t) = T_{i,\mathrm{a}} - T_{i,\mathrm{b}} \cos(w_i(t - \delta_i)), t \in (t_1 \sim t_2) \tag{1}$$

where i is the number of boundary surfaces, $T_{i,\mathrm{a}}$ and $T_{i,\mathrm{a}}$ are the average temperature and amplitude of the fluid medium, respectively, w_i and δ_i are the frequency and phase of the function, respectively, and t_1 and t_2 are the sunrise and sunset time, respectively.

As there is no internal heat source in the temperature model, the heat conduction equation is expressed as Equation (2). The boundary conditions of surfaces BC1 and BC2 are considered by Equations (3) and (4), and Equation (5) gives the initial condition of this problem:

$$\alpha \frac{\partial^2 T(x,t)}{\partial x^2} = \frac{\partial T(x,t)}{\partial t} \text{ in } 0 < x < d,\ t_1 < 0 < t_2 \tag{2}$$

$$\text{BC1}:\ -k\frac{\partial T(x,t)}{\partial x} + h_1 T(x,t) = h_1 f_1(t)\ ,\ x = 0 \tag{3}$$

$$\text{BC2}:\ k\frac{\partial T(x,t)}{\partial x} + h_2 T(x,t) = h_2 f_2(t)\ ,\ x = d \tag{4}$$

$$\text{IC}:\ T(x,t_1) = T_0 \tag{5}$$

where α is the thermal diffusivity ($m^2 \cdot s^{-1}$), k is the thermal conductivity (W $m^{-1} \cdot K^{-1}$), h_1 and h_2 are heat transfer coefficients at different surfaces ($W \cdot m^{-2} \cdot K^{-1}$), and $T(x, t)$ is the temperature (°C) at an arbitrary point x at any time t.

To make the initial temperature $T(x, t_1)$ be equal to zero, the following temperature variable is introduced:

$$\theta(x,t) = T(x,t) - T_0 \tag{6}$$

Then, Equations (3)–(5) are transformed as:

$$\text{BC1}:\ -k\frac{\partial \theta(x,t)}{\partial x} + h_1 \theta(x,t) = h_1 (f_1(t) - T(x,t_1)) \tag{7}$$

$$\text{BC2}:\ k\frac{\partial \theta(x,t)}{\partial x} + h_2 \theta(x,t) = h_2 (f_2(t) - T(x,t_1)) \tag{8}$$

$$\text{IC}:\ \theta(x,t_1) = 0 \tag{9}$$

To solve the aforementioned heat conduction problem, the integral transform pair for the function $\theta(x, t)$ with respect to the x variable is constructed based on the integral transformation method [27]:

$$\theta(x,t) = \sum_{n=1}^{\infty} \frac{X(\beta_n, x')}{N(\beta_n)} \overline{\theta}(\beta_n, t) \tag{10}$$

$$\overline{\theta}(\beta_n, t) = \int_0^d X(\beta_n, x')\theta(x', t)\mathrm{d}x' \tag{11}$$

where $N(\beta_n)$ is the norm and $X(\beta_n, t)$ is the eigenfunction. There are infinite norms and eigenfunctions for the eigenvalues β_n. By the application of the transformation (Equations (10) and (11)), the general solution of Equation (2) is in the following form [27]:

$$\theta(x,t) = \sum_{n=1}^{\infty} \frac{X(\beta_n, x')}{N(\beta_n)} \mathrm{e}^{-\alpha\beta_n{}^2 t} \int_0^t \mathrm{e}^{\alpha\beta_n{}^2 t'} A(\beta_n, t')\mathrm{d}t' \tag{12}$$

where

$$N(\beta_n) = \int_0^d [X(\beta_n, x')]^2 \mathrm{d}x' \tag{13}$$

$$A(\beta_n, t') = \frac{\alpha}{k}[X(\beta_n, 0)h_1 f_1(t') + X(\beta_n, d)h_2 f_2(t')] \tag{14}$$

For the boundary value problem with a double-convective boundary condition, the eigenfunction and transcendental function are expressed as follows [27]:

$$X(\beta_n, x) = \beta_n \cos(\beta_n, x) + H_1 \sin(\beta_n, x) \tag{15}$$

$$\tan(\beta_n d) = \frac{(H_1 + H_2)\beta_n}{\beta_n{}^2 - H_1 H_2} \tag{16}$$

According to temperature variable $\theta(x, t)$, the function of ambient fluid temperature (Equation (1)) becomes:

$$f_i(t) = \Delta T_{f,i} + T_{f,i}(t) i = 1,2 \quad (17)$$

where $\Delta T_{f,i} = T_{i,\mathrm{a}} - T_0$ and $T_{f,i}(t) = -T_{i,\mathrm{b}} \cos(w_i(t - \delta_i))$.

The expression (Equation (17)) is introduced into the general solution (Equation (12)). Then, the definite integral can be evaluated, and the solution is expressed as:

$$\theta(x,t) = \sum_{i=1}^{2} \sum_{n=1}^{\infty} \frac{X(\beta_n,x)}{N(\beta_n)} \frac{h_i X(\beta_n,x_i)}{k\beta_n{}^2} \cos(w_i\varphi_{i,n}) \left\{ T_{f,i}(t-\varphi_{i,n}) + \frac{\Delta T_{f,i}}{\cos(w_i\varphi_{i,n})} - \left[T_{f,i}(t_1-\varphi_{i,n}) + \frac{\Delta T_{f,i}}{\cos(w_i\varphi_{i,n})} \right] \mathrm{e}^{-\alpha\beta_n{}^2(t-t_1)} \right\} \quad (18)$$

By substituting Equation (6) into Equation (18), the analytical solution of the one-dimensional original heat conduction problem is obtained, as shown in Equation (19):

$$T(x,t) = T_0 + \sum_{i=1}^{2} \sum_{n=1}^{\infty} C_{i,n} X(\beta_n,x) [f_{i,n}(t) - \mathrm{e}^{-\alpha\beta_n{}^2(t-t_1)} f_{i,n}(t_1)] \quad (19)$$

where it is defined that:

$$f_{i,n}(t) = \frac{\Delta T_{f,i}}{\cos(w_i\varphi_{i,n})} + T_{f,i}(t-\varphi_{i,n}) \quad (20)$$

$$C_{i,n} = \frac{H_i \cos(w_i\varphi_{i,n}) X(\beta_n, x_i)}{N(\beta_n)\beta_n{}^2} \quad (21)$$

$$\cos(w_i\varphi_{i,n}) = \frac{\alpha\beta_n{}^2}{\sqrt{(\alpha\beta_n{}^2)^2 + w_i{}^2}} \quad (22)$$

In Equation (19), the temperature $T(x, t)$ mainly consists of a superposition expression of the boundary temperature term. The analytical solution is a parametric formula, which benefits discussing the effect of the dimensional or environmental parameters. When the initial temperature T_0 is determined, the calculation accuracy of the temperature field is only related to the series expression. The larger the series expansion term n, the higher the accuracy.

3.2. Temperature Distribution Decomposition

In the analytical solution, the temperature $T(x, t)$ depends on the initial temperature term and the boundary temperature term. According to the linear superposition principle, the temperature distribution caused by boundary layers at the ith boundary can be obtained with decomposing Equation (19). As shown in Figure 8, given an arbitrary thermal field of a slab section with two boundaries, the total thermal field is the sum of these three contributions (see Equation (23)).

$$T = T_{\mathrm{IC}} + T_{\mathrm{BC1}} + T_{\mathrm{BC2}} \quad (23)$$

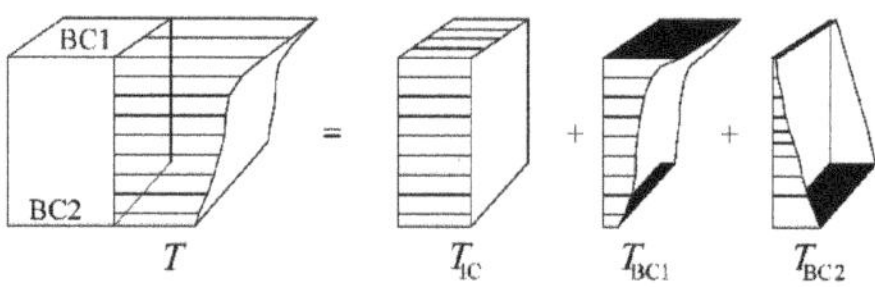

Figure 8. Temperature decomposition of concrete slab.

The initial temperature component T_{IC} is the temperature distribution of the slab at the initial time. The temperature gradient is quite small for a thin slab. The initial temperature distribution is assumed to be linear. Then, the component T_{IC} is the area weighted average value of the thermal field:

$$T_{\mathrm{IC}} = T_0(x, 0) \tag{24}$$

The boundary temperature components T_{BC1} and T_{BC2} are generated by boundary surfaces BC1 and BC2, respectively. The component $T_{\mathrm{BC}i}$ (x, t) consists of the periodic function term and exponential attenuation term, as shown in Equation (25):

$$T_{\mathrm{BC}i}(x,t) = \sum_{n=1}^{\infty} C_{i,n} X(\beta_n, x) \left[f_{i,n}(t) - \mathrm{e}^{-\alpha \beta_n{}^2 (t-t_1)} f_{i,n}(t_1) \right] \tag{25}$$

The temperature decomposition has the advantage of investigating the different actions of the solar radiation and the air temperature on the vertical temperature distribution of slab track. Thereby, the effect of different boundary conditions on the thermal field can be obtained.

3.3. The Method of Dealing with Meteorological Parameters

The site environmental conditions are considered and the procedure for dealing with meteorological parameters is illustrated. The main steps of the analytical prediction method for the concrete slab temperature distribution are summarized in Figure 9.

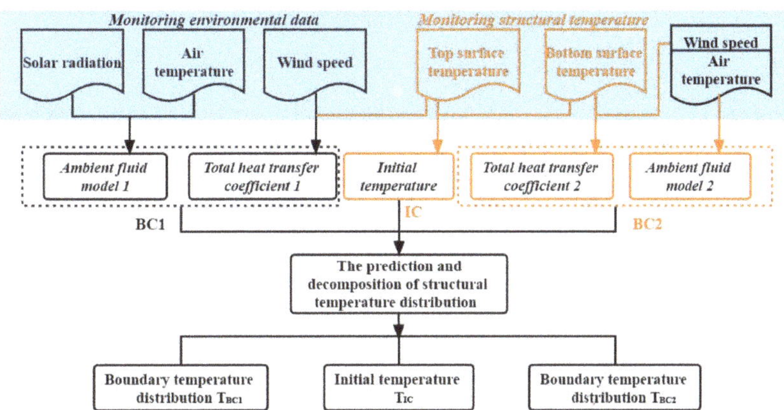

Figure 9. Flowchart of the analytical method using meteorological parameters.

Stage 1: Obtain the site meteorological data for every hour on a sunny and cloudless day, mainly including three meteorological parameters, namely, solar radiation, wind speed, and air temperature. The treatment of the meteorological data is described in Section 3.3.1.

In particular, the air temperature variation is not wide in the continuous sunny weather condition, which is beneficial for the establishment of the ambient fluid model 2 and the initial temperature distribution. Besides, the regular solar radiation can better establish the ambient fluid model 1. The wind speed is mainly used to calculate the heat transfer coefficient, which is not an affecting parameter.

Stage 2: Establish the boundary condition functions f_1(t) and f_1(t) of the concrete slab according to the meteorological parameters in Section 3.3.2. The mathematical model is a cosine function (Equation (1)).

Stage 3: Measure the top and bottom surface temperatures of the slab to develop the initial condition (IC) described in Section 3.3.3. For the thin concrete slab, the initial condition is a constant initial temperature.

Stage 4: Calculate the total heat transfer coefficient by the formula of the radiation heat transfer and convection heat transfer in Section 3.3.4. The coefficient of the radiation heat transfer is calculated approximately using Equation (28).

Stage 5: Apply the analytical solution to obtain the temperature distribution in Section 3.1.

Stage 6: Use the temperature decomposition method to obtain the temperature components in Section 3.2.

3.3.1. Equivalent Radiation Temperature

In the daytime, the structure absorbs heat from the solar radiation and the atmospheric radiation. Meanwhile, the structure releases heat to the exterior environment through the convection and longwave radiation. The main processes of heat exchange are split in radiation, convection, and conduction, as shown in Figure 10.

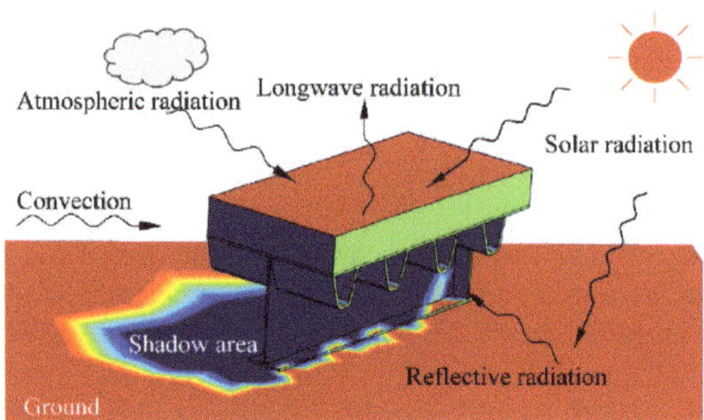

Figure 10. Mechanisms of heat transfer.

Taking the heat exchange at the top surface (BC1) of the concrete slab as an example, the general equation for net heat transfer is written as:

$$\begin{aligned} q &= \gamma q_s - h_c(T_u - T_a) - \varepsilon C_0({T_u}^4 - {T_a}^4) \\ &= \gamma q_s - (h_c + h_r)(T_u - T_a) \end{aligned} \tag{26}$$

where q_s is the total solar radiation on a plane, which is gained from the measured solar radiation on a horizontal plane or the empirical models of solar radiation, γ is the solar absorptivity coefficient of the concrete slab and is taken as 0.5 [14], ε is the emissivity of the surface, C_0 is the Stefan–Boltzmann constant (5.67×10^{-8} W·m^{-2}·K^{-4}), T_u and T_a are the surface and fluid temperatures (K), respectively, and h_c is the coefficient of the convection heat transfer (W·m^2·K^{-1}). h_c is a function of wind speed v (m·s^{-1}) and is expressed as [14,15]:

$$h_c = 4v + 5.7 \tag{27}$$

The variable h_r is the coefficient of the radiation heat transfer (W·m^2·K^{-1}) calculated by the radiation heat transfer Equation (28).

$$h_r = \varepsilon C_0\left(T_u^2 + {T_a}^2\right)(T_u + T_a) \tag{28}$$

The net heat conduction equation (Equation (26)) is simplified to the boundary condition of the third type by substituting into Equations (27) and (28), which is expressed as:

$$q = -h\left[T_u - \left(T_a + \frac{\gamma q_s}{h}\right)\right] \tag{29}$$

where h is the total heat transfer coefficient (W·m^{-2}·K^{-1}) and $h = h_c + h_r$.

The combined action of the solar radiation and atmospheric temperature on concrete slab surface is considered as the equivalent radiation temperature [19]. The assumed temperature T_e is obtained from Equation (29), giving:

$$T_e = T_a + \frac{\gamma q_s}{h} \tag{30}$$

3.3.2. The Boundary Condition

Figure 11a shows the time history curves of the equivalent radiation temperature T_e and solar radiation q_s from 21 to 27 July 2020 at the test site. The curve of the equivalent radiation temperature is proportional to that of solar radiation, and the change trends are consistent. The variable T_e changes regularly with time, which makes it convenient to be fitted by the cosine function (Equation (1)) during the solar time. In addition, the continuous warming weather condition makes the uniform temperature distribution in the concrete slab; thereby, the heating data on the third day were selected to determine the govern equations of the temperature model, as shown in Figure 11b.

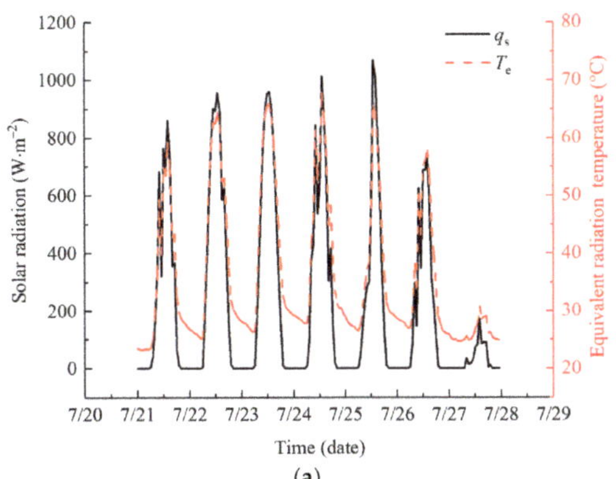

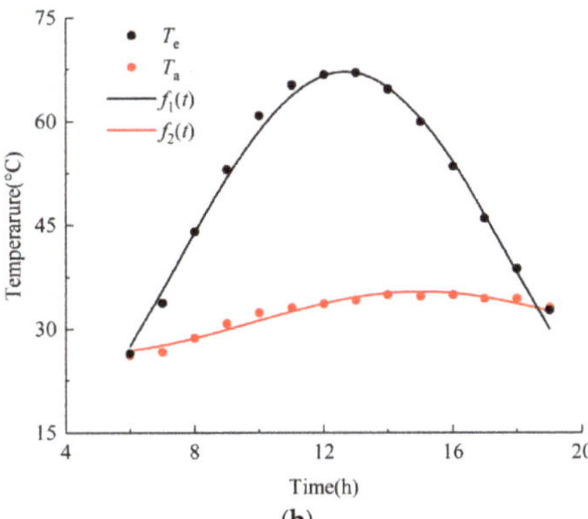

Figure 11. Time history curves of environmental parameters. (**a**) Measured data of the equivalent radiation temperature and solar radiation from 21 to 27 July 2020; (**b**) Fitting results of the equivalent radiation temperature and air temperature on 23 July 2020.

In Figure 11b, the temperature variation of the parameter T_e exposed to the solar radiation is wider. The fitting results of the equivalent radiation temperature and air temperature have a good degree of correlation, with an index R^2 of 0.99 and 0.98, respectively. The fitting formulas $f_1(t)$ and $f_2(t)$ at the boundaries are, respectively:

$$f_1(t) = 32.9 - 32.53\cos\left[\left(\frac{2\pi}{23.43}(t - 0.891)\right)\right], R^2 = 0.99 \quad (31)$$

$$f_2(t) = 31 + 4.5\cos\left[\left(\frac{2\pi}{20.5}(t - 15)\right)\right], R^2 = 0.98 \quad (32)$$

3.3.3. The Initial Condition

The sensitivity for the initial condition may result in the instability of the calculation for the time period in which the temperature gradient is high [18]. To avoid the unexpected oscillations in solutions, the initial temperature distribution in concrete structure was assumed to be uniform and equal to the atmospheric temperature at initial time. Emerson et al. [28] pointed out that the initial temperature can be approximated by the atmospheric temperature at 8:00, which was verified for the temperature of concrete beams and composite beams. The calculated result is more accurate when the initial condition is determined by the measurement of the structural temperature [29,30].

Figure 12 shows the initial temperature distributions through the depth of the slab at 6:00 for a week. These temperature profiles present various nonlinear temperature distributions and temperature differences. The expression of the temperature profile is necessary when applying analytical methods. Based on the above assumptions of the initial condition, a uniform temperature along the depth was calculated by the measured temperatures on 23 July. The maximum error between the calculated temperature and the measured one was 0.72 °C. Thereby, the calculated temperature (30 °C) was taken as the initial condition of the temperature model.

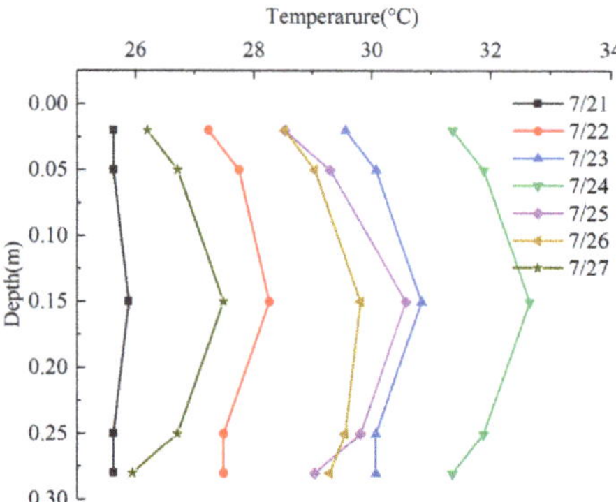

Figure 12. Initial temperature distributions at 6:00 for a week.

3.3.4. The Total Heat Transfer Coefficient

The influence of total heat transfer coefficient on the predicted temperature has been studied and it was shown that the result with the daily average coefficient instead of the time-change coefficient involved a maximum error of 6% [31]. The daily averages of the total heat transfer coefficient on 23 July were calculated using Equation (29), which were h_1 = 15.8 W·m^{-2}·K^{-1} and h_2 = 11.4 W·m^{-2}·K^{-1}.

4. Results and Discussion

4.1. Comparison with the Numerical Solution

The numerical solution of the testing specimen (Figure 3) was calculated using the COMSOL software. A three-dimensional thermal field model was established by considering the actual heat transfer behaviors, including the atmospheric radiation, solar radiation, free convection, and reflected radiation, as shown in Figure 10. The measured meteorological parameters including the solar radiation, ambient temperature, and wind speed, were inputted into the boundary conditions of the temperature model. The measured temperature profile on 23 July was taken as the initial condition. Besides, the ground model was established to calculated the reflected radiation. The material properties and thermal coefficients [21,22,31] are listed in Tables 1 and 2.

Table 1. Properties of concrete and steel.

Material Property	Concrete	Steel
Density, ρ (Kg·m^{-3})	2800	7850
Specific heat capacity, c (J·Kg^{-1}·K^{-1})	880	475
Thermal conductivity, k (W m^{-1}·K^{-1})	1.8	47

Table 2. Thermal coefficients adopted in the temperature model.

Thermal Coefficient	Concrete	Steel
Shortwave absorptivity, γ (W m^{-1}·K^{-1})	0.5	0.9
Longwave absorptivity, γ_1 (W m^{-1}·K^{-1})	0.82	0.88
Emissivity, ε	0.82	0.88

The simulation results of the solar radiation absorbed by surfaces are shown in Figure 13. Figure 13a,c shows the distribution of solar radiation absorbed by surfaces at 8:00 and 10:00, respectively. The horizontal solar radiation intensity increases from 540 W/m^2 to 785 W/m^2 with time variation, and the intensity of the shading area without the solar radiation is 0 W/m^2. The shading zone matches well that of the testing specimen (Figure 13b,d). This indicates that the simulation method of the thermal field model when considering site conditions is correct.

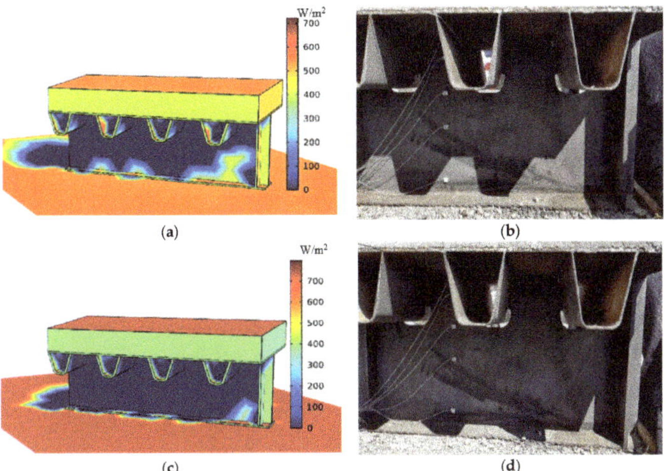

Figure 13. Simulation results of the solar radiation absorbed by surfaces at different times. (**a**) Simulation of the absorbed radiation at 8:00; (**b**) Shading zone at 8:00; (**c**) Simulation of the absorbed radiation at 10:00; (**d**) Shading zone at 10:00.

Figure 14 shows the time–temperature curve of the analytical solution (n = 9) and numerical solution at different depths. The calculated temperatures at various depths converge to the initial temperature of 30 °C at 6:00. The maximum errors at the top and bottom surface are 2.4 °C and 1.3 °C, respectively. Meanwhile, the error within the concrete slab is less than 0.8 °C. The nonhomogeneous boundary condition makes the nonconvergence of the analytical solution at boundary surfaces. Thereby, the analytical temperature at the boundary surface requires larger superposition terms n for generating a higher prediction accuracy.

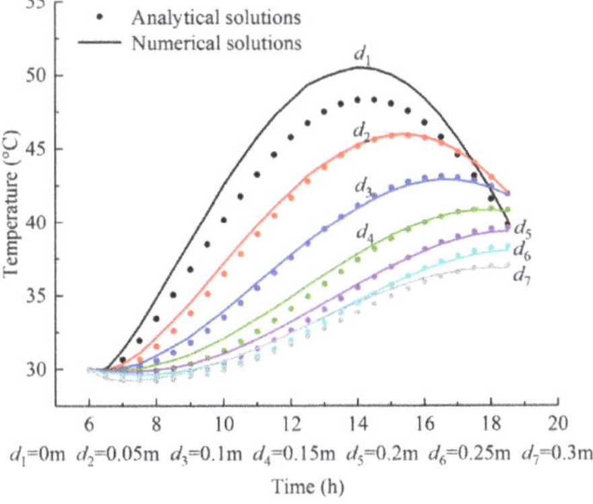

Figure 14. Comparison between numerical solutions and analytical solutions (n = 9).

Table 3 lists the maximum error (ME) between the analytical solution and the numerical solution, which is calculated using Equation (33). It can be seen in Table 3 that the convergence at $n = 1$ is rather slow, with an accuracy of 11% reached after four terms. The error after summing the first nine terms is less than 5%. The calculated temperature at $d_5 = 0.2$ m converged to within 3% (1.3 °C) of the numerical value after summing only the first five terms of the series. In general, as term n increases, the convergence of the series summation becomes much more rapid. This is readily explained by the exponential term of Equation (25), which depends on the power of $-\alpha\lambda_n^2(t - t_1)$. Clearly this term decreases exponentially with increasing time and increasing eigenvalues, thereby causing the rapid convergence of the series.

$$ME = \begin{cases} \max(T(n,x,t) - T_{\mathrm{m}}(x,t)/T_{\mathrm{m}}(x,t) \times 100\%)\ T(x,t) > T_{\mathrm{m}}(x,t) \\ \min(T(n,x,t) - T_{\mathrm{m}}(x,t)/T_{\mathrm{m}}(x,t) \times 100\%)\ T(x,t) < T_{\mathrm{m}}(x,t) \end{cases}, n = 1,2,3\cdots 15;\ 0 < x = d < 0.3;\ 6 < t < 19 \quad (33)$$

where $T(n, x, t)$ is the analytical solution and $T_{\mathrm{m}}(x, t)$ is the numerical solution.

Table 3. Error statistical results.

N	Maximum Error						
	$d_1 = 0$ m	$d_2 = 0.05$ m	$d_3 = 0.1$ m	$d_4 = 0.15$ m	$d_5 = 0.2$ m	$d_6 = 0.25$ m	$d_7 = 0.3$ m
1	30.46%	16.99%	4.60%	5.59%	11.25%	11.80%	6.96%
2	20.24%	3.6%	6.55%	6.90%	4.92%	6.91%	10.17%
3	13.79%	2.95%	2.26%	4.11%	2.03%	2.41%	2.66%
4	10.73%	3.37%	3.69%	4.24%	2.95%	3.83%	4.54%
5	8.56%	2.63%	3.01%	2.59%	1.96%	2.21%	2.55%
6	7.26%	3.09%	3.31%	2.61%	2.35%	2.42%	2.96%
7	6.26%	2.31%	1.84%	1.27%	1.46%	2.17%	2.42%
8	5.61%	2.45%	1.97%	1.32%	1.54%	1.95%	2.54%
9	4.34%	1.93%	1.64%	1.15%	1.22%	1.52%	2.1%

The calculation error at the position $d_7 = 0.3$ m was used for error statistics, which are shown in the boxplot of Figure 15. In Figure 15, the maximum error line clearly fluctuates over a range of $n = 1$ to $n = 8$. With the increase in terms n, the fluctuation range of the line is relatively small. When n is an odd number, it is beneficial for reducing the error. However, we note that every other term, namely, the even n terms, are adverse for convergence. This is readily explained by the positive or negative value of the eigenfunction $X(\beta_n,x)$ in the analytical equation, which depends on the quadrant of the eigenvalue β_n. For the even n terms, β_n is in the third quadrant, which produces the positive eigenfunction. β_n changes in the first and third quadrants in turn with the change in the parity of n. Thus, the maximum error line clearly fluctuates in first eight terms. Furthermore, the error fluctuation is smaller with greater increases in the terms. This indicates that the errors tend to be stabilized quickly and change linearly.

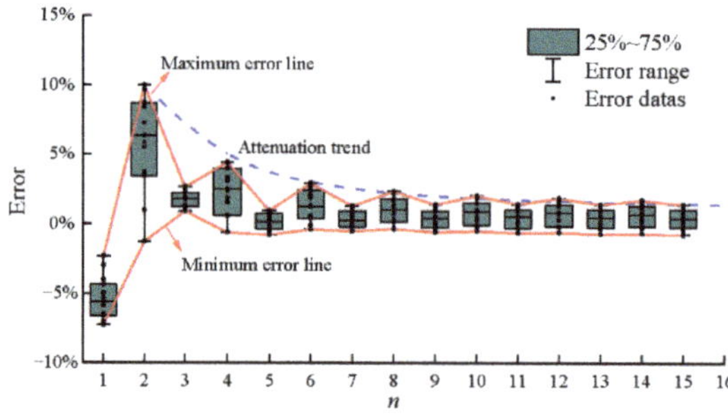

Figure 15. Error results ($d_7 = 0.3$ m).

4.2. Comparison with Empirical Formula and Experimental Data

The empirical result of the slab temperature was calculated to present the improvement of the analytical solution. The empirical formula in the one-dimensional temperature model was developed without considering the convection heat transfer at the shading surface. Thereby, the slab temperature is directly related to the equivalent radiation temperature at the radiation surface. The empirical result is calculated by modifying the amplitude and phase of the equivalent radiation temperature. The empirical formula is as follows [15]:

$$T_f(x,t) = f_1(t)' + \gamma\left[f_1(t-\delta) - f_1(t)'\right],\ t_1 < t < t_2,\ 0 < x < d \tag{34}$$

$$\gamma = \frac{\exp\left(-\sqrt{\pi/24\alpha}x\right)}{\sqrt{1+\sqrt{\pi k^2/6\alpha {h_u}^2}+\pi k^2/12\alpha {h_u}^2}} \tag{35}$$

$$\delta = -\tan^{-1}\frac{1}{1+\sqrt{24\alpha {h_u}^2/\pi k^2}} - \sqrt{\pi/24\alpha}x \tag{36}$$

where $f_1(t)'$ is the daily average value of the equivalent radiation temperature $f_1(t)$, γ is the amplitude correction factor, and δ is the phase correction factor.

Figure 16 shows the analytical, empirical, and measured temperatures at different measuring points on 23 July 2020. It can be observed that the analytical result ($T(x, t)$) matches well with the measured temperature (H2~H5). The errors between the calculated temperature ($T(x, t)$) and the measured temperature are very small during the three hours before sunset and after sunrise. However, the error becomes larger when the solar radiation is stronger at 10:00~15:00. The maximum error rate of the analytical solution occurs at the measuring point H5, at 5.5% (2.2 °C) in Figure 16a. Only while considering the influence of the top surface on the temperature field, the maximum error rate between the empirical result ($T_f(0.02, 12)$) and the measured temperature (H5) is 12.8% (5.5 °C) in Figure 16a. This can be explained by the fact that the daily average of total heat transfer coefficient deviates too much from that coefficient, changing with time. By contrast, the analytical method with appropriate assumptions can provide a higher accuracy method for the temperature of the slab track.

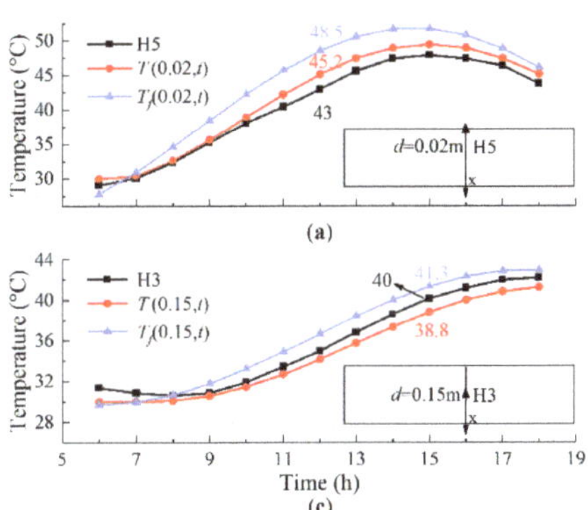

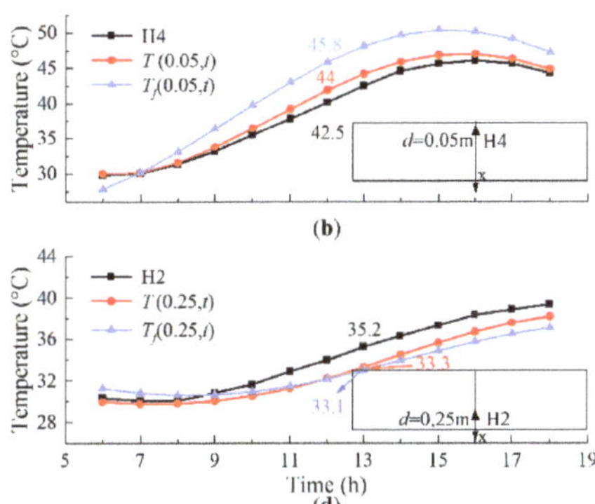

Figure 16. Results of the analytical, empirical, and measured temperatures on 23 July 2020. (**a**) H5; (**b**) H4; (**c**) H3; (**d**) H2.

4.3. Decomposition of the Concrete Slab Surface Temperature

The temperature components of the concrete slab are a function of variables i, x, and t, which can be calculated at an arbitrary point x and at any time t using Equation (25). Figure 17 shows the time history curve of temperature components at the top slab surface. The curve of component T_{BC1} is generated through heat exchange with the solar radiation and air temperature at the top surface (BC1), and the curve of component T_{BC2} is generated by convection with the air temperature $f_2(t)$ at the bottom surface (BC2).

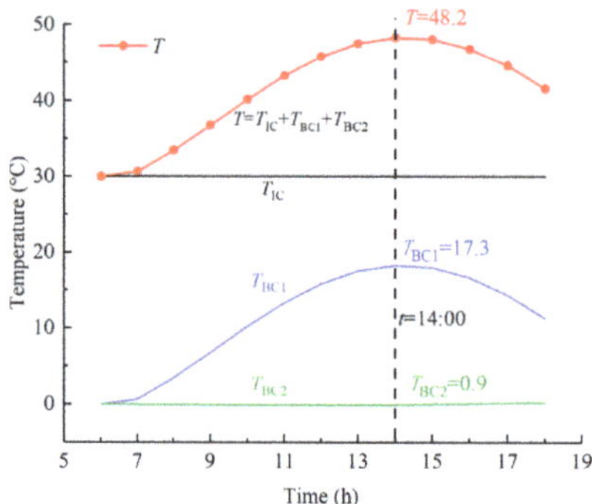

Figure 17. Time history curves of temperature components.

In the solar radiation duration, the change trend of the surface temperature T is consistent with that of the component T_{BC1}. When time is t = 14, the calculated temperature of the top surface is 48.2 °C. Then, by substituting the temperature value into Equations (29) and (25), the instantaneous net heat flux at the top surface is 253 W·m^{-2}, and the increment of the top surface temperature contributed by the component T_{BC1} is 17.3 °C, as shown in Figure 17. In addition, the bottom surface of the slab belongs to the shading surface, where there is only convection heat transfer and a small amount of radiation heat transfer. The limited heat flow (t = 14, q = 34 W·m^{-2}) contributes little to the increment in the top surface temperature. The result from the curve of T_{BC2} shows that the average contribution of the component T_{BC2} is only about 0.5 °C in the daytime.

Figure 18 shows the hourly percentages of the components T_{BC1} and T_{BC2} in the heating process of the top surface. According to statistics, the contribution percentage of the bottom surface temperature accounts on average for only 5% in the increment of the top surface temperature. The result shows that the temperature change of the top surface is mainly affected by the temperature components T_{BC1}. For instance, when the top surface temperature increases by 20 °C, the bottom surface temperature only contributes about 1 °C to the increment in the top surface temperature.

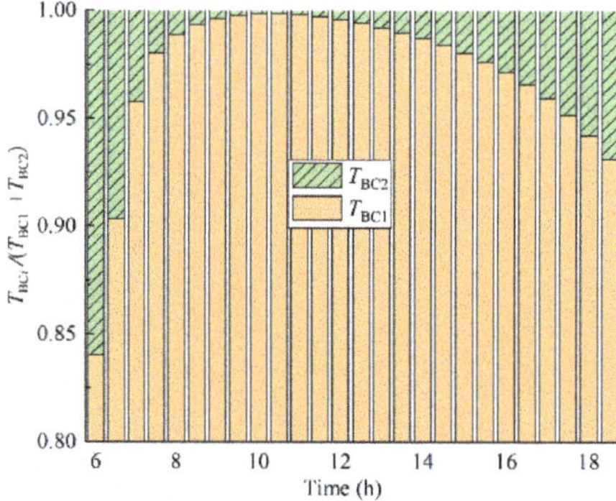

Figure 18. Hourly percentages of components T_{BC1} and T_{BC2} at the top surface.

The comparison of the change characteristics of the temperature components T_{BC1} and T_{BC2} shows that the temperature variation in the top surface is related to the component

T_{BC1} and is independent of component T_{BC2}. Using the decomposition method, we can also obtain the similar change trend of temperature components at different depths.

4.4. Decomposition of the Temperature Distribution of the Concrete Slab

Figure 19 illustrates the temperature distribution of the temperature component T_{BC1} through the depth of the slab at different times. The temperature profile of the slab is approximately a straight line at 6:00. With the more heat absorbed by the surface BC1, the profile becomes nonlinear over the solar time. The largest temperature variation occurs on the top surface and reaches a maximum of 17.6 °C at 15:00, while the bottom surface temperature is 3.7 °C. After that time, the top surface temperature begins to drop. However, due to the low thermal conductivity of concrete, the bottom surface temperature steadily increases and reaches 5.2 °C at 18:00. The maximum positive temperature difference is 14.7 °C at 15:00 under the thermal action of solar radiation.

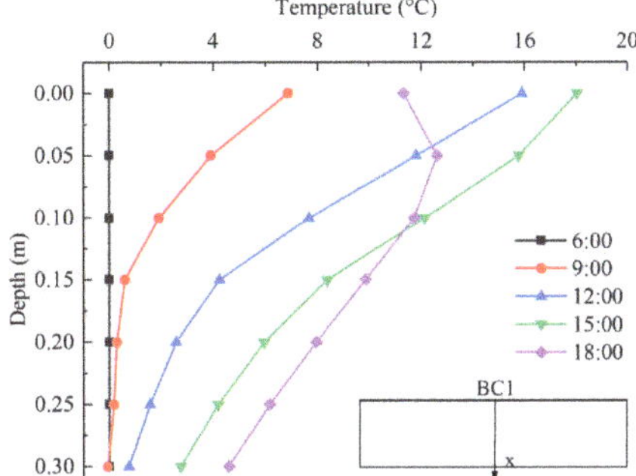

Figure 19. Temperature distributions of component T_{BC1} at different times.

Figure 20 shows the temperature distribution of component T_{BC2} at different times. Only with the convective and reflected radiation heat transfer at the bottom surface (BC2), the temperature difference varies slightly in the daytime. In addition, one can observe that the negative temperature difference occurs at 6:00~9:00. Due to the weak solar radiation during early sunrise hours, the surface BC2 remains heat loss, and the temperature at a depth of 0.1~0.3 m drops. With the limited amount of heat loss, the negative temperature difference is very small, at just −0.5 °C. The temperature difference values become gradually positive with an increment of the combined action of the convection and reflected radiation on the surface BC2. The maximum positive temperature difference is 2.2 °C at 18:00 under the thermal action of air temperature.

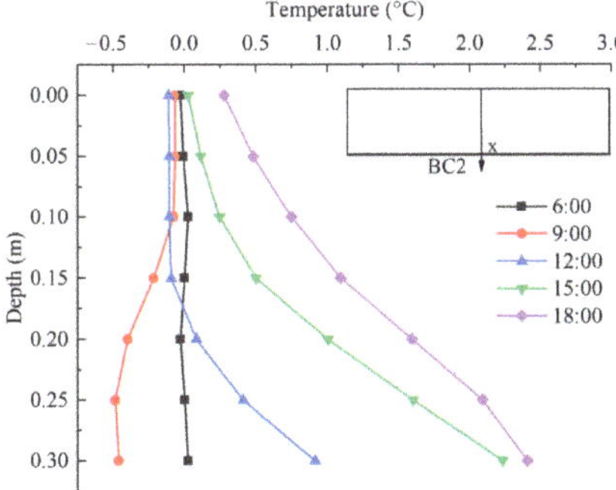

Figure 20. Temperature distributions of component T_{BC2} at different times.

The comparison of the temperature distribution of the temperature component shows that the solar radiation has a great influence on the nonlinear temperature distribution of the concrete slab. Due to the different amount of the heat exchange at the boundary surface, the temperature profiles of the temperature components T_{BC1} and T_{BC2} are not symmetric along the depth of the concrete slab.

5. Conclusions

With a special focus on the influence of environmental conditions on the temperature distribution of slab tracks, an analytical method is proposed to calculate and decompose the temperature distribution of slab tracks. A one-dimensional temperature model was solved using the integral transformation method. An experimental program on a concrete slab track structure was conducted to validate the reliability and accuracy of the developed methodology in this study. Based on the results, the following conclusions are drawn:

(1) The proposed method is convenient to predict the real-time temperature distribution of concrete slab tracks using meteorological parameters, and shows a high accuracy and a rapid convergence speed.
(2) The relationship between the temperature of the slab track and meteorological parameters is established through the proposed analytical solution. Based on the temperature decomposition method, the temperature distribution of slab tracks affected by solar radiation and atmospheric temperature can be calculated separately.
(3) A method for dealing with meteorological parameters is proposed. The combined action of solar radiation and atmospheric temperature on the boundary surface is considered as a fluid medium, which is the expression of a cosine function.
(4) Solar radiation is the main reason for the nonlinear temperature distribution in slab tracks during the daytime. By contrast, the convection heat transfer caused by air has little effect, and the temperature change in the slab surface resulting from the atmospheric temperature accounts for only 5% in the hot weather condition.

In this study, an analytical method of the one-dimensional thermal field in a slab track is proposed. To improve the applicability of the method, a multidimensional temperature model of slab tracks in different environmental conditions need to be investigated. In future works, the transverse temperature distribution can be considered in temperature measurement, and the effect of extreme weather conditions on slab tracks can be further studied.

Author Contributions: Conceptualization, G.D.; methodology, G.D. and Q.Z.; software, Q.Z.; validation, Q.Z. and Y.T.; formal analysis, Q.Z. and Y.T.; writing—original draft preparation, Q.Z.; writing—review and editing, G.D. and Y.T.; project administration, Q.Z. and Y.T. All authors have read and agreed to the published version of the manuscript.

Funding: The research was funded by the Science and Technology Research and Development Key Program of the China Railway Corporation, grant number 2017G006-N and the Fundamental Research Funds for the Central Universities of Central South University, grant number 2020zzts151.

Institutional Review Board Statement: Not applicable.

Informed Consent Statement: Not applicable.

Data Availability Statement: The data that support the findings of this study are available upon request from the authors.

Acknowledgments: The authors would like to thank the Science and Technology Research and Development Key Program of the China Railway Corporation, the Fundamental Research Funds for the Central Universities of Central South University, and the Southeast Coastal Railway Fujian Co., Ltd.

Conflicts of Interest: The authors declare no conflict of interest.

References

1. Matias, S.R.; Ferreira, P.A. Railway slab track systems: Review and research potential. *Struct. Infrastruct. Eng.* **2020**, *16*, 1635–1653. [CrossRef]
2. Yang, R.S.; Li, J.L.; Kang, W.X.; Liu, X.Y.; Cao, S.H. Temperature characteristics analysis of the ballastless track under continuous hot weather. *J. Transp. Eng. Part A Syst.* **2017**, *143*, 04017048. [CrossRef]
3. Cai, X.P.; Luo, B.C.; Zhong, Y.L.; Zhang, Y.R.; Hou, B.W. Arching mechanism of the slab joints in CRTS II slab track under high temperature conditions. *Eng. Fail. Anal.* **2019**, *98*, 95–108. [CrossRef]
4. Zhong, Y.; Gao, L.; Zhang, Y. Effect of daily changing temperature on the curling behavior and interface stress of slab track in construction stage. *Constr. Build. Mater.* **2018**, *185*, 638–647. [CrossRef]
5. Robertson, I.; Masson, C.; Sedran, T.; Barresi, F.; Caillau, J.; Keseljevic, C.; Vanzenberg, J.M. Advantages of a new ballastless track form. *Constr. Build. Mater.* **2015**, *92*, 16–22. [CrossRef]
6. Li, Y.; Chen, J.; Jiang, Z.; Cheng, G.; Shi, X. Thermal performance of the solar reflective fluorocarbon coating and its effects on the mechanical behavior of the ballastless track. *Constr. Build. Mater.* **2021**, *291*, 123260.
7. Jiang, H.; Zhang, J.; Zhou, F.; Wang, Y. Optimization of PCM coating and its influence on the temperature field of CRTS II ballastless track slab. *Constr. Build. Mater.* **2020**, *236*, 117498. [CrossRef]
8. Liu, D.; Zhang, W.; Tang, Y.; Jian, Y.; Gong, C.; Qiu, F. Evaluation of the uniformity of protective coatings on concrete structure surfaces based on cluster analysis. *Sensors* **2021**, *21*, 5652.
9. Zhang, Y.; Zhou, L.; Mahunon, A.D.; Zhang, G.; Peng, X.; Zhao, L.; Yuan, Y. Mechanical performance of a ballastless track system for the railway bridges of high-speed lines: Experimental and numerical study under thermal loading. *Materials* **2021**, *14*, 2876.
10. Zhou, R.; Zhu, X.; Ren, W.X.; Zhou, Z.X.; Yao, G.W.; Ma, C.; Du, Y.L. Thermal evolution of CRTS II slab track under various environmental temperatures: Experimental study. *Constr. Build. Mater.* **2022**, *326*, 126699. [CrossRef]
11. Zhou, L.; Wei, T.; Zhang, G.; Zhang, Y.; Mahunon, A.D.G.; Zhao, L.; Guo, W. Experimental study of the influence of extremely repeated thermal loading on a ballastless slab track-bridge structure. *Appl. Sci.* **2020**, *2*, 461. [CrossRef]
12. Zhou, L.; Yuan, Y.H.; Zhao, L.; Mahunon, A.D.G.; Zhou, L.; Hou, W. Laboratory investigation of the temperature-dependent mechanical properties of a CRTS-II ballastless track-bridge structural system in summer. *Appl. Sci.* **2020**, *10*, 5504. [CrossRef]
13. Zhao, L.; Zhou, L.Y.; Zhang, G.C.; Wei, T.Y.; Mahunon, A.D.G.; Jiang, L.Q.; Zhang, Y.Y. Experimental study of the temperature distribution in CRTS II ballastless tracks on a high-speed railway bridge. *Appl. Sci.* **2020**, *10*, 1980. [CrossRef]
14. Ou, Z.; Li, F. Analysis and prediction of the temperature field based on in-situ measured temperature for CRTS-II ballastless track. *Energy Procedia* **2014**, *61*, 1290–1293.
15. Liu, X.; Li, J.; Kang, W.; Liu, X.; Yang, R. Simplified calculation of temperature in concrete slabs of ballastless track and influence extreme weather. *J. Southwest Jiaotong Univ.* **2017**, *52*, 1037–1045. (In Chinese)
16. Zeng, R.; Zhang, J.; Hu, W.; Chen, J. Study on the evolution characteristics of temperature field in CRTS-III ballastless track slab. *Railw. Stand. Des.* **2022**, *66*, 7. (In Chinese)
17. Riding, K.A.; Poole, J.L.; Schindler, A.K.; Juenger, M.C.G.; Folliard, K.J. Evaluation of temperature prediction methods for mass concrete member. *ACI Mater. J.* **2006**, *103*, 357–365.
18. Chen, J.; Wang, H.; Zhu, H. Analytical approach for evaluating temperature field of thermal modified asphalt pavement and urban heat island effect. *Appl. Therm. Eng.* **2017**, *113*, 739–748. [CrossRef]
19. Chong, W.; Tramontini, R.; Specht, L.P. Application of the Laplace Transform and Its Numerical Inversion to Temperature Profile of a Two-Layer Pavement under Site Conditions. *Numer. Heat Trans. Part A Appl.* **2009**, *55*, 1004–1018. [CrossRef]
20. Wang, D. Analytical approach to predict temperature profile in a multilayered pavement system based on measured surface temperature data. *J. Transport. Eng.* **2012**, *138*, 674–679. [CrossRef]
21. Yang, Y.; Lu, H.; Tan, X.; Chai, H.K.; Wang, R.; Zhang, Y. Fundamental mode shape estimation and element stiffness evaluation of girder bridges by using passing tractor-trailers. *Mech. Syst. Signal Processing* **2022**, *169*, 108746. [CrossRef]
22. Yang, Y.; Zhang, Y.; Tan, X. Review on Vibration-Based Structural Health Monitoring Techniques and Technical Codes. *Symmetry* **2021**, *13*, 1998. [CrossRef]
23. Yang, Y.; Ling, Y.; Tan, X.; Wang, S.; Wang, R. Damage identification of frame structure based on approximate Metropolis–Hastings algorithm and probability density evolution method. *J. Struct. Stab. Dyn.* **2022**, *22*, 2240014. [CrossRef]
24. Riding, K.A.; Poole, J.L.; Schindler, A.K.; Juenger, M.C.G.; Folliard, K.J. Temperature boundary condition models for concrete bridge members. *Aci. Mater. J.* **2007**, *104*, 379–387.
25. Lawson, L.; Ryan, K.L.; Buckle, I.G. Bridge Temperature Profiles Revisited: Thermal analyses based on recent meteorological data from Nevada. *J. Bridge Eng.* **2020**, *25*, 04019124. [CrossRef]
26. Liu, D.; Chen, H.; Tang, Y.; Liu, C.; Cao, M.; Gong, C.; Jiang, S. Slope Micrometeorological Analysis and Prediction Based on an ARIMA Model and Data-Fitting System. *Sensors* **2022**, *22*, 1214. [CrossRef]
27. Hahn, D.W.; ÖziŞik, M.N. *Heat Conduction*, 3rd ed.; Wiley: Hoboken, NJ, USA, 2012; pp. 32–48.
28. Emerson, M. *The Calculation of the Distribution of Temperature in Bridges*; TRRL Rep. No. LR 561; Transport and Road Research Laboratory: Wokingham, UK, 1973.
29. Wang, A.; Zhang, Z.; Lei, X.; Xia, Y.; Sun, L. All-Weather thermal simulation methods for concrete maglev bridge based on structural and meteorological monitoring data. *Sensors* **2021**, *21*, 5789. [CrossRef]

30. Lei, X.; Fan, X.T.; Jiang, H.W.; Zhu, K.N.; Zhan, H.Y. Temperature field boundary conditions and lateral temperature gradient effect on a PC box-girder bridge based on real-time solar radiation and spatial temperature monitoring. *Sensors* **2020**, *20*, 5261. [CrossRef]
31. Kehlbeck, F.; Liu, X. *Effect of Solar Radiation on Bridge Structure*; Liu, X., Ed.; Chinese Railway Publishing Company: Beijing, China, 1981.

Review

A Review on Rail Defect Detection Systems Based on Wireless Sensors

Yuliang Zhao [1,*], Zhiqiang Liu [1], Dong Yi [1], Xiaodong Yu [1], Xiaopeng Sha [1], Lianjiang Li [1], Hui Sun [2], Zhikun Zhan [1,3] and Wen Jung Li [2,*]

1 School of Control Engineering, Northeastern University at Qinhuangdao, Qinhuangdao 066004, China
2 Department of Mechanical Engineering, City University of Hong Kong, Hong Kong SAR, China
3 School of Electrical Engineering, Yanshan University at Qinhuangdao, Qinhuangdao 066104, China
* Correspondence: zhaoyuliang@neuq.edu.cn (Y.Z.); wenjli@cityu.edu.hk (W.J.L.)

Abstract: Small defects on the rails develop fast under the continuous load of passing trains, and this may lead to train derailment and other disasters. In recent years, many types of wireless sensor systems have been developed for rail defect detection. However, there has been a lack of comprehensive reviews on the working principles, functions, and trade-offs of these wireless sensor systems. Therefore, we provide in this paper a systematic review of recent studies on wireless sensor-based rail defect detection systems from three different perspectives: sensing principles, wireless networks, and power supply. We analyzed and compared six sensing methods to discuss their detection accuracy, detectable types of defects, and their detection efficiency. For wireless networks, we analyzed and compared their application scenarios, the advantages and disadvantages of different network topologies, and the capabilities of different transmission media. From the perspective of power supply, we analyzed and compared different power supply modules in terms of installation and energy harvesting methods, and the amount of energy they can supply. Finally, we offered three suggestions that may inspire the future development of wireless sensor-based rail defect detection systems.

Keywords: rail defects detection; wireless sensing system; railway sensors

Citation: Zhao, Y.; Liu, Z.; Yi, D.; Yu, X.; Sha, X.; Li, L.; Sun, H.; Zhan, Z.; Li, W.J. A Review on Rail Defect Detection Systems Based on Wireless Sensors. *Sensors* **2022**, *22*, 6409. https://doi.org/10.3390/s22176409

Academic Editors: Phong B. Dao, Tadeusz Uhl, Liang Yu, Lei Qiu and Minh-Quy Le

Received: 8 June 2022
Accepted: 19 August 2022
Published: 25 August 2022

1. Introduction

During rail service, defects are produced due to material degradation, wheel–rail stress, thermal stress, residual stress [1], and other reasons. If small defects are not discovered and repaired in time, they will be aggravated [2] and in turn cause rail breakage [3] and even serious accidents such as train derailment [4]. Main railway track defects include surface defects, inner defects [5], and component (fastener) defects [6]. With the continuous increase in the railway transportation speed, density, and load [7] there has also been an increase in accidents caused by rail defects. For example, in a mere 10 days in August 2017, four train derailments occurred in India, causing very serious losses [8]. Therefore, the detection of rail defects and the life cycle management [9] of rails has become extremely important. There are also research works focusing on the other part of the railways. For example, Kaewunruen et al. [10] carried out a related study and investigation on the stress of railway sleepers. Setsobhonkul et al. [11] assessed the life cycle of railway bridge transitions exposed to extreme climatic events. Melo et al. showed that the interaction of defects between different parts of the rail can cause different severe consequences, and investigated the related methods for predicting the deterioration of the rail [12].

In the early days, the detection of railway track defects mainly relied on manual detection. Manual inspection is performed by well-trained inspectors who regularly walk along the railway line to identify rail defects. However, manual inspection is inefficient and costly and sometimes even threatens the safety of inspectors [13]. Since the world's first railway ultrasonic inspection vehicle was put into use in 1959, manual inspection methods have been gradually replaced by large inspection vehicles. For example, in 2004, E. Deutschl et al. [14] designed a vision-based rail surface defect detection system which can automatically detect rail defects. Large-scale rail defect detection vehicles mainly use defects detection devices on the train bogies to detect rail defects [15]. The inspections are conducted once every few months but are not for real-time rail status monitoring [16]. This method occupies the rail, resulting in the inability to transport passengers or cargo during the inspection. Furthermore, neither manual detection nor defect detection vehicles can detect rail damage at the first time when an accident (e.g., derailment and rail breakage) occurs. This means that it is not possible to promptly locate the injury and fix it before damage is caused, and this cannot meet the daily inspection needs of modern railways, especially for high-speed railways [17]. M. Vohra et al. [18] invented a robot-based infrared sensor rail defects detection system. However, the robot-based detection systems are still unable to achieve real-time detection since they usually have larger locomotion ranges than detection ranges. The development of wireless communication and self-organizing networks (GSM, ZigBee ad hoc networks, etc.) enables the information collected by sensing devices to be transmitted to terminal in real time and with good realiabitly [19]. This makes the wireless sensor network a perfect option for the real-time detection of rail defects. E. Aboelela et al. [20] established a wireless sensor network model for railway safety, which laid the foundation for the application of wireless networks in railway track detection. After that, a lot of works [2,21–24] on the wireless sensor-based rail defect detection systems (WSRDDS) emerged to research the availability of using different sensing methods, wireless commnication, power supply, and data processing in this area. However, there is no systmatic review covering all related aspects of the WSRDDS. From the perspective of data acquisition of the WSRDDS, we focus on the key elements, inlcluding sensing methods, wireless communication, and power supply, in this paper to give an overview.

For a WSRDDS, the sensor is the core and should be to be considered first. Different defects require different types of sensors for detecting. To have a long lifetime, a regenerative power supply is required for the wireless sensor. Furthermore, a reliable sensor network architecture should be designed for the data transmission of sensor readings. The main components of the whole system are shown in Figure 1. With the rapid development of high-speed railways, our requirements for the maintenance of rail infrastructure status are increasingly demanded. How to detect rail defects comprehensively, reliably, and in real-time has become extremely important.

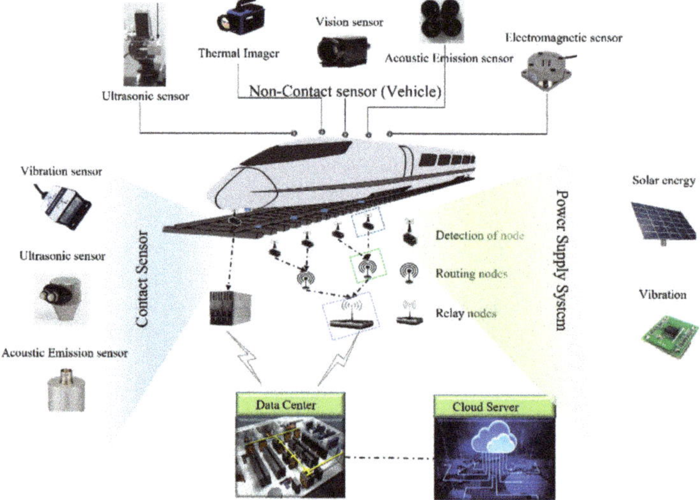

Figure 1. A wireless sensor-based rail defect detection system.

2. Sensing Method

Since the first rail ultrasonic inspection vehicle was put into use in 1953, various inspection methods have gradually been proposed for rail defect detection [25]. These detection technologies can be summarized as contact detection and non-contact detection based on whether there is physical contact between sensor and rail. Contact sensor detection technology includes: vibration [26], ultrasonic [27,28], and acoustic emission technology [29]. Non-contact sensor detection technology includes: ultrasonic [30], thermal imaging [31], vision [32], electromagnetic wave diffusion [33], etc. In the contact detection methods, the detection sensors are usually installed on the abdomen of the rail [34]. For the non-contact detection methods, the detection sensors are often installed on a large rail detection vehicle [35] or a smart car [36] to detect rail defects. The selection of sensors is highly dependent on the defect types. For example, visual inspection is suitable for detecting rail surface defects, ultrasonic and electromagnetic wave diffusion are suitable for detecting internal rail defects, and thermal imaging is suitable for detecting rail subsurface defects. This section selects a variety of typical sensing methods for the introduction of the sensing mechanism and detectable defect types of these methods. At the end of this section, summarizes the main differences between these methods are summarized.

2.1. Vibration

When the train passes the railroad track, it causes vibration of the railroad track [37,38]. There is a significant difference in vibration signals between healthy rails and defective rails. Defective rails have flatter peaks and troughs in the vibration acceleration signal compared to healthy rails [39]. Q. Wei et al. [16] showed that the instantaneous energy distribution is an effective defect feature. For example, among three defects (rail corrugation, ail head sag, and rail surface stripping) the intra-class cross-correlation coefficient of the instantaneous energy distribution is greater than 0.7, while the inter-class ross-correlation coefficient is below 0.45. Therefore, the vibration signal characteristics of different types of rail defects can be extracted through multiple experiments. Finally, classification algorithms can be used to identify and classify rail defects based on these features. M. Sun et al. [40] applied the sequential backward selection (SBS) method to select important feature parameters, and the support vector machine method to recognize and classify the rail defects. This study compares the accuracy of classification before and after using the SBS method and proves that optimizing the parameter set can improve the accuracy of the classification.

MEMS accelerometers are widely used in rail detection due to their small size, low price, and high accuracy [41]. M. David et al. [42] compared MEMS sensors with geophones in 2016. The results prove that MEMS sensors are suitable for track defect detection. Z. Zhan et al. [5] developed a wireless sensor system for rail fastener detection, which can reliably identify fasteners with a looseness coefficient greater than 60%. In addition, strain gauges can also be used to detect missing or broken fasteners. J. J. Zhao et al. [43] demonstrated a linear relationship between the strain voltage and tightness of fasteners, finding that the tighter the fastener, the smaller the strain voltage. We summarized the existing techniques in the literature shown in Table 1.

Table 1. Comparison of vibration testing methods.

Methods	Types of Detected Defects	Algorithm	Results	Comments
MEMS accelerometers	Rail fastener [5]	Finite element method	Reliable identification of fasteners with a looseness factor greater than 60%	Small size, low price, high accuracy
	/ [42]	The high- and low-pass filter	This study proves that MEMS sensors are suitable for rail defect detection.	
	Rail head sag, rail surface stripping, height joint [40].	Peak-finding algorithm	The accuracy rate of the classification of rail defect types can reach 93.8%.	
Strain gauge	Rail fastener [43]	Sequential backward selection	Demonstrated a linear relationship between strain voltage and fastener tightness.	Small size, low price, low accuracy
	Rail fastener [44]	Support vector machines	Demonstrated a linear relationship between strain voltage and fastener tightness.	

2.2. Acoustic Emission

Different from other detection methods, the acoustic emission (AE) method is suited to investigate the dynamic behavior of materials and structures [7]. The dynamic expansion process of rail defects releases transient elastic waves. The acoustic emission (AE) sensor method works based on this phenomenon [45] (as shown in Figure 2). It is more sensitive to the forming and expanding of defect but less influenced by the structural geometry. Furthermore, this method can achieve a detection range as far as 30 m [46]. This method can estimate the dynamic characteristics of defects and is an ideal choice for online continuous monitoring [43]. This method can detect railhead defects, inner defects, welding defects, and surface defects.

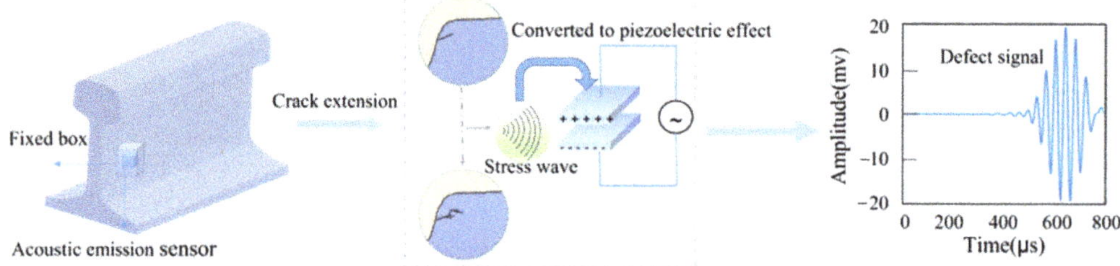

Figure 2. AE sensor detection.

H. Jian et al. [47] demonstrated that the acoustic emission frequencies of defective rails are mainly located in the 100–150 KHz and 150–200 KHz frequency bands, and a small part is located in the 380–430 kHz frequency band. In 2013, A. G. Kostryzhev et al. [48] found that the spectral characteristics of the acoustic emission signal depend on the extended mode of the defect. That is, long duration and low-frequency signals come from ductile fractures; short duration and high-frequency signals come from brittle fractures. In 2015, the K.S.C. Kuang team of the Department of Civil and Environmental Engineering of the National University of Singapore [46] found that the railhead side is the best location for inspection. At the same time, the research team used the wavelet transform-based modal analysis location (WTMAL) method to locate defects. The error is less than ±0.30 m in a high-noise environment, and the average working range reaches 30.0 m. However, the defect acoustic emission signal is often interfered with by strong noise. To solve this problem, X. Zhang et al. [29] presented a joint optimization method based on long short-term memory (LSTM) network and k-means clustering to cluster noise signals, and the results showed that most of the noise signals can be reduced. To suppress the influence of noise and ensure proper time resolution, the research team further studied the characteristic frequency of the time window for defect detection [43]. This research has greatly promoted the application of AE sensors in the detection of rail defects. Based on the AE sensor, the dynamic expansion process of the inner defects of the rail can be detected in real-time. However, this method is susceptible to interference from external sound waves (trains and nature). We summarized the existing techniques in the literature shown in Table 2.

Table 2. Comparison of acoustic emission methods.

Methods	Types of Detected Defects	Algorithm	Results	Comments
AE	Rail-head defects [46]	Hilbert transform Wavelet transform	The error of the location of rail defects is less than 0.3 m. Detection distance can reach 30 m.	Long detection distance
AE	/[49]	Signal adapted wavelet in the frame of a two-band analysis/synthesis system	The wavelet designed by the proposed method has superior performance in expressing the defect AE signal, and can outperform the most suitable existing wavelet.	The designed wavelet shows good robustness against noise, which has profound meaning for rail defect detection in practical applications.
AE	Rail fatigue defect [48]	Single-hit waveform and power spectrum analysis	High duration, low frequency signals result from ductile fractures. Low duration, high frequency signals result from brittle fractures.	It is demonstrated that the AE signal associated with defect propagation depends on the fracture mode.
AE	Rail defect, small bearing defect, and worse bearing defect [47]	Cepstrum analysis	This study verifies that AE signals can detect bearing/rail defects.	

2.3. *Ultrasonic*

The ultrasonic sensor detects the rail defect by analyzing the sound waves reflected from the rail [50]. The prerequisite for the use of ultrasonic sensors for detection is that sound waves must be excited inside the rail. The excitation can be realized by either piezoelectric elements (as shown in Figure 3a) or by lasers (as shown in Figure 3b) and so on. This method has a high detection rate for the inside of the rail (particularly in the railhead and waist) [7]. In this subsection, we summarize the existing techniques in the literature, shown in Table 3.

The focus angle and focus depth of ordinary ultrasound probes are fixed, so the coverage rate of this method on the guide rail is relatively low [30,51]. To overcome the shortcomings of ordinary ultrasonic probes, Zhang et al. [51] proposed a high-speed phased array ultrasonic testing technology. This technology can generate multi-angle beams and receive defect echo signals from all channels, which greatly improves the detection speed and detection range. C. Ling et al. [52] combined traditional probes with phased array probes to detect defects on 60 kg/m rails, and the detection accuracy can reach 6 mm. Acoustic guided waves can cover the entire rail cross section and have a longer propagation distance [53]. Therefore, the efficiency of ultrasonic guided wave detection of rail defects is much greater than the ultrasonic waves. However, different guided wave modes have different sensitivities to defects in different parts of the rail, which greatly increases the complexity of detection [54]. H. Shi et al. [55] studied the mode of the guided wave propagating in the rail at a frequency of 35 KHZ. The study showed that there are a total of 20 guided wave modes propagating in the rail at this frequency. After experimental verification, it proves that modes 7, 3, and 1 are suitable for detecting defects at the rail head, waist, and seat, respectively. Kaewunruen et al. [56] used ultrasonic measurement technology to achieve accurate drawing of the three-dimensional profiles of the deep-sinking defect of the rail. This has important implications for on-site inspections.

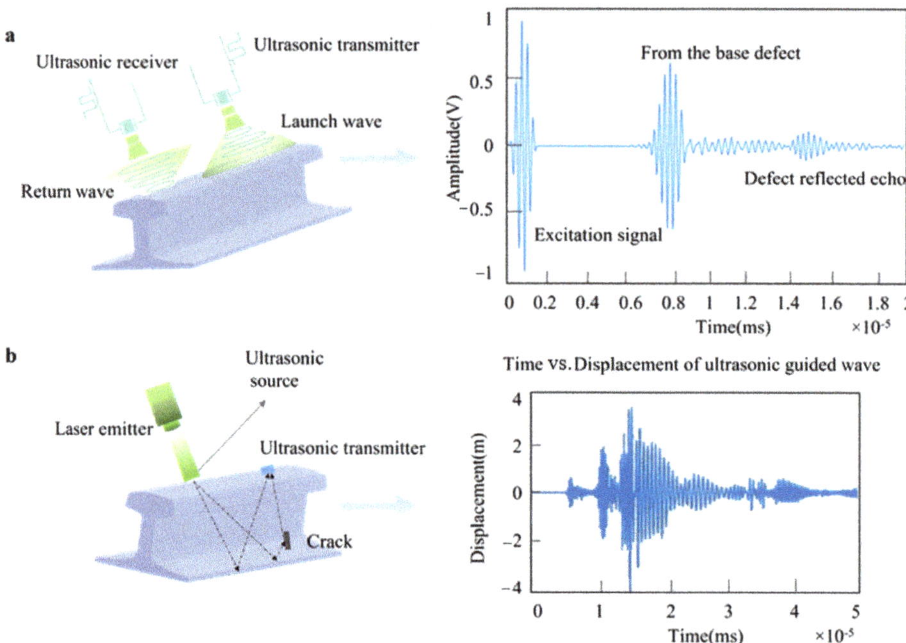

Figure 3. (**a**) Piezoelectric transducer excitation. (**b**) Laser excitation.

Piezoelectric-based ultrasonic technology is applied in almost all the available studies, but the application of this technology is greatly limited by a variety of shortcomings, such as dependence on acoustic coupling agents and requirements for surface pretreatment of the measured object [57]. The laser-based ultrasonic detection technology can excite various modes of ultrasonic waves inside the rail [27,58]. The method has higher accuracy than conventional ultrasonic in the non-destructive detection of small defects in railway tracks.

The ultrasonic sensor emits ultrasound waves that have strong penetration capability and can detect defects in the head and waist of the rail. Under laser excitation, it can further detect defects near the surface and on the foot of the rail. However, when the detection frequency increases, it may easily lose a lot of defect information, and its accuracy is low in detecting very small cracks. We summarized the existing techniques in the literature shown in Table 3.

Table 3. Comparison of ultrasonic testing methods.

Methods			Algorithm or Simulation	Types of Detectable Defects	Results	Summarize
Detection Method	Ordinary ultrasound	Multi-angle ultrasonic probe [59]	PCA and LSSVM	Different types of defects in rail head, rail waist and rail foot	Classification recognition accuracy: 92%. Identify seven types of rail defects.	Ordinary ultrasonic waves are usually single-modal at low frequencies, and cannot achieve high-sensitivity omnidirectional detection of all parts of the rail (track surface, underground, and interior).
		Combination of wheeled ultrasonic probes [60]	LSTM-based deep learning model		Average f1-score: 95.5%. Maximum detection speed: 22 m/s.	
	Phased array ultrasonic	Combination of the conventional probe and phased array probe [51]	/	Defects around bolt holes, vertical defects and transverse imperfections in the rail head, waist and foundation area	Ultrasonic beam coverage rate up to 80%	The rails can be inspected more comprehensively and the inspection efficiency is improved. Multiple angles monitoring the same area.
		Phased array with transverse wedge block(railhead), transverse and longitudinal wave probes (rail waist and rail foot) [61]	/	Different types of defects in rail head, rail waist and rail foot	Effectively covers the railhead, rail foot, and rail waist	
		Combination of the conventional probe and phased array probe [52]	/	Different types of defects in rail head, rail waist and rail foot	The detection accuracy can reach 6 mm.	
	Ultrasonic guided wave	High voltage pulse sequences [62]	/	/	Coverage up to 1000 m	The efficiency of ultrasonic guided wave detection of rail defects is much greater than the ultrasonic waves.
		Sine wave modulated by the Hanning window with a frequency of 35 kHz [55]	Phase control and time delay technology.	Rail head, rail waist and rail foot	Enhance expected mode and suppress interference mode. The optimal excitation direction and excitation node of the modes are calculated.	
Excitation source	Laser ultrasonic	High energy laser pulses [58]	Finite element simulations	Rail foot	The best detection position is 300 mm in front of the defect position. The best detection frequency is 20 KHZ.	Can cover the head, web, and foot parts of the rail
		Non-ablative laser source [63]	Analysis of Variance. Monte-Carlo simulations.	Head surface defects, horizontal defects, vertical longitudinal split defects, star defects at colt holes and diagonal defect in waist.	The position of the sensor has a greater impact on detection accuracy. The research results can find the best detection position of the sensor.	
		Hybrid laser/air coupling sensor system [35]	Wavelet transform and outlier analysis.	Surface defects(Transverse defects and alongitudinal defects)	Inner defects and surface defects of the rail can be distinguished.	
		Two staggered beams of laser [27]	Finite element simulations.	Irregular scratches on rail surface	The error is about 0.014%.	
	Electromagnetic ultrasonic	/	Finite element analysis [57]	Rail base	Able to detect common defects in rail bases	No couplant required

2.4. Electromagnetic

The motion-induced eddy current (MIEC) is generated on the surface of the rail by the relative motion between the rail and the detection device under the high-speed electromagnetic non-destructive testing [64]. Therefore, it is possible to determine whether defects are present in the rail by analyzing the changes in the inner and surface magnetic fields of the rail [50]. Electromagnetic wave diffusion detection methods are applied on the rail defects detection based on the change, mainly include magnetic flux leakage (MFL) [65] and eddy-current inspection (ECI) [66].

Magnetic flux leakage sensor consists of an excitation source and a detection sensor. Based on the magnetized excitation source, it can be divided into alternating magnetic field, DC magnetic field, and permanent magnet. The sensor first magnetizes the rail under test to saturation through the excitation source [67]. When defects such as cracks or pits appear on the surface of the rail, the evenly distributed lines of magnetic field inside the rail bend to deform and spread outside of the rail (as shown in Figure 4b), forming a leakage magnetic field on the surface of the defect area [68]. For traditional AC magnetic field excitation technology, the excitation signal is usually a single-frequency sinusoidal signal, which cannot accurately extract rail defect information. P. Wang [69] solved this problem by introducing the periodic square wave pulse technology. For high-speed magnetic flux leakage detection, the collected MFL signals often contain complex noise. The increase of detection frequency has an approximately linear relationship with the decrease of the magnetic flux leakage signal. K. Ji et al. [70] proposed an improved adaptive filtering method that can effectively remove noise. L. Yang et al. [71] proposed a high-speed MFL detection technique based on multi-level magnetization to effectively suppress the influence of magnetic after-effects on rail defect detection.

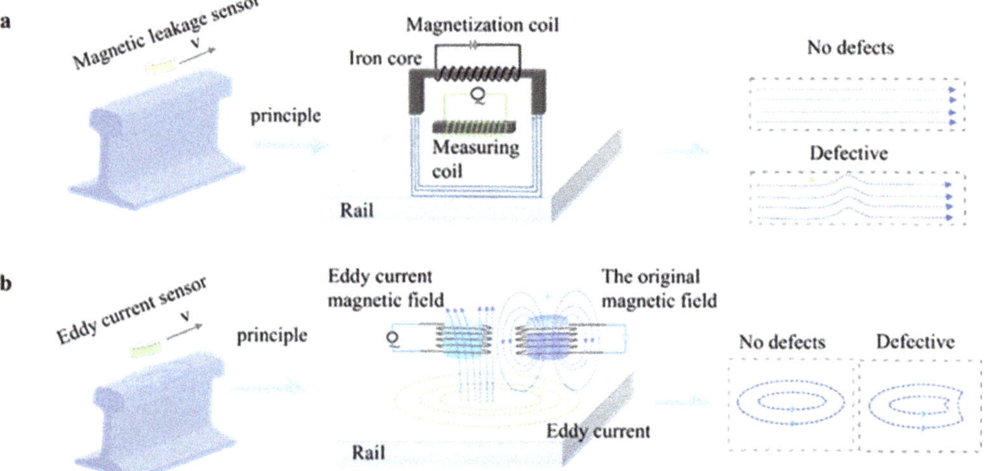

Figure 4. (**a**) Eddy-current detection. (**b**) Magnetic flux leakage detection.

The eddy-current inspection sensor consists of an excitation coil and an induction coil (Figure 4a). Eddy currents cause secondary changes in the strength and distribution of the magnetic field, which leads to changes in the impedance of the detection coil [50]. If no defects are present in the inspection area, the impedance of the detection coil remains constant. If there are defects on and near the surface of the rail, it causes the surface magnetic field to fluctuate and the impedance of the detection coil to change. In this way, rail defects can be detected by analyzing the changes in detection coil impedance. However, many problems are also encountered in high-speed eddy-current testing. For example, the detection signal varies depending on the location of the sensor and the depth of the detection depending on the detection speed. F. Yuan et al. [64] used the DC electromagnetic detection method to study the optimal detection position of rail defects, and the study showed that the optimal detection position is near the inner edge of the excitation coil against the probe movement direction. The team [72] further demonstrated that the PEC detection signal increases with the detection speed, and when the detection speed is constant, the detection signal positively correlates with the defect width and defect depth.

For electromagnetic detection, the velocity effect can affect the amplitude of the signal, and the signal is subject to greater external interference. Therefore, a well-designed algorithm is needed to offset these effects. Compared with ultrasonic inspection, the electromagnetic inspection can detect near-surface defects. We summarize the existing techniques in the literature, as shown in Table 4.

Table 4. Comparison of electromagnetic testing methods.

Methods		Algorithm or Simulation	Types of Detectable Defects	Research Content and Results
Eddy current	Pulsed eddy current [72]	3D transient model	Different installation positions can detect rail defects in different parts.	• The team studied the relationship between the pulsed eddy current detection signal and the velocity of different defect depths and widths.
	Direct current [64]	2D Finite element method	Different installation positions can detect rail defects in different parts.	• The optimal detection position is determined.
	AC bridge techniques [73]	Digital lock-in amplifier algorithm	Four typical types of rail defects (transverse defects, compound fissure, crushed head, detail fracture)	• The effect of solving the lift-off effect is better.
	Differential eddy-current (EC) sensor system [33]	• Low-pass filter • Rotation of EC signal (To extract maximum information and have better visualization)	The degree of looseness of fasteners	• Can detect fastener features 65 mm above the track • The type of missing fixture can be detected by analyzing the characteristics of the fastener.
Magnetic flux leakage	Pulsed magnetic flux leakage [69]	2D transient analysis model under	Vertical and oblique defects	• With the sensor array, not only the magnetic field distribution of the defect can be detected, but also the edge effect caused by the magnetic yoke can be eliminated. • The introduction of periodic square wave pulses solves the problem that single-frequency sinusoidal signals cannot effectively extract rail defect information.
	Multistage magnetization [71]	Finite element method	Rail inner defects	• Magnetic aftereffects are effectively inhibited in high-speed MFL detection.
	Direct current [68]	2D simulation model	Oblique defect and rectangle defect	• Analyzed the influence of speed on magnetic flux leakage signal (At high speed, the magnitude of the flux leakage signal is smaller, but more stable.)

Table 4. *Cont.*

Methods	Algorithm or Simulation	Types of Detectable Defects	Research Content and Results
Magnetic flux leakage [70]	Improved adaptive filtering	Different types of defects in rail surface	• The noise intensity of the MFL signal is reduced by up to about 80%. • The generalization ability of the algorithm is better, and the filtering effect becomes more significant as the speed increases.
Combination of permanent magnets and yoke [74]	3-D FEM simulations	Different types of defects in rail surface	• The MFL signals from the subsurface defect will be more affected by the weakly magnetized regions compared to the surface defect. • The increase in speed reduces the magnetization of the rail.

2.5. Thermal Imaging

When an excitation source such as an eddy current is used to excite the rail, a local heating effect is generated inside the rail [75]. The method for analyzing this effect is called thermal imaging detection.

Pulse thermal imaging based on eddy currents involves two thermal processes in the measurement: Joule heating caused by eddy currents and thermal diffusion inside the material [76,77]. In the heating stage, the presence of surface defects affects the eddy current density distribution and leads to changes in temperature. The geometry of angular defects also affects the temperature difference between the groove edges, which leads to changes in the thermal diffusion mode. Therefore, if the spatial and transient temperature distribution can be obtained, they can be used to detect and characterize the inner defects of the sample [31,78]. This method can accurately detect defects with a width larger than 100 μm [36]. A single-channel blind source separation method for eddy-current pulsed thermography image sequence processing was proposed to extract abnormal patterns and strengthen the comparison of defects [79]. To verify the influence of the sensor's shape on detection, Y. Wu et al. designed sensors of different shapes for comparison [75]. The results showed that a sensor with a round core structure can only detect partial defects. The arc-shaped and U-shaped sensors can detect almost all defects, and the arc-shaped ones have a higher thermal contrast than the U-shaped ones. However, the excitation of rails by ordinary coils cause angle and instability problems. To solve these problems, J. Peng et al. [76] proposed a Helmholtz coil with a larger and more stable detection range than a linear coil (as shown in Figure 5b) and achieved greatly improved detection efficiency.

There are some other methods for generating thermal effects (as shown in Figure 5a). In 2018, R. Usamentiaga et al. [80] applied optical stimulation to thermally stimulate the rail. The study has shown that when the camera is installed at 1 m from the track, defects of 1 cm^2 can be detected.

This technology can detect subsurface defects that cannot be directly discovered on the surface of the rail, greatly improving the ability to detect rail defects. We summarize the existing techniques in the literature, as shown in Table 5.

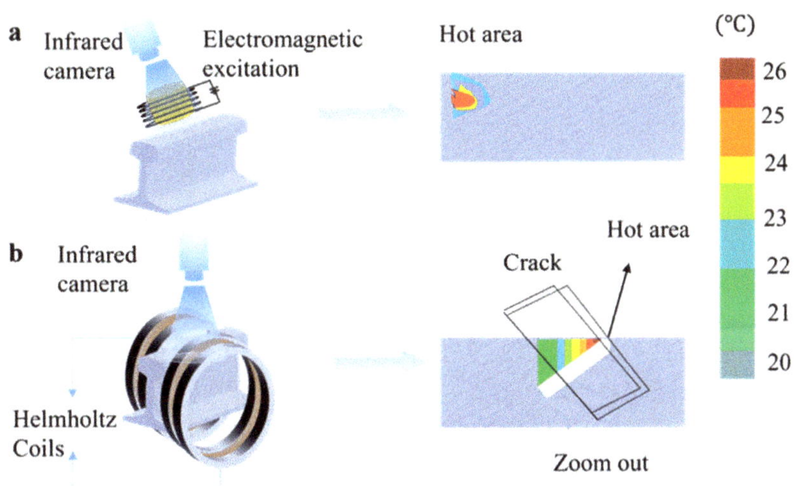

Figure 5. (**a**) Electromagnetic excitation. (**b**) Helmholtz coil excitation.

Table 5. Comparison of thermal imaging testing methods.

Thermal Stimulation		Algorithm	Types of Detectable Defects	Results	Comments
Eddy current	Eddy-current pulsed thermography [79]	Single-channel blind source separation	Thermal fatigue defects	The method can automatically detect rail defects in both the time and the spatial domains.	• The research innovatively discovered the changing process of the mixing vector in the heating and cooling phases.
	Helmholtz coils [76]	Finite element method	Rolling contact fatigue (RCF) defects	Solved the problem that the excitation of ordinary coils on the rails would cause unstable detection areas	• This method provides a larger detection area than linear coils.
	Various shapes of sensors [75]	Inverse Fourier transformation (deblurring method)	RCF defects and micro-defect	Verify the detection effect of various shape sensors	• The research is helpful to design sensors with better detection performance.
	Easyheat 224 system with induction heater [81]	Normalized difference vegetation index (NDVI)	RCF defects	The proposed method can have a good correction for the emissivity.	• Good for correcting ECPT emissivity
Laser	Two halogen lamps [80]	/	Rolled-in material defect	Defects of 1 cm^2 can be detected.	• The study compared multiple methods to enhance the defect signal-to-noise ratio.

Table 5. *Cont.*

Thermal Stimulation	Algorithm	Types of Detectable Defects	Results	Comments
Pulsed air-flow thermography [82]	Subtract the first image in the sequence from the last image acquired in the heating sequence when removing the background.	Rail surface defects	The study proved that the pulsed air-flow thermography method used in the experiment is effective for detecting rail defects.	• The method needs further improvement.
High-frequency continuous sine-wave current [83]	Metric learning modules	Fatigue defects	The method proposed in this study can not only reduce the influence of interference factors but also expand the feature space distance between defective samples and normal samples.	• Using an open set of supervision frameworks, it is easy to add new defect samples. • Good anti-interference performance
Apply uniform heat flux for a time [84]	pulse phase thermography (PPT)	Lateral surface defects	After thermal stimulation for the same time, the cooling rate of shallow defects is faster than that of deep defects.	• The study proved the feasibility of active infrared thermography for detecting rail defects.

2.6. Visual

Visual inspection technology is one of the most important methods in current rail defect detection. An automatic visual inspection system usually consists of a light source, a camera, or other image acquiring devices [85,86] (as shown in Figure 6). The visual inspection system has been widely used in the defect detection of rail facilities along the railway line [50]. According to the visual inspection algorithms, the existing methodologies are categorized into two groups: traditional image processing and deep learning.

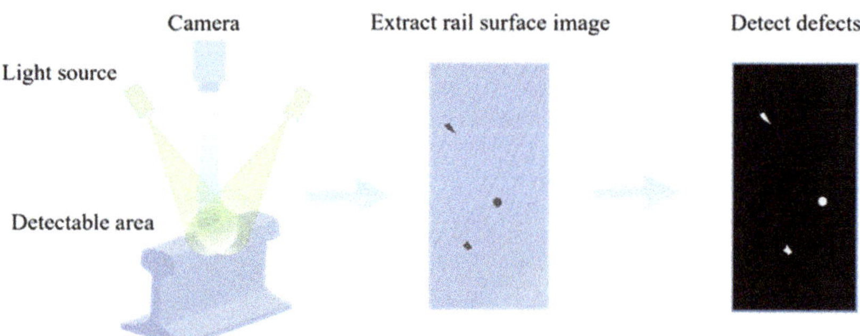

Figure 6. An auxiliary vision system based on a light source.

The traditional image processing process includes two main steps: extracting effective rail surface images and identifying defects. L. Guo et al. [85] applied Hough transform to extract the image of the effective rail surface, and then applied the improved Sobel algorithm and area filter algorithm to detect rail defects with a minimum area of 0.0068 cm^2. O. Yaman et al. [87] applied the Otsu segmentation method to extract the image of the rail surface. Next, feature signals are obtained by calculating the variance value through the rail surface image. By analyzing these signals, and combining fuzzy logic to determine the defect type, the success rate can reach 72.05%. Gan et al. [88] proposed a coarse-to-fine extractor for rail defect detection. The method locates the abnormality of rail defects by rough extraction, and then further extracts the defect information. However, the computational complexity of this method is too high.

Due to the extremely complex characteristics of rail surface defects, the use of ordinary image processing techniques cannot achieve good detection results. Z. Liang et al. [89] compared the SegNet (a deep convolutional network architecture for semantic segmentation) algorithm with artificial and automatic threshold segmentation algorithms. The results showed that the accuracy of the deep learning algorithm is 100%, which is much higher than that of ordinary image processing algorithms (77.8% for manual threshold segmentation and 55.6% for automatic threshold segmentation). For the first time, Li et al. [90] combined the U-Net graph segmentation network with the saliency cues method of damage location and applied on the damage detection of high-speed railway rails, with an accuracy rate of 99.76%. L. Zhuang et al. [91] proposed a cascading rail surface flaw identifier. The method detects the presence of defect based on DenseNet-169, and then performs defect classification for the defective rails with a feature joint learning module (FJLM) and a feature reduction module (FRM).

Visual inspection can effectively detect the surface defects of the rail. However, it does not provide any information about the inner defects of the rails. We summarize the existing techniques in the literature, as shown in Table 6.

Table 6. Comparison of visual inspection methods.

Algorithm		Results	Comments	Summarize
Traditional algorithm	Hough transform and improved Sobel algorithm [85]	Minimum detection area: 0.0068 cm^2	• Fast processing speed • Harder to apply to complex situations	Weak generalization ability and low accuracy
	Otsu segmentation and fuzzy logic [87]	The success rate of identifying defect types: 72.05%	• Types of defects can be identified.	
	Coarse-to-fine model [88,92]	CTFM outperforms state-of-the-art methods in terms of pixel-level indices and defect-level indices.	• Effectively suppress the influence of noise points • The proposed computational requires high computational resources.	
Deep learning	SegNet [89]	Detection accuracy: 100%	• outperform ordinary image processing algorithms	Strong generalization ability and high accuracy
	SCueU-Net [90]	Detection accuracy: 99.76%	• Overcome the interference of image noise and solve the current problem of low detection efficiency	
	MOLO [93]	This algorithm improves the accuracy 3–5% more than the YOLOv3 algorithm.	• Image features are extracted using MobileNetV2 as the backbone network. At the same time, the multi-scale prediction and the loss calculation method of YOLOv3 are used. • The network structure is relatively simple, which balances detection accuracy and detection speed.	
	Cascading rail surface flaw identifier [91]	The detection accuracy rate of defect type: 98.2%	• Better processing performance for complex scenes • Accurately identify multiple types of defects	

2.7. Other Detection Methods

There are other rail defect detection methods such as structured light detection, fiber grating detection, and infrared detection.

Q. Mao et al. [94] proposed a fastener detection method based on a structured light sensor. They used a decision tree classifier to classify fastener defects and achieved an overall accuracy no less than 99.8%, indicating that this method can offer a promising way to detect fasteners. Compared with two-dimensional vision, the structured light sensor can obtain a three-dimensional point cloud of fasteners, thereby obtaining more detailed fastener information.

A laser-based non-contact sensor can be an effective tool for detecting rail defects. Generally, the sensor consists of two infrared modules: a transmitter and a receiver. The transmitter emits infrared rays, and the receiver receives the pulses reflected from the rails. By analyzing the time when the pulses are reflected from the rails, the geometric parameters of the rails can be tracked, and a high-resolution map with three-dimensional objects can be generated. This method is effective in detecting surface and welding defects on the rail [18,95].

2.8. Technology Comparison

The vibration sensors based on MEMS technology have been widely used. This type of sensor features high accuracy, low price, small size, and convenient installation while having the ability to detect various rail defects through data analysis. The detection accuracy of visual inspection technology is high, but this method can only detect the surface defects. The ultrasonic-based detection technology can detect the inner defects of the rail, but the detection depth is not less than 5 mm [65], which cannot be used for defect detection on the near-surface ($\leq$5 mm) of the rail. Magnetic flux leakage and eddy-current detection technology have high detection accuracy for near-surface defects of rails. The above methods can only detect the static defects of the rails. The acoustic emission detection method is suitable for studying the dynamic expansion process of rail defects. Therefore, the combination of multiple methods can realize the simultaneous monitoring of multiple different defects. In Table 7, a horizontal comparison is provided to distinguish between the different methods.

Table 7. Comparison of rail defect detection methods.

<table>
<tr><th colspan="3">Detection Method</th><th>Types of Detectable Defects</th><th colspan="2">Detection Performance</th><th>Influence of Environment on Detection Performance</th></tr>
<tr><td colspan="3">Vibration accelerometer</td><td>• The degree of looseness of fasteners [5,96]
• Inner [2] and surface [47] defects of the rail</td><td colspan="2">• Can detect the degree of looseness of fasteners [5]
• Small size, easy installation, wide detection range [42]</td><td>Temperatures that are too low will reduce the sensitivity of the sensor.</td></tr>
<tr><td rowspan="3">Ultrasonic</td><td rowspan="2">Ordinary ultrasonic [51,52,61]</td><td>Conventional probe</td><td rowspan="2">• Railhead inner defects
• Rail foot defects
• Rail waist defects</td><td>• Single angle and low efficiency</td><td rowspan="2">In high-speed inspection systems, rail defects with a depth of less than 4 mm are often undetectable [76].</td><td rowspan="3">When the temperature changes, it will affect the speed of the sound wave in the rail, so the localization of the defect will have an impact.</td></tr>
<tr><td>Phased array probe</td><td>• Multi-angle detection
• Better ultrasonic beam coverage
• Higher efficiency than traditional ultrasonic testing</td></tr>
<tr><td colspan="2">Electromagnetic ultrasonic</td><td>• Rail inner defects
• surface defects</td><td colspan="2">• High precision
• No complaint required [57,97]</td></tr>
</table>

Table 7. *Cont.*

Detection Method		Types of Detectable Defects	Detection Performance	Influence of Environment on Detection Performance
	Laser ultrasonic	• Rail inner defects [35] • Surface defects [98] and subsurface defects [58]	• Good penetration ability • Can cover the entire track for testing [63]	
AE		• Subsurface defects [48]	• Suitable for studying the dynamic expansion process of rail defects [48] • Acoustic emission signals are easily submerged by high-frequency vehicle speed signals [48].	Other noises will affect the detection results.
Electromagnetic	MFL	• Surface and shallow surface defects [69,74]	• Highly susceptible to the environment (white noise and power frequency interference in the environment) [71] • Easily affected by lift-off [64,74] • As the detection rate increases, the depth of detection of rail defects decreases [64]	The temperature will drift the detection results of the eddy-current sensor, and the two are negatively correlated.The increase in temperature will cause the magnetic permeability to decrease.
	ECI	• Rail inner defects [99] • surface and subsurface defects [64]	• Easily affected by lift-off [64]. • Can effectively detect subsurface defects.	
Thermal imaging		• Subsurface defects [76] and surface defects [82]	• It can characterize the shape and size of rail defects [76,80].	Contamination present on the Rail surface will attenuate the signal.
Vision		• Missing fastener fixture [100] • Surface defects [93]	• Can only detect surface defects • High detection accuracy • Mature detection algorithm • Affected by the surface condition (dirt occlusion, others)	Contaminants such as snowflakes and leaves can block rail defects, making visual inspection methods unable to detect rail defects.

3. Wireless Transmission

In the rail defect detection system, especially in the contact detection technology, many sensors need to be installed on the rail. If wired transmission is used, too many lines need to be placed for data transmission. The rapid advance in wireless communication and self-organizing networks (GSM, ZigBee, others) makes it more reliable and convenient to wirelessly transmit information to terminals [19]. Therefore, reliable wireless transmission is more suitable for rail defect detection than wired transmission [101]. E. Aboelela et al. [20] established a wireless sensor network system for monitoring railway safety which has laid the foundation for the application of wireless networks in railway detection. In the last 10 years, wireless sensor networks have gradually replaced wired monitoring along railway lines. This changes provide enormous convenience for real-time monitoring of railway facilities [101]. Because wireless transmission networks are responsible for data exchange between wireless sensors and terminals, they must be carefully designed to prevent transmission errors, delays, network interruptions, and data loss or damage [19].

3.1. Transmission Node Settings

A good network topology can reduce communication interference, extend the network's service life, and improve communication efficiency [24]. In a wireless sensor system for rail defect detection, the nodes are usually set up in the following three ways, as illustrated in Figure 7 and compared in Table 8. Each method contains three nodes: the terminal node, routing node, and coordinating node. The terminal node collects data, and the routing node can forward information and assist the coordinator in maintaining the network. The coordinating node is the central hub of the entire network for transmission data.

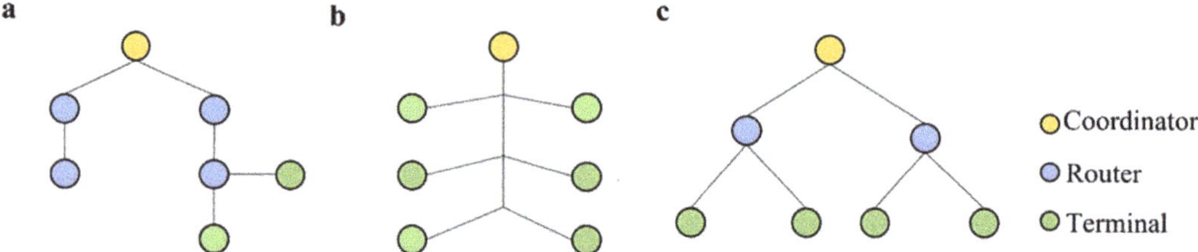

Figure 7. Network topology. (**a**) Tree topology. (**b**) Line topology. (**c**) Star topology.

Table 8. Comparison of network topology.

Network Topology	Advantages	Disadvantages	References
Star topology	• Short network delay time • Simple structure • Easy to maintain	• Low line utilization • The central node load is too heavy.	[43,102]
Tree topology	• Simple structure • Easy to maintain • Easy to expand	• The dependence of each node on the root is too large.	[20,24]
Line topology	• Simple structure • Low cost • Easy to expand	• Low reliability • Difficulty in fault diagnosis and isolation	[103]

3.2. Transmission Media

In rail wireless sensor systems, the transmission from node to node and from node to base station is usually a short distance, so the wireless communication can be achieved by various technologies such as Bluetooth, Wi-Fi, and ZigBee. In 2007, Aw et al. [104] developed a method that uses Bluetooth to transmit rail detection information. This method, however, has become obsolete due to its weak anti-interference ability and short transmission distance [105]. Zigbee offers limited bandwidth when used for rail condition monitoring [106]. To overcome this disadvantage, M. Tolani et al. [23] designed a two-layer transmission network composed of power-efficient ZigBee nodes as the first layer and bandwidth-efficient WLAN as the second layer. In recent years, Global System for Mobile Communications (GSM) [107] has emerged as a powerful tool for mobile communications [108] and has been used in wireless sensor systems for rail defect detection. Its main advantages are low costs and global availability. Jiaying et al. [102] applied GSM technology to the detection of the environment surrounding the rail, and the results showed that it offers good transmission performance.

3.3. Information Transmission

Sustainable running is an essential goal in the design of a wireless sensor network. Therefore, it is important to minimize the energy consumption of the system [109], which can be realized by two main methods: optimizing the transmission protocol and optimizing the hardware design.

In the wireless transmission media access control (MAC) protocol, the energy consumption mainly comes from collision, eavesdropping, and idle monitoring [110]. The schedule-based protocol is collision-free, thus reducing energy waste due to collisions. However, they lack adaptability and scalability to adapt to changes in node density or traffic load. Contention-based protocols have good scalability but cannot avoid wasting energy due to collisions, overhearing, and idle listening [111]. Energy consumption can be reduced by filtering useless data and reducing idle listening time. GM Shafiullah et al. [112] proposed a new protocol named E-BMA. The protocol minimizes the idle time during competition and can achieve improved energy efficiency at low and medium traffic. A. Philipose et al. [113] proposed an improved media access control (MAC) protocol. In this protocol, each node is awakened only when it needs to work, which reduces the energy consumption of sensor nodes.

Another method is to reduce energy consumption by optimizing sensor hardware. This method adopts a sleep strategy when the system is not working, so as to minimize the energy consumption of the system. M. F. Islam et al. [2] proposed a lazy pole strategy, in which data is sent only when the vibration sensed by the sensor node is different from a pre-defined pattern. H. Zhang et al. [114] adopted a synchronous sleep and wake-up strategy to make idle nodes to sleep and shut down most hardware to greatly reduce energy consumption.

Therefore, for future designs of wireless sensor networks for rail defect detection systems, these two methods can be combined to minimize energy consumption.

4. Power Supply

It is extremely important to make a wireless sensor self-powered to reduce the cost of maintenance and have a long lifetime. The energy generated by the surrounding environment such as rail vibration [115], solar energy [116], and other types can be stored in a rechargeable battery [117] for powering the wireless sensing detection system. The power generation methods of the rail wireless sensor detection system at present are shown in Figure 8.

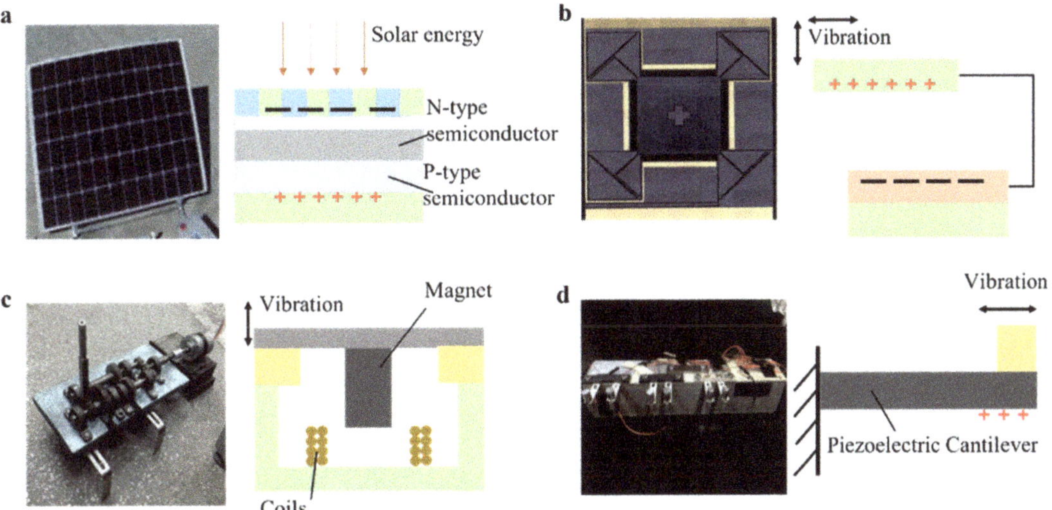

Figure 8. Principles of power generation. (**a**) Principle of solar energy harvester [34]. (**b**) Principle of electrostatic harvester [118]. (**c**) Principle of electromagnetic harvester [119]. (**d**) Principle of piezoelectric energy harvester [120].

4.1. Solar

Solar power generation methods include photovoltaic power generation (light energy is converted into electrical energy) and thermal power generation (thermal energy is converted into electrical energy) [121]. Photovoltaic power generation can achieve a power density as high as 10–15 mW/cm^2 [116], which a enough to power wireless sensors. Solar thermal power generation is a technology that uses solar concentrators to convert solar radiant energy into thermal energy and then into electrical energy. However, solar thermal power generators cannot be used on a large scale in wireless sensors as they are highly susceptible to weather conditions and other environmental impacts, and it is difficult to find a suitable location to install them.

4.2. Vibration

Vibration power generation converts the kinetic energy of rail vibration into electrical energy. Typical vibration power generators include electromagnetic, piezoelectric, and electrostatic generators [106]. An electromagnetic generator converts the orbital vibration into the relative motion between the permanent magnet and the coil [119] and converts it further into extremely low-frequency electrical energy [122]. For the first time, X. Zhang et al. [119] applied supercapacitors to vibration energy harvesting systems. The system amplifies the small vibrations of the track and store energy from rapidly changing transient currents. For piezoelectric power generation, when pressure is applied to a piezoelectric material, a potential difference is generated on the surface of the piezoelectric material [123,124]. J. Wang et al. [125] studied a theoretical model of using the patch and stacked piezoelectric transducers to collect piezoelectric energy from railway systems. The electrostatic generator needs to be driven by an external voltage, so it features high output impedance and high voltage and is not readily applicable for sensing devices. Table 9 summarizes the main differences of the related studies. These technologies focus on converting environmental energy into electrical energy [120].

A solar power generator is highly dependent on external conditions. It requires stable light conditions, which make it difficult to find an appropriate location for sensor installation. A vibration-based piezoelectric energy collector has the advantages of a simple structure and a small size. However, this method requires a large vibration amplitude of the rail. An electromagnetic energy harvester is much less demanding on the amplitude of rail vibration, but it is susceptible to external electromagnetic interference.

Table 9. Comparison of energy harvesters based on vibration principle.

Energy Harvesting Device	Application Conditions	Installation Location	Voltage	Power	Reference
Piezoelectric energy harvester	2.5 mph (the speed of the train) The resistor connected in the PZT0 (a single piezoelectric energy harvester) was 9.9 KΩ	Harvester Track Rail pads	40 V (the maximum voltage)	0.18 mW (the maximum power)	[120]
Magnetic levitation oscillator	105 km/h (the speed of the train) (one-car train)	Harvester	2.3 V (peak–peak output voltage)	/	[106]
Galfenol magnetostictive device	60 km/h (the speed of the train) 60 m (The train is far from the sensor of 60 m.)	Harvester	0.15 V (The voltage varies with the distance between the train and the sensor, when the distance is shorter, the voltage is larger, and the longer the distance, the smaller the voltage.)	When the terminal voltage is about 0.56 V, the power is maximum.	[115]

Table 9. *Cont.*

Energy Harvesting Device	Application Conditions	Installation Location	Voltage	Power	Reference
A patch-type piezoelectric transducer	30 m/s (the speed of the train)	Base Harvester	4.82 V (at the beginning of a valid signal)	0.19 mW (at the beginning of a valid signal)	[125]
Drum transducer	0.15 m/s (running speed) 120 kg (the weight of a fully-loaded train)	Harvester	50–70 V (peak open-circuit voltage)	100 mW	[123]
Electromagnetic energy harvesting system	6 mm (amplitude) 1 Hz and 2 Hz (frequencies)	fixture Harvester	6.45 V (the output peak–peak voltage)	0.0912 J	[119]
Magnetic levitation harvester	low-frequency (3–7 Hz) Rail displacement	Harvester	2.32 V (the output peak–peak voltage)	119 mW	[126]

5. Summary and Future Work

This paper reviews the existing conventional rail detection technologies such as vibration, ultrasonic, electromagnetic detection, and visual detection and makes comparisons between them and briefly introduces the wireless transmission method and power generation methods for the WSRDDS.

There is still some potential improvement for existing systems. We suggest the following optimizations to make these systems more reliable, intelligent, and powerful in detecting rail defects.

(1) Rail defect feature signals can be extracted to build a complete database of rail defect and fastener defect features. This database can be used to automatically classify rail defects and determine the degree of damage to other track components.
(2) For a single detection technology, it is difficult to detect all the information from the rails. Combining a variety of sensors can achieve all-round and high-precision detection of rail defects.
(3) Building a comprehensive monitoring system for rail defects based on big data management and information mining technology is a good direction for achieving all-round and high-precision detection of rail infrastructure.

Author Contributions: Conceptualization, Z.L., D.Y. and Y.Z.; methodology, Z.L. and X.Y.; investigation, Z.L., Z.Z. and W.J.L.; resources, Z.L., X.S. and L.L.; writing—original draft preparation, Z.L.; writing—review and editing, Z.L. and H.S.; project administration, Y.Z.; funding acquisition, Y.Z. All authors have read and agreed to the published version of the manuscript.

Funding: This work was supported by the National Natural Science Foundation of China (Grant No. 61873307), the Hebei Natural Science Foundation (Grant No. F2020501040, F2021203070, F2022501031), the Fundamental Research Funds for the Central Universities under Grant N2123004, the Administration of Central Funds Guiding the Local Science and Technology Development (Grant No. 206Z1702G), the HMRF-Health and Medical Research Fund under Project 17181811, in part of the TBRS-RGC Theme-based Research Scheme under Project T42-717/20-R, and CRF-Collaborative Research Fund under Project C7174-20G.

Institutional Review Board Statement: Not applicable.

Informed Consent Statement: Not applicable.

Data Availability Statement: Not applicable.

Conflicts of Interest: We declare that we do not have any commercial or associative interests that represent any conflicts of interest in connection with the work submitted.

References

1. Ishida, M.; Akama, M.; Kashiwaya, K.; Kapoor, A. The Current Status of Theory and Practice on Rail Integrity in Japanese Railways-Rolling Contact Fatigue and Corrugations. *Fatigue Fract. Eng. Mater. Struct.* **2003**, *26*, 909–919. [CrossRef]
2. Islam, M.F.; Maheshwari, S.; Kumar, Y. Energy Efficient Railway Track Security Using Vibration Sensing Network. In Proceedings of the 2nd International Conference on Signal Processing and Integrated Networks, SPIN 2015, Noida, India, 19–20 February 2015; pp. 973–978. [CrossRef]
3. Wang, P. Longitudinal Force Measurement in Continuous Welded Rail with Bi-Directional FBG Strain Sensors. *Smart Mater. Struct.* **2015**, *25*, 15019. [CrossRef]
4. Sharma, K.; Maheshwari, S.; Solanki, R.; Khanna, V. Railway Track Breakage Detection Method Using Vibration Estimating Sensor Network: A Novel Approach. In Proceedings of the 2014 International Conference on Advances in Computing, Communications and Informatics, ICACCI 2014, Delhi, India, 24–27 September 2014; pp. 2355–2362. [CrossRef]
5. Zhan, Z.; Sun, H.; Yu, X.; Yu, J.; Zhao, Y.; Sha, X.; Chen, Y.; Huang, Q.; Li, W.J. Wireless Rail Fastener Looseness Detection Based on MEMS Accelerometer and Vibration Entropy. *IEEE Sens. J.* **2020**, *20*, 3226–3234. [CrossRef]
6. Kaewunruen, S.; Osman, M.H.B.; Rungskunroch, P. The Total Track Inspection. *Front. Built Environ.* **2019**, *4*, 84. [CrossRef]
7. Zhang, X.; Feng, N.; Wang, Y.; Shen, Y. An Analysis of the Simulated Acoustic Emission Sources with Different Propagation Distances, Types and Depths for Rail Defect Detection. *Appl. Acoust.* **2014**, *86*, 80–88. [CrossRef]
8. Yi, J. Four Times in 10 Days, Why Do Indian Trains Always Derail. *Urban Mass Transit* **2017**, *20*, 197–199.
9. Goto, K.; Matsumoto, A.; Ishida, M.; Chen, H.; Kaewunruen, S. Editorial: UK-Japan Symposium on Highspeed Rails. *Front. Built Environ.* **2020**, *6*, 54. [CrossRef]
10. Kaewunruen, S.; Ngamkhanong, C.; Sengsri, P.; Ishida, M. On Hogging Bending Test Specifications of Railway Composite Sleepers and Bearers. *Front. Built Environ.* **2020**, *6*, 592014. [CrossRef]
11. Setsobhonkul, S.; Kaewunruen, S.; Sussman, J.M. Lifecycle Assessments of Railway Bridge Transitions Exposed to Extreme Climate Events. *Front. Built Environ.* **2017**, *3*, 35. [CrossRef]
12. De Melo, A.L.O.; Kaewunruen, S.; Papaelias, M.; Bernucci, L.L.B.; Motta, R. Methods to Monitor and Evaluate the Deterioration of Track and Its Components in a Railway In-Service: A Systemic Review. *Front. Built Environ.* **2020**, *6*, 118. [CrossRef]
13. Feng, H.; Jiang, Z.; Xie, F.; Yang, P.; Shi, J.; Chen, L. Automatic Fastener Classification and Defect Detection in Vision-Based Railway Inspection Systems. *IEEE Trans. Instrum. Meas.* **2014**, *63*, 877–888. [CrossRef]
14. Deutschl, E.; Gasser, C.; Niel, A.; Werschonig, J. Defect Detection on Rail Surfaces by a Vision Based System. In Proceedings of the IEEE Intelligent Vehicles Symposium, Parma, Italy, 14–17 June 2004; pp. 507–511. [CrossRef]
15. Zhong, Y.; Ma, Y.; Li, P.; Xiong, L.; Yan, F. The Existing GTC-80 Rail Inspection Vehicle Automatic System Upgrade and Transformation. *Chin. Railw.* **2018**, *6*, 98–102.
16. Wei, Q.; Zhang, X.; Wang, Y.; Feng, N.; Shen, Y. Rail Defect Detection Based on Vibration Acceleration Signals. In Proceedings of the Conference Record—IEEE Instrumentation and Measurement Technology Conference, Minneapolis, MN, USA, 6–9 May 2013; pp. 1194–1199. [CrossRef]
17. Min, Y.; Xiao, B.; Dang, J.; Yue, B.; Cheng, T. Real Time Detection System for Rail Surface Defects Based on Machine Vision. *Eurasip J. Image Video Processing* **2018**, *2018*, 3. [CrossRef]
18. Vohra, M.; Gabhane, S.K. Efficient Monitoring System for Railways for Crack Detection. In Proceedings of the International Conference on I-SMAC (IoT in Social, Mobile, I-SMAC 2018, Palladam, India, 30–31 August 2018, Analytics and Cloud); pp. 676–681. [CrossRef]
19. Hodge, V.J.; O'Keefe, S.; Weeks, M.; Moulds, A. Wireless Sensor Networks for Condition Monitoring in the Railway Industry: A Survey. *IEEE Trans. Intell. Transp. Syst.* **2015**, *16*, 1088–1106. [CrossRef]
20. Aboelela, E.; Edberg, W.; Papakonstantinou, C.; Vokkarane, V. Wireless Sensor Network Based Model for Secure Railway Operations. In Proceedings of the IEEE International Performance, Computing, and Communications Conference, Phoenix, AZ, USA, 10–12 April 2006; pp. 623–628. [CrossRef]
21. Grudén, M.; Westman, A.; Platbardis, J.; Hallbjörner, P.; Rydberg, A. Reliability Experiments for Wireless Sensor Networks in Train Environment. In Proceedings of the European Microwave Week 2009: Science, Progress and Quality at Radiofrequencies, Conference Proceedings—2nd European Wireless Technology Conference, EuWIT 2009, Rome, Italy, 28–29 September 2009; pp. 37–40.
22. Grover, J. Anjali Wireless Sensor Network in Railway Signalling System. In Proceedings of the 2015 5th International Conference on Communication Systems and Network Technologies, CSNT 2015, Gwalior, India, 4–6 April 2015; pp. 308–313. [CrossRef]
23. Tolani, M.; Sunny; Singh, R. K.; Shubham, K.; Kumar, R. Two-Layer Optimized Railway Monitoring System Using Wi-Fi and ZigBee Interfaced Wireless Sensor Network. *IEEE Sens. J.* **2017**, *17*, 2241–2248. [CrossRef]

24. Zhou, D.; Shi, T.; Lv, X.; Bai, W. A Research on Banded Topology Control of Wireless Sensor Networks along High-Speed Railways. In Proceedings of the Chinese Control Conference, Hangzhou, China, 28–30 July 2015; pp. 7736–7740. [CrossRef]
25. Falamarzi, A.; Moridpour, S.; Nazem, M. A Review on Existing Sensors and Devices for Inspecting Railway Infrastructure. *Jurnal Kejuruteraan* **2019**, *31*, 1–10.
26. Zhang, X.; Jia, L.; Wei, X.; Ru, N. Railway Track Condition Monitoring Based on Acceleration Measurements. In Proceedings of the 2015 27th Chinese Control and Decision Conference, CCDC 2015, Qingdao, China, 23–25 May 2015; pp. 923–928. [CrossRef]
27. Zhong, Y.; Gao, X.; Luo, L.; Pan, Y.; Qiu, C. Simulation of Laser Ultrasonics for Detection of Surface-Connected Rail Defects. *J. Nondestruct. Eval.* **2017**, *36*, 70. [CrossRef]
28. Suhas, B.N.; Bhagavat, S.; Vimalanand, V.; Suresh, P. Wireless Sensor Networks Based Monitoring of Railway Tracks. In Proceedings of the 2018 International CET Conference on Control, Communication, and Computing, IC4 2018, Thiruvananthapuram, India, Thiruvananthapuram, India ; pp. 187–192. [CrossRef]
29. Zhang, X.; Wang, K.; Wang, Y.; Shen, Y.; Hu, H. Rail Crack Detection Using Acoustic Emission Technique by Joint Optimization Noise Clustering and Time Window Feature Detection. *Appl. Acoust.* **2020**, *160*, 107141. [CrossRef]
30. Yılmaz, H.; Öztürk, Z. Investigation of Rail Defects Using an Ultrasonic Inspection Method: A Case Study of Aksaray-Airport Light Rail Transit Line in Istanbul. *Urban Transp. XXI* **2015**, *1*, 687–698. [CrossRef]
31. Zhu, J.; Tiany, G.; Min, Q.; Wu, J. Comparison Study of Different Features for Pocket Length Quantification of Angular Defects Using Eddy Current Pulsed Thermography. *IEEE Trans. Instrum. Meas.* **2019**, *68*, 1373–1381. [CrossRef]
32. Ren, W.Z.; Min, Y.Z.; Tao, J.; Hu, J. Research on Embedded Rail Surface DefectDetection System Based on Multi Core DSP. In Proceedings of the 2018 Chinese Automation Congress, CAC 2018, Xi'an, China, 30 November–2 December 2018; pp. 4107–4112. [CrossRef]
33. Chandran, P.; Rantatalo, M.; Odelius, J.; Lind, H.; Famurewa, S.M. Train-Based Differential Eddy Current Sensor System for Rail Fastener Detection. *Meas. Sci. Technol.* **2019**, *30*, 125105. [CrossRef]
34. Zhang, M.; Qi, S.; Zhang, X.; Zhao, Y.; Sha, X.; Liu, L. Multi-Modal Wireless Sensor Platform for Railway Monitoring. In Proceedings of the 9th IEEE International Conference on Cyber Technology in Automation, Control and Intelligent Systems, CYBER 2019, Suzhou, China, 29 July–2 August 2019; pp. 1658–1662. [CrossRef]
35. Rizzo, P.; Cammarata, M.; Bartoli, I.; di Scalea, F.L.; Salamone, S.; Coccia, S.; Phillips, R. Ultrasonic Guided Waves-Based Monitoring of Rail Head: Laboratory and Field Tests. *Adv. Civ. Eng.* **2010**, *2010*, 291293. [CrossRef]
36. Netzelmann, U.; Walle, G.; Ehlen, A.; Lugin, S.; Finckbohner, M.; Bessert, S. NDT of Railway Components Using Induction Thermography. *AIP Conf. Proc.* **2016**, *1706*, 150001. [CrossRef]
37. Jiang, B. Design of Railway Vibration Detection System Based on ARM and Acceleration Sensor. Master's Thesis, Lanzhou Jiaotong University, Lanzhou, China, 2015.
38. Ng, A.K.; Martua, L.; Sun, G. Dynamic Modelling and Acceleration Signal Analysis of Rail Surface Defects for Enhanced Rail Condition Monitoring and Diagnosis. In Proceedings of the 4th International Conference on Intelligent Transportation Engineering, ICITE 2019, Singapore, 5–7 September 2019; pp. 69–73. [CrossRef]
39. Li, B.; Chen, X.; Wang, Z.; Tan, S. Vibration Signal Analysis for Rail Flaw Detection. In Proceedings of the 2019 11th CAA Symposium on Fault Detection, SAFEPROCESS 2019, Xiamen, China, 5–7 July 2019; pp. 830–835. [CrossRef]
40. Sun, M.; Wang, Y.; Zhang, X.; Liu, Y.; Wei, Q.; Shen, Y.; Feng, N. Feature Selection and Classification Algorithm for Non-Destructive Detecting of High-Speed Rail Defects Based on Vibration Signals. In Proceedings of the Conference Record—IEEE Instrumentation and Measurement Technology Conference 2014, Montevideo, Uruguay, 12–15 May 2014; pp. 819–823. [CrossRef]
41. Humbe, A.A.; Karmude, S.A. Analysis of Mechanical Vibration and Fault Detection of Railway Track Using Lab View System. *Ijireeice* **2019**, *7*, 9–15. [CrossRef]
42. Milne, D.; Pen, L.L.; Watson, G.; Thompson, D.; Powrie, W.; Hayward, M.; Morley, S. Proving MEMS Technologies for Smarter Railway Infrastructure. *Procedia Eng.* **2016**, *143*, 1077–1084. [CrossRef]
43. Zhao, J.; Wang, B.; Niu, W.; Li, X.; Zhang, B.; Wang, Y. Detection System of Fasteners State Based on ZigBee Networks. *MATEC Web Conf.* **2015**, *35*, 3005. [CrossRef]
44. Wang, Y.; Li, X.; Zhao, J. The Design and Research of Rail Fastener State Detection System. *Hardw. Circuits* **2015**, *39*, 22–34.
45. Bruzelius, K.; Mba, D. An Initial Investigation on the Potential Applicability of Acoustic Emission to Rail Track Fault Detection. *NDT E Int.* **2004**, *37*, 507–516. [CrossRef]
46. Kuang, K.S.C.; Li, D.; Koh, C.G. Acoustic Emission Source Location and Noise Cancellation for Crack Detection in Rail Head. *Smart Struct. Syst.* **2016**, *18*, 1063–1085. [CrossRef]
47. Jian, H.; Lee, H.R.; Ahn, J.H. Detection of Bearing/Rail Defects for Linear Motion Stage Using Acoustic Emission. *Int. J. Precis. Eng. Manuf.* **2013**, *14*, 2043–2046. [CrossRef]
48. Kostryzhev, A.G.; Davis, C.L.; Roberts, C. Detection of Crack Growth in Rail Steel Using Acoustic Emission. *Ironmak. Steelmak.* **2013**, *40*, 98–102. [CrossRef]
49. Hao, Q.; Zhang, X.; Wang, K.; Shen, Y.; Wang, Y. A Signal-Adapted Wavelet Design Method for Acoustic Emission Signals of Rail Cracks. *Appl. Acoust.* **2018**, *139*, 251–258. [CrossRef]
50. Li, Q.; Zhong, Z.; Liang, Z.; Liang, Y. Rail Inspection Meets Big Data: Methods and Trends. In Proceedings of the 2015 18th International Conference on Network-Based Information Systems, NBiS 2015, Taipei, Taiwan, 2–4 September 2015; pp. 302–308. [CrossRef]

51. Zhang, Y.; Gao, X.; Peng, C.; Wang, Z.; Li, X. Rail Inspection Research Based on High Speed Phased Array Ultrasonic Technology. Proceedings of 2016 IEEE Far East NDT New Technology and Application Forum, FENDT 2016, Nanchang, China, 22–24 June 2016; pp. 181–184. [CrossRef]
52. Ling, C.; Chen, L.; Guo, J.; Gao, X.; Wang, Z.; Li, J. Research on Rail Defect Detection System Based on FPGA. In Proceedings of the 2016 IEEE Far East NDT New Technology and Application Forum, FENDT 2016, Nanchang, China, 22–24 June 2016; pp. 195–200. [CrossRef]
53. Rose, J.L.; Avioli, M.J.; Mudge, P.; Sanderson, R. Guided Wave Inspection Potential of Defects in Rail. *NDT E Int.* **2004**, *37*, 153–161. [CrossRef]
54. Xu, X.; Zhuang, L.; Xing, B.; Yu, Z.; Zhu, L. An Ultrasonic Guided Wave Mode Excitation Method in Rails. *IEEE Access* **2018**, *6*, 60414–60428. [CrossRef]
55. Shi, H.; Zhuang, L.; Xu, X.; Yu, Z.; Zhu, L. An Ultrasonic Guided Wave Mode Selection and Excitation Method in Rail Defect Detection. *Appl. Sci.* **2019**, *9*, 1170. [CrossRef]
56. Kaewunruen, S.; Ishida, M. In Situ Monitoring of Rail Squats in Three Dimensions Using Ultrasonic Technique. *Exp. Tech.* **2015**, *40*, 1179–1185. [CrossRef]
57. Wang, S.J.; Chen, X.Y.; Jiang, T.; Kang, L. Electromagnetic Ultrasonic Guided Waves Inspection of Rail Base. In Proceedings of the FENDT 2014—Proceedings, 2014 IEEE Far East Forum on Nondestructive Evaluation/Testing: New Technology and Application, Increasingly Perfect NDT/E, Chengdu, China, 20–23 June 2014; pp. 135–139. [CrossRef]
58. Pathak, M.; Alahakoon, S.; Spiryagin, M.; Cole, C. Rail Foot Flaw Detection Based on a Laser Induced Ultrasonic Guided Wave Method. *Meas. J. Int. Meas. Confed.* **2019**, *148*, 106922. [CrossRef]
59. Li, Y.; Yao, F.; Jiao, S.; Huang, W.; Zhang, Q. Identification and Classification of Rail Damage Based on Ultrasonic Echo Signals. In Proceedings of the Chinese Control Conference, CCC 2020, Shenyang, China, 27–29 July 2020; pp. 3077–3082. [CrossRef]
60. Luo, X.; Hu, Y.Q.; Liu, Y.; Huang, M.; Chu, W.; Lin, J. A Novel Text-Style Sequential Modeling Method for Ultrasonic Rail Flaw Detection. In Proceedings of the 2020 IEEE Vehicle Power and Propulsion Conference, VPPC 2020—Proceedings 2020, Gijon, Spain, 18 November–16 December 2020; pp. 1–4. [CrossRef]
61. Su, X.; Zhang, X. High—Speed Railway Monitoring System Based on Wireless Sensor Network. *J. Terahertz Sci. Electron. Inf. Technol.* **2019**, *17*, 239–242.
62. Wei, X.; Yang, Y.; Yu, N. Research on Broken Rail Real-Time Detection System for Ultrasonic Guided Wave. In Proceedings of the 2017 19th International Conference on Electromagnetics in Advanced Applications, ICEAA 2017, Verona, Italy, 11–15 September 2017; pp. 906–909. [CrossRef]
63. Benzeroual, H.; Khamlichi, A.; Zakriti, A. Reliability of Rail Transverse Flaw Detection by Means of an Embedded Ultrasonic Based Device. *MATEC Web Conf.* **2018**, *191*, 5. [CrossRef]
64. Yuan, F.; Yu, Y.; Liu, B.; Li, L. Investigation on Optimal Detection Position of DC Electromagnetic NDT in Crack Characterization for High-Speed Rail Track. In Proceedings of the I2MTC 2019—2019 IEEE International Instrumentation and Measurement Technology Conference, Proceedings 2019, Auckland, New Zealand, 20–23 May 2019. [CrossRef]
65. Wang, P.; Gao, Y.; Tian, G.; Wang, H. Velocity Effect Analysis of Dynamic Magnetization in High Speed Magnetic Flux Leakage Inspection. *NDT E Int.* **2014**, *64*, 7–12. [CrossRef]
66. Rajamäki, J.; Vippola, M.; Nurmikolu, A.; Viitala, T. Limitations of Eddy Current Inspection in Railway Rail Evaluation. *Proc. Inst. Mech. Eng. Part F J. Rail Rapid Transit* **2018**, *232*, 121–129. [CrossRef]
67. Gao, J.; Du, G.; Wei, H. The Research of Defect Detection Test System Based on Magnetic Flux Leakage. In Proceedings of the 6th International Forum on Strategic Technology, IFOST 2011, Harbin, China, 22–24 August 2011; pp. 1225–1229. [CrossRef]
68. Chen, Z.; Xuan, J.; Wang, P.; Wang, H.; Tian, G. Simulation on High Speed Rail Magnetic Flux Leakage Inspection. In Proceedings of the Conference Record—IEEE Instrumentation and Measurement Technology Conference 2011, Hangzhou, China, 10–12 May 2011; pp. 760–764. [CrossRef]
69. Wang, P.; Xiong, L.; Sun, Y.; Wang, H.; Tian, G. Features Extraction of Sensor Array Based PMFL Technology for Detection of Rail Cracks. *Meas. J. Int. Meas. Confed.* **2014**, *47*, 613–626. [CrossRef]
70. Ji, K.; Wang, P.; Jia, Y.; Ye, Y.; Ding, S. Adaptive Filtering Method of MFL Signal on Rail Top Surface Defect Detection. *IEEE Access* **2021**, *9*, 87351–87359. [CrossRef]
71. Yang, L.; Geng, H.; Gao, S. Study on High-Speed Magnetic Flux Leakage Testing Technology Based on Multistage Magnetization. *Yi Qi Yi Biao Xue Bao/Chin. J. Sci. Instrum.* **2018**, *39*, 148–156. [CrossRef]
72. Yuan, F.; Yu, Y.; Liu, B.; Tian, G. Investigation on Velocity Effect in Pulsed Eddy Current Technique for Detection Cracks in Ferromagnetic Material. *IEEE Trans. Magn.* **2020**, *56*, 6201008. [CrossRef]
73. Liu, Z.; Koffman, A.D.; Waltrip, B.C.; Wang, Y. Eddy Current Rail Inspection Using AC Bridge Techniques. *J. Res. Natl. Inst. Stand. Technol.* **2013**, *118*, 140–149. [CrossRef]
74. Piao, G.; Li, J.; Udpa, L.; Udpa, S.; Deng, Y. The Effect of Motion-Induced Eddy Currents on Three-Axis MFL Signals for High-Speed Rail Inspection. *IEEE Trans. Magn.* **2021**, *57*, 6200211. [CrossRef]
75. Wu, Y.; Gao, B.; Zhao, J.; Liu, Z.; Luo, Q.; Shi, Y.; Xiong, L.; Tian, G.Y. Induction Thermography for Rail Nondestructive Testing under Speed Effect. In Proceedings of the 2018 IEEE Far East NDT New Technology and Application Forum, FENDT 2018, Xiamen, China, 6–8 July 2018; pp. 185–189. [CrossRef]

76. Peng, J.; Tian, G.Y.; Wang, L.; Zhang, Y.; Li, K.; Gao, X. Investigation into Eddy Current Pulsed Thermography for Rolling Contact Fatigue Detection and Characterization. *NDT E Int.* **2015**, *74*, 72–80. [CrossRef]
77. Gao, Y. Research on Nondestructive Detection of Rail Cracks in Multiphysical Electromagnetic and Thermal Imaging. Ph.D. Thesis, Nanjing University of Aeronautics and Astronautics, Nanjing, China, 2018.
78. Gibert, X.; Patel, V.M.; Chellappa, R. Robust Fastener Detection for Autonomous Visual Railway Track Inspection. In Proceedings of the 2015 IEEE Winter Conference on Applications of Computer Vision, WACV 2015, Waikoloa, HI, USA, 5–9 January 2015; pp. 694–701. [CrossRef]
79. Gao, B.; Bai, L.; Woo, W.L.; Tian, G.Y.; Cheng, Y. Automatic Defect Identification of Eddy Current Pulsed Thermography Using Single Channel Blind Source Separation. *IEEE Trans. Instrum. Meas.* **2014**, *63*, 913–922. [CrossRef]
80. Usamentiaga, R.; Sfarra, S.; Fleuret, J.; Yousefi, B.; Garcia, D. Rail Inspection Using Active Thermography to Detect Rolled-in Material. In Proceedings of the 14th Quantitative InfraRed Thermography Conference, Berlin, Germany, 25–29 June 2018; pp. 845–852. [CrossRef]
81. Gao, Y.; Tian, G.Y.; Wang, P.; Wang, H. Emissivity Correction of Eddy Current Pulsed Thermography for Rail Inspection. In Proceedings of the 2016 IEEE Far East NDT New Technology and Application Forum, FENDT 2016, Nanchang, China, 22–24 June 2016; pp. 108–112. [CrossRef]
82. Lu, X.; Tian, G.; Wu, J.; Gao, B.; Tian, P. Pulsed Air-Flow Thermography for Natural Crack Detection and Evaluation. *IEEE Sens. J.* **2020**, *20*, 8091–8097. [CrossRef]
83. Zhang, X.; Gao, B.; Shi, Y.; Woo, W.L.; Li, H. Memory Linked Anomaly Metric Learning of Thermography Rail Defects Detection System. *IEEE Sens. J.* **2021**, *21*, 24720–24730. [CrossRef]
84. Ramzan, B.; Malik, S.; Ahmad, S.M.; Martarelli, M. Railroads Surface Crack Detection Using Active Thermography. In Proceedings of the 18th International Bhurban Conference on Applied Sciences and Technologies, IBCAST 2021, Islamabad, Pakistan, 12–16 January 2021; pp. 183–197. [CrossRef]
85. Guo, L.; Zhang, J.; Chen, Z.; Sun, L.; Ge, J.; Lü, K.; Dai, G. Automatic Detection for Defects of Railroad Track Surface. *Appl. Mech. Mater.* **2013**, *278–280*, 856–860. [CrossRef]
86. Fu, S.; Jiang, Z. Research on Image-Based Detection and Recognition Technologies for Cracks on Rail Surface. In Proceedings of the 2019 International Conference on Robots and Intelligent System, ICRIS 2019, Haikou, China, 15–16 June 2019; pp. 98–101. [CrossRef]
87. Yaman, O.; Karakose, M.; Akin, E. A Vision Based Diagnosis Approach for Multi Rail Surface Faults Using Fuzzy Classificiation in Railways. In Proceedings of the 2nd International Conference on Computer Science and Engineering, UBMK 2017, Antalya, Turkey, 5–8 October 2017; pp. 713–718. [CrossRef]
88. Gan, J.; Li, Q.; Wang, J.; Yu, H. A Hierarchical Extractor-Based Visual Rail Surface Inspection System. *IEEE Sens. J.* **2017**, *17*, 7935–7944. [CrossRef]
89. Liang, Z.; Zhang, H.; Liu, L.; He, Z.; Zheng, K. Defect Detection of Rail Surface with Deep Convolutional Neural Networks. In Proceedings of the World Congress on Intelligent Control and Automation (WCICA), Changsha, China, 4–8 July 2018; pp. 1317–1322. [CrossRef]
90. Lu, J.; Liang, B.; Lei, Q.; Li, X.; Liu, J.; Liu, J.; Xu, J.; Wang, W. SCueU-Net: Efficient Damage Detection Method for Railway Rail. *IEEE Access* **2020**, *8*, 125109–125120. [CrossRef]
91. Zhuang, L.; Qi, H.; Zhang, Z. The Automatic Rail Surface Multi-Flaw Identification Based on a Deep Learning Powered Framework. *IEEE Trans. Intell. Transp. Syst.* **2021**, *23*, 12133–12143. [CrossRef]
92. Yu, H.; Li, Q.; Tan, Y.; Gan, J.; Wang, J.; Geng, Y.A.; Jia, L. A Coarse-to-Fine Model for Rail Surface Defect Detection. *IEEE Trans. Instrum. Meas.* **2019**, *68*, 656–666. [CrossRef]
93. Yuan, H.; Chen, H.; Liu, S.; Lin, J.; Luo, X. A Deep Convolutional Neural Network for Detection of Rail Surface Defect. In Proceedings of the 2019 IEEE Vehicle Power and Propulsion Conference, VPPC 2019—Proceedings 2019, Hanoi, Vietnam, 14–17 October 2019; pp. 2019–2022. [CrossRef]
94. Mao, Q.; Cui, H.; Hu, Q.; Ren, X. A Rigorous Fastener Inspection Approach for High-Speed Railway from Structured Light Sensors. *ISPRS J. Photogramm. Remote Sens.* **2018**, *143*, 249–267. [CrossRef]
95. Divya, V. Crack Detection for Railway Tracks and Accident Prevention. *Int. J. Res. Appl. Sci. Eng. Technol.* **2017**, *V*, 448–450. [CrossRef]
96. Wei, J.; Liu, C.; Ren, T.; Liu, H.; Zhou, W. Online Condition Monitoring of a Rail Fastening System on High-Speed Railways Based on Wavelet Packet Analysis. *Sensors* **2017**, *17*, 318. [CrossRef]
97. Tian, G.Y.; Gao, B. Review of Railway Rail Defect Non-Destructive Testing and Monitoring. *J. Instrum.* **2016**, *37*, 1763–1780. [CrossRef]
98. Jiang, Y.; Wang, H.; Tian, G.; Chen, S.; Zhao, J.; Liu, Q.; Hu, P. Non-Contact Ultrasonic Detection of Rail Surface Defects in Different Depths. Proceedings of 2018 IEEE Far East NDT New Technology and Application Forum, FENDT 2018, Xiamen, China, 6–8 July 2018; pp. 46–49. [CrossRef]
99. Song, Z.; Yamada, T.; Shitara, H.; Takemura, Y. Detection of Damage and Crack in Railhead by Using Eddy Current Testing. *J. Electromagn. Anal. Appl.* **2011**, *3*, 546–550. [CrossRef]
100. Wei, X.; Yang, Z.; Liu, Y.; Wei, D.; Jia, L.; Li, Y. Railway Track Fastener Defect Detection Based on Image Processing and Deep Learning Techniques: A Comparative Study. *Eng. Appl. Artif. Intell.* **2019**, *80*, 66–81. [CrossRef]

101. Chen, R.; Shi, T.; Lv, X. Transmission Performance Analysis of Wireless Sensor Networks under Complex Railway Environment. In Proceedings of the 29th Chinese Control and Decision Conference, CCDC 2017, Chongqing, China, 28–30 May 2017; pp. 2970–2974. [CrossRef]
102. Duan, J.; Shi, T.; Lv, X.; Li, Z. Optimal Node Deployment Scheme for WSN-Based Railway Environment Monitoring System. In Proceedings of the 28th Chinese Control and Decision Conference, CCDC 2016, Yinchuan, China, 28–30 May 2016; pp. 6529–6534. [CrossRef]
103. Lv, X.; Li, J.; Shi, T.; Jia, X. Topology Analysis Based on Linear Wireless Sensor Networks in Monitoring of High-Speed Railways. In Proceedings of the 28th Chinese Control and Decision Conference, CCDC 2016, Yinchuan, China, 28–30 May 2016; pp. 1797–1802. [CrossRef]
104. Germaine, J.T.; Whittle, A.J. Low Cost Monitoring System to Diagnose Problematic Rail Bed: Case Study at a Mud Pumping Site. Ph.D. Thesis, Massachusetts Institute of Technology, Cambridge, MA, USA, 2007; p. 203.
105. Hernandez, A.; Valdovinos, A.; Perez-Diaz-De-Cerio, D.; Valenzuela, J.L. Bluetooth Low Energy Sensor Networks for Railway Applications. In Proceedings of the IEEE Sensors, Glasgow, UK, 29 October–1 November 2017; pp. 1–3. [CrossRef]
106. Gao, M.; Wang, P.; Wang, Y.; Yao, L. Self-Powered ZigBee Wireless Sensor Nodes for Railway Condition Monitoring. *IEEE Trans. Intell. Transp. Syst.* **2018**, *19*, 900–909. [CrossRef]
107. Nallathambi, M.M. Remote Sensor Networks for Condition Monitoring: An Application on Railway Industry. In Proceedings of the 2017 IEEE International Conference on Electrical, Instrumentation and Communication Engineering (ICEICE), Karur, India, 27–28 April 2017. [CrossRef]
108. Kljaic, Z.; Cipek, M.; Mlinaric, T.J.; Pavkovic, D.; Zorc, D. Utilization of Track Condition Information from Remote Wireless Sensor Network in Railways-A Mountainous Rail Track Case Study. In Proceedings of the 27th Telecommunications Forum, TELFOR 2019, Belgrade, Serbia, 26–27 November 2019. [CrossRef]
109. Punetha, D.; Tripathi, D.M.; Kumar, A. A Wireless Approach with Sensor Network for Real Time Railway Track Surveillance System. *Int. J. Eng. Trends Technol.* **2014**, *9*, 426–429. [CrossRef]
110. Philipose, A.; Rajesh, A. Investigation on Energy Efficient Sensor Node Placement in Railway Systems. *Eng. Sci. Technol. Int. J.* **2016**, *19*, 754–768. [CrossRef]
111. Munadi, R.; Sulistyorini, A.E.; Fauzi, F.U.S.; Adiprabowo, T. Simulation and Analysis of Energy Consumption for S-MAC and T-MAC Protocols on Wireless Sensor Network. In Proceedings of the APWiMob 2015—IEEE Asia Pacific Conference on Wireless and Mobile, Bandung, Indonesia, 27–29 August 2015; pp. 142–146. [CrossRef]
112. Shafiullah, G.M.; Azad, S.A.; Ali, A.B.M.S. Energy-Efficient Wireless Mac Protocols for Railway Monitoring Applications. *IEEE Trans. Intell. Transp. Syst.* **2013**, *14*, 649–659. [CrossRef]
113. Philipose, A.; Rajesh, A. Performance Analysis of an Improved Energy Aware MAC Protocol for Railway Systems. In Proceedings of the 2nd International Conference on Electronics and Communication Systems, ICECS 2015, Coimbatore, India, 26–27 February 2015; pp. 233–236. [CrossRef]
114. Zhang, H.; Jiang, H. Research and Application on WSNs of Monitoring High-Speed Rail Infrastructure Based on ZigBee. *Railw. Comput. Appl.* **2013**, *22*, 44–47.
115. Chomsuwan, K.; Srisuthep, N.; Pichitronnachai, C.; Toshiyuki, U. Energy Free Railway Monitoring with Vibrating Magnetostrictive Sensor for Wireless Network Sensor. In Proceedings of the International Conference on Sensing Technology, ICST 2018, Sydney, NSW, Australia, 4–6 December 2017; pp. 1–5. [CrossRef]
116. Sharma, H.; Haque, A.; Jaffery, Z.A. Solar Energy Harvesting Wireless Sensor Network Nodes: A Survey. *J. Renew. Sustain. Energy* **2018**, *10*, 023704. [CrossRef]
117. Shang, Q.; Guo, H.; Liu, X.; Zhou, M. A Wireless Energy and Thermoelectric Energy Harvesting System for Low Power Passive Sensor Network. In Proceedings of the 2020 IEEE MTT-S International Wireless Symposium, IWS 2020—Proceedings 2020, Shanghai, China, 20–23 September 2020; pp. 2020–2022. [CrossRef]
118. Kim, S.; Bang, S.; Chun, K. Temperature Effect on the Vibration-Based Electrostatic Energy Harvester. In Proceedings of the IEEE Region 10 Annual International Conference, Proceedings/TENCON 2011, Sanur, Bali, Indonesia, 21–24 November 2011; pp. 1317–1320. [CrossRef]
119. Zhang, X.; Zhang, Z.; Pan, H.; Salman, W.; Yuan, Y.; Liu, Y. A Portable High-Efficiency Electromagnetic Energy Harvesting System Using Supercapacitors for Renewable Energy Applications in Railroads. *Energy Convers. Manag.* **2016**, *118*, 287–294. [CrossRef]
120. Li, J.; Jang, S.; Tang, J. Implementation of a Piezoelectric Energy Harvester in Railway Health Monitoring. In Proceedings of the Sensors and Smart Structures Technologies for Civil, Mechanical, and Aerospace Systems 2014, San Diego, CA, USA, 9–13 March 2014; Volume 9061, p. 90612. [CrossRef]
121. Alva, G.; Liu, L.; Huang, X.; Fang, G. Thermal Energy Storage Materials and Systems for Solar Energy Applications. *Renew. Sustain. Energy Rev.* **2017**, *68*, 693–706. [CrossRef]
122. Kalaagi, M.; Seetharamdoo, D. Electromagnetic Energy Harvesting Systems in the Railway Environment: State of the Art and Proposal of a Novel Metamaterial Energy Harvester. In Proceedings of the 13th European Conference on Antennas and Propagation, EuCAP 2019, Krakow, Poland, 31 March–5 April 2019; pp. 7–11.
123. Tianchen, Y.; Jian, Y.; Ruigang, S.; Xiaowei, L. Vibration Energy Harvesting System for Railroad Safety Based on Running Vehicles. *Smart Mater. Struct.* **2014**, *23*, 125046. [CrossRef]

124. Zhao, X.; Wei, G.; Li, X.; Qin, Y.; Xu, D.; Tang, W.; Yin, H.; Wei, X.; Jia, L. Self-Powered Triboelectric Nano Vibration Accelerometer Based Wireless Sensor System for Railway State Health Monitoring. *Nano Energy* **2017**, *34*, 549–555. [CrossRef]
125. Wang, J.; Shi, Z.; Xiang, H.; Song, G. Modeling on Energy Harvesting from a Railway System Using Piezoelectric Transducers. *Smart Mater. Struct.* **2015**, *24*, 105017. [CrossRef]
126. Gao, M.; Wang, P.; Cao, Y.; Chen, R.; Cai, D. Design and Verification of a Rail-Borne Energy Harvester for Powering Wireless Sensor Networks in the Railway Industry. *IEEE Trans. Intell. Transp. Syst.* **2017**, *18*, 1596–1609. [CrossRef]

 sensors

MDPI

Article

Mechanic-Electric-Thermal Directly Coupling Simulation Method of Lamb Wave under Temperature Effect

Xiaofei Yang, Zhaopeng Xue, Hui Zheng, Lei Qiu * and Ke Xiong

Research Center of Structural Health Monitoring and Prognosis, State Key Laboratory of Mechanics and Control of Mechanical Structures, Nanjing University of Aeronautics and Astronautics, Nanjing 210016, China
* Correspondence: lei.qiu@nuaa.edu.cn

Abstract: Lamb Wave (LW)-based structural health monitoring method is promising, but its main obstacle is damage assessment in varying environments. LW simulation based on piezoelectric transducers (referred to as PZTs) is an efficient and low-cost method. This paper proposes a multi-physics simulation method of LW propagation with the PZTs under temperature effect. The effect of temperature on LW propagation is considered from two aspects. On the one hand, temperature affects the material parameters of the structure, the adhesive layers and the PZTs. On the other hand, it is considered that the thermal stress caused by the inconsistency of thermal expansion coefficients among the structure, the adhesive layers, and the PZTs affect the piezoelectric constant of the PZTs. Based on the COMSOL Multiphysics, the mechanic–electric–thermal directly coupling simulation model under temperature effect is established. The simulation model consists of two steps. In the first step, the thermal-mechanic coupling is carried out to calculate the thermal stress, and the thermal stress effect is introduced into the piezoelectric constant model. In the second step, mechanic–electric coupling is carried out to simulate LW propagation, which considers the piezoelectric effect of the PZTs for the LW excitation and reception. The simulation results at −20 °C to 60 °C are obtained and compared to the experiment. The results show that the A_0 and S_0 mode of simulation signals match well with the experimental measurements. Additionally, the effect of temperature on LW propagation is consistent between simulation and experiment; that is, the amplitude increases, and the phase velocity decreases with the increment of temperature.

Keywords: structural health monitoring; Lamb Wave; multiphysics simulation; temperature effect; thermal stress

Citation: Yang, X.; Xue, Z.; Zheng, H.; Qiu, L.; Xiong, K. Mechanic-Electric-Thermal Directly Coupling Simulation Method of Lamb Wave under Temperature Effect. *Sensors* **2022**, 22, 6647. https://doi.org/10.3390/s22176647

Academic Editor: Simon Laflamme

Received: 3 August 2022
Accepted: 30 August 2022
Published: 2 September 2022

1. Introduction

Structural Health Monitoring (SHM) technology has the advantages of real-time monitoring, reducing detection times and improving detection efficiency and condition-based maintenance. It has been widely studied and applied in the field of aerospace [1–3]. The Lamb Wave (LW)-based SHM method is a promising one, which has the advantages of high sensitivity and large area monitoring through piezoelectric transducers (PZTs) network. The method's fundamental principle is to directly attach PZTs to a monitored structure and acquire LW signals via the excitation and sensing of these PZTs. The damage can be measured by analyzing the LW features that the damage has changed. Therefore, it shows great application potential in structural damage assessment [4–6]. However, in recent decades, the development of the LW-based SHM method from theoretical and basic research to engineering application has been quite slow [7–9].

Time-varying conditions are one of the main obstacles to the development of LW-based SHM [10]. When LW propagates in aircraft structures, it will inevitably be affected by time-varying conditions, including temperature, moisture, load, and aerodynamic noise. The extracted damage features are affected by these environmental factors, which reduce the reliability of damage monitoring. A variety of damage monitoring methods have been developed, such as baseline free method, environmental compensation method and probabilistic model method, to minimize the impact of time-varying conditions [11–15]. In order to further study and verify these damage monitoring methods in aerospace applications, a large number of LW signals must be obtained under time-varying conditions.

Although the LW signals under time-varying conditions can be obtained by experiment and simulation, respectively, it is expensive and time-consuming to obtain LW signals via experiment—especially, those experiments carried out on real aircraft structures under in-service conditions. However, the LW simulation under time-varying conditions is an effective and low-cost method, which can not only study the LW propagation in an aircraft structure but also can be used to verify the related damage monitoring methods.

In recent decades, Finite Element Analysis (FEA)-based LW simulation methods [16–22] have been intensively explored for simulating LW propagation in metallic and composite structures. However, the LW simulation method under time-varying conditions considering the PZTs attached to structures is rarely reported.

The varying temperature condition is a significant one. It has been studied that the temperature can cause significant changes in the velocity and amplitude of LW signals acquired by the PZTs [23,24]. The influence of temperature on LW propagation is studied via theoretical studies and numerical analysis [25–28]. The results show that the main reason for the variations of the LW velocity and amplitude under the temperature effect is due to the change in structural materials and the piezoelectric properties, including the piezoelectric constant and dielectric constant. The effect of temperature change on the LW is investigated over a temperature range of −200 °C to 204 °C via ABAQUS [29]. Since the influence of temperature on material parameters is not considered, temperature change in the range of −200 °C to 93 °C has no effect on the displacement responses. Attarian et al. [30] experimentally investigated that thermal cycling reduces the sensitivity of damage diagnosis because the properties of adhesive layers have changed and how the influence of the adhesive layers is not negligible. So far, the influence of thermal stress caused by the inconsistency of thermal expansion coefficients among different materials on LW propagation has not been studied.

To simulate the temperature influence on LW, Lonkar et al. [31] studied the piezo-enabled spectral element analysis method containing a piezoelectric model for LW propagation. In the simulation model, the piezoelectric constant and dielectric constant of the piezoelectric sheet and the shear modulus, Poisson's ratio and elastic modulus of the adhesive layers are considered as a function of temperature. Palazotto et al. [32] developed a 2D simulation model containing a piezoelectric sheet and investigated the influence of temperature on LW on aluminum plates using ABAQUS simulation software. The influence of temperature on the elastic modulus and Poisson's ratio of aluminum plate is considered in the simulation model. The results show that LW propagation velocity decreases with the increment of temperature. Yule et al. [33] conducted a 2D-guided wave simulation based on COMSOL Multiphysics software, considering the influence of temperature on material parameters, and the influence law of temperature on LW was obtained. However, these studies neglected the influence of thermal stress caused by the inconsistency of thermal expansion coefficients among different materials on LW propagation. The piezoelectric constant of PZT is very sensitive to thermal stress. Thermal stress leads to the change of the piezoelectric constant of PZT, which will affect LW propagation. Therefore, thermal stress needs to be considered.

This paper proposes a multiphysics simulation method for LW propagation with the PZTs under temperature effect. The simulation model includes the changes in material parameters of the structure, the adhesive layers, and the PZTs with temperature. In particular, the thermal stress that affects the piezoelectric constant due to the inconsistency of the thermal expansion coefficients among the three is considered. The FEA model of LW propagation with the PZTs under temperature effect is constructed based on the COMSOL Multiphysics computational platform. The LW signals under the temperature effect can be derived directly from the simulation model. The results show that the waveform of simulation signals matches well with the experimental measurements, which indicates that the simulation method is feasible.

2. Simulation Mechanisms of Temperature Influence on LW

Take the plate-like structure as an example to illustrate the simulation mechanisms of LW under the temperature effect [34], as shown in Figure 1. Two PZTs are attached to the structure with an adhesive and are used as an LW actuator and LW sensor, respectively. A voltage excitation waveform is provided to the actuator to generate the LW signal. The actuator deforms and transmits the stress to the adhesive layers and the structure due to the inverse piezoelectric effect. When LW propagates to the sensor in the structure, it is received and converted into a voltage signal due to the direct piezoelectric effect. It is proposed to simplify the problem of temperature affecting LW propagation into two parts. One part is that temperature affects the material parameters of the structure, adhesive layers, and PZTs. The other part is that thermal stress affects the piezoelectric constant of the PZTs due to the inconsistency of thermal expansion coefficients among the structure and the adhesive layers, and the PZTs. The two parts are superimposed for LW propagation simulation under the temperature effect. In this Section, two parts of the influence mechanisms are introduced, and the numerical model used in this paper is established.

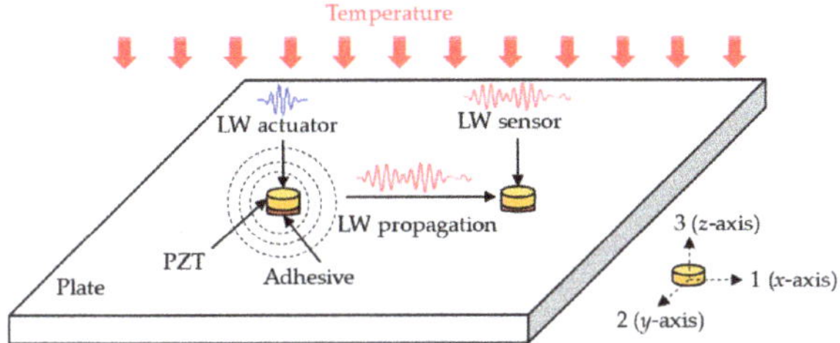

Figure 1. Schematic diagram of LW propagation with the PZTs under temperature effect.

2.1. Excitation–Propagation–Sensing Model

2.1.1. Excitation Model

When the PZT is employed as an excitation element, the PZT's principle for LW excitation is the inverse piezoelectric effect. The piezoelectric constitutive equations are presented in the following Equation (1).

$$\begin{aligned} \boldsymbol{e} &= \boldsymbol{d}^{\mathrm{T}}\boldsymbol{E} + \boldsymbol{s\sigma} \\ \boldsymbol{D} &= \boldsymbol{\varepsilon}\mathbf{E} + \boldsymbol{d\sigma} \end{aligned} \tag{1}$$

where $\boldsymbol{e}$ and $\boldsymbol{E}$ refer to strain vector and electric field vector, respectively, $\boldsymbol{s}$ is the elastic compliance matrix, $\boldsymbol{\sigma}$ is the stress vector, $\boldsymbol{\varepsilon}$ is the dielectric constant matrix, and $\boldsymbol{d}$ and $\boldsymbol{D}$ refer to the piezoelectric coefficient matrix and electric displacement vector, respectively.

According to Giurgiutiu's study [35], the strain transfer model between the actuator and the structure is derived by converting the input voltage to the mechanical strain and then by calculating the shear-lag model between the top and lower adhesive layer interfaces. The schematic diagram of LW excitation is shown in Figure 2.

Where l_{act} and t_{act} are the diameter and thickness of the PZT, respectively, t_{bond} is the thickness of the adhesive layer, t_{plate} is the thickness of the plate structure, G_{bond} is the shear strength of the adhesive layer, Y_{act}^{E} is Young's modulus of the PZT, and E_{plate} is elastic modulus of the structure.

In this paper, we use a circular PZT with d_{31} = d_{32} and excite the LW by applying a voltage excitation signal in the 3-direction. Only the piezoelectric constant d_{31} needs to be considered in m/V. The driving strain at the bottom of the excitation sensor along the x-direction is expressed in the following Equation (2), where V_{in} is the excitation voltage.

$$\varepsilon_{act}(t) = -\frac{d_{31}V_{in}(t)}{t_{act}} \tag{2}$$

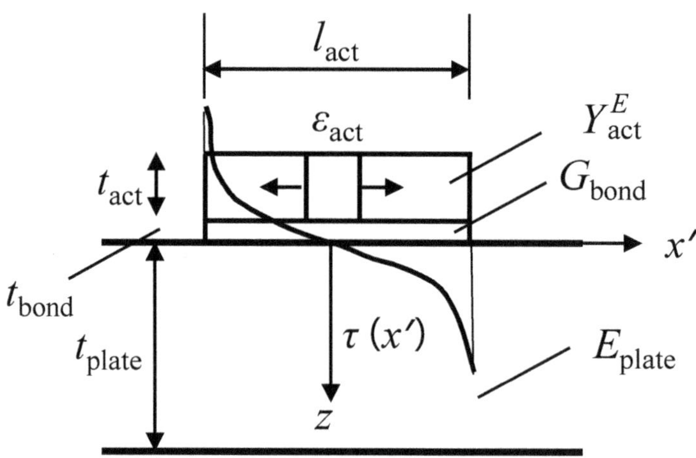

Figure 2. Schematic diagram of LW excitation [35].

Based on the shear-lag model, the stress generated by the PZT coupled to the structure through the adhesive layer is shown in Equation (3).

$$\tau(x) = -\frac{G_{bond}}{t_{bond}} \times \frac{l_{act}\varepsilon_{act}}{\Gamma \cosh \Gamma} \sinh\left[\Gamma\left(\frac{2x}{l_{act}}\right)\right] \tag{3}$$

where Γ is the shear-lag coefficient, and the expression is shown in Equation (4). When $\alpha = 1$, shear excitation is used to excite S mode. When $\alpha = 3$, bending excitation is used to excite A mode.

$$\Gamma^2 = \frac{G_{bond}l_{act}^2}{t_{bond}}\left(\frac{1}{Y_{act}^E t_{act}} + \frac{\alpha}{E_{plate}t_{plate}}\right) \tag{4}$$

2.1.2. Propagation Model

For the plate structure with a free surface, the displacement and stress in the isotropic plate can be simplified into the LW frequency equation in symmetric and antisymmetric mode without consideration for in-plate stress, as shown in Equations (5) and (6).

$$\frac{\tan(qh)}{\tan(ph)} = -\frac{4k^2pq}{(q^2-k^2)^2} \tag{5}$$

$$\frac{\tan(qh)}{\tan(ph)} = -\frac{(q^2-k^2)^2}{4k^2pq} \tag{6}$$

where p and q are given by

$$p^2 = \left(\frac{\omega^2}{c_L^2} - k^2\right),\ q^2 = \left(\frac{\omega^2}{c_T^2} - k^2\right) \tag{7}$$

where h is half the thickness of the plate, k is wave number, ω is angel frequency, c_L and c_T are the velocities of the longitudinal wave and transverse wave, respectively, as shown in Equations (8) and (9).

$$c_L = \sqrt{\frac{Y_{plate}^{E}\left(1-\nu_{plate}\right)}{\rho_{plate}(1+\nu_{plate})(1-2\nu_{plate})}} \tag{8}$$

$$c_T = \sqrt{\frac{Y_{plate}^{E}}{2\rho_{plate}(1+\nu_{plate})}} \tag{9}$$

2.1.3. Sensing Model

Assuming that there is no loss in LW propagation, the strain at the sensor is obtained according to the basic equation in the adhesive shear-lag model, and the voltage output can be obtained according to the direct piezoelectric effect [26], as shown in Equation (10).

$$V_{out}(t) = d_{31}^{act} C_{act}(\Gamma) C_{sen}(\Gamma) \left[\frac{d_{31}}{e_{33} s_{13}(1-\nu_{act})}\right]_{sen} V_{in}(t) \tag{10}$$

where ν_{act} is the Poisson's ratio of the PZT, e_{33} is the dielectric constant of the PZT, s_{13} is the elastic coefficient of the PZT, and $C(\Gamma)$ is a function of the Γ shear-lag parameter, as shown in Equations (11) and (12).

$$C_{act}(\Gamma) = \frac{G_{bond}(1+\nu_{act})R}{t_{bond}[(\Gamma R)I_0(\Gamma R) - (1-\nu_{act})I_1(\Gamma R)]} I_1(\Gamma R) \tag{11}$$

$$C_{sen}(\Gamma) = \frac{\iint \left(\overset{\frown}{\varepsilon}{}_{rr}^{sen} + \overset{\frown}{\varepsilon}{}_{\theta\theta}^{sen}\right) r dr d\theta \,|_{\tau_0=1}}{\pi R^2} \tag{12}$$

where R is the radius of the PZT. $I(\Gamma R)$ is the Bessel function.

The LW theoretical model and the related temperature factors that affect the LW propagation velocity and response amplitude are summarized in Table 1. On the one hand, the temperature affects the material parameters of the structure, the adhesive layer and the PZT. On the other hand, the thermal expansion coefficients of the three are inconsistent, so the PZT is affected by thermal stress, which changes the mechanic-to-electric conversion characteristics. This is equivalent to the change produced by the external load on the PZT. Since the PZT is sensitive to stress, the thermal stress caused by thermal expansion is not negligible, so the change in temperature is accompanied by the load effect.

Table 1. Summary of Lamb Wave theory model and key parameters influenced by temperature.

LW Propagation Characteristics	Expressions	Temperature Effects
Propagation velocity	$c_L = \sqrt{\frac{Y_{plate}^{E}(1-\nu_{plate})}{\rho_{plate}(1+\nu_{plate})(1-2\nu_{plate})}}$ $c_T = \sqrt{\frac{Y_{plate}^{E}}{2\rho_{plate}(1+\nu_{plate})}}$	Temperature affects the propagation velocity of LW by influencing the elastic modulus Y_{plate}^{E}, density ρ_{plate}, and Poisson's ratio ν_{plate} of the structure.
Response amplitude	$V_{out}(t) = d_{31}^{act} C_{act}(\Gamma) C_{sen}(\Gamma) \left[\frac{d_{31}}{e_{33} s_{13}(1-\nu_{act})}\right]_{sen} V_{in}(t)$ $\Gamma^2 = \frac{G_{bond}\, l_{act}^2}{t_{bond}} \left(\frac{1}{Y_{act}^{E} t_{act}} + \frac{\alpha}{E_{plate}\, t_{plate}}\right)$	Temperature affects the LW amplitude by influencing piezoelectric coefficient d_{31} including the effects of thermal stress, the dielectric constant e_{33} and elastic flexibility coefficient s_{13} of the PZT, shear modulus G_{bond} and shear-lag constant of the adhesive layer.

2.2. Material Parameters under Temperature Effect

In this paper, we take 2024 aluminum alloy as an example. The elastic modulus and Poisson's ratio of 2024 aluminum material measured by laser ultrasound was published by Sandia National Laboratories [36]. Elastic modulus decreases approximately linearly with the increasing temperature, and Poisson's ratio increases approximately linearly with the increasing temperature, as shown in Equations (13) and (14).

$$E_{\text{plate}}(\Delta T) = 73.5 - 0.06 \times \Delta T \tag{13}$$

$$v(\Delta T) = 0.344 + 5.13 \times 10^{-5} \times \Delta T \tag{14}$$

According to the parameter manual of a typical two-component epoxy adhesive, its working temperature range is −55 °C to 250 °C, and its storage modulus at room temperature is 3.61 Gpa with a thermal expansion rate of $54 \times 10^{-6}/°C$. Barakat et al. [37] gave the curve of storage modulus with temperature. The numerical model for the variation of adhesive shear modulus with temperature is shown in Equation (15).

$$E_{\text{bond}}(\Delta T) = 3.61 - 0.01 \times \Delta T \tag{15}$$

Compared to metallic structures, piezoelectric materials contain both force and electrical properties, so the parameters affected by temperature are more complex. By reviewing the relevant literature, NASA reported some measurement results about the variation of PZT parameters with temperature [38]. From the reported results, the impedance, mechanic-electric coupling coefficient, and dielectric loss of PZT-5A can be approximately constant in the range of −55 °C to 100 °C, which can be neglected in the simulation. As the temperature increases, the dielectric constant and piezoelectric constant show a linear increasing trend, which needs to be considered in the simulation.

The numerical model of the piezoelectric constant and dielectric constant of PZT-5A was studied in the temperature range of −20 °C to 60 °C [34], as shown in Equation (16) and Equation (17), in agreement with the findings reported by NASA [38]. This numerical model is used for simulation, but it does not take into account the thermal stress effect.

$$d_{31}(\Delta T) = -167.7 - 0.194 \times \Delta T \tag{16}$$

$$e_{33}(\Delta T) = 2155 + 4.12 \times \Delta T \tag{17}$$

2.3. Piezoelectric Constants under Thermal Stress Effect

The thermal expansion coefficient of the PZT, adhesive layer, and aluminum structure are $3 \times 10^{-6}/°C$, $54 \times 10^{-6}/°C$, and $23 \times 10^{-6}/°C$, respectively. The thermal stress is not negligible for the PZT because the difference between the three thermal expansion coefficients is large, and the piezoelectric sensor is sensitive to changes in stress.

For most materials, warming causes expansion and cooling causes contraction, and tiny cells generate thermal stress due to the constraints of adjacent cells and boundary conditions. It is common to assume a linear relationship between strain and temperature, as shown in Equation (18), which forms a set of intrinsic relationships between strain and temperature. The coefficient of thermal expansion is a fundamental parameter of the material, and the effect of thermal stress can be approximated as a PZT subjected to a static load.

$$\varepsilon x = \varepsilon y = \varepsilon z = \alpha \Delta T \tag{18}$$

When a PZT is subjected to an external load, its polarization state changes. The internal electric dipole moments are aligned in the direction of the polarization field and are confined by the domain walls, thus changing the mechanic-electric transition characteristics of the PZT.

In the application of excitation-response of LW, the mechanic-electric transition characteristics of piezoelectric materials are usually considered to be linear. However, in fact, the piezoelectric material itself is nonlinear, and its mechanic-electric characteristics are usually reflected as a hysteresis curve, which can be approximated as linear due to the small voltage and displacement of the excitation and response of LW.

Qiu et al. [39] studied the variation law of LW amplitude and propagation velocity under static load. It was pointed out that LW propagation velocity is affected by load due to the acoustoelastic effect, and the structure has a nonlinear change in its stress-strain intrinsic relationship under external load. Since the effect of thermal stress on the structure is ignored when discussing thermal stress, the acoustoelastic effect is not considered in this paper. The amplitude is affected by the load because the piezoelectric constant of the PZT is changed. This paper focuses on the effect of thermal stress on the LW amplitude. The nonlinear numerical relationship model between the load and the piezoelectric constant d_{31} is summarized as shown in Equation (19), where σ is the actual stress caused by the load in MPa.

$$d_{31}(\Delta\sigma) = d_{31} + d_{31} \times \left(-1.1 \times 10^{-5} \times \Delta\sigma^2 + 4.2 \times 10^{-3} \times \Delta\sigma\right) \quad (19)$$

In summary, the numerical model of the influence of static load on the piezoelectric constant studied by scholar Qiu is used to simulate the influence of thermal stress on the piezoelectric constant. The numerical models of the influence of temperature on the piezoelectric constant d_{31} and the influence of thermal stress on the piezoelectric constant d_{31} are superimposed as the numerical model of the temperature-influenced LW propagation simulation, including thermal stress, as shown in Equation (20).

$$d_{31}\left(\Delta T, \Delta\sigma_T\right) = -167.7 - 0.194 \times \Delta T - 167.7 \times \left(-1.1 \times 10^{-5} \times \Delta{\sigma_T}^2 + 4.2 \times 10^{-3} \times \Delta\sigma_T\right) \quad (20)$$

where ΔT is relative to the reference temperature 20 °C, in °C, $\Delta\sigma_T$ is the thermal stress, in MPa.

3. Simulation Method of LW under Temperature Effect

3.1. Architecture of the Multiphysics Simulation Method

Three kinds of physics need to be coupled with each other to simulate the LW under temperature effect. One is mechanic-electric coupling controlled by a piezoelectric instantaneous equation, used to simulate LW excitation and sensing. The other is solid mechanics controlled by fluctuation equations, which is used to simulate LW propagation. The temperature field is only a factor affecting LW propagation; that is, it is weakly coupled with the solid mechanics field and electrostatic field.

This paper adopts the COMSOL Multiphysics platform. Electrostatic and solid mechanics can be directly coupled. In addition, the thermal stress effect, piezoelectric effect, and temperature effect of PZTs can be integrated into the simulation model. The multiphysics simulation architecture of LW propagation with PZTs under temperature effect is shown in Figure 3.

In taking an aluminum plate as the research object, nine PZTs are arranged on an aluminum plate with an adhesive to simulate LW under temperature effect, as shown in Figure 4. The working condition of the simulation is shown in Table 2, described in detail as follows.

Table 2. Working condition of LW propagation simulation under temperature effect.

Structure	Geometry	Excitation Signal Frequency	Temperature
2024 Aluminum plate	500 mm × 500 mm × 2 mm (length × width × thickness)	150 kHz, 200 kHz	−20 °C to 60 °C

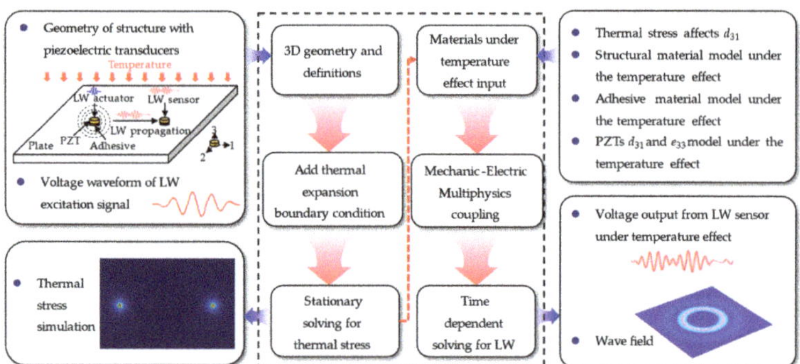

Figure 3. Architecture of multiphysics simulation method of LW under temperature effect.

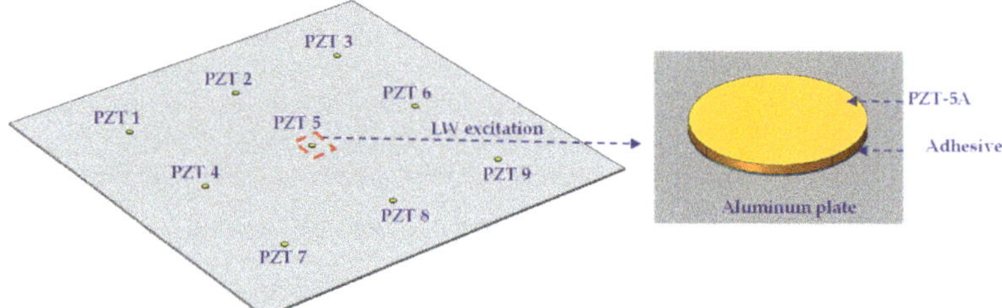

Figure 4. LW propagation under temperature effect simulation model.

3.1.1. D Geometry and Definitions

In this paper, the 3D geometry used includes the aluminum plate, the PZTs, and the adhesive layers. The size of the plate is 500 mm × 500 mm × 2 mm. The diameter and thickness of the PZTs are 8 mm and 0.48 mm, respectively. Nine PZTs are arranged symmetrically. The distance between two PZTs is 150 mm. The diameter and thickness of the adhesive layers are 8 mm and 0.08 mm, respectively.

There are two parts to define. One part is to define the voltage excitation signal that is modulated by a Hanning window, as shown in Equation (21). Parameters are set to A = 35 V, f = 200 kHz and N = 5. The other part is to define simulation temperature, reference temperature, and numerical model coefficients for material parameters under the influence of temperature.

$$Ex = A \cdot [1 - \cos(2\pi f t/N)] \cdot \sin(2\pi f t) \cdot [t < (N/f)] \tag{21}$$

where Ex is the excitation signal, A is the amplitude, f is the central frequency, t is the wave propagating duration, and N is the number of cycles within the signal window.

3.1.2. Material Parameters Numerical Model of Temperature Effect

According to the study in Sections 2.2 and 2.3, the temperature influence on LW propagation can be equated to the effect of temperature on the material parameters of the structure, PZTs, and adhesive layers and the effect of thermal stress load on the material parameters of piezoelectric ceramics due to the inconsistent coefficient of thermal expansion. The numerical models of each part used in this paper are summarized, as shown in Table 3.

3.1.3. Multiphysics Coupling under Temperature Effect

In solid mechanics, the default linear elastic materials, free and initial values are assigned to the aluminum structure, the adhesive layers, and the PZTs, while piezoelectric material is only assigned to the PZTs, and low reflection boundary is only assigned to the aluminum structure to suppress boundary reflection. In electrostatics, the upper surface of only one PZT is set as electric potential to apply a defined voltage signal, and the lower surface of all PZTs is set as ground. The voltage response signals of all PZTs can be received by defining the upper surfaces of all PZTs as boundary probes. The COMSOL will directly couple two physical fields in the calculation process by adding a piezoelectric effect to mulphysical fields and selecting solid mechanics and electrostatic.

Table 3. Material properties of aluminum plate, adhesive, and PZT (PZT-5A).

Material	Parameter	Value
2024 Aluminum plate	Elastic modulus	$E_{plate}(\Delta T)(\text{GPa}) = 73.5 - 0.06\Delta T$
	Poisson's ratio	$v(\Delta T) = 0.344 + 5.13 \times 10^{-5}\Delta T$
	Density	2700 (kg/m^3)
	Coefficient of thermal expansion	23.1×10^{-6}(/K)
Adhesive	Shear modulus	$G_{bond}(\Delta T)(\text{GPa}) = 3.61 - 0.01\Delta T$
	Poisson's ratio	0.3
	Density	1110 (kg/m^3)
	Coefficient of thermal expansion	54×10^{-6} (/K)
PZT-5A	Piezoelectric constant	$d_{31}(\Delta T, \Delta\sigma_T) = -167.7 - 0.194 \times \Delta T$ $-167.7(-1.1 \times 10^{-5} \times \Delta\sigma_T{}^2 + 4.2 \times 10^{-3} \times \Delta\sigma_T)$
	Relative permittivity	$e_{33}(\Delta T) = 4.12 \times \Delta T + 2155$
	Coefficient of thermal expansion	3×10^{-6} (/K)

For the calculation of thermal stress, it is necessary to add the thermal expansion to the nodes of linear elastic material and piezoelectric material, and then set the reference temperature to T0 and set the target temperature to Tem. It is considered that the temperature of structure, adhesive layers, and PZTs are the same, so it is not necessary to solve the solid heat transfer problem.

In solid mechanics physics, it is necessary to add rigid motion suppression to aluminum structure when calculating thermal stresses. The reason is that if there is no displacement constraint, the analysis will not converge without a unique solution, and the model will indicate "no solution found". In the stationary study, it is necessary to find an equilibrium solution where the object is free to deform but not free to move or rotate, so the reaction forces must balance with each other. If no constraint is provided, the unbalanced forces will move or rotate the object, and the stationary solver cannot converge. COMSOL provides rigid motion suppression boundary conditions that can handle the missing displacement constraints and obtain the correct thermal stress results.

3.1.4. Finite Element Meshes

Yang et al. [40] gave the relationship between the finite element size and the LW wavelength. The maximum mesh size recommended in the literature is 1/10 to 1/6 of the minimum wavelength. For LW at 200 kHz, the phase velocities of S_0 mode and A_0 mode on a 2 mm thick aluminum plate are 5382 m/s and 1731 m/s, respectively, and the corresponding wavelengths are 27 mm and 9 mm, respectively. Therefore, the maximum mesh size of the aluminum plate shall be less than 1.5 mm. Considering that the thickness of PZTs and adhesive layers are 0.48 mm and 0.08 mm, the mesh size of the PZTs and the adhesive layers are set to 1 mm and 0.5 mm, respectively. Due to the small mesh size, the model contains 1,890,000 domain elements, 553,000 boundary elements, and 3800 edge elements, with 5,200,000 degrees of freedom.

3.1.5. Stationary and Time-Dependent Solver Settings

The solver includes the stationary study for thermal stress simulation and the time-dependent study for LW propagation. First, the stationary study is used to calculate the thermal stress of piezoelectric elements caused by inconsistent thermal expansion coefficient in step 1. Then, the thermal stress value is input into the parameter table. Finally, the time-dependent study is used to calculate LW propagation in step 2.

In the thermal stress simulation, the electrostatics and piezoelectric effect are disabled. In the LW propagation simulation, the thermal expansion and rigid motion suppression related to thermal stress simulation are disabled. If not disabled, the piezoelectric element will not only generate the LW signals but also be subjected to the thermal stress of the structure. At the beginning of the time-dependent study, the thermal stress at the coupling part between the piezoelectric element and the structure will act as a transient excitation to generate a wide-band voltage response signal.

When the simulation solver runs once, LW signals at one temperature level can be obtained. After several runs, LW signals at all temperature levels can be obtained. The simulation time step is set to 1×10^{-7} s, and the time range in the time-dependent study is from 0 s to 1.5×10^{-4} s.

3.2. Simulation Results

For LW at 200 kHz, the typical von Mises stress wave fields with the PZTs under temperature effect are shown in Table 4. It can be seen that both the S_0 mode and A_0 mode are excited normally. The S_0 mode propagates faster than the A_0 mode, and the amplitude of the S_0 mode is weaker than that of the A_0 mode. It can be seen that the amplitude increases with the increment of temperature by comparing the color bar of maximum stress at different temperatures. In order to better study the effect of temperature on the LW phase, the region in the red box is selected and enlarged. The wave fields at $t = 3 \times 10^{-5}$ s are given in Figure 5. It can be seen that the phase is delayed with the increment of temperature. The third wave packet just reaches the white line at 20 °C, exceeds the white line at −20 °C, and does not reach the white line at 60 °C.

Table 4. Typical wave fields of LW propagation with the PZTs under temperature effect.

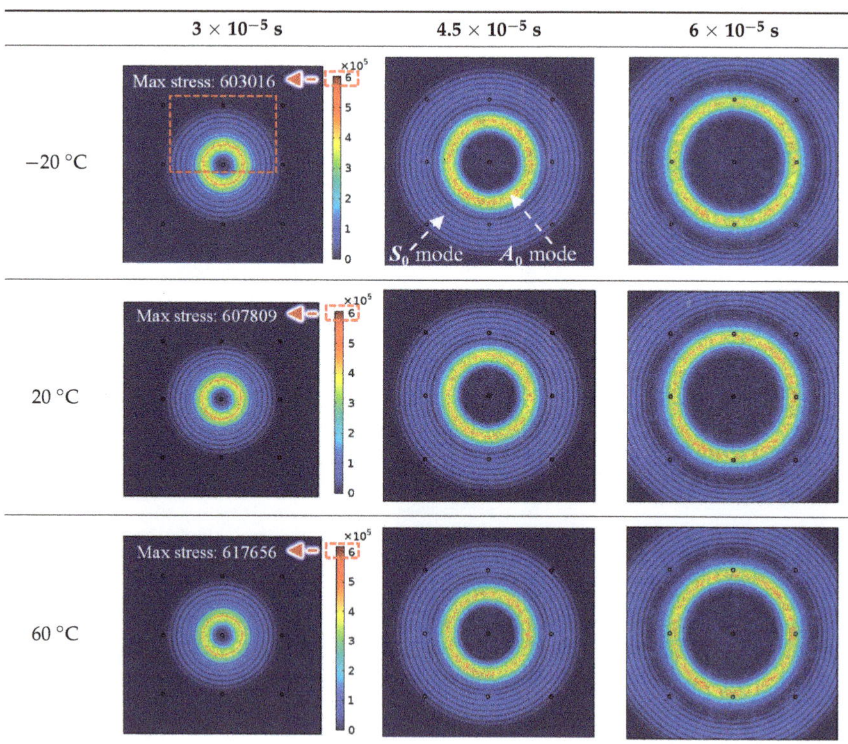

	3×10^{-5} s	4.5×10^{-5} s	6×10^{-5} s
−20 °C	Max stress: 603016	S_0 mode, A_0 mode	
20 °C	Max stress: 607809		
60 °C	Max stress: 617656		

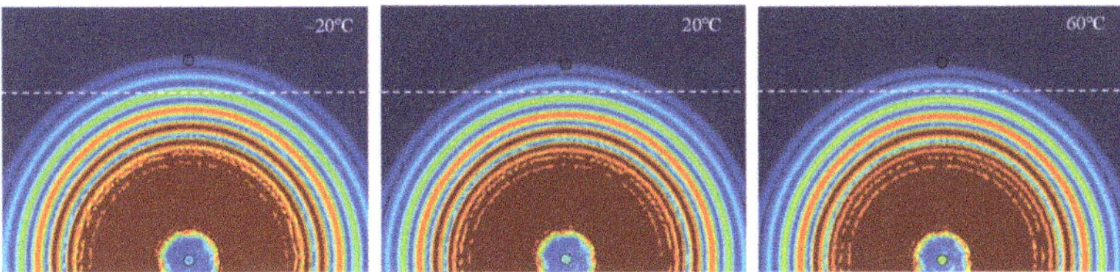

Figure 5. The effect of temperature on LW phase.

The simulated LW signals under different temperatures are given in Figure 6, which are the LW signals of channel 5–6 when the center frequency of the excitation signal is 200 kHz. An enlarged view of S_0 mode is given to observe the changes in phase and amplitude better. It can be seen that the amplitude increases, and the phase delays with the increment of temperature.

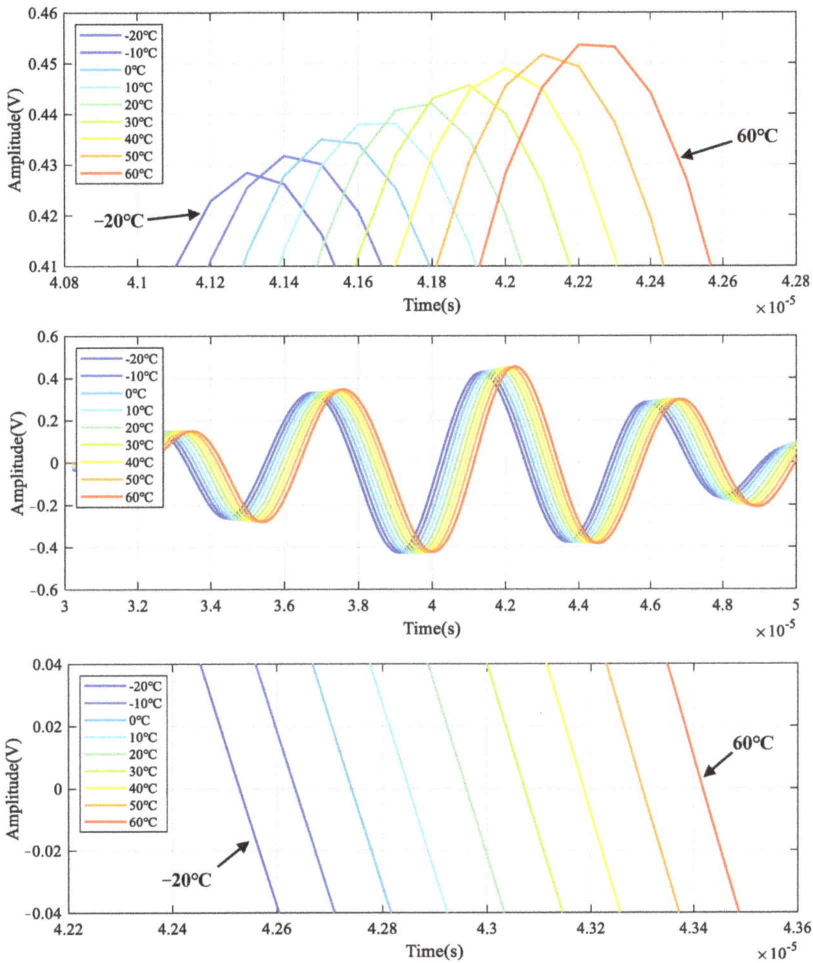

Figure 6. Typical LW response signal results of simulation under temperature effect.

4. Experimental Verification of the Simulation Method

In order to verify whether the above simulation model can correctly and effectively simulate the LW propagation under the influence of temperature, experiments were conducted using an aluminum plate of the same size and material. The same adhesive and the PZTs were arranged at the same position. The LW signals at different temperatures were obtained. The differences between the experimental signals and the simulated signals were compared in terms of amplitude and propagation velocity to verify the correctness of the simulation method.

4.1. Experimental Setup

The geometric dimensions and material parameters of the aluminum plate, the adhesives, and the PZTs used in the experiment are the same as those of the simulation. PZT 5 is used to excite LW. The distance between two PZTs is 150 mm, as shown in Figure 7. The aluminum plate is fixed on an environmental test chamber THV1070W, which is used to provide the required temperature environment. The integrated SHM system [41] is used to excite and obtain LW signals. The experimental system of LW propagation in the aluminum plate under the influence of temperature is shown in Figure 8.

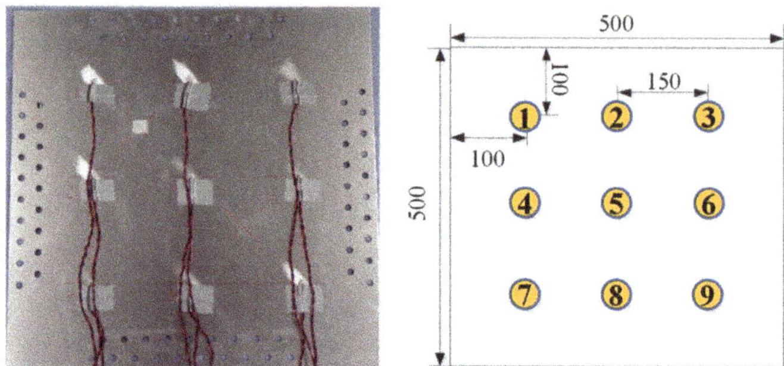

Figure 7. The aluminum plate with the PZTs and placement of the PZTs network.

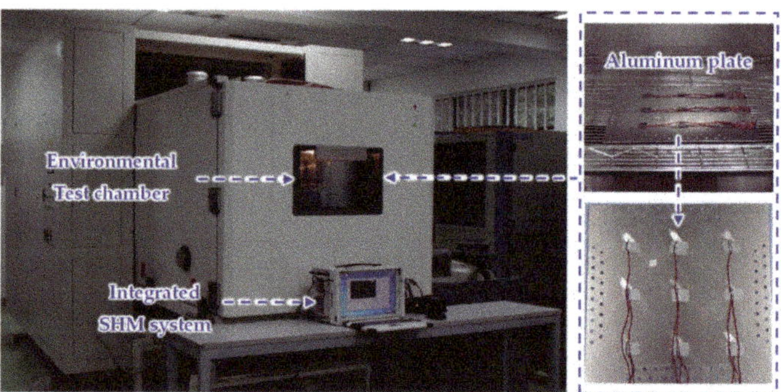

Figure 8. Experimental system of temperature influence on LW.

The whole range of temperature is from −20 °C to 60 °C, with the increment of 5 °C from −20 °C to 40 °C and the increment of 2 °C from 40 °C to 60 °C. The whole experimental process is about 6 h. There is no continuous signal acquisition in the process of temperature rise. The signal is acquired only when the temperature reaches the desired value. The excitation signal is a five-cycle sine burst modulated by a Hanning window with an amplitude of ±70 V. The central frequency of the LW excitation signal is 150 kHz and 200 kHz, respectively. The sampling rate is 10 MSamples/s. The signal acquired in the experiment is amplified by voltage, so the experimental signal cannot be directly compared with the simulation signal, and amplitude normalization is needed.

4.2. Experimental Results

The LW signals at 200 kHz under different temperatures are given in Figure 9. The first wave packet is the signal crosstalk, which can be ignored in the analysis. An enlarged view of the S_0 mode is given to observe the changes in phase and amplitude better. The amplitude increases, and the phase delays with the increase of temperature. Figure 10 shows the relationship between signal amplitude and temperature. The amplitude increases linearly with the increase in temperature.

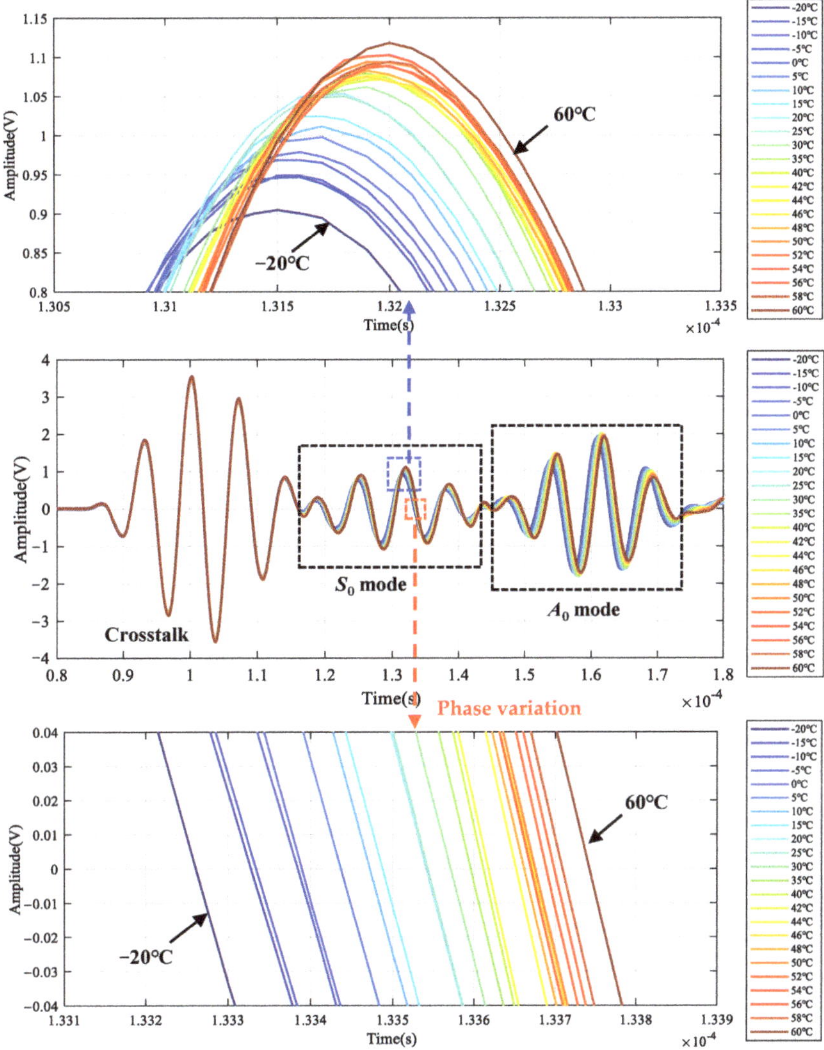

Figure 9. LW signals of central frequency 200 kHz at all temperatures.

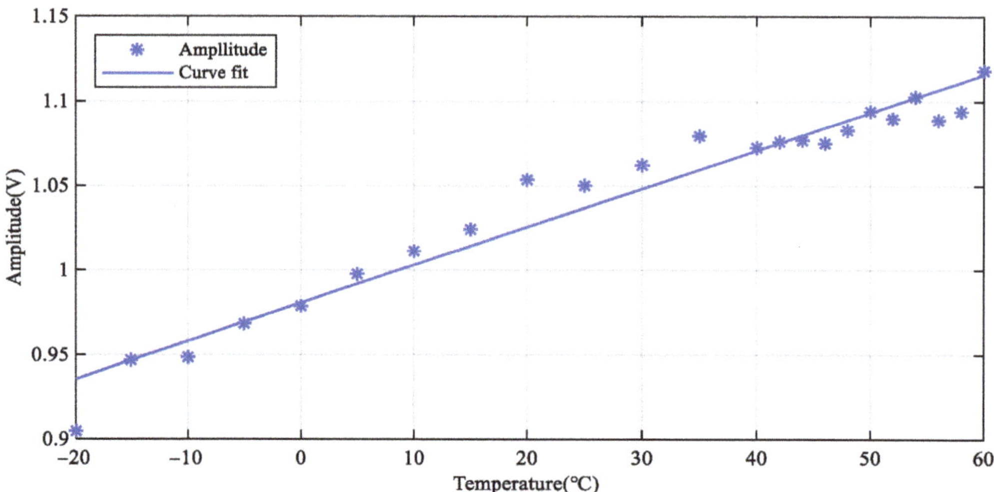

Figure 10. Amplitude change of LW signals at all temperatures.

4.3. Comparison between Simulation and Experiment

The comparison between simulation signals and experimental signals of channel 5–6 under the temperature of −20 °C, 20 °C, and 60 °C are shown in Figure 11. The experimental signals are amplified by a charge amplifier, but the simulation signals are not. So, in order to better compare the simulation signals with the experimental signals, the complex continuous Shannon wavelet transform is used for filtering, and then the amplitude of S_0 mode is normalized. It can be found that the waveform of S_0 mode matches well, while the amplitude and phase of A_0 mode have small errors. The reason for the error may be that the wavelength of A_0 mode is less than S_0; therefore, the mesh size of A_0 mode needs to be smaller to ensure sufficient accuracy.

Equations (22) and (23) are used to measure the changes in signal amplitude and phase at different temperatures [42]. The data fit cross zero was performed.

$$\Delta \text{Amp} = \frac{\text{Amp}_{\text{Tem}} - \text{Amp}_{\text{T0}}}{\text{Amp}_{\text{T0}}} \times 100\% \tag{22}$$

$$\Delta c_{\text{p}} = \frac{-c_{\text{p}}^2}{l_{\text{p}}} \Delta t \tag{23}$$

where Amp_{Tem} is the amplitude of the LW signals at different temperatures, Amp_{T0} is the amplitude of the LW signals at −20 °C, c_{p} is the phase velocity, l_{p} is the distance of LW propagation, and Δt is the time shift of the constant phase of the LW signals.

Figure 12 shows the quantitative variations of amplitude and phase velocity of S_0 mode between simulation and experiment. Experiment 1 and experiment 2 represent channel 5–6 and channel 5–8, respectively. It can be seen that the amplitude increases, and the phase velocity decreases with the increment of temperature. However, there may be differences in the influence of temperature on material parameters between simulation and experiment, resulting in the amplitude variations rate of the experiment being greater than that of the simulation. The phase velocity variations match well between the simulation and experiment.

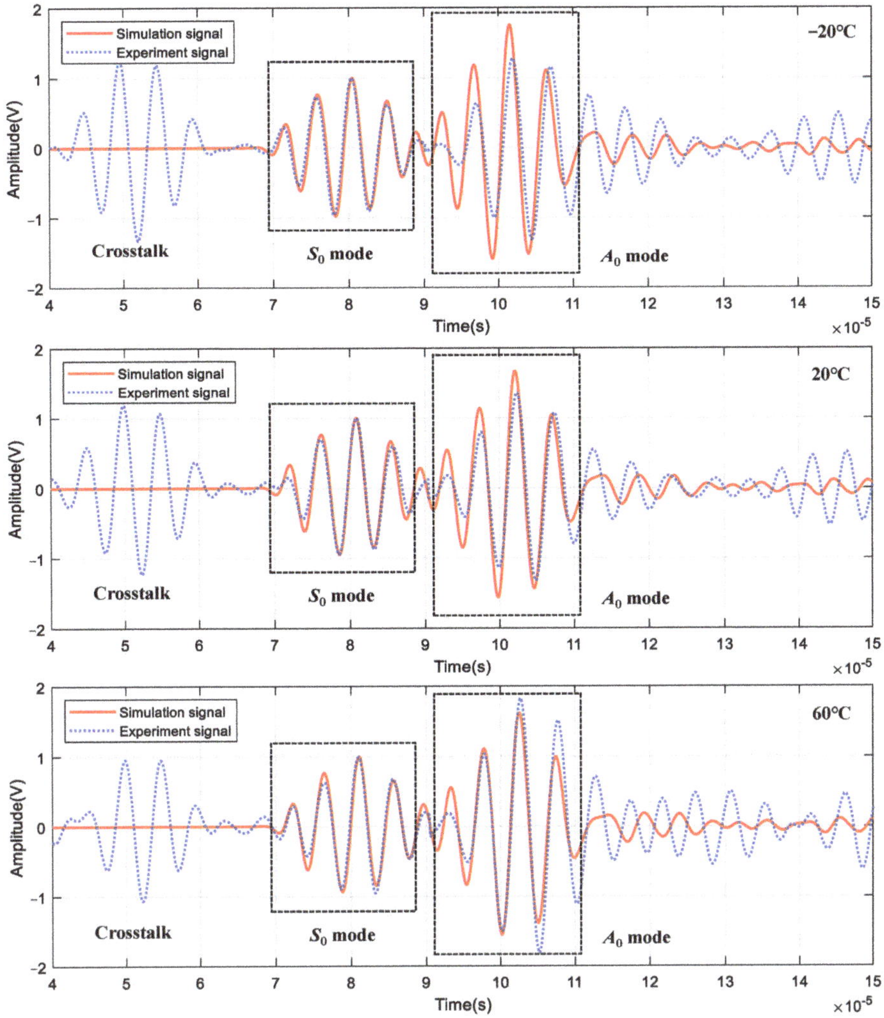

Figure 11. Comparison of LW signals between simulation and experiment.

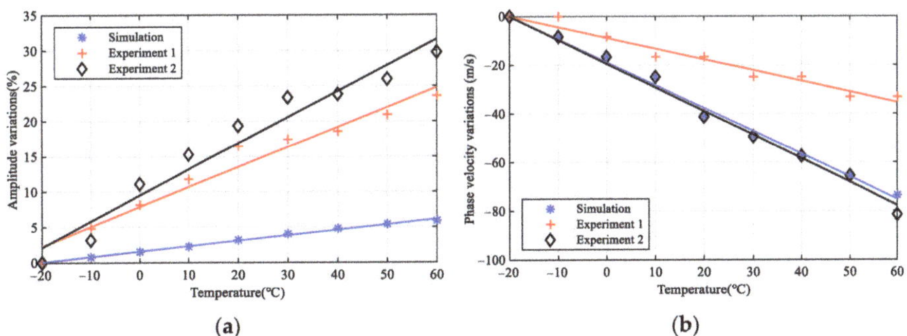

Figure 12. Comparison for variations of S_0 mode between simulation and experiment. (**a**) Variations of amplitude; (**b**) variations of phase velocity.

5. Conclusions

The paper aims to provide a contribution to the modeling of the LW propagation with the PZTs under temperature effect. Temperature mainly affects the propagation of LW from two aspects. On the one hand, temperature affects the material parameters. On the other hand, the influence of thermal stress on the piezoelectric constant is due to the inconsistent thermal expansion coefficient. The stationary study for thermal stress and the time-dependent study for LW propagation are established. The simulation results at −20 °C to 60 °C are obtained and compared with the experimental results. The results show that the waveform of S_0 mode matches well, while the amplitude and phase of A_0 mode have small errors. In addition, the influence of temperature on the LW between simulation and experiment is also consistent; that is, the amplitude increases, and the phase velocity decreases with the increase of temperature. However, there may be differences in the influence of temperature on material parameters between simulation and experiment, resulting in the amplitude change rate of the experiment being greater than that of the simulation.

However, there still exists some issues that will be studied in future work.

(1) LW propagation with the PZTs under load conditions will be considered and combined with the method proposed in this paper to realize the multiphysics simulation of LW propagation under temperature and load conditions.

(2) Complex structures and composite structures are widely used in the aerospace field. Therefore, it is very important to study the multiphysics simulation method of LW propagation in these structures under time-varying conditions.

Author Contributions: Funding acquisition, L.Q.; Methodology, L.Q. and X.Y.; Validation, Z.X. and H.Z.; Software, X.Y. and Z.X.; Writing—original draft, X.Y.; Writing—review and editing, X.Y., Z.X. and K.X. All authors have read and agreed to the published version of the manuscript.

Funding: This work is sponsored by the National Natural Science Foundation of China (Nos. 51921003, and 51975292), the Outstanding Youth Foundation of Jiangsu Province of China (No. BK20211519), the Research Fund of State Key Laboratory of Mechanics and Control of Mechanical Structures (Nanjing University of Aeronautics and Astronautics) (No. MCMS-I-0521K01), the Fundamental Research Funds for the Central Universities (No. 1001-XAC21022), the Priority Academic Program Development of Jiangsu Higher Education Institutions of China, the Fund of Prospective Layout of Scientific Research for Nanjing University of Aeronautics and Astronautics.

Institutional Review Board Statement: Not applicable.

Informed Consent Statement: Not applicable.

Data Availability Statement: Not applicable.

Conflicts of Interest: The authors declare no conflict of interest.

Nomenclature

Symbol	Implication
e	Strain vector
d	Piezoelectric coefficient matrix
E	Electric field vector
s	Elastic compliance matrix
σ	Stress vector
D	Electric displacement vector
ε	Dielectric constant matrix
l_{act}	Diameter of PZT
t_{act}	Thickness of PZT
t_{bond}	Thickness of the adhesive layer
t_{plate}	Thickness of the structure
G_{bond}	Shear strength of the adhesive layer
Y_{act}^{E}	Young's modulus of the PZT
E_{plate}	Elastic modulus of the structure

d_{31}	Piezoelectric constant
V_{in}	Excitation voltage
Γ	Shear-lag coefficient
k	Wavenumber
ω	Angel frequency of the LW
c_{L}	Velocity of the longitudinal wave
c_{T}	Velocity of the transverse wave
ρ_{plate}	Density of the structure
ν_{plate}	Poisson's ratio of the structure
e_{33}	Dielectric constant of PZT
s_{13}	Elastic coefficient of PZT
V_{out}	Response voltage
$C(\Gamma)$	Function of the Γ shear-lag parameter
$I(\Gamma R)$	Bessel function
ΔT	Relative to the reference temperature
$\Delta\sigma_{\text{T}}$	Thermal stress
Ex	Excitation signal
A	Amplitude of excitation signal
f	Central frequency of excitation signal
t	Wave propagating duration
N	Number of cycles within the signal window
Amp_{Tem}	Amplitude of LW signal at corresponding temperature
Amp_{T0}	Amplitude of LW signal at -20 °C
c_{p}	Phase velocity of LW signal
l_{p}	Distance of LW propagation
Δt	Time shit of the constant phase of LW signal

References

1. Boller, C.; Chang, F.K.; Fujino, Y. *Encyclopedia of Structural Health Monitoring*; John Wiley & Sons: New York, NY, USA, 2009.
2. Bekas, D.G.; Sharif-Khodaei, Z.; Aliabadi, M.H.F. An innovative diagnostic film for structural health monitoring of metallic and composite structures. *Sensors* **2018**, *18*, 2084. [CrossRef] [PubMed]
3. Yuan, S.F.; Ren, Y.Q.; Qiu, L.; Mei, H.F. A multi-response-based wireless impact monitoring network for aircraft composite structures. *IEEE Trans. Ind. Electron.* **2016**, *63*, 7712–7722. [CrossRef]
4. Chen, J.; Yuan, S.F.; Jin, X. On-line prognosis of fatigue cracking via a regularized particle filter and guided wave monitoring. *Mech. Syst. Signal Process.* **2019**, *131*, 1–17. [CrossRef]
5. Wang, Y.; Qiu, L.; Luo, Y.J.; Ding, R. A stretchable and large-scale guided wave sensor network for aircraft smart skin of structural health monitoring. *Struct. Health Monit.* **2021**, *20*, 861–876. [CrossRef]
6. Wandowski, T.; Malinowski, P.; Ostachowicz, W. Elastic wave mode conversion phenomenon in glass fiber-reinforced polymers. *Int. J. Struct. Integr.* **2019**, *10*, 337–355. [CrossRef]
7. Yuan, S.F.; Chen, J.; Yang, W.B.; Qiu, L. On-line crack prognosis in attachment lug using Lamb wave-deterministic resampling particle filter-based method. *Smart Mater. Struct.* **2017**, *26*, 085016. [CrossRef]
8. Ren, Y.Q.; Tao, J.Y.; Xue, Z.P. Design of a large-scale piezoelectric transducer network layer and its reliability verification for space structures sensors. *Sensors* **2020**, *20*, 4344. [CrossRef]
9. Xu, Q.H.; Yuan, S.F.; Huang, T.X. Multi-dimensional uniform initialization gaussian mixture model for spar crack quantification under uncertainty. *Sensors* **2021**, *21*, 1283. [CrossRef]
10. Gorgin, R.; Luo, Y.; Wu, Z.J. Environmental and operational conditions effects on lamb wave based structural health monitoring systems: A review. *Ultrasonics* **2020**, *105*, 106114. [CrossRef]
11. Su, Z.Q.; Zhou, C.; Hong, M.; Cheng, L.; Wang, Q.; Qing, X.L. Acousto-ultrasonics-based fatigue damage characterization: Linear versus nonlinear signal features. *Mech. Syst. Signal Process.* **2014**, *45*, 225–239. [CrossRef]
12. Qiu, L.; Fang, F.; Yuan, S.F. Improved density peak clustering-based adaptive Gaussian mixture model for damage monitoring in aircraft structures under time-varying conditions. *Mech. Syst. Signal Process.* **2019**, *126*, 281–304. [CrossRef]
13. Ren, Y.Q.; Qiu, L.; Yuan, S.F.; Fang, F. Gaussian mixture model and delay-and-sum based 4D imaging of damage in aircraft composite structures under time-varying conditions. *Mech. Syst. Signal Process.* **2020**, *135*, 106390. [CrossRef]
14. Singh, P.; Keyvanlou, M.; Sadhu, A. An improved time-varying empirical mode decomposition for structural condition assessment using limited sensors. *Eng. Struct.* **2021**, *232*, 111882. [CrossRef]
15. Qiu, L.; Yuan, S.F.; Chang, F.K.; Bao, Q.; Mei, H.F. On-line updating Gaussian mixture model for aircraft wing spar damage evaluation under time-varying boundary condition. *Smart Mater. Struct.* **2014**, *23*, 125001. [CrossRef]
16. Shen, Y.F.; Giurgiutiu, V. Effective non-reflective boundary for Lamb waves: Theory, finite element implementation, and applications. *Wave Motion* **2015**, *58*, 22–41. [CrossRef]

17. Ge, L.Y.; Wang, X.W.; Wang, F. Accurate modeling of PZT-induced Lamb wave propagation in structures by using a novel spectral finite element method. *Smart Mater. Struct.* **2014**, *23*, 95018. [CrossRef]
18. Hafezi, M.H.; Alebrahim, R.; Kundu, T. Peri-ultrasound for modeling linear and nonlinear ultrasonic response. *Ultrasonics* **2017**, *80*, 47–57. [CrossRef]
19. Radecki, R.; Su, Z.Q.; Cheng, L.; Packo, P.; Staszewski, W.J. Modelling nonlinearity of guided ultrasonic waves in fatigued materials using a nonlinear local interaction simulation approach and a spring model. *Ultrasonics* **2018**, *84*, 272–289. [CrossRef]
20. Shen, Y.F.; Giurgiutiu, V. Combined analytical FEM approach for efficient simulation of Lamb wave damage detection. *Ultrasonics* **2016**, *69*, 116–228. [CrossRef]
21. Gravenkamp, H.; Prager, J.; Saputra, A.A.; Song, C. The simulation of Lamb waves in a cracked plate using the scaled boundary finite element method. *J. Acoust. Soc. Am.* **2012**, *132*, 1358–1367. [CrossRef]
22. Kumar, A.; Kapuria, S. Finite element simulation of axisymmetric elastic and electroelastic wave propagation using local-domain wave packet enrichment. *J. Vib. Acoust.* **2022**, *144*, 021011. [CrossRef]
23. Ajay, R.; Carlos, E.S.C. Effects of elevated temperature on guided-wave structural health monitoring. *J. Intell. Mater. Syst. Struct.* **2008**, *19*, 1383–1398.
24. Radecki, R.; Staszewski, W.J.; Uhl, T. Impact of changing temperature on Lamb wave propagation for damage detection. *Key Eng. Mater.* **2014**, *588*, 140–148. [CrossRef]
25. Dodson, J.C.; Inman, D.J. Thermal sensitivity of Lamb waves for structural health monitoring applications. *Ultrasonics* **2013**, *53*, 677–685. [CrossRef] [PubMed]
26. Roy, S.; Lonkar, K.; Janapati, V.; Chang, F.K. A novel physics-based temperature compensation model for structural health monitoring using ultrasonic guided waves. *Struct. Health Monit.* **2014**, *13*, 321–342. [CrossRef]
27. Francesco, L.; Salamone, S. Temperature effects in ultrasonic Lamb wave structural health monitoring systems. *J. Acoust. Soc. Am.* **2008**, *124*, 161–174.
28. Marzani, A.; Salamone, S. Numerical prediction and experimental verification of temperature effect on plate waves generated and received by piezoceramic sensors. *Mech. Syst. Signal Process.* **2012**, *30*, 204–217. [CrossRef]
29. Esfarjani, S.M. Evaluation of effect changing temperature on lamb-wave based structural health monitoring. *J. Mech. Energy Eng.* **2020**, *3*, 329–336. [CrossRef]
30. Attarian, V.A.; Cegla, F.B.; Cawley, P. Long-term stability of guided wave structural health monitoring using distributed adhesively bonded piezoelectric transducers. *Struct. Health Monit.* **2014**, *13*, 265–280. [CrossRef]
31. Lonkar, K.P. Modeling of Piezo-Induced Ultrasonic Wave Propagation for Structural Health Monitoring. Ph.D. Thesis, Stanford University, Palo Alto, CA, USA, 2013.
32. Han, S.J.; Palazotto, A.N.; Leakeas, C.L. Finite-element analysis of Lamb wave propagation in a thin aluminum plate. *J. Aerospace Eng.* **2018**, *22*, 185–197. [CrossRef]
33. Yule, L.; Zaghari, B.; Harris, N.; Hill, M. Modelling and validation of a guided acoustic wave temperature monitoring system. *Sensors* **2021**, *21*, 7390. [CrossRef] [PubMed]
34. Liu, A.Q. Research on the Propagation and Simulation of Lamb Wave under Temperature Effect. Master's Thesis, Nanjing University of Aeronautics and Astronautics, Nanjing, China, 2020.
35. Giurgiutiu, V. Tuned Lamb wave excitation and detection with piezoelectric wafer active sensors for structural health monitoring. *J. Intell. Mater. Syst. Struct.* **2005**, *16*, 291–305. [CrossRef]
36. Brammer, J.A.; Percival, C.M. Elevated-temperature elastic moduli of 2024 aluminum obtained by a laser-pulse technique. *Exp. Mech.* **1970**, *10*, 245–250. [CrossRef]
37. Barakat, S. The Effects of Low Temperature and Vacuum on the Fracture Behavior of Organosilicate Thin Films. Masters's Thesis, University of Waterloo, Waterloo, ON, Canada, 2011.
38. Lee, H.J.; Saravanos, D.A. *The Effect of Temperature Dependent Material Nonlinearities on the Response of Piezoelectric Composite Plates*; NASA Technical Memorandum; NASA: Brook Park, OH, USA, 1997; pp. 1–20. Available online: https://ntrs.nasa.gov/citations/19980017194. (accessed on 2 May 2022).
39. Qiu, L.; Yan, X.X.; Lin, X.D.; Yuan, S.F. Multiphysics simulation method of lamb wave propagation with piezoelectric transducers under load condition. *Chin. J. Aeronaut.* **2019**, *32*, 1071–1086. [CrossRef]
40. Yang, C.H.; Ye, L.; Su, Z.Q.; Bannister, M. Some aspects of numerical simulation for Lamb wave propagation in composite laminates. *Compos. Struct.* **2006**, *75*, 267–275. [CrossRef]
41. Qiu, L.; Yuan, S.F.; Wang, Q.; Sun, Y.J.; Yang, W.W. Design and experiment of PZT network-based structural health monitoring scanning system. *Chin. J. Aeronaut.* **2009**, *22*, 505–512.
42. Nader, G.; Carlos, E.; Silva, N.; Adamowski, J.C. Effective damping value of piezoelectric transducer determined by experimental techniques and numerical analysis. *ABCM Symp. Ser. Mechatron.* **2004**, *1*, 271–279.

MDPI

Review

Review of Machine-Learning Techniques Applied to Structural Health Monitoring Systems for Building and Bridge Structures

Alain Gomez-Cabrera and Ponciano Jorge Escamilla-Ambrosio *

Centro de Investigación en Computación, Instituto Politécnico Nacional, Mexico City 07738, Mexico
* Correspondence: pescamilla@cic.ipn.mx

Abstract: This review identifies current machine-learning algorithms implemented in building structural health monitoring systems and their success in determining the level of damage in a hierarchical classification. The integration of physical models, feature extraction techniques, uncertainty management, parameter estimation, and finite element model analysis are used to implement data-driven model detection systems for SHM system design. A total of 68 articles using ANN, CNN and SVM, in combination with preprocessing techniques, were analyzed corresponding to the period 2011–2022. The application of these techniques in structural condition monitoring improves the reliability and performance of these systems.

Keywords: structural health monitoring; machine learning; physics-based model; data-based model; building structures

Citation: Gomez-Cabrera, A.; Escamilla-Ambrosio, P.J. Review of Machine-Learning Techniques Applied to Structural Health Monitoring Systems for Building and Bridge Structures. *Appl. Sci.* **2022**, *12*, 10754. https://doi.org/10.3390/app122110754

Academic Editors: Liang Yu, Phong B. Dao, Lei Qiu, Tadeusz Uhl and Minh-Quy Le

Received: 8 September 2022
Accepted: 14 October 2022
Published: 24 October 2022

1. Introduction

Structural health monitoring (SHM) is a field of science that focuses its efforts on evaluating and monitoring the integrity of a structure of interest [1]. Structural health monitoring systems are based on the design of sensing systems and structural models to evaluate machines and structures.

Although SHM systems are not a new field of research, computational advances in sensing hardware and the computational power of embedded devices drive the generation of reliable data for developing models based on classification and prediction data, including machine-learning algorithms in SHM systems. Moreover, sensors, such as accelerometers, are inexpensive compared to other sensors and can be effectively deployed in a sensing system to implement vibration-based SHM systems [2]. Accelerometers, or when combining them with other sensors, are the dominant sensing approaches for these SHM applications [3]. Because vibration-based systems date back to the late 1970s [4], technological advances represent a field of opportunities to improve existing solutions in the field of damage identification.

Previous studies have been conducted in the field of SHM using machine-learning (ML) techniques. However, the importance of this study is to compare the application of artificial neural networks (ANNs), convolutional neural networks (CNNs) and support vector machine (SVM) techniques considering the input data, feature selection techniques, the structure of interest, data size, the level of damage identification and the accuracy of the ML model. In addition to the fact that each of the above ML techniques has been used for similar tasks in structural damage and system identification, some of them perform better when the data comes from data generated by multi-sensor data fusion or when the data are processed with damage-sensitive feature extraction techniques, such as Hilbert–Huang transform (HHT) or wavelet packet transform (WPT). In addition, this study could provide a starting point for the selection of ML techniques and signal processing techniques for future SHM ML-based solutions where structural configuration or data features have similarity with previous studies that achieved good structural damag or system identification performance.

This paper is structured as follows. First, a brief description of the concepts related to SHM systems is presented in Section 2. In Section 3, the most common feature selection algorithms are presented, and a review of related works is given. In Section 4, the two main branches of SHM system models are presented, namely, the physics-related model (Section 4.1) and the data-driven models (Section 4.2). In Section 4.2, three ML techniques and their application in SHM systems are explored: ANN, CNN and SVM. At the end of Section 4, a comparison between the two main model branches is established in Section 4.3. Section 5 includes a brief review of the use of signal processing techniques to mitigate uncertainties effects from the environment and noise. In Section 6, a summary of the most important features of the application of damage identification techniques are presented in a list. Finally, Section 7 contains conclusions and future work.

2. Damage Classification in SHM

SHM covers several application areas and the assets monitored range from small components to huge civil structures and complex machines. Building SHM systems focus on measuring changes in the physical parameters to assess the current state of the structure and, in some cases, predict the building's response to future seismic excitations. To make these predictions, it is necessary to identify the natural frequencies of the buildings [5]. In the case of buildings, the structure is subjected to the effects of static and dynamic loads, so the complexity of the analysis presents a challenge in giving an accurate model that includes all these known and unknown effects.

SHM systems are composed of several hardware and software elements. An overview of the main components of SHM systems as defined by Farrar and Worden [6] are:

1. Operational assessment: The aspect related to damage conceptualization and operational conditions.
2. Data acquisition: The sensor system design and data preprocessing.
3. Feature extraction: The selection of sensitive damage features according to the damage identification capabilities of the desired SHM system.
4. Statistical model development: The design and implementation of the physics-based or data-based model.

Yuan et al. [1] explored the data acquisition aspect (1) of recent proposals for SHM systems, showing several features of accelerometer sensing systems that are attractive for the structural monitoring and evaluation of SHM systems. This study focuses on areas three and four of SHM systems, analyzing proposals of the physics-based models and the data-based models presented in the literature that belong to these areas. Figure 1 shows these areas in SHM systems, and the methods, techniques and algorithms involved.

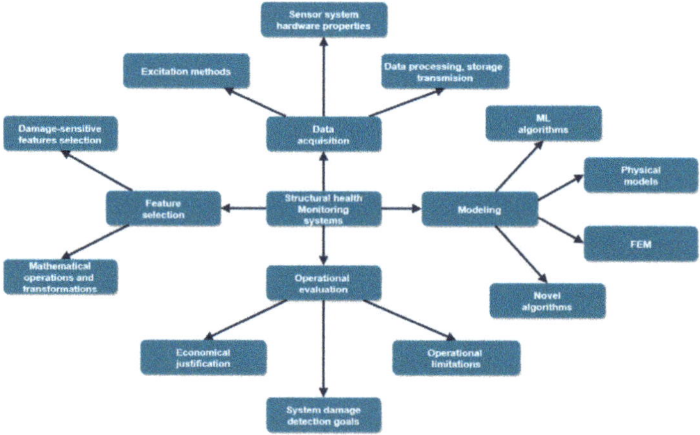

Figure 1. Diagram of structural health monitoring systems areas.

The core of SHM systems is their ability to perform damage identification. Damage is the change in the material's physical properties due to progressive deterioration or as a result of a single event on a structure. This change can detract from the behavior or integrity of a structure. Damage characterization can be conceived in several ways depending on the objectives of the SHM system, and damage states can be defined in terms of extent, severity, remaining operational time, thresholds, and damage index standards. Rytter [7] also presents a damage identification classification in SHM systems, as shown in Table 1.

Table 1. Damage characterization levels.

Damage Characterization Level	Description
I: Detection	The SHM system can decide if there is any damage to the structure of interest.
II: Localization	The SHM system can determine the existence and location of damage in the structure of interest.
III: Assessment	The SHM system can estimate the extent of damage in the structure of interest.
IV: Prediction	The SHM system can estimate the remaining lifetime of the structure.

3. Feature Selection Algorithms

In the field of SHM, several signal processing techniques can be used to extract damage-sensitive features and detect, locate and estimate structural damage in the machines and structures of interest. Some of the techniques used in previous works in this field include ANFIS [8,9], wavelet-based techniques [10], frequency–response functions (FRF) [11], Kalman filter-based techniques [12] and soft computing-based techniques [13].

The selection of damage-sensitive features by manual means represents a problem, as it requires expertise to identify candidate damage-sensitive signals. Algorithms for producing these features have been explored in the literature and two approaches to the generation of these features are widely proposed. One common approach is to propose novel features based on signal processing techniques. Another is to let a machine model the task of automatically extracting some sensitive features from data hidden on the surface. Composite damage indicators can also be constructed from the combination of damage-sensitive features. For example, Lubrano Lobicanco et al. [14] used residual drift ratios and lateral stiffness to define damage levels and quantify the amount of damage to structures due to seismic events.

A comprehensive review of some damage-sensitive feature construction algorithms can be found in the review of Civera and Surace [15]. Among the most recent techniques applied to the problem of extracting useful information from data for structural damage identification are principal component analysis (PCA), the wavelet transform (WT), along with its variants, and the application of empirical modal decomposition (EMD) in combination with the Hilbert–Huang transform (HHT).

3.1. Principal Component Analysis

Principal component analysis (PCA) is a statistical method mainly applied to reduce data dimensionality, which keeps relevant information from the original data. PCA application in ML models improve training time and model performance because redundant information in training data is removed. Initial data are projected into new variables, named principal components, with the most data variance. PCA use in SHM systems reduces large sensor-gathered data processing times, thereby enhancing the system response for structural changes or events resulting from structural damage.

Table 2 shows a compilation of recent works incorporating the PCA technique on the data for subsequent use by a machine-learning model. As can be seen, PCA is a popular technique in recent years due to the advantages that it offers. Although PCA is a powerful tool for ML applications, the PCA computation calculates the eigenvalues and eigenvectors of the data covariance matrix, which is an expensive operation for large data dimensions.

3.2. Wavelet Transform

The wavelet transform (WT) decomposes a signal into a series expansion where the number of wavelet functions is multiplied by a set of coefficients representing the original signal. Similar to Fourier transform (FT), WT expresses a signal in more specific located-in-time functions, but in the case of WT, the original signal can be non-periodic. In particular, WT analysis of a signal can bring frequency information and its position in time. Wavelet-based techniques include the discrete wavelet transform (DWT), the wavelet packet transform (WPT), the continuous wavelet transform (CWT), and the stationary wavelet transform (SWT). Table 3 shows the combination of WT-based techniques with ML models for structural damage identification. In data-based models, WT techniques serve as dimensionality reduction techniques, noise removal filters, and damage-sensitive feature generators, thereby improving SHM systems' accuracy and robustness.

3.3. Empirical Mode Decomposition

Empirical mode decomposition (EMD) is one of the most used signal decomposition techniques in SHM. Several proposals exist in structural damage identification, modal parameters identification and structural prognosis. The advantages of using EMD over nonlinear and non-stationary signals, such as structural responses due to mechanical excitation, are the time–frequency information of the analyzed signal in all the signal lengths. Another advantage of the application of EMD in comparison to other signal decomposition techniques is that EMD is signal-adaptative.

Over time, variations have been proposed to tackle the limitations and issues of EMD applications, such as noise and the well-known mode-mixing problem. Ensemble EMD (EEMD), multivariate EMD (MEMD) and time-varying filter-based EMD (TVF-EMD) are used in combination with ML models to achieve better accuracy in structural damage and system identification, as can be seen in Table 4.

A systematic comparative study between the above-mentioned EMD variants can be found in [16], proving EMD to be a robust signal processing technique in structural damage identification, system identification and anomaly detection. Regardless of the benefits of using EMD variations, including mode-mixing minimization and noise removal, the computation of these algorithms can be prohibitively expensive in SHM real-time applications.

Table 2. PCA review.

Publication	Structure	Excitation Source	Signal Processing and/or Feature Extraction Technique	Dataset Size	Purpose	Year
[17]	Instrumented bridge	Operational conditions	FFT	43,200,000 data points	Improve anomaly detection accuracy and reduce the size of SHM transmitted data	2021
[18]	Steel-supported experimental beam model	Impact hammer	FRF	16,384 data points	Study the relation between PCA-generated data to modal analysis and frequencies to establish a physical interpretation of PCA-compressed data	2020
[19]	The Phase II IASC-ASCE SHM benchmark structure FE model	Simulation	MSD	480,000 data points	Propose a new method to extract damage-sensitive features that are invariant to environmental effects.	2020

Table 2. *Cont.*

Publication	Structure	Excitation Source	Signal Processing and/or Feature Extraction Technique	Dataset Size	Purpose	Year
[20]	Simply supported steel truss bridge experimental model	Impact hammer	FRF, PCA	1200 samples of 512 data points	Propose a combination of PCA and FRF to obtain a new damage index for structural damage classification	2020
[21]	Benchmark concrete beam numerical model	Simulation	ARM, LDA, QDA, NB, DT	8002 data points	Present a comparative study between PCA and AR models to extract damage-sensitive features and several classification methods	2019
[22]	Truss bridge FE model and two-story frame FE models	Simulation	FRF, LUT	Not specified	Propose a combination of PCA and LUT to train an ML model for structural damage detection and localization	2017
[23]	Truss bridge and two-story frame FE models	Simulation	FRF, 2D-PCA, ICA	49 samples (bridge) and 75 samples (two-story frame)	Propose a structural damage detection, localization and severity estimation using a PCA-extracted damage index and an ANN	2017
[24]	Twin-tower steel structure experimental model	Shaking table	FRF	Not specified	Propose a scalogram-based damage detection and localization system using PCA and FRF	2017
[25]	Steel plate experimental model	PZT burst signal	Golay filter algorithm	8 samples of 10,000 data points	Study the noise filtering capabilities of PCA and Golay algorithms to train an ML model for structural damage identification	2016
[26]	Thin aluminum plate experimental model	PZT burst signal	ARM	1000 samples	Propose methodology to detect structural changes based on statistical inference	2016
[27]	Three-story frame aluminum structure experimental model	Bumper	AANN, FA, MSD, SVD, SVDD, KPCA	17 samples of 4096 data points	Propose a comparative study between kernel algorithms for structural damage detection purposes	2016
[28]	Three-story frame aluminum structure experimental model	Shaking table	ARM	1700 samples	Compare the performance of four PCA-based algorithms for structural damage detection under environmental conditions	2015
[29]	Grid FE model and steel grid structure experimental model	Simulation	PCA, WGN	4000 samples	Propose an ANN SHM system for damage detection due to temperature changes and noise perturbations	2015
[30]	Small aluminum plate experimental model	PZT burst signal	Chi-square goodness-of-fit test	600 samples	Propose a statistical analysis for structural damage detection using PCA-compressed data	2014

Table 2. *Cont.*

Publication	Structure	Excitation Source	Signal Processing and/or Feature Extraction Technique	Dataset Size	Purpose	Year
[31]	12-DOF mass–spring-damped experimental model	Impact hammer	PCP	20,000 samples	Propose a denoising method using PCP for SHM systems	2014
[32]	Building structure experimental model and Sydney Harbour-instrumented bridge	Electrodynamic shaker and operational conditions	RP, FT, SVM, PAA	3240 samples of 8192 data points (building), 6370 samples of 3600 data points (bridge)	Propose a comparative study between kernel algorithms for structural damage detection purposes	2014
[33]	Aluminum plate experimental model	Piezoelectric transducers attached to the plate	SOM	750 samples	Propose an SHM model for structural damage detection and classification	2013
[34]	Instrumented bridge	Numerical simulation	HHT, EMD	500 samples	Propose an environmental noise removal method to improve ML model damage detection capabilities	2013
[35]	10-story low-rise building FE model	Simulation	FRF, PCA	240 samples	Propose a structural damage detection method using PCA for data dimensionality reduction and noise removal	2013
[36]	Aluminum plate experimental model	PZT burst signal	CDLM	300 samples	Present a comparative study between five different methods for structural damage detection and localization	2011
[37]	Aluminum plate experimental model	PZT burst signal	Chi-square distribution	100 samples	Propose a new structural damage detection methodology based on the probability distribution of PCA projections	2011
[38]	Supported steel beam experimental models	Impact hammer	FRF	8192 data points	Propose a structural damage identification method using residual FRF data in combination with ANN	2011

Table 3. WT review.

Publication	Structure	Machine-Learning Algorithm	Dataset Size	Accuracy	Purpose	Year
[39]	Three-story building structure experimental model and steel frame experimental model	DCNN	2280 samples (building) and 2000 samples (steel frame)	ACC = 99% (building), ACC = 99.610% (steel frame)	Generate 2D training samples using CWT from accelerations scalograms	2021
[40]	Aluminum plate FE model and aluminum plate experimental model	CNN	300,000 data points	ACC = 99.9%	Wavelet transform was applied to acceleration signals to obtain wavelet coefficient matrix (WCM) in order to train a CNN	2019

Table 3. *Cont.*

Publication	Structure	Machine-Learning Algorithm	Dataset Size	Accuracy	Purpose	Year
[41]	Offshore platform FE model and scaled experimental model	TF, WPD, PCA	120 samples of 10,000 data points (FE model), 48 samples of 17,500 data points (scaled model)	ACC = 78.125% (FE model), ACC = 84.375% (instrumented model)	WPE and low-order principal components were applied to structural response for SVM model training	2018
[42]	IASC–ASCE SHM benchmark four-story steel structure FE model	WPD	99 samples	ACC = 91.75%	WPE was used to extract damage-sensitive features from accelerations signal to train a damage detection SVM model	2015
[43]	IASC–ASCE SHM benchmark four-story steel structure FE model	LS-SVM, PSO, HSA, SF, WPT	16 samples of 40,000 data points	ACC = 100% (PSHS), ACC = 98.2% (Harmony), ACC = 98.68% (PSO)	Damage features were extracted using WPE in order to train an ANN and an LS-SVM model	2014
[44]	IASC–ASCE SHM benchmark four-story steel structure FE model	ANN	621 samples	ACC = 95.49% (damage case 1), ACC = 96.78% (Damage case 2)	WPT was used to remove noise and interferences from structural responses signal to build a training dataset for an ANN damage detection model	2012
[45]	Four-story building FE model	BP-NN, WPT, Battle–Lemarie decomposition, WGN	183 samples	ACC = 90% (single sensor), ACC = 100% (multiple sensor)	WPRE was used to extract damage-sensitive features from acceleration data in order to train an ANN damage detection model	2011

Table 4. EMD review.

Publication	Structure	Machine-Learning Algorithm	Accuracy	Purpose	Year
[46]	Mass–spring system numerical model and instrumented bridge	MSD, RARMX	Error graph	EMD method and IMF were used to build damage feature vectors	2020
[47]	Fourteen-bay steel truss bridge experimental model	ANN, CEEMD, HHT, GWN	RMSE = 5.156×10^{-4}	A combination of CEEMDAN and HHT was proposed to extract four key damage-sensitive features to train an ANN model for damage detection, localization and severity estimation	2020
[48]	Simply supported steel Warren truss instrumented bridge	Neuro-fuzzy ANN	ACC = 80%	EMD method was used to reduce data dimensionality of SHM system for structural damage assessment	2020
[48]	Hanxi bridge FE model	FastICA	Correlation = 0.84	EEMD and PCA were used to extract independent deflection components from structural responses and determine the presence of damage	2018
[49]	Manavgat cable-stayed bridge FE model	SVM, THT	ACC = 95.43%	HHT and THT were used to extract damage-sensitive features to train an SVM damage detection model	2018

3.4. Hilbert–Huang Transform

Hilbert–Huang transform (HHT) was designed to represent a non-stationary signal in time and frequency. Intrinsic mode function (IMF) is used in combination with the Hilbert algorithm to locate signal frequency information over a specific time of a given signal. Like EMD, HHT is adaptive for a given signal, and it does not require a prior selection of a function or parameter to compute it. The amplitude and instantaneous phase of a signal at a specific time can be obtained by applying HHT for system identification (e.g., estimating natural frequencies and damping) and damage identification purposes by detecting structural response changes.

Several applications of HHT in damage identification, anomaly detection and system identification in different structures can be found in [50]. Despite the effectivity shown in HHT-proposed SHM systems, better performance can be achieved when HHT is combined with ML algorithms and data fusion techniques. Moreover, some drawbacks of applying HHT in SHM systems can be mitigated by combining other signal decomposition techniques and data-based models, such as ML models. Table 5 shows some proposals found in the literature that combine HHT and ML algorithms in the damage identification applications of structures.

Table 5. HHT review.

Publication	Structure	Machine-Learning Algorithm	Accuracy	Purpose	Year
[51]	IASC–ASCE SHM benchmark four-story steel structure FE model	CNN	ACC = 92.36% (25 db SNR)	HHT was applied to structural response data to obtain time–frequency graphs and the marginal spectrums for CNN model training	2021
[52]	Five-story offshore platform experimental model	HHT, MEEMD, SVM	ACC = 62.5%, MSE = 0.90	HHT and EMD were applied to obtain damage-sensitive features from vibration signals and then used to train an SVM damage detection model	2021
[47]	Fourteen-bay steel truss bridge experimental model	CEEMD, HHT, WGN, FFMLP	MSE = 2.65×10^{-7}	A combination of CEEMDAN and HHT was proposed to extract four key damage-sensitive features to train an ANN model for damage detection, localization and severity estimation	2020
[53]	Three-story steel moment-resisting frame FE model	EMD, ANN	ACC = 98.8%	A combination of CEEMDAN and HHT was proposed to estimate first mode shapes and structural natural frequencies and compare them against reference values to determine the presence of damage	2019
[54]	12-story-reinforced concrete frame experimental model	RBFNN	ACC = 100%	HHT was studied for structural modal parameter identification and damage diagnosis	2014

4. SHM System Models

4.1. Physics-Based SHM Systems

In the case of physical asset monitoring, there are two main branches of modeling: physics-based modeling, also known as physical-law modeling, and data-driven modeling. Physics-based modeling aims to describe phenomena by formulating mathematical models that integrate interdisciplinary knowledge to generate models that replicate observed behavior. Models are commonly presented in differential equations whose complexity increases as more factors become involved. Several terms and parameters must be defined to fully describe the system phenomena. Initial and boundary conditions must be identified to obtain physics equation solutions, and the computational cost associated with this operation can be very time-consuming for complex phenomena.

In SHM systems, these models are implemented to assess the condition of an asset under operating conditions to monitor changes that may indicate the presence of damage and shorten the remaining useful life of the asset. Finite element modeling (FEM) software implements well-known analyses, such as modal analysis, and allows the simulation of different structures and initial and boundary conditions straightforwardly. FEM software includes physical law models integrated into software libraries to perform damage analysis on a virtualized model of structures efficiently.

The complexity of modeling building structures under seismic excitation is caused by the intervening factors that can modify the behavior of these structures and by the difficulty of correctly defining their physical properties. Several works focus on estimating model parameters and uncertainties to improve the model of the structure. For example, Xu et al. [55] estimated the parameters of the structures based on linear and nonlinear regression analyses. These structures' linear and nonlinear parameters, such as elastic stiffness and yield displacement, are obtained for a three-story structure. Gomes et al. [56] addressed an inverse identification problem using numerical models and a genetic algorithm.

Model parameters obtained from experimental and recorded data can increase the model's accuracy compared to the actual measured results. However, environmental and operating conditions do not remain constant throughout the life of the structure. In addition, physical law models have drawbacks that limit their applications in some SHM systems. The time to solve the equations in a real-time SHM system impacts the response time to ensure safety and the reduction of economic losses in response systems for seismic protocols and evacuation procedures. In order to reduce the time costs of performing calculations, optimization algorithms applied to the problem of SHM are encouraged and proposed in the literature [57].

Table 6 lists physics-based proposals for SHM systems in buildings under vibration excitation for multi-story structures. In this table, two types of proposal contributions are shown: the identification of system parameters and damage. According to Farrar's classification, the level of damage identification for the practical proposals is also presented.

Table 6. N-story building structure health monitoring systems based on physical model techniques.

Publication	Structure	Damage Indicator	Algorithm or Analysis Method	Damage Identification and Level		Year
[58]	An eight-story physical building model in FEM software	Stiffness reduction	Vibration-based damage methods	Damage detection and localization	II	2018
[59]	A 14-story physical building prototype under vibration table	Modal frequencies	Operational modal analysis	Modal identification	N/A	2017
[60]	A five-story physical building prototype under vibration table	Stiffness reduction	Novel damage localization algorithm based on wave propagation	Damage detection and localization	II	2020
[61]	A 51-story building with accelerometers and tilt sensors	Modal frequencies	Modal parameters estimation through Bayesian algorithm combined with FFT	Modal identification	N/A	2019
[62]	A 12-story frame structure	Stiffness reduction	Hysteresis loop analysis method	Damage detection	I	2017
[63]	An 86-story physical building in FEM software	Modal frequencies	Wave-based damage detection based on propagation analysis	Modal identification	N/A	2018
[64]	A three-story frame structure	Inter-story displacement	Two novel damage indices based on the displacement of the structure	Damage detection and localization	II	2015

4.2. Data-Based SHM Systems

Machine-learning techniques are a subset of the field of artificial intelligence. Due to their statistical nature, their vision is to address problems of interest in pattern recognition identification and classification tasks. SHM from the ML point of view is a classification problem in which at least two states are compared in SHM systems employing ML techniques: damaged and undamaged states.

A machine-learning model extracts information in the form of features from a given data set and classifies those data features. These data-driven models require large amounts of information to train the model and avoid the overfitting problem. The generalization problem depends on the amount of available data and significant diversity of this training data to avoid overfitting and ensure a reasonable level of generalization. In an idealized situation, the data set should include the samples of the possible range of excitation that can be applied to the structure. In addition, data quality improvement using signal processing techniques, such as normalization and noise filtering, is desirable for the generation of the data set. ML algorithms are applied in the damage identification process and in analyzing the anomalous data obtained from the sensors [8,65], thereby improving data quality. In addition, signal processing techniques, such as WT and HHT also improve data quality and are applied in SHM systems [9,66].

ML techniques can be divided into supervised learning (for regression and classification tasks), unsupervised learning (anomaly detection and clustering) and reinforcement learning. The most popular ML techniques implemented in the construction of SHM solutions are support vector machines (SVMs) and convolutional neural networks (CNNs) [10,67]. In the case of neural network techniques, damage-sensitive feature selection plays a crucial role in the performance of the SHM system. SVM optimally classifies features in linear and nonlinear problems. CNN, a subset of neural network (NN) methods, include convolution operations in the hidden layer of neural networks to classify data, usually in image format. Other techniques, such as PCA, improve the features of the training data set by making them uncorrelated.

The selection of a ML technique is guided by the limitations of each technique and the requirements of SHM in terms of damage identification level and operating conditions. Identifying the modal systems of building structures can also be performed using deep neuronal networks (DNNs) [11,68]. Table 7 summarizes the ML-based proposals for SHM systems in buildings under vibration excitation for multi-story structures.

Table 7. N-story building structure health monitoring systems based on machine-learning techniques.

Publication	Structure	Data Type Used for Training	Machine-Learning Technique	Damage Identification and Level		Year
[69]	A four-story physical building prototype under vibration table	Acceleration response data from a physical prototype	ANN	Damage existence and localization	II	2016
[70]	An eight-story physical building mathematical model	Artificially generated dataset from an algorithm	FCN	Damage existence and localization	II	2020
[71]	A three-story physical building simulated model	Simulation-generated dataset from OpenSeesMD software	ANN	Damage detection and localization	II	2017
[72]	30 buildings including 3, 5 and 7 stories with different structural parameters	Simulation-generated dataset from Raumoko3D software	ANN	Damage detection and localization	II	2017
[73]	An instrumented main steel frame	Experimental simulation from a physical prototype with modal shaker excitation	CNN	Damage detection and localization	II	2016
[74]	A three-story physical building-simulated model	Simulation-generated dataset from OpenSeesMD software	SVM	Damage existence, localization and severity	III	2019
[75]	A three-story steel frame structure	Intensity-based features	SVM	Damage detection	I	2019
[76]	A seven-story steel structure	Simulation-generated dataset	ANN	Damage existence, localization and severity	III	2018
[77]	A five-story steel structure	Simulation-generated dataset	ANN	Damage detection and localization	II	2008

4.2.1. Artificial Neural Networks (ANNs)

Inspired by the working of brain neurons, artificial neural networks are the most well-known ML algorithm applied for classification and regression problems. ANN architectures consist of an input layer, several hidden layers and an output layer. Several hidden layers can be stacked between the input and output layer according to the complexity of the ANN. Figure 2 shows a typical architecture of a forward neural network, showing input neurons, successive hidden layers and output neurons. In general, ANN power resides in the interconnection of its neurons and a set of weights associated with each interconnection. ANN learning algorithms update network weights to minimize an error function. BP-ANN is a widely used ANN where the backpropagation algorithm (BP) performs the network's training.

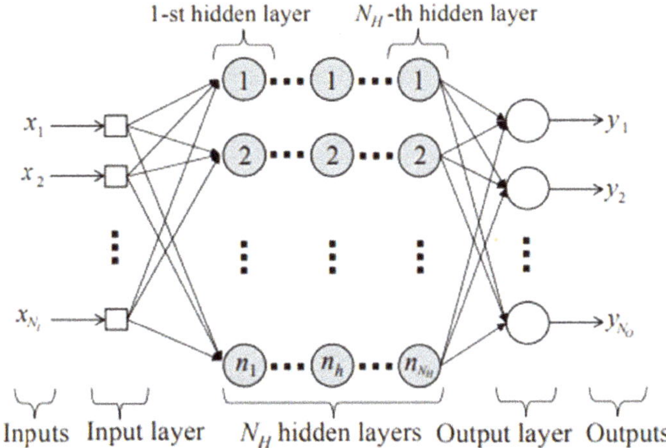

Figure 2. Multi-layer hidden layer feedforward ANN [78].

In many SHM systems, FE structural models generate training datasets for ANN weight calculation. In these datasets, damaged and undamaged training examples are provided. Natural frequencies and changes in the structural properties are associated with structural damage. Table 8 groups several SHM proposed systems using ANN and signal processing techniques and modal analysis for damage detection, localization and severity estimation.

Tan et al. [79] proposed a method in which the modal strain energy was used as a damage-sensitive feature and an ANN to locate and quantify structural damage in a series of damaged scenarios. A structural damage identification method using RFR functions in combination with PCA was described by Padil et al. [20] to reduce the input data size and minimize the effects of measurement uncertainties and interferences to reduce the model error. Mousavi et al. [47] proposed a damage identification methodology using the CEEDAM technique to extract damage-sensitive features from the measured data and employing an ANN to perform the classification task in various damage scenarios. The bridge damage presence, severity and location were obtained using the methodology mentioned above.

In [80], Finotti et al. presented a comparison between ANN and SVM techniques in damage identification capability using data from ten statistical indicators from acceleration measurements. These indicators were proposed to avoid manual selection and calculation of modal structural parameters that may introduce undesirable inaccuracies or errors in damage identification performance. Chang et al. [76] proposed a post-seismic evaluation method based on an ANN. This ANN was trained with simulation-generated data from a previously constructed FE model with structural parameters identified by the stochastic identification of the system subspace. Although the accuracy of damage identification was based on the identification of system parameters, it was possible to efficiently locate damage scenarios in one or several elements of the structure.

In the study conducted by Morfidis and Kostinakis [72], combinations of 14 seismic parameters together with an MLP were explored for structural damage prediction. Their analysis observed that with at least five seismic parameters combined with an MLP network, the seismic damage state for an R/C building can be accurately predicted. Correlations between seismic parameters and MLP network training algorithms were also observed and analyzed. Using natural frequencies and modal shapes as input to the network, Padil et al. [81] defined a damage identification system using an ANN. A relationship between input and output parameters was established by defining a probability of damage existence to classify damage states. Numerical and experimental samples were used to validate the proposed method regarding modeling errors and noise interference.

Jin et al. [82] proposed an artificial neural network with an extended Kalman filter to remove temperature effects in SHM systems. This EKF was also used to establish confidence variation intervals and thresholds for structural changes that were suggestive of damage. It was shown that temperature changes induced changes in natural frequencies that can lead to incorrect damage identification if these temperature effects are not considered in the training samples. Smarsly et al. [69] provided a decentralized sensor fault detection system consisting of a set of ANNs that perform anomaly detection in an SHM wireless sensor network. An anomaly detection ANN was incorporated at each sensor node to perform local anomaly detection. Each ANN utilized sensor data redundancy throughout the sensor network using the correlation of sensor measurements in structural response.

Kourehli [83] designed an ANN-based SHM system using incomplete FEM modal data and natural frequencies for damage location and damage severity estimation using only the first two natural frequencies. Using stiffness reduction as a damage indicator, damage identification was successfully achieved in an 8-DOF spring–mass system, a supported beam, and a three-story flat frame. Natural frequencies are commonly used as damage-sensitive features in SHM systems, and the importance of eliminating or reducing interference from temperature changes was noted by Gu et al. [29]. Using a multilayer ANN, they discriminated between changes in natural frequencies due to structural damage and changes in these natural frequencies due to temperature effects. A damage index based on Euclidean distance was employed to quantify the difference between an undamaged structural state under temperature variations and a possible structurally damaged state.

Ng [78] proposed an ANN for structural damage localization and severity estimation using a Bayesian class selection method, improving the selection of optimal ANN hyperparameters (optimal transfer function and the number of neurons). Validation tests were performed using the Phase II IASC-ASCE SHM reference structure data, where ANN estimates stiffness reduction values as an indicator of damage location and severity for six damage cases. Xie et al. [84] presented a bridge damage management system using a BP-ANN algorithm that utilizes four hundred sensors randomly distributed over the bridge. A comparative study was performed employing the proposed BP-ANN algorithm, an SVM, a decision tree (DT) and a logistic regression model, showing that ANN achieves the highest accuracy in damage identification for the proposed SHM system, even under noisy operating conditions. Goh et al. [85] designed a two-step damage identification method. In the first step, a prediction of the modal shape of the unmeasured structure was performed using an ANN. The predicted modal shape was compared with a cubic spline interpolation prediction method to verify the prediction accuracy. Next, a second ANN trained with limited structural response measurement points was used to perform structural damage localization and damage severity. Shu et al. [86] proposed another ANN-BP for bridge damage identification using the statistical parameters of displacement and acceleration measurements as training input for ANN. This proposal showed that measurement noise negatively impacts damage identification performance and should be eliminated or minimized. For this approach, single and multiple damage cases were used for validation testing.

Table 8. ANN review.

Publication	Structure	Excitation Source	Used Techniques and/or Algorithms	Number of Neurons	Number of Hidden Layers	Training Algorithm	Activation Function	Error	Dataset Size	Purpose	Year
[79]	A multiple steel girder composite bridge experimental model	Electrodynamic shaker	FFT, FRF	4 ANN architectures: 10, 15, 20, 25 and 30	1	LMBP	Sigmoid	MSE = 0.000029 (30 neurons)	114 samples	Propose ANN architectures for damage detection and localization in structures	2020
[20]	A simply supported steel truss bridge FE model	Simulation	FRF, PCA	20	1	LMBP	Tangent-sigmoid	MSE = 0.015213	1200 random damage cases	Identify presence of damage considering several uncertainties	2020
[47]	A fourteen-bay steel truss bridge experimental model	Electrodynamic shaker	CEEMD, HHT, GWN	34	1	BP	Log-sigmoid	MSE = 2.65×10^{-7}	A time series of 14,000 data points	Identify presence, location and severity of bridge damage	2020
[80]	Simply supported beam FE model and railway bridge experimental models	Simulation (beam) and operational conditions (bridge)	SSaI, ARN	10	1	LMBP	Sigmoid	DCR = 93.54% (beam), DCR = 87.09% (bridge)	5400 samples (beam), 4500 samples (bridge)	Detect structural changes based on statistical indicators	2019
[76]	A twin-tower building experimental model	Shaking table	SSI	140	2	BP	Log-sigmoid	Error graph	279,936 data points	Identify presence, location and severity of building damage	2018
[87]	A carbon fiber-reinforced plastic plate	Simulation	FT	627	1	BP	Sigmoid	ACC = 100%	40 samples with 150 data points	Identify presence and type of plate damage	2018
[72]	A set of 30 reinforced concrete buildings numerical models	Simulation	14 ground motion parameters, NTHA	30	1	LMBP, SCG	Tangent-sigmoid	MSE = 0.045	3900 vectors of size 18×1	Propose damage-sensitive features for improving ANN structural damage prediction	2017
[81]	A single-span steel frame numerical model	Impact hammer	SRF	20	1	LMBP	Tangent-sigmoid	MSE = 0.0026	2400 samples	Identify presence of structural damage using noisy ANN training data	2017
[82]	A Meriden bridge FE model	Simulation	EKF	6	1	BP	Tangent	MAE = 0.0572	6480 samples	Propose a damage detection method using ANN and EKF in structures under temperature changes	2016
[88]	An instrumented aircraft panel and an instrumented wind turbine	PZT burst signal and operational conditions	Ensemble classifier	1 (panel ANN) and 27 (turbine ANN)	1 (panel), 2 (turbine)	Not specified	Not specified	MSE = 0.0098 (panel), MSE = 0.000194 (turbine)	110 data samples (aircraft) and 3450 data samples (turbine)	Propose an ensemble design method for ANN hyperparameter selection	2016

Table 8. *Cont.*

Publication	Structure	Excitation Source	Used Techniques and/or Algorithms	Number of Neurons	Number of Hidden Layers	Training Algorithm	Activation Function	Error	Dataset Size	Purpose	Year
[69]	A four-story frame structure experimental model	Not specified	Cooley–Tukey FFT algorithm	24	1	BP	Not specified	RMS = 0.807	100 samples	Propose a detection method for sensor faults in response data of SHM systems	2016
[83]	3 numerical models: a simply supported beam, three-story plane frame and an 8-DOF spring–mass system	Simulation	ARN	6 ANN architectures: 30, 40, 46, 53, 44 and 51	1	LMBP, GDM	Tangent	MSE = 3.44×10^{-3} (beam), MSE = 4.77×10^{-3} (plane frame), MSE = 9.12×10^{-4} (spring–mass system)	2304 samples (beam), 2592 samples (three-story frame) and 2187 samples (DOF spring system)	Propose a structural damage detection ANN using natural frequencies and incomplete structure mode shapes	2015
[29]	A simply supported steel beam girder FE model	Simulation and a piston	PCA, GWN	2 ANN architectures: 35 and 17	1	LMBP	Tangent-sigmoid	MSE = 0.1137	4000 samples	Use of an ANN to associate changes in modal frequencies to structural damage due temperature changes	2015
[78]	The Phase II IASC-ASCE SHM benchmark structure FE model	Simulation	Bayesian ANN design algorithm	17	1	Not specified	Tangent-sigmoid	ACC = 97%	226 samples	A Bayesian model class selection method was proposed to select optimal ANN hyperparameters	2014
[84]	An instrumented bridge	Operational conditions	N/A	674	1	BP	Sigmoid	ACC = 98%	8000 samples	Propose a damage detection method that achieves high accuracy and more robust performance in the presence of noise	2013
[85]	A two-span-reinforced concrete slab FE model	Simulation	SRF	Not specified	1	SCG	Tangent-sigmoid	MSE = 0.00342	3000 samples	Propose an ANN to predict unmeasured mode shape values for damage detection	2013
[86]	A Banafjäl Bridge FE model	Simulation	GWN	4 ANN architectures: 19 and 23 (single damage case); 19 and 25 (multiple damage case)	2	BP	Not specified	ACC = 79%	416 samples (single damage case) and 900 samples (multiple damage case)	Propose a BP-ANN for detection, localization and extent of structural damage using statistical indicators	2013
[35]	A 10-story low-rise building FE model	Simulation	FRF, PCA	2 ANN architectures: 25 and 37	2	BP	Log-sigmoid	ACC = 98.77%	240 samples	Propose an ANN for detection, localization and extent of structural damage using RF as damage index and PCA	2013

Table 8. *Cont.*

Publication	Structure	Excitation Source	Used Techniques and/or Algorithms	Number of Neurons	Number of Hidden Layers	Training Algorithm	Activation Function	Error	Dataset Size	Purpose	Year
[89]	An aluminum beam experimental model	Mechanical roller	CC	36	2	BP	Sigmoid	ACC = 87.7%	250 samples	This paper reported a NN technique to select damage-sensitive frequency ranges and diagnose structural damage	2012
[90]	A building FE model	Simulation	FEMA damage index	49	1	BP	Sigmoid	MSE = 1.6×10^{-10}	835 samples	Propose an SHM system based on ANN to predict the building damage index	2012
[91]	A four-story building FE model	Simulation	FRF, PNN	16	1	BP	Not specified	ACC = 100%	30 samples+J24	Propose a PNN and a BP-NN for detection, localization and extent of structural damage using RF	2011
[45]	A four-story building FE model	Simulation	WPT, GWN, Battle–Lemarie decomposition	2 ANN architectures: 39 (single sensor, damage extend) and 90 (multi sensor fusion, damage extend)	1	BP	Not specified	ACC = 90% (single sensor), ACC = 100% (multiple sensor)	183 samples	Propose a feature fusion and a neural network model for structural damage	2011

Using a combination of PCA and FRF frequency–response functions, Bandara et al. [35] extracted damage-sensitive features from the structural response data to train an ANN-based damage identification system. The damage location and severity of the structure were represented by damage index classes in the output of the proposed ANN. Several noise levels were added to the acceleration measurements to show the denoising capabilities of PCA on the ANN training data and the reduction of data size to improve training time and computational cost. Mardiyono et al. [90] presented a BP-ANN architecture as part of a post-earthquake SHM response system using the FEMA damage index framework to classify the structural condition of buildings. A complete building SHM framework was developed, including a warning system module, an intelligent module, a monitoring module, and a data acquisition module. The complete function of the framework included accelerometer data collection, data processing and feeding to ANN, damage index prediction and alert notification in case of structural compromise. Wang et al. [91] presented a building damage identification method using a BP-ANN and a probabilistic neural network (PNN). In this method, the damage location process consists of three stages in which the damage location is refined at each stage. In the three stages, the damage was detected in each history, the extent of damage in the history was predicted and the compromised structural members were identified, respectively. Accuracy results indicated that the BP-ANN performed better in estimating the severity of structural damage, and the PNN suggested the damage location more accurately. Liu et al. [45] proposed structural damage identification where energy components were extracted from acceleration measurements and fused with data fusion techniques. An ANN classifier was used to identify, locate and estimate structural damage. The combination of WTERP and data fusion techniques improved the accuracy of ANN damage identification, and WPT-based component energies were shown to be an effective damage-sensitive feature.

4.2.2. Convolutional Neural Networks (CNNs)

Convolutional neural networks (CNNs) are maybe the most representative deep algorithm and have been adopted in several application fields of data-based classification and regression problems. A CNN is composed, essentially, of three types of layers according to its function: convolutional layers, pooling layers and fully connected layers. Convolutional layers automatically extract features from the input data through a convolution between input data and a user-defined matrix named filter. Pooling layers reduce the data size, and fully connected layers perform data classification tasks. Several layers can be stacked in the network design architecture, thereby increasing network complexity at the cost of incrementing computation training time and the amount of resources used.

CNN networks' inputs consist of n-dimensional training samples, where n can be from 1 to N. Figure 3 shows a convolutional network with an input vector of dimension 1D (128×1) for a multiclass classification task. These inputs are, in most cases, images according to the classification or regression problem. In structural damage identification, proposals using CNN architectures use 1D acceleration records or 2D data that consist of images or matrices using reshaping methods, signal processing techniques or feature extraction methods. One advantage of using CNN models in SHM systems is that convolutional layers perform the selection of damage-sensitive features automatically.

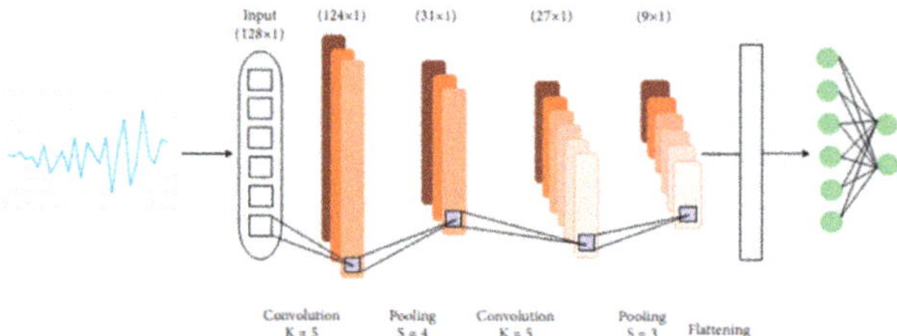

Figure 3. A CNN network example using raw acceleration input data [92]. Input data can be reshaped in 1D or higher dimension vector fashion.

Table 9 lists several proposals for SHM applications that employ CNNs. In these proposals, structural responses are captured as acceleration measurements are used. A complete and exhaustive review of SHM systems using CNN and mainly image data can be consulted in [93].

Some relevant limitations of the applications of CNN arise in SHM systems for structural assessment. A large amount of training data are usually required in CNN training to achieve the generalization property of the model, and the structural data of damage states are unavailable in most cases. Moreover, the structural data of damage states in large civil structures are uncommon. A possible solution is to generate damage structural state data from FE models, which is the usual approach in several SHM proposals, as shown in Table 6 However, FE-generated data rely on the accuracy of the FE models, and this also depends on the correct value of the structural parameters that are used for the model. Some of these parameter values are unknown or change over time, so they should be estimated or calculated from experimental data.

Wang et al. [51] proposed a structural damage identification method using a time–frequency plot of the acceleration signal. The marginal spectrum was used as the input to a CNN, and the hyperparameters of this network were optimized using PSO, thereby achieving 10% better accuracy than a CNN without PSO. Oh and Kim [94] explored the application of two objective function techniques for the optimal hyperparameter selection of a structural damage identification system using CNN. Their results showed a computational cost reduction of 40%. Sony et al. [95] designed a 1D-CNN to perform multiclass damage identification using bridge vibration data. The random search technique performed hyperparameter tuning to obtain the optimal network architecture. Oh and Kim [96] also proposed a hyperparameter search technique in CNN. These hyperparameters included the subsampling size and the number and size of kernels using a genetic algorithm and a multi-objective optimization technique.

Damage-sensitive features were extracted from the automatic signal using a CNN architecture in the damage identification strategy presented by Rosafalco et al. [70], which obtained good performance in damage identification and localization. In [97] Zhang et al. proposed a physics-guided CNN architecture for structural response modeling. Structural physics provide a set of constraints for CNN training, which avoid overfitting problems and reduce the size of the training dataset. de Rezende et al. [98] presented a 1D-CNN network with an electromechanical impedance-based methodology to achieve damage identification under various environmental temperature conditions, proposing a robust structural damage identification scheme against temperature effects. Sarawgi et al. [99] designed a distributed SHM damage identification system using 1D-CNN and a four-step methodology. In the first step, a multiclass classification of the vibration data is performed, and the second step consists of training the CNN with patterns of damaged and undamaged data. The damage location is identified in the third step. The first three steps are performed on each node, and in the final step, a combined data set from all nodes are used to train a global damage identification model.

Focusing on joint damage identification, Sharma and Sen [100] proposed a joint damage identification method based on using a 1D-CNN with robustness to signal noise. Single and multiple damage cases were successfully classified. Using the modal shapes and curvature differences as input to the CNN, Zhong et al. [101] proposed a damage identification system based on CNN, detecting that the modal shapes contain damage-sensitive features, which are sensitive to all degrees of damage, thereby allowing high detection accuracy. Additionally, it was shown that the parameters of the first-order modal signal have helpful information for the localization of structural damage. Seventekidis et al. [102] presented the idea of building an optimal FE model to generate training data for a CNN. A comparative case was analyzed comparing this generated optimal data set against a FE data set, showing that the latter is unreliable for CNN training given model uncertainties. This optimal FE model was obtained by initial structure measurements and, subsequently, the use of the covariance matrix adaptation evolution strategy.

Ibrahim et al. [2] evaluated several machine-learning approaches using low-cost accelerometers and noise filtering techniques for structural evaluation after excitation events, including SVM, k-nearest neighbors and CNN. This work demonstrated that the CNN-based detection method in this structure outperformed both SVM and the k-nearest neighbors' method using noise-contaminated raw measurements.

In a study by Duan and Zhu [103], the Fourier amplitude spectrum was used with a CNN to perform damage state structure identification. This work showed that the use of FAS outperforms traditional CNN proposals using time–history acceleration responses and that the proposed FAS-CNN SHM system was robust against noise effects. Gulgec et al. [104] proposed a damage identification and location method that used a CNN architecture and was trained by analytical simulations. The presented method consisted of two steps for damage identification: the first to detect the presence of damage as a binary classification problem and the second as a regression problem to estimate the damage location by predicting the damage boundaries in the structure. Using the stiffness histogram of the measured structural response and stiffness degradation as an indicator of damage, Zhou et al. [105] proposed a deep learning network to provide a real-time alarm in case of structure compromise. Zhang and Wang [106] proposed a 1D-CNN for structural state identification using raw time-domain data. GANs were used to generate training data and improve the robustness and effectiveness of structural damage identification.

To avoid the manual selection and elaboration of damage-sensitive features from the raw measurements, Yu et al. [107] proposed a CNN architecture to identify and localize structural damage. The input to the proposed network consisted of frequency-domain features obtained by FFT on the raw measurement signals. Compared with other ML techniques, including general regression neural network (GRNN) and ANFIS, the proposed network outperformed these in error classification. Tang et al. [108] proposed a structural damage identification system where real-world anomaly detection acceleration data are transformed into dual-channel time–frequency images. The anomaly detection patterns were correctly identified with an overall average accuracy of 93.5%. They highlight the importance and effect of balancing on the training dataset. Wu and Jahanshahi [109] presented a structural response prediction method using a DCNN in three structures: a linear SDOF, a nonlinear SDOF, and a three-story MDOF steel frame. This approach was compared with an MLP network regarding noise signal robustness, and the convolutional kernel analysis showed a frequency signature feature extraction conducted by the convolutional layers.

Khodabandehlou et al. [110] presented a novel approach using 2D convolutional networks from the raw measurements of a scaled bridge model. This network was able to classify the bridge's structural condition into four severe damage states, achieving 100% accuracy for the validation test. Avci et al. [111] presented a decentralized 1D-CNN architecture for structural damage identification. In this SHM system, each sensor performs damage identification locally using a 1D-CNN per sensor. Damage identification and localization were achieved through the classification network of each sensor, and the need for data transmission and aggregation is reduced because each sensor analyzes its local measurement data. Azimi and Pekcan [112] designed a CNN SHM system for damage identification and localization using compressed acceleration response histograms as network input. Deep transfer learning techniques were applied to this structural damage classification problem and proved feasible for damage identification in similar structures using discrete acceleration histograms.

Table 9. CNN review.

Publication	Structure	Excitation Source	Signal Preprocessing Techniques	Dataset Size	Convolution Layers Size	Pooling Layers Size	Activation Function	Accuracy	Purpose	Year
[51]	The Phase II IASC-ASCE SHM benchmark structure FE model	Simulation	HHT	1000 samples of size 16 × 4000	CNN 1D: 10 × 1, 10 × 1, 6 × 1 (3 layers), CNN 2D: 14 × 14, 5 × 5 (2 layers)	CNN 1D: 3 × 1, 6 × 1 (2 layers), CNN 2D: 2 × 2, 3 × 3 (2 layers)	Softmax	ACC = 100%	Propose a novel damage identification method using HHT and CNN with improved accuracy in comparison with CNN and SVM methods	2021
[94]	A steel beam numerical model	Numerical model forces	GWN	2000 samples of size 10 × 10	CNN 1: 5 × 1, 3 × 1 (2 layers), CNN 2: 4 × 4, 2 ×2 (2 layers)	CNN 1: 4 × 1, 2 × 1 (2 layers), CNN 2: 2 × 2, 1 × 1 (2 layers)	Sigmoid	RMS = 2.5	Propose a method for CNN hyperparameters selection in SHM systems	2021
[95]	Z24 bridge experimental model	Electrodynamic shaker	Statistical scaling	1,231 time series with 65,530 data points	16 × 16 (1 layer)	N/A	Softmax	ACC = 0.85 (pier settlement), ACC = 0.66 (tendon rupture)	Propose a 1D-CNN for multiclass structural damage detection using limited datasets	2021
[96]	A three-story building frame experimental model	Shaking table	N/A	2000 samples of size 10 × 10	8 × 8, 8 × 8, 7 × 7 (3 layers)	8, 8, 7 (3 layers)	Sigmoid	RMS ≈ 5.6	Propose a method for CNN hyperparameters selection in SHM systems	2021
[70]	An eight-story building numerical model	Lateral and vertical loads applied at each story	GWN	9216 samples	8 × 1, 5 × 1, 3 × 1 (3 layers)	N/A	ReLU and Softmax	ACC = 99.3%	Propose an FCN architecture for damage detection and localization in structures	2020
[97]	A six-story hotel instrumented building	Historically recorded response	Butterworth HPF	11 samples with 7200 data points	4 × 1, 4 × 1 (2 layers)	N/A	ReLU	Confidence = 93% (worst case)	Propose a CNN architecture to develop a surrogate model for modeling the seismic response of building structures	2020
[98]	A three-aluminum-beam experimental model	Mass structure addition and temperature changes	N/A	900 samples with 888 data points	237 × 1 (1 layer)	2 × 1 (1 layer)	ReLU and Softmax	ACC = 97%	Propose a combination of CNN and EMI methods for structural damage prediction	2020
[99]	The Phase I IASC-ASCE SHM benchmark structure FE model	Simulation	N/A	6 samples of size 1000 × 400	41 × 1, 41 × 1 (2 layers)	41 × 1, 41 × 1 (2 layers)	tanH	ACC = 96.11%	Propose a 1D-CNN for predicting damage from vibration data of structures	2020
[100]	A three-story shear frame FE model	Simulation	GWN	19,800 samples	128 × 1, 64 × 1, 32 × 1, and 16 × 1 (4 layers)	2 × 1, 2 × 1, 2 × 1 and 2 × 1 (4 layers)	Softmax	ACC > 81.8%	Propose CNN-based approach to identify and locate damage in structural joints	2020

Table 9. *Cont.*

Publication	Structure	Excitation Source	Signal Preprocessing Techniques	Dataset Size	Convolution Layers Size	Pooling Layers Size	Activation Function	Accuracy	Purpose	Year
[101]	A steel truss FE model	Simulation	N/A	5326 samples of size 15 × 3	3 × 3, 2 × 2, 1 × 1, 1 × 1 (4 layers)	N/A	ReLU and Softmax	ACC > 90% (50% damage degree)	Propose a CNN method to locate damaged rods using first-order mode shapes and mode curvature differences	2020
[102]	A linear steel beam FE model	Electrodynamical shaker	N/A	7500 samples	12 × 12, 10 × 10, 8 × 8 (3 layers)	4 × 4, 5 × 5, 6 × 6 (layers)	ReLU	91.75% and 92.18%	Present a novel SHM damage detection method based on a CNN trained with FE and experimental data	2020
[2]	Four- and eight-story building FE models	Simulation	HPF	5000 samples	50 × 50, 10 × 10 (2 layers)	20 × 20, 2 × 2 (2 layers)	ReLU and Softmax	ACC > 90% (several noise levels)	Propose a CNN method using raw data for structural damage detection	2019
[103]	A tied-arch bridge FE model	Simulation	FDD, FFT	10,000 samples	3 × 7 (2 layers)	4 × 1 (1 layer)	ReLU and Sigmoid	ACC = 99.92 (worst case)	Propose a method for structural damage identification using Fourier amplitude spectra of bridge data	2019
[104]	A structural connection FE model	Simulation	GWN	60,000 samples	3 × 3 (3 layers)	2 × 2 (2 layers)	tanH	ACC = 92.5% (6 cm crack), ACC = 95% (8 cm crack)	Propose a CNN for damage detection and localization in structures	2019
[105]	A three-story full-scale building experimental model	Shaking table	Bouc–Wen model, Baber–Noori hysteretic model	35,400 samples	10 × 1 (1 layer)	N/A	Sigmoid	ACC = 97.2%	Propose a DLN to predict the damage index (DI) of stiffness degradation for structures during earthquakes	2019
[106]	A steel Warren truss bridge scale experimental model	Impact hammer	N/A	Not specified	7 × 1, 5 × 1, 3 × 1 (3 layers)	3 × 1, 3 × 1, 3 × 1 (3 layers)	ReLU	ACC = 100%	Propose a new CNN framework to detect and locate damage in different structural crack scenarios	2019
[107]	A five-level benchmark building experimental model	Simulation	GWN, FFT, PSD	2832 samples of length 5	1000 × 1, 30 × 3, 10 × 3 (3 layers)	3 × 1, 3 × 1, 3 × 1 (3 layers)	ReLU and Softmax	RMSE = 0.0163	Propose a novel method based on DCNN for damage identification and location in building structures	2018
[108]	A long-span cable-stayed instrumented bridge	Operational conditions	FFT	333,792 samples of size 100 × 100 × 2	41 × 41 × 2 (1 layer)	2 × 2 (1 layer)	ReLU and Softmax	ACC = 93.5%	Propose a novel method based on CNN using time series that are visualized in the time domain and frequency domain	2018

Table 9. *Cont.*

Publication	Structure	Excitation Source	Signal Preprocessing Techniques	Dataset Size	Convolution Layers Size	Pooling Layers Size	Activation Function	Accuracy	Purpose	Year
[109]	A three-story steel frame experimental model	Shaking table	N/A	600 samples	5 × 8 (7 layers)	N/A	ReLU and tanH	RMS = 0.41 (worst noise case)	This study presents a DCNN approach to estimate the dynamic response of structures	2018
[110]	A highway bridge experimental model	Shaking table	N/A	48 samples of size 122 × 70	4 × 4, 8 × 8, 16 × 16, 32 × 32, 32 × 32 (5 layers)	2 × 2 (5 layers)	ReLU and tanH	ACC = 98.437%	Propose a two-dimensional CNN-based SHM system to identify structural damage states	2018
[111]	A steel structure experimental model	Electrodynamic shaker	Signal normalization	245,760 samples	80 × 1 (2 layers)	N/A	Not specified	Error graph	Propose a CNN SHM-based approach using raw signal without preprocessing or signal extraction techniques to detect structural damage	2018
[113]	An instrumented main steel frame	Electrodynamic shaker	N/A	6 samples of size	42 × 1 (1 layer)	2 × 1 (1 layer)	Not specified	Average error = 0.54%.	Propose a damage detection and localization method using an adaptive 1D-CNN	2017
[112]	The Phase II IASC-ASCE SHM benchmark structure FE model	Simulation	GWN	262,144 samples	3 × 1 (9 layers)	2 × 1 (3 layers)	ReLU and Softmax	ACC = 90–100% (several damage cases)	Present a novel CNN-based approach that uses acceleration data to estimate structural responses through transfer learning (TL)-based techniques	2016

4.2.3. Support Vector Machines (SVMs)

A support vector machine (SVM) is a popular machine-learning algorithm used for classification and regression problems. Its working principle is essential to maximize the distance value between a set of vectors, named support vectors, and a defined hyperplane. The maximizing goal in SVM can be seen as an optimization problem, and several approaches can be used to solve this problem. An exciting feature of the SVM decision boundary is that this boundary is the best possible boundary between a set of data classes for a given set of data and an SVM formulation. Table 10 shows that several SVM structural approaches for damage identification have high accuracy. However, SVM models' training stage is usually computationally expensive and unsuitable for real-time SHM systems that rely on post-training updates and ML model improvement. A common practice to reduce this computational cost is to combine SVM models with signal processing techniques such as PCA, WT and HHT, or damage-sensitive feature extraction methods such as RFR. Figure 4 shows an example of an SVM-based structural damage identification system.

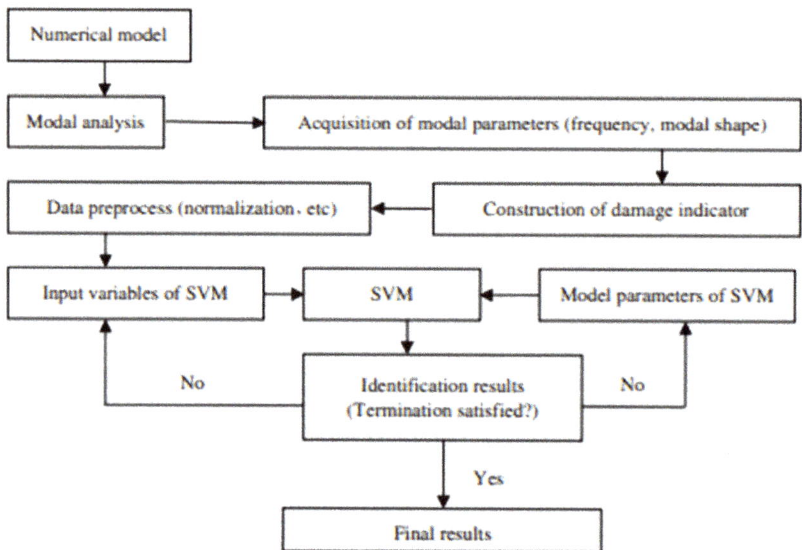

Figure 4. An example of an SVM-based structural damage identification system [114].

Cuong-Le et al. [115] proposed a combination of PSO and SVM for structural damage identification, location and severity. Four ML algorithms, including ANN, DNN, ANFIS and SVM, were trained and compared to predict damaged elements and damage severity. The proposed PSO-SVM combination achieved the highest accuracy among the compared techniques and correctly predicted all damage locations in the validation test. In a proposal by Agrawal and Chakraborty [116], Bayesian optimization (BO) was used in the hyperparameter search of an SVM, in which the SVM was used to perform damage identification. A comparison between BO and PSO in hyperparameter optimization search showed that BO was an efficient method that accelerated the optimization problem. Diao et al. [52] designed an SHM damage identification system, in which MEEMD and HHT were applied to acceleration measurements to extract damage characteristics through Hilbert spectrum energy. The SVM trained with Hilbert energy identified, located and estimated the severity of the damage in the structure, and the theoretical and experimental validation of the proposed method was provided.

Table 10. SVM review.

Publication	Structure	Excitation Source	Signal Processing Technique	Type of Kernel	Classification Error	Dataset Size	Purpose	Year
[115]	Truss bridge structure FE model and two-story frame structure FE model	Simulation	PSO	GRBF	RMSE = 0.0461 (truss structure), RMSE= 0.0621 (frame structure)	1000 samples	Propose an ML-based method to detect and locate structural damage using noisy measurements	2021
[116]	The Phase II IASC-ASCE SHM benchmark structure FE model	Simulation	PSO	GRBF	ACC = 100%	1768 samples	Propose a method to train an SVM classifier to optimally detect structural damage	2021
[52]	A five-story offshore platform experimental model	Shaking table	HHT, MEEMD	GRBF	ACC = 62.5%, MSE = 0.90	50,000 data points	HHT was used to extract damage-sensitive features from raw signals in order to detect, locate and estimate the severity of structural damage	2021
[51]	12-story building numerical model and laboratory-scaled steel bridge experimental model	Simulation and a rubber hammer	Deep autoencoder, GWN	GRBF	ACC = 97.4% (building), ACC = 91.0% (steel bridge)	15,600 samples (building model). 5810 samples (steel bridge)	Propose an SHM system for structural damage detection using an autoencoder to achieve high accuracy in several damage cases	2021
[117]	Two planar steel frame structure numerical models of 18 elements and 49 elements	Simulation	DEA	Newly proposed kernel function	ACC = 90.05% (double damage case 49 elements)	2500 samples	Propose a two-step method for structural damage location and extend it through the combination of SVM (a first step) and DEA (a second step)	2020
[118]	Steel frame structure experimental model and wooden house FE model	Shaking table and simulation	DNN	GRBF	ACC = 94.7% (steel frame), ACC = 90% (wooden structure)	6600 samples (steel frame structure), 100 samples (wooden house)	An SVM-based SHM was proposed to automatically detect several damage patterns from training data to improve the damage detection generalization capabilities of an SHM system	2020
[80]	Simply supported steel beam FE model and instrumented bridge	Electrodynamic shaker	SF	GRBF	ACC = 85.87%	19 samples of 2,400,000 data points (beam), 504 samples (bridge)	Ten statistical parameters of acceleration signals are used as damage-sensitive features to train an SHM system	2019

Table 10. *Cont.*

Publication	Structure	Excitation Source	Signal Processing Technique	Type of Kernel	Classification Error	Dataset Size	Purpose	Year
[75]	3D-reinforced concrete moment frame FE model	Simulation	Cumulative intensity measures	GRBF	ACC = 83.1%	6240 samples	Propose a structural damage detection and localization SHM system using an SVM model	2019
[74]	3D-reinforced concrete moment frame FE model	Simulation	GWN	GRBF	ACC = 92.48%	5400 samples	A framework was presented for near real-time prediction of damage existence, location and severity of building damage	2018
[119]	31-bar planar truss FE model and double-layer grid FE model	Simulation	LS-SVM	Not specified	MSE = 0.00051 (planar truss), RMSE = 0.0129 (double layer grid)	11,625 data points	Present a comparative study of the application of several ML techniques for damage detection of structures	2018
[120]	Three-story plane steel frame FE model and 8-DOF spring–mass system experimental model	Simulation and electrodynamic shaker	LS-SVM	GRBF	MSE = 3.243×10^{-7} (building), MSE = 7.303×10^{-9} (plate)	1944 samples (building), 2333 samples (plate), 2187 samples (8-DOF system)	Propose structural damage detection, localization and severity using an LS-SVM approach	2018
[41]	Offshore platform FE model and an instrumented scaled model	Simulation and shaking table	TFA, WPD, PCA	GRBF	ACC = 78.125% (FE model), ACC = 84.375% (instrumented model)	120 samples of 10,000 data points (FE model), 48 samples of 17,500 data points (instrumented model)	WPD and PCA were used as damage-sensitive feature extractors to train an SVM model for robust structural damage detection against measurement noise	2018
[121]	Three-story frame structure experimental model	Electrodynamic shaker	GA, PSO, grid search techniques, RE	GRBF	ACC = 100%	510 samples	Three optimization techniques were used to optimally select SVM hyperparameters using acceleration signals to perform structural damage detection tasks	2017
[122]	IASC-ASCE SHM benchmark dome truss FE model	Simulation	WPRE, HM	Newly proposed kernel function	MAE = 1.02×10^{-2} (benchmark), MAE = 1.02×10^{-2} (dome truss)	162 samples of 54,000 data points	Propose a new SVM kernel in order to increase the speed and the accuracy of LS-SVM for structural damage detection	2016

Table 10. *Cont.*

Publication	Structure	Excitation Source	Signal Processing Technique	Type of Kernel	Classification Error	Dataset Size	Purpose	Year
[123]	Three numerical models: simply supported beam, two-story plane frame, 8-DOF spring–mass system	Simulation	IRS, CSA, LS-SVM	GRBF	MSE = 3.0204×10^{-10} (beam), MSE = 7.2822×10^{-4} (plane frame), MSE = 4.2270×10^{-10} (8-DOF spring–mass system)	2187 samples (beam), 1944 samples two-story plane frame), 2315 samples (8-DOF spring–mass system)	Propose a new SHM damage detection system using incomplete mode shapes and structural natural frequencies	2016
[27]	Three-story frame aluminum structure experimental model	Bumper	AANN, FA, MSD, SVD, SVDD, KPCA	GRBF	ACC = 96.64% (one class SVM)	17 samples of 4096 data points	Four kernel SVM algorithms are tested to predict and detect structural damage in a building structure	2016
[42]	The Phase II IASC-ASCE SHM benchmark structure FE model	Simulation	WPD	GRBF	ACC = 91.75%	99 samples	Propose a structural building damage detection system based on an SVM using data fusion techniques	2015
[124]	A supported beam FE model	Simulation	BP-NN	Not specified	ACC = 100% (single crack), ACC = 99.9% (double crack)	Not specified	Propose a structural damage detection and localization method combining an SVM and BP-NN in beam structures	2014
[32]	Building structure experimental model and Sydney Harbour-instrumented bridge	Electrodynamic shaker and operational conditions	PCA, RP, PAA	GRBF	ACC = 97% (supervised SVM), ACC = 71% (unsupervised SVM)	3240 samples of 8192 data points (building), 6370 samples of 3600 data points (bridge)	Data dimensionality reduction techniques were used to generate training datasets for an SVM model for structural damage detection in buildings and bridge structures	2014
[43]	The Phase II IASC-ASCE SHM benchmark structure FE model	Simulation	LS-SVM, PSO, HSA, SF, WPT	GRBF	ACC = 100% (PSHS), ACC = 98.2% (Harmony), ACC = 98.68% (PSO)	16 samples of 40,000 data points	Present a comparative study between an ANN model and LS-SVM model using hyperparameter optimization techniques to achieve optimal structural damage detection	2014
[125]	Five-story, nine-story and twenty-one-story shear structures FE models and five-story steel experimental model	Simulation and electrodynamic shaker	Damage location indicators	Not specified	ACC = 100%	Not specified	Propose a structural damage detection and localization SHM system using an SVM model	2013

Table 10. *Cont.*

Publication	Structure	Excitation Source	Signal Processing Technique	Type of Kernel	Classification Error	Dataset Size	Purpose	Year
[126]	Three-story building structure experimental model	Magnetorheological dampers	ARM, WARM, DEF	Linear kernel and GRBF	RMSE = 0.0380 (AR, damaged system), RMSE = 0.0549 (WAR, damaged system)	100 samples	Propose SHM system using a combination of ARM and WARM to extract damage features from acceleration data and an SVM for damage detection purposes	2013
[114]	Simply supported bridge numerical model	Simulation	GA	GRBF	ACC= 100% (damage localization), ACC= 98.16% (triple damaged extend)	Not specified	A GA algorithm was used to select the optimal hyperparameters for an SVM structural damage classification model	2011

Seyedpoor and Nopour [117] proposed a two-step method using a differential evolution algorithm (DEA) and an SVM for damage identification in a steel structure. In the first step, the location of the potential damage is determined using an SVM, and in the second step, the severity of the damage and a more accurate location is obtained using the DEA. The results obtained showed high accuracy in the capabilities of the damage identification system. Kohiyama et al. [118] proposed the integration of a DNN and SVMs for the SHM damage identification framework. SVMs were used in the output layer of the DNN to detect data associated with an unlearned structural damage pattern. The added value of the proposed framework is the correct classification of unlearned damage patterns that may lead to misclassification, thereby improving the generalization feature of the classification model for damage identification.

Sajedi and Liang [75] proposed a damage identification SVM system to determine the presence and location of damage using intensity-based features as damage-sensitive features. This SHM system achieved over 83% accuracy in damage classification.

Intensity-based damage features were also used in a system proposed by Sajedi and Liang [74], in which an SVM classifier performed the classification task. The optimal damage-sensitive features and hyperparameters of the SVM were obtained using Bayesian optimization techniques. The proposed system showed robustness against signal distortion and reliability in detecting damage to building structures.

Ghiasi et al. [119] presented a comprehensive comparative study among several ML algorithms, including BP-NN, LS-SVMs, ANFIS, RBFN, LMNN, ELM, GP, multivariate adaptive regression spline (MARS), random forests and Kriging, for damage identification classification, location and severity estimation tasks. The results obtained indicate that the Kriging and LS-SVM models better predict damage location and severity than the other ML models.

Kourehli [120] presented a two-step unmeasured mode prediction method using two LS-SVMs and limited sensor measurements. The first step estimates missing modal shapes through an LS-SVM. In the second step, when complete modal system information is available, a second LS-SVM performs structural damage identification. The effects of noise and modeling error interference were analyzed, showing that the proposed system is robust against these effects. In the damage identification proposal presented by Diao et al. [41], transmissibility functions, WPEV and PCA were calculated on the acceleration response data and fed into an SVM damage classification model. WPEV and PCA were used as damage-sensitive feature extractors to construct the training samples of the ML model. As an improvement of the proposed damage identification scheme, the training data size was greatly reduced with the proposed strategy. The SVM classifier performed damage localization, and damage severity was obtained from the SVM regression result.

Gui et al. [121] presented a comparative study of optimization techniques combined with SVM models. Grid-search, PSO, and GA were used to optimize the SVM hyperparameters in structural damage identification, and the selected damage features included selected autoregressive and residual error features. The application of the above techniques showed improved SVM prediction performance, and the GA with AR features shows the best classification results. Ghiasi and Noori [122] proposed a new method for structural damage identification using an SVM with a newly proposed kernel. A Littlewood–Paley wavelet kernel was presented to improve the accuracy of the SVM for the SHM damage classification task, and the social harmony search algorithm was used to optimize the hyperparameters of the SVM. The results improved damage identification accuracy with the proposed kernel compared to other kernels combined.

An LS-SVM-based damage identification method was proposed by Kourehli [123] using incomplete modal data and natural frequencies as training samples and an iterated improved reduction system method. Coupled simulated annealing (CSA) was used to determine the optimal LS-SVM fitting parameters combined with a 10-fold cross-validation method. The experimental validation of the method was performed by applying it to three different structures, and the results showed an acceptable performance of the LS-SVM in damage identification.

Santos et al. [27] conducted a study in which four kernel-based ML models for damage identification tasks under various noise conditions were compared to analyze their performance. Among the kernel-based ML models, one-class support vector machine, support vector data description, kernel principal component analysis and greedy kernel principal component analysis were selected and used for damage identification. Compared with the previous studies performed by other ML models, kernel-based damage identification showed better accuracy results and better generalization ability.

Zhou et al. [42] presented an a posteriori probability support vector machine based on Dempster–Shafer evidence theory combined with a multi-sensor data fusion strategy. Signal energy features were extracted from vibration data using WPT and then processed by a posteriori probability support vector machine (PPSVM). The experimental results demonstrated that the proposed method was a robust structural damage identification method due to the fusion of information sensors and the application of the Dempster–Shafer evidence theory. Yan et al. [124] proposed a beam structural damage identification method for crack detection using a BP-ANN model and an SVM. The training data were generated through a FE beam model, and the strain differences between the models were considered damage indicators. The two ML models were used and compared, and the results indicated that both ML models performed well in crack identification for single and two-crack damage cases.

In the SHM damage identification approach presented by Khoa et al. [32], random projection (RP) was used to reduce the size of the vibration data, thereby reducing the computational processing time in structural monitoring. Data anomalies or structural damage were detected through one-class SVM and SVM models using the reduced bridge data. The results showed that the combination of SVM and RP improves the computational speed without affecting the detection accuracy.

Ghiasi et al. [43] presented a comparative study between ANN and LS-SVM for structural damage identification. The wavelet power spectrum was used for feature extraction, and the particle swarm harmony search (PSHS) algorithm was used for the hyperparameter selection of LS-SVM and ANN. The increased model performance of both ML techniques was achieved with the PSHS optimization method in terms of accuracy, and the combination of PSHS and LS-SVM was better than ANN with PSHS.

HoThu and Mita [125] designed an SVM-based method using only the first natural frequencies for damage identification and localization in shear structures as input data. The experimental results showed that the proposed system performed damage localization in a five-story laboratory model with response data from two accelerometers: one for the basement and the other from the roof of the structure. Kim et al. [126] presented a novel SHM framework for damage identification in smart structures. DWT, autoregressive models, and SVM were combined in this proposal, in which damage-sensitive feature extraction is performed from wavelet-based AR time series models. The detection results showed that the proposed system effectively detected structural damage in a three-story structure. Liu and Jiao [114] proposed a genetic algorithm to select the optimal hyperparameters of the SVM. Figure 4 shows the diagram with the proposed structural damage detection methodology based on the combination of a SVM and a GA.Mode shape and frequency ratios were used as input data for the SVM model. In the results, the combination of the GA-SVM damage identification method showed an accuracy of 98.16% for the case of single, double and triple damage elements, outperforming other methods, such as RBF networks and GA-BP networks.

4.2.4. Unsupervised Learning Algorithms

Unlike the supervised machine-learning methods discussed in the previous sections, unsupervised learning-based methods use training data samples without the need for class labels. Intuitively, these types of methods avoid the problem of defining possible multiple structural damage cases, including multiple structural damage scenarios. By applying unsupervised methods, two types of detections can be performed from the data obtained by the sensor network measurements of SHM systems: novelty detection and anomaly detection.

To address the drawbacks of supervised machine-learning approaches applied to structural damage identification, including sample labeling and the need for data from multiple damage scenarios, Wang and Cha [51] presented an unsupervised ML approach that uses an autoencoder to extract features from the raw measurement data. An SVM was trained with the extracted features, allowing it to detect even small damages (<10% stiffness reduction) in a structural member of a steel bridge.

Entezami and Shariatmadar [127] proposed two new damage indices for damage localization and quantification from time series modeling using an AR model. In this approach, thresholds are defined to detect possible structural damage without establishing damage patterns. Daneshvar and Sarmadi [128] explored an unsupervised anomaly-based method for short- and long-term SHM using an anomaly score and also setting an alarm threshold to establish damage. The core of the method implemented unsupervised feature selection through a one-class nearest neighbor rule.

de Almeida Cardoso et al. [129] proposed a damage detection framework based on outlier detection in the acquired data. The data descriptors used in this proposal included expected data value, data variability, data symmetry, and data flatness. In this proposal, a k-medoid clustering technique was used to detect outlier data using a proposed symbolic distance-voting scheme. Eltouny and Liang [130] developed a post-earthquake damage detection and localization approach that combines intensive cumulative measurements with an unsupervised learning algorithm. This approach consists of two stages: in the first stage, a feature extraction process is performed to extract the cumulative measures and in the second stage, joint probabilities are used as damage indicators obtained through the kernel multivariate maximum entropy method.

This review will emphasize supervised learning methods such as ANN, CNN and SVM in combination with preprocessing techniques such as WT, EMD, HHT and PCA. Despite the advantages of using unsupervised techniques in the problem of structural damage identification, the use of supervised techniques is the most adopted branch in the literature.

4.3. Model Type Comparative in Building SHM Systems

Table 11 analyzes the advantages and disadvantages of applying the physics-based model and the data-based model based on reviewing the proposals mentioned above.

Recent proposals implement hybrid approaches that improve SHM damage identification models and integrate physics-based and data-driven modeling solutions. One of the most common strategies is to build artificial datasets used to train ML models from FEM-generated data. Conversely, FEM parameters can be estimated from the output of an ML regression model and improve a structural virtual model.

Table 11. Advantages and disadvantages of a physics-based model and data-based models in SHM building applications.

Model Approach	Advantages	Disadvantages
Physics-based SHM	• The model parameters have a straightforward physical interpretation. Stiffness changes and displacements are consistent as damage-sensitive features. • It can reach all the levels of damage identification if the parameters are defined or estimated in the building structure. • The effect of the variation of the parameters can be estimated in the final result of the model. Parameter variation allows the simulation of different scenarios, and the structural safety thresholds can be established.	• The calculations of the solution of the model equations may not be feasible for real-time SHM applications, where the complexity of the structure requires long processing times. • The uncertainties and changing parameters may reduce the accuracy of the output model. Therefore, an estimation using other methods, such as model-based techniques, is encouraged.
Data-based SHM	• Noise and environmental effects on the data collected by the sensors can be minimized for the classification performed by the ML model. • Damage identification can be performed even if the parameters of the structure are unknown or cannot be estimated.	• The solution process is hidden from the user, so the rationale for successful damage classification is not explicit. • The training data set must be large enough to avoid overfitting, especially in algorithms such as CNN. In addition, obtaining the training samples of damage states is in most cases limited to an artificially generated dataset from simulations. • Computational training times can be costly for some ML algorithms (SVM, for example). Real-time SHM monitoring systems require faster methods such as NN solutions.

5. Uncertainties Effect Minimization in SHM Systems

Within the training and operation process of SHM systems, model uncertainties and monitoring system anomalies adversely affect the performance and the damage detection system capability. These uncertainties can be found in the data that are used for training and in the data that are processed during system operation.

In the training phase of the machine-learning model, it is common to use synthetic training data generated from FEM models of the structure of interest. The simulation of the structural response facilitates the task of generating data under different operating and structural damage conditions. The initial model of the structure of interest must accurately reflect the structural behavior and response, as it is the starting point for comparison with future states to determine the presence of structural damage. To minimize the uncertainties of the structural FEM model, several solutions have been proposed in recent years to obtain a model with reliable parameters and several solutions have been proposed for the optimal way to update FEM models in the SHM area.

Lee and Cho [131] proposed a direct FEM update strategy based on the identification of modal parameters and their comparison with the predicted structural response from measurements on a bridge structure. Jafarkhani [132] presented an evolutionary strategy in combination with a preliminary damage detection update triggering process and an ARMA model. An iterative FEM model updating process is used in Pachon's approach [133], where a sensitivity analysis determines the most significant structural parameters that have influence on the overall behavior of the structure. A survey of different FEM model updating methods in the area of SHM can be found in [134]. In the case of bridge-type structures, an analysis of the different model updating techniques can be found in Sharry's work in [135].

On the other hand, during SHM systems operation it is necessary to identify the impact of environmental and operational effects on structural parameters to minimize their effects on damage-sensitive features, especially in these structures under harsh or extreme conditions. For example, high sensor sensitivity measurements without noise removal techniques may only be permissible when environmental and operational conditions remain unchanged and this scenario is very uncommon in the implementations of SHM systems.

Yuen and Kuok [136] conducted a study to analyze the environmental effects (temperature and humidity) on the modal frequencies of the structure. Through one year of measurements in a 22-story building, Yuen identifies a correlation between the mentioned environmental conditions and the modal frequencies of the structure. The normalization of the sensing system data can reduce the effects of environmental noise and sensor failures. Some of these effects may be correlated with damage and should be eliminated to avoid a false-positive error in damage identification.

Cross et al. [137] employed cointegration and (PCA) techniques to remove operational and environmental effects and extract damage-sensitive features. Damikoukas et al. [138] proposed an identification method based on sensor measurements with noise to obtain the stiffness and damping matrices of a structure model. A final model including these parameters obtained from the noise measurements predicts building response to seismic excitation. Using SVM, relevant vector machines (RVM) and cointegration techniques, Coletta [139] implemented an SHM to monitor the structural condition of a historic building. This SHM system only considers the identified frequencies of the given structure and not the environmental effects of performing structural monitoring. Temperature variation is another environmental factor that can impact the performance of an SHM system and its ability to accurately detect damage, as the sensors can be affected by these variations. Huang et al. [140] proposed a new damage identification system that employs a new algorithm, which includes the effect of temperature variations and their correlation with structural properties. A genetic algorithm (GA) is also used to solve the optimization of the process.

Other approaches in the area of unsupervised learning that consider the effect of uncertainties in measured data have been proposed for anomaly detection systems that are resistant to uncertainty in sensor data. Yan et al. [141] studied the use of symmetric Kullback–Leibler (SKL) distance and transmissibility function to perform structural anomaly detection by comparing a baseline state and potential damage scenarios. Bayesian inference and a Monte Carlo discordance test are used as a statistical screening scheme to deal with measurement uncertainties. To improve the probabilistic function of the anomaly detection model, Mei et al. [142]. used the Bhattacharyya distance of TF. The performance of this proposal outperforms other proposals using Mahalanobis distance-based methods and have better resistance against the effects of uncertainties. Sarmadi and Karamodin [143] presented a Mahalanobis-squared distance-based method that combines the one-class kNN rule and an adaptive distance measure. In this study, the clustering process eliminates the effects of varying environmental conditions on the anomaly detection process.

6. Discussion and Summary

Within the advantages and limitations of physics-based SHM systems, we summarize the following highlights when choosing this approach in the design of a structural damage identification or anomaly detection system.

6.1. Physics-Based SHM Highlights

- Physics-based SHM systems produce a high level of accuracy of structural response prediction; however, they require domain expertise to define an FEM model with correct boundary conditions, constraints and structural properties.
- For long-term monitoring systems and in operation, a strategy for updating the FEM model should be defined in order to have a model that reflects and simulates the

current state of the structure. This strategy should define the update method, which parameters to update and when to update the structural parameters.

- To define the initial FEM model of the structure of interest, a stage of identification of the structural parameters must be considered, using estimation algorithms that include wear effects, operating conditions, the uncertainty of material properties, as well as the possibility of underlying structural damage. Using a data-driven method for the identification of these parameters, through a regression task by a machine-learning algorithm is recommended.
- For real-time SHM applications based on physical models, optimization algorithms in the solution of the numerical computation problem of an FEM model must be implemented to guarantee the responsiveness and availability of the SHM system to an event or in the continuous monitoring task.

Derived from the analysis of the works shown above in Tables 8–10, the most relevant observations of the application of ML techniques are presented below.

6.2. ANN SHM Highlights

- The ANN-based SHM systems showed good damage identification results using the modal information extracted from the structural acceleration responses with less training time compared to the proposed CNN and SVM, and simplicity in implementation.
- The selection of damage-sensitive features include the use of natural frequencies, modal shapes, damage index based on modal strain energy, FRF estimation, IMF or a combination of them.
- The removal of environmental effects and noise is also encouraged by signal processing techniques, such as filters (Kalman filter), and transforms, such as WT, EMD and HHT. Indeed, natural frequencies are susceptible to effects by temperature changes [23].
- The need for a large number of sensors is a drawback for ANN-based SHM system implementations. Analytical or data-driven estimation of unmeasured locations in the structure can enrich the input data to improve structural damage identification, as shown in [76]. The use of a sensor network allows the application of data fusion techniques that mitigate redundant signal information and anomalous sensor data.

6.3. CNN SHM Highlights

- The optimization techniques for the design of CNN architectures for SHM systems should be considered. Some popular proposals include optimization algorithms such as PSO and HSA. The design of the CNN architecture in SHM applications is the focus of attention since there is no method to select the hyperparameters of the CNN architecture.
- Two main dimension variants are used in the proposed CNN-based SHM systems: 1D-CNN and 2D-CNN. In the proposals with 1D input in SHM CNN, marginal HHT spectra, raw response time series windows, mode curvature differences, first-order mode shapes and Fourier amplitude spectra (FAS) have been used as input vectors. For 2D input matrices in SHM CNN, time–frequency plots, time series data rearranged into square matrices, multi-sensor matrix series data, heat map matrices and discrete response histograms have been used.
- A large amount of training data are needed to train CNN architectures. FE-generated data can be used, but this generated data can be unreliable if the FE model has uncertainties. Data enrichment can be performed to make the data reliable for training purposes, as shown in [92]. Data augmentation techniques and GAN-generated data can also be used and it is also necessary to balance the training datasets.
- CNN shows reliable performance, especially on noisy datasets [94] using raw measurements, compared to SVM [2] and MLP models [99]. In addition, CNN pre-preserves dominant frequency signal features, eliminating high-frequency noise components [99].

6.4. SVM SHM Highlights

- To ensure a good level of generality in structural damage identification capabilities, sufficient examples of damage scenarios should be provided in the model training step. The data generated by the FEM could be useful in scenarios where the damage state of the building is not available. Classification errors may occur when unlearned pattern scenarios are fed to the SVM model.

- It is essential to keep the number of SVM features to a computationally feasible number due to the computational processing cost. To reduce this cost, dimensionality reduction techniques such as PCA, random projection and boosting techniques are encouraged.
- The performance of the final SVM depends critically on the damage-sensitive features selected in the model input. Popular selected damage features include ARM features, wavelet-based energy features, and statistical features. If denoising techniques are not applied, the final performance may also be affected [45].
- There is a high dependence between the selected hyperparameters of the SVM and its performance, so the use of optimization techniques for hyperparameter tuning, such as PSO, PSH, CSA and GA, is recommended. More recent optimization techniques include sunflower optimization [144].
- The model training time using SVM approaches might be inadequate for model update stages in dynamic SHM building models and real-time building monitoring systems.
- SVM approaches show high capability for small structural damage identification [104] in combination with feature selection techniques.

7. Conclusions

This study presented a review of recent solutions in the field of SHM systems for multi-story buildings and bridges, addressing areas of improvement, limitations and opportunities. The limitations of physics-based and data-driven models in SHM in buildings and bridges were also discussed. This review showed the popularity of SHM systems based on the ML techniques of SVM, NN, and CNN due to the benefits of these algorithms in classification capabilities. The advantages and disadvantages of applying these ML techniques were also discussed in order to provide valuable insight for researchers.

In general, a fair comparison between ML models and feature extraction techniques implies a similar definition of damage scenarios (i.e., loss of stiffness, bolt loosening or mass addition) to provide a reasonable comparison. The occurrence of multiple damage locations within the structure is also an underexplored area in damage scenarios.

Real-time monitoring with in-service structure systems also needs to be developed. In addition, anomalous sensor data must be taken into account in real-time applications. Real-world structures are more complex than some idealized FE models and must involve a higher degree of uncertainty than those idealized models, so more damage scenarios must be studied to verify damage identification. FE model updating techniques and ML model retraining techniques should be explored for long-term SHM systems in the construction stages of the operational phase. Sensor placement plays a crucial role in the performance of ML-based SHM systems, so more studies should be conducted analyzing this fact.

The implications due to constraints imposed by the operational and environmental effects on the model need to be explored in implementation scenarios. For example, the periodicity, in which data should be collected from the monitored system to adjust or update the model due to the degradation of physical properties or operational conditions. Signal processing techniques, such as data fusion, normalization and compression, could improve data quality. In addition, with the development of increasingly powerful devices with higher computing power, signal processing can be performed in the fog computing paradigm. Fog devices can speed up the process of noise filtering, data compression and data fusion in SHM systems to assess the state of the structure after a seismic event. Hybrid approaches that integrate physics-based and data-based models are encouraged to address some of the drawbacks of stand-alone modeling, and this is the approach taken by some of the most recent developments in the SHM field. The need for an SHM system that synergistically integrates the advances reported in the literature in data generation, sensing, processing, model parameter updating and real-time monitoring, including the rapid assessment of post-seismic event response, is envisioned.

Author Contributions: Conceptualization, A.G.-C. and P.J.E.-A.; investigation, A.G.-C.; writing—original draft preparation, A.G.-C.; writing—review and editing, P.J.E.-A.; supervision, P.J.E.-A. All authors have read and agreed to the published version of the manuscript.

Funding: This work was funded in part by Consejo Nacional de Ciencia y Tecnología (CONACYT) and by Instituto Politécnico Nacional (IPN) under Grant SIP-20221495.

Institutional Review Board Statement: Not applicable.

Informed Consent Statement: Not applicable.

Data Availability Statement: Not applicable.

Conflicts of Interest: The authors declare no conflict of interest.

Abbreviations

Definition	Abbreviation	Definition	Abbreviation
Added random noise	ARN	Linear discriminant analysis	LDA
Auto-associative neural network	AANN	Mahalanobis-squared distance	MSD
Autoregressive model	ARM	Mean absolute error	MAE
Back-propagation	BP	Mean-squared error	MSE
Back-propagation neural network	BP-NN	Modal assurance criterion	MAC
Camage classification rates	DCR	Modified ensemble empirical mode decomposition	MEEMD
Complete ensemble empirical mode decomposition	CEEMD	Naïve bayes	NB
Contribution damage localization methods	CDLM	Nonlinear time–history analysis	NTHA
Convolutional neural network	CNN	Particle swarm optimization	PSO
Coupled simulated annealing	CSA	Piecewise aggregate approximation	PAA
Cross-correlation coefficient	CC	Poly-reference least-square complex frequency domain	p-LSCF
Damage-sensitive energy feature	DEF	Power spectral densities	PSD
Decision tree	DT	Principal component analysis	PCA
Deep convolutional neural network	DCNN	Principal component pursuit	PCP
Deep neural network	DNN	Probabilistic neural network	PNN
Differential evolution algorithm	DEA	Quadratic discriminant analysis	QDA
Electromechanical impedance	EMI	Radial basis function neural network	RBFNN
Extended Kalman filter	EKF	Random projection	RP
Factor analysis	FA	Recursive algorithm autoregressive-moving average with the exogenous inputs	RARMX
Fast Fourier transform	FFT	Residual error	RE
Feed-forward multilayered perceptron	FFMLP	Scaled conjugate gradient algorithm	SCG
Fourier transform	FT	Self-organizing maps	SOM
Frequency-domain decomposition	FDD	Signal statistical indicators	SSaI
Frequency response functions	FRF	Singular value decomposition	SVD
Gaussian radial basis function	GRBF	Statistical features	SF
Gaussian white noise	GWN	Stiffness reduction factor	SRF
Generative adversarial network	GAN	Stochastic subspace identification	SSI
Gradient descent momentum	GDM	Support vector data description	SVDD
Harmony memory	HM	Support vector machine	SVM
Harmony search algorithm	HSA	Teager–Huang transform	THT
High-pass filter	HPF	Trainable look-up tables	LUT
Hilbert–Huang transform	HHT	Transmissibility function analysis	TFA
Imperial competitive algorithm	ICA	Unsupervised image transformation model	UITM
Improved reduction system	IRS	Wavelet package relative energy	WPRE
Kernel principal component analysis	KPCA	Wavelet packet decomposition	WPD
Least-square support vector machine	LS-SVM	Wavelet packet transform	WPT
Levenberg–Marquardt back-propagation	LMBP	Wavelet-based autoregressive model	WARM

References

1. Yuan, F.-G.; Zargar, S.A.; Chen, Q.; Wang, S. Machine learning for structural health monitoring: Challenges and opportunities. *Sens. Smart Struct. Technol. Civ. Mech. Aerosp. Syst.* **2020**, *11379*, 1137903. [CrossRef]
2. Ibrahim, A.; Eltawil, A.; Na, Y.; El-Tawil, S. A Machine Learning Approach for Structural Health Monitoring Using Noisy Data Sets. *IEEE Trans. Autom. Sci. Eng.* **2020**, *17*, 900–908. [CrossRef]
3. Sivasuriyan, A.; Vijayan, D.; Górski, W.; Wodzyński, Ł.; Vaverková, M.; Koda, E. Practical Implementation of Structural Health Monitoring in Multi-Story Buildings. *Buildings* **2021**, *11*, 263. [CrossRef]
4. Kong, X.; Cai, C.-S.; Hu, J. The State-of-the-Art on Framework of Vibration-Based Structural Damage Identification for Decision Making. *Appl. Sci.* **2017**, *7*, 497. [CrossRef]
5. Valinejadshoubi, M.; Bagchi, A.; Moselhi, O. Structural health monitoring of buildings and infrastructure. *Struct. Health Monit.* **2016**, *1*, 50371.
6. Farrar, C.R.; Worden, K. *Structural Health Monitoring: A Machine Learning Perspective*; Wiley: Hoboken, NJ, USA, 2012.
7. Rytter, A. Vibrational Based Inspection of Civil Engineering Structures. Ph.D. Thesis, Aalborg University, Aalborg, Denmark, 1993.
8. Escamilla-Ambrosio, P.; Liu, X.; Lieven, N.; Ramírez-Cortés, J.J.r. ANFIS-Wavelet Packet Transform Approach to Structural Health Monitoring. *Ratio* **2010**, *10*, 1.
9. Escamilla-Ambrosio, P.J.; Liu, X.; Ramírez-Cortés, J.M.; Rodríguez-Mota, A.; Gómez-Gil, M.D.P. Multi-Sensor Feature Extraction and Data Fusion Using ANFIS and 2D Wavelet Transform in Structural Health Monitoring. In *Structural Health Monitoring—Measurement Methods and Practical Applications*; InTech: Houston, TX, USA, 2017.
10. Escamilla-Ambrosio, P.J.; Liu, X.; Lieven, N.A.J.; Ramirez-Cortes, J.M. Wavelet-fuzzy logic approach to structural health monitoring. In Proceedings of the 2011 Annual Meeting of the North American Fuzzy Information Processing Society, El Paso, TX, USA, 18–20 March 2011.
11. Liu, X.; Lieven, N.; Escamilla-Ambrosio, P. Frequency response function shape-based methods for structural damage localisation. *Mech. Syst. Signal Process.* **2009**, *23*, 1243–1259. [CrossRef]
12. Liu, X.; Escamilla-Ambrosio, P.; Lieven, N. Extended Kalman filtering for the detection of damage in linear mechanical structures. *J. Sound Vib.* **2009**, *325*, 1023–1046. [CrossRef]
13. Escamilla-Ambrosio, P.J.; Lieven, N. Soft Computing Feature Extraction for Health Monitoring of Rotorcraft Structures. In Proceedings of the 2007 IEEE International Fuzzy Systems Conference, London, UK, 23–26 July 2007.
14. Lubrano Lobianco, A.; Del Zoppo, M.; Di Ludovico, M. Seismic damage quantification for the SHM of existing RC structures. In *Lecture Notes in Civil Engineering*; Springer International Publishing: Cham, Switzerland, 2021; pp. 177–195.
15. Civera, M.; Surace, C. A Comparative Analysis of Signal Decomposition Techniques for Structural Health Monitoring on an Experimental Benchmark. *Sensors* **2021**, *21*, 1825. [CrossRef]
16. Barbosh, M.; Singh, P.; Sadhu, A. Empirical mode decomposition and its variants: A review with applications in structural health monitoring. *Smart Mater. Struct.* **2020**, *29*, 093001. [CrossRef]
17. Moallemi, A.; Burrello, A.; Brunelli, D.; Benini, L. Model-based vs. Data-driven approaches for anomaly detection in structural health monitoring: A case study. In Proceedings of the 2021 IEEE International Instrumentation and Measurement Technology Conference (I2MTC), Glasgow, UK, 17–20 May 2021.
18. Lin, Y.-Z.; Nie, Z.-H.; Ma, H.-W. Mechanism of Principal Component Analysis in Structural Dynamics under Ambient Excitation. *Int. J. Struct. Stab. Dyn.* **2020**, *20*, 2050136. [CrossRef]
19. Kumar, K.; Biswas, P.K.; Dhang, N. Time series-based SHM using PCA with application to ASCE benchmark structure. *J. Civ. Struct. Health Monit.* **2020**, *10*, 899–911. [CrossRef]
20. Padil, K.H.; Bakhary, N.; Abdulkareem, M.; Li, J.; Hao, H. Non-probabilistic method to consider uncertainties in frequency response function for vibration-based damage detection using Artificial Neural Network. *J. Sound Vib.* **2020**, *467*, 115069. [CrossRef]
21. Entezami, A.; Shariatmadar, H.; Mariani, S. Structural Health Monitoring for Condition Assessment Using Efficient Supervised Learning Techniques. *Proceedings* **2019**, *42*, 17. [CrossRef]
22. Khoshnoudian, F.; Talaei, S. A New Damage Index Using FRF Data, 2D-PCA Method and Pattern Recognition Techniques. *Int. J. Struct. Stab. Dyn.* **2017**, *17*, 1750090. [CrossRef]
23. Khoshnoudian, F.; Talaei, S.; Fallahian, M. Structural Damage Detection Using FRF Data, 2D-PCA, Artificial Neural Networks and Imperialist Competitive Algorithm Simultaneously. *Int. J. Struct. Stab. Dyn.* **2017**, *17*, 1750073. [CrossRef]
24. Loh, C.-H.; Huang, Y.-T.; Hsueh, W.; Chen, J.-D.; Lin, P.-Y. Visualization and Dimension Reduction of High Dimension Data for Structural Damage Detection. *Procedia Eng.* **2017**, *188*, 17–24. [CrossRef]
25. Vitola, J.; Vejar, M.A.; Burgos, D.A.T.; Pozo, F. Data-Driven Methodologies for Structural Damage Detection Based on Machine Learning Applications. In *Pattern Recognition–Analysis and Applications*; InTech: Houston, TX, USA, 2016.
26. Pozo, F.; Arruga, I.; Mujica, L.; Ruiz, M.; Podivilova, E. Detection of structural changes through principal component analysis and multivariate statistical inference. *Struct. Health Monit.* **2016**, *15*, 127–142. [CrossRef]
27. Santos, A.; Figueiredo, E.; Silva, M.F.M.; Sales, C.S.; Costa, J.C.W.A. Machine learning algorithms for damage detection: Kernel-based approaches. *J. Sound Vib.* **2016**, *363*, 584–599. [CrossRef]

28. Santos, A.D.F.; Silva, M.F.M.; Sales, C.S.; Costa, J.C.W.A.; Figueiredo, E. Applicability of linear and nonlinear principal component analysis for damage detection. In Proceedings of the 2015 IEEE International Instrumentation and Measurement Technology Conference (I2MTC), Pisa, Italy, 11–14 May 2015.
29. Gu, J.; Gul, M.; Wu, X. Damage detection under varying temperature using artificial neural networks. *Struct. Control Health Monit.* **2017**, *24*, e1998. [CrossRef]
30. Mujica, L.E.; Ruiz, M.; Pozo, F.; Rodellar, J.; Guemes, A. A structural damage detection indicator based on principal component analysis and statistical hypothesis testing. *Smart Mater. Struct.* **2013**, *23*, 25014. [CrossRef]
31. Yang, Y.; Nagarajaiah, S. Blind denoising of structural vibration responses with outliers via principal component pursuit. *Struct. Control Health Monit.* **2014**, *21*, 962–978. [CrossRef]
32. Khoa, N.L.; Zhang, B.; Wang, Y.; Chen, F.; Mustapha, S. Robust dimensionality reduction and damage detection approaches in structural health monitoring. *Struct. Health Monit.* **2014**, *13*, 406–417. [CrossRef]
33. Tibaduiza, D.A.; Mujica, L.E.; Rodellar, J. Damage classification in structural health monitoring using principal component analysis and self-organizing maps. *Struct. Control Health Monit.* **2012**, *20*, 1303–1316. [CrossRef]
34. Zhang, H.; Guo, J.; Xie, X.; Bie, R.; Sun, Y. Environmental effect removal based structural health monitoring in the internet of things. In Proceedings of the 2013 Seventh International Conference on Innovative Mobile and Internet Services in Ubiquitous Computing, Taichung, Taiwan, 3–5 July 2013.
35. Bandara, R.; Chan, T.; Thambiratnam, D. The three-stage artificial neural network method for damage assessment of building structures. *Aust. J. Struct. Eng.* **2013**, *14*, 13–25. [CrossRef]
36. Tibaduiza, D.A.; Mujica, L.E.; Rodellar, J. Comparison of several methods for damage localization using indices and contributions based on PCA. *J. PhysicsConf. Ser.* **2011**, *305*, 012013. [CrossRef]
37. Mujica, L.E.R.M.P.F.; Rodellar, J. Damage detection index based on statistical inference and PCA. *Struct. Health Monit.* **2011**, *2011*, 1–18. Available online: http://hdl.handle.net/2117/15362 (accessed on 23 August 2022).
38. Li, J.; Dackermann, U.; Xu, Y.-L.; Samali, B. Damage identification in civil engineering structures utilizing PCA-compressed residual frequency response functions and neural network ensembles. *Struct. Control Health Monit.* **2011**, *18*, 207–226. [CrossRef]
39. Chen, Z.; Wang, Y.; Wu, J.; Deng, C.; Hu, K. Sensor data-driven structural damage detection based on deep convolutional neural networks and continuous wavelet transform. *Appl. Intell.* **2021**, *51*, 5598–5609. [CrossRef]
40. Ewald, V.; Groves, R.M.; Benedictus, R. Deep SHM: A deep learning approach for structural health monitoring based on guided Lamb wave technique. In Proceedings of the SPIE, Denver, CO, USA, 27 March 2019; Volume 10970, p. 109700H-1-16.
41. Diao, Y.; Men, X.; Sun, Z.; Guo, K.; Wang, Y. Structural Damage Identification Based on the Transmissibility Function and Support Vector Machine. *Shock Vib.* **2018**, *2018*, 4892428. [CrossRef]
42. Zhou, Q.; Zhou, H.; Zhou, Q.; Yang, F.; Luo, L.; Li, T. Structural damage detection based on posteriori probability support vector machine and Dempster–Shafer evidence theory. *Appl. Soft Comput.* **2015**, *36*, 368–374. [CrossRef]
43. Ghiasi, R.; Torkzadeh, P.; Noori, M. Structural damage detection using artificial neural networks and least square support vector machine with particle swarm harmony search algorithm. *Int. J. Sustain. Mater. Struct. Syst.* **2014**, *1*, 303. [CrossRef]
44. Shi, A.; Yu, X.-H. Structural damage detection using artificial neural networks and wavelet transform. In Proceedings of the 2012 IEEE International Conference on Computational Intelligence for Measurement Systems and Applications (CIMSA) Proceedings, Tianjin, China, 2–4 July 2012.
45. Liu, Y.-Y.; Ju, Y.-F.; Duan, C.-D.; Zhao, X.-F. Structure damage diagnosis using neural network and feature fusion. *Eng. Appl. Artif. Intell.* **2011**, *24*, 87–92. [CrossRef]
46. Chen, C.; Wang, Y.; Wang, T.; Yang, X. A Mahalanobis Distance Cumulant-Based Structural Damage Identification Method with IMFs and Fitting Residual of SHM Measurements. *Math. Probl. Eng.* **2020**, *2020*, 6932463. [CrossRef]
47. Mousavi, A.A.; Zhang, C.; Masri, S.F.; Gholipour, G. Structural Damage Localization and Quantification Based on a CEEMDAN Hilbert Transform Neural Network Approach: A Model Steel Truss Bridge Case Study. *Sensors* **2020**, *20*, 1271. [CrossRef]
48. Vagnoli, M.R.-P.R.; Andrews, J. A machine learning classifier for condition monitoring and damage detection of bridge infrastructure. *Train. Reducing Uncertain. Struct. Saf.* **2018**, *1*, 53.
49. Pan, H.; Azimi, M.; Yan, F.; Lin, Z. Time-Frequency-Based Data-Driven Structural Diagnosis and Damage Detection for Cable-Stayed Bridges. *J. Bridg. Eng.* **2018**, *23*, 04018033. [CrossRef]
50. Chen, B.; Zhao, S.-L.; Li, P.-Y. Application of Hilbert-Huang Transform in Structural Health Monitoring: A State-of-the-Art Review. *Math. Probl. Eng.* **2014**, *2014*, 317954. [CrossRef]
51. Wang, X.; Zhang, X.; Shahzad, M.M. A novel structural damage identification scheme based on deep learning framework. *Structures* **2021**, *29*, 1537–1549. [CrossRef]
52. Diao, Y.; Jia, D.; Liu, G.; Sun, Z.; Xu, J. Structural damage identification using modified Hilbert–Huang transform and support vector machine. *J. Civ. Struct. Health Monit.* **2021**, *11*, 1155–1174. [CrossRef]
53. Vazirizade, S.M.; Bakhshi, A.; Bahar, O.; Nozhati, S. Online Nonlinear Structural Damage Detection Using Hilbert Huang Transform and Artificial Neural Networks. *Sci. Iran.* **2019**, *26*, 1266–1279. [CrossRef]
54. Han, J.; Zheng, P.; Wang, H. Structural modal parameter identification and damage diagnosis based on Hilbert-Huang transform. *Earthq. Eng. Eng. Vib.* **2014**, *13*, 101–111. [CrossRef]
55. Xu, C.; Chase, J.G.; Rodgers, G.W. Physical parameter identification of nonlinear base-isolated buildings using seismic response data. *Comput. Struct.* **2014**, *145*, 47–57. [CrossRef]

56. Gomes, G.F.; de Almeida, F.A.; da Cunha, S.S.; Ancelotti, A.C. An estimate of the location of multiple delaminations on aeronautical CFRP plates using modal data inverse problem. *Int. J. Adv. Manuf. Technol.* **2018**, *99*, 1155–1174. [CrossRef]
57. Pereira, J.L.J.; Francisco, M.B.; da Cunha, S.S., Jr.; Gomes, G.F. A powerful Lichtenberg Optimization Algorithm: A damage identification case study. *Eng. Appl. Artif. Intell.* **2020**, *97*, 104055. [CrossRef]
58. Frigui, F.; Faye, J.; Martin, C.; Dalverny, O.; Peres, F.; Judenherc, S. Global methodology for damage detection and localization in civil engineering structures. *Eng. Struct.* **2018**, *171*, 686–695. [CrossRef]
59. López, J.O.; Reyes, L.V.; Oyarzo-Vera, C. Structural health assessment of a R/C building in the coastal area of Concepción, Chile. *Procedia Eng.* **2017**, *199*, 2214–2219. [CrossRef]
60. Morales-Valdez, J.; Alvarez-Icaza, L.; Escobar, J.A. Damage Localization in a Building Structure during Seismic Excitation. *Shock Vib.* **2020**, *2020*, 8859527. [CrossRef]
61. Zhang, F.; Yang, Y.; Xiong, H.; Yang, J.; Yu, Z. Structural health monitoring of a 250-m super-tall building and operational modal analysis using the fast Bayesian FFT method. *Struct. Control Health Monit.* **2019**, *26*, e2383. [CrossRef]
62. Zhou, C.; Chase, J.G.; Rodgers, G.W.; Huang, B.; Xu, C. Effective Stiffness Identification for Structural Health Monitoring of Reinforced Concrete Building using Hysteresis Loop Analysis. *Procedia Eng.* **2017**, *199*, 1074–1079. [CrossRef]
63. Sun, H.; Al-Qazweeni, J.; Parol, J.; Kamal, H.; Chen, Z.; Büyüköztürk, O. Computational modeling of a unique tower in Kuwait for structural health monitoring: Numerical investigations. *Struct. Control. Health Monit.* **2019**, *26*, e2317. [CrossRef]
64. Isidori, D.; Concettoni, E.; Cristalli, C.; Soria, L.; Lenci, S. Proof of concept of the structural health monitoring of framed structures by a novel combined experimental and theoretical approach. *Struct. Control Health Monit.* **2016**, *23*, 802–824. [CrossRef]
65. Bao, Y.; Chen, Z.; Wei, S.; Xu, Y.; Tang, Z.; Li, H. The State of the Art of Data Science and Engineering in Structural Health Monitoring. *Engineering* **2019**, *5*, 234–242. [CrossRef]
66. Amezquita-Sanchez, J.P.; Adeli, H. Signal Processing Techniques for Vibration-Based Health Monitoring of Smart Structures. *Arch. Comput. Methods Eng.* **2014**, *23*, 1–15. [CrossRef]
67. Flah, M.; Nunez, I.; Ben Chaabene, W.; Nehdi, M.L. Machine Learning Algorithms in Civil Structural Health Monitoring: A Systematic Review. *Arch. Comput. Methods Eng.* **2020**, *28*, 2621–2643. [CrossRef]
68. Bao, Y.; Li, H. Machine learning paradigm for structural health monitoring. *Struct. Health Monit.* **2021**, *20*, 1353–1372. [CrossRef]
69. Smarsly, K.; Dragos, K.; Wiggenbrock, J. Machine learning techniques for structural health monitoring. In Proceedings of the 8th European Workshop on Structural Health Monitoring (EWSHM 2016), Bilbao, Spain, 5–8 July 2016; pp. 5–8.
70. Rosafalco, L.; Manzoni, A.; Mariani, S.; Corigliano, A. Fully convolutional networks for structural health monitoring through multivariate time series classification. *Adv. Model. Simul. Eng. Sci.* **2020**, *7*, 38. [CrossRef]
71. Vazirizade, S.M.; Nozhati, S.; Zadeh, M.A. Seismic reliability assessment of structures using artificial neural network. *J. Build. Eng.* **2017**, *11*, 230–235. [CrossRef]
72. Morfidis, K.; Kostinakis, K. Seismic parameters' combinations for the optimum prediction of the damage state of R/C buildings using neural networks. *Adv. Eng. Softw.* **2017**, *106*, 1–16. [CrossRef]
73. Abdeljaber, O.; Avci, O.; Kiranyaz, S.; Gabbouj, M.; Inman, D.J. Real-time vibration-based structural damage detection using one-dimensional convolutional neural networks. *J. Sound Vib.* **2017**, *388*, 154–170. [CrossRef]
74. Sajedi, S.O.; Liang, X. A data-driven framework for near real-time and robust damage diagnosis of building structures. *Struct. Control Health Monit.* **2020**, *27*, e2488. [CrossRef]
75. Sajedi, S.O.; Liang, X. Intensity-Based Feature Selection for Near Real-Time Damage Diagnosis of Building Structures. In Proceedings of the IABSE Congress, New York, NY, USA, 4–6 September 2019.
76. Chang, C.-M.; Lin, T.-K.; Chang, C.-W. Applications of neural network models for structural health monitoring based on derived modal properties. *Measurement* **2018**, *129*, 457–470. [CrossRef]
77. González, M.P.; Zapico, J.L. Seismic damage identification in buildings using neural networks and modal data. *Comput. Struct.* **2008**, *86*, 416–426. [CrossRef]
78. Ng, C.-T. Application of Bayesian-designed artificial neural networks in Phase II structural health monitoring benchmark studies. *Aust. J. Struct. Eng.* **2014**, *15*, 27–36. [CrossRef]
79. Tan, Z.X.; Thambiratnam, D.P.; Chan, T.H.T.; Gordan, M.; Razak, H.A. Damage detection in steel-concrete composite bridge using vibration characteristics and artificial neural network. *Struct. Infrastruct. Eng.* **2019**, *16*, 1247–1261. [CrossRef]
80. Finotti, R.P.; Cury, A.A.; Barbosa, F.D.S. An SHM approach using machine learning and statistical indicators extracted from raw dynamic measurements. *Lat. Am. J. Solids Struct.* **2019**, *16*, 1–17. [CrossRef]
81. Padil, K.H.; Bakhary, N.; Hao, H. The use of a non-probabilistic artificial neural network to consider uncertainties in vibration-based-damage detection. *Mech. Syst. Signal Process.* **2017**, *83*, 194–209. [CrossRef]
82. Jin, C.; Jang, S.; Sun, X.; Li, J.; Christenson, R. Damage detection of a highway bridge under severe temperature changes using extended Kalman filter trained neural network. *J. Civ. Struct. Health Monit.* **2016**, *6*, 545–560. [CrossRef]
83. Kourehli, S.S. Damage Assessment in Structures Using Incomplete Modal Data and Artificial Neural Network. *Int. J. Struct. Stab. Dyn.* **2015**, *15*, 1450087. [CrossRef]
84. Xie, X.; Guo, J.; Zhang, H.; Jiang, T.; Bie, R.; Sun, Y. Neural-network based structural health monitoring with wireless sensor networks. In Proceedings of the 2013 Ninth International Conference on Natural Computation (ICNC), Shenyang, China, 23–25 July 2013; pp. 57–66. [CrossRef]

85. Goh, L.D.; Bakhary, N.; Rahman, A.A.; Ahmad, B.H. Prediction of Unmeasured Mode Shape Using Artificial Neural Network for Damage Detection. *J. Teknol.* **2013**, *61*, 57–66. [CrossRef]
86. Shu, J.; Zhang, Z.; Gonzalez, I.; Karoumi, R. The application of a damage detection method using Artificial Neural Network and train-induced vibrations on a simplified railway bridge model. *Eng. Struct.* **2013**, *52*, 408–421. [CrossRef]
87. Geng, X.; Lu, S.; Jiang, M.; Sui, Q.; Lv, S.; Xiao, H.; Jia, Y.; Jia, L. Research on FBG-Based CFRP Structural Damage Identification Using BP Neural Network. *Photon-Sensors* **2018**, *8*, 168–175. [CrossRef]
88. Dworakowski, Z.; Stepinski, T.; Dragan, K.; Jablonski, A.; Barszcz, T. Ensemble ANN Classifier for Structural Health Monitoring. In *Artificial Intelligence and Soft Computing*; Springer International Publishing: Cham, Switzerland, 2016; pp. 81–90.
89. Min, J.; Park, S.; Yun, C.-B.; Lee, C.-G.; Lee, C. Impedance-based structural health monitoring incorporating neural network technique for identification of damage type and severity. *Eng. Struct.* **2012**, *39*, 210–220. [CrossRef]
90. Mardiyono, M.; Suryanita, R.; Adnan, A. Intelligent Monitoring System on Prediction of Building Damage Index using Neural-Network. *TELKOMNIKA (Telecommun. Comput. Electron. Control.)* **2012**, *10*, 155–164. [CrossRef]
91. Wang, B.; Ni, Y.-Q.; Ko, J. Damage detection utilising the artificial neural network methods to a benchmark structure. *Int. J. Struct. Eng.* **2011**, *2*, 229. [CrossRef]
92. Chamangard, M.; Amiri, G.G.; Darvishan, E.; Rastin, Z. Transfer Learning for CNN-Based Damage Detection in Civil Structures with Insufficient Data. *Shock Vib.* **2022**, *2022*, 3635116. [CrossRef]
93. Azimi, M.; Eslamlou, A.D.; Pekcan, G. Data-Driven Structural Health Monitoring and Damage Detection through Deep Learning: State-of-the-Art Review. *Sensors* **2020**, *20*, 2778. [CrossRef] [PubMed]
94. Oh, B.K.; Kim, J. Multi-Objective Optimization Method to Search for the Optimal Convolutional Neural Network Architecture for Long-Term Structural Health Monitoring. *IEEE Access* **2021**, *9*, 44738–44750. [CrossRef]
95. Sony, S.; Gamage, S.; Sadhu, A.; Samarabandu, J. Multiclass Damage Identification in a Full-Scale Bridge Using Optimally Tuned One-Dimensional Convolutional Neural Network. *J. Comput. Civ. Eng.* **2022**, *36*, 04021035. [CrossRef]
96. Oh, B.K.; Kim, J. Optimal architecture of a convolutional neural network to estimate structural responses for safety evaluation of the structures. *Measurement* **2021**, *177*, 109313. [CrossRef]
97. Zhang, R.; Liu, Y.; Sun, H. Physics-guided convolutional neural network (PhyCNN) for data-driven seismic response modeling. *Eng. Struct.* **2020**, *215*, 110704. [CrossRef]
98. de Rezende, S.W.F.; de Moura, J.d.R.V., Jr.; Neto, R.M.F.; Gallo, C.A.; Steffen, V., Jr. Convolutional neural network and impedance-based SHM applied to damage detection. *Eng. Res. Express* **2020**, *2*, 035031. [CrossRef]
99. Sarawgi, Y.; Somani, S.; Chhabra, A. Dhiraj Nonparametric Vibration Based Damage Detection Technique for Structural Health Monitoring Using 1D CNN. In *Communications in Computer and Information Science*; Springer: Singapore, 2020; pp. 146–157.
100. Sharma, S.; Sen, S. One-dimensional convolutional neural network-based damage detection in structural joints. *J. Civ. Struct. Health Monit.* **2020**, *10*, 1057–1072. [CrossRef]
101. Zhong, K.; Teng, S.; Liu, G.; Chen, G.; Cui, F. Structural Damage Features Extracted by Convolutional Neural Networks from Mode Shapes. *Appl. Sci.* **2020**, *10*, 4247. [CrossRef]
102. Seventekidis, P.; Giagopoulos, D.; Arailopoulos, A.; Markogiannaki, O. Structural Health Monitoring using deep learning with optimal finite element model generated data. *Mech. Syst. Signal Process.* **2020**, *145*, 106972. [CrossRef]
103. Duan, Y.; Chen, Q.; Zhang, H.; Yun, C.B.; Wu, S.; Zhu, Q. CNN-based damage identification method of tied-arch bridge using spatial-spectral information. *Smart Struct. Syst. Int. J.* **2019**, *23*, 507–520.
104. Gulgec, N.S.; Takáč, M.; Pakzad, S.N. Convolutional Neural Network Approach for Robust Structural Damage Detection and Localization. *J. Comput. Civ. Eng.* **2019**, *33*, 04019005. [CrossRef]
105. Zhou, C.; Chase, J.G.; Rodgers, G.W. Degradation evaluation of lateral story stiffness using HLA-based deep learning networks. *Adv. Eng. Informatics* **2019**, *39*, 259–268. [CrossRef]
106. Zhang, T.; Wang, Y. Deep learning algorithms for structural condition identification with limited monitoring data. In Proceedings of the International Conference on Smart Infrastructure and Construction 2019 (ICSIC), Cambridge, UK, 8–10 July 2019.
107. Yu, Y.; Wang, C.; Gu, X.; Li, J. A novel deep learning-based method for damage identification of smart building structures. *Struct. Health Monit.* **2019**, *18*, 143–163. [CrossRef]
108. Tang, Z.; Chen, Z.; Bao, Y.; Li, H. Convolutional neural network-based data anomaly detection method using multiple information for structural health monitoring. *Struct. Control Healyh Monit.* **2019**, *26*, e2296. [CrossRef]
109. Wu, R.-T.; Jahanshahi, M.R. Deep Convolutional Neural Network for Structural Dynamic Response Estimation and System Identification. *J. Eng. Mech.* **2019**, *145*, 04018125. [CrossRef]
110. Khodabandehlou, H.; Pekcan, G.; Fadali, M.S. Vibration-based structural condition assessment using convolution neural networks. *Struct. Control Health Monit.* **2018**, *26*, e2308. [CrossRef]
111. Avci, O.; Abdeljaber, O.; Kiranyaz, S.; Hussein, M.; Inman, D.J. Wireless and real-time structural damage detection: A novel decentralized method for wireless sensor networks. *J. Sound Vib.* **2018**, *424*, 158–172. [CrossRef]
112. Azimi, M.; Pekcan, G. Structural health monitoring using extremely compressed data through deep learning. *Comput. Civ. Infrastruct. Eng.* **2020**, *35*, 597–614. [CrossRef]
113. Avci, O.; Abdeljaber, O.; Kiranyaz, S.; Inman, D. Structural Damage Detection in Real Time: Implementation of 1D Convolutional Neural Networks for SHM Applications. In *Structural Health Monitoring & Damage Detection*; Springer International Publishing: Cham, Switzerland, 2017; Volume 7, pp. 49–54.

114. Liu, H.-B.; Jiao, Y.-B. Application Of Genetic Algorithm-Support Vector Machine (Ga-Svm) For Damage Identification Of Bridge. *Int. J. Comput. Intell. Appl.* **2011**, *10*, 383–397. [CrossRef]
115. Cuong-Le, T.; Nghia-Nguyen, T.; Khatir, S.; Trong-Nguyen, P.; Mirjalili, S.; Nguyen, K.D. An efficient approach for damage identification based on improved machine learning using PSO-SVM. *Eng. Comput.* **2021**, *38*, 3069–3084. [CrossRef]
116. Agrawal, A.K.; Chakraborty, G. On the use of acquisition function-based Bayesian optimization method to efficiently tune SVM hyperparameters for structural damage detection. *Struct. Control Health Monit.* **2021**, *28*, e2693. [CrossRef]
117. Seyedpoor, S.M.; Nopour, M.H. A two-step method for damage identification in moment frame connections using support vector machine and differential evolution algorithm. *Appl. Soft Comput.* **2020**, *88*, 106008. [CrossRef]
118. Kohiyama, M.; Oka, K.; Yamashita, T. Detection method of unlearned pattern using support vector machine in damage classification based on deep neural network. *Struct. Control Healthy Monit.* **2020**, *27*, e2552. [CrossRef]
119. Ghiasi, R.; Ghasemi, M.R.; Noori, M. Comparative studies of metamodeling and AI-Based techniques in damage detection of structures. *Adv. Eng. Softw.* **2018**, *125*, 101–112. [CrossRef]
120. Kourehli, S.S. Prediction of unmeasured mode shapes and structural damage detection using least squares support vector machine. *Struct. Monit. Maint.* **2018**, *5*, 379–390.
121. Gui, G.; Pan, H.; Lin, Z.; Li, Y.; Yuan, Z. Data-driven support vector machine with optimization techniques for structural health monitoring and damage detection. *KSCE J. Civ. Eng.* **2017**, *21*, 523–534. [CrossRef]
122. Ghiasi, R.; Torkzadeh, P.; Noori, M. A machine-learning approach for structural damage detection using least square support vector machine based on a new combinational kernel function. *Struct. Health Monit.* **2016**, *15*, 302–316. [CrossRef]
123. Kourehli, S.S. LS-SVM Regression for Structural Damage Diagnosis Using the Iterated Improved Reduction System. *Int. J. Struct. Stab. Dyn.* **2016**, *16*, 1550018. [CrossRef]
124. Yan, B.; Cui, Y.; Zhang, L.; Zhang, C.; Yang, Y.; Bao, Z.; Ning, G. Beam Structure Damage Identification Based on BP Neural Network and Support Vector Machine. *Math. Probl. Eng.* **2014**, *2014*, 850141. [CrossRef]
125. HoThu, H.; Mita, A. Damage Detection Method Using Support Vector Machine and First Three Natural Frequencies for Shear Structures. *Open J. Civ. Eng.* **2013**, *3*, 104–112. [CrossRef]
126. Kim, Y.; Chong, J.W.; Chon, K.H.; Kim, J. Wavelet-based AR–SVM for health monitoring of smart structures. *Smart Mater. Struct.* **2012**, *22*, 015003. [CrossRef]
127. Entezami, A.; Shariatmadar, H. An unsupervised learning approach by novel damage indices in structural health monitoring for damage localization and quantification. *Struct. Health Monit.* **2018**, *17*, 325–345. [CrossRef]
128. Daneshvar, M.H.; Sarmadi, H. Unsupervised learning-based damage assessment of full-scale civil structures under long-term and short-term monitoring. *Eng. Struct.* **2022**, *256*, 114059. [CrossRef]
129. Cardoso, R.D.A.; Cury, A.; Barbosa, F.; Gentile, C. Unsupervised real-time SHM technique based on novelty indexes. *Struct. Control Health Monit.* **2019**, *26*, e2364. [CrossRef]
130. Eltouny, K.; Liang, X.J.a.p.a. A nonparametric unsupervised learning approach for structural damage detection. *arXiv* **2020**, arXiv:2009.13692.
131. Lee, Y.-J.; Cho, S. SHM-Based Probabilistic Fatigue Life Prediction for Bridges Based on FE Model Updating. *Sensors* **2016**, *16*, 317. [CrossRef]
132. Jafarkhani, R.; Masri, S.F. Finite Element Model Updating Using Evolutionary Strategy for Damage Detection. *Comput. Civ. Infrastruct. Eng.* **2011**, *26*, 207–224. [CrossRef]
133. Pachón, P.; Castro, R.; García, M.E.P.; Compán, V.; Puertas, E.E. Torroja's bridge: Tailored experimental setup for SHM of a historical bridge with a reduced number of sensors. *Eng. Struct.* **2018**, *162*, 11–21. [CrossRef]
134. Moravej, H.; Jamali, S.; Chan, T.; Nguyen, A. Finite element model updating of civil engineering infrastructures: A literature review. In Proceedings of the 8th International Conference on Structural Health Monitoring of Intelligent Infrastructure, Brisbane, Australia, 5–8 December 2017; pp. 1–12.
135. Sharry, T.; Guan, H.; Nguyen, A.; Oh, E.; Hoang, N. Latest Advances in Finite Element Modelling and Model Updating of Cable-Stayed Bridges. *Infrastructures* **2022**, *7*, 8. [CrossRef]
136. Yuen, K.-V.; Kuok, S.-C. Ambient interference in long-term monitoring of buildings. *Eng. Struct.* **2010**, *32*, 2379–2386. [CrossRef]
137. Cross, E.; Manson, G.; Worden, K.; Pierce, S. Features for damage detection with insensitivity to environmental and operational variations. *Proc. R. Soc. A Math. Phys. Eng. Sci.* **2012**, *468*, 4098–4122. [CrossRef]
138. Damikoukas, S.; Chatzieleftheriou, S.; Lagaros, N.D. Direct identification of reduced building models based on noisy measurements for performance based earthquake engineering. *J. Build. Eng.* **2021**, *34*, 101776. [CrossRef]
139. Coletta, G.; Miraglia, G.; Pecorelli, M.; Ceravolo, R.; Cross, E.; Surace, C.; Worden, K. Use of the cointegration strategies to remove environmental effects from data acquired on historical buildings. *Eng. Struct.* **2019**, *183*, 1014–1026. [CrossRef]
140. Huang, M.-S.; Gül, M.; Zhu, H.-P. Vibration-Based Structural Damage Identification under Varying Temperature Effects. *J. Aerosp. Eng.* **2018**, *31*, 04018014. [CrossRef]
141. Yan, W.-J.; Chronopoulos, D.; Yuen, K.-V.; Zhu, Y.-C. Structural anomaly detection based on probabilistic distance measures of transmissibility function and statistical threshold selection scheme. *Mech. Syst. Signal Process.* **2022**, *162*, 108009. [CrossRef]
142. Mei, L.-F.; Yan, W.-J.; Yuen, K.-V.; Beer, M. Structural novelty detection with Laplace asymptotic expansion of the Bhattacharyya distance of transmissibility and Bayesian resampling scheme. *J. Sound Vib.* **2022**, *540*, 117277. [CrossRef]

143. Sarmadi, H.; Karamodin, A. A novel anomaly detection method based on adaptive Mahalanobis-squared distance and one-class kNN rule for structural health monitoring under environmental effects. *Mech. Syst. Signal Process.* **2020**, *140*, 106495. [CrossRef]
144. Magacho, E.G.; Jorge, A.B.; Gomes, G.F. Inverse problem based multiobjective sunflower optimization for structural health monitoring of three-dimensional trusses. *Evol. Intell.* **2021**, *14*, 1–21. [CrossRef]

MDPI

Article

Application of Fiber Optic Sensing System for Predicting Structural Displacement of a Joined-Wing Aircraft

Yang Meng [1], Ying Bi [2], Changchuan Xie [1,*], Zhiying Chen [1] and Chao Yang [1]

[1] School of Aeronautics Science and Engineering, Beihang University, Beijing 100191, China
[2] Institute of Engineering Thermophysics, Chinese Academy of Sciences, Beijing 100190, China
* Correspondence: xiechangc@buaa.edu.cn

Abstract: This work aims to achieve real-time monitoring of strains and structural displacements for the target Joined-Wing aircraft. To this end, a Fiber Optic Sensing System (FOSS) is designed and deployed in the aircraft. The classical modal method, which is used for Strain-to-Displacement Transformation (SDT), is improved to adapt to different boundary conditions by introducing extra constraint equations. The method is first verified by numerical studies on a cantilever beam model and the high-fidelity finite element model of the Joined-Wing aircraft. Ground static tests are then carried out to further demonstrate the capability of the developed FOSS and SDT algorithm in practical application. The results have shown that the improved modal method is able to predict structural deformation under different boundary conditions by using only free–free modes. In addition, the errors between the predicted displacement and the reference in the ground test are within 10%, which proves the FOSS has reasonable accuracy and the potential for future flight tests.

Keywords: joined-wing; fiber optic sensing; deformation measurement; modal method

Citation: Meng, Y.; Bi, Y.; Xie, C.; Chen, Z.; Yang, C. Application of Fiber Optic Sensing System for Predicting Structural Displacement of a Joined-Wing Aircraft. *Aerospace* **2022**, *9*, 661. https://doi.org/10.3390/aerospace9110661

Academic Editor: Liang Yu

Received: 21 September 2022
Accepted: 25 October 2022
Published: 27 October 2022

1. Introduction

The Joined-Wing configuration is usually considered to have a substantial increase in design space and be able to provide more options in aerodynamics, flight dynamics, aeroelasticity, propulsions, etc. [1]. In the 2000s, the U.S. Air Force proposed the new generation of the Intelligence, Surveillance and Reconnaissance (ISR) platform, which is called "SensorCraft" [2]. The Joined-Wing configuration is one of the three basic platform shapes that was considered. The large overall size of "SensorCraft" makes flexibility of great significance. Previous studies have investigated and highlighted the effects of structural deformation or structural geometric nonlinearity on the aeroelastic features of the Joined-Wing aircraft [3,4]. Real-time measurement of structural deformation is of great significance in Structural Health Monitoring (SHM), structural control, safety assessment and methodology validation.

In the past few years, the Chinese Academy of Sciences has started up a project to design and manufacture a Joined-Wing platform for civil applications. This work aims to provide an onboard strain and displacement measurement technique for the platform. Benefiting from numerous advantages such as light weight, small size, multiplexing capability and resistance to electromagnetic disturbance, fiber optic sensors have become the first choice for Structural Health Monitoring (SHM) in aeronautical applications [5]. Compared to the existing shape-sensing techniques of using electrical strain sensors, accelerometers, cameras, inclinometers or laser scanners, fiber optic sensing offers an extremely promising alternative with the capability to track the shape continuously and dynamically [6]. Among various strain-sensing technologies in fiber optic sensing, Fiber Bragg Gratings (FBG) are the most widely used optical fiber sensors and have extensive applications. Despite of the limitation in distributed sensing capability, FBG sensors have prominent advantages such as a large sensing length, a low cost, high strain sensing accuracy, high reliability and so on.

Compared with the distributed sensing based on Rayleigh or Brillouin scattering [7], FBG sensors are still the first choice in engineering applications at this stage. After the famous mishap of the "Helios" [8], shape-sensing methodology as well as the Fiber Optic Sensing System (FOSS) have been investigated at NASA Dryden Flight Research Center (DFRC). In present work, a FOSS including both software and hardware is designed to be adaptive to the target Joined-Wing platform. FBG sensors are used for strain measurement.

Three main methods for Strain-to-Displacement Transformation (SDT) have been investigated. The first one is called the Ko displacement theory [9,10], which computes the displacement by double integration of measured strains based on linear Euler beam model. The method was then verified by tests on a swept plate [11], Ikahana wing [12], a composite aircraft wing [13], the business jet Global 7500 wing [14], etc. Meng [15] extended the Ko displacement theory to nonlinear two-dimensional problems. The major advantage of this method is that it does not require any additional modeling, which means the only required physical information of the structure is the geometric parameter. Another method is called the inverse Finite Element Method (iFEM), which is proposed and developed by Tessler and Gherlone et al. [16–20]. The iFEM depends on a least-squares variational principle to predict the three-dimensional deformation field using strain measurements. Comparative studies on a composite wing box conducted by Ghelone have shown that the iFEM is more accurate than the Ko theory and the classical modal method [21,22]. However, a barrier of the iFEM in practical applications is that the element library should be enriched to satisfy the accurate modeling requirements [18]. Many other researchers utilize the modal method to construct the relationship between the measured strains and the structural displacements. FOSS and Haugse [23] first used modal test results to develop SDT. For simple structures, the strain and displacement modes can be computed analytically [24,25]. In practical problems, high-fidelity finite element models are usually used to calculate modal results numerically. The modal method has been verified and applied to wind turbine blades [26] and aerospace structures [27,28]. The major advantage of the modal method is that it can reconstruct the accurate three-dimensional displacement field with few strain sensors, which makes it applicable to complex structures. As mentioned in [11], the main drawback of the modal method is that it requires prior knowledge of a numerically structural model that matches the real structures, which can be very challenging. Nevertheless, the modal method can be a promising method for SDT if the high-fidelity finite element model is available.

To the best of the authors' knowledge, few studies associated with the shape-sensing problem have dealt with the Joined-Wing aircraft. Different from the single main wing in the traditional configuration, the Joined-Wing aircraft usually has the front wing and aft wing, which makes the deformation more complicated. In addition, previous studies usually considered fixed boundary conditions. Nevertheless, the aircraft can experience different boundary conditions during the whole flight process. It is troublesome to switch modal shapes according to boundary conditions. An intuitive way is to compute the structural displacement by using the modal information obtained by one calculation. To this end, this work will extend the classical modal approach to adapt to various boundary conditions by introducing extra constraint equations. Only free–free modes are needed in the improved modal method. This paper is organized as follows. In Section 2, the theoretical background including the main principle of Fiber Optic Sensing Technology, and a detailed description of the improved modal method is provided. In Section 3, the Joined-Wing aircraft is first introduced, followed by the description of the FOSS and sensor arrangement. In Section 4, a cantilever beam model is first investigated to validate the ability of the proposed method in dealing with problems under different boundary conditions. Then, the FOSS on the Joined-Wing aircraft is numerically and experimentally studied, respectively. Finally, several concluding remarks are highlighted in Section 5.

2. Theoretical Background

2.1. Principle of Fiber Optic Sensing

Fiber Bragg Grating (FBG) sensors are created by exposing an optical fiber to an ultraviolet interference pattern, which produces a periodic change in the core index of refraction. The periodic changes further cause a reflection when the light in the waveguide is of a wavelength while other wavelengths are transmitted in the optic fiber. This wavelength is called Bragg wavelength, λ_B, which is determined by the period of core index modulation, Λ, and the effective core index of refraction, n_0:

$$\lambda_B = 2n_0\Lambda \tag{1}$$

Both strain and temperature can change Λ and n_0, which further cause a variation in λ_B. Assuming the fiber is optically isotropic, the variation of the Bragg wavelength, $\delta\lambda_B$, can be expressed by

$$\frac{\delta\lambda_B}{\lambda_B} = \alpha_\varepsilon\varepsilon + \alpha_T\Delta T \tag{2}$$

where α_ε is the strain-optic coefficient, α_T is the temperature-optic coefficient, ε is the strain along the fiber-grating-axis direction, ΔT is the variation of temperature. For common silicon fibers, α_ε and α_T equal to 0.78×10^{-6} and 6.67×10^{-6}, respectively. However, these two coefficients need to be recalibrated due to the uncertainties in the installation process. According to Equation (2), the strain and temperature can be obtained by identification of the variation of Bragg wavelength as follows:

$$\begin{aligned} \varepsilon &= \frac{\lambda_B^{strain} - \lambda_{B,0}^{strain}}{\alpha_\varepsilon\lambda_{B,0}^{strain}} - \frac{\alpha_T\Delta T}{\alpha_\varepsilon} \\ \Delta T &= \frac{\lambda_B^{temp} - \lambda_{B,0}^{temp}}{\alpha_T\lambda_{B,0}^{temp}} \end{aligned} \tag{3}$$

In Equation (3), FBG sensors are divided into two types, one of which can reflect both temperature variations, and the other one can only reflect temperature variations. The former one is denoted by the superscript "*strain*". Sensors directly attached to the structure surface naturally fall into this category. The second one is denoted by the superscript "*temp*". In this work, we first put the FBG sensor into a tube, which contains a short steel needle that is quite rigid. Then, the tube is bonded to the structure surface. As the steel needle is quite rigid compared to the tested structure, the wavelength variation of the FBG sensor in the tube can be approximately considered to be only affected by the temperature, thereby realizing temperature compensation.

2.2. Strain-to-Displacement Transformation (SDT)

Following the classical modal method, several primary strain modes and displacement modes are firstly calculated from modal analysis of high-fidelity Finite Element Models (FEM). The measured strain is then used to solve the least-square solutions of the coefficients of the best linear combination of strain shape functions. The same coefficients are used to determine the deformed shape by a linear combination of the displacement shape functions. Here, we denote the strain shape function of the ith mode as $\boldsymbol{\varphi}_i$, the displacement shape function of the ith mode as $\boldsymbol{\Phi}_i$. The structural displacement in the physical degrees of freedom, $\boldsymbol{u}$, can be written as a linear combination of m displacement modes:

$$\boldsymbol{u} = \sum_{i=1}^{m} \boldsymbol{\Phi}_i q_i = \boldsymbol{\Phi}\boldsymbol{q} \tag{4}$$

where $\boldsymbol{\Phi} = \begin{bmatrix} \boldsymbol{\Phi}_1 & \boldsymbol{\Phi}_2 & \cdots & \boldsymbol{\Phi}_m \end{bmatrix}$ is the matrix of displacement mode, $\boldsymbol{q} = \begin{bmatrix} q_1 & q_2 & \cdots & q_m \end{bmatrix}^T$ is the column vector of generalized coordinates. In the same way, the strain vector, $\boldsymbol{\varepsilon}$, can be written as a combination of strain modes

$$\boldsymbol{\varepsilon} = \sum_{i=1}^{m} \boldsymbol{\varphi}_i q_i = \boldsymbol{\varphi q} \tag{5}$$

where $\boldsymbol{\varphi} = \begin{bmatrix} \boldsymbol{\varphi}_1 & \boldsymbol{\varphi}_2 & \cdots & \boldsymbol{\varphi}_m \end{bmatrix}$ is the matrix of strain mode. The generalized coordinates can be solved from a set of measured strains by least squares as

$$\boldsymbol{q} = \left(\boldsymbol{\varphi}^T \boldsymbol{\varphi}\right)^{-1} \boldsymbol{\varphi}^T \boldsymbol{\varepsilon} \tag{6}$$

Substituting Equation (6) into Equation (4), a strain-to-displacement (STD) transformation is obtained and we have

$$\boldsymbol{u} = \boldsymbol{\Phi}\left(\boldsymbol{\varphi}^T \boldsymbol{\varphi}\right)^{-1} \boldsymbol{\varphi}^T \boldsymbol{\varepsilon} \tag{7}$$

With appropriate mode selection and strain measurement, Equation (7) can provide relatively accurate prediction of structural deformation in the majority of test cases.

From the above equations, one may find a major deficiency of the classical modal method is that the boundary condition, which is important for the modal results, has not been deliberately considered. If the 6 rigid-body degrees of freedom (DOFs) are not fully constrained, there must be zero strain modes, which causes a singular solution of Equation (6). In another words, a strain state does not uniquely correspond to a displacement state due to rigid-body motions. This is why previous studies usually used constrained structural modes for deformation reconstruction. In practical problems, an undesirable phenomenon is that the aircraft is often in the situation of moving boundaries. Different boundary conditions may lead to different modal results, which further make the shape-sensing problem difficult for the whole process of flight, including sliding, flying and landing. Specifically, the boundary conditions of an aircraft in flight are usually considered to be free–free, while to be supported at the landing gear when sliding or parking on the ground.

In this work, we propose a more generalized SDT algorithm by introducing extra boundary degrees of freedom, in which only the free–free modes are needed for various situations. To this end, the matrices of displacement modes $\boldsymbol{\Phi}$ and strain modes $\boldsymbol{\varphi}$ with free boundary condition are divided into two parts:

$$\boldsymbol{\Phi} = \begin{bmatrix} \boldsymbol{\Phi}_r & \boldsymbol{\Phi}_e \end{bmatrix} \tag{8}$$

$$\boldsymbol{\varphi} = \begin{bmatrix} \boldsymbol{\varphi}_r & \boldsymbol{\varphi}_e \end{bmatrix} \tag{9}$$

Here, the subscripts r and e represent the rigid-body modes and the elastic modes, respectively. Accordingly, the generalized coordinates can be written as $\boldsymbol{q} = \begin{bmatrix} \boldsymbol{q}_r & \boldsymbol{q}_e \end{bmatrix}^T$. It is apparent that the strain modes $\boldsymbol{\varphi}_r$ are comprised of zero vectors. The generalized coordinates corresponding to elastic modes can be estimated in manner the same as Equation (6):

$$\boldsymbol{q}_e = \left(\boldsymbol{\varphi}_e^T \boldsymbol{\varphi}_e\right)^{-1} \boldsymbol{\varphi}_e^T \boldsymbol{\varepsilon} \tag{10}$$

Consider (1) that some nodes are constrained, and (2) that its DOFs are part of a set denoted as A-set; (3) that the remaining DOFs are part of a set denoted as B-set; and (4) that all DOFs are a set denoted as N-set. In addition, if $\boldsymbol{\Xi}_{NA}$ is the Boolean matrix that locates the a-set DOFs in the n-set and $\boldsymbol{\Xi}_{NB}$ is the equivalent for the b-set, then the desired $\boldsymbol{q}_r$ should satisfy the following constraint equations:

$$\boldsymbol{u}_A = \boldsymbol{\Xi}_{NA} \boldsymbol{\Phi}_r \boldsymbol{q}_r + \boldsymbol{\Xi}_{NA} \boldsymbol{\Phi}_e \boldsymbol{q}_e = 0 \tag{11}$$

It should be noted that the number of r is equal to the number of A-set, which means the number of equations should be equal to the number of unknowns. Therefore, we have

$$q_r = -(\Xi_{NA}\Phi_r)^{-1}\Xi_{NA}\Phi_e q_e \tag{12}$$

Combining Equations (10) and (12), the generalized coordinates can be calculated by

$$q = \begin{bmatrix} -(\Xi_{NA}\Phi_r)^{-1}\Xi_{NA}\Phi_e(\varphi_e^T\varphi_e)^{-1}\varphi_e^T\varepsilon \\ (\varphi_e^T\varphi_e)^{-1}\varphi_e^T\varepsilon \end{bmatrix} \tag{13}$$

Then, the displacement can be obtained by Equation (4). Using the above equations, we do not need to calculate the modes corresponding to each boundary condition. Instead, only the free–free modes are used to determine structural deformation under arbitrary essential boundary conditions.

3. System Development

In this section, we will firstly present a brief introduction of the tested model, i.e., a Joined-Wing aircraft. Then, the components and workflow of the Fiber Optic Sensing System (FOSS) for collecting and storing Bragg wavelengths are introduced.

3.1. Model Description: A Joined-Wing Aircraft

As shown in Figure 1, a Joined-Wing aircraft with diamond wings is designed and manufactured. The aircraft has a span length of approximately 60 m and a fuselage length of approximately 25 m. The main wing can be divided into front wing, aft wing and outboard wing. The front wing and the outboard wing are connected by two nacelles, which are further connected with the aft wing. The main structure of the aircraft is made of composite materials. The FOSS to be introduced next is mainly used to collect the strain information on the wing spar and predict the three-dimensional deformation of the entire wing.

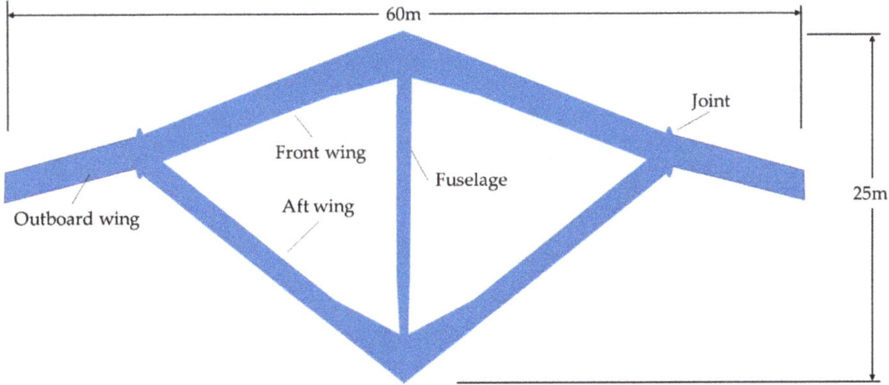

Figure 1. Top view of the Joined-Wing aircraft.

3.2. FOSS Design

The overall scheme of the FOSS, including the hardware system and software system, is shown in Figure 2. The hardware system is divided into two independent systems for collecting and storing sensor data of the left and right wings. Each of the system consists of a FBG interrogation module, an onboard computer, a GPS and a memory module. The requirement of such an interrogation module would depend on the strain measurement along the front and aft wings of the 60-m wingspan. Figure 3a shows the interrogation system, the main parameters of which include an overall size of 234 mm × 172 mm × 28 mm, a sampling rate of 1000 Hz, 16 fiber channels, 1.04 kg weight and Ethernet interface. The environmental requirements were to meet a maximum wavelength shift of 5 pm under −10 °C to 50 °C operating temperature range, and a sine wave vibration in a range of 10–150 Hz with overload of 2 g. The characteristic wavelengths obtained by the interrogation system are transmitted to the onboard computer through UDP protocol. Meanwhile, the time series from GPS along with the fiber sensor data will be reorganized and stored in the memory module. The time consistency of test data from difference devices is guaranteed by the time series provided by GPS. Figure 3b shows the onboard computer, which has a 1.5 GHz 4-core CPU, a maximum RAM of 8 GB and multi-type interfaces including USB, HDMI, LAN, etc. The operating system is based on LINUX. A 64 GB Micro SD Card is used to store the sensor data. The onboard computer and a cooling fan are enclosed in an aluminum box with a size of 89 mm × 74 mm × 31 mm. The total weight is approximately 0.156 kg. Additionally, the main parameters of the used FBG sensor are listed in Table 1.

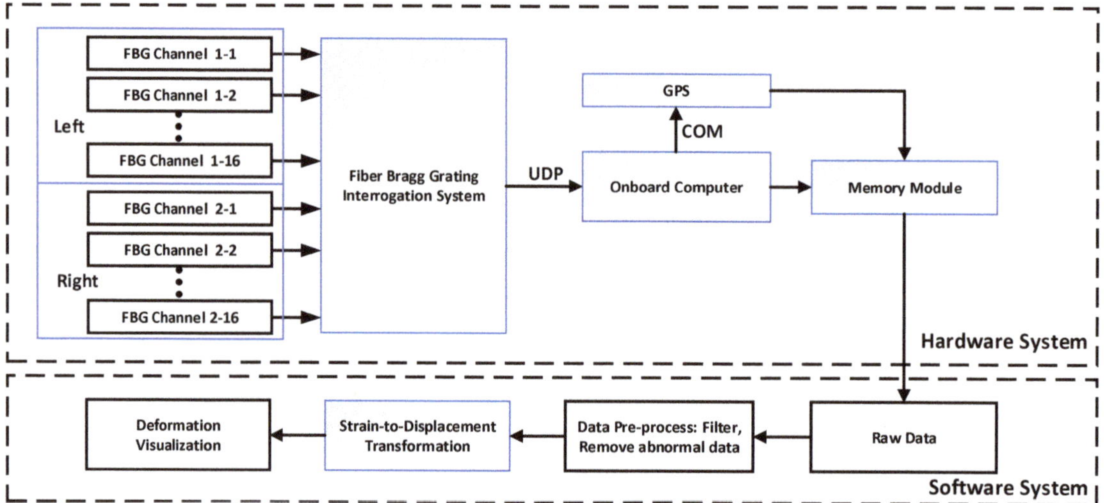

Figure 2. Overall design of FOSS.

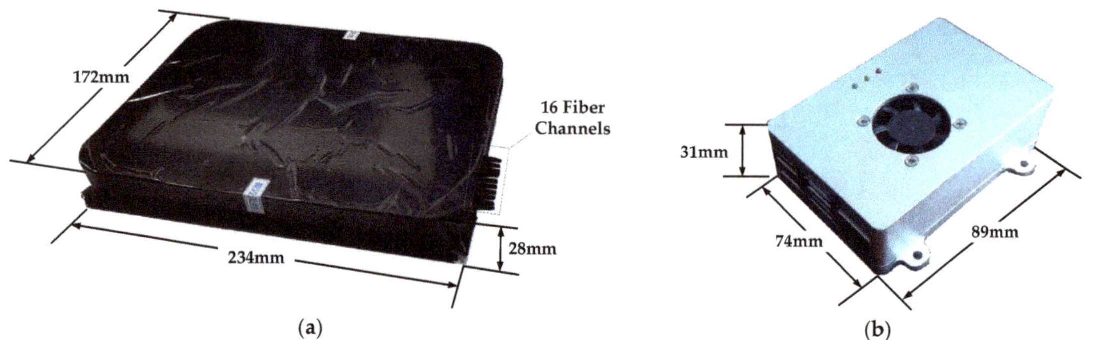

Figure 3. (**a**) FBG Interrogation System; (**b**) onboard computer.

Table 1. The main parameters of FBG sensor.

Parameter	Value
3 dB bandwidth	≤0.3 nm
Reflectivity	≥90%
Side Lobe Suppression (SLS)	≥15 dB
Grating length	10 mm
Temperature sensitivity (temperature-optic coefficient)	6.67 $K^{-1}{\cdot}10^{-6}$
Strain sensitivity (strain-optic coefficient)	7.8 $\mu\varepsilon^{-1}{\cdot}10^{-7}$

The software system was designed to transform the preprocessed sensing data into the structural deformation and display the final results. In general, the raw data are not directly used because of high-frequency noise and abnormal data. Therefore, the raw data need to be filtered, and the abnormal data need to be removed. After that, the desired structural deformations can be reconstructed by the STD transformation. Finally, the displacement response of arbitrary point on the structure or the whole structure configuration can be plotted by the visualization module.

In this work, the simplest and most effective filtering strategy was adopted, i.e., mean filtering. Denote the raw wavelength data as $\lambda(\tau)$, then the filtered time series $\lambda(t)$ can be expressed as:

$$\lambda(t) = \frac{1}{N}\sum_{t \in S} \lambda(\tau) \tag{14}$$

in which S is the filter window with the size of N and the center at t. As mentioned before, interrogation system has a sampling rate of 1000 Hz, which is usually 5 Hz for GPS. By default, the window size N is set to 200.

The criterion for judging the abnormal data depends on the wavelength bandwidth and the strain range. The total wavelength bandwidth of a fiber channel is approximately 40 nm. To be specific, the interrogator is limited to identify Bragg wavelength within a range of 1529 nm to 1569 nm. The maximum number of FBGs multiplexed within an optic fiber is designed to be eight, which means the bandwidth occupied by one FBG sensor is at least 5nm. Therefore, the variation of the Bragg wavelength, $\delta\lambda_B$, is ±2.5 nm. Assuming the original Bragg wavelength λ_B is approximately 1550 nm, the observed one can be (1550 ± 2.5) nm. According to Equation (2), the strain range can cover approximately ±2000 $\mu\varepsilon$ without consideration of temperature change. The strain range can basically meet the requirements of this work. Based on the above considerations, the wavelength variation exceeding 2.5 nm is considered as abnormal, i.e.,

$$|\lambda_B - \lambda_{B0}| > 2.5 \tag{15}$$

and needs to be removed.

3.3. Sensor Arrangement

In this section, we will take the half model of the Joined-Wing aircraft as an example to illustrate the sensor layout and installation. Figure 4 shows the sensor layout of the right wing. As shown in Figure 4a, there are 16 fiber channels, each of which is multiplexed with different numbers of FBG sensors. For example, the optical fibers numbered 1–4 are used for strain and temperature measurements of the inner section of the front wing, as shown in Figure 4b. Each fiber contains six FBG sensors, which are represented by rectangular blocks, and the blocks with black border are used for temperature compensation. The solid ones are attached to the upper edge of the wing spar, while the hollow ones are attached to the lower edge. Figure 4c shows a detail view of FBG sensor installation on the surface of the spar. Due to the breakdown interfaces, the right wing is further divided into the six sections as shown in Figure 4a. The front wing and the aft wing consist of two sections and three sections, respectively.

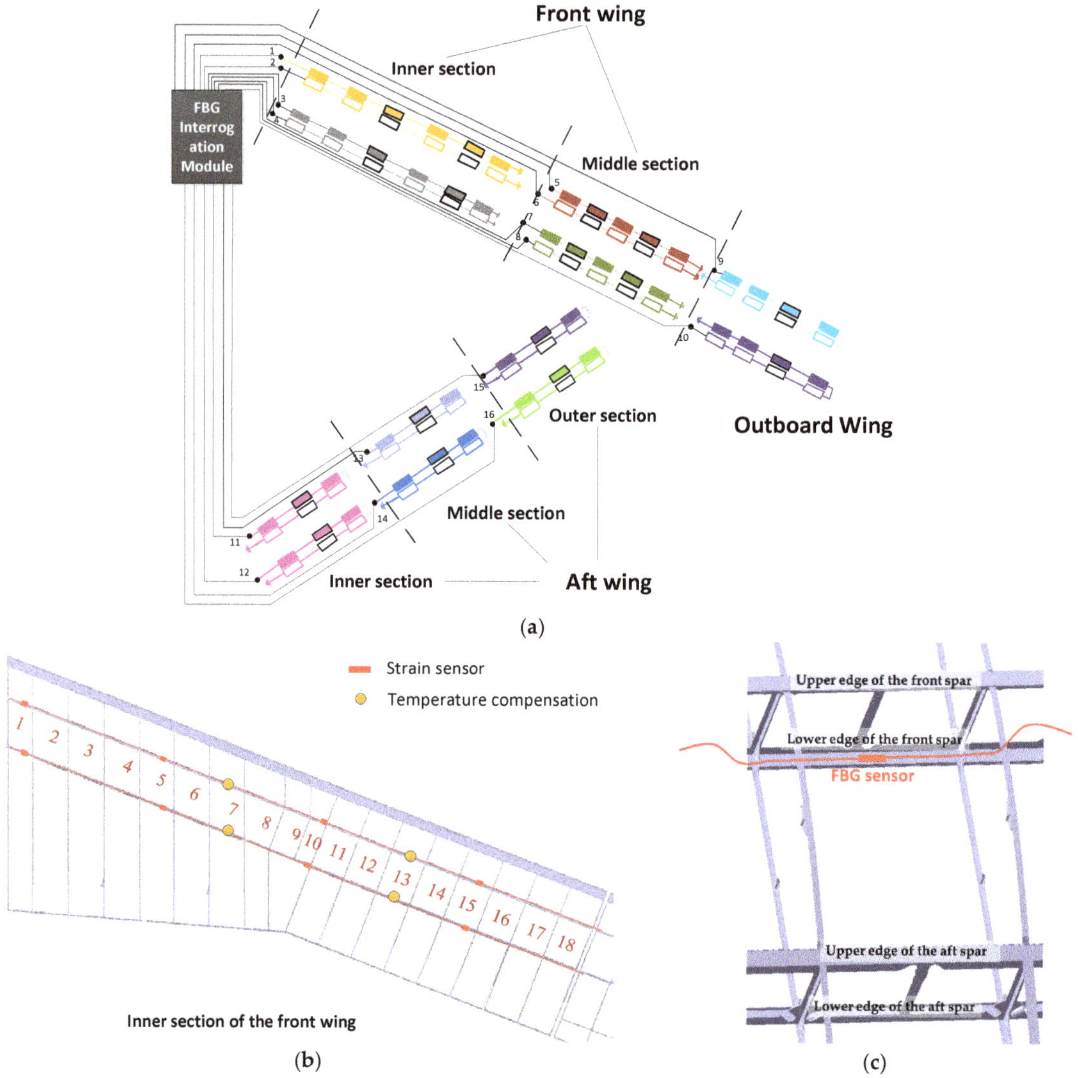

Figure 4. Sensor layout. (**a**) The overall sensor arrangement of the half model; (**b**) sensor arrangement of the inner section of the front wing; (**c**) detail view of FBG sensor installation on the surface of the spar.

Figure 5 is a snapshot of the wing interior. Both the electric cables and the optical fibers are bundled with the skeleton inside the wing. The FBG sensors are attached to the spar surfaces using epoxy adhesive. The temperature compensation is realized by the FBG sensor in a tube with a rigid steel needle inside.

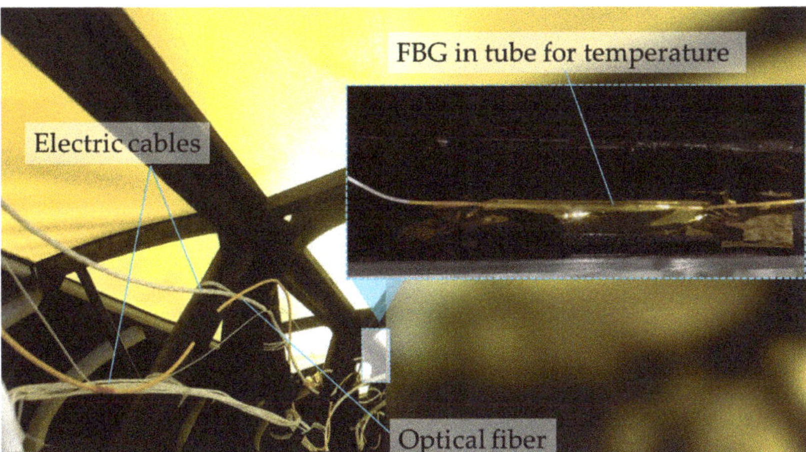

Figure 5. Wing interior and sensor attachment on the wing spar.

4. Results and Discussion

4.1. Validation on a Cantilever Beam Model

A cantilever beam supported at the midpoint was first investigated to demonstrate the improved modal method. The beam has a rectangular cross-section with a width of 0.035 m and a height of 0.0015 m. The elastic modulus is 210 GPa. The material density is 7750 kg/m^3. As shown in Figure 6, a concentrated force is applied at the beam tip, with its vertical components being 0.1 N. The beam is solved in MSC NASTRAN with the discretization of 50 elements.

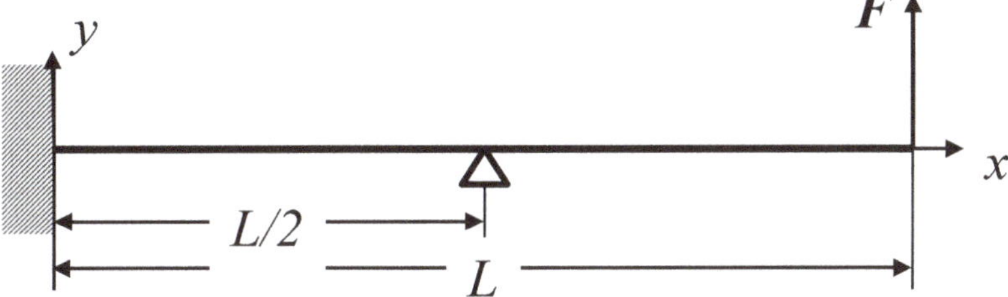

Figure 6. Cantilever beam model with a tip force.

Four types of boundary conditions are considered for linear modal analysis. The first one is consistent with Figure 6, in which the beam is fixed at the root and supported at the midpoint. The second and third ones retain fixed and supported constraints, respectively. The fourth one is totally free without any constraint. The mathematical expressions of the boundary conditions are given in Equation (16).

$$\begin{cases} \text{BC1} : y(0) = y'(0) = 0, y(L/2) = 0 \\ \text{BC2} : y(0) = y'(0) = 0 \\ \text{BC3} : y(L/2) = 0 \\ \text{BC4} : free - free \end{cases} \tag{16}$$

Mode shapes obtained through the modal analysis can be found in Appendix A. As expected, a rigid-body rotational mode around the support point occurs in the third case. In the last case, one may find an axial mode and two rigid-body modes, which are superpositions of y-direction motion and rotational motion around z-axis.

Figure 7 presents the comparisons of displacement directly computed from FEM and those obtained from SDT under different conditions. In all cases, the first 20 modes with no distinction between rigid-body modes and elastic modes are used for SDT. Under BC1 and BC2, the displacements are computed by Equation (7) consistent with the classical modal approach. Under BC3 and BC4, the displacements are computed by the improved modal method due to the existence of rigid-body modes. Additionally, the predicted displacements show differences when only elastic modes are considered under BC3 and BC4. The differences are perfectly compensated by superposition of rigid-body motions. Finally, one can observe that the predicted displacements are highly consistent with the FEM results, no matter what boundary conditions are adopted.

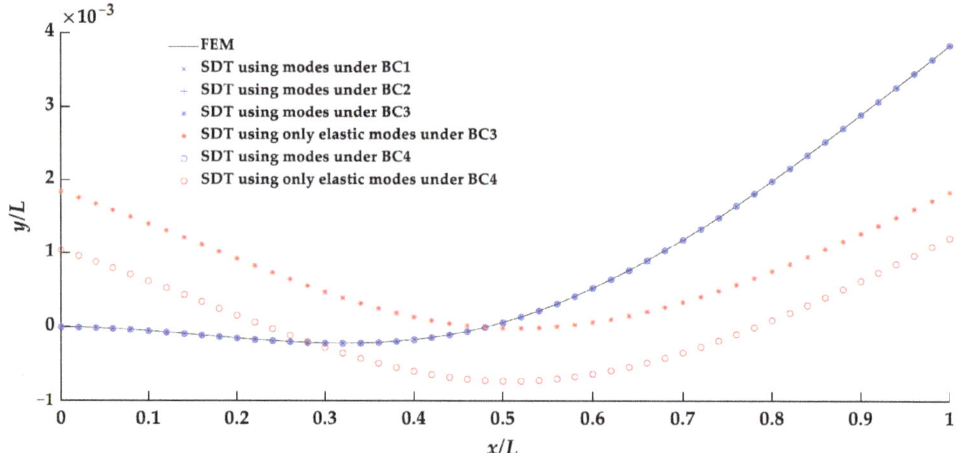

Figure 7. Comparisons of calculated displacements from FEM and those obtained from SDT.

4.2. Numerical Studies on the Joined-Wing Aircraft

The FE model was constructed from the original CAD model in MSC NASTRAN, as shown in Figure 8a. Most of the structures including the wing spars and skin were modeled using composite plate elements. The fuselage skeleton uses a large number of bar elements. The FE model has approximately 370,000 nodes and 340,000 elements. A number of 64 massless rod elements with negligible stiffness were attached to the spar surface of both right and left wing, respectively. These elements do not change the stiffness and inertial properties of the original aircraft model. They are distributed at the positions in accordance with the sensor layout in Figure 4, serving as virtual strain sensors to provide strain information at these locations. Figure 8b shows a schematic view of a rod element.

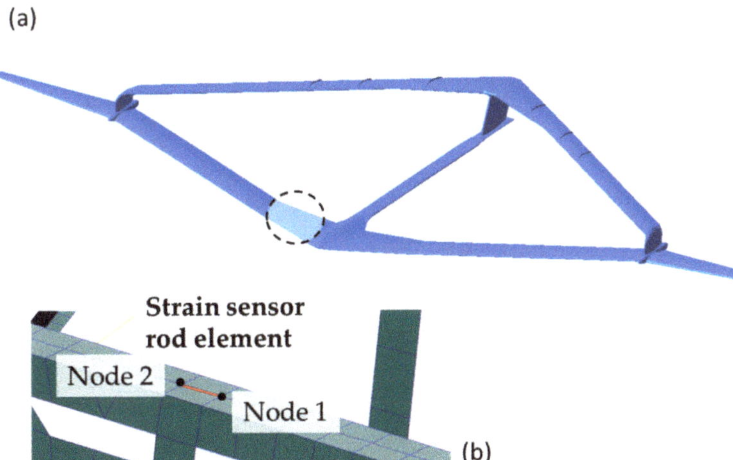

Figure 8. (**a**) FE model and (**b**) zoom on the front spar of the front wing and a rod element to simulate strain sensor.

Modal analysis was then conducted to obtain mode frequencies and mode shapes of both displacement modes and strain modes. The strain modes are provided by the virtual strain sensor elements. The boundary condition is set to free–free without constraints. The first 30 modes are used for SDT. The first six modes are rigid-body modes. Several primary elastic mode shapes are plotted in Figure 9. The first symmetric vertical bending mode has a very low frequency of 0.354 Hz. Different from the common-configuration aircraft, a distinctive modal feature of the joined-wing configuration is the coupling of the front wing and the aft wing, which can be found in Figure 9b,c.

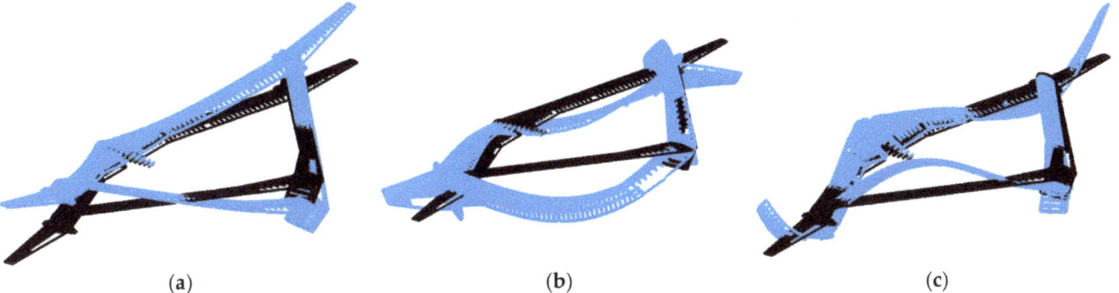

Figure 9. Several primary elastic mode shapes: (**a**) 7th mode—symmetric vertical bending, $f_1 = 0.354$ Hz; (**b**) 13th mode—anti-symmetric vertical bending coupled with torsion, $f_7 = 1.101$ Hz; (**c**) 30th mode—symmetric vertical bending coupled with torsion, $f_{30} = 3.234$ Hz.

In the linear static solver of MSC NASTRAN, the six degrees of freedom (DOFs) of the grid point coincident with the mass center are constrained. One times gravity loads along the anti-gravity direction are applied to make the wing bend upward. The free–free modes were used for SDT. Figure 10 shows the comparison of an undeformed and deformed configuration under such loading conditions. The displacements calculated directly by FEM are in very good agreement with those reconstructed by SDT. The maximum displacement in z-axis direction is up to 4.71 m, which is approximately 14.1% of the half wingspan. Due to the complexity of the FE model with up to two million DOFs, it is too time-consuming to reconstruct the deformation using the modal information of all nodes. Here, we select

1188 grid points on the wing structure for SDT. Three rotational DOFs in the displacement modal vector are ignored. In this way, the matrix of displacement mode $\boldsymbol{\Phi}$ in Equation (4) has 3564 rows and 30 columns. The matrix of strain mode $\boldsymbol{\varphi}$ in Equation (5) has 196 rows and 30 columns. Although only the deformation of the selected grid points is shown here, the position of any other point inside the whole aircraft can be conveniently calculated by surface spline interpolation [29]. Figure 11 presents the interpolated lifting surface, which is used to conveniently and efficiently evaluate the overall deformation of the aircraft.

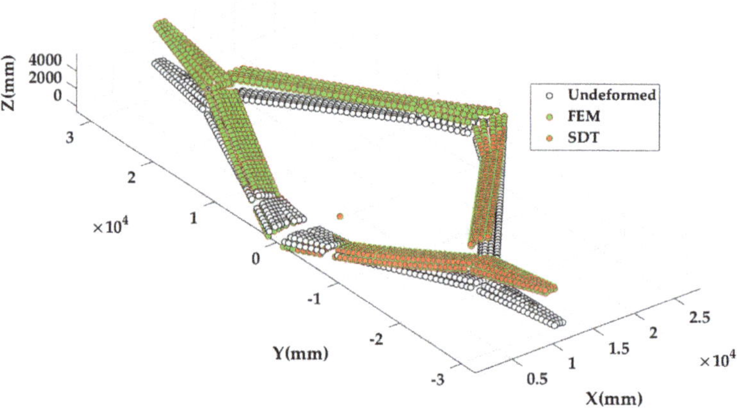

Figure 10. Positions of selected grid points in undeformed configuration and deformed configuration under gravity loads along the anti-gravity direction.

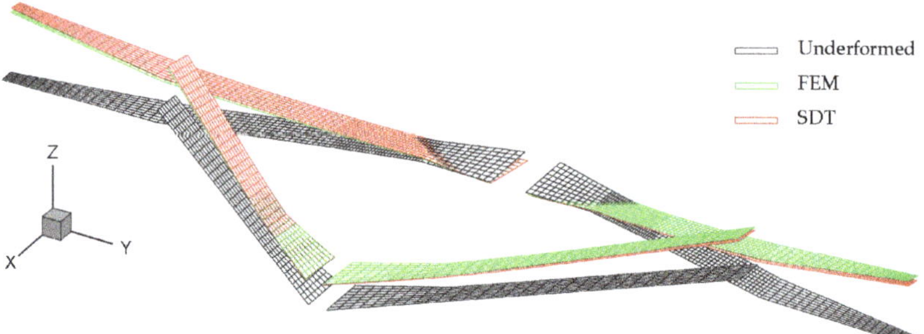

Figure 11. Interpolated lifting surfaces for evaluation of overall deformation.

To quantitatively investigate the deformation reconstruction accuracy, the following two kinds of Root Mean Square Errors (RMSE) are used,

$$RMSE_1 = \sqrt{\sum_{i=1}^{N}(z_i - \hat{z}_i)^2 \Big/ \sum_{i=1}^{N} z_i^2} \tag{17}$$

$$RMSE_2 = \sqrt{(z_{i_{\max}} - \hat{z}_{i_{\max}})^2 \Big/ z_{i_{\max}}^2} \tag{18}$$

In Equation (17), z is the displacement in z-axis direction calculated by FEM, and $\hat{z}$ is the displacement calculated by SDT. The subscript i is the index number of the grid point, and N is the total number of grid points. In Equation (18), $i_{\max}$ represents the node number with the maximum displacement. These two equations provide a global and local evaluation of deformation reconstruction accuracy, respectively. Finally, the global error $RMSE_1$ is 2.86%, and the local error $RMSE_2$ is 4.47%. The results further show the accuracy of SDT in reconstructing the wing deformation.

4.3. Ground Test on the Joined-Wing Aircraft

The system's performance was then evaluated in a ground quasi-static test in which the wing was lifted by two lifting platforms under the nacelles at the joint. Figure 12 shows the schematic of the ground test. The lifting attitude z can be accurately controlled by the lifting platform. Before the test, the Joined-Wing aircraft was on the ground supported by two front main landing gears, two auxiliary landing gears and a rear landing gear. The optical fiber sensing data in this state were recorded as the initial reference value. At the beginning of the test, the lifting attitudes of both right and left side were set to 785 mm. After holding on for a certain time, the lifting altitude may drop by approximately 100 mm within approximately 10 s. The above maintenance and decent operations will be repeated several times until the altitude drops to zero, which means the aircraft returns to its initial reference state. Table 2 lists the lifting altitudes over time and the specific operations, which are recorded as H and D (Hold and Drop). The ground test lasted nearly 2000 s. The altitudes were reduced from 785 mm to 0 mm in 7 operations.

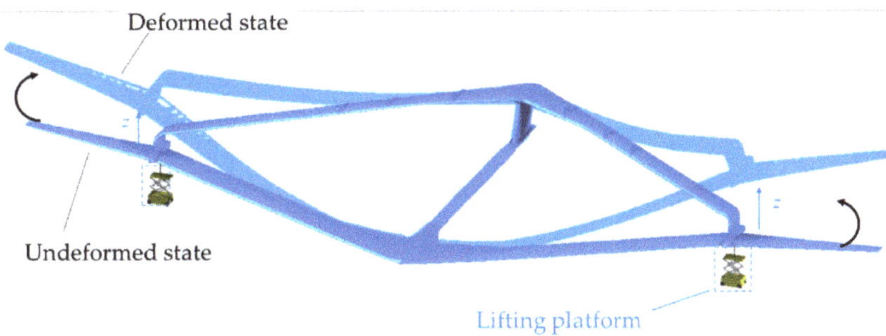

Figure 12. A schematic of ground test.

Table 2. Lifting altitudes of the lifting platform and operations at different time.

Time (s)	Lifting Altitude z	Operation (H-Hold/D-Drop)
0	785	H
265	785	D
275	690	H
375	690	D
385	590	H
468	590	D
478	490	H
578	490	D
588	390	H
645	390	D
655	290	H
763	290	D
773	190	H
1294	190	D
1304	90	H
1854	90	D
1864	0	H

Figure 13 plots the temperature curve recorded by FBG sensors. Since the test is in a relatively closed indoor environment, we assume that the temperature is independent of the location. Therefore, the temperature curve reflects the average of collected data from all FBG sensors for temperature compensation. According to the records of the onboard temperature sensor with limited precision, the temperature increased from 22 °C to 23 °C during the test. The test was conducted from 2:00 pm to 3:00 pm, during which time the temperature rose slowly. The results of temperature sensor and FBG sensor show that the temperature change during the test is less than 1 °C, which means the temperature effect on the following strain data acquisition is almost negligible.

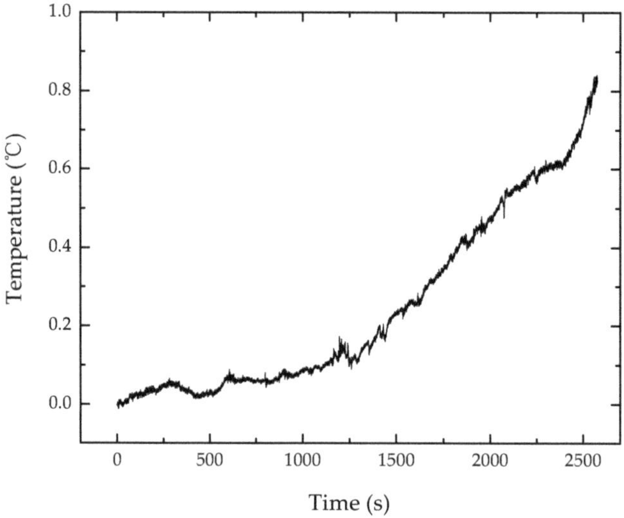

Figure 13. Temperature during the ground test recorded by FBG sensors.

Figure 14a plots the displacement responses at the support point during the ground test. The deformations on both sides of the wing are approximately considered symmetrical, specifically synchronous and of the same size. Thus, the displacement plotted in this figure is the average of the z-direction deformation at the left and right support points. The test results are consistent with those listed in Table 2. As a comparison, wing displacements were also obtained from FEM. To simulate the ground test, we constrained the translational DOFs of four nodes at the front of fuselage and one node at the rear of the fuselage. These nodes are consistent with the positions of landing gears. As for the nodes corresponding to the support point, translational DOFs except the vertical motion are constrained. Nodal forces are applied to the support point towards positive z-direction to simulate the deformation during the test. The results show that the numerical simulation accurately reflects the displacement response in the test. The difference between SDT and TEST results is plotted in Figure 14b. The absolute displacement error is within $\pm$100 mm. One can also calculate the relative error in the manner similar to Equation (17) by replacing i and N with time variables corresponding to the horizontal axis in the figure. The calculated error is 6.6%. The error mainly comes from two aspects. One is the model-related error. For such a complex engineering model, the deviation between the FE model and the realistic one is inevitable. Additionally, the sensor placement and the mode selection will affect the reconstruction accuracy. Future efforts can be made to improve the accuracy by optimizing the sensor positions and selected mode as in Ref. [22]. The second one is the test-related error. In the test, the strain measured by the FBG sensor is affected by the adhesive layer and cannot exactly reflect the strain on the structural surface. It is challenging to analyze and predict the strain transfer to the fiber core reliably, which is beyond the scope of this work.

A practical way is to consider the error caused by strain transfer as uncertainties satisfying normal distribution, and then to achieve more reliable results by filtering methods.

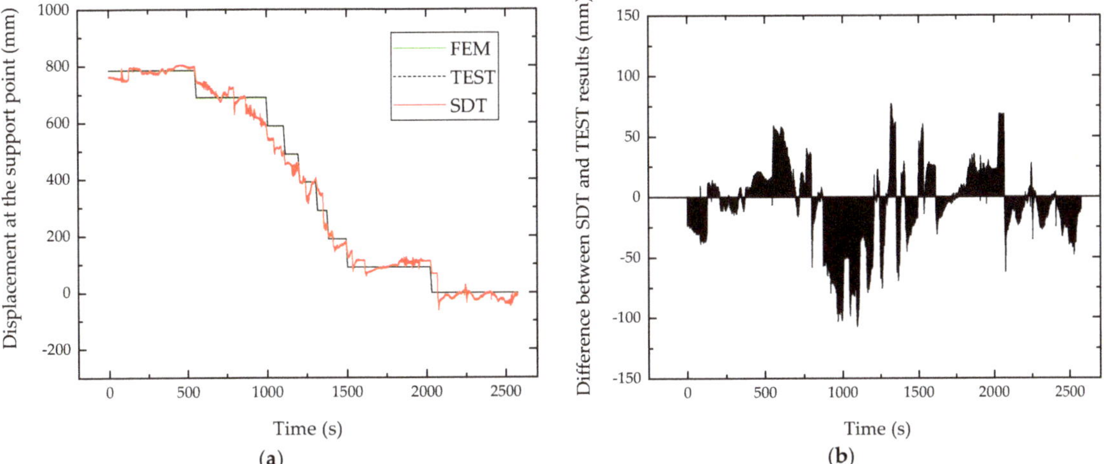

Figure 14. (**a**) Displacement responses at the support point; (**b**) difference between SDT and TEST results.

Figure 15 shows the deformed configuration obtained from FEM and SDT at the beginning of the test. The overall deformation obtained by direct numerical simulation is in good agreement with that reconstructed by measured strains. Taking FEM results as reference, the global error $RMSE_1$ at each moment was calculated and plotted in Figure 16. The maximum error is approximately 7%. The average error is 2.62%. The results show that the FOSS and SDT algorithm in this work can accurately measure the deformation of the Joined-Wing aircraft during the ground test.

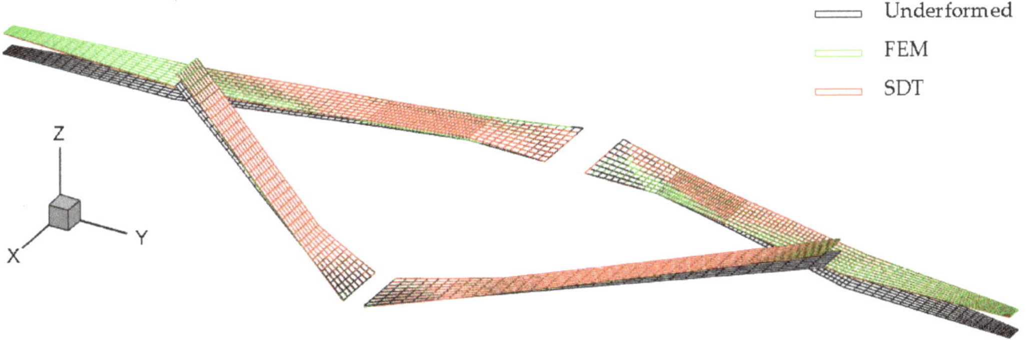

Figure 15. Deformed configuration obtained from FEM and SDT at the beginning of the test.

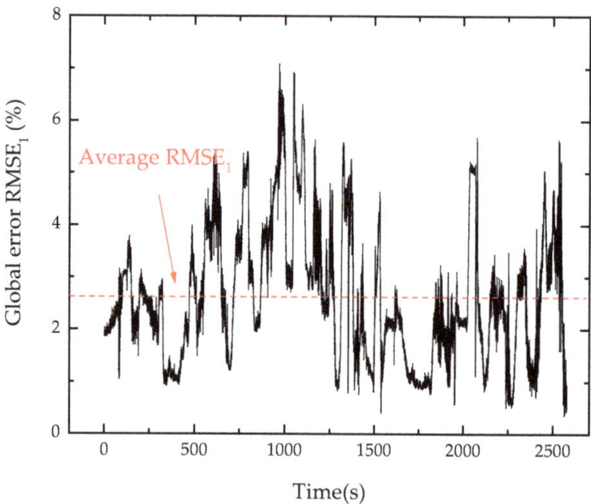

Figure 16. Global error response of SDT.

5. Conclusions

This work presents a complete framework for predicting structural displacement of a Joined-Wing aircraft by using Fiber Optic Sensing techniques. Several concluding remarks can be drawn as follows:

(1) A FOSS with hardware and software subsystems is designed and installed on the target Joined-Wing aircraft. The system was then verified by the ground test.
(2) The classical modal method is modified to adapt to various boundary conditions, which is common in practical applications. The improved SDT algorithm was then verified by numerical studies on a cantilever beam model and the Joined-Wing aircraft.
(3) Both the numerical and experimental results show that the proposed SDT algorithm can accurately predict the overall configuration of the aircraft or deformations of a particular point. In the ground test, the relative error of the displacement at the support point is 6.6%. The global error of the overall deformation is less than 7%, and the average error is only 2.62%.

In general, the FOSS was proved to have reasonable accuracy and the potential for future flight tests.

Author Contributions: Conceptualization, Y.M. and C.X.; Data curation, Y.B.; Methodology, Y.M. and Z.C.; Software, Y.M.; Supervision, C.Y.; Validation, Y.B. and C.X.; Writing—original draft, Y.M.; Writing—review and editing, Y.B. and C.X. All authors have read and agreed to the published version of the manuscript.

Funding: This research received no external funding.

Data Availability Statement: Not applicable.

Acknowledgments: We would appreciate Ziqiang Wang for his support in hardware system development and ground test.

Conflicts of Interest: The authors declare no conflict of interest.

Appendix A. Mode Shapes of the Cantilever Beam Model with Different Boundary Conditions

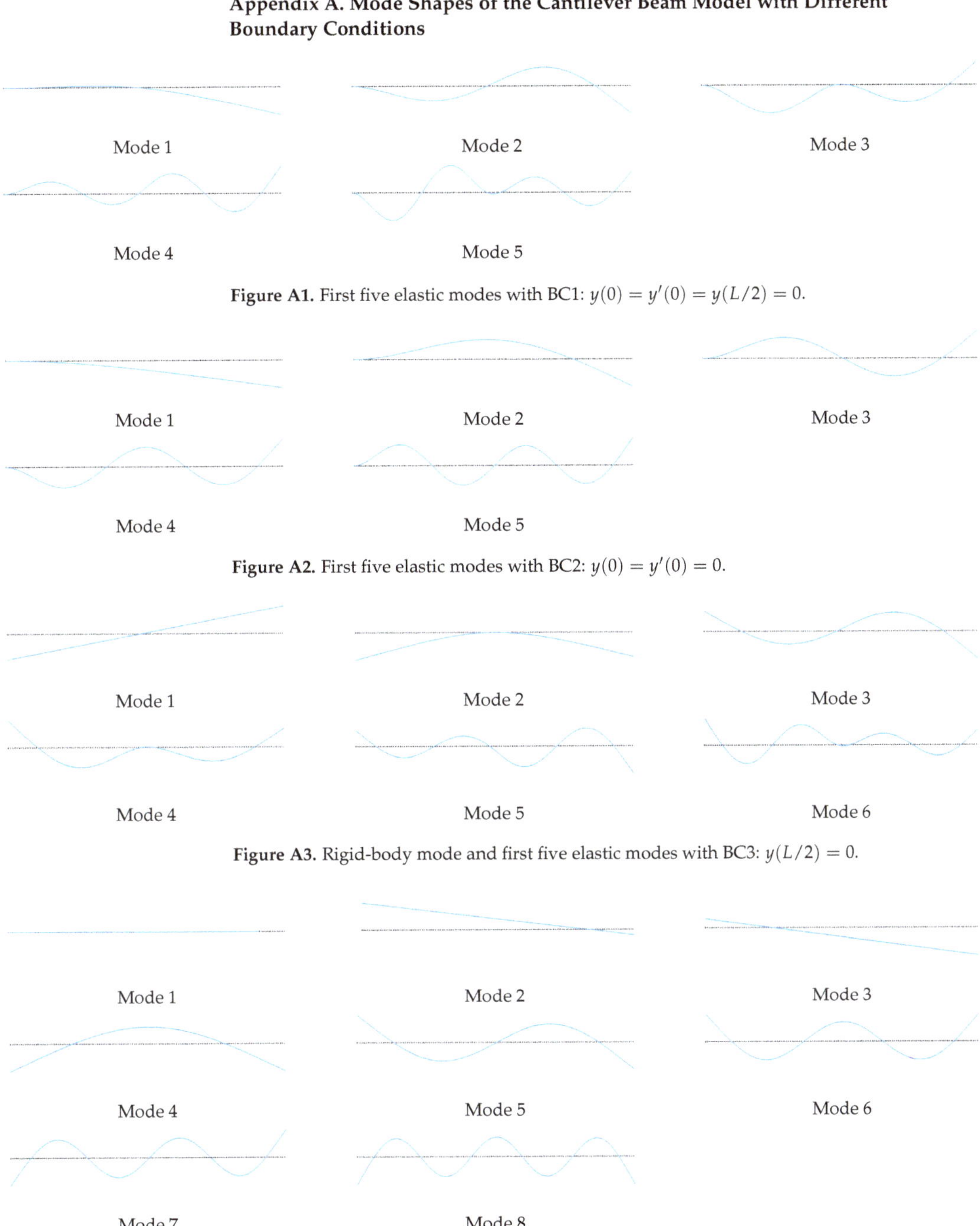

Figure A1. First five elastic modes with BC1: $y(0) = y'(0) = y(L/2) = 0$.

Figure A2. First five elastic modes with BC2: $y(0) = y'(0) = 0$.

Figure A3. Rigid-body mode and first five elastic modes with BC3: $y(L/2) = 0$.

Figure A4. Rigid-body modes and first five elastic modes with BC4: free–free.

References

1. Cavallaro, R.; Demasi, L. Challenges, Ideas, and Innovations of Joined-Wing Configurations: A Concept from the Past, an Opportunity for the Future. *Prog. Aerosp. Sci.* **2016**, *87*, 1–93. [CrossRef]
2. Tilmann, C.P.; Flick, P.M.; Martin, C.A.; Love, M.H. High-Altitude Long Endurance Technologies for SensorCraft. In Proceedings of the RTO Paper MP-104-P-26, RTO AVT Symposium on Novel and Emerging Vehicle and Vehicle Technology Concepts, Brussels, Belgium, 7–11 April 2003.
3. Su, W. *Coupled Nonlinear Aeroelasticity and Flight Dynamics of Fully Flexible Aircraft*; University of Michigan: Ann Arbor, MI, USA, 2008.
4. Cavallaro, R.; Iannelli, A.; Demasi, L.; Razon, A.M. Phenomenology of nonlinear aeroelastic responses of highly deformable joined wings. *Adv. Aircr. Spacecr. Sci.* **2015**, *2*, 125–168. [CrossRef]
5. Ma, Z.; Chen, X. Fiber Bragg Gratings Sensors for Aircraft Wing Shape Measurement: Recent Applications and Technical Analysis. *Sensors* **2018**, *19*, 55. [CrossRef] [PubMed]
6. Floris, I.; Adam, J.M.; Calderón, P.A.; Sales, S. Fiber Optic Shape Sensors: A comprehensive review. *Opt. Lasers Eng.* **2021**, *139*, 106508. [CrossRef]
7. Masoudi, A.; Newson, T.P. Contributed Review: Distributed optical fibre dynamic strain sensing. *Rev. Sci. Instrum.* **2016**, *87*, 011501. [CrossRef] [PubMed]
8. Noll, T.E.; Brown, J.M.; Perez-Davis, M.E.; Ishmael, S.D.; Tiffany, G.C.; Gaier, M. *Investigation of the Helios Prototype Aircraft Mishap*; Volume I Mishap Report; NASA: Washington, DC, USA, 2004.
9. Ko, W.L.; Richards, W.L.; Tran, V.T. *Displacement Theories for In-Flight Deformed Shape Predictions of Aerospace Structures*; NASA Dryden Flight Research Center: Edwards, CA, USA, 2007.
10. Ko, W.L.; Fleischer, V.T. *Further Development of Ko Displacement Theory for Deformed Shape Predictions of Nonuniform Aerospace Structures*; NASA Dryden Flight Research Center: Edwards, CA, USA, 2009.
11. Derkevorkian, A.; Masri, S.F.; Alvarenga, J.; Boussalis, H.; Bakalyar, J.; Richards, W.L. Strain-Based Deformation Shape-Estimation Algorithm for Control and Monitoring Applications. *AIAA J.* **2013**, *51*, 2231–2240. [CrossRef]
12. KO, W.L.; Richards, W.L.; Fleischer, V.T. *Applications of KO Displacement Theory to the Deformed Shape Predictions of the Doubly-Tapered Ikhana Wing*; NASA Dryden Flight Research Center: Edwards, CA, USA, 2009.
13. Nicolas, M.J.; Sullivan, R.W.; Richards, W.L. Large Scale Applications Using FBG Sensors: Determination of In-Flight Loads and Shape of a Composite Aircraft Wing. *Aerospace* **2016**, *3*, 18. [CrossRef]
14. Klotz, T.; Pothier, R.; Walch, D.; Colombo, T. Prediction of the business jet Global 7500 wing deformed shape using fiber Bragg gratings and neural network. *Results Eng.* **2020**, *9*, 100190. [CrossRef]
15. Meng, Y.; Xie, C.C.; Wan, Z.Q. Deformed Wing Shape Prediction using Fiber Optic Strain Data. In Proceedings of the International Forum on Aeroelasticity and Structural Dynamics, Como, Italy, 25–28 June 2017.
16. Tessler, A.; Spangler, J.L. A least-squares variational method for full-field reconstruction of elastic deformations in shear-deformable plates and shells. *Comput. Methods Appl. Mech. Eng.* **2005**, *194*, 327–339. [CrossRef]
17. Tessler, A. *Structural Health Monitoring Using High-Density Fiber Optic Strain Sensor and Inverse Finite Element Methods*; NASA Langley Research Center TM-214871: Hampton, VA, USA, 2007.
18. Gherlone, M.; Cerracchio, P.; Mattone, M.; Di Sciuva, M.; Tessler, A. Shape sensing of 3D frame structures using an inverse Finite Element Method. *Int. J. Solids Struct.* **2012**, *49*, 3100–3112. [CrossRef]
19. Gherlone, M.; Cerracchio, P.; Mattone, M.; Di Sciuva, M.; Tessler, A. An inverse finite element method for beam shape sensing: Theoretical framework and experimental validation. *Smart Mater. Struct.* **2014**, *23*, 045027. [CrossRef]
20. Tessler, A.; Roy, R.; Esposito, M.; Surace, C.; Gherlone, M. Shape Sensing of Plate and Shell Structures Undergoing Large Displacements Using the Inverse Finite Element Method. *Shock Vib.* **2018**, *2018*, 8076085. [CrossRef]
21. Gherlone, M.; Cerracchio, P.; Mattone, M. Shape sensing methods: Review and experimental comparison on a wing-shaped plate. *Prog. Aerosp. Sci.* **2018**, *99*, 14–26. [CrossRef]
22. Esposito, M.; Gherlone, M. Composite wing box deformed-shape reconstruction based on measured strains: Optimization and comparison of existing approaches. *Aerosp. Sci. Technol.* **2020**, *99*, 105758. [CrossRef]
23. Foss, G.C.; Haugse, E.D. Using Modal Test Results to Develop Strain to Displacement Transformations. In *SPIE The International Society for Optical Engineering*; SPIE: Bellingham, WA, USA, 1995; p. 112.
24. A Davis, M.; Kersey, A.D.; Sirkis, J.S.; Friebele, E.J. Shape and vibration mode sensing using a fiber optic Bragg grating array. *Smart Mater. Struct.* **1996**, *5*, 759–765. [CrossRef]
25. Kang, L.-H.; Kim, D.-K.; Han, J.-H. Estimation of dynamic structural displacements using fiber Bragg grating strain sensors. *J. Sound Vib.* **2007**, *305*, 534–542. [CrossRef]
26. Kim, H.; Han, J.; Bang, H. Real-time deformed shape estimation of a wind turbine blade using distributed fiber Bragg grating sensors. *Wind. Energy* **2014**, *17*, 1455–1467. [CrossRef]
27. Freydin, M.; Rattner, M.K.; Raveh, D.E.; Kressel, I.; Davidi, R.; Tur, M. Fiber-Optics-Based Aeroelastic Shape Sensing. *AIAA J.* **2019**, *57*, 5094–5103. [CrossRef]
28. Martins, B.L.; Kosmatka, J.B. Health Monitoring of Aerospace Structures via Dynamic Strain Measurements: An Experimental Demonstration. In Proceedings of the AIAA Scitech 2020 Forum, Orlando, FL, USA, 6–10 January 2020. [CrossRef]
29. Chang, C.X.; Chao, Y.; Xie, C.; Yang, C. Surface Splines Generalization and Large Deflection Interpolation. *J. Aircr.* **2007**, *44*, 1024–1026. [CrossRef]

Article

Prediction of Traffic Vibration Environment of Ancient Wooden Structures Based on the Response Transfer Ratio Function

Cheng Zhang [1], Nan Zhang [1,*], Yunshi Zhang [2,3] and Xiao Liu [1]

1 School of Civil Engineering, Beijing Jiaotong University, Beijing 100044, China
2 China Electronic Engineering Design Institution Co., Ltd., Beijing 100142, China
3 Beijing Engineering Research Center for Micro-Vibration Environmental Control, Beijing 100048, China
* Correspondence: nzhang@bjtu.edu.cn

Abstract: Traffic−induced vibration is increasingly affecting people's lives, which necessitates scrutiny of the environmental vibrations caused by traffic. This paper proposed a vibration prediction method suitable for the ancient wooden structures subjected to traffic−induced vibrations based on the multi−point response transfer ratio function. The accuracy of the proposed approach was also checked by comparing the predicted results with the measured results in the context of both the time domain and frequency domain. Subsequently, the environmental vibrations due to heavy−duty trucks passing at various speeds were measured, and the measurements were utilized as the input vibration excitation to assess the structural vibration of the Feiyun Pavilion. The structural safety was evaluated according to the "Technical specifications for protecting historic buildings against man−made vibration". In order to meet the structural safety requirements of the Feiyun Pavilion, it is strongly recommended to limit the type and speed of vehicles in the nearby area.

Keywords: environmental vibrations; traffic−induced vibration; vibration prediction; response transfer ratio (RTR); ancient wooden structures

Citation: Zhang, C.; Zhang, N.; Zhang, Y.; Liu, X. Prediction of Traffic Vibration Environment of Ancient Wooden Structures Based on the Response Transfer Ratio Function. *Sensors* **2022**, *22*, 8414. https://doi.org/10.3390/s22218414

Academic Editors: Phong B. Dao, Tadeusz Uhl, Liang Yu, Lei Qiu and Minh-Quy Le

Received: 27 August 2022
Accepted: 28 October 2022
Published: 2 November 2022

1. Introduction

With the development of cities, environmental vibrations and noises produced by traffic are attracting more and more attention within the international scientific community [1]. The environmental vibrations caused by traffic will not only affect people's lives and works, but also the usage of some precision instruments [2,3]. Additionally, the structural performance of buildings will deteriorate under the action of long−term traffic vibration, particularly for ancient buildings with wooden structures [4,5]. Consequently, environmental vibrations due to traffic should be further examined in the near future.

The prediction of environmental vibrations generated by traffic loading is one of the leading research directions of the understudied problem. The commonly used prediction methods for examining traffic−induced vibration mainly include theoretical analyses [6–8], numerical simulations [9–12], field tests [13–18], and empirical prediction formulas [19,20]. Although the theoretical analysis method leads to the exact solution in most cases, numerous assumptions and simplifications are made to the solution process; therefore, it cannot wholly reflect the actual situations of the induced vibration process. On the other hand, numerical methods have been extensively employed with the progress of computer technology; however, it is challenging to determine various parameters and complex structural modeling of a complex system problem. Commonly, the empirical prediction formula method requires a large number of measured data as the premise, and the accuracy of the prediction results will be affected by the judgment of factors.

Field tests have essentially focused on the dynamic performance analysis of several typical high−rise ancient wooden structures, while ancient wooden structures with different structural features have not been properly investigated. At the same time, most of the vibration sources are strong but short−lasting dynamic loadings such as earthquakes, while there is still a lack of research on long−lasting but micro−amplitude vibrations loadings such as traffic−induced vibration.

Therefore, this paper aimed to utilize field tests to propose a response−to−transfer ratio (RTR) vibration prediction method for ancient wooden structures under traffic loads based on the transfer function. The RTR function is the ratio of the output responses between systems. For a complex system with multiple subsystems, the overall RTR function of the system can be obtained through the RTR of each subsystem.

Similar to the transfer function vibration prediction method, the RTR vibration prediction method proposed in this paper does not need to establish the finite element model and has high calculation accuracy. At the same time, it is different from the vibration attenuation prediction method, which can only predict the magnitude of vibration energy but cannot obtain the spectral characteristics at the predicted point. The RTR vibration prediction method operates in the frequency domain. Therefore, it can accurately predict the frequency component of the vibration at the prediction point. For ancient wooden structures, the vibration of some frequencies (such as natural frequency) will cause more serious damage to the structure, so it is necessary to predict the vibration within a specific frequency band.

We took Feiyun Pavilion as the case study to verify the properties of the RTR. Then, the correctness of the RTR vibration prediction method was verified based on the field−measured data. Considering that the vibration system is a complex one with multiple subsystems, in order to reduce interference from noise vibration, the multi−point RTR was used for vibration prediction. In addition, the structural safety under the action of extreme traffic loads according to relevant codes was also evaluated.

2. Prediction Method of the Response Transfer Ratio Function

2.1. The Transfer Function and the Response Transfer Ratio Function

The transfer function is defined as the ratio of the Laplace transform of the linear system response (output) to the Laplace transform of the excitation (input) under the rest initial conditions [21]:

$$T(s) = \frac{Y(s)}{X(s)} = \frac{L\{y(t)\}}{L\{x(t)\}} \tag{1}$$

where $T(s)$ denotes the transfer function of the linear system, $Y(s)$ and $L\{y(t)\}$ in order are the response of the system and the Laplace transform of the output, $X(s)$ and $L\{x(t)\}$ represent the excitation of the system and the Laplace transform of the input, respectively. Generally, the transfer function requires that the system can be represented by a linear time−invariant system and must be applied in the presence of the rest initial condition. It describes the differential relationships between the input and output of the system.

For complex systems with multiple subsystems, the existing interactions between the subsystems make it difficult to get the excitation of each subsystem; Therefore, it is difficult to solve the transfer function of each subsystem by exploiting the output and input. In contrast, the response of each subsystem is usually easy to get. To distinguish from the traditional transfer function, the output ratio between each pair subsystems is defined as the response transfer ratio (RTR) function:

$$H_n = \frac{R_n}{R_m} \tag{2}$$

In Equation (2), H_n represents the RTR function between the $n-$th subsystem and the $m-$th subsystem. As shown in Figure 1, R_n and R_m in order denote the output response of the $n-$th and $m-$th subsystems. Similar to the transfer function, the RTR function can describe the dynamic performance of the linear system.

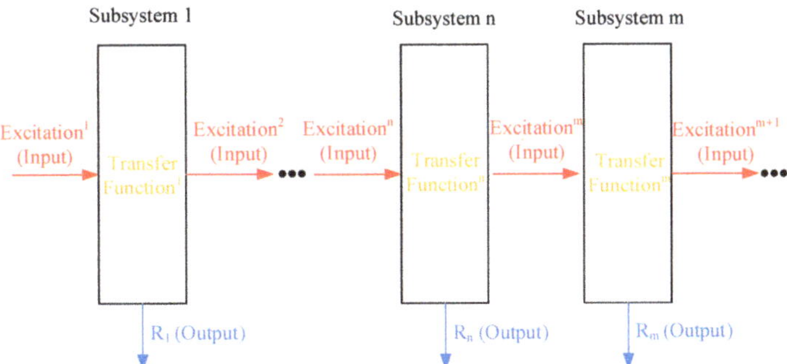

Figure 1. The complex systems with multiple subsystems.

For the multi−degrees−of−freedom system, the external excitation load of the system is assumed to be a simple harmonic load, that is:

$$\boldsymbol{M}\ddot{\boldsymbol{X}} + \boldsymbol{C}\dot{\boldsymbol{X}} + \boldsymbol{K}\boldsymbol{X} = \boldsymbol{P}sin(\omega t) \tag{3}$$

where $\boldsymbol{M}$, $\boldsymbol{C}$, and $\boldsymbol{K}$ in order are the mass, damping, and stiffness matrices of the system. The parameters $\ddot{\boldsymbol{X}}$, $\dot{\boldsymbol{X}}$, and $\boldsymbol{X}$ represent the acceleration vector, velocity vector, and displacement vector of the system, respectively, $\boldsymbol{P}$ denotes the vector of the external force applied to the system, ω is the circular frequency of the external force, and t is the time factor.

By employing the orthogonality properties of vibration modes, it is obtainable:

$$M_n\ddot{X}_n + C_n\dot{X}_n + K_nX_n = P_nsin(\omega t) \tag{4}$$

A harmonic solution to Equation (4) can be sought in the following form:

$$X_n = A_nsin(\omega_nt - \varphi_n) \tag{5}$$

where: $A_n = \frac{P_n}{K_n}\frac{1}{\sqrt{\left(1-\frac{\omega}{\omega_n}\right)^2+\left(\frac{2\xi_n\omega}{\omega_n}\right)^2}}$, $\varphi_n = tan^{-1}\frac{2\xi_n\frac{\omega}{\omega_n}}{1-\frac{\omega^2}{\omega_n^2}}$, and $\omega_n = \sqrt{\frac{K_n}{M_n}}$; P_n and ω_n in order are the generalized load and natural frequency associated with the $n-$th vibration mode; M_n and K_n represent the generalized mass and stiffness corresponding to the $n-$th mode, respectively, and ξ_n denotes the generalized damping ratio. By employing the superposition of modes in view of Equation (5), we can arrive at:

$$\boldsymbol{X} = \sum_{n=1}^{N}\boldsymbol{\phi}_n^TX_n = \sum_{n=1}^{N}\left[\boldsymbol{\phi}_n^TA_nsin(\omega t - \varphi_n)\right] \tag{6}$$

where $\boldsymbol{\phi}_n$ denotes the vector pertinent to the $n-$th vibration mode, X_n represents the generalized mode participation coefficient, while A_n and φ_n are their corresponding constants.

According to Equation (6), in the multi−degrees−of−freedom system under the action of single frequency vibration excitation, the steady−state response of each degree−of−freedom is still evaluated based on a single frequency, and the vibration frequency is the same as the excitation one.

When the input load in Equation (4) varies in constant λ times, the resulting solution is readily resulted by:

$$X_n = \lambda A_nsin(\omega t - \varphi_n) \tag{7}$$

$$\boldsymbol{X} = \sum_{n=1}^{N}\boldsymbol{\phi}_n^TX_n = \sum_{n=1}^{N}\left[\boldsymbol{\phi}_n^T\lambda A_nsin(\omega t - \varphi_n)\right] \tag{8}$$

According to Equation (8), when the ratio of the input load remains unchanged in various frequency bands, the ratio of the output acceleration also remains unchanged in the corresponding frequency band.

Therefore, for the multi−point RTR, input a simple harmonic force $\boldsymbol{P} = A\sin\omega t$ at the loading point, the RTR function between the points P_{n-1} and P_n is represented by H_{n-1}. As a result, the RTR function between the vibration source replacement point P_1 and the vibration prediction point P_n can be expressed by:

$$H(\omega) = \prod_{m}^{n-1} H_m(\omega) \tag{9}$$

where H_m denotes the RTR between point P_{m+1} and point P_m, n represents the number of transfer points on the transfer path. The calculation diagram of the multipoint RTR has been demonstrated in Figure 2.

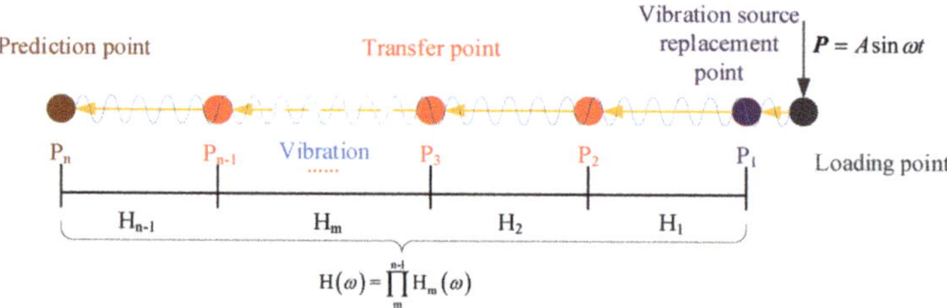

Figure 2. Schematic representation of the main procedure of the multipoint RTR.

2.2. Vibration Prediction Based on the RTR

For the environmental vibration caused by traffic, when a vehicle passes the road, the roadside vibration acceleration, $x(t)$, and the acceleration at the prediction point, $y(t)$, can be recorded simultaneously. Considering the calculation efficiency and accuracy, we then proceed in dividing the accelerations $x(t)$ and $y(t)$ according to the frequency bandwidth of the one−third octave. The result $X(t,f)$ and $Y(t,f)$ of such a division is the corresponding acceleration time−history data in each frequency band, the so−called octave time−history data in the present study. Specific processes for octave time−history data are as follows: 1. Take the Fourier transform of the acceleration data from the time domain to the frequency domain; 2. According to the one−third octave band, band−pass filtering is performed in turn to select the acceleration data in each frequency band; 3. Finally, the inverse Fourier transform is performed on the selected acceleration data in each frequency band to obtain the acceleration time−history data in the corresponding frequency band, which is called octave time−history data.

Subsequently, the ratio of maximum value of the octave time−history data, obtainable from the corresponding frequency band $MAX(X(t,f))$ and $MAX(Y(t,f))$, are defined as the amplitude−RTR. Mathematically, it is stated by:

$$H(f)_{MAX} = \frac{MAX(X(t,f))}{MAX(Y(t,f))} \tag{10}$$

Further, the ratio of the acceleration root−mean−square (RMS) of the octave time−history data $RMS(X(t,f))$ and $RMS(Y(t,f))$ in the corresponding frequency band is defined as the RMS−RTR, which is calculated by:

$$H(f)_{RMS} = \frac{RMS(X(t,f))}{RMS(Y(t,f))} \tag{11}$$

Using hammer excitation at the same excitation point, the roadside vibration acceleration time−history $x'(t)$ and the acceleration time−history $y'(t)$ at the prediction point can be simultaneously recorded. According to the one−third octave calculation method, the octave time−history data $X'(t,f)$ and $Y'(t,f)$ are computed, and similar to Equations (10) and (11), the RTR function of the roadside−prediction point acted upon by the hammering excitation is evaluated according to Equations (12) and (13):

$$H\prime(f)_{MAX} = \frac{MAX(X'(t,f))}{MAX(Y'(t,f))} \tag{12}$$

$$H\prime(f)_{RMS} = \frac{RMS(X'(t,f))}{RMS(Y'(t,f))} \tag{13}$$

It is assumed that the RTR function calculated by the hammer excitation test can be exploited to replace the transfer ratio function of the roadside−prediction point under the action of the traffic excitation. Thereby,

$$H(f)_{MAX} \approx H\prime(f)_{MAX} \tag{14}$$

$$H(f)_{RMS} \approx H\prime(f)_{RMS} \tag{15}$$

The roadside−prediction point RTR functions $H'(f)_{RMS}$ and $H'(f)_{MAX}$ can be measured by the hammer excitation and then combined with the traffic−induced roadside acceleration octave time−history data $X_{pre}(t,f)$. Subsequently, the output octave time−history data at the prediction point can be calculated through the following relations:

$$Y_{pre}(t,f) = X_{pre}(t,f)/H(f)_{MAX} \approx X_{pre}(t,f)/H\prime(f)_{MAX} \tag{16}$$

$$Y_{pre}(t,f) = X_{pre}(t,f)/H(f)_{RMS} \approx X_{pre}(t,f)/H\prime(f)_{RMS} \tag{17}$$

Finally, by superimposing the octave time−history data $Y_{pre}(t,f)$ associated with each frequency band, the acceleration time−history data $y_{pre}(t)$ at the predicted point can be evaluated as follows:

$$y_{pre}(t) = \sum_{f} Y_{pre}(t,f) \tag{18}$$

The proposed traffic−induced vibration prediction method has been flowcharted in Figure 3.

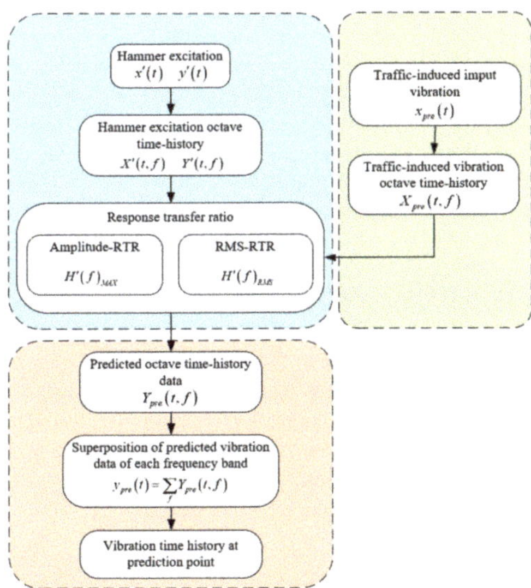

Figure 3. The flowchart of traffic environmental vibration prediction based on the measured RTR.

The vibration prediction method based on the RTR is as follows: Firstly, the response transfer ratio between the roadside and the prediction point is obtained through the artificial excitation vibration test; Then, collect the roadside environmental vibration caused by traffic load; Finally, the measured traffic−induced vibration is used as the excitation to calculate the vibration response of the prediction point by calculated RTR.

2.3. Field Test of the RTR Function

Theoretically, the transfer function is only related to the tested object, representing its inherent attribute, and it does not change with different external incentives. In order to verify whether the proposed RTR function also satisfies this characteristic, we designed two groups of hammer excitation experiments with a hammer weight of 30 kg. One group was oriented to control the hammer to fall from different heights, and only the change of the excitation energy was allowed without altering the excitation frequency. The other group dropped the hammer at the same height, and placed rubber, wood, and steel blocks at the landing point of the hammer. Therefore, the latter group was aimed at changing the input load spectrum characteristics, without altering the excitation energy. The test conditions are also presented in Table 1.

Table 1. Working conditions of the drop weight test: (**a**) Different excitation heights, (**b**) Different cushion blocks.

(a)

Test Name	Working Condition					Measuring Point Location	
	Condition 1	Condition 2	Condition 3	Condition 4	Condition 5	Vibration Source Replacement Point R	First Floor Measuring Point
Drop weight height (cm)	50	55	60	65	70	R	A−P1 B−P1 C−P1 D−P1

(b)

Test Name	Working Condition				Measuring Point Location	
	Condition 1	Condition 2	Condition 3	Condition 4	Vibration Source Replacement Point R	First Floor Measuring Point
Cushion block	–	Wood block	Rubber block	Steel block	R	A−P1 B−P1 C−P1 D−P1

As part of research series, this paper takes the Feiyun Pavilion as the case study. The Feiyun Pavilion (see Figure 4) is a purely wooden building in the Yuan (1271–1368 AD) and Ming Dynasty (1368–1683 AD) styles. It is located within the Dongyue temple in Wanrong County, Yuncheng City, Shanxi Province, China. The entire building is mainly made of wood, and the structural connection uses mortise and tenon joints without any metal components. The pavilion has three floors on the outside and five floors inside. The total height of the building is about 23 m.

The layout of the measurement points is demonstrated in Figures 5 and 6. The excitation point of the drop weight is on the road, which is ten meters away from the south side of the Feiyun Pavilion. The vibration source replacement point (R) is arranged at 1m perpendicular to the road, which is to collect the output response of road traffic subsystem. On the first floor of the Feiyun Pavilion near the four through columns, first floor measuring points (A−P1, B−P1, C−P1, D−P1) are set up and employed to measure the output acceleration of the system under the action of hammering excitation. In order to measure the RTR in three directions, all measuring points should be equipped with acceleration sensors in horizontal east−west, horizontal north−south direction, and vertical direction. The sampling frequency is set equal to 512 Hz.

Figure 4. The Feiyun Pavilion.

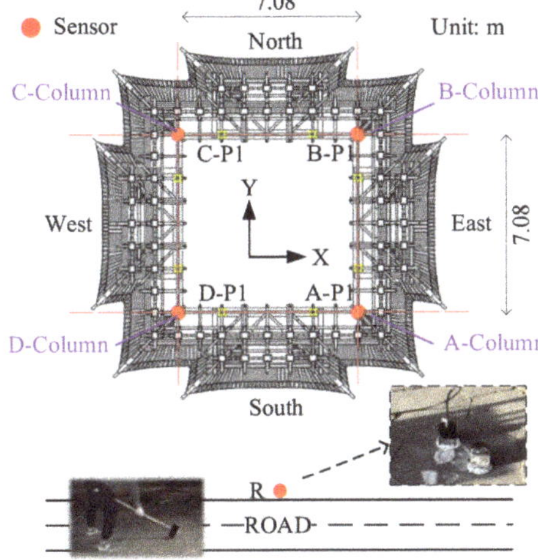

Figure 5. The horizontal layout of the measurement points. (Field Test of the RTR Function).

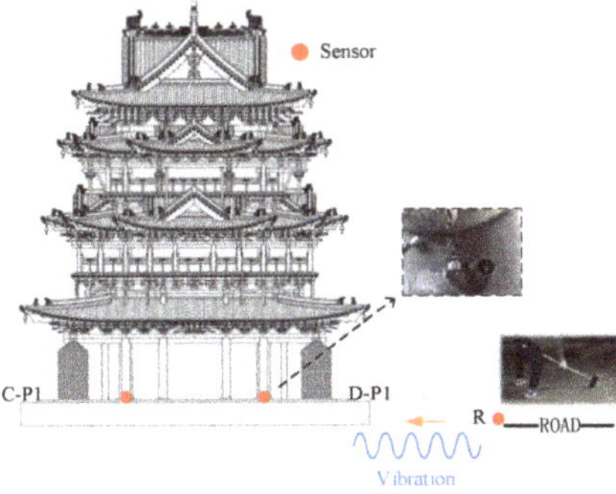

Figure 6. The vertical layout of the measurement points. (Field Test of the RTR Function).

The equipment utilized in the test includes an INV3020C synchronous data acquisition system with 28 channels, and 15 uniaxial (10 horizontal and 5 vertical) 941b acceleration sensors. Before the test, the acceleration sensors are appropriately calibrated for consistency and sensitivity.

2.3.1. Variation of the RTR Function with Excitation Vibration Energy

The artificial excitation vibration was applied to the system by using a free−falling weight. By controlling the hammer falling from heights of 50–70 cm with an increment of 5 cm, we examined whether the RTR of the system would change due to the same spectral characteristics but with various vibrational energies. Considering that the environmental vibration caused by traffic is mainly low−frequency vibration, we mainly analyzed the vibration below 80 Hz.

According to Equations (12) and (13), the RTRs of the system subjected to different levels of the input energy were calculated. For this purpose, at least five sets of valid hammer vibration data were collected for each working condition, and the RTR from the vibration source replacement point (R) to the first floor measuring point (A−P1, B−P1, C−P1, D−P1) was calculated due to each hammering excitation. Subsequently, the average value of five groups of RTRs in the same direction under the same working condition was taken as the RTR in this direction under this working condition. Taking the horizontal east−west direction (X direction) as an example, the amplitude−RTR and the RMS−RTR under different hammering excitations were evaluated, and the obtained results are graphed in Figures 7 and 8.

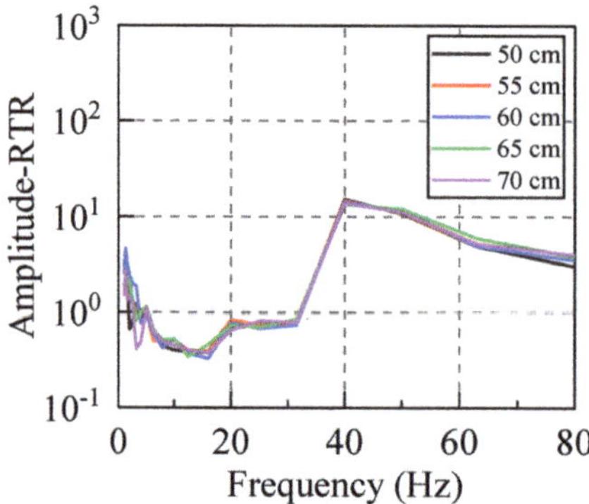

Figure 7. Amplitude−RTR. (Different excitation heights).

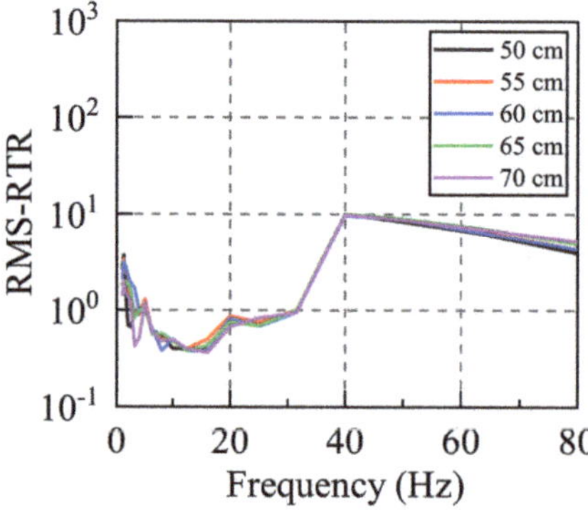

Figure 8. RMS−RTR. (Different excitation heights).

Firstly, the RTR functions calculated by the two distinct methods were compared. It can be seen that although their calculation bases are completely different, the discreteness of the two approaches is small for frequencies below 80 Hz, and the variation laws in terms of the frequency are the same. The RTR functions, which are calculated by the same method, are also compared. The RTR varies slightly when the input vibration energy is significantly different, and the consistency is satisfactory in the frequency range of 5–63 Hz. Therefore, both methods can appropriately calculate the RTR function of the system, and its value is independent of the input vibration energy.

2.3.2. Variation of the RTR Function with Excitation Vibration Frequency

In order to input the same vibration energy to the system, we let the hammer have a free fall from 60 cm. In addition, steel plate, wood plate, and rubber plate are placed at the hammering point in order to apply vibrations with the same excitation energy but different spectral characteristics to the system. The RTR function between vibration source replacement point to the first floor measuring point was calculated according to Equations (12) and (13). Taking the X direction as an example, the calculation results for this case are illustrated in Figures 9 and 10.

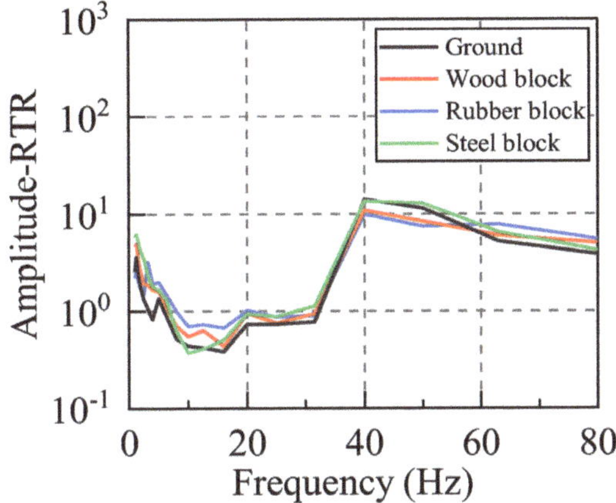

Figure 9. Amplitude−RTR. (Different cushion blocks).

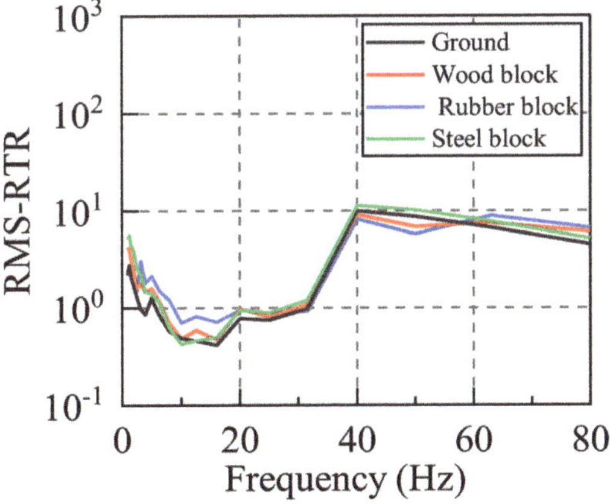

Figure 10. RMS−RTR. (Different cushion blocks).

By comparing the RTR acted upon by the excitation with different spectrum characteristics, it can be seen that the calculation results of the RTR in the presence of various working conditions are relatively consistent. Therefore, the different input excitation spectrum has little influence on the RTR of the same system, and this characteristic is more prominent for frequencies in the range of 5–63 Hz. The RTR is independent of the input vibration spectral characteristics.

Through the calculation and analysis of the RTR under various working conditions, it can be inferred that the RTR is only affected by the dynamic characteristics of the structural system. It is the inherent attribute of the structural system, and it will not change for different input excitations. It implies that the application of the hammering excitation to calculate the RTR of the system can be utilized as an appropriate replacement of the RTR under the action of the traffic excitation. This indicates that the Equations (14) and (15) are reasonable.

3. Vibration Prediction Based on the Measured RTR Function

3.1. Introduction to the On−Site Dynamic Test

As demonstrated in Figure 11, Feiyun Pavilion is only 10 m away from Houtu road in the south and close to Feiyun Bei road in the east. The long−term wind, rain erosion, and the impact of traffic vibration have made damages to Feiyun Pavilion up to a certain extent. Therefore, it is necessary to predict the vibration of the Feiyun Pavilion under the action of traffic loading.

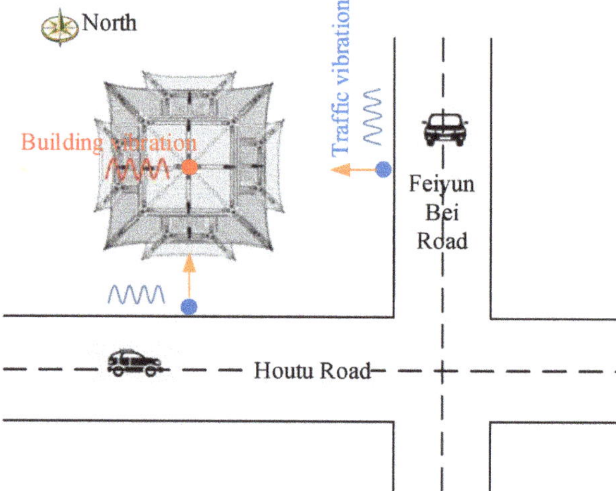

Figure 11. Schematic representation of the traffic environment around the Feiyun Pavilion.

Considering the whole media of the vibration transmission is composed of the road, soil, foundation, and superstructure. The measuring points of such a media are particularly arranged as presented in Figures 12 and 13. This is somehow similar to the measurement point arrangement in the "Field test of the RTR function". In addition to arranging the vibration source replacement point (R) and the first floor measuring points (A−P1, B−P1, C−P1, and D−P1), it is also necessary to arrange the third floor measuring points (A−P2, B−P2, C−P2, and D−P2) at the top of the four corner columns. The third floor measuring points are also exploited to collect the vibration at the highest point of the structure. At the same time, the vibration data of the third floor measuring point can also evaluate the structural safety of the Feiyun Pavilion according to the relevant specifications.

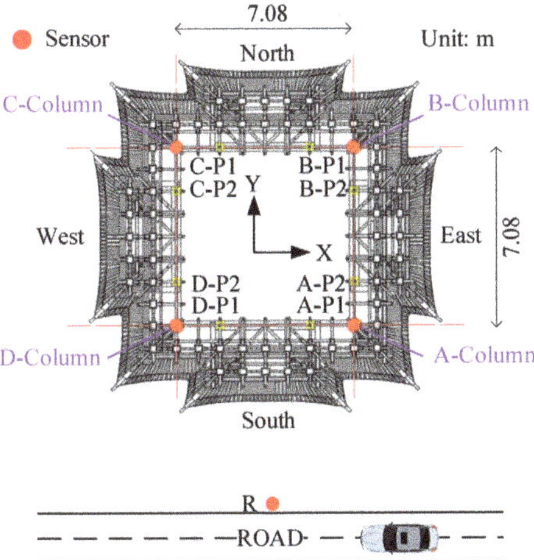

Figure 12. The layout plan of the measurement points.

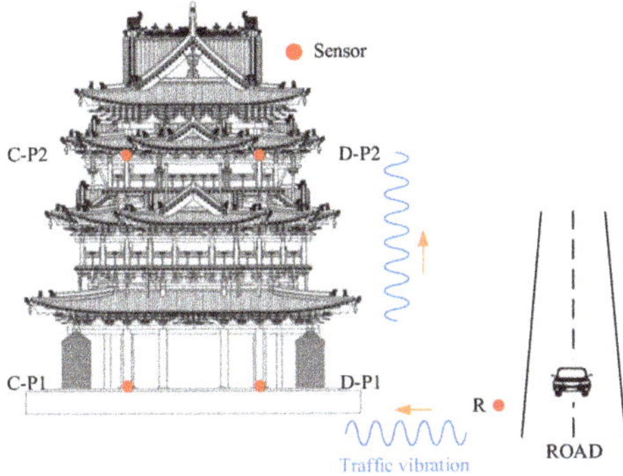

Figure 13. The vertical layout of the measurement points.

Due to the increase of the measuring points, we choose the following equipment to collect vibration data: INV3020C synchronous data acquisition system with 28 channels and 27 uniaxial (18 horizontal and 9 vertical) 941b acceleration sensors.

3.2. The On−Site Test of the Multi−Point RTR

Let us take the measuring point in column D as an example. Firstly, the free−fall hammer is utilized to input excitation vibration to the system from a height of 60 cm, on the Houtu road. Then, the multi−point RTR function between the vibration source replacement point (R), the first floor measuring point (D−P1), and the third floor measuring point (D−P2) were calculated according to the acceleration data. After that, under the action of the traffic−induced vibration, the measured environmental vibration at point R was employed as the excitation, and the multi−point RTR calculated by the hammering test was employed to predict the traffic−induced vibration at the third floor measuring point (D−P2). Finally, the effectiveness of the proposed vibration prediction method was verified by comparing the measured vibration of the third floor measuring point (D−P2) with the predicted vibration.

According to Equation (9), the collected acceleration data were calculated to obtain the multi−point (R to D−P2) RTR. The plotted results are demonstrated in Figure 14.

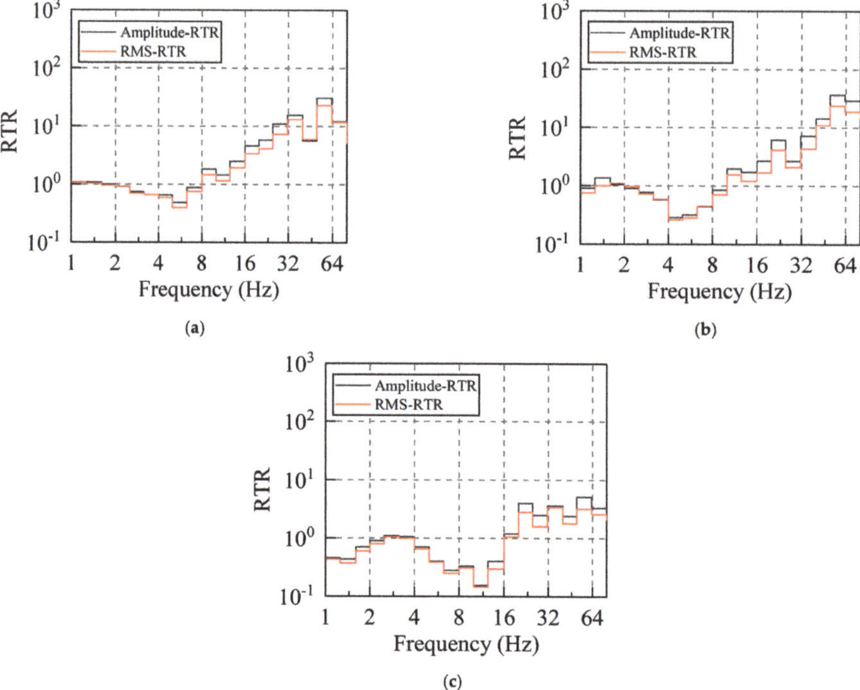

Figure 14. The measured RTR: (**a**) vertical direction; (**b**) east–west direction; and (**c**) south–west direction.

3.3. Structural Vibration Prediction due to Traffic–Induced Vibration

In order to ensure that a sufficient amount of traffic vibration data were collected, the sampling duration and frequency in order were set to 1200 s and 512 Hz. Due to the traffic–induced vibration, the partial acceleration data of the third floor measuring point (D–P2) in the horizontal east–west direction (X direction) has been presented in Figure 15.

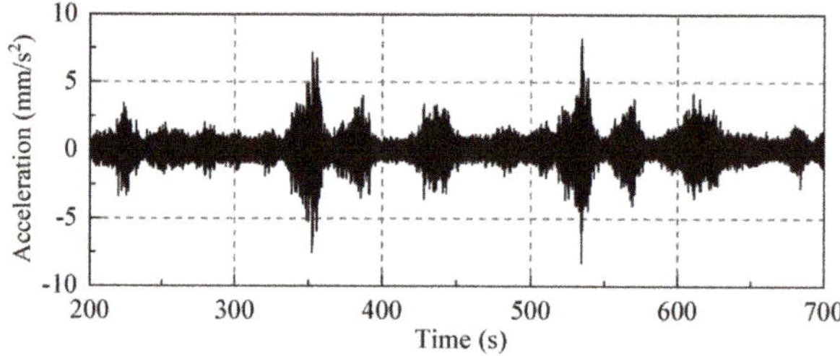

Figure 15. The time–history of acceleration of the third floor measuring point (D–P2) in the east–west direction.

A representative traffic–induced vibration data segment (520s–570s) was selected for fast Fourier transform (FFT) analysis. The vibration data in the frequency–domain were then analyzed, as shown in Figure 16.

The vibration at the vibration source replacement point (R) is mainly low frequency vibration below 80 Hz, of which 5 Hz–60 Hz represents its excellent frequency band. When the vibration is transmitted from the vibration source replacement point (R) to the third floor measuring point (D–P2), its spectrum characteristics have considerably altered: the vibration with a frequency above 40 Hz is greatly attenuated, and the vibration is mainly low−frequency vibration with frequencies in the range of 5−40 Hz. Therefore, this frequency interval is defined as the prediction and analysis frequency band. In the vibration prediction, the acceleration for the frequency interval 5–40 Hz will be mainly calculated.

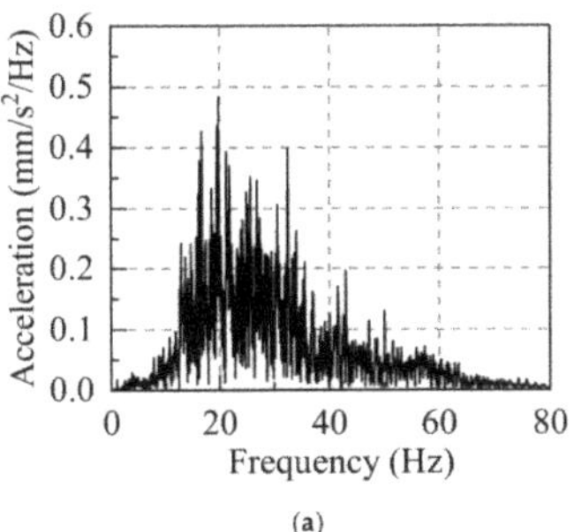

(a)

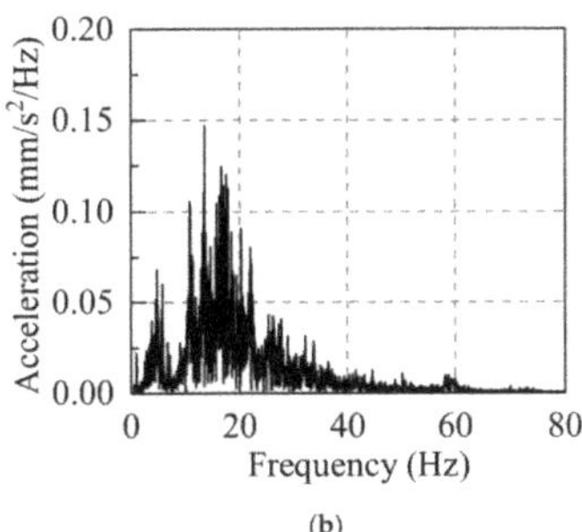

(b)

Figure 16. Acceleration spectrums: (**a**) spectrum of the vibration source replacement point (R); and (**b**) spectrum of the third floor measuring point (D−P2).

According to the RTR, the traffic−induced vibration data at the third floor measuring point (D−P2) can be predicted. The time−history data of the prediction results are presented in Figure 17:

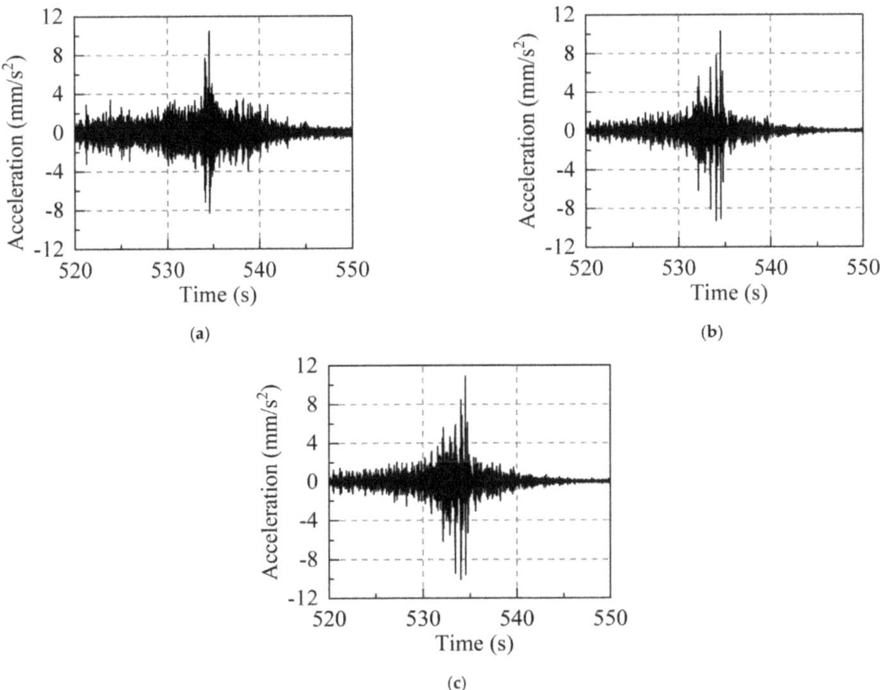

Figure 17. Acceleration time−history of the third floor measuring point (D−P2): (**a**) traffic−induced acceleration in third floor measuring point (D−P2); (**b**) third floor measuring point (D−P2) acceleration time−history prediction based on the amplitude−RTR, and (**c**) third floor measuring point (D−P2) acceleration time−history prediction based on the RMS−RTR.

3.4. Evaluate Prediction Accuracy

Because the phase difference between the measuring points is ignored in the prediction process, the prediction accuracy is generally evaluated by statistical indicators.

The square of the root−mean−square (RMS) can be employed to quantify the average vibration energy at each measurement point [22]. Hence, this factor is capable of evaluating the vibration response produced by traffic vibrations. The RMS value of a discrete−time signal is defined by:

$$a_{RMS} = \sqrt{\frac{\sum\limits_{i=1}^{N} a^2(i)}{N}} \tag{19}$$

where a denotes the measured acceleration, and N represents the number of data points analyzed.

The computed results of the maximum and RMS at the third floor measuring point (D−P2) during the analysis period have been presented in Table 2.

Table 2. Comparison between the predicted and measured acceleration data at the third floor measuring point (D−P2).

	Maximum Acceleration in Time−Domain (mm/s^2)	RMS of Acceleration in Time−Domain (mm/s^2)
Measured value	10.47	0.97
Predicted value of amplitude−RTR	10.28	0.91
Predicted value of RMS−RTR	10.89	0.94

By comparing the measured data and the predicted one, it can be seen that the amplitude−RTR can be utilized to predict the vibration of the third floor measuring point (D−P2) subjected to traffic−induced vibration. The difference between the maximum acceleration and the real value is reported as 1.8%, and the relative discrepancy between the predicted acceleration RMS and the real value is about 6.1%. Employing the RMS−RTR to predict the vibration of the third floor measuring point, the generated relative error between the maximum acceleration and the real value is 4.5%, and the resulting error between the RMS and the real value is about 3.1%. The main reasons for the produced error are as follows: (1) The signal−to−noise ratio of the acceleration data due to the traffic−induced vibration is poor; (2) The environment of the on−site test is complex, so the measured signal can be easily disturbed; (3) Only the excellent frequency band is predicted, while the vibration energy in other frequency bands is overlooked.

The accuracy of the prediction results was also evaluated in the frequency−domain. The spectrum, 1/3−octave of the measured data, and predicted data of the third floor measuring point are calculated. The predicted results have been now illustrated in Figure 18. The frequency−domain analysis revealed that both the predicted and measured results can have good consistency in the predicted analysis frequency band.

In both the time−domain and frequency−domain, the application of the measured RTR function for examining the vibration of ancient wooden structures exhibited a high prediction accuracy due to the traffic−induced vibration.

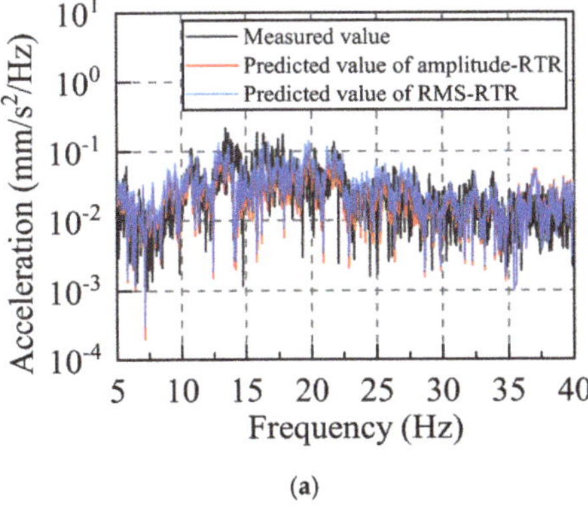

(a)

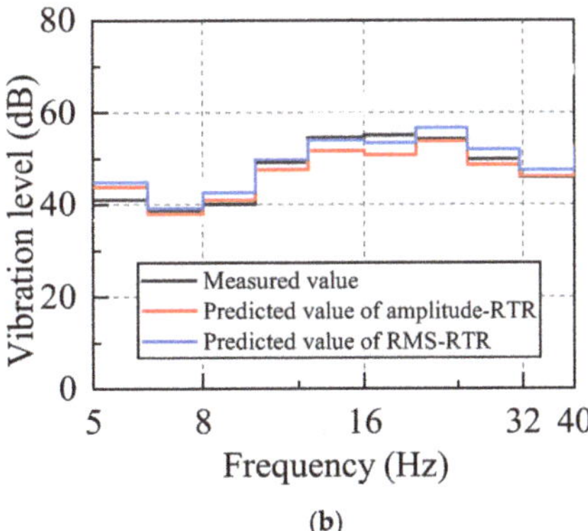

(b)

Figure 18. The frequency−domain calculation results of the measured and predicted values: (**a**) Power spectra density, (**b**) 1/3−octave.

4. Prediction and Evaluation of Structural Safety of the Feiyun Pavilion due to Extreme Traffic Loading

In order to explore the vibration of the Feiyun Pavilion under the action of extreme traffic loading and evaluate its structural safety according to the prediction results, we performed the excitation test of heavy−duty trucks (the vehicle's weight is 40 tons) on a road similar to the Houtu road. The acceleration data (horizontal north−south, horizontal east−west, and vertical components) were collected at the vibration source replacement point when the vehicle speeds are 30, 40, and 50 km/h. The sampling frequency is 512 Hz and the recording time period is 30 s. The horizontal east−west acceleration data at the vibration source replacement point have been plotted in Figure 19.

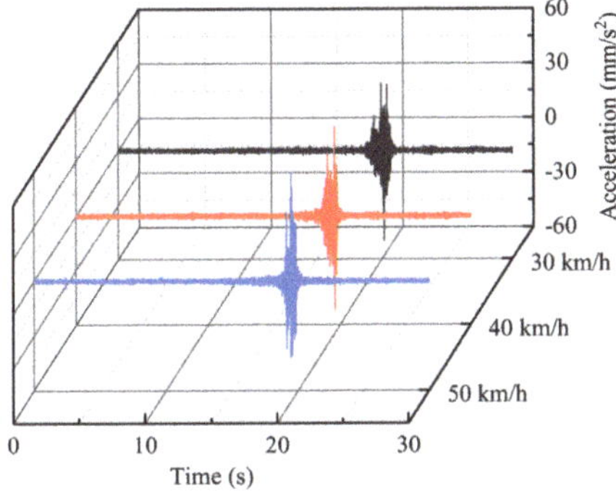

Figure 19. The acceleration of the vibration source replacement point (horizontal east−west direction).

According to the "Technical specifications for protection of historic buildings against man−made vibration" [23], the allowable vibration of ancient structures should be controlled by the horizontal vibration velocity at the highest point of the structure. Therefore, the limit of horizontal vibration velocity at the highest point of the Feiyun Pavilion is 0.18 mm/s.

According to the proposed vibration prediction method, the vibration acceleration of third floor measuring point (D−P2) caused by a heavy−duty truck can be obtained, based on the amplitude−RTR and RMS−RTR (Figure 20). The predicted vibration velocity of third floor measuring point (Figure 21) was obtained by integrating the predicted acceleration data in the frequency−domain. The maximum predicted vibration velocity of the third floor measuring point have been provided in Table 3.

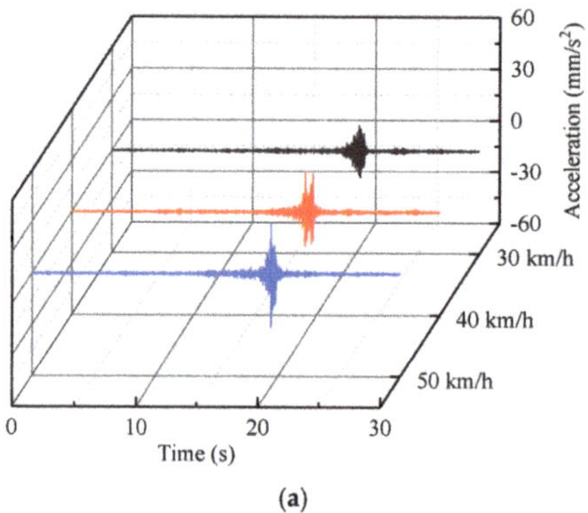

(a)

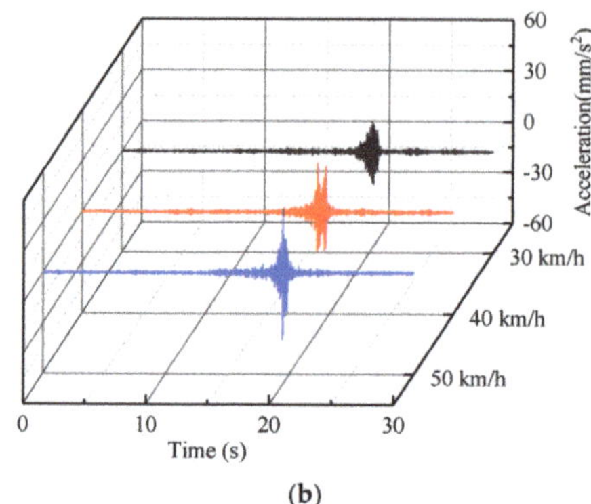

(b)

Figure 20. Predicted acceleration time−history of the third floor measuring point (D−P2) under the heavy−duty truck (horizontal east−west direction): (**a**) third floor measuring point (D−P2) acceleration time−history prediction based on the amplitude−RTR; and (**b**) third floor measuring point (D−P2) acceleration time−history prediction based on the RMS−RTR.

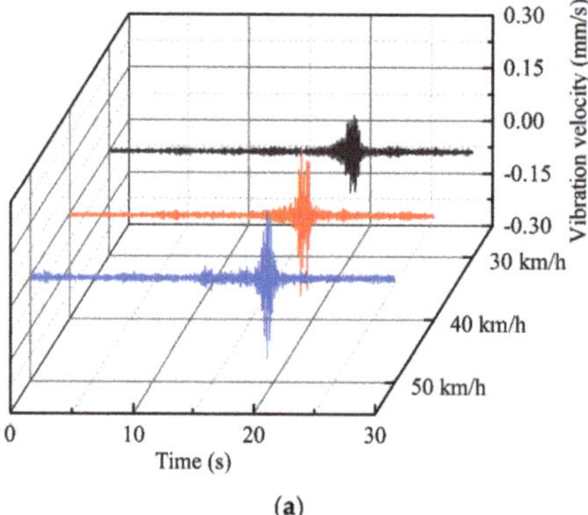

(a)

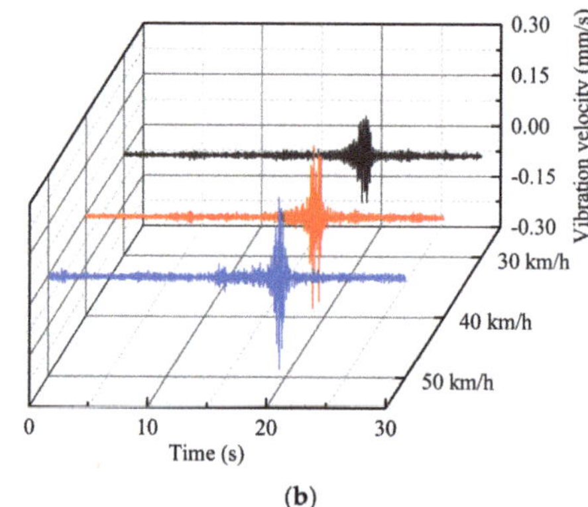

(b)

Figure 21. Predicted velocity time−history of the third floor measuring point (D−P2) under the heavy−duty truck (horizontal east−west direction): (**a**) third floor measuring point (D−P2) vibration velocity time−history prediction based on the amplitude−RTR; and (**b**) third floor measuring point (D−P2) vibration velocity time−history prediction based on the RMS−RTR.

Table 3. The maximum predicted vibration velocity of the third floor measuring point (D−P2) under the action of heavy−duty trucks.

Vehicle Speed (km/h)	The Maximum of Vibration Velocity Based on Amplitude−RTR		The Maximum of Vibration Velocity Based on RMS−RTR	
	East−West Direction (mm/s)	North−South Direction (mm/s)	East−West Direction (mm/s)	North−South Direction (mm/s)
30	0.12	0.30	0.14	0.31
40	0.23	0.69	0.27	0.75
50	0.24	2.03	0.28	2.13

The calculation results revealed in Table 3 show that even when the heavy−duty truck passes through the Feiyun pavilion at a minimum test speed of 30 km/h, the structural vibration has exceeded the specification limit, which will pose a threat to its structural safety. Therefore, it is suggested to limit the type and speed of vehicles in the Feiyun Pavilion areas for protecting the ancient wooden structures.

5. Conclusions

(1) The multipoint RTR was derived, and showed that the RTR function is the inherent property of the structure and does not alter with excitation load energy and frequency;
(2) Based on the RTR function, a vibration prediction method suitable for the ancient wooden structures subjected to traffic−induced vibration was proposed. By comparing with the measured data, the prediction results represented a good accuracy in both the time and frequency domains;
(3) The structural vibration of the Feiyun Pavilion due to extreme traffic loads was predicted, and the corresponding structural safety was evaluated according to the "Technical specifications for protection of historic buildings against man−made vibration". The calculation results reveal that in the Feiyun Pavilion area, it is necessary to restrict the type and speed of vehicles to protect the ancient wooden structures from traffic−induced vibrations.

Author Contributions: C.Z. and Y.Z. came up with the concept, C.Z. did the field test and analyzed the data, and edited the draft of manuscript. N.Z. conducted the literature review, checked the computations, wrote the draft of the manuscript, replied to reviewers' comments, and revised the final version. X.L. verified the simulated data and polished the article. All authors have read and agreed to the published version of the manuscript.

Funding: This research received no external funding.

Institutional Review Board Statement: Not applicable.

Informed Consent Statement: Not applicable.

Data Availability Statement: The data presented in this study are available upon request from the corresponding author.

Conflicts of Interest: We declared that we have no financial and personal relationships with other people or organizations that can inappropriately influence our work, there is no professional or other personal interest of any nature or kind in any product service and company that could be construed as influencing the review of "Prediction of traffic vibration environment of ancient wooden structures based on the response transfer ratio function".

References

1. Cao, Y.M.; Wang, F.X.; Zhang, Y.S. Ansys Method and vibration characteristics of field vibrations induced by high-speed trains. *J. China Railw. Soc.* **2017**, *39*, 118–124.
2. Liu, W.F.; Liu, W.N.; Nie, Z.L.; Wu, Z.Z.; Li, K.F. Prediction of effects of vibration induced by running metro trains on sensitive instruments. *J. Vib. Shock* **2013**, *32*, 18–23.
3. Zhang, P.F.; Lei, X.Y.; Gao, L.; Liu, Q.J. Study on the ground vibration and its impact on precision instruments induced by freight train. *J. Railw. Sci. Eng.* **2013**, *10*, 108–113.
4. Hu, W.B.; Yang, J.; Wu, Y.H.; Meng, Z.B. The calculation of ancient wooden pillars horizontal velocity under the traffic load. *J. Xi'an Univ. Archit. Technol. Nat. Sci. Ed.* **2019**, *51*, 315–320.
5. Jia, X.P. Study on Vibration Caused by Subway and Its Impact for Buildings. Master's Thesis, Tongji University, Shanghai, China, 2008.
6. Xia, H.; Cao, Y.M.; Roeck, G.D. Theoretical modeling and characteristic analysis of moving-train induced ground vibrations. *J. Vib. Eng.* **2009**, *329*, 819–832. [CrossRef]
7. Lombaert, G.; Degrande, G. Ground-borne vibration due to static and dynamic axle loads of InterCity and high-speed trains. *J. Sound Vib.* **2009**, *319*, 1036–1066. [CrossRef]
8. Zou, C.; Moore, J.A.; Sanayei, M.; Wang, Y. Impedance model for estimating train-induced building vibrations. *Eng. Struct.* **2018**, *172*, 739–750. [CrossRef]
9. Ma, M.; Lui, W.N.; Ding, D.Y. Influence of metro train-induced vibration on Xi'an Bell Tower. *J. Beijing Jiaotong Univ.* **2010**, *34*, 88–90.
10. Ma, M.; Liu, W.N.; Qian, C.Y.; Deng, G.H.; Li, Y.D. Study of the train-induced vibration impact on a historic Bell Tower above two spatially overlapping metro lines. *Soil Dyn. Earthq. Eng.* **2016**, *81*, 58–74. [CrossRef]
11. Kouroussis, G.; Parys, L.V.; Conti, C.; Verlinden, O. Prediction of ground vibrations induced by urban railway traffic: An analysis of the coupling assumptions between vehicle, track, soil, and buildings. *Int. J. Acoust. Vib.* **2013**, *18*, 163–172. [CrossRef]

12. Coulier, P.; Lombaert, G.; Degrande, G. The influence of source–receiver interaction on the numerical prediction of railway induced vibrations. *J. Sound Vib.* **2014**, *333*, 2520–2538. [CrossRef]
13. Yan, W.M.; Nie, H.; Ren, M.; Feng, J.H.; Zhang, Y.W.; Chen, J.Q. In situ experiment and analysis of ground surface vibration induced by urban subway transit. *J. Railw. Sci. Eng.* **2006**, *3*, 1–5.
14. Agata, S.; Anna, J.G.; Robert, J. The idea of using Bayesian networks in forecasting impact of traffic-induced vibration transmitted through the ground on residential buildings. *Geosciences* **2019**, *9*, 339–352.
15. Giacomo, Z.; Michele, B.; Gianni, B. Experimental analysis of the traffic-induced-vibration on an ancient lodge. *Struct. Control Health Monit.* **2022**, *29*, e2900.
16. Ivo, H.; Marijan, B.; Stjepan, L. Analysis of Tram Traffic-Induced Vibration Influence on Earthquake Damaged Buildings. *Buildings* **2021**, *11*, 590. [CrossRef]
17. Tao, Z.; Wang, Y.; Sanayei, M.; Moore, J.A.; Zou, C. Experimental study of train-induced vibration in over-track buildings in a metro depot. *Eng. Struct.* **2019**, *198*, 109473. [CrossRef]
18. Zou, C.; Moore, J.A.; Sanayei, M.; Tao, Z.; Wang, Y. Impedance Model of Train-Induced Vibration Transmission Across a Transfer Structure into an Over Track Building in a Metro Depot. *J. Struct. Eng.* **2022**, *148*, 04022187. [CrossRef]
19. Verbraken, H.; Lombaert, G.; Degrande, G. Verification of an empirical prediction method for railway induced vibrations by means of numerical simulations. *J. Sound Vib.* **2011**, *330*, 1692–1703. [CrossRef]
20. Vogiatzis, K. Protection of the cultural heritage from underground metro vibration and ground-borne noise in Athens centre: The case of the Kerameikos Archaeological Museum and Gazi Cultural Centre. *Int. J. Acoust. Vib.* **2012**, *17*, 59–72. [CrossRef]
21. Chen, J.M. *Principles of Automatic Control*; National Defense Industry Press: Beijing, China, 2014.
22. Zhang, C.; Zhang, N.; Wang, J.; Yao, J.B. Vibration reduction and isolation performance of a platform foundation and column base of an ancient wooden structure based on the energy transfer analysis. *J. Low Freq. Noise Vib. Act. Control* **2022**, *41*, 14613484221082635. [CrossRef]
23. Ministry of Housing and Urban Rural Development of the People's Republic of China. *Technical Specifications for Protection of Historic Buildings against Man-Made Vibration*; China Building Industry Press: Beijing, China, 2008.

applied sciences

Review

Developments in 3D Visualisation of the Rail Tunnel Subsurface for Inspection and Monitoring

Thomas McDonald *, Mark Robinson and Gui Yun Tian

Newcastle University Centre of Excellence for Mobility and Transport, Newcastle upon Tyne NE1 7RU, UK
* Correspondence: t.mcdonald@newcastle.ac.uk

Featured Application: The review presented in this work has practical application to the conception, development and refinement of new technologies and visualisation frameworks pertaining to railway tunnel subsurface inspection. Subsequent application to the development of prototype self-sustaining digital twin tunnels also presents opportunity. In both cases, practical end user benefit would be improvement to the clarity and comprehensiveness of subsurface inspection datasets, better informing targeted maintenance strategy planning.

Abstract: Railway Tunnel SubSurface Inspection (RTSSI) is essential for targeted structural maintenance. 'Effective' detection, localisation and characterisation of fully concealed features (i.e., assets, defects) is the primary challenge faced by RTSSI engineers, particularly in historic masonry tunnels. Clear conveyance and communication of gathered information to end-users poses the less frequently considered secondary challenge. The purpose of this review is to establish the current state of the art in RTSSI data acquisition and information conveyance schemes, in turn formalising exactly what constitutes an 'effective' RTSSI visualisation framework. From this knowledge gaps, trends in leading RTSSI research and opportunities for future development are explored. Literary analysis of over 300 resources (identified using the 360-degree search method) informs data acquisition system operation principles, common strengths and limitations, alongside leading studies and commercial tools. Similar rigor is adopted to appraise leading information conveyance schemes. This provides a comprehensive whilst critical review of present research and future development opportunities within the field. This review highlights common shortcomings shared by multiple methods for RTSSI, which are used to formulate robust criteria for a contextually 'effective' visualisation framework. Although no current process is deemed fully effective; a feasible hybridised framework capable of meeting all stipulated criteria is proposed based on identified future research avenues. Scope for novel analysis of helical point cloud subsurface datasets obtained by a new rotating ground penetrating radar antenna is of notable interest.

Keywords: railways; tunnel; subsurface; inspection; visualisation; ground penetrating radar; 360GPR; structural health monitoring; building information modelling; extended reality

Citation: McDonald, T.; Robinson, M.; Tian, G.Y. Developments in 3D Visualisation of the Rail Tunnel Subsurface for Inspection and Monitoring. *Appl. Sci.* **2022**, *12*, 11310. https://doi.org/10.3390/app122211310

Academic Editors: Phong B. Dao, Tadeusz Uhl, Liang Yu, Lei Qiu and Minh-Quy Le

Received: 16 September 2022
Accepted: 1 November 2022
Published: 8 November 2022

1. Introduction

Railway tunnels provide critical transport links for passengers and freight through terrain otherwise impassable to trains, facilitating time-efficient navigation through mountains, under waterbodies and bypassing human-made obstructions (e.g., buildings, utilities, mass-transit routes). As confined high-traffic subterranean infrastructure, tunnels are inherently hostile and dangerous environments, suffering perpetual degradation from both environmental and human factors (e.g., shifting landmass, extreme weather, above-ground construction) [1–5] which seed discernible damage to the intrados—the innermost surface of the tunnel arch [6]—surface and subsurface. In the UK, unlike comparatively modern highway and metro tunnels, railway tunnels frequently date back to the Victorian era. These historic masonry structures are inherently weaker than their modern concrete counterparts, meaning complex degradation can rapidly develop in the vicinity of seeded damage. Therefore, detection and hazard-level evaluation of all structural features (assets, defects) during Railway Tunnel Inspection (RTI) surveys is essential to inform targeted maintenance, ensuring continued safe and efficient operation. Use of Non-Destructive Inspection/Evaluation (NDI/E) techniques for Rail Tunnel SubSurface Inspection (RTSSI) is of paramount importance in modern surveys; however, they are not infallible due to accuracy and clarity limitations. Consequently, undetected seeding and growth of concealed subsurface defects can complicate or even scrub maintenance attempts, irrespectively posing serious safety risks that can endanger life. Timely reminders include undetected microfracture growth which caused catastrophic failure of the Gerrards Cross Tunnel (UK, 2005) [7,8] and two violent crown failures which partially collapsed 18 m of the Yangshang Tunnel (China, 2017) [9].

Collectively these dangers highlight urgent need for a comprehensive, reliable, repeatable, time-efficient and clear RTSSI visualisation framework, based on NDI techniques, for accurate intrados subsurface feature detection and evaluation. In this work, we review the effectiveness of current RTSSI-relatable visualisation frameworks, focusing on the increasing capabilities of realistic 3D surveys and discussion of future research opportunities. Our aims are to highlight common critical limitations of current strategies and propose viable, pertinent improvements. Following an overview of research methodology (Section 2) we explore leading NDI methods in RTI for RTSSI (Section 3), before deliberating the issues of applying heuristic comparisons (Section 4). From this, we formulate criteria for effective RTSSI visualisation frameworks (Section 5), then appraise the scope of current connected research efforts (Section 6). A discussion of trends and identified findings is finally presented (Section 7).

2. Materials and Methods

We analyse journal references, practical studies and commercial systems pertaining to RTSSI-relatable visualisation frameworks. For this purpose, we partition notion of an RTSSI visualisation framework into two sequential phases: (1) Data Acquisition Approaches; (2) Information Conveyance Schemes. A subtle remark, note that information is not itself raw data but the meaning from raw data. 'The tree holds 5 apples' is raw data, but knowing we expect it to hold 20 gives meaning to the data (i.e., the harvest is poor). Context and analysis turn raw data into useful information.

No prior review work encountered considers both described phases in detail. In fact, we found only two literary reviews directly related to RTSSI [10,11]. Both principally analyse literature concerning phase (1), of which [10] is the only dedicated review article. However, being published in 2015, it now lacks current relevance due to technological advancements. Discussions of ROBO-SPECT (RS) are the only RTSSI-specific scheme we do not consider outdated, but consideration will only be paid its most recent publications. The 2015 review provides a comparative 'baseline' for discerning small updates on already established methods from genuinely novel RTSSI innovations. Our review focuses on the latter as this provides greater benefit to current researchers and practitioners, with [10] providing historic reference. By contrast, we note although [11] is the more recent publication, it relies on many references previously provided in [10]. Our observations clearly necessitate creation of a novel review addressing both phases of RTSSI visualisation framework development, spanning research projects and commercial systems documented between 2015 and 2021.

Our review procedure is illustrated in Figure 1 and draws upon knowledge from over 300 information resources. We formally cite 255 primary resources (e.g., reviews, articles, proceedings, etc.). Upwards of 60 secondary resources (e.g., internal reports, cooperate multimedia, web-resources, etc.) provide further insight, however are not formally referenceable Owing to the fundamental dependency phase (2) has on phase (1), materials considered have frequently contributed to discussion of both areas. As a conservative estimate, the distribution of total literature is approximately 67% across phase (1) and

55% across phase (2), with 33% providing contextualisation. Maximal crossover is approximately 69%. Academic publications are obtained by 360-degree searches of journal articles and conference proceedings through databases including IEEE Xplore, Springer, MDPI and Google Scholar. Materials concerning commercial technology solutions are primarily sourced from the relevant corporate organisations' website, system manuals and any associated technical papers. Keywords recurringly searched include: "Railway Tunnels", "Structural Health Monitoring", "Subsurface", "Defects", "Computer Vision", "Visualisation", "Human Computer Interaction", "LiDAR", "Photogrammetry", "Ground Penetrating Radar" and "Extended Reality". For a detailed breakdown of key references, see Appendix A.

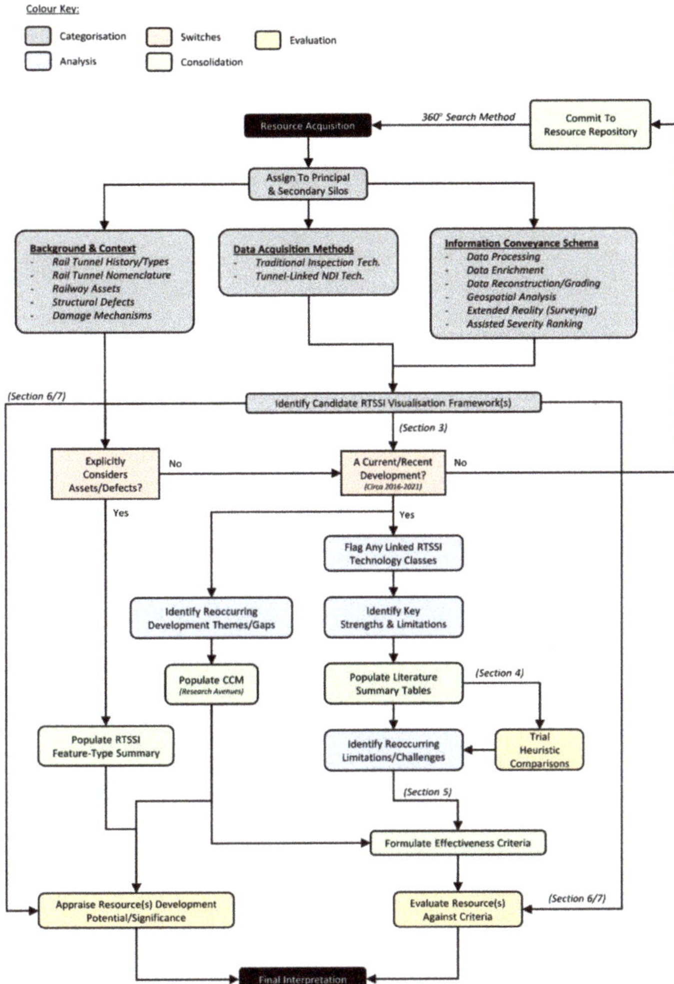

Figure 1. Flowchart of reviewing procedure prescribes handling, analysis and management of resources, alongside providing a rigorous procedure for appraising the effectiveness and scope of frameworks considered.

3. RTI Data Acquisition Methods

To gauge current state of the art in RTSSI-relatable visualisation frameworks, we first consider current leading subsurface data acquisition methods. Note that data acquisition must logically precede information conveyance in all conceivable frameworks by chronological reasoning.

3.1. Visual Methods

Visual assessment is the longest-established NDI method for RTI and is still widely adopted today, particularly across the UK and Chinese rail networks [12,13]. We subdivide methods into two classes: (i) Traditional and (ii) Modernised. Traditional evaluation is exclusively based on engineers' learnt association between visual indicators (e.g., workmanship inconsistencies, material fatigue hallmarks) and fault likelihood. Problematically, engineers infrequently share similar extents of practical experience, resulting in high subjectivity. Crosschecks and multi-pass surveys can partially reduce accuracy and consistency variations but take significantly longer to implement at increased resource cost, closure times and rail worker risk. Handwritten notetaking ambiguity, incompleteness and inherent susceptibility to human error also present issues for later analysis. They entice misinterpretation, causing unnecessary delays and disruption. However, being low expense and reasonably accurate (if performed by more experienced engineers), coupled with human-aptitude at informed predications from non-structural information (e.g., history of construction practices); traditional methods can time-efficiently localise visibly degraded quadrants requiring repair.

Modernised methods mostly utilise Close-Range Photogrammetry (CRP) to provide referenceable intrados imagery. Units commonly employ RGB optical cameras mounted on moving platforms for stability and time-efficiency. These include: Pushcarts/Rail-Trolley (RT), Road-Rail Vehicle (RRV) and Robotic Traction Unit (RTU). Merging resultant overlapping orthophotos via mosaicing [14,15] allows tunnels to be 'unwrapped'—permitting analysis in 2D—although we more frequently find studies adopt 3D CRP topography model reconstruction via 'Structure from Motion' (SfM) algorithms [16–19].

A noteworthy recent innovation includes 'Digital Imaging for Condition Asset Monitoring System' (DIFCAM) [20]; an RRV-mounted optical array deigned to reduce crew sizes and inspection durations. Although 2014 marks DIFCAM's last major study [21], scope of its successor project DIFCAM Evolution [22] discusses subsurface imaging and automated defect recognition technology integration. However, due Visual assessment is the longest-established NDI method for RTI and is still widely a lack of available details or recent publication activity, we reside this to speculation only. Of comparable interest, [23] presents a 'Moving Tunnel Profile Measurement' system (MTPM-1) which deploys a novel rotating camera for CRP that tracks a translating laser target to achieve swift 3D capture of a 100 m tunnel in 3 min. Use of a more lightweight camera is necessary for smoother rotation and reduction of prevalent lens-distortion.

Overall, visual methods provide extensive surface inspection prospects but are impractical for subsurface inspection since tunnel intrados' are opaque, except where defects have already exposed the subsurface. For a summary of defect types see Section 6.1.2 and consult 'Ring Separation and Debonding'. We believe proposed revisions of MTPM-1 show promise and would further benefit from fusion with automatous RTU locomotion described in [24] to facilitate 24/7 remote deployment.

3.2. Acoustic Methods

Subsurface features modify the characteristics of propagating soundwaves. Acoustic methods pulse predefined waveforms into the tunnel intrados and analyse resultant distortion and delay to identify audible indicators of defects. Acoustic methods can be subdivided into Ultrasonic Testing (UST) and Infrasonic Testing (IST). In UST, reductions in travelling pulse velocity correspond to elastic deformation of defected regions [25,26];

contrastingly for IST, defects are indicated by high resonant frequency components in returning pulses [27].

We only encountered two research groups directly applying UST to tunnel subsurface inspection. In [28], UST is extremely time-inefficient, requiring 9–25 min to scan 1m of tunnel wall and necessitating use of a preliminary GPR scan (Section 3.6) to localise suspected features. Likewise, despite robotic automation, UST scans performed by tunnel profiler ROBOSPECT achieve comparably inefficient durations of one hour to scan 6m [29] and are optimised for surface level crack and spall detection only [30,31]. Evidently, UST can be considered ill-suited for RTSSI, where surveys must be swift to minimise periods of tunnel closure.

IST proves more useful for RTSSI. Traditionally, hammer-strike emissions are performed by experienced human operatives who detect audible defect indicators 'by ear' alone, but those remaining are few, approaching retirement and are not being replaced. Faster robotic schemes are now preferential, boasting improved high level access achieved by mounting hammers to robotic arms [32–35] on Variable Guide Frames [36] and UAVs [37,38]. We notably uncovered a unique non-contact infrasonic UAV system [39] successfully inducing hammer-strike reminiscent flexural vibrations in infrastructure at distances of up to 5 m, for which application to remote-RTSSI presents an interesting research venture.

Detrimentally, inherent reliance on human interpretation of audio-spectra (which do not physically resemble subsurface features they convey) critically limits the insight non-specialist end-users can draw from IST without costly training or additional contextual metadata (e.g., maps of striking locations).

3.3. Laser Methods

Terrestrial Laser Scanning (TLS), also termed LiDAR (Light Detection And Ranging), utilises directed lasers to scan the visible tunnel intrados, generating dense 3D point clouds (Figure 2a) at up to 1×10^6 datapoints per second [40–43]. Visible light impulses reflect with variable intensity informing relative distances. However, datapoints lack classification labels and do not penetrate the subsurface. This makes segmentation of tunnel features challenging [44], but does permit direct insight into subsurface condition (e.g., profile distortions indicate abnormal strains) [45]. Pursuit of TLS integration with counterpart penetrating NDI methods marks an emerging avenue of long-term research. We note that the development of a standardised, efficient and reliable method to perform the essential alignment of multiple point cloud datasets—to form a unified digital environments—will be a key milestone for innovators to achieve before practical deployment becomes mainstream RTSSI practice.

Returning to standalone laser methods, we found TLS-RTI studies and commercial contractors most commonly deploy FARO® FOCUS scanning modules [46–49] (Figure 2b) or the Z+F Profiler® 9012 [50–52] to assimilate RGB optical photography for improved end-user navigational ease in recovered point clouds. Noteworthy innovations include an automated deformation detection assembly [11], which utilises a novel Circular Laser Scanning System (CLSS), highlighting the practicality of adopting circular sensing arrays that complement natural tunnel curvature.

Notable innovation is showcased in the Tunnel Monitoring and Measurement System (TMMS) developed by [52]. The prototype visualisation framework utilises a Z+F Profiler® 9012 mounted on a bespoke rail trolley (Figure 2c) to pass RGB LiDAR tunnel point clouds and a 'roaming video' feed of intrados condition to an engineer's tablet PC. The developed hardware bares strong similarities with a similar mobile TLS apparatus used in [49] employing a FARO X330 scanner. Validation trials in China's Zhengzhou Metro network demonstrate practical deployment capability but also relay that primary functions of ingress and cross-sectional deformation detection suffer noteworthy accuracy and stability reduction when applied to non-circular tunnel profiles (e.g., horseshoe, elliptical, etc.). TMMS therefore flags the importance adaptability in the design of new RTSSI solutions for wide-scale deployment, particularly on older rail networks (e.g., UK) which adopt multiple 'standard' tunnel cross-section variants.

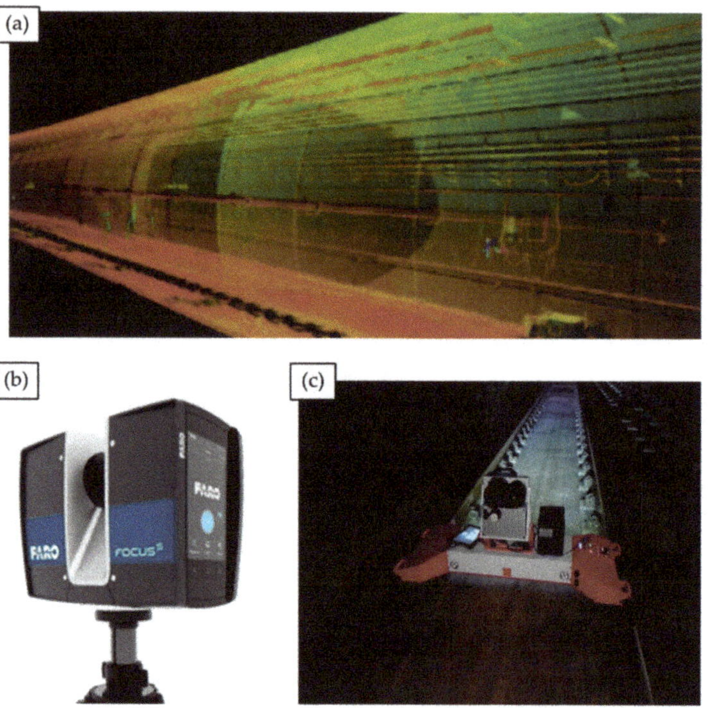

Figure 2. TLS for tunnel inspection. (**a**) 3D point cloud returned from a TLS metro tunnel survey [52]; (**b**) A FARO® FOCUS 350 scanning module, commonly deployed for infrastructure surveys; (**c**) TMMS rail-trolley transports a Z+F9012 to capture a TLS point cloud of a Zhengzhou metro tunnel.

3.4. Thermographic Methods

Subsurface faults modify thermal emission patterns of nearby interior tunnel surfaces, causing abnormal variations. Visualising temperature distribution profiles (Thermometry) facilitates localisation of suspected near-surface features (Thermography) [53,54], but recovery of specific attributes defers to higher quality UST or localised GPR imaging. Active Thermography (ACT) heats surfaces using halogen lamps [55], air guns [56] or inductive-heating elements [57] to induce exaggerated thermal responses. Abandoned testing by [58] and remarks of [59] affirm that heating element operation for RTSSI would incur impractical cost and could debond masonry, explaining its literary absence. We find use of infrared camera arrays for passive Infrared Thermography (IRT) more commonplace, owing to swifter and less costly implementation. Leading systems identify both air and water filled voids, with individual scans displayable as 2D panoramic imagery [53] or pioneering 3D mesh overlays on digital structural models rendered using TOSCA-FI [60] (Figure 3) or Augmented Reality [61] (Section 6.2). Despite recent work, Thermography still exhibits persistent limitations [54] undermining direct application to RTSSI:

- Results are highly sensitive to ambient temperature conditions which diminishes anomaly contrast (e.g., daily and seasonal variation);
- Thermal insulation and heat-resistant coatings used for tunnel temperate regulation and fire resilience can skew results. High thermal dissipation can easily restrict penetration d < 30 mm [62];

- Subsurface water content variation (e.g., increased permeation following rainfall or snow) can mask or exaggerate thermal profiles of faults;
- Enclosed, curved tunnel geometry restricts available viewing angles and confine results to 2D, even in a 3D mesh overlay, making inference of feature depth and physical form very challenging even for experienced operatives.

Figure 3. TOSCA-FI Software Platform: 2D heatmap overlays on a 3D digital bridge model [60].

3.5. Gravity Methods

Gravity Surveys (GS) use portable gravimeters [63], placed at regularly spaced sampling locations, to measure subtle variation in gravity surrounding railway tunnels [64]. Anomalies observed in returned Complete Bouguer Anomaly (CBA) curves inform subsurface material composition [65] and indirectly, structural health assessment. Regional trends in subsurface density conveyed by CBA curves can vary across scales comparable to the tunnel itself, granting extensive inspection coverage. Likewise, localised negative field displacements can indicate the presence of irregular low density regions, strong indicators of voids and deformation zones [66–68]. However, few common defects exhibit substantially large density variations (compared to their surrounding landmass) that would noticeably influence a CBA curve, which despite informing the general nature of the subsurface, does not comprehensively nor clearly visualise subsurface features themselves. Moreover, localising large features relative to the tunnel (i.e., in front, behind, left, right) is further complicated by the structure's cylindrical profile. This makes modelling the corresponding gravity field a multi-solution problem, introducing significant uncertainty and greatly increasing involved computation efforts [69].

3.6. Radar Methods

Ground Penetrating Radar (GPR) directs radio pulse emissions at the tunnel intrados, which penetrate and partially backscatter off strong dielectric gradients in the subsurface associated with features of interest [10,70–73]. Pulsed Radar (PR) samples consecutively emit wideband waveforms to measure backscatter in the time domain. Step-Frequency Continuous Wave (SFCW) radar incrementally sweeps an emission sinusoid through a predefined frequency band; the Fourier Spectrum of the returning signal is directly ascertained in the frequency domain by frequency-wise inspection of return signal strength [74]. In RTI, systems fall under three categories:

- **Trolley-Mounted** [75–77] (Figure 4a)—Units commonly feature interchangeable air-coupled antenna of differing frequencies to facilitate trade-off between penetration depth and output image resolution [28]. However, motorisation is infrequent, scans are unidirectional (typically railbed only) and offer no protection to operatives;
- **Handheld** [78–81]—Compact ground-coupled scanners guided by hand can achieve real-time scanning of curved tunnel sidewalls and crown. Typically restricted by limited penetrative depth (d < 50 cm), coverage speed (under m^2 h^{-1}) and gantry requirement to reach high surfaces make units impractical for full RTI;
- **Vehicle-Mounted** [82–86] (Figure 4b)—Multidirectional fixed antenna units attached to locomotives, rolling stock or RRVs. Although capable of data capture at speeds ranging from to 50–30 km h^{-1}, fixed directionality guarantees blind spot and air-coupling reduces achievable penetrative depth.

Figure 4. A selection of leading GPR systems. (**a**) Ballast fouling inspection trolley with selectable-frequency GPR antenna module [77]. (**b**) Loco-mounted fixed directional air-coupled GPR antennas.

Radargrams (B-scans) traditionally convey survey output, which although encoding multiple feature characteristics (e.g., depth, extent, orientation, heterogeneity, etc.) [70,73], demand extensive digital filtering [87–91] and may still contain abundant false-artifacts (e.g., ringing effects from rails, airwaves from overhead surfaces) [92–94], making them notoriously unintuitive. Increasing clarity via orthogonal intersection [95,96] (Figure 5a) and parallel stacking [97,98] (Figure 5b) of B-scans to form C-scans is now well-established practice in many commercially available GPR processing software packages [99–103]. Collectively, we denote this pseudo-3DGPR. We stipulate 'pseudo' to emphasise the inherent information loss resulting from 2D projections of 3D tomography. Beneficially, C-scan datasets can exploit time-slicing [104–106] (Figure 6a,b), in situ transparency filtering [107] and false-colouration [108] to improve conveyance of 3D forms. More recent investigations into true-3D volumetric reconstruction [98,109–117] (Figure 6c) show promise for advances towards practically viable fully immersive GPR-based subsurface inspection surveys (undertaken in fully digitised virtual survey environments) [80,118–122]. Whilst conceivable and under trial, achieving mainstream commercial deployment will require blind spot alleviation through adoption of rotary scan motion complementary to tunnel curvature, possibly similar to the superposition of concentric cylindrical 'look-ahead' radargrams pioneered by the TULIPS system [123] for tunnel excavation monitoring. Pursuit of blind spot elevation is therefore of critical importance if comprehensive RTSSI profiles are to be captured.

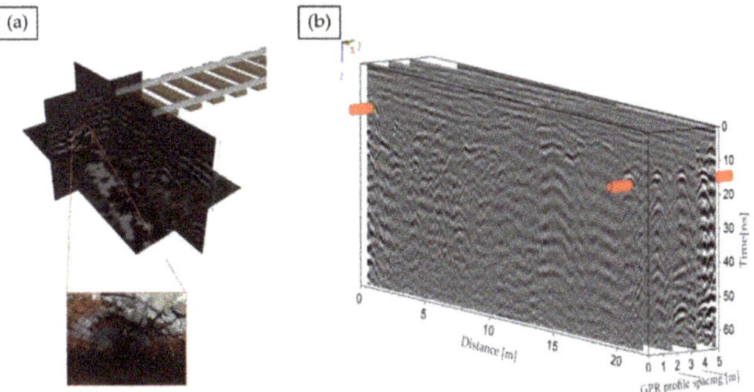

Figure 5. Principal methodology for combining planar radargrams to form pseudo-3DGPR visuals. (**a**) Orthogonal Intersection: B-scans meet at 90° helps focus attention on the central region. (**b**) Parallel Stacking: Aligning B-scans as slices of a cuboidal volume helps identify reoccurring targets [97].

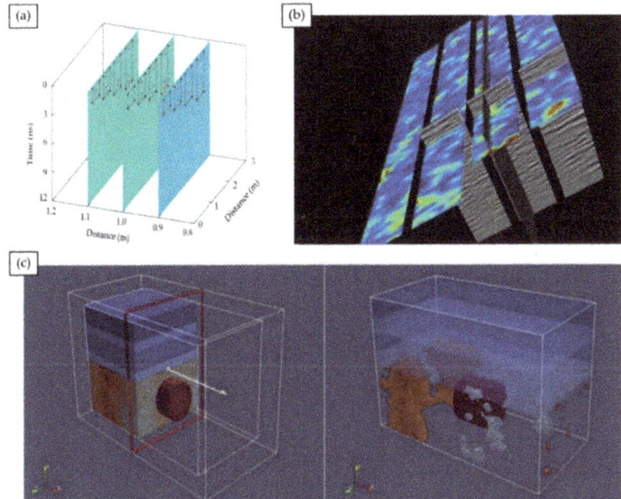

Figure 6. Mechanisms for improved contextualisation of 3D GPR datasets. Time-slicing performed (**a**) horizontally [100] or (**b**) vertically [102] can improve perception of relative depths and 3D feature shape. (**c**) Volumetric reconstruction increases feature faithfulness to reality, reducing human memory dependency, but processing remains highly involved [103].

3.7. Robotic Methods

Robotic systems reduce necessary human involvement in RTI, thereby beneficially reducing human error during data acquisition (e.g., mis-recordings due to subjectivity or lapses in concentration). Scope for varying degrees of autonomy further reduces dependency on onsite human presence, thereby increasing crew safety and cutting overhead costs. However, we must remember robotic methods share the limitations of their constituent sensors and also exhibit their own unique set of challenges (e.g., collision avoidance, recovery, stabilisation, power management, miniaturisation).

3.7.1. Unmanned Aerial Vehicles

Unmanned Aerial Vehicles (UAVs) have become increasingly popular for tunnel inspection owing to their low cost designs, programmability and exceptional manoeuvrability, which has motivated in excess of $4 billion global investment in UAV technology development for infrastructure inspection [124]. However, practical performance of current UAVs remains limited by poor onboard charge retention [125]; stabilisation challenges from near-wall turbulence and common dependency on GPS. Note that being subterranean, Global Positioning Systems (GPS) typically struggle to operate reliably in tunnels. [126,127]. Although, we did find considerable recent research applying collision-aversion protocol [128,129] and 'smart pathfinding' [130–132] (e.g., PLUTO [133]) to develop autonomous UAVs [134–137]. However, backup pilots remain necessary which add costs and safety-risks [137]. Furthermore, no commercially available autonomous UAV has yet to be developed specifically for RTSSI, despite similar systems existing for hydroelectric penstock surface level inspection [138].

Being airborne, UAVs could quickly transport RTSSI sensors where articulated booms cannot reach, for instance UAV-SWIRL hovers inside vertical ventilation shafts [125,139]. However, most systems still favour Optical Photometry and LiDAR sensing [140–142], permitting only implicit subsurface measurements. Of novel importance, we discuss several significant exceptions developed since 2015. These include development of new UAV-mounted GPR prototypes [143–145]; we found one commercial system [146] capable of 10 m penetration, however it is unclear if this incorporates UAV altitude.

In addition, hybrid locomotion UAVs now encompass:

- **Fixed Anchor-Point Docking** [147–150]—Sustains surface contact for IST and UST but requires highly involved pre-installation of anchors;
- **Pivoting RTUs** [151] (Figure 7)—Tracks enable uninterrupted contact and continuous one-way surface coverage but increase weight and power drain, limiting survey completeness;
- **Negative Pressure Wall-Climbers** [152,153]—Faster than track-based RTUs but require large flat contact surfaces, hence curved tunnel geometry risks UAV slip and hazardous control loss;
- **Fully Actuated Configurations** [154,155]—Provides best all-round solution, providing unrestricted multidirectional movement even on curved surfaces, but is liable to near-surface turbulence.

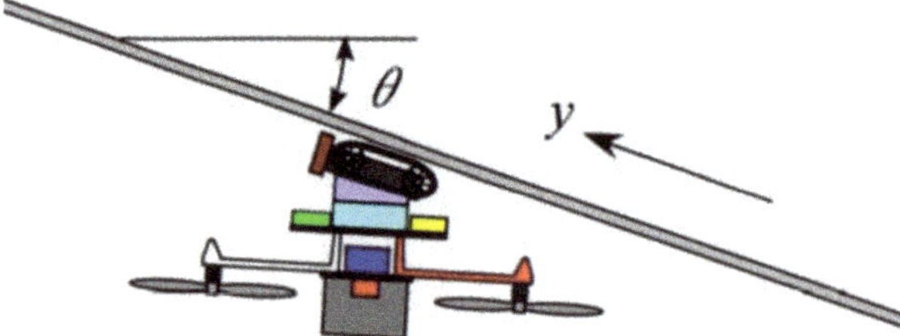

Figure 7. A recent novel innovation in hybrid locomotion UAVs. The pivoting traction crawler UAV performs IST on angled infrastructure surfaces inaccessible to engineers without gantries [151].

3.7.2. Adaptive Robots

We consider adaptive tunnel inspection robots to be devices capable of automatic geometry, operation or locomotion mechanism modification that combats demanding environmental conditions. As developments found were primarily proof-of-concept prototypes, current systems lack direct applicability to RTSSI without significant refinement efforts.

Nonetheless, we shall consider how such systems could be of practical future benefit in RTSSI. Foremost, adaption permits infiltration of inaccessible survey areas (e.g., drainage pipe interiors, capped shafts), increasing survey coverage. Moreover, units can swiftly traverse complex terrain (e.g., steps, rail tracks, damaged surfaces, angled walls) without human interaction, inviting remote inspection innovation potential.

Reconfigurable UAVs [156,157] fold (Figure 8a) to pass though narrow channels before unfolding (Figure 8b) to survey unknown void-like environments, which could be applied to preliminary surveys of hidden shafts via small diameter drill holes in capping facades. However, with more moving parts, damage likelihood during transit or execution is increased, potentially trapping systems behind walls incurring excess repair, replacement or recovery costs. Self-disassembly [158] could provide an easier route towards recovery.

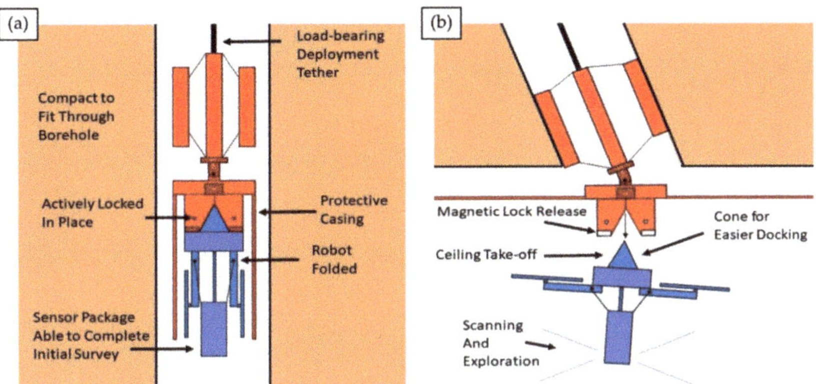

Figure 8. Folding UAV PROMETHEUS for subterranean inspection [156] can enter boreholes (**a**) to explore inaccessible voids (**b**) of potential benefit for probing hidden shafts in RTSSI.

Burrowing inspection devices [159] could create subsurface channels, then deploy 'snake robots' [160,161] fitted with endoscopes or fibrous sensing elements to directly image subsurface condition, detect ground movement [162] or moisture content [163]. However, burrowing is destructive and could exacerbate damage to already defective quadrants of intrados. Alternate use of modular configurations [164,165], compact step-climbers [111,166,167] or deformable 'soft robots' [3,168,169] fitted with NDI sensors could traverse small but pre-existing subsurface channels (e.g., pipes, vents, data cables) avoiding destructive burrowing. Soft robots uniquely could contort to bypass obstructions for multidirectional inspection of clogged drainage pipes. However, extensive development remains necessary to form a coherent self-arrangement of modular robots [170–174] capable of emulating established NDI techniques.

3.8. BIM-Integration

Building Information Modelling (BIM)represents a new paradigm for large structure lifecycle information management [175,176]. Current survey outputs represent one-way information exchanges between the physical tunnel environment and reconstructed digital models. By contrast, Digital Twin Tunnel (DTT) BIMs would facilitate two-way information exchange from any point in time during its perpetual update cycle. In two-way exchange, state changes in physical tunnel prompt reactive changes in the digital tunnel informing future maintenance, which cause further state changes in the physical tunnel and so on and so forth [177].

We find multiple recent experimental rail tunnel-BIM studies exist [178–182], typically deploying laser methods to profile and categorise trackside assets (Figure 9) but only [183] directly approaches RTSSI, developing a prototype AI-assisted BIM for ingress detection (developed on the Amber Inspection Could). Problematically, none currently exhibit adequate automation to be considered idealised DTTs. By inference, visualisation quality and overheads would clearly benefit from the significantly increased data pools and optimised network architectures anticipated [184]. However, challenges remain. Existing BIM architectures frequently lack specialisation to account for unique RTSSI challenges, such as complex terrain deformations and changeable subsurface geological conditions [182,185]. Reoccurring incompleteness of feeder data from NDI methods further limit BIM efficacy for RTSSI, despite recent improvements in multi-label datasets recovery [186–189].

Evidently, BIM integration for RTSSI will be essential for developing the first self-sustaining DTT [188], but insufficient without complimentary improvements to survey completeness.

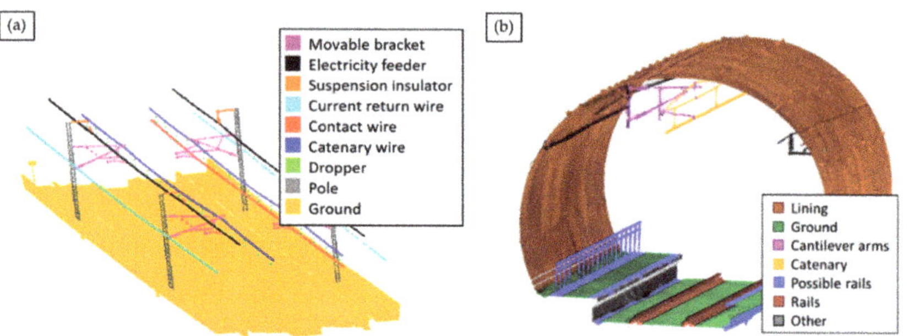

Figure 9. BIM for RTI. 'As-is' BIMs [181] frequently adopt LiDAR point cloud segmentation (via RANSAC and Supported Vector Machine methods) to label trackside assets both (**a**) outside and (**b**) within the tunnel environment [41].

3.9. Other Methods

Aforementioned NDI methods are most commonly deployed in routine RTSSI surveys based on encountered literature, motivating distinction from (i) more antiquated methods (e.g., invasive, inefficient, overly localised), (ii) less established experimental practices and (iii) schemes for real-time subsurface monitoring. In (i), we group: Borehole/Drill Core Sampling [190,191], Electrical Resistivity Tomography (ERT) [192–195], Endoscopic Probing [196,197] and Schmidt Hammer Strength Testing [53,198]. In group (ii), we gather: Radiography/Muon Tomography [199,200] and multiple additional prototype robotic RTSSI systems [14,166,201–204]. Group (iii) accounts for Time Domain Reflectometry schemes [205,206] and other Embedded Sensors [207].

4. Heuristic Comparisons of RTI Methods

Having discussed key attributes of leading RTI methods for RTSSI in isolation, we now direct the reader to our more comprehensive summary provided in Appendix A. It is tempting to directly compare advantages and disadvantages, 'ranking' methods to find an 'optimal' choice—a process we'll term 'heuristic comparison'. Noting that railway networks must balance inspection funding, duration and result quality, whilst researchers similarly prefer to invest effort in developments that return greatest impact, both for practical surveys and advances to the research field. This optimisation problem initially appears well-posed but this is not the case.

We find heuristic comparisons lack natural scaling. We may regard 'scaling' as a fixed-reference, quantifiable metric for comparing the importance of two characteristics. We draw parallels with use of numbered scales on questionnaires gauging attitudes (e.g., perceived risk between different dangers) [208]), therefore frequently suffer from:

1. Ambiguity maintaining a consistent comparative ground throughout;
2. Contextual variation between significance of comparative grounds.

Consistency ambiguity typically arises during first evolution of an argument (e.g., spoken discussion in planning meetings):

> *"Let's compare the accuracy of subsurface 3D visuals produced by LiDAR and pseudo-3DGPR. The latter are clearly more accurate because LiDAR can only indirectly visualise the subsurface (cross-sectional deformation). But the former is more accurate because deformation appears as point cloud deformation, whereas physical features are not actually hyperbolae-shaped as they're shown in pseudo-3DGPR".*

Note that both comparisons are valid and concern accuracy, but lack a definitive conclusion. The comparison ground for 'accuracy' subtly shifts from data-type to data-faithfulness. We argue the origin is vague definition of what constitutes an 'accurate' visual in the comparison posed. By contrast, contextual variation is more obvious:

> *"The spatial resolution of gravitational surveys would be inferior to Thermography for detecting small voids in an operating tunnel, but superior for strata mapping during construction".*

Evidently, a more robust rationale is required.

5. Criteria for an 'Effective' RTSSI Visualisation Framework

So far, we have found heuristic arguments unsatisfactory for appraisal of RTI methods in a RTSSI context. Significant disparity exists between respective operating principles, deployment methodologies and output conveyance; not least in the inherent multifaceted and context-dependant grounds for suitability and performance comparison. Two specific examples would include: (i) the nature/variety of detectable features and (ii) assessment timescale.

The basis for our criteria is twofold.

First, we recognise each RTI method discussed exhibits at least one critical limitation (Table 1). Logically, an effective RTSSI visualisation framework would not share any, meaning an innovation directly addressing either would have significant impact on the current RTI hardware market. This motivates our novel formulation of well-defined but sufficiently general criteria for an 'effective' RTSSI visualisation framework. We illustrate the benefits visually using a conceptual network diagram (Figure 10).

Table 1. Main current issues facing NDI methods.

<table>
<tr><th>Method</th><th colspan="2">Critical Limitations for RTSSI</th></tr>
<tr><td>Visual
Laser</td><td colspan="2">Lack necessary penetrative capability to directly visualise subsurface.</td></tr>
<tr><td>Therm.
Acoustic
Radar</td><td>Information loss (2D projections of 3D features) limits survey faithfulness.</td><td>Skew from ambient temperature variations.
Inefficient implementation.
Curvature induces blind spots.
Visuals lack interpretive clarity.</td></tr>
<tr><td>Gravity</td><td>Struggles to resolve localised features.</td><td rowspan="2">Onsite supervision still required.
Majority of systems are still concepts or early prototypes.
Architectures frequently lack
sufficient optimisation to react to RTSSI data dynamically.</td></tr>
<tr><td>Robotic
BIM-Int.</td><td>Systems share the limits of their ancillary sensors.</td></tr>
</table>

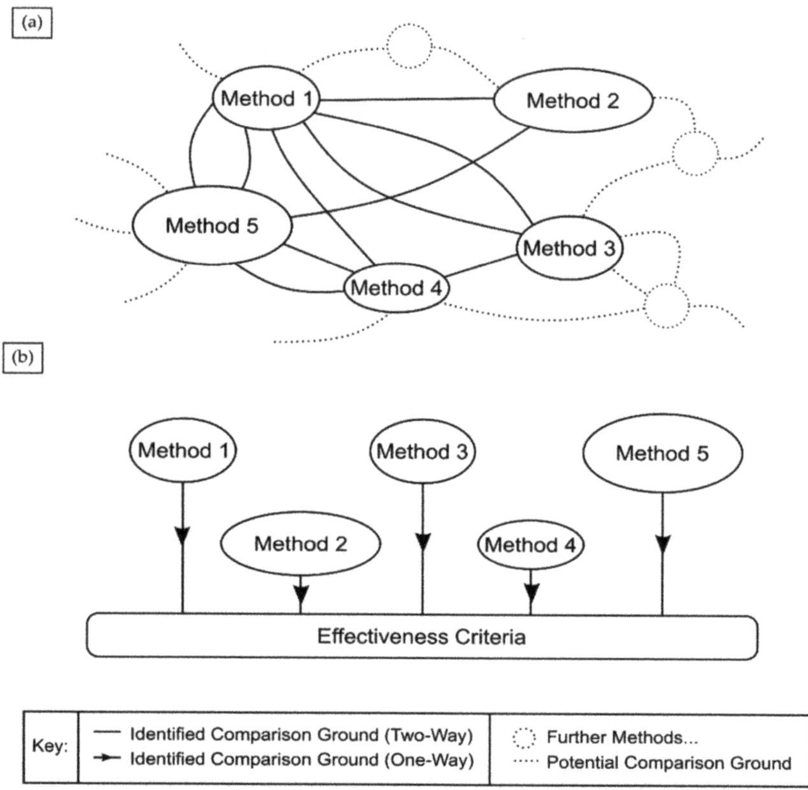

Figure 10. Using criteria greatly simplifies comparison networks. (**a**) Heuristic comparisons are irregular and multi-directional networks. (**b**) Criteria inclusion collapses the network to be regularised and unidirectional.

Second, our creation of a Category Connection Matrix (CCM) from encountered literature (Table 2)—inspired by the hybrid workflow matrix presented in [209])—highlights emerging research trends, which we use to infer future research avenues in RTSSI. The categories extracted inform where research attention is currently most directed. Furthermore, collectively, motivation for all relevant studies is to contribute to producing the most effective RTSSI visualisation framework possible. Thus, each identified research category must align with at least one criterion.

From this, we formulate our proposed criteria for an 'effective' RTSSI visualisation framework, which considers:

Data Acquisition:

1. **Completeness**—Uninterrupted scan coverage should be achieved to record the full extent of the influential regions of the tunnel subsurface. Current methods either lack deep penetrative capability or exhibit bind spots due high localisation of scans or geometry curvature.
2. **Duration**—Survey execution should balance acquisition speeds with recovered data quality (i.e., resolution, distortions) ensuring inspections and repairs cause minimal network disruption. A rapid low-quality scan limits inspection closure, but misinformed repairs take longer to fix and vice versa.

Information Conveyance:

1. **Accessibility/Interpretive Clarity**—A railway network end-user who is not a specialist in the utilised RTSSI technique(s) (e.g., planner) should independently be able to understand and make informed decisions based on visualisation output (consider radargrams the antithesis to this, containing considerable but mostly incomprehensible information).
2. **Faithfulness**—Inconsistency between the physical subsurface geometry undergoing inspection and corresponding representation within the visualisation medium should be kept to a minimum. An example of unfaithful conveyance is how overhead structures can confusingly appear as below-ground features (airwaves) in radargrams [93,210].
3. **Interactivity**—Visualisations should react intuitively to end-user engagement in ways that make surveys more ergonomic, efficient and versatile. Again consider radargrams; time-slicing in pseudo-3DGPR conveys depth more ergonomically than viewing am isolated B-scan.

Understandably, developing fully effective frameworks will take time, but we can make informed predictions. In consulting phase (1) literature; current GPR technology offers greatest versatility for RTSSI. Hence, we anticipate earliest industrial impact will likely stem from its unification with pre-existing conveyance innovations such as interpolative [111,116] and AI-assisted 3D feature recovery (e.g., DepthNet [114]).

Table 2. Category Connection Matrix (CCM).

Categories	Identified Research Avenues							
Method	A	B	C	D	E	F	G	H
Visual	•	◐	◐	◐	◐	◐	◐	•
Acoustic	○	○	○	•	◐	○	○	○
Laser	◐	◐	◐	•	○	◐	◐	◐
Therm.	○	○	◐	•	◐	○	○	○
Gravity	○	○	◐	○	○	○	○	○
Radar	○	◐	◐	○	◐	◐	○	◐
Robotic	•	◐	○	•	○	○	○	○
BIM-Int.	○	○	○	○	◐	○	◐	◐

Avenue Codes: A: Autonomous Tunnel Surveys; B: Alternatives to Fixed-Direction Sensor Arrays; C: Surface-Subsurface Tunnel Survey Fusion; D: Automated Tunnel Feature Detection; E: Tunnel Subsurface Feature Severity Ranking; F: Volumetric Tunnel Feature Reconstruction; G: BIM/DDT Development; H: XR/RTI Integration. Icons: (○): Indicates a literature gap due to critical method limitations or a currently unexplored research avenue. (◐): Indicates works connected to an RTSSI research avenue exist but are indirectly related, signifying opportunity for new research via novel idea synthesis either amalgamated from or inspired by present literature. (•): Indicates works connected to a RTSSI research avenue exist and are directly related, signifying relevant practical research is proposed, presently underway or considered surplus to requirement.

6. Steps towards Criteria Fulfilment

Relevant research is already underway targeted at fulfilling our outlined criteria for an effective RTSSI visualisation framework.

Regarding the completeness criterion, we draw attention to the significant recent development of rotating, air-launched GPR antenna by Railview Ltd. (UK) [211] (Figure 11). Compared to fixed array GPR systems, including the Zetica Advanced Rail Radar (ZARR) [83,212] and the IDS SafeRailSystem (SRS) [84], the helical scanning trajectory of rotating antenna more closely mirrors naturally curved tunnel geometry, facilitating more comprehensive 360-degree RTSSI imaging, at competitive depths.

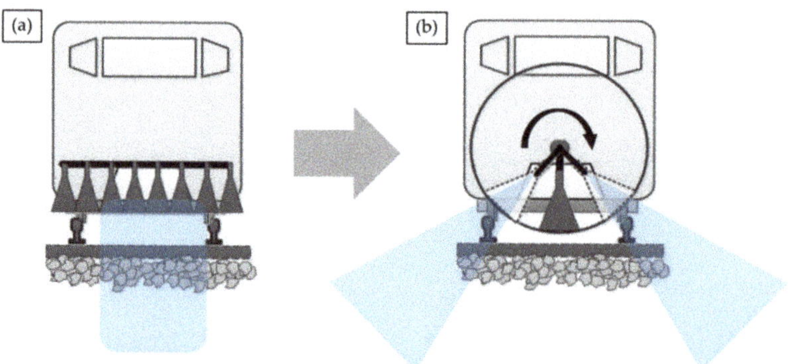

Figure 11. Unlike conventional fixed direction antenna (**a**) which typically only image the railbed; rotating GPR antenna capture 360-degree subsurface profiles including tunnel sidewalls, haunch and crown (**b**).

Lastly, we consider the conveyance criteria. Helical scans directly capture 3D RTSSI geometry in situ as 360GPR datasets, more akin to laser-methods than pseudo-3DGPR (requiring B-scan stacking or intersection). Ergo, forthcoming analysis of 360GPR presents an interesting opportunity for new research into volumetric feature reconstruction for RTSSI.

Thus, 360GPR has scope to form a visualisation framework meeting at least three of our five effectiveness criteria, feasibly disrupting the current RTSSI hardware market. At this point, owing to the larger proportion of relevant literature focused on addressing our conveyance criteria, we discuss the main innovations this review encountered into feature identification within subsurface datasets (Section 6.1) and dynamic interaction between visualisations and end-users (Section 6.2).

6.1. Automated Feature Detection and Evaluation

If an end-user cannot clearly interpret subsurface data, the inspection yields little useful insight into tunnel structural health or targeted maintenance. Detecting degradation indicators is critical for localising damage, whilst characteristic evaluation (e.g., location, extent, maximal depth, etc.) informs repair urgency. Searching RTSSI visuals manually is impractical: tunnel datasets are cumbersome; defect types are wide-ranging through a tunnel's operational lifespan; human cognition speeds are slow and our evaluation is subjective. Unintuitive data visualisations only compound the issue (e.g., radargrams), explaining why research tackling Automated Feature Detection/Evaluation (AFD/E) accounts for over 1/3 of CCM-featured research connections and encompasses Convolutional Neural Network (CNN) [213,214] and Deep Learning (DL) [49,215,216] detectors, alongside severity ranking schemes [217,218]. Scope of feature variety and complexity current AFD can simultaneously identify with accuracy drew our attention. Restricting our consideration to subsurface studies, a DL image grid workflow flags four distinct features [96] (manhole, cavity, pipe, heterogeneous soil background), yielding widest feature detection scope, albeit not concurrently. Studies successfully achieving simultaneous detection of realistically complex defect configurations [219,220] likewise favoured 2D GPR imagery but discriminated two types maximum [81]. However, with exception to [96], test environments featured only assets or defects. This implies GPR-based research is currently leading developments in AFD and seemingly the upper limits of DL feature detector capability have yet to be fully explored. Thus, any study classifying over four mixed type (assets, defects), variety (shaft, void, pipes) or complexity of features would mark a significant advance in RTSSI-AFD.

AFE, namely defect severity ranking, proves less researched. Contrary to our initial expectations of exclusively dictionary-based schemes, of studies found, most now adopt contextual evaluation via fuzzy logic devices [221,222] and probabilistic analysis [33,223]. We infer more complex evaluation grounds are being considered in parallel when grading repair urgency, if not yet for RTSSI. For example, many nearby cracks in close proximity can be of greater concern than one occurring in isolation. Collectively, this suggests development of a robust contextual severity ranking scheme for RTSSI would be of worthwhile pursuit in future research.

ADF/E is clearly transitioning from proof-of-concept simulations to practical deployment tests. In RTSSI, schemes will need to discriminate tunnel assets from more hazardous defects, therefore training demands programmer knowledge of common features. For already aged masonry tunnels, we found no consolidated summary, concerning given a forecast 30–50% increase in rail-traffic demand by 2050 [224]. We therefore now present our own bespoke consolidated summary.

6.1.1. Common Assets in Masonry Railway Tunnels

An 'asset' denotes any useful or valuable item associated with railway network operation, encompassing employees, track, signalling, buildings, utilities and civils (structures, earthworks) [225]. Tunnel assets fall under civils, with any unprotected structurally significant entity designated a critical element.

Hidden Critical Element (HCE) is unobservable from at least one side [226]. Locating HCEs has presented considerable challenges for Network Rail (UK) in RTSSI. We identify that for current inspection methods, greatest challenge is presented by detection of blind (concealed but disenable) and hidden (concealed and indiscernible) shafts:

a. **Ventilation Shafts**—Hollow columns extending from tunnel crown to the surface. They facilitate air circulation and were originally used to remove material during construction [227].
b. **Maintenance Shafts**—To allow simultaneous excavation of multiple faces, many shafts would be sunk along proposed tunnel routes [228]. Typically infilled or converted to ventilation shafts, capping frequently conceals them for aesthetics (processes rarely recorded in writing). Being unreinforced, many have deformed or partially collapsed.

This is most true for Wales and Western Regions of the UK, following Network Rail's failure 'to deliver on a commitment to identify all hidden tunnel shafts by the end of 2016–2017' [229] and more recent delays tackling HCE examination schedules in 2019–2020 due to pandemic impacts [230].

As cavities are prime sources of water infiltration, failure to identify hidden shafts significantly increases risk of accelerated compromise to surrounding structure. Therefore, innovations towards hidden shaft detection present opportunity for highest new research impact in RTSSI.

Training detectors to discriminate hidden shafts from other features motivates continued summary of other common masonry railway tunnel assets:

c **Overhead Line Equipment (OLE)**—Furthermore, dubbed 'traction wires' or the 'catenary', these high voltage electrical pickup lines power electric locomotives via onboard pantograph connectors. Systems can be integrated into older tunnels during electrification works. Forms include tensioned metallic cables mounted to the crown and the Rigid Overhead Conductor Rail System (ROCS) [231], which provide more efficient operation in low-clearance tunnels. Structural weakening can result from necessary drilling during install, whilst strong electromagnetic fields generated by the power feed can interfere with data acquisition systems and present line of sight obstruction during haunch or crown inspections.

d **Portals**—Reinforced surfaces surrounding tunnel entrances combating outward deformation induced by shear stresses from continuous shifting of landmass encircling the tunnel [232]. Exposed to the elements, portal rigidity deteriorates, risking collapse if cracks and displacement are not detected early. Reinforcement schemes include buttresses, ground anchors, and steel mesh coverage fixed with soil nails [233].

e **Refuges**—Small arched recesses within the tunnel lining to protect railway workers from locomotives.

f **Buried Utilities**—These can include both metallic and plastic water drainage pipes [234], electrical wiring and telecom cables.

g **Trackside Objects**—These include signage, signals, electrical junction boxes and CCTV units.

h **Culverts**—Small passages allowing watercourses to pass under railway tracks, including underground rivers [235]. Old masonry culverts particularly can be weakened by solution and hydraulic action resulting in partial section collapse, deforming the railbed above and causing water backlog which floods tunnels.

6.1.2. Common Defects in Masonry Railway Tunnels

Defects constitute any imperfections in the material or form of a structure. Indicative of degradation and increased failure likelihood under strain. Swift detection in an enclosed tunnel environment is critical. We provide a brief overview of causes and hallmarks for common defects in masonry railway tunnels, then reflect on the efficacy of their detection in modern surveys, highlighting any opportunities for future research:

a. **Arch Barrel and Cross Sectional Deformations** (Figure 12a)—Shifting tunnel landmass induces changeable tensile and compressive forces within arch barrels. This can trigger sidewall bulging and buckling; haunch distortion; tunnel floor bowing or side-offset of the crown.

b. **Cracks and Fracturing** (Figure 12b)—Localised shear forces and vibrations from rolling stock can split and displace masonry. Damage ranges from hairline cracks lacking obvious signs of displacement, to large open fractures exhibiting significant displacement.

c. **Water Ingress** (Figure 12c)—Rain infiltrates tunnels through shafts and groundwater propagates through dissolved subsurface joints and fissures (karsts) [236]. Leaching of mortar begins as water percolates between masonry before flowing down the sidewall. Lubrication of masonry joints leads to movement of the structure, resulting in lining deformation. Owing to joint length, significant quantities of ingress can accumulate, scouring supports and flooding tunnels without proper drainage or saturated catchpits, as ingress often carries dissolved ochre (acquired during subsurface percolation) which forms crystalised limonite deposits [197] that block drains. Out-flow down sidewalls forms noticeable white streaks emanating from the region of breech, therefore can be used as indicators of ingress source. However, establishing subsurface mortar leaching extent is reliant on NDI methods for RTSSI.

d. **Open Joints and Perished Mortar** (Figure 12d)—Characterised by the deterioration and eventual absence of mortar between brickwork. As brick tunnels can date back over 150 years, mortar naturally begins to deteriorate from reactions with moisture in the bricks, air and subsurface [227,231,237]. This process is accelerated by wash from nearby ingress, vibrations induced by both rolling stock and air-pressure waves from passing locomotives.

e. **Loose and Missing Brickwork** (Figure 12e)—Early onset of spalling and failed patching repairs can result in loosened or missing brickwork, indicated by brickwork rubble on the railbed. Individually, gaps pose no substantial loss of structural rigidity, but can provide opportunity for larger defect growth if not addressed quickly. Furthermore, if present in tunnel haunch or crown, falling rubble can damage rolling stock, or cause serious injury to ground crews working below.

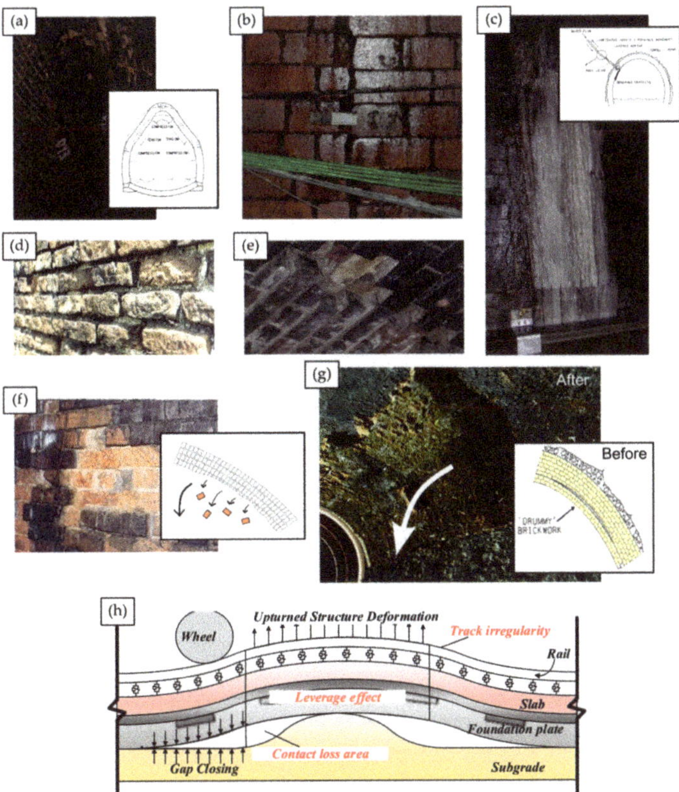

Figure 12. Common defects in masonry railway tunnels: (**a**) Haunch Deformation; (**b**) Cracks; (**c**) Water Ingress; (**d**) Perished Mortar; (**e**) Missing/Loose Brickwork; (**f**) Spalling; (**g**) Ring Separation and Debonded Wall-Section; (**h**) Frost Heave [238].

f **Spalling** (Figure 12f)—Perishing masonry on the tunnel intrados in vicinity of the haunch or crown can be dislodged by gravity, leaving the next layer of brickwork exposed. Repetition gradually creates rough-profiled recessed quadrants. Bricks in newly exposed layers lack tarnishing from exposure to soot and locomotive exhaust fumes, thus a second hallmark is more vibrant brickwork colouration.

g **Ring Separation and Debonding** (Figure 12g)—Arch barrels contain multiple layers of concentric brickwork rings. Literature encountered discussed brick tunnels ranging from 3–15 layers [237,239,240], evidencing significant possible contextual variation. Ingress, deterioration of intrados mortar and poor quality workmanship can all cause neighbouring rings to separate within the wall. Gravity pulls innermost layers downward, causing debonding from layers behind forming slit voids (subsurface hairline fractures). Over time, large voids begin to grow. If near-surface, separating rings may briefly cause visible cracking of intrados mortar, allowing detection. However, separations deeper than a single ring are completely invisible to an in-tunnel observer. With time, large sections of rings can debond, causing arcs of brickwork to fall away as slabs. Such 'delamination' events can deform rails or damage the railbed presenting a derailment risk. Depending on size, debonding can significantly weaken substantial volumes of surrounding masonry.

h **Railbed Faults** (Figure 12i)—Subgrade layers of ballast facilitate even distribution of rail-traffic weight, ensuring rails remains level on uneven ground to prevents derailments. Displacement induced by ballast fouling [75–77] and frost heave [238,241,242] can damage rails and offsets train weight distribution, increasing in-tunnel derailment risk and subsequent likelihood of major network disruptions.

i **Drainage Faults**—Tunnels contain integrated pipework and catchment pits to safely remove excess water and silt. Flooding may occur if pipes become blocked, rupture due to freezing and expansion, or become overwhelmed by intense weather events [243]. Improvement works can also fail. For example, 18 bolts supporting a water catchment tray in Balcombe Tunnel (2011) decoupled due to resin failure [244]. Sag reduced tunnel clearance from 0.87 m to 0.3 m posing a dangerous obstruction to rail-traffic.

Upon reflection, we first note recent studies focus primarily on (and achieve) detection of three main defects: (i) surface-visible cracks via CRP; (ii) voids via Acoustics, Thermography or GPR; (iii) water ingress via Thermography and GPR. By extension, with suitable modification to enable multi-directional scanning, similar hardware could reasonably detect (i) open joints and perished mortar, spalling, missing brickwork; (ii) ring separation and debonding and (iii) drainage faults in future applications.

Secondly, we observe that currently cross-sectional deformation can only be directly imaged by laser methods as they boast 3D point cloud data capture. We note that traditional visual inspection also detects deformation, but notetaking is inherently far less accurate than imaging. Although no encountered literature directly utilised laser methods to detect railbed faults, we may reasonably assume cross-sectional imaging would also provide usable insight with relative ease.

Finally, we report that no data acquisition method is single-handedly able to detect each identified common RTSSI defect. This agrees with our findings in Section 3. Discounting Thermography since it cannot directly measure target depth, we remark that GPR has capability to detect the widest range of defect types (five of nine). Note that direct arrival waveforms are liable to mask small surface level defects, hence are not included in this statistic.

To improve this ratio, we believe future research into multi-sensory data acquisition systems presents high potential impact. A unit amalgamating RGB-CRP (surface imaging), LiDAR (cross-sectional profiling) and 360GPR (subsurface imaging) could feasibly detect all defects listed.

6.2. Extended Reality for Dynamic Survey Interaction

We define Dynamic Survey Interaction (DSI) to encompass any information conveyance technique that intuitively responds to end-user triggers (e.g., gestures, camera proximity, movement speed, metadata, field-of-view) [245–247], thereby increasing clarity of information within a survey and ergonomics of use. DSI attributes fundamentally reduce application complexity, boosting data usage efficiency and its accessibility to non-specialists, which had made them common features in Extended Reality (XR) interfaces.

Here, for clarity, we restrict our consideration of XR to just two subsets: Augmented/ Mixed Reality (AR/MR) and Virtual Reality (VR). Note there is a distinction between AR and MR, AR is effectively passive information overlay, whereas MR is active, allowing the physical-environment to influence the digital environment and vice versa. In both cases survey data is conveyed through a head-mounted-display, which in AR/MR superimposes assistive digital information over the users' view-field, whereas in VR it provides full immersion in a digitally rendered environment [247,248]. As an emerging exploratory medium, academics and industry are now exploring practical applications of XR for DSI, including use for infrastructure inspections.

We encountered multiple noteworthy AR/MR visualisation developments outside phase (1). In [117], a rendering pipeline utilising back-projection of air-coupled 3D multistatic GPR data and Jerman Enhancement Filtering is presented but lacks an interface. Contrastingly, [249] demonstrates a prototype Unity3D-built AR pavement-subsurface visualiser for IOS based on 'Reality-Capture' modelling. However, performance of all subsurface AR inspection tools found appears unstudied in real railway tunnels. We anticipate large volumes of subsurface data necessary for practical surveys will forgo local storage on commonly utilised tablets, requiring wireless relay from remote data-hubs. Herein we speculate the frequent absence of underground wireless communication networks in older railway tunnels may be responsible for research stagnation.

By comparison, several synchronisation schemes have been developed for real-time VR-BIM data exchange (e.g., BVRS [250]) and trailed in real tunnel construction projects (e.g., The Shenzhen-Zhongshan Immersed Tunnel [251]). For operating tunnels, literature concerning disaster situation training [252,253] dominates, whilst we found inspection studies to be scarce.

All but one VR visualisation framework was encountered (across [218,254]) tailored for RTI. The Enhanced Photorealistic Immersive (EPI) Survey Platform is developed in UE4 from SfM. Processed CRP data feeds a novel interactive dashboard to provide an extensive range of DSI attributes.

Techniques include: (i) defect highlighting filter toggles; (ii) a mini-map of in-model user location and (iii) proximity-triggered defect information modules detailing TCMI grading. The TCMI (Tunnel Condition Marking Index) ranges from 0 to 100, where 100 denotes a defect free aspect of the tunnel. Sadly dependency on visual CPR data does not facilitate subsurface inspection, undermining direct application to RTSSI.

Applying recent innovations in XR for DSI to RTSSI presents a promising direction for future research, which could significantly increase the clarity and accessibility to non-RTSSI specialist roles in tunnel management (e.g., asset engineers, environmental managers, operations risk advisors, etc.) [255–257].

The next key milestone facing AR/MR-RTSSI system deployment will be development of a dedicated wireless subterranean communication network supporting real-time information exchange with rail network data-hubs. We believe use of IoT/WSN Wireless Mesh mine communication nodes [257–259] could present a feasible solution and interesting research opportunity. For VR systems, we believe the lack of comprehensive subsurface datasets discussed in Section 5 explains the absence of research into a dedicated RTSSI application, whilst the visualisation framework presented in [218] showcases current state of the art in DSI for tunnels. We theorise future successes in 360GPR visualisation would open a practical research avenue into VR-based RTSSI.

7. Discussion

Method-centred subdivision of state of the art literature, spanning both data acquisition approaches and conveyance schemes (circa 2015–2021) reveals considerable recent advances in the capabilities of RTSSI-linked visualisation frameworks. Our review addresses two key knowledge gaps and presents three promising considerations for future research.

Appendix A summarises our deconstruction of leading NDI techniques for RTSSI, establishing that a multitude of valid comparative grounds exist between methods. Their variable importance subject to survey context undermines heuristic direct comparisons of method performance, suitability or efficacy by balancing advantages against disadvantages, resulting in ambiguity and inconclusiveness. This justifies need for a robust definition of an 'effective' RTSSI visualisation framework to address the knowledge gap. Our formulation of explicit criteria from five key research gaps was based on common shortcomings identified between considered methods. Respective criterions consider the (i) completeness and (ii) duration of survey data acquisition; alongside the (iii) interpretive clarity, (iv) faithfulness and (v) interactivity of information conveyance.

We overall find that despite recent innovations, a fully effective RTSSI visualisation framework has yet to be developed.

Creation of our CCM facilitated inference of eight key avenues for future research. Grading connected developments by relevance drew distinction between emerging and established trends within the literature. Initial thoughts towards achieving complete, time-efficient RTSSI surveys highlights pioneering analysis of 360GPR datasets from a novel rotary, air-launched GPR antenna. Already meeting three criteria, we anticipate development of a suitable information conveyance scheme will present a prime research opportunity, with feasible scope for creating the first fully effective RTSSI visualisation framework within the next decade. This would establish 360GPR as a mainstream and potentially preferential technique amongst current RTSSI methods.

We note over 1/3 of research categories concerning information conveyance are aligned to automated target detection and defect severity ranking. However, their practical application on 3D datasets containing realistic quantities, varieties and complexities of subsurface features remains largely unexplored. We believe successful trails on authentic RTSSI datasets present an upcoming milestone for future research efforts to achieve if devised methods are to eventually aid real surveys.

Thoughts on detection scheme training for recognition of subsurface features in masonry tunnels flagged a further knowledge gap. Namely, this review was unable to find a combined nor comprehensive list of common masonry-related assets and defects within recent literature. Focus primarily centred on modern concrete tunnels. To address this, we presented our own bespoke consolidated summary for masonry tunnels.

Our exploration of emerging trends in dynamic interaction with current visualisation frameworks found studies addressing dataset interfacing with XR hardware featured 63% more prevalently than BIM/DTT system design. Most notable interaction potential was demonstrated by a VR tunnel surface survey platform presented in [218], which justified by end-user trails, quantifiably evidences the high levels of intuitiveness both 3D rendered environments and contextual dashboard modules can achieve. As with most encountered schemes, optimisation is for surface surveys only and no preview facility is provided during data acquisition. Arguably, this could also be beneficial; encouraging off site analysis of survey data in safer environments reduces crew risk.

Nonetheless, based on this review we believe VR presents the most versatile and intuitive tunnel survey interaction medium presently available, therefore would provide an ideal basis for future RTSSI visualisation framework developments. Application to a hybridisation of CRP, LiDAR and 360GPR datasets poses an interesting research opportunity and potential industrial solution for simultaneous surface and subsurface RTSSI surveys.

Supplementary Materials: The following supporting information can be downloaded at: https://www.mdpi.com/article/10.3390/app122211310/s1, for the literature summary tables referenced in Sections 3.1–3.8 and Appendix A, we direct the reader to consult the supplementary file <Literature_Summary_Tables.xlsx> uploaded alongside this manuscript.

Author Contributions: Conceptualization, T.M., M.R. and G.Y.T.; methodology, T.M.; software, N/A; validation, T.M., M.R. and G.Y.T.; formal analysis, T.M.; investigation, T.M.; resources, T.M.; data curation, T.M.; writing—original draft preparation, T.M.; writing—review and editing, T.M., M.R. and G.Y.T.; supervision, M.R. and G.Y.T.; project administration, T.M.; funding acquisition, T.M. All authors have read and agreed to the published version of the manuscript.

Funding: This work was supported by the UK Engineering and Physical Sciences Research Council (EPSRC) Grant EPT5179141 for Newcastle University.

Institutional Review Board Statement: Not applicable.

Informed Consent Statement: Not applicable.

Data Availability Statement: Not applicable.

Acknowledgments: The authors would like to thank Railview Ltd. (UK) for technical consultation on RTSSI hardware specification and provision of photographs presented in Figure 12.

Conflicts of Interest: The authors declare no conflict of interest. The funders had no role in the design of the study; in the collection, analyses, or interpretation of data; in the writing of the manuscript; or in the decision to publish the results.

Abbreviations

(ACT) Active Thermography; (AD) Adaptive; (AFD/E) Automated Feature Detection/Evaluation; (AR/MR) Augmented Reality/Mixed Reality; (BIM) Building Information Modelling; (CBA) Complete Bouguer Anomaly; (CCM) Category Connection Matrix; (CLSS) Circular Laser Scanning System; (CNN) Convolutional Neural Network; (CRP) Close-Range Photogrammetry; (DIFCAM) Digital Imaging for Condition Asset Monitoring System; (DL) Deep Learning; (DSI) Dynamic Survey Interaction; (DTT) Digital Twin Tunnel; (EPI) Enhanced Photorealistic Immersive; (ERT) Electrical Resistivity Tomography; (FA) Fully Autonomous; (GPR) Ground Penetrating Radar; (GPS) Global Positioning System; (GS) Gravity Surveys; (HCE) Hidden Critical Element; (IRT) Infrared Thermography; (IST) Infrasonic Testing; (LiDAR) Light Detection And Ranging; (MTPM) Moving Tunnel Profile Measurement; (NDI/E) Non-Destructive Inspection/Evaluation; (OLE) Overhead Line Equipment; (PR) Pulsed Radar; (ROCS) Rigid Overhead Conductor Rail System; (RRV) Road-Rail Vehicle; (RS) ROBO-SPECT; (RT) Rail-Trolley; (RTI) Railway Tunnel Inspection; (RTSSI) Railway Tunnel Subsurface Inspection; (RTU) Robotic Traction Unit; (SA) Semi-Autonomous; (SFCW) Step-Frequency Continuous Wave; (SfM) Structure From Motion; (SRS) SafeRailSystem; (TLS) Terrestrial Laser Scanning; (UAV) Unmanned Aerial Vehicle; (UST) Ultrasonic Testing; (VR) Virtual Reality; (XR) Extended Reality; (ZARR) Zetica Advanced Rail Radar.

Appendix A

For NDI methods discussed in Sections 3.1–3.8, we provide comprehensive summary tables of key literature analysed in this review, both for completeness and to evidence the multitude of valid comparative grounds on which heuristic comparisons may be based. The Literature Summary Tables are included as Supplementary Material.

Abbreviations adopted in the tables carry over from each section of the review. Below we define an additional visual scale to rank the interpretive clarity of systems presented, alongside a global legend of additional shorthand notation used exclusively in the Literature Summary Tables.

Table A1. Literature Summary Tables: Interpretive clarity scale.

Symbol	Clarity	Necessary Training
○○○	High	Training Not Required.
○○●	Mid-High	Some Require Light Training.
○●●	Low-Mid	Most Require Moderate Training.
●●●	Low	Extensive Training Essential.

Table A2. Literature Summary Tables: Shorthand notation.

Column Field	Type	Shorthand
Status	Concept	C
	Prototype	P
	Commercial System	CS
Motion	Static	ST
	Handheld	HH
	On-Rail	OR
	Airborne	AB
	Crawler-Unit	CU
	Robotic Arm	RA
	Adaptive Traction Unit	ATU
	Pneumatic Suction Feet	PSF
	Tunnel Boring Machine	TBM

Table A2. *Cont.*

Column Field	Type	Shorthand
Duration	Seconds	S
	Minutes	M
	Hours	H
	Days	D
	Weeks	W
Key Target Types	Cross Sectional Deformation	CSD
	Hot/Cold Spots	H/C
	Groundwater Flow	GF
	Buried Utilities	BU
	Ballast Fouling	BF
	Trackside Assets	TA
	Power Distribution	PD
	Voids/Debonding	V/D
	Ore-Deposits	OD
Additional Symbols	Not Applicable	N/A
	Information Unavailable	-
	Important Note	*

References

1. Jones, C. Transportation planning in an era of inequality and climate change. *Fordham Urban Law J.* **2017**, *44*, 1005–1009.
2. Wan, M.; Standing, J.; Potts, D.; Burland, J. Measured short-term subsurface ground displacements from EPBM tunnelling in London clay. *Geotechnique* **2017**, *67*, 748–779. [CrossRef]
3. Zhang, X.; Jiang, Y.; Sugimoto, S. Seismic damage assessment of mountain tunnel: A case study on the Tawarayama tunnel due to the 2016 Kumamoto earthquake. *Tunn. Undergr. Space Technol.* **2018**, *71*, 138–148. [CrossRef]
4. Attard, L.; Debono, C.; Valentino, G.; Di Castro, M.; Osborne, J.; Scibile, L.; Ferre, M. A comprehensive Virtual Reality System for Tunnel Surface Documentation and Structural Health Monitoring. In Proceedings of the 2018 IEEE International Conference on Imaging Systems and Techniques (IST), Krakow, Poland, 16–18 October 2018; pp. 1–6.
5. Palin, E.; Stipanovic Oslakovic, I.; Gavin, K.; Quinn, A. Implications of climate change for railway infrastructure. *Wiley Interdiscip. Rev. Clim. Chang.* **2021**, *12*, 1–47. [CrossRef]
6. Bickerdike, G. Forgotten Relics of an Enterprising Age. Organisation Website. 2017. Available online: http://www.forgottenrelics.co.uk/glossary/index.html (accessed on 22 November 2021).
7. Shillito, C. The Fiery Jack. Railway & Canal Historical Society 2007. Volume 774. Available online: https://rchs.org.uk/wpcontent/uploads/2020/02/Journal-200-Dec-2007.pdf#page=40 (accessed on 27 July 2021).
8. Prakhya, G.; Hopkin, I.; Hansford, B. Construction of a Concrete Segmental Arch Bridge Over a Railway. In *Institution of Civil Engineers-Bridge Engineering;* Thomas Telford Ltd.: London, UK, 2019; Volume 172, pp. 226–240. [CrossRef]
9. Liu, F.; Ma, T.; Tang, C.; Liu, X.; Chen, F. A case study of collapses at the Yangshang tunnel of the coal transportation channel from the western inner Mongolia to the central China. *Tunn. Undergr. Space Technol.* **2019**, *92*, 103063. [CrossRef]
10. Montero, R.; Victores, J.; Martinez, S.; Jardón, A.; Balaguer, C. Past, present and future of robotic tunnel inspection. *Autom. Constr.* **2015**, *59*, 99–112. [CrossRef]
11. Farahani, B.; Barros, F.; Sousa, P.; Cacciari, P.; Tavares, P.; Futai, M.; Moreira, P. A coupled 3D laser scanning and digital image correlation system for geometry acquisition and deformation monitoring of a railway tunnel. *Tunn. Undergr. Space Technol.* **2019**, *91*, 102995. [CrossRef]
12. Bahadori-Jahromi, A.; Rotimi, A.; Roxan, A. Sustainable conditional tunnel inspection: London underground, UK. *Infrastruct. Asset Manag.* **2018**, *5*, 22–31. [CrossRef]
13. Liu, S.; Wang, Q.; Luo, Y. A review of applications of visual inspection technology based on image processing in the railway industry. *Transp. Saf. Environ.* **2019**, *1*, 185–204. [CrossRef]
14. Stent, S.; Girerd, C.; Long, P.; Cipolla, R. A Low-Cost Robotic System for the Efficient Visual Inspection of Tunnels. In Proceedings of the International Symposium on Automation and Robotics in Construction, Oulu, Finland, 15–18 June 2015; IAARC Publications: Oulu, Finland, 2015; Volume 32. Available online: http://mi.eng.cam.ac.uk/~{}cipolla/archive/Publications/inproceedings/2015-ISARC-tunnel-inspection.pdf (accessed on 13 July 2021).
15. Attard, L.; Debono, C.; Valentino, G.; Di Castro, M. Image Mosaicing of Tunnel Wall Images Using High Level Features. In Proceedings of the 10th International Symposium on Image and Signal Processing and Analysis, Ljubljana, Slovenia, 18–20 September 2017; pp. 141–146.
16. Tannant, D. Review of photogrammetry-based techniques for characterization and hazard assessment of rock faces. *Int. J. Georesour. Environ. IJGE* **2015**, *1*, 76–87. [CrossRef]

17. Krisada, C.; Tae-Kyun, K.; Fabio, V.; Roberto, C.; Kenichi, S. Distortion-free image mosaicing for tunnel inspection based on robust cylindrical surface estimation through structure from motion. *J. Comput. Civ. Eng.* **2016**, *30*, 04015045. [CrossRef]
18. Jenkins, M.D.; Buggy, T.; Morison, G. An Imaging System for Visual Inspection and Structural Condition Monitoring of Railway Tunnels. In Proceedings of the 2017 IEEE Workshop on Environmental, Energy, and Structural Monitoring Systems (EESMS), Milan, Italy, 24–25 July 2017; pp. 1–6.
19. Xue, Y.; Zhang, S.; Zhou, M.; Zhu, H. Novel SfM-DLT method for metro tunnel 3D reconstruction and visualization. *Undergr. Space* **2021**, *6*, 134–141. [CrossRef]
20. Aleksieva, N.; Hermosilla Carrasco, C.; Brown, A.; Dean, R.; Carolin, A.; Täljsten, B.; García-Villena, F.; Morales-Gamiz, F. *Inspection and Monitoring Techniques for Tunnels and Bridges*; Technical Report, IN2TARCK2; Research into Enhanced Tracks, Switches and Structures: Málaga, Spain, 2019. [CrossRef]
21. Mccormick, N.; Kimkeran, S.; Najimi, A.; Jonas, D. Assessing the Condition of Railway Assets Using DIFCAM: Results from Tunnel Examinations. In Proceedings of the 6th IET Conference on Railway Condition Monitoring (RCM 2014), Birmingham, UK, 17–18 September 2014; pp. 1–6. [CrossRef]
22. Npl Management Limited. DIFCAM Evolution. Organisation Website. 2021. Available online: https://gtr.ukri.org/projects?ref=971711 (accessed on 13 July 2021).
23. Xue, Y.; Zhang, S. A Fast Metro Tunnel Profile Measuring Method Based on Close-Range Photogrammetry. In Proceedings of the International Conference on Information Technology in Geo-Engineering, Guimaraes, Portugal, 29 September–2 October 2019; Springer: Cham, Switzerland, 2019; pp. 57–69. [CrossRef]
24. Leonidas, E.; Xu, Y. The Development of an Automatic Inspection System Used for the Maintenance of Rail Tunnels. In Proceedings of the 2018 24th International Conference on Automation and Computing (ICAC), Newcastle Upon Tyne, UK, 6–7 September 2018; pp. 1–6. [CrossRef]
25. IOWA State University. The Speed of Sound in Other Materials. Organisation Website. 2021. Available online: https://www.ndeed.org/Physics/Sound/speedinmaterials.xhtml (accessed on 15 June 2021).
26. TWI Ltd. What Is Ultrasonic Testing and How Does It Work? Organisation Website. 2021. Available online: https://www.twiglobal.com/technical-knowledge/faqs/ultrasonic-testing (accessed on 17 June 2021).
27. Whitlow, R.; Haskins, R.; Mccomas, S.; Crane, C.; Howard, I.; Mckenna, M. Remote bridge monitoring using infrasound. *J. Bridge Eng.* **2019**, *24*, 04019023. [CrossRef]
28. White, J.; Wieghaus, K.; Karthik, M.; Shokouhi, P.; Hurlebaus, S.; Wimsatt, A. Nondestructive testing methods for underwater tunnel linings: Practical application at Chesapeake channel tunnels. *J. Infrastruct. Syst.* **2017**, *23*, B4016011. [CrossRef]
29. Menendez, E.; Victores, J.; Montero, R.; Martínez, S.; Balaguer, C. Tunnel structural inspection and assessment using an autonomous robotic system. *Autom. Constr.* **2018**, *87*, 117–126. [CrossRef]
30. Protopapadakis, E.; Stentoumis, C.; Doulamis, N.; Doulamis, A.; Loupos, K.; Makantasis, K.; Kopsiaftis, G.; Amditis, A. Autonomous robotic inspection in tunnels. *ISPRS Ann. Photogramm. Remote Sens. Spat. Inf. Sci.* **2016**, *5*, 167–174. [CrossRef]
31. ROBO-SPECT. Robotic System with Intelligent Vision and Control for Tunnel Structural Inspection and Evaluation. Organisation Website. 2021. Available online: http://www.robo-spect.eu/index.php/project (accessed on 18 July 2021).
32. Watanabe, A.; Even, J.; Morales, L.; Ishi, C. Robot-Assisted Acoustic Inspection of Infrastructures—Cooperative Hammer Sounding Inspection. In Proceedings of the 2015 IEEE/RSJ International Conference on Intelligent Robots and Systems (IROS), Hamburg, Germany, 28 September–2 October 2015; pp. 5942–5947. [CrossRef]
33. Jamshidi, A.; Faghih Roohi, S.; Núñez, A.; Babuska, R.; De Schutter, B.; Dollevoet, R.; Li, Z. Probabilistic defect-based risk assessment approach for rail failures in railway infrastructure. *IFAC-PapersOnLine* **2016**, *49*, 73–77. [CrossRef]
34. Fujii, H.; Yamashita, A.; Asama, H. Defect Detection with Estimation of Material Condition Using Ensemble Learning for Hammering Test. In Proceedings of the 2016 IEEE International Conference on Robotics and Automation (ICRA), Stockholm, Sweden, 16–21 May 2016; pp. 3847–3854. [CrossRef]
35. Louhi Kasahara, J.Y.; Yamashita, A.; Asama, H. Acoustic inspection of concrete structures using active weak supervision and visual information. *Sensors* **2020**, *20*, 629. [CrossRef] [PubMed]
36. Nakamura, S.; Yamashita, A.; Inoue, F.; Inoue, D.; Takahashi, Y.; Kamimura, N.; Ueno, T. Inspection test of a tunnel with an inspection vehicle for tunnel lining concrete. *J. Robot. Mechatron.* **2019**, *31*, 762–771. [CrossRef]
37. Moreu, F.; Ayorinde, E.; Mason, J.; Farrar, C.; Mascarenas, D. Remote railroad bridge structural tap testing using aerial robots. *Int. J. Intell. Robot. Appl.* **2018**, *2*, 67–80. [CrossRef]
38. Lattanzi, D.; Miller, G. Review of robotic infrastructure inspection systems. *J. Infrastruct. Syst.* **2017**, *23*, 1–15. [CrossRef]
39. Sugimoto, T.; Sugimoto, K.; Uechi, I.; Utagawa, N.; Kuroda, C. Efficiency Improvement of Outer Wall Inspection by Noncontact Acoustic Inspection Method Using Sound Source Mounted Type UAV. In Proceedings of the 2019 IEEE International Ultrasonics Symposium (IUS), Glasgow, UK, 6–9 October 2019; pp. 2091–2094. [CrossRef]
40. Arastounia, M. Automated as-built model generation of subway tunnels from mobile LIDAR data. *Sensors* **2016**, *16*, 1486. [CrossRef] [PubMed]
41. Soilán, M.; Sánchez-rodríguez, A.; Del Río-barral, P.; Perez-Collazo, C.; Arias, P.; Riveiro, B. Review of laser scanning technologies and their applications for road and railway infrastructure monitoring. *Infrastructures* **2019**, *4*, 58. [CrossRef]
42. Fröhlich, Z. Z+F profiler®6007 Duo. Online Document. 2014. Available online: https://www.zf-laser.com/fileadmin/editor/Broschueren/Broschuere_PROFILER_6007_duo_E_compr.pdf (accessed on 16 July 2021).

43. Cui, H.; Ren, X.; Mao, Q.; Hu, Q.; Wang, W. Shield Subway Tunnel Deformation Detection Based on Mobile Laser Scanning. *Autom. Constr.* **2019**, *106*, 102889. [CrossRef]
44. Gézero, L.; Antunes, C. Automated three-dimensional linear elements extraction from mobile LIDAR point clouds in railway environments. *Infrastructures* **2019**, *4*, 46. [CrossRef]
45. Kemp, D. 3D Crossrail Tunnel Scan Unwrapped into 2D for First Time. Online Document. 2016. Available online: https://www.constructionnews.co.uk/tech/3d-crossrailtunnel-scan-unwrapped-into-2d-for-first-time-03-08-2016/#Tunnel_slice (accessed on 19 July 2021).
46. Tan, K.; Cheng, X.; Ju, Q. Combining mobile terrestrial laser scanning geometric and radiometric data to eliminate accessories in circular metro tunnels. *J. Appl. Remote Sens.* **2016**, *10*, 030503. [CrossRef]
47. Mccrory, K. Case Study of Llandudno Junction Station Survey. Online Document. 2020. Available online: https://scantechinternational.com/case_study/llandudno-junction-station (accessed on 16 July 2021).
48. Scantech International Ltd. Railway Surveys. Organisation Website. 2021. Available online: https://scantechinternational.com/sectors/railway-surveys (accessed on 24 July 2021).
49. Cheng, X.; Hu, X.; Tan, K.; Wang, L.; Yang, L. Automatic detection of shield tunnel leakages based on terrestrial mobile LIDAR intensity images using deep learning. *IEEE Access* **2021**, *9*, 55300–55310. [CrossRef]
50. Fröhlich, Z. Case Study: Train Mounted Laser Survey of Birmingham New Street Area Resignalling Phase 7. Online Document. 2016. Available online: https://www.zf-laser.com/fileadmin/editor/Case_studies/Case_Study_omnicom_E_comp.pdf (accessed on 16 July 2021).
51. Heinz, E.; Mettenleiter, M.; Kuhlmann, H.; Holst, C. Strategy for determining the stochastic distance characteristics of the 2D laser scanner Z+F Profiler 9012a with special focus on the close range. *Sensors* **2018**, *18*, 2253. [CrossRef]
52. Sun, H.; Xu, Z.; Yao, L.; Zhong, R.; Du, L.; Wu, H. Tunnel monitoring and measuring system using mobile laser scanning: Design and deployment. *Remote Sens.* **2020**, *12*, 730. [CrossRef]
53. Yamazaki, F.; Ueda, H.; Liu, W. Basic study on detection of deteriorated RC structures using infrared thermography camera. *Eng. J.* **2018**, *22*, 233–242. [CrossRef]
54. Farahani, B. Innovative Methodology for Railway Tunnel Inspection. Ph.D. Thesis, Faculty of Engineering, University of Porto, Porto, Portugal, 2019. Available online: https://www.researchgate.net/publication/336406410_Innovative_Methodology_for_Railway_Tunnel_Inspection (accessed on 12 July 2021).
55. Ishikawa, M.; Koyama, M.; Kasano, H.; Ogasawara, N.; Yamada, Y.; Hatta, H.; Fukui, R.; Nishitani, Y.; Utsunomiya, S. Inspection of Concrete Structures Using the Active Thermography Method with Remote Heating Apparatuses. In Proceedings of the 15th Asia Pacific Conference for Non-Destructive Testing (APCNDT2017), Singapore, 13–17 November 2017.
56. Lu, X.; Tian, G.; Wu, J.; Gao, B.; Tian, P. Pulsed air-flow thermography for natural crack detection and evaluation. *IEEE Sens. J.* **2020**, *20*, 8091–8097. [CrossRef]
57. Liu, Z.; Gao, B.; Tian, G. Natural crack diagnosis system based on novel l-shaped electromagnetic sensing thermography. *IEEE Trans. Ind. Electron.* **2020**, *67*, 9703–9714. [CrossRef]
58. Konishi, S.; Kawakami, K.; Taguchi, M. Inspection method with infrared thermometry for detect void in subway tunnel lining. *Procedia Eng.* **2016**, *165*, 474–483. [CrossRef]
59. Afshani, A.; Akagi, H. Investigate the Detection Rate of Defects in Concrete Lining Using Infrared-Thermography Method. In Proceedings of the 7th Japan-China Geotechnical Symposium, Sanya, China, 16–18 March 2018; pp. 1005–1016. Available online: https://www.researchgate.net/publication/324890164_Investigate_the_detection_rate_of_defects_in_concrete_lining_using_infrared-thermography_method (accessed on 14 July 2021).
60. Olmi, R.; Palombi, L.; Durazzani, S.; Poggi, D.; Renzoni, N.; Costantino, F.; Durazzani, S.; Frilli, G.; Raimondi, V. Integrating thermographic images in a user-friendly platform to support inspection of railway bridges. *Proceedings* **2019**, *27*, 12. [CrossRef]
61. Liu, F.; Seipel, S. Infrared-visible image registration for augmented reality-based thermographic building diagnostics. *Vis. Eng.* **2015**, *3*, 16. [CrossRef]
62. Afshani, A.; Kawakami, K.; Konishi, S.; Akagi, H. Study of infrared thermal application for detecting defects within tunnel lining. *Tunn. Undergr. Space Technol.* **2019**, *86*, 186–197. [CrossRef]
63. Scintrex. *CG-3/3M Autograv Automated Gravity Meter Operator Manual*, 5th ed.; Scintrex: Vaughan, ON, Canada, 1995. Available online: https://scintrexltd.com/support/product-manuals/cg3-manual/ (accessed on 27 November 2021).
64. Butler, D. Detection and Characterization of Cavities, Tunnels, and Abandoned Mines. Online Document. 2008. Available online: https://digital.lib.usf.edu//content/SF/S0/05/54/94/00001/K26-05045-Butler--ICEEG_Presentation_on_Cavities_and_Tunnels.pdf (accessed on 14 November 2020).
65. Fores, B.; Champollion, C.; Lesparre, N.; Pasquet, S.; Martin, A.; Nguyen, F. Variability of the water stock dynamics in karst: Insights from surface-to-tunnel geophysics. *Hydrogeol. J.* **2021**, *29*, 2077–2089. [CrossRef]
66. Blecha, V.; Mašín, D. Observed and calculated gravity anomalies above a tunnel driven in clays–implication for errors in gravity interpretation. *Near Surf. Geophys.* **2013**, *11*, 569–578. [CrossRef]
67. Zahorec, P.; Papčo, J.; Vajda, P.; Szabó, S. High-precision local gravity survey along planned motorway tunnel in the Slovak karst. *Contrib. Geophys. Geod.* **2019**, *49*, 207–227. [CrossRef]

68. Bloedau, E. Assessment of Change to Gravity Field due to Underground Railroad Tunnel Construction. Ph.D. Thesis, University of Stuttgart, Stuttgart, Germany, 2021. Available online: https://elib.uni-stuttgart.de/handle/11682/11296 (accessed on 14 June 2021).
69. Han, R.; Li, W.; Cheng, R.; Wang, F.; Zhang, Y. 3D high-precision tunnel gravity exploration theory and its application for concealed inclined high-density ore deposits. *J. Appl. Geophys.* **2020**, *180*, 104119. [CrossRef]
70. Alani, A.; Tosti, F. GPR Applications in Structural Detailing of a Major Tunnel Using Different Frequency Antenna Systems. *Constr. Build. Mater.* **2018**, *158*, 1111–1122. [CrossRef]
71. Lai, W.; Derobert, X.; Annan, P. A review of ground penetrating radar application in civil engineering: A 30-year journey from locating and testing to imaging and diagnosis. *NDTE Int.* **2018**, *96*, 58–78.
72. Sensors & Software. What Is Ground Penetrating Radar (GPR)? Online Document. 2020. Available online: https://www.sensoft.ca/blog/what-is-gpr/ (accessed on 17 November 2020).
73. Solla, M.; Pérez-Gracia, V.; Fontul, S. A Review of GPR Application on Transport Infrastructures: Troubleshooting and Best Practices. *Remote Sens.* **2021**, *13*, 672. [CrossRef]
74. Shrestha, S.; Arai, I. Signal processing of ground penetrating radar using spectral estimation techniques to estimate the position of buried targets. *EURASIP J. Adv. Signal Process.* **2003**, *2003*, 970543. [CrossRef]
75. Anbazhagan, P.; Dixit, P.; Bharatha, T. Identification of type and degree of railway ballast fouling using ground coupled GPR antennas. *J. Appl. Geophys.* **2016**, *126*, 183–190. [CrossRef]
76. Cafiso, S.; Capace, B.; D'Agostino, C.; Delfino, E.; Di Graziano, A. Application of NDT to Railway Track Inspections. In Proceedings of the 3rd International Conference on Traffic and Transport Engineering (ICTTE), Lucerne, Switzerland, 6–10 July 2016.
77. Ciampoli, L.; Calvi, A.; D'Amico, F. Railway ballast monitoring by GPR: A test-site investigation. *Remote Sens.* **2019**, *11*, 2381. [CrossRef]
78. Proceq. Proceq GPR Live. Proceq, Screening Eagle Technologies AG Ringstrasse 28603 Schwerzenbach Zürich Switzerland. 2017. Available online: https://www.screeningeagle.com/en/product-family/proceq-ground-penetrating-radars (accessed on 3 October 2020).
79. GSSI. StructureScan Mini XT. Online Document. 2017. Available online: https://www.geophysical.com/wp-content/uploads/2018/01/GSSI-StructureScanMiniXTBrochure.pdf (accessed on 21 July 2021).
80. Proceq. Portable Ground Penetrating Radar—Proceq GP8000. Organisation Website. 2021. Available online: https://www.screeningeagle.com/en/products/proceq-gp8000-portable-concrete-gpr-radar (accessed on 12 June 2021).
81. Dawood, T.; Zhu, Z.; Zayed, T. Deterioration mapping in subway infrastructure using sensory data of GPR. *Tunn. Undergr. Space Technol.* **2020**, *103*, 103487. [CrossRef]
82. 3D-RADAR. GEOSCOPE MK IV: High-Speed 3D GPR with High-Resolution and Deep Penetration. Online Document. 2019. Available online: http://3d-radar.com/wp-content/uploads/2019/10/3DRadar_GeoScope_ProductSheet_2019.pdf (accessed on 18 June 2021).
83. Zetica Rail. Zetica—Advanced Rail Radar (ZARR) Solution to Augment Inspection Trains. Organisation Website. 2021. Available online: https://zeticarail.com/systems-software/zarr/ (accessed on 4 July 2021).
84. IDS GeoRadar. SRS SafeRailSystem: Safe Railway Ballast Inspections with Ground Penetrating Radar. Organisation Website. 2021. Available online: https://idsgeoradar.com/products/ground-penetratingradar/srs-saferailsystem (accessed on 4 July 2021).
85. Zan, Y.; Li, Z.; Su, G.; Zhang, X. An innovative vehicle-mounted GPR technique for fast and efficient monitoring of tunnel lining structural conditions. *Case Stud. Nondestruct. Test. Eval.* **2016**, *6*, 63–69. [CrossRef]
86. Xiong, H.; Su, G.; Zhang, C.; Li, B.; Wei, W. A train-mounted GPR System for Operating Railway Tunnel Inspection. In Advances in Transdisciplinary Engineering Series. In Proceedings of the ISMR 2020 7th International Symposium on Innovation & Sustainability of Modern Railway, Nanchang, China, 23–25 October 2020; Volume 14. Available online: https://ebooks.iospress.nl/volume/ismr-2020-proceedings-of-the-7th-international-symposium-on-innovation-amp-sustainability-of-modern-railway (accessed on 7 July 2020).
87. Han, X.; Jin, J.; Wang, M.; Jiang, W.; Gao, L.; Xiao, L. A review of algorithms for filtering the 3D point cloud. *Signal Process. Image Commun.* **2017**, *57*, 103–112. [CrossRef]
88. Rial, F.; Uschkerat, U. Improving SCR of Underground Target Signatures from Air-Launched GPR Systems Based on Scattering Center Extraction. In Proceedings of the 2017 18th International Radar Symposium (IRS), Prague, Czech Republic, 28–30 June 2017; pp. 1–10. [CrossRef]
89. Xiang, Z.; Rashidi, A.; Ou, G. States of practice and research on applying GPR technology for labelling and scanning constructed facilities. *J. Perform. Constr. Facil.* **2019**, *33*, 03119001. [CrossRef]
90. Lyu, Y.; Wang, H.; Gong, J. GPR detection of tunnel lining cavities and reverse-time migration imaging. *Appl. Geophys.* **2020**, *17*, 1–7. [CrossRef]
91. Bugarinović, V.; Pajewski, L.; Ristić, A.; Vrtunski, M.; Govedarica, M.; Borisov, M. On the introduction of canny operator in an advanced imaging algorithm for real-time detection of hyperbolas in ground-penetrating radar data. *Electronics* **2020**, *9*, 541. [CrossRef]

92. Li, C.; Xing, S.; Lauro, S.; Su, Y.; Dai, S.; Feng, J.; Cosciotti, B.; Di Paolo, F.; Mattei, E.; Xiao, Y.; et al. Pitfalls in GPR data interpretation: False reflectors detected in lunar radar cross sections by Chang'e-3. *IEEE Trans. Geosci. Remote Sens.* **2018**, *56*, 1325–1335. [CrossRef]
93. Johnston, G. The Basics of Interpreting GPR Data—Part 2. Webinar. 2018. Available online: https://www.sensoft.ca/trainingevents/webinars/interpreting-gpr-data-part2/ (accessed on 16 November 2020).
94. Kilic, G.; Eren, L. Neural network based inspection of voids and karst conduits in hydroelectric power station tunnels using GPR. *J. Appl. Geophys.* **2018**, *151*, 194–204. [CrossRef]
95. Wang, X.; Sun, S.; Wang, J.; Yarovoy, A.; Neducza, B.; Manacorda, G. Real GPR signal processing for target recognition with circular array antennas. In Proceedings of the 2016 URSI International Symposium on Electromagnetic Theory (EMTS), Espoo, Finland, 14–18 August 2016; pp. 818–821. [CrossRef]
96. Kim, N.; Kim, S.; An, Y.; Lee, J. A novel 3D GPR image arrangement for deep learning-based underground object classification. *Int. J. Pavement Eng.* **2019**, *22*, 740–751. [CrossRef]
97. Šarlah, N.; Podobnikar, T.; Ambrožič, T.; Mušič, B. Application of kinematic GPR-TPS model with high 3D georeference accuracy for underground utility infrastructure mapping: A case study from urban sites in Celje, Slovenia. *Remote Sens.* **2020**, *12*, 1228. [CrossRef]
98. Sjödin, R. Interpolation and Visualization of Sparse GPR Data. Master's Thesis, Umea University, Department of Physics, Umea, Sweden, 2020. Available online: https://www.diva-portal.org/smash/record.jsf?pid=diva2:1431027&dswid=8251 (accessed on 30 July 2021).
99. GoldenSoftwareLLC2015. Voxler®4: 3D Well & Volumetric Data Visualization. GoldenSoftwareLLC2015. 2015. Available online: https://downloads.goldensoftware.com/guides/Voxler4UserGuide.pdf (accessed on 12 December 2020).
100. Zhang, S.; Zhang, L.; He, W.; Ling, T.; Deng, Z.; Fu, G. Three-dimensional quantitative recognition of filler materials ahead of a tunnel face via time-energy density analysis of wavelet transforms. *Minerals* **2022**, *12*, 234. [CrossRef]
101. MALA. MALA Vision User Manual. GUIDELINE GEO, Hemvärnsgatan 9SE-171 54 Solna, Stockholm VAT: SE 556606-1155-01. 2021. Available online: https://www.guidelinegeo.com/product/mala-vision/ (accessed on 2 December 2021).
102. Coli, M.; Ciuffreda, A.L.; Marchetti, E.; Morandi, D.; Luceretti, G.; Lippi, Z. 3D HBIM model and full contactless GPR tomography: An experimental application on the historic walls that support Giotto's mural paintings, Santa Croce Basilica, Florence—Italy. *Heritage* **2022**, *5*, 132. [CrossRef]
103. Hou, F.; Liu, X.; Fan, X.; Guo, Y. DL-aided underground cavity morphology recognition based on 3D GPR data. *Mathematics* **2022**, *10*, 2806. [CrossRef]
104. Núñez-Nieto, X.; Solla, M.; Prego, F.J.; Lorenzo, H. Assessing the Applicability of GPR Method for Tunnelling Inspection: Characterization and Volumetric Reconstruction. In Proceedings of the 2015 8th International Workshop on Advanced Ground Penetrating Radar (IWAGPR), Florence, Italy, 7–10 July 2015; pp. 1–4. [CrossRef]
105. Simi, A.; Manacorda, G. The NETTUN Project: Design of a GPR Antenna for a TBM. In Proceedings of the 2016 16th International Conference on Ground Penetrating Radar (GPR), Hong Kong, China, 13–16 June 2016; pp. 1–6. [CrossRef]
106. Garcia-Garcia, F.; Valls-Ayuso, A.; Benlloch-Marco, J.; Valcuende-Paya, M. An optimization of the work disruption by 3D cavity mapping using GPR: A new sewerage project in Torrente (Valencia, Spain). *Constr. Build. Mater.* **2017**, *154*, 1226–1233. [CrossRef]
107. Kadioglu, S.; Kadioğlu, Y. Determining buried remains under the ala gate road of Anavarza ancient city in the southern of Turkey with interactive transparent 3D GPR data imaging. *Int. Multidiscip. Sci. Geoconf. Sgem* **2019**, *19*, 773–779.
108. Grasmueck, M.; Viggiano, D. PondView: Intuitive and Efficient Visualization of 3D GPR Data. In Proceedings of the 2018 17th International Conference on Ground Penetrating Radar (GPR), Rapperswil, Switzerland, 18–21 June 2018; pp. 1–6. [CrossRef]
109. Agrafiotis, P.; Lampropoulos, K.; Georgopoulos, A.; Moropoulou, A. 3D modelling the Invisible Using Ground Penetrating radar. In *The International Archives of the Photogrammetry, Remote Sensing and Spatial Information Sciences*; Ktisis: Nafplio, Greece, 2017; Volume XLII/W3, pp. 33–37. [CrossRef]
110. Tong, Z.; Gao, J.; Zhang, H. Recognition, location, measurement, and 3D reconstruction of concealed cracks using convolutional neural networks. *Constr. Build. Mater.* **2017**, *146*, 775–787. [CrossRef]
111. Chen, K.; Kamezaki, M.; Katano, T.; Kaneko, T.; Azuma, K.; Ishida, T.; Seki, M.; Ichiryu, K.; Sugano, S. Compound locomotion control system combining crawling and walking for multi-crawler multi-arm robot to adapt unstructured and unknown terrain. *Robomech J.* **2018**, *5*, 2. [CrossRef]
112. Neubauer, W.; Bornik, A.; Wallner, M.; Verhoeven, G. Novel Volume Visualisation of GPR Data Inspired by Medical Applications. In New Global Perspectives on Archaeological Prospection. In Proceedings of the 13th International Conference on Archaeological Prospection, Sligo, Ireland, 28 August–1 September 2019.
113. Liu, Y.; Qiao, J.; Han, T.; Li, L.; Xu, T. A 3D image reconstruction model for long tunnel geological estimation. *J. Adv. Transp.* **2020**, *2020*, 8846955. [CrossRef]
114. Feng, J.; Yang, L.; Wang, H.; Song, Y.; Xiao, J. GPR-Based Subsurface Object Detection and Reconstruction Using Random Motion and DepthNet. In Proceedings of the 2020 IEEE International Conference on Robotics and Automation (ICRA), Paris, France, 31 May 2020–31 August 2020; pp. 7035–7041. [CrossRef]
115. Feng, J.; Yang, L.; Biao, J.; Xiao, J. Robotic inspection and 3D GPR-based reconstruction for underground utilities. *arXiv* **2021**, arXiv:2106.01907.

116. Dinh, K.; Gucunski, N.; Tran, K.; Novo, A.; Nguyen, T. Full-resolution 3D imaging for concrete structures with dual-polarization GPR. *Autom. Constr.* **2021**, *125*, 103652. [CrossRef]
117. Pereira, M.; Burns, D.; Orfeo, D.; Zhang, Y.; Jiao, L.; Huston, D.; Xia, T. 3D multistatic ground penetrating radar imaging for augmented reality visualization. *IEEE Trans. Geosci. Remote Sens.* **2020**, *58*, 5666–5675. [CrossRef]
118. Wu, S.; Hou, L.; Zhang, G. Integrated Application of Bm and Extended Reality Technology: A review, Classification and Outlook. In Proceedings of the International Conference on Computing in Civil and Building Engineering, São Paulo, Brazil, 18–20 August 2020; Springer: Berlin/Heidelberg, Germany, 2020; pp. 1227–1236. [CrossRef]
119. Karaaslan, E.; Bagci, U.; Catbas, F. Artificial intelligence assisted infrastructure assessment using mixed reality systems. *Transp. Res. Rec.* **2019**, *2673*, 413–424. [CrossRef]
120. Childs, J.; Orfeo, D.; Burns, D.; Huston, D.; Xia, T. Enhancing Ground Penetrating Radar with Augmented Reality Systems for Underground Utility Management. In *Virtual, Augmented, and Mixed Reality (XR) Technology for Multi-Domain Operations*; International Society for Optics and Photonics: Bellingham, WA, USA, 2020; Volume 11426, p. 1142608. [CrossRef]
121. Jin, R. Developing a Mixed-Reality Based Application for Bridge Inspection and Maintenance. In Proceedings of the 20th International Conference on Construction Applications of Virtual Reality (CONVR 2020), Middlesbrough, UK, 30 September–2 October 2020.
122. Hu, D.; Hou, F.; Blakely, J.; Li, S. Augmented Reality Based Visualization for Concrete Bridge Deck Deterioration Characterized by Ground Penetrating Radar. In *Construction Research Congress 2020: Computer Applications*; American Society of Civil Engineers: Reston, VA, USA, 2020; pp. 1156–1164. [CrossRef]
123. Wei, L.; Magee, D.; Cohn, A. An anomalous event detection and tracking method for a tunnel look-ahead ground prediction system. *Autom. Constr.* **2018**, *91*, 216–225. [CrossRef]
124. ESCAP. Inspection and Monitoring of Railway Infrastructure Using Aerial Drones. Online Document. 2019. Available online: https://www.unescap.org/sites/default/files/TARWG_4E_Inspectionandmonitoring.pdf (accessed on 1 July 2021).
125. Tan, C.H.; Shaiful, D.S.B.; Ang, W.J.; Win, S.K.H.; Foong, S. Design optimization of sparse sensing array for extended aerial robot navigation in deep hazardous tunnels. *IEEE Robot. Autom. Lett.* **2019**, *4*, 862–869. [CrossRef]
126. Jordan, S.; Moore, J.; Hovet, S.; Box, J.; Perry, J.; Kirsche, K.; Lewis, D.; Tse, Z. State-of-the-art technologies for UAV inspections. *IET Radar Sonar Navig.* **2018**, *12*, 151–164. [CrossRef]
127. Galtarossa, L.; Navilli, L.; Chiaberge, M. Visual-Inertial Indoor Navigation Systems and Algorithms for UAV Inspection Vehicles. In *Industrial Robotics*; IntechOpen: London, UK, 2020; pp. 1–16. [CrossRef]
128. Azevedo, F.; Oliveira, A.; Dias, A.; Almeida, J.; Moreira, M.; Santos, T.; Ferreira, A.; Martins, A.; Silva, E. Collision Avoidance for Safe Structure Inspection with Multirotor UAV. In Proceedings of the 2017 European Conference on Mobile Robots (ECMR), Paris, France, 6–8 September 2017; pp. 1–7. [CrossRef]
129. Quan, Q.; Fu, R.; Li, M.; Wei, D.; Gao, Y.; Cai, K. Practical distributed control for VTOL UAVs to pass a tunnel. *arXiv* **2021**, arXiv:2101.07578.
130. Petrlík, M.; Báča, T.; Heřt, D.; Vrba, M.; Krajník, T.; Saska, M. A robust UAV system for operations in a constrained environment. *IEEE Robot. Autom. Lett.* **2020**, *5*, 2169–2176. [CrossRef]
131. Moletta, M. Path Planning for Autonomous Aerial Robots in Unknown Underground Zones Optimized for Vertical Tunnels Exploration. Master's Thesis, KTH Royal Institute of Technology in Stockholm, School of Electrical Engineering and Computer Science (EECS), Stockholm, Sweden, 2020. Available online: https://www.diva-portal.org/smash/record.jsf?pid=diva2:1499089&dswid=5052 (accessed on 18 March 2021).
132. Elmokadem, T.; Savkin, A. A method for autonomous collision-free navigation of a quadrotor UAV in unknown tunnel-like environments. *Robotica* **2021**, *40*, 1–27. [CrossRef]
133. Falcone, A.; Vaccarino, G. Primary Level UAV for Tunnel Inspection: The PLUTO Project. SEMANTIC SCHOLAR. 2020. Available online: https://www.semanticscholar.org/paper/Primary-Level-UAVfor-Tunnel-Inspection:-the-PLUTO-Falcone-Vaccarino/32a694d6dbe4f7dba61181c54d8681fbe7503245 (accessed on 18 July 2020).
134. Özaslan, T.; Shen, S.; Mulgaonkar, Y.; Michael, N.; Kumar, V. Inspection of Penstocks and Featureless Tunnel-Like Environments Using Micro UAVs. In *Field and Service Robotics: Results of the 9th International Conference*; Mejias, L., Corke, P., Roberts, J., Eds.; Springer Tracts in Advanced Robotics Book Series; Springer International Publishing: Cham, Switzerland, 2015; Volume 105, pp. 123–136. [CrossRef]
135. Sakuma, M.; Kobayashi, Y.; Emaru, T.; Ravankar, A. Mapping of Pier Substructure Using UAV. In Proceedings of the 2016 IEEE/SICE International Symposium on System Integration (SII), Sapporo, Japan, 13–15 December 2016; pp. 361–366. [CrossRef]
136. Wu, W.; Qurishee, M.; Owino, J.; Fomunung, I.; Onyango, M.; Atolagbe, B. Coupling Deep Learning and UAV for Infrastructure Condition Assessment Automation. In Proceedings of the 2018 IEEE International Smart Cities Conference (ISC2), Kansas City, MO, USA, 16–19 September 2018; pp. 1–7. [CrossRef]
137. Dorafshan, S.; Maguire, M.; Hoffer, N.; Coopmans, C. Challenges in Bridge Inspection Using Small Unmanned Aerial Systems: Results and Lessons Learned. In Proceedings of the 2017 International Conference on Unmanned Aircraft Systems (ICUAS), Miami, FL, USA, 13–16 June 2017; pp. 1722–1730. [CrossRef]
138. Hovering Solutions Ltd. Case Studies: Penstock Inspections and Mapping by Using Autonomous Flying Robots. *Organisation Website.* 2020. Available online: http://www.hoveringsolutions.com/aboutus/penstocks-mapping (accessed on 19 April 2021).

139. Tan, C.; Ng, M.; Shaiful, D.; Win, S.; Ang, W.; Yeung, S.; Lim, H.; Do, M.; Foong, S. A smart unmanned aerial vehicle (UAV) based imaging system for inspection of deep hazardous tunnels. *Water Pract. Technol.* **2018**, *13*, 991–1000. [CrossRef]
140. Hovering Solutions Ltd. Case Studies: London Crossrail Tunnels Are Scanned Using Drones. *Organisation Website.* 2017. Available online: http://www.hoveringsolutions.com/about-us/crossrailtunnels-3d-mapping-using-drones (accessed on 19 April 2021).
141. Pahwa, R.; Chan, K.; Bai, J.; Saputra, V.; Do, M.; Foong, S. Dense 3D Reconstruction for Visual Tunnel Inspection Using Unmanned Aerial Vehicle. In Proceedings of the 2019 IEEE/RSJ International Conference on Intelligent Robots and Systems (IROS), Macau, China, 3–8 November 2019; pp. 7025–7032. [CrossRef]
142. Cwiakala, P.; Gruszczynski, W.; Stoch, T.; Puniach, E.; Mrochen, D.; Matwij, W.; Matwij, K.; Nedzka, M.; Sopata, P.; Wojcik, A. UAV applications for determination of land deformations caused by underground mining. *Remote Sens.* **2020**, *12*, 1733. [CrossRef]
143. Garcia-Fernandez, M.; Alvarez-Lopez, Y.; Gonzalez-Valdes, B.; Arboleya-Arboleya, A.; Rodriguez-Vaqueiro, Y.; Heras, F.L.; Pino, A. UAV-Mounted GPR for NDT Applications. In Proceedings of the 2018 15th European Radar Conference (EuRAD), Madrid, Spain, 26–28 September 2018; pp. 2–5. [CrossRef]
144. Garcia-Fernandez, M.; Alvarez-Lopez, Y.; Heras, F.L.; Gonzalez-Valdes, B.; Rodriguez-Vaqueiro, Y.; Pino, A.; Arboleya-Arboleya, A. GPR System Onboard a UAV for Non-Invasive Detection of Buried Objects. In Proceedings of the 2018 IEEE International Symposium on Antennas and Propagation USNC/URSI National Radio Science Meeting, Boston, MA, USA, 8–13 July 2018; pp. 1967–1968. [CrossRef]
145. Lamsters, K.; Karušs, J.; Krievāns, M.; Ješkins, J. High-resolution surface and bed topography mapping of Russell Glacier (SW Greenland) using UAV and GPR. *ISPRS Ann. Photogramm. Remote Sens. Spat. Inf. Sci.* **2020**, *2*, 757–763. [CrossRef]
146. MALA. MALA Geodrone 80 Technical Specification. *GUIDELINEGEO, Hemvarnsgatan 9SE-171 54 Solna, StockholmVAT: SE 556606-1155-01.* 2021. Available online: https://wwwguidelinegeoc.cdn.triggerfish.cloud/uploads/2020/01/MALA-GeoDrone-80-Technical-Specification-2020-04-27.pdf (accessed on 2 December 2021).
147. Delamare, Q. Algorithms for Estimation and Control of Quadrotors in Physical Interaction with Their Environment. Ph.D. Thesis, University Rennes, Rennes, France, 2019. Available online: https://tel.archives-ouvertes.fr/tel-02410023 (accessed on 18 July 2021).
148. Delamare, Q.; Giordano, P.; Franchi, A. Toward aerial physical locomotion: The contact-fly-contact problem. *IEEE Robot. Autom. Lett.* **2018**, *3*, 1514–1521. [CrossRef]
149. Sanchez-Cuevas, P.; Ramon-Soria, P.; Arrue, B.; Ollero, A.; Heredia, G. Robotic system for inspection by contact of bridge beams using UAVs. *Sensors* **2019**, *19*, 305. [CrossRef]
150. Kocer, B.; Tjahjowidodo, T.; Pratama, M.; Seet, G. Inspection-while-flying: An autonomous contact-based nondestructive test using UAV-tools. *Autom. Constr.* **2019**, *106*, 102895. [CrossRef]
151. Iwamoto, T.; Enaka, T.; Tada, K. Development of testing machine for tunnel inspection using multi-rotor UAV. *J. Phys. Conf. Ser.* **2017**, *842*, 012068. [CrossRef]
152. PRODRONE Co., Ltd. PD6-CI-L. Organisation Website. 2021. Available online: https://www.prodrone.com/products/pd6-ci-l/ (accessed on 19 July 2021).
153. Mahmood, S.; Bakhy, S.; Tawfik, M. Propeller-Type Wall-Climbing Robots: A Review. *IOP Conf. Ser. Mater. Sci. Eng.* **2021**, *1094*, 012106. [CrossRef]
154. Ikeda, T.; Yasui, S.; Fujihara, M.; Ohara, K.; Ashizawa, S.; Ichikawa, A.; Okino, A.; Oomichi, T.; Fukuda, T. Wall Contact by Octo-rotor UAV with one DoF Manipulator for Bridge Inspection. In Proceedings of the 2017 IEEE/RSJ International Conference on Intelligent Robots and Systems (IROS), Vancouver, BC, Canada, 24–28 September 2017; pp. 5122–5127. [CrossRef]
155. Jiang, G.; Voyles, R.; Choi, J. Precision Fully-Actuated UAV for Visual and Physical Inspection of Structures for Nuclear Decommissioning and Search and Rescue. In Proceedings of the 2018 IEEE International Symposium on Safety, Security, and Rescue Robotics (SSRR), Philadelphia, PA, USA, , 6–8 August 2018; pp. 1–7. [CrossRef]
156. Mosaddek, A.; Kommula, H.; Gonzalez, F. Design and Testing of a Recycled 3D Printed and Foldable Unmanned Aerial Vehicle for Remote Sensing. In Proceedings of the 2018 International Conference on Unmanned Aircraft Systems (ICUAS), Dallas, TX, USA, 12–15 June 2018; pp. 1207–1216. [CrossRef]
157. Brown, L.; Clarke, R.; Akbari, A.; Bhandari, U.; Bernardini, S.; Chhabra, P.; Marjanovic, O.; Richardson, T.; Watson, S. The design of Prometheus: A reconfigurable UAV for subterranean mine inspection. *Robotics* **2020**, *9*, 95. [CrossRef]
158. Ecker, G.; Zagar, B.; Schwab, C.; Saliger, F.; Schachinger, T.; Stur, M. Conceptualising an Inspection Robot for Tunnel Drainage Pipes. *IOP Conf. Ser. Mater. Sci. Eng.* **2020**, *831*, 12016. [CrossRef]
159. Naclerio, N.; Karsai, A.; Murray-Cooper, M.; Ozkan-Aydin, Y.; Aydin, E.; Goldman, D.; Hawkes, E. Controlling subterranean forces enables a fast, steerable, burrowing soft robot. *Sci. Robot.* **2021**, *6*, eabe2922. [CrossRef] [PubMed]
160. Xiao, X.; Murphy, R. A review on snake robot testbeds in granular and restricted manoeuvrability spaces. *Robot. Auton. Syst.* **2018**, *110*, 160–172. [CrossRef]
161. Liu, J.; Tong, Y.; Liu, J. Review of snake robots in constrained environments. *Robot. Auton. Syst.* **2021**, *141*, 103785. [CrossRef]
162. Ghazali, M.; Mohamad, H. Monitoring Subsurface Ground Movement Using Fibre Optic Inclinometer Sensor. *IOP Conf. Ser. Mater. Sci. Eng.* **2019**, *527*, 012040. [CrossRef]

163. Ciocca, F.; Bodet, L.; Simon, N.; Karaulanov, R.; Clarke, A.; Abesser, C.; Krause, S.; Chalari, A.; Mondanos, M. Towards the Wetness Characterization of Soil Subsurface Using Fibre Optic Distributed Acoustic Sensing. AGU Fall Meeting Abstracts 2017, Volume 2017, (H21A–1423). Available online: https://agu.confex.com/agu/fm17/meetingapp.cgi/Paper/270785 (accessed on 19 July 2021).
164. Guzman, R.; Navarro, R.; Beneto, M.; Carbonell, D. Robotnik—Professional Service Robotics Applications with ROS. In *Robot Operating System (ROS)*; Springer: Berlin/Heidelberg, Germany, 2016; pp. 253–288. [CrossRef]
165. Brunete, A.; Ranganath, A.; Segovia, S.; de Frutos, J.; Hernando, M.; Gambao, E. Current trends in reconfigurable modular robots design. *Int. J. Adv. Robot. Syst.* **2017**, *14*, 1729881417710457. [CrossRef]
166. Zou, M.; Bai, H.; Wang, Y.; Yu, S. Mechanical design of a self-adaptive transformable tracked robot for cable tunnel inspection. In Proceedings of the 2016 IEEE International Conference on Mechatronics and Automation, Harbin, China, 7–10 August 2016; pp. 1096–1100. [CrossRef]
167. Bruzzone, L.; Fanghella, P.; Quaglia, G. Experimental performance assessment of MANTIS 2, hybrid leg-wheel mobile robot. *Int. J. Autom. Technol.* **2017**, *11*, 396–403. [CrossRef]
168. Calderón, A.; Ugalde, J.; Zagal, J.; Pérez-Arancibia, N. Design, Fabrication and Control of a Multi-Material-Multi-Actuator Soft Robot Inspired by Burrowing Worms. In Proceedings of the 2016 IEEE International Conference on Robotics and Biomimetics (ROBIO), Qingdao, China, 3–7 December 2016; pp. 31–38. [CrossRef]
169. Ahmadzadeh, H.; Masehian, E.; Asadpour, M. Modular robotic systems: Characteristics and applications. *J. Intell. Robot. Syst.* **2016**, *81*, 317–357. [CrossRef]
170. Zhang, X.; Pan, T.; Heung, H.; Chiu, P.; Li, Z. A biomimetic Soft Robot for Inspecting Pipeline with Significant Diameter Variation. In Proceedings of the 2018 IEEE/RSJ International Conference on Intelligent Robots and Systems (IROS), Madrid, Spain, 1–5 October 2018; pp. 7486–7491. [CrossRef]
171. Kopperger, E.; List, J.; Madhira, S.; Rothfischer, F.; Lamb, D.; Simmel, F. A self-assembled nanoscale robotic arm controlled by electric fields. *Science* **2018**, *359*, 296–301. [CrossRef] [PubMed]
172. Amir, Y.; Abu-Horowitz, A.; Werfel, J.; Bachelet, I. Nanoscale robots exhibiting quorum sensing. *Artif. Life* **2019**, *25*, 227–231. [CrossRef] [PubMed]
173. Dong, J.; Wang, M.; Zhou, Y.; Zhou, C.; Wang, Q. DNA-based adaptive plasmonic logic gates. *Angew. Chem.* **2020**, *132*, 15148–15152. [CrossRef]
174. Romanishin, J. Creating Modular Robotic Systems Which Can Reconfigure Themselves in Order to Create New Robots. Organisation Website. 2018. Available online: https://www.csail.mit.edu/research/m-blocksmodular-robotics (accessed on 20 July 2021).
175. NBS Enterprises Ltd. What Is Building Information Modelling (BIM)? *Organisation Website.* 2021. Available online: https://www.thenbs.com/knowledge/what-is-buildinginformation-modelling-bim (accessed on 12 July 2021).
176. Kupriyanovsky, V.; Pokusaev, O.; Klimov, A.; Volodin, A. BIM on the way to IFC5-alignment and development of IFC semantics and ontologies with UML and OWL for road and rail structures, bridges, tunnels, ports, and waterways. *Int. J. Open Inf. Technol.* **2020**, *8*, 69–78.
177. Soilán, M.; Nóvoa, A.; Sánchez-Rodríguez, A.; Riveiro, B.; Arias, P. Semantic segmentation of point clouds with PointNet and KPConv architectures applied to railway tunnels. *ISPRS Ann. Photogrammetry. Remote Sens. Spat. Inf. Sci.* **2020**, *V-2-2020*, 281–288. [CrossRef]
178. Nuttens, T.; De Breuck, V.; Cattoor, R.; Decock, K.; Hemeryck, I. Using BIM models for the design of large rail infrastructure projects: Key factors for a successful implementation. *Int. J. Sustain. Dev. Plan.* **2018**, *13*, 77–89. [CrossRef]
179. ERA LEARN: Eurostars 2. Project: Operation Oriented Tunnel Inspection System. Organisation Website. 2018. Available online: https://www.era-learn.eu/networkinformation/networks/eurostars-2/eurostars-cut2013off-9/operation-oriented-tunnel-inspection-system (accessed on 17 July 2021).
180. Sorge, R.; Buttafoco, D.; Debenedetti, J.; Menozzi a Cimino, G.; Maltese, F.; Tiberi, B. BIM Implementation—Brenner Base Tunnel Project. In *Tunnels and Underground Cities: Engineering and Innovation Meet Archaeology, Architecture and Art*; CRC Press: Boca Raton, FL, USA, 2019; pp. 3122–3131. Available online: https://www.researchgate.net/publication/332517232_BIM_implementation_-_Brenner_Base_Tunnel_project (accessed on 17 July 2021).
181. Cheng, Y.; Qiu, W.; Duan, D. Automatic creation of as-is building information model from single-track railway tunnel point clouds. *Autom. Constr.* **2019**, *106*, 102911. [CrossRef]
182. Tijs, K. Digital Tunnel Twin: Enriching the Maintenance and Operation of Dutch Tunnels. Master's Thesis, Delft University of Technology, Civil Engineering, Construction Management and Engineering, Delft, The Netherlands, 2020. Available online: http://resolver.tudelft.nl/uuid:9cbf5ecf-66ce-4dde-9306-16bd8ccfdb9d (accessed on 1 July 2021).
183. Schneider, O.; Prokopová, A.; Modetta, F.; Petschen, V. The Use of Artificial Intelligence for a Cost-Effective Tunnel Maintenance. In *Tunnels and Underground Cities: Engineering and Innovation Meet Archaeology*; CRC Press: Boca Raton, FL, USA, 2019; pp. 3050–3059. Available online: https://hagerbach.ch/fileadmin/user_upload/ch323_OliverSchneider.pdf (accessed on 23 July 2021).
184. Kapogiannis, G.; Mlilo, A. Digital Construction Strategies and BIM in Railway Tunnelling Engineering. In *Tunnel Engineering-Selected Topics*; IntechOpen: London, UK, 2019. Available online: https://www.intechopen.com/chapters/68102 (accessed on 25 November 2021).

185. Song, Z.; Shi, G.; Wang, J.; Wei, H.; Wang, T.; Zhou, G. Research on management and application of tunnel engineering based on BIM technology. *J. Civ. Eng. Manag.* **2019**, *25*, 785–797. [CrossRef]
186. Monica, R.; Aleotti, J.; Zillich, M.; Vincze, M. Multi-Label Point Cloud Annotation by Selection of Sparse Control Points. In Proceedings of the 2017 International Conference on 3D Vision (3DV), Qingdao, China, 10-12 October 2017; pp. 301–308. [CrossRef]
187. Xu, C.; Wu, B.; Wang, Z.; Zhan, W.; Vajda, P.; Keutzer, K.; Tomizuka, M. Squeezesegv3: Spatially-Adaptive Convolution for Efficient Point-Cloud Segmentation. In Proceedings of the European Conference on Computer Vision, Virtual, 23–28 August 2020; Springer: Berlin/Heidelberg, Germany, 2020; pp. 1–19. [CrossRef]
188. Kaewunruen, S.; Peng, S.; Phil-Ebosie, O. Digital twin aided sustainability and vulnerability audit for subway stations. *Sustainability* **2020**, *12*, 7873. [CrossRef]
189. Singh, V.; Willcox, K. Engineering design with digital thread. *AIAA J.* **2018**, *56*, 4515–4528. [CrossRef]
190. Shi, P.; Zhang, D.; Pan, J.; Liu, W. Geological investigation and tunnel excavation aspects of the weakness zones of Xiang'an subsea tunnels in China. *Rock Mech. Rock Eng.* **2016**, *49*, 4853–4867. [CrossRef]
191. Zhou, S.; Tian, Z.; Di, H.; Guo, P.; Fu, L. Investigation of a loess-mudstone landslide and the induced structural damage in a high-speed railway tunnel. *Bull. Eng. Geol. Environ.* **2020**, *79*, 2201–2212. [CrossRef]
192. Ghezzi, A.; Schettino, A.; Pierantoni, P.P.; Conyers, L.; Tassi, L.; Vigliotti, L.; Schettino, E.; Melfi, M.; Gorrini, M.; Boila, P. Reconstruction of a segment of the UNESCO world heritage hadrian villa tunnel network by integrated GPR, magnetic–paleomagnetic, and electric resistivity prospections. *Remote Sens.* **2019**, *11*, 1739. [CrossRef]
193. Moghaddam, S.; Azadi, A.; Sadeghi, E. Detection of Landslide Geometry Using ERT, a Case Study: The Tunnel of Kermanshah-Khosravy Railway. In Proceedings of the 19th Iranian Geophysical Conference, Online, 9–16 November 2020; Iranian National Geophysical Society: Tehran, Iran, 2020; pp. 68–71. Available online: http://www.nigsconference.ir/article_4094.pdf (accessed on 30 November 2021).
194. ABEM. User Manual Terrameter LS 2. Guideline Geo Abem Mala, Abem Instrument AB, Löfströms Allé 6A, S-172 66 Sundbyberg, Sweden. 2017. Available online: https://wwwguidelinegeoc.cdn.triggerfish.cloud/uploads/2017/08/Terrameter-LS-2-User-Manual-2017-08-14-1.pdf (accessed on 3 December 2021).
195. Lataste, J.; Bruneau, J. Geophysical Investigations of a Landslide to Interpret the Distortion of a Railway Tunnel. In Proceedings of the NSG2021 27th European Meeting of Environmental and Engineering Geophysics, Bordeaux, France, 29 August–2 September 2021; EarthDoc: Bunnik, The Netherlands; Volume 2021, pp. 1–5. [CrossRef]
196. Rhayma, N.; Talon, A.; Breul, P.; Goirand, P. Mechanical investigation of tunnels: Risk analysis and notation system. *Struct. Infrastruct. Eng.* **2016**, *12*, 381–393. [CrossRef]
197. Zhou, Y.; Zhang, X.; Wei, L.; Liu, S.; Zhang, B.; Zhou, C. Experimental study on prevention of calcium carbonate crystallizing in drainage pipe of tunnel engineering. *Adv. Civ. Eng.* **2018**, *2018*, 9430517. Available online: https://www.hindawi.com/journals/ace/2018/9430517/ (accessed on 25 July 2021).
198. Futai, M.; Cacciari, P.; Monticeli, J.; Cantarella, V. Study of an Old Railway Rock Tunnel: Site Investigation, Laboratory Tests, Weathering Effects and Computational Analysis. In Proceedings of the 19th International Conference on Soil Mechanics and Geotechnical Engineering, Seoul, Korea, 17–21 September 2017; ISSMGE, COEX Convention Centre: Seoul, Korea, 2021. Available online: https://www.issmge.org/publications/publication/study-ofan-old-railway-rock-tunnel-site-investigation-laboratorytests-weathering-effects-and-computational-analysis (accessed on 25 July 2021).
199. Thompson, L.; Stowell, J.; Fargher, S.; Steer, C.; Loughney, K.; O'sullivan, E.; Gluyas, J.; Blaney, S.; Pidcock, R. Muon tomography for railway tunnel imaging. *Phys. Rev. Res.* **2020**, *2*, 023017. [CrossRef]
200. Han, R.; Yu, Q.; Li, Z.; Li, J.; Cheng, Y.; Liao, B.; Jiang, L.; Ni, S.; Yi, Z.; Liu, T.; et al. Cosmic muon flux measurement and tunnel overburden structure imaging. *J. Instrum.* **2020**, *15*, P06019. [CrossRef]
201. Di Castro, M.; Tambutti, M.L.B.; Ferre, M.; Losito, R.; Lunghi, G.; Masi, A. I-TIM: A robotic System for Safety, Measurements, Inspection and Maintenance in Harsh Environments. In Proceedings of the 2018 IEEE International Symposium on Safety, Security, and Rescue Robotics (SSRR), Philadelphia, PA, USA, 6–8 August 2018; pp. 1–6. [CrossRef]
202. Shi, C.; Che, H.; Hu, H.; Wang, W.; Xu, X.; Li, J. Research on Laser Positioning System of a Underground Inspection Robot Based on Signal Reflection Principle. In Proceedings of the 2019 3rd International Conference on Robotics and Automation Sciences (ICRAS), Wuhan, China, 1–3 June 2019; pp. 58–62. [CrossRef]
203. Vithanage, R.; Harrison, C.; Desilva, A. Importance and applications of robotic and autonomous systems (RAS) in railway maintenance sector: A review. *Computers* **2019**, *8*, 56. [CrossRef]
204. Lincseek2021. Rail-Mounted Robot. Organisation Website. 2021. Available online: http://en.launchdigital.net/product.aspx?t=25 (accessed on 2 July 2021).
205. Ziegler, M.; Loew, S. Investigations in the New TBM-Excavated Belchen Highway Tunnel. In *Program, Design and Installations (Part 1)*; Techreport, ETH Zürich: Zürich, Switzerland, 2017. Available online: https://www.researchgate.net/publication/333653106_Investigations_in_the_new_TBM-excavated_Belchen_highway_tunnel - Program_design_and_installations_Part_1 (accessed on 25 November 2021).
206. Zhang, T.; Shi, B.; Zhang, C.; Xie, T.; Yin, J.; Li, J. Tunnel Disturbance Events Monitoring and Recognition with Distributed Acoustic Sensing (DAS). In Proceedings of the 11th Conference of Asian Rock Mechanics Society, Asian Rock Mechanics Society,

School of Earth Sciences and Engineering, Nanjing University, Beijing, China, 21–25 October 2021; IOP Publishing: Nanjing, China, 2021; Volume 861, p. 042034. [CrossRef]
207. Lienhart, W.; Buchmayer, F.; Klug, F.; Monsberger, C. Distributed Fibre-Optic Sensing Applications at the Semmering Base Tunnel, Austria. In *Institution of Civil Engineers–Smart Infrastructure and Construction*; Institute of Engineering Geodesy and Measurement Systems, Graz University of Technology: Graz, Austria; ICE Publishing: London, UK, 2020; Volume 172, pp. 148–159. [CrossRef]
208. Hulse, L.; Xie, H.; Galea, E. Perceptions of autonomous vehicles: Relationships with road users, risk, gender and age. *Saf. Sci.* **2018**, *102*, 1–13. [CrossRef]
209. El Masri, Y.; Rakha, T. A scoping review of non-destructive testing (NDT) techniques in building performance diagnostic inspections. *Constr. Build. Mater.* **2020**, *265*, 120542. [CrossRef]
210. Zou, L.; Yi, L.; Sato, M. On the use of lateral wave for the interlayer debonding detecting in an asphalt airport pavement using a multistatic GPR system. *IEEE Trans. Geosci. Remote Sens.* **2020**, *58*, 4215–4224. [CrossRef]
211. McDonald, T.; Robinson, M.; Tian, G. Spatial Resolution Enhancement of Rotational-Radar Subsurface Datasets Using Combined Processing Method. In JPCS Conference Series. In Proceedings of the 10th International Conference on Mathematical Modelling in Physical Sciences, Online, 6–9 September 2021; Vlachos, D., Ed.; IC-MSQUARE; IOP Publishing: Bristol, UK, 2021. [CrossRef]
212. ZeticaRail. ZARR Zetica Advanced Rail Radar. Online Document. 2017. Available online: https://zeticarail.com/wpcontent/uploads/2017/02/English-International-Flyer.pdf (accessed on 13 July 2021).
213. Han, J.; Cho, Y.; Lee, H.; Yang, H.; Jeong, W.; Moon, Y. Crack Detection Method on Surface of Tunnel Lining. In Proceedings of the 2019 34th International Technical Conference on Circuits/Systems, Computers and Communications (ITC-CSCC), JeJu, Korea, 23–26 June 2019; pp. 1–3. [CrossRef]
214. Hou, S.; Dong, B.; Wang, H.; Wu, G. Inspection of surface defects on stay cables using a robot and transfer learning. *Autom. Constr.* **2020**, *119*, 103382. [CrossRef]
215. Liu, B.; Ren, Y.; Liu, H.; Xu, H.; Wang, Z.; Cohn, A.; Jiang, P. GPRInvNet: Deep learning-based ground-penetrating radar data inversion for tunnel linings. *IEEE Trans. Geosci. Remote Sens.* **2021**, *59*, 8305–8325. [CrossRef]
216. Zhang, P.; Chen, R.; Dai, T.; Wang, Z.; Wu, K. An AIoT-based system for real-time monitoring of tunnel construction. *Tunn. Undergr. Space Technol.* **2021**, *109*, 103766. [CrossRef]
217. Gagarin, N.; Mekemson, J.; Goulias, D. Second-Generation Analysis Approach for Condition Assessment of Transportation Infrastructure Using Step-Frequency (SF) Ground-Penetrating-Radar (GPR) Array System. In *Bearing Capacity of Roads, Railways and Airfields*, 1st ed.; Andreas, L., Al-Qadi, I., Scarpas, T., Eds.; CRC Press: Boca Raton, FL, USA, 2017; pp. 1573–1581. Available online: https://doi.org/10.1201/9781315100333-209/secondgeneration-analysis-approach-condition-assessmenttransportation-infrastructure-using-step-frequency-sfground-penetrating-radar-gpr-array-system-gagarinmekemson-goulias (accessed on 23 July 2021). [CrossRef]
218. Insa-Iglesias, M.; Jenkins, M.; Morison, G. 3D visual inspection system framework for structural condition monitoring and analysis. *Autom. Constr.* **2021**, *128*, 1–9. [CrossRef]
219. Makantasis, K.; Protopapadakis, E.; Doulamis, A.; Doulamis, N.; Loupos, C. Deep Convolutional Neural Networks for Efficient vision Based Tunnel Inspection. In Proceedings of the 2015 IEEE International Conference on Intelligent Computer Communication and Processing (ICCP), Cluj-Napoca, Romania, 3–5 September 2015; pp. 335–342. [CrossRef]
220. Nasrollahi, M.; Bolourian, N.; Hammad, A. Concrete Surface Defect Detection Using Deep Neural Network Based on LIDAR Scanning. In Proceedings of the (CSCE 2019) 7th International Construction Conference Jointly with the Construction Research Congress (CRC 2019), Montreal, QC, Canada, 12–15 June 2019. Available online: https://www.researchgate.net/publication/335276365_Concrete_Surface_Defect_Detection_Using_Deep_Neural_Network_Based_on_LiDAR_Scanning (accessed on 30 July 2021).
221. Arbabsiar, M.; Farsangi, M.; Mansouri, H. Fuzzy logic modelling to predict the level of geotechnical risks in rock tunnel boring machine (TBM) tunnelling. *Min.-Geol.-Pet. Bull.* **2020**, *35*, 1–14. Available online: https://hrcak.srce.hr/ojs/index.php/rgn/article/view/9979 (accessed on 27 July 2021). [CrossRef]
222. El-Khateeb, L.; Mohammed Abdelkader, E.; Al-Sakkaf, A.; Zayed, T. A hybrid multi-criteria decision making model for defect-based condition assessment of railway infrastructure. *Sustainability* **2021**, *13*, 7186. [CrossRef]
223. Sajid, S.; Taras, A.; Chouinard, L. Defect detection in concrete plates with impulse-response test and statistical pattern recognition. *Mech. Syst. Signal Process.* **2021**, *161*, 107948. [CrossRef]
224. Islam, D.; Jackson, R.; Zunder, T.; Burgess, A. Assessing the impact of the 2011 EU transport white paper—A rail freight demand forecast up to 2050 for the EU27. *Eur. Transp. Res. Rev.* **2015**, *7*, 22. [CrossRef]
225. Network Rail Asset Management Policy January 2018. Online Document. 2018. Available online: https://www.networkrail.co.uk/wpcontent/uploads/2019/10/Asset-Management-Policy-2018.pdf (accessed on 25 July 2021).
226. Network Rail Establishing. Condition of Hidden Critical Elements. Online Document. 2019. Available online: https://www.networkrail.co.uk/wpcontent/uploads/2019/06/Challenge-Statement-Bridges-HCEHidden-Critical-Elements.pdf (accessed on 12 November 2021).
227. Pragnell, H. Early British Railway Tunnels: The Implications for Planners, Landowners and Passengers between 1830 and 1870. Ph.D. Thesis, University of York, York, UK, 2016. Available online: https://etheses.whiterose.ac.uk/16826/1/Railwaytunnelsrecovered3.pdf (accessed on 29 July 2021).

228. Taylor, P. Search for Hidden Construction Shafts within the Welsh Railway Tunnels. In Proceedings of the XVII ECSMGE-2019 Geotechnical Engineering Foundation of the Future, Reykjavik, Iceland, 1–6 September 2019. Available online: https://www.ecsmge-2019.com/uploads/2/1/7/9/21790806/0215-ecsmge-2019_taylor.pdf (accessed on 25 July 2021).
229. Office of Rail and Road (ORR). *Network Rail Monitor: Quarters 1–2 of Year 4 of CP5*; Techreport; ORR: London, UK, 2017. Available online: https://www.orr.gov.uk/sites/default/files/om/networkrail-monitor-2017-18-q1-2.pdf (accessed on 12 November 2021).
230. Office of Rail and Road (ORR). *Annual Assessment of Network Rail—April 2020 to March 2021*; Techreport; ORR: London, UK, 2021. Available online: https://www.orr.gov.uk/sites/default/files/2021-07/annualassessment-of-network-rail-2020-21.pdf (accessed on 12 November 2021).
231. Fletcher, N.; Brown, M.; Sadek, T. Great Western Railway Electrification, UK: Patchway Tunnels. In *Institution of Civil Engineers-Civil Engineering*; Thomas Telford Ltd.: London, UK, 2020; Volume 173, pp. 37–45. [CrossRef]
232. Khan, R.; Emad, M.; Jo, B. Tunnel portal construction using sequential excavation method: A case study. *MATEC Web Conf.* **2017**, *138*, 04002. [CrossRef]
233. Spinks, J. Strengthening of Heritage Tunnel Portals. In OPUS. In Proceedings of the 12th Australia New Zealand Conference on Geomechanics, Wellington, New Zeeland, 22–25 February 2015; Ramsey, G., Ed.; SIMSG ISSMGE: Wellington, New Zealand, 2021; Volume 1. Available online: https://www.issmge.org/publications/publication/strengthening-of-heritage-tunnel-portals (accessed on 25 July 2021).
234. POLYPIPE. Rail Construction Solutions. Online Document. 2014. Available online: https://www.polypipe.com/sites/default/files/Rail_Construction_Solutions_Dec2014.pdf (accessed on 28 July 2021).
235. Zhang, P.; Huang, Z.; Liu, S.; Xu, T. Study on the control of underground rivers by reverse faults in tunnel site and selection of tunnel elevation. *Water* **2019**, *11*, 889. [CrossRef]
236. Hui, H.; Bowen, Z.; Yanyan, Z.; Chunmei, Z.; Yize, W.; Zeng, G. The mechanism and numerical simulation analysis of water bursting in filling karst tunnel. *Geotech. Geol. Eng.* **2018**, *36*, 1197–1205. [CrossRef]
237. Atkinson, C.; Paraskevopoulou, C.; Miller, R. Investigating the rehabilitation methods of victorian masonry tunnels in the UK. *Tunn. Undergr. Space Technol.* **2021**, *108*, 103696. [CrossRef]
238. Gao, L.; Zhao, W.; Hou, B.; Zhong, Y. Analysis of influencing mechanism of subgrade frost heave on vehicle-track dynamic system. *Appl. Sci.* **2020**, *10*, 8097. [CrossRef]
239. Parrott, J.; Lahra, J. Masonry Arch Bridges and Tunnels Repair and Strengthening: A Case Study. Online Document. 2014. Available online: https://bridgerestoration.co.uk/wpcontent/uploads/2019/10/Underpass-strengthening.pdf (accessed on 12 November 2021).
240. Chen, H.; Yu, H.; Smith, M. Physical model tests and numerical simulation for assessing the stability of brick-lined tunnels. *Tunn. Undergr. Space Technol.* **2016**, *53*, 109–119. [CrossRef]
241. Akagawa, S.; Hori, M.; Sugawara, J. Frost heaving in ballast railway tracks. *Procedia Eng.* **2017**, *189*, 547–553. [CrossRef]
242. Luo, Y.; Chen, J. Research status and progress of tunnel frost damage. *J. Traffic Transp. Eng.* **2019**, *6*, 297–309. [CrossRef]
243. BBC News. Flood-Prone Crick Railway Tunnel Repairs 'Will Reduce Delays'. *BBC News.* 2020. Available online: https://www.bbc.co.uk/news/ukengland-northamptonshire-56354714 (accessed on 22 November 2021).
244. RAIB. *Partial Failure of a Structure inside Balcombe Tunnel, West Sussex 23 September 2011. Tech. Report R132013-130815, GOV.UK*; The Wharf Stores: Derby, UK, 2014. Available online: https://www.gov.uk/raib-reports/partial-failure-of-astructure-inside-balcombe-tunnel-west-sussex (accessed on 22 November 2021).
245. Nielsen, J. *Progressive Disclosure. Organisation Website*. 2006. Available online: https://www.nngroup.com/articles/progressive-disclosure/ (accessed on 9 November 2020).
246. Vi, S.; da Silva, T.; Maurer, F. User Experience Guidelines for Designing HMD Extended Reality Applications. In *Human-Computer Interaction—INTERACT 2019*; Lamas, D., Loizides, F., Nacke, L., Petrie, H., Winckler, M., Zaphiris, P., Eds.; University of Calgary: Calgary, AB, Canada; Springer: Cham, Switzerland, 2019; Volume 11749, pp. 319–341. [CrossRef]
247. Fast-Berglund, A.; Gong, L.; Li, D. Testing and validating extended reality (XR) technologies in manufacturing. *Procedia Manuf.* **2018**, *25*, 31–38. [CrossRef]
248. Chuah, S. *Why and Who Will Adopt Extended Reality Technology? Literature Review, Synthesis, and Future Research Agenda*; Elsevier: Amsterdam, The Netherlands, 2018; pp. 1–55. [CrossRef]
249. Hansen, L.; Wyke, S.; Kjems, E. Combining Reality Capture and Augmented Reality to Visualise Subsurface Utilities in the Field. In *ISARC Proceedings of the International Symposium on Automation and Robotics in Construction*; IAARC Publications: Waterloo, ON, Canada, 2020; Volume 37, pp. 703–710. Available online: https://vbn.aau.dk/ws/portalfiles/portal/401875244/Combining_Reality_Capture_and_Augmented_Reality_to_Visualise_Subsurface_Utilities_in_the_Field.pdf (accessed on 2 July 2021).
250. Du, J.; Zou, Z.; Shi, Y.; Zhao, D. Simultaneous data exchange between BIM and VR for collaborative decision making. *Comput. Civ. Eng.* **2017**, 1–8. [CrossRef]
251. Wang, F.; Sui, H.; Kong, W.; Zhong, H. Application of BIM + VR technology in immersed tunnel construction. *IOP Conf. Ser. Earth Environ. Sci.* **2021**, *798*, 012019. [CrossRef]
252. Cosma, G.; Ronchi, E.; Nilsson, D. Way-finding lighting systems for rail tunnel evacuation: A virtual reality experiment with oculus rift®. *J. Transp. Saf. Secur.* **2016**, *8* (Suppl. 1), 101–117. [CrossRef]

253. Arias, S.; La Mendola, S.; Wahlqvist, J.; Rios, O.; Nilsson, D.; Ronchi, E. Virtual reality evacuation experiments on way-finding systems for the future circular collider. *Fire Technol.* **2019**, *55*, 2319–2340. [CrossRef]
254. Insa-Iglesias, M.; Jenkins, M.; Morison, G. An Enhanced Photorealistic Immersive System Using Augmented Situated Visualization within Virtual Reality. In Proceedings of the 2021 IEEE Conference on Virtual Reality and 3D User Interfaces Abstracts and Workshops (VRW), Lisbon, Portugal, 27 March–1 April 2021; pp. 514–515. [CrossRef]
255. Network Rail. Our Business Areas. Organisation Website. 2021. Available online: https://www.networkrail.co.uk/careers/our-business-areas/ (accessed on 19 July 2021).
256. Network Rail. Careers. Organisation Website. 2021. Available online: https://www.networkrail.co.uk/careers/careers-search/ (accessed on 19 July 2021).
257. MTI. Tunnelmesh—100% Wireless Connectivity for Tunnels. Organisation Website. 2021. Available online: https://mti-technology.co.uk/tunnelmesh/ (accessed on 23 November 2021).
258. MTI. UK Mine Communication System. Organisation Website. 2020. Available online: https://mti-technology.co.uk/mine-communicationtunnelmesh/ (accessed on 23 November 2021).
259. Singh, A.; Singh, U.K.; Kumar, D. IoT in mining for sensing, monitoring and prediction of underground mines roof support. In Proceedings of the 4th International Conference on Recent Advances in Information Technology (RAIT), Dhanbad, India, 15–17 March 2018; pp. 1–5. [CrossRef]

 *sensors*

Article

FEM Simulation-Based Adversarial Domain Adaptation for Fatigue Crack Detection Using Lamb Wave

Li Wang [1,2], Guoqiang Liu [2], Chao Zhang [1], Yu Yang [2,*] and Jinhao Qiu [1,*]

[1] State Key Laboratory of Mechanics and Control of Mechanical Structures, Nanjing University of Aeronautics and Astronautics, Nanjing 210016, China
[2] Structural Damage Monitoring Laboratory, Aircraft Strength Research Institute of China, Xi'an 710065, China
* Correspondence: yangyu@cae.ac.cn (Y.Y.); qiu@nuaa.edu.cn (J.Q.)

Abstract: Lamb wave-based damage detection technology shows great potential for structural integrity assessment. However, conventional damage features based damage detection methods and data-driven intelligent damage detection methods highly rely on expert knowledge and sufficient labeled data for training, for which collecting is usually expensive and time-consuming. Therefore, this paper proposes an automated fatigue crack detection method using Lamb wave based on finite element method (FEM) and adversarial domain adaptation. FEM-simulation was used to obtain simulated response signals under various conditions to solve the problem of the insufficient labeled data in practice. Due to the distribution discrepancy between simulated signals and experimental signals, the detection performance of classifier just trained with simulated signals will drop sharply on the experimental signals. Then, Domain-adversarial neural network (DANN) with maximum mean discrepancy (MMD) was used to achieve discriminative and domain-invariant feature extraction between simulation source domain and experiment target domain, and the unlabeled experimental signals samples will be accurately classified. The proposed method is validated by fatigue tests on center-hole metal specimens. The results show that the proposed method presents superior detection ability compared to other methods and can be used as an effective tool for cross-domain damage detection.

Keywords: fatigue crack detection; lamb waves; finite element method; domain-adversarial neural network; maximum mean discrepancy; metal structures

Citation: Wang, L.; Liu, G.; Zhang, C.; Yang, Y.; Qiu, J. FEM Simulation-Based Adversarial Domain Adaptation for Fatigue Crack Detection Using Lamb Wave. *Sensors* **2023**, *23*, 1943. https://doi.org/10.3390/s23041943

Academic Editor: Gilbert-Rainer Gillich

Received: 22 December 2022
Revised: 1 February 2023
Accepted: 6 February 2023
Published: 9 February 2023

1. Introduction

As a unique non-destructive evaluation (NDE) system, structural health monitoring (SHM) has shown great potential in reducing maintenance cost, extending service life, and ensuring structural integrity [1,2]. Several SHM techniques have been implemented for damage detection in the past few years, such as strain-based SHM [3], electromechanical impedance-based SHM [4], smart coating-based SHM [5], Lamb waves-based SHM [6], etc. Due to the long-distance propagation and small attenuation of Lamb waves in structures, Lamb waves-based damage detection technology has received extensive concerns.

However, the complexities involved with Lamb wave due to its multi-modal and dispersive nature make the signals analysis quite strenuous [7], and its physics modeling to predict the output and identifying the damage [8,9] is a difficult and prohibitive task. Conventional Lamb wave-based damage detection methods are to extract predesigned damage features of Lamb wave in time and frequency domain and identify structural damages by comparing damage features with their thresholds [10,11]. Due to the effect of structures geometry on Lamb wave, the damage feature threshold needs to be adjusted according to different structures, which often presents a less robustness and poor knowledge generalization performance in real life structures with complicated geometry. These methods also need a reasonable selection of damage features, which highly rely on expert experience. To circumvent these limitations, many damage detection methods based on machine learning have been developed for automatic damage detection without a specific threshold. Atashipour et al. [12] proposed an automatic damage identification approach for steel beams based on Lamb wave and artificial neural network (ANN). Damage character points based on continuous wavelet transform were extracted first, then a multilayer ANN supervised by error-back propagation algorithm was trained to automatic detect damage. Li et al. [13] used Hilbert transform, power spectral density, fast Fourier transform, and wavelet fractal dimension to extract multi-features from time domain, frequency domain, and fractal dimension of Lamb wave. Following that, a machine learning method based on support vector machine (SVM) was used to fuse multi-features and further identify damage. Yang et al. [14] developed an integrated damage identification method based on least margin, which integrates multiple machine-learning models and outputs the fused damage identification result by polling all models' decisions. Twelve damage features and seven machine learning methods, including k-nearest neighbor (KNN), radial basis function support vector machine (RBF-SVM), Gaussian process (GP), decision tree (DTree), neural network (NN), Gaussian naive Bayes (GNB) and quadratic discriminant analysis (QDA), were applied to predict the damage identification results.

Instead of manual feature extraction for machine learning, various damage detection methods based on deep learning have been developed to automatically feature extraction and damage detection. Lee et al. [15] adopted a deep autoencoder (DAE) to capture hidden representation and effective tracking of signal variations, and the reconstruction error was used to diagnosis fatigue damage in composites structures. Chen et al. [16] and Wu et al. [17] converted the Lamb wave signals into a two-dimensional time-frequency spectrogram with the continuous wavelet transform, then input them into a 2D convolutional neural network (CNN) to classify damage. To minimize loss of information in conversion of time signals to image, Pandey et al. [7] used 1D CNN to detect damage directly using original Lamb wave signals of an aluminum plate, in which Lamb wave response signals obtained with FEM simulations were used as training samples and experimental data were used for testing. Sampath et al. [18] incorporated the long short-term memory (LSTM) with trispectrum-based higher-order spectral analysis to propose a novel hybrid method for reliable fatigue crack detection under noisy environments. The DL model based on LSTM was used to eliminate the random noise by reconstructing the original Lamb wave signals, and trispectrum-based higher-order spectral analysis method was adopted to extract the nonlinear components considered as an indication of fatigue cracks. Yang et al. [19] used the temporal distributed conventional neural network (TDCNN) to extract less expertise-dependent features, in which the long short-term memory (LSTM) was used to associate features of data fragments.

However, these data-driven intelligent damage detection methods require sufficient labeled data to train the intelligent model for good performance, and the training and testing data must follow the same distribution. In industrial scenarios, collecting sufficient labeled data for training is usually expensive and impractical, and usually just a large amount of unlabeled data can be obtained. Meanwhile, many available training data are obtained by simulations or from simple structure forms, which may not follow the same distribution with data from the practical complex conditions. When the training data are non-labeled or the distributions are mismatched, the performance of data-driven intelligent damage detection methods may drop sharply.

Transfer learning is an effective knowledge generalization tool and can transfer the knowledge learned from an abundant labeled source domain into a new but related target domain [20]. By combining the hidden features of learning ability and deep learning and knowledge transfer ability of transfer learning, deep transfer learning has been widely studied. Nowadays, deep domain adaptation has a dominant position in deep transfer learning, which can be summarized into three categories: discrepancy-based, adversarial discrimination-based, and adversarial generation-based deep domain adaptation methods [21]. The main idea of discrepancy-based deep domain adaptation methods is to minimize the distribution discrepancy between source domain and target domain to get domain-invariant features, in which MMD [22–24], Kullback–Leibler (KL) divergence [25–27], and Wasserstein distance [28,29], et al., can be used as the metrics of inter-domain distribution divergence. Deep domain confusion (DDC) [23] and deep adaptation network (DAN) [24] are the classical discrepancy-based deep domain adaptation methods. Adversarial discrimination-based deep domain adaptation methods aim to extract fault-discriminative and domain-invariant features through adversarial training with the gradient reversal layer (GRL) [30]. DANN [31] is a classical adversarial discrimination-based deep domain adaptation method of feed-forward architectures, combining domain adaptation and deep feature learning within one training process. Adversarial generation-based deep domain adaptation methods aim to minimize the discrepancy between the source data and target data with the adversarial training. The widely used adversarial generation-based deep domain adaptation models include generation adversarial network (GAN) [32], Wasserstein-GAN (WGAN) [33], Wasserstein-GAN with gradient penalty (WGAN-GP) [34], et al. Many deep domain adaptation models for structural health monitoring using Lamb wave have been developed. Alguri et al. [35] proposed a transfer learning framework for Lamb waves' full wave field reconstruction, in which autoencoder was used to learn the general propagation of signals, then the learned knowledge was combined with sparse spatial measurements to reconstruct full wavefield. Zhang et al. [36] applied the joint distribution adaptation (JDA) to adapt both the marginal distribution and conditional distribution of the Lamb waves from aluminum plate and composite plate, then used the LSTM network to learn the damage indexes for damage probability imaging. Zhang et al. [37] proposed a muti-task deep transfer learning methods by transferring the high-level shared features of damage level detection task to damage location task. Wang et al. [38] used the MMD-based deep adaptation network for learning transferable features to make the classifier trained on the labeled source data achieve comparable performance on the unlabeled target data. Even though several deep transfer learning methods have been developed for Lamb wave-based damage monitoring, fatigue crack detection with deep transfer learning model for Lamb wave has not yet been fully studied.

In order to automatically detect fatigue crack and further improve the detection accuracy, in this paper, a deep domain adaptation method based on FEM simulation and MMD-DANN was proposed for damage detection using Lamb wave. FEM simulations are employed to obtain the simulated response signals under different conditions. Therefore, the insufficient labeled data of real-world can be possibly solved. However, the variabilities of real response signals during the fatigue crack growth have not been represented in the simulated response signals. For engineering structures, the real response signals are affected by complex uncertainties, like structure manufacturing procedure, environmental variables, crack geometries, multi-sensors performance, and the sensor installation process. Thus, the classifier model of simulated signals cannot be directly applied to the real structures. By fusing the distribution discrepancy metric and the adversarial discrimination training to minimize the domain disparity of the simulated source data and experimental target data, the MMD-DANN model was developed to learn damage-discriminative and domain-invariant feature representations. Then, the classifier model of simulated signals can be directly transferred to experimental signals with comparative ability.

The outline of this paper is as follows. The details of the proposed fatigue crack detection method are specified in Section 2, including the network architecture of MMD-DANN model, training of MMD-DANN model, and detection procedure of target domain. In Section 3, the proposed method is demonstrated through fatigue test data of center-hole metal specimens. Conclusions are drawn in Section 4.

2. Proposed MMD-DANN-Based Fatigue Crack Detection Method

By combining the distribution discrepancy metric of MMD and the adversarial discrimination training of DANN model, an unsupervised deep domain adaption method based on MMD-DANN model was proposed to detect fatigue crack in this paper, which bridges the source and target domains in an isomorphic latent feature space and performs a superior diagnosis performance on the unlabeled target data.

In the proposed method, the simulated Lamb wave response signals obtained from the undamaged case and multiple damaged cases are assigned as the labeled source domain data. The experimental Lamb wave response signals obtained from the fatigue test are assigned as unlabeled target domain data. Assume that the source and target domain data are $\mathcal{D}_s = \{x_s^i, y_s^i\}_{i=1}^m$ and $\mathcal{D}_t = \{x_t^j\}_{j=1}^n$, in which $\mathcal{D}_s$ are the m labeled source samples and $\mathcal{D}_t$ are the n unlabeled target samples. $x_s, y_s \in \{1, \ldots, K\}$ are the simulated response signals and the corresponding labels for K types of damage categories in the source domain.x_t are the unlabeled experimental response signals in the target domain.

2.1. Network Architecture of MMD-DANN Model

The architecture of the proposed MMD-DANN model is illustrated in Figure 1, with all hyperparameters used in the paper given besides the layers. It consists of a deep feature extractor G_f, a deep label predictor G_y, and a domain classifier G_d. As shown in Figure 1, the labeled source and unlabeled target Lamb wave signals are first input into the feature extractor G_f to extract the multi-dimensional features vector. The label predictor G_y takes the extracted features of the labeled source data as input and predicts the class labels. The domain classifier G_d takes the extracted features of the source and target data as input and predicts the domain labels. The distribution discrepancy of the extracted features is reduced by multi-layer domain adaptation and the adversarial training between the feature extractor and domain classifier.

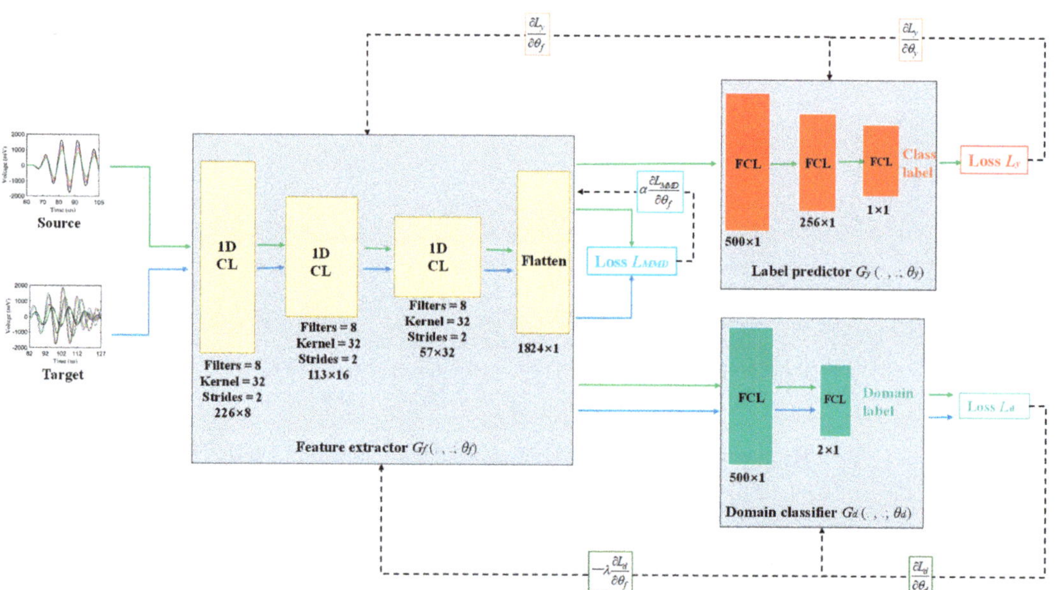

Figure 1. The architecture of the proposed MMD-DNN model.

The feature extractor G_f is composed of three 1-D convolutional layers (CLs) and one flattened layer. The features of the input response signals are extracted through three layers of convolutional computation, and the resulted features map is compressed down to multi-dimensional features vector through a flattened layer. The rectified linear unit (ReLU) was used as an activation function of every CL to improve computational efficiency [39], and a batch normalization (BN) layer was used after every CL to normalize the data and reduce the internal covariate shift [40]. The weights of G_f of the source and target domain are shared.

The label predictor G_y is made up of three fully connected layers (FCLs). The high-dimensional features vector of the source domain extracted by G_f is input to G_y and further compressed through two FCLs with ReLU activation function. The sigmoid activation function is used in the third FCL to predict the class label. The weights of G_y of the source and target domain are also shared.

The domain classifier G_d is composed of two FCLs. The high-dimensional features vector of the source and target domain extracted by G_f are used as input and further compressed through one FCL with ReLU activation function. The softmax function is used in the second FCL to predict the domain label. The weights of G_d of the source and target domain are also shared.

2.2. Training of MMD-DANN Model

According to the three outputs of MMD-DANN model, three losses are constructed, including the label prediction loss L_y based on the outputs of the label predictor G_y, the domain classification loss L_d based on the outputs of domain classifier G_d, and the domain adaptation loss L_{MMD} based on the outputs of the feature extractor G_f, as illustrated in Figure 1. Three losses are detailed as follows:

The label predictor G_y takes the extracted features of the labeled source data as input and outputs the predicted labels. During the training stage, for the good performance of the label predictor, the label predictor G_y aims to minimize the label prediction loss between the predicted label and the actual label for the training source data in a supervised way. The parameters θ_f, θ_y of both the feature extractor G_f and the label predictor G_y are optimized at the same time. The label prediction loss L_y of the label predictor G_y based on the cross-entropy loss function is defined as:

$$L_y(\mathcal{D}_s) = -\mathrm{E}_{(x_s,y_s)\sim\mathcal{D}_s} \sum_{k=1}^{K} 1_{[k=y_s]} \log\left(G_y\left(G_f(x_s)\right)\right) \tag{1}$$

where $\mathrm{E}_{(xs,\ ys)\sim\mathcal{D}s}$ means calculates the expectation of the samples from $\mathcal{D}_s$. $1_{[K\,=\,ys]}$ is an indicator function, if $K = y_s$, its value equals 1, else it equals 0.

Simultaneously, the domain classifier G_d takes the extracted features of the source and target domain data as input and outputs the predicted domain labels. During the training stage, for the good performance of the domain classifier, the domain classifier G_d aims to minimize the domain classification loss of two domains in a supervised way. The domain label of the source domain data is assigned as 0, and the domain label of the target domain data is assigned as 1. The parameters θ_f, θ_d of both the feature extractor G_f and the domain classifier G_d are optimized at the same time. The domain classification loss L_d of the domain classifier G_d based on the cross-entropy loss function is defined as:

$$L_d(\mathcal{D}_s, \mathcal{D}_t) = -\mathrm{E}_{x_s\sim\mathcal{D}_s}\left[\log\left(G_d\left(G_f(x_s)\right)\right)\right] - \mathrm{E}_{x_t\sim\mathcal{D}_t}\left[\log\left(1 - G_d\left(G_f(x_t)\right)\right)\right] \tag{2}$$

To further reduce the distribution discrepancy between the two domains and improve the domain adaptation of the feature extractor, the distribution discrepancy of the output features vector between two domains is measured and incorporated into the model training, which is defined as the domain adaptation loss L_{MMD}. MMD is a common distance metric in deep domain adaptation to measure the distribution discrepancy between two datasets. MMD of $\mathcal{D}_s$ and $\mathcal{D}_t$ after the feature extractor G_f can be expressed as:

$$L_{MMD}(\mathcal{D}_s, \mathcal{D}_t) = \sup_{G_f\in\mathcal{F}} \left(\mathrm{E}_{\mathcal{D}_s}[G_f(x_s)] - \mathrm{E}_{\mathcal{D}_t}[G_f(x_t)]\right) \tag{3}$$

where $\mathcal{F}$ represents the reproducing kernel Hilbert space (RKHS). sup(·) is the supremum of the input. By replacing the population expectations with empirical expectations, a biased empirical estimate of MMD can be obtained and written as:

$$L_{MMD}(\mathcal{D}_s, \mathcal{D}_t) = \sup_{G_f \in \mathcal{F}} \left(\frac{1}{m} \sum_{i=1}^{m} G_f\left(x_s^i\right) - \frac{1}{n} \sum_{j=1}^{n} G_f\left(x_t^j\right) \right) \tag{4}$$

By means of the kernel mean embedding of distribution, RKHS is induced by the characteristic kernels, such as Gaussian and Laplace kernels [38]. The empirical estimate of the squared MMD is defined as:

$$L_{MMD}{}^2(\mathcal{D}_s, \mathcal{D}_t) = \frac{1}{m^2} \sum_{i,j=1}^{m} k\left(x_s^i, x_s^j\right) - \frac{2}{mn} \sum_{i,j=1}^{m,n} k\left(x_s^i, x_t^j\right) + \frac{1}{n^2} \sum_{i,j=1}^{n} k\left(x_t^i, x_t^j\right) \tag{5}$$

where $k(.,.)$ is the characteristic kernel. In order to avoid the difficulty of selecting the kernel function, MK-MMD assumes that the optimal kernel can be obtained linearly from multiple kernels. The characteristic kernel associated with the feature map f, $k(x_s^i, x_t^j) = <f(x_s^i), f(x_t^j)>$, is defined as the convex combination of N kernels $\{k_u\}$:

$$\mathcal{K} \triangleq \left\{ k = \sum_{u=1}^{N} \beta_u k_u : \sum_{u=1}^{N} \beta_u = 1, \beta_u \geq 0, \forall u \right\} \tag{6}$$

where $\{\beta_u\}$ are the constraints on coefficients and are imposed to guarantee that the derived multi-kernel k is characteristic [39]. Multiple Gaussian kernels with different radial basis function (RBF) bandwidths are widely used as a nonparametric method.

In the forward-training process, the training of $\boldsymbol{G_f}$ can make the features discriminative. However, in order to make the extracted features domain-invariant, $\boldsymbol{G_f}$ should furthermore maximize the domain classification loss, which is run in the opposite direction to the training of $\boldsymbol{G_d}$. In order to implement the adversarial training between $\boldsymbol{G_f}$ and $\boldsymbol{G_d}$, GRL is inserted between the feature extractor and the domain classifier. In the forward propagation-based training, GRL acts as an identity transformer. However, during the back propagation-based training, the GRL multiplies the gradient by a certain negative constant $-\lambda$, leading the domain classification loss negative feedback to $\boldsymbol{G_f}$. A detailed discussion of GRL can be found in Ref. [30].

Therefore, by incorporating the MK-MMD loss into the adversarial training, the discriminative and domain-invariant features can be learned by the feature extractor. The total loss function of the feature extractor $\boldsymbol{G_f}$ is expressed as:

$$L_f(\mathcal{D}_s, \mathcal{D}_t) = L_y(\mathcal{D}_s) - \lambda L_d(\mathcal{D}_s, \mathcal{D}_t) + \alpha L_{MMD}(\mathcal{D}_s, \mathcal{D}_t) \tag{7}$$

where α is the trade-off hyperparameter of MK-MMD loss.

Based on the above loss function of $\boldsymbol{G_f}$, $\boldsymbol{G_y}$, and $\boldsymbol{G_d}$, the training is performed. The corresponding parameters θ_f, θ_y, θ_d of $\boldsymbol{G_f}$, $\boldsymbol{G_y}$, and $\boldsymbol{G_d}$ are updated as follows:

$$\theta_f \leftarrow \theta_f - \mu \frac{\partial L_f}{\partial \theta_f} \tag{8}$$

$$\theta_y \leftarrow \theta_y - \mu \frac{\partial L_y}{\partial \theta_y} \tag{9}$$

$$\theta_d \leftarrow \theta_d - \mu \frac{\partial L_d}{\partial \theta_d} \tag{10}$$

where μ is the learning rate. In this paper, the stochastic gradient descent (SGD) algorithm with 0.9 momentum and an annealing learning rate is used to optimize the model parameters [30]. The pseudo code of the proposed MMD-DANN model is presented in Appendix A.

2.3. Detection Procedure of Target Domain

When the training is complete, the unlabeled target domain data can be classified using the trained feature extractor and the trained label predictor. The target data are first input into the feature extractor to extract features vector, then fed forward into the label predictor, and the sigmoid activation function predicts the class labels. The detection procedure of the proposed MMD-DANN model is presented in Figure 2 and can be described as follows:

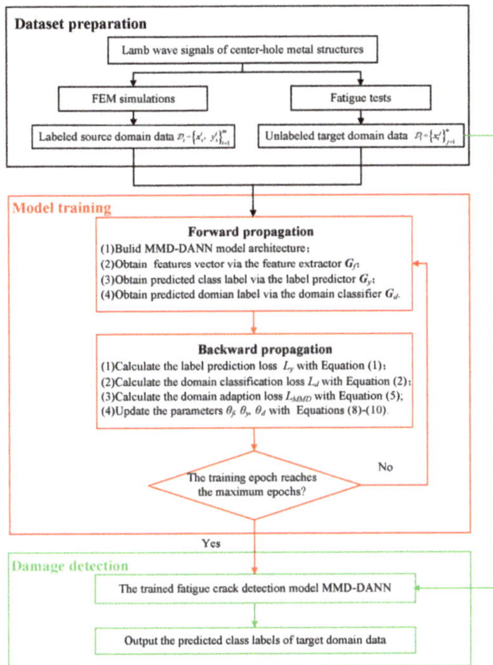

Figure 2. Flowchart of the proposed fatigue crack detection method based on MMD-DANN model.

First, the Lamb wave response signals of center-hole metal structures are obtained by FEM simulations and fatigue tests. The simulated response signals under different conditions are considered as the labeled source domain data, and the experimental response signals are considered as the unlabeled target domain data.

Then, in the step of model training, the labeled source domain data and unlabeled target domain data are fed forward into the MMD-DANN model. The extracted features vector, the predicted class labels, and the predicted domain labels are obtained. Three loss functions are calculated and propagated backward to update the model parameters until the training of the proposed MMD-DANN model is finished.

Finally, the unlabeled target domain data are used as testing samples to input to the trained model, and the damage detection results are obtained.

3. Experimental Validation

The proposed unsupervised domain adaptation method is validated on the fatigue crack detections of a center-hole metal structure. We aimed to detect the fatigue crack of the metal structures by transferring the learning knowledge from the abundant labeled source domain into an unlabeled target domain, in which the source domain data are assumed as the simulated response signals under different conditions and the target domain data are assumed as the experimental response signals under uncertain conditions. The simulated and experimental dataset are used to demonstrate the transfer results of the proposed MMD-DANN model and the detection accuracy of the experimental dataset.

3.1. Simulated Dataset and Data Preprocessing

The dataset consists of Lamb wave response signals of metal structures manufactured using 7050 aluminum of 3 mm thickness with a center-hole 25 mm diameter from simulation and experiment. Material properties of the center-hole metal specimen are shown in Table 1. Four PZT sensors P51 were installed on every specimen to monitor the healthy conditions of both sides of the center-hole. Two sensing paths, A1-S1 and A2-S2, were formed, in which A1 and A2 serve as actuators and S1 and S2 serve as sensors. The dimensions of PZT sensors are 8 mm in diameter and 0.45 mm in thickness. In order to demonstrate the detection performance on small fatigue cracks of the proposed method, only fatigue cracks under 8 mm are studied. Specimen geometry and PZT sensors placement are shown in Figure 3. The material and structure form considered as specimen are common in the aircraft. The FE model and the physical fatigue test are built with the same specimen geometrical dimensions and PZT sensors placement.

Table 1. Material properties of center-hole metal specimens.

ρ/(kg/m^3)	E/GPa	ν
2700	70	0.33

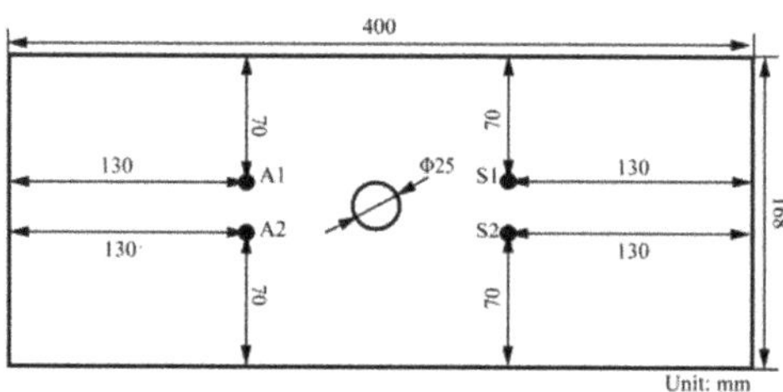

Figure 3. Specimen geometry and PZT sensors placement.

Considering the symmetries of structure geometry and sensors placement, FEM simulations are performed to acquire the response Lamb wave signals of only A1-S1 sensing path. The PZT sensor is modeled by an electromechanical coupling plate element, and the positive and negative piezo-conductive effects of actuators and sensors can be realized [41]. In order to simulate the uncertainty of the actual crack morphology, fatigue crack is modeled with a notch of 0.05 mm width, different length, and different orientations. Hanning window tone burst signals with a five-cycle frequency of 230 kHz is used as excitation signals. The central frequency is referred to the experiment. A fixed time increment of 0.1 µs is consistent with sampling rate of the experiment, which is sufficient to capture the interested time period of signals. The output time of field results for sensors is set as 4 ms to ensure that the number of the output data point is consistent with the experiment. The global element size is 1 mm, and the local encrypted element size of crack and hole are 0.5 mm.

The availability of the FE model is validated by comparing the simulated group velocity and the analytical group velocity for the same frequency. The simulated group velocity of S_0 mode is estimated as 5381 m/s, and the simulated group velocity of A_0 mode is estimated as 2955 m/s. The respective errors for group velocity of S_0 and A_0 mode are about 1.93% and 2.96%. In this paper, the response signals of S_0 mode are used to analyze. A two-dimensional FE model of the center-hole specimen with 5 mm fatigue crack and 90° orientation is created in ABAQUS/Explicit with one edge fixed and another edge loaded with 3 kN, as shown in Figure 4. The holding load is the same as the experiment. To illustrate the effect of crack on the wave propagation process, a screenshot of the simulated wave propagation of the center-hole specimen with 5 mm fatigue crack and 90° orientation is shown in Figure 5. The displacement at 85 μs can indicates the wave reflections by the damage and the top boundary, which further demonstrates the availability of the simulations.

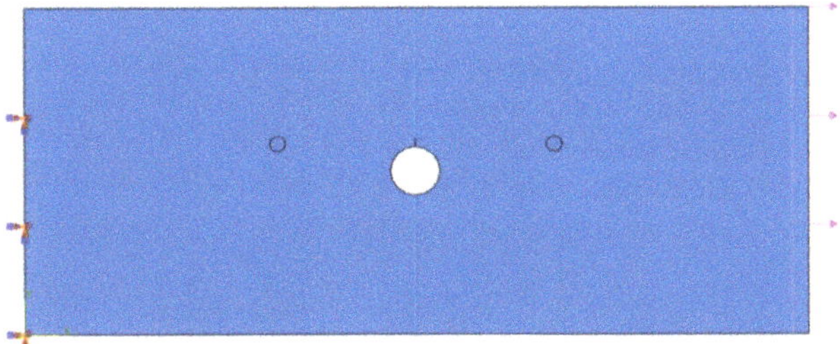

Figure 4. FE model of center-hole metal specimen using ABAQUS/Explicit.

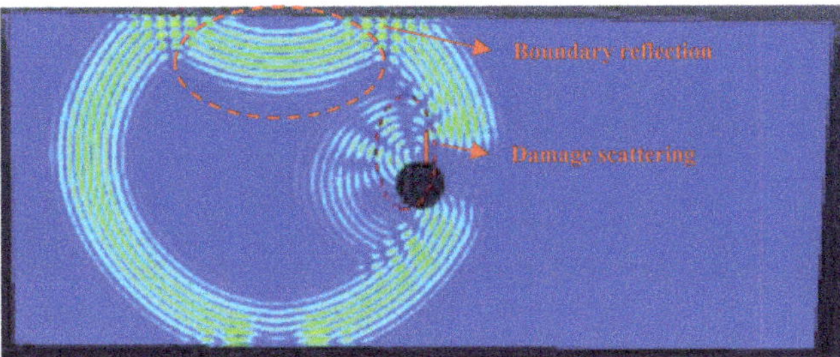

Figure 5. Total displacement nephogram of simulated wave propagation at 85 μs.

Simulation was performed for an undamaged case followed by multiple damaged cases. Damaged cases are induced at sixteen different crack lengths (from 0.5 mm to 8 mm at a step of 0.5 mm) with five different orientations (80°, 85°, 90°, 95°, 100°). Simulation was performed for one damage case at a time. Considering that direct wave packet contains the most effective structural information in the response signals, a fixed-length rectangular window [60 μs 105 μs] was employed to extract a direct wave packet for S_0 mode. Consequently, 81 simulated samples were 80 samples from damaged cases and the remaining 1 sample is undamaged, with each sample consisting of 451 data points which corresponded to the time instances of the extracted direct wave packet. Figure 6 shows the simulated Lamb wave response signals of different crack lengths for the A1-S1 sensing path. Significant phase right-shifting and amplitude decreasing can be observed with the undamaged and damaged conditions, which means that the simulated signals can reflect the variations of structural conditions.

In order to consider the effect of dispersions of specimens and sensors performance on the signals, data augment technology was used to introduce the fluctuations of amplitude and phase into the signals to expand the simulated samples. Firstly, the signals are converted into the analytical signals with Hilbert transform. Then, the amplitude and phase of the analytical signals are multiplied by a scaling coefficient to generate the virtual simulated samples. For the undamaged sample, the scaling coefficient for the amplitude varies from 0.82 to 1.20 at a step of 0.02, and the scaling coefficient for the phase varies from 0.82 to 1.21 at a step of 0.03. The scaling of the amplitude and phase are cyclic preceded. For the damaged samples, the scaling coefficient for the amplitude varies from 0.80 to 1.25 at a step of 0.15, and the scaling coefficient for the phase varies from 0.80 to 1.25 at a step of 0.15. The scaling of the amplitude and phase are meanwhile preceded.

That is, a total of 600 simulated samples are obtained with 280 undamaged samples and 320 damaged samples.

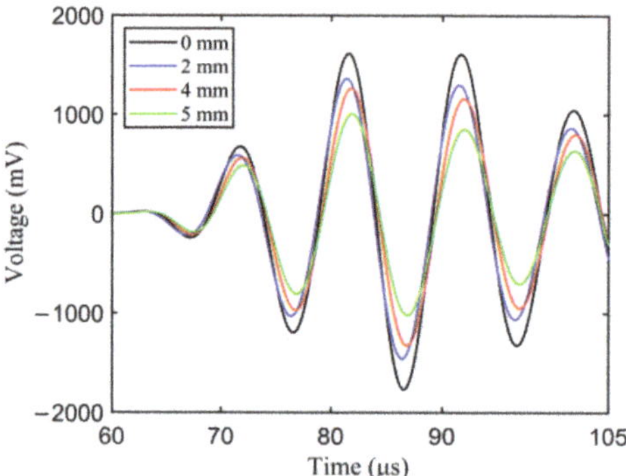

Figure 6. Simulated Lamb wave response signals of different crack length.

3.2. Experimental Dataset and Data Preprocessing

Fatigue tests on eight center-hole metal specimens were performed to obtain the experimental response signals, labeled from T1–T8. The experimental setup is shown in Figure 7. For a constant amplitude axial tensile cyclic load with load ratio $R = 0.1$, the maximum load of 40 kN and loading frequency of 8 Hz was applied to the specimens using a hydraulic MTS machine to introduce fatigue cracks in the specimens. The initiation and growth of fatigue cracks were measured offline with a charge-coupled device (CCD) camera. Crack lengths are measured from optical microscopic images of CCD camera. Representative examples of actual fatigue crack through the microscopic are shown in Figure 8. At the early stages of the fatigue test, only one crack initialized and grew. At the medium stage of the fatigue test, two cracks grew simultaneously.

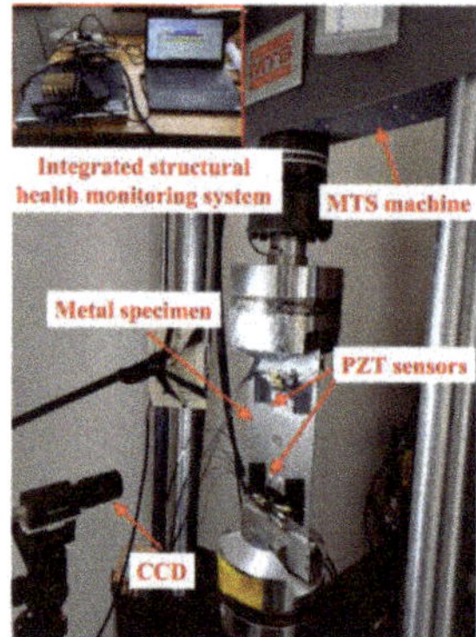

Figure 7. Experimental setup for center-hole metal specimens.

An integrated structural health monitoring system was utilized to generate and acquire online Lamb waves. During the fatigue test, Lamb waves were periodically acquired when the load was held at 3 kN. The load holding was to ensure the identical boundary condition for every signal acquisition. The temperature variation at the laboratory was maintained below 1 °C to minimize temperature effects on acquired Lamb waves [42]. The excitation signal is a five-cycle tone burst modulated by a Hanning window with the sampling rate set to 10 MHz and the sampling length set to 4000.

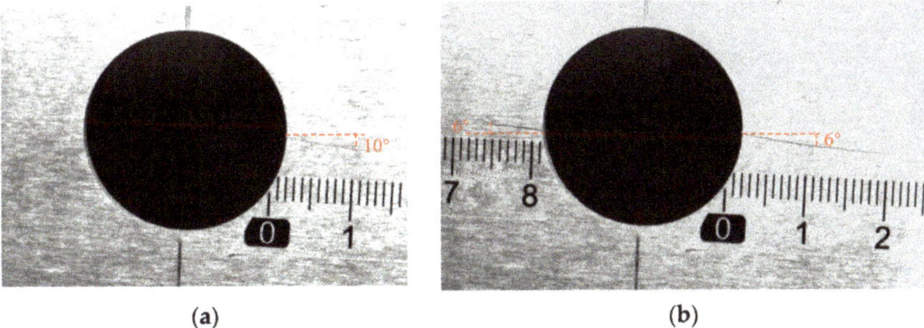

Figure 8. Cracks observed from the microscope of specimen T1 at: (**a**) 59,673 load cycles and (**b**) 70,766 load cycles.

By analyzing the waveform envelope and the arrival wave packet overlap under 150 kHz: 20 kHz: 270 kHz excitation frequencies, typical response signals of A1-S1 sensing path under 230 kHz are given in Figure 9. The separated first arrival wave packet can be observed without overlapping with the crosstalk signals and boundary refection signals, and the experimental group velocity of S_0 mode is approximately estimated as 5106 m/s. Therefore, the central frequency of excitation signal was confirmed as 230 kHz to obtain separated direct wave packet. A fixed-length rectangular window [82 μs 127 μs] was employed to extract the direct wave packet. The non-coupling of A1-S1 and A2-S2 sensing paths is demonstrated by comparing the response signals of A2-S2 sensing path only with one side crack and that with two sides cracks. Thus, the response signals of each sensing path are regarded as one independent sample. Figure 10 plots the experimental Lamb wave response signals of different length for specimens T1–T3. Due to the dispersions of specimens, sensor performance, and installation process, the response signals amplitude and damage sensitivities for different specimens are different, and are manifested as different changes in amplitude and phases under the same crack length growth. Further comparison of the simulated and experimental response signals shows that the direct wave packet can be extracted earlier in simulated data, and the simulated data have higher damage sensitivity.

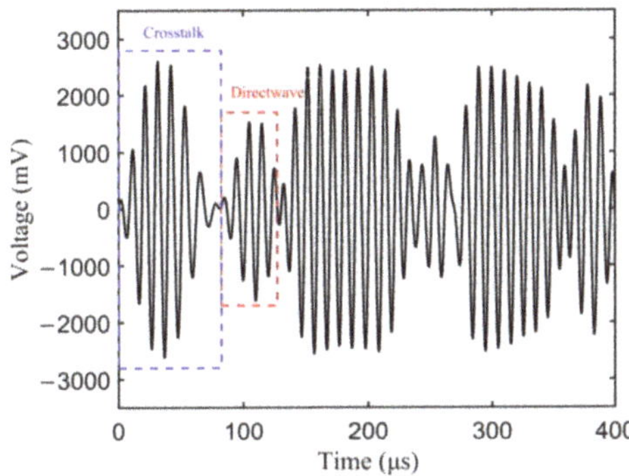

Figure 9. Typical response signals of A1-S1 sensing path under 230 kHz excitation frequency.

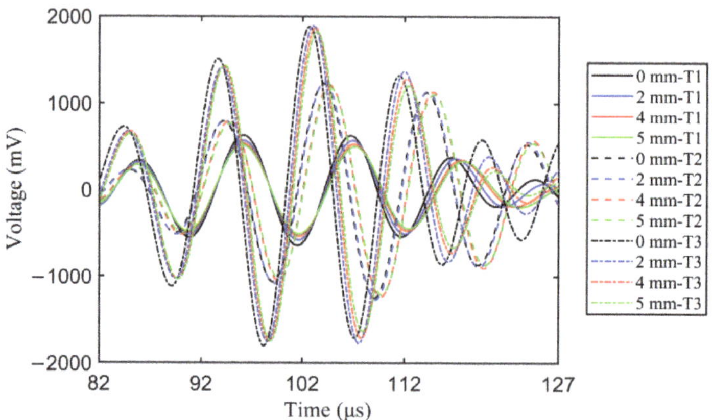

Figure 10. Experimental Lamb wave response signals of different crack lengths for specimens T1–T3.

The experimental dataset of eight center-hole metal specimens consists of one hundred and forty undamaged samples and eighty damaged samples, with each sample consisting of 451 data points, in which the undamaged and damaged samples are extremely unbalanced. In order to balance the undamaged and damaged samples to further improve the model performance, data augment technology is also applied to introduce the fluctuations of amplitude and phase into the response signals to expand the experimental samples. For the undamaged samples, the scaling coefficient for the amplitude varies at (0.90, 1.20), and the scaling coefficient for the phase varies at (0.90, 1.20). Meanwhile, the scaling of the amplitude and phase are preceded. For the damaged samples, the scaling coefficient for the amplitude varies from 0.80 to 1.25 at a step of 0.15, and the scaling coefficient for the phase varies from 0.80 to 1.25 at a step of 0.15. The scaling of the amplitude and phase are meanwhile preceded. That is, a total of 600 experimental samples are obtained with 280 undamaged samples and 320 damaged samples.

Descriptions of the simulated and experimental dataset are shown in Table 2. The dataset is randomly divided as training and testing datasets at a ratio of 4:3. Out of 600 cases of simulated data and experimental data, 210 undamaged cases and 240 damaged cases are used for training while 70 undamaged cases and 80 damage cases are used for testing. In addition, max-min normalization is a necessary step to convert the different scales of the simulated and experimental dataset into a common scale, which enables the unbiased contribution from the output of every response signal.

Table 2. Introduction to datasets.

Dataset	Healthy Conditions	Number of Samples
Simulation	Undamaged	280
	Damaged	320
Experiment	Undamaged	320
	Damaged	280

3.3. Transfer Results of MMD-DANN Model

In the proposed MMD-DANN model, λ, and α are important trade-off parameters which seriously affect the transfer performance of the model. Thus, the studies of the trade-off parameters are implemented first. The classification accuracy is related to distribution divergence between two domains. Thus, MMD of the extracted features output by the second fully connected layer of the label predictor is used to evaluate the transfer performance with different trade-off parameters, which is the highest-level feature before classification.

In order to suppress noisy signals from the domain classifier at the early stages of the training procedure [31], the parameter λ gradually changed from 0 to 1 using the formula $2/(1 + e^{-10p}) - 1$ instead of a fixed value, in which p is the training progress linearly changing from 0 to 1. With the training process progressing, the trade-off parameter λ gradually increases. The parameter α is selected from {0.01, 0.05, 0.1, 0.5, 1, 5, 10, 50}. In order to improve the optimization of SGD during the training, an annealing learning rate is adopted using the formula $\mu = \eta_0/(1 + 10p)^{0.75}$, in which η_0 is the initialized learning rate. The SGD method with an initialized learning rate 1×10^{-3} and mini-batch 64 was used to train the model. Every training is carried out for five trails and the average us obtained to reduce the effect of randomness. All training is performed using python on the Inter(R) Xeon(R) Gold 6462R CPU.

Figure 11a shows MMD of the extracted features with different trade-off parameter α. The classification accuracy of different parameter α on the experimental dataset is given in Figure 11b. It can be seen that MMD of the extracted features and the corresponding classification accuracy have a nonmonotonic trend with the parameter α, but the smaller MMD corresponds to the higher classification accuracy. MMD takes the minimum value when α is set as 5 and 50, and the classification accuracy takes the maximum value when α is set as 5, and the corresponding classification accuracy of the experimental dataset is 98.12%. Therefore, MMD-DANN model with the trade-off parameter α of 5 is trained to detect fatigue cracks.

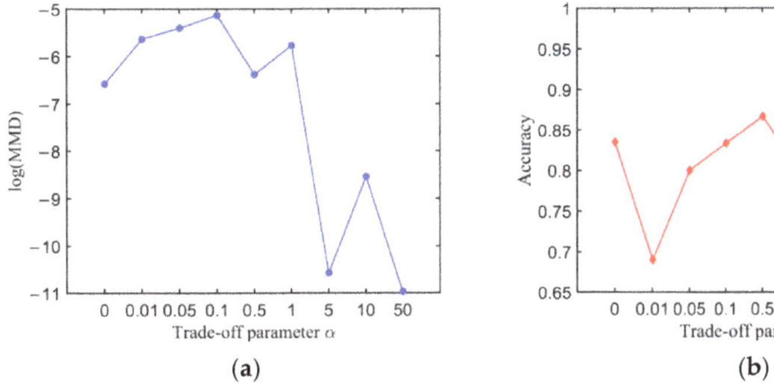

Figure 11. Transfer results of MMD-DANN model with different trade-off parameter α: (**a**) MMD of the extracted features and (**b**) classification accuracy.

To further demonstrate the superiority of the progressing strategy for trade-off parameter λ, the transfer results of the model with a fixed parameter λ are compared. The parameter λ is selected from {0.1, 0.5, 1, 5, 10}. Figure 12 shows MMD of the extracted features and classification accuracy with different parameter λ when α is set to 5. The parameter λ has an obvious effect on the transfer results of the model. Once the parameter λ is set as a large value, such as being larger than 1, the classification accuracy of the model sharply drops to a poor level. A safe way is to select a small parameter λ to balance the domain classification loss and the domain adaptation loss in the loss function. Especially, the classification accuracy takes the maximum value when λ is set as 1 and α is set as 5, and the corresponding classification accuracy of the experimental dataset is 94.33%, which is still lower than when λ is set as a changing value and α is set as 5. The progressive training strategy of the trade-off parameter λ significantly improves the classification performance and simplifies the parameter-selecting. Finally, the MMD-DANN model, with a changing trade-off parameter λ and a fixed trade-off parameter α of 5, was trained to detect fatigue cracks.

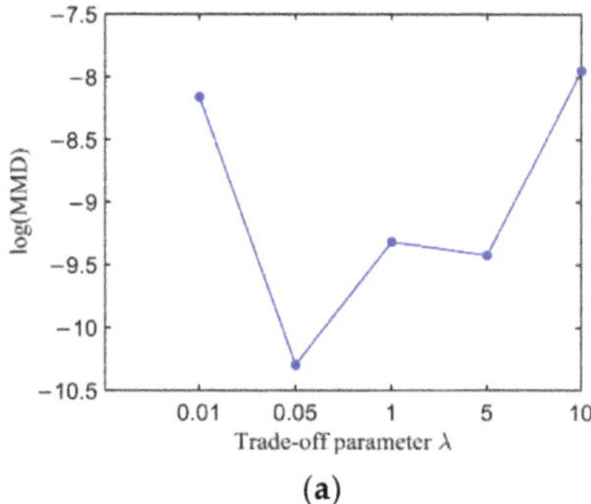

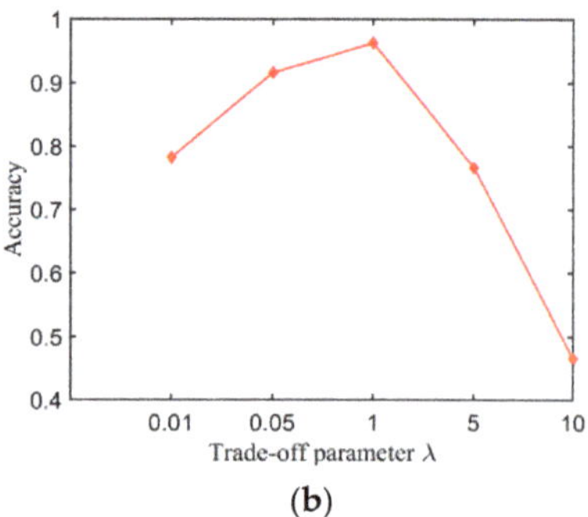

Figure 12. Transfer results of MMD-DANN model with different trade-off parameter λ: (**a**) MMD of the extracted features and (**b**) classification accuracy.

3.4. Comparisons with Other Methods

To further validate the transfer results and transfer performances of the proposed method, we compared our method with other methods, including 1D-CNN, WGAN-GP, DDC, DAN, and DANN.

For comparison, 1D-CNN consists of a feature extractor and a label predictor, in which the architectures of the feature extractor and the label predictor are the same as MMD-DANN model. A 1D-CNN model trained on the labeled simulated data was used to classify the unlabeled domain data without domain adaptation knowledge. The architecture of CNN classifier in WGAN-GP model is the same as the 1D-CNN model. WGAN-GP is a classical adversarial generation-based deep domain adaptation method made up of a generator, a discriminator, and the gradient penalty. By the adversarial training between the generator and the discriminator, a CNN classifier trained on the synthetic domain can be directly transferred to the target domain. The architecture of CNN classifier in WGAN-GP model is the same as a 1D-CNN model. DDC is a common unsupervised domain adaptation method made up of a fixed CNN, an adaptation layer, and MMD, whose architecture analogous MMD-DANN model is without a domain classifier. The position to place the adaptation layer in is decided by comparing every condition. By introducing more adaptation layers and MK-MMD, DAN is developed, which is made up of a feature extractor, a label predictor, and MK-MMD. The trade-off parameter α of MK-MMD in the loss function for DAN model is searched from {0.1, 0.5, 1, 5, 10, 50}. The optimal transfer result of DAN model is calculated as 0.5. DANN model is the simplified version of MMD-DANN model, without MK-MMD in the loss function. The training dataset of 1D-CNN, WGAN-GP, DDC, DAN, DANN, and the proposed model are 70% of the labeled simulated domain data and 70% of the unlabeled target domain data. The testing dataset of 1D-CNN, WGAN-GP, DDC, DAN, DANN, and the proposed model are the experimental domain data. Every model carried out six trainings. The network architecture of 1D-CNN, WGAN-GP, DDC, DAN, and DANN with all hyperparameters used in the paper is shown in Appendix B.

To observe the transfer results specifically, Table 3 denotes MMD for six models, in which MMD for CNN model is calculated from the source and target data, equal to initial MMD without domain adaptation. Based on the domain adaptation mechanisms of other five models, the transfer result for WGAN-GP model can be expressed as MMD of the synthetic and target data, and the transfer results for DDC, DAN, DANN, and the proposed model can be expressed as MMD of the extracted features vector output by the feature extractor. Compared with six models, the MMD of raw data is the highest, and the MMD of five deep domain adaptation methods is smaller, indicating that the deep domain adaptation methods are effective tools to reduce the distribution discrepancy. The proposed method has the smallest MMD, implying that the proposed method has the best learning ability of domain adaptation.

Table 3. Transfer results of six models.

Models	Data	MMD
1D-CNN	Raw source data and target data	0.008052
WGAN-GP	Synthetic data and target data	0.003773
DDC	Extracted features vector of source data and target data	0.003083
DAN	Extracted features vector of source data and target data	0.002932
DANN	Extracted features vector of source data and target data	0.001397
Proposed	Extracted features vector of source data and target data	0.0001998

To evaluate the performance of models comprehensively, false alarm rate and missing alarm rate were introduced to enrich the evaluation of the classification accuracy of the proposed MMD-DANN and other methods. The false alarm rate is defined as the proportion of undamaged samples predicted as damaged samples in total predicted damaged samples. The missing alarm rate is defined as the proportion of damaged samples predicted as undamaged samples in total damaged samples.

The average classification accuracy, false alarm rate, and missing alarm rate on the experimental dataset are detailed in Table 4. The classification accuracy of five trails in every method is shown in Figure 13. The average classification accuracy of the proposed MMD-DANN model on the experimental dataset is 98.12%, which is the highest among six methods. The average false alarm rate and missing alarm rate of the proposed MMD-DANN model are 3.08% and 1.56%, which are also the smallest among the six methods.

Table 4. Detection results for experimental dataset with multiple methods.

Methods	Accuracy (%)	False Alarm Rate (%)	Missing Alarm Rate (%)
1D-CNN	58.53	30.69	33.33
WGAN-GP	66.93	23.88	30.28
DDC	75.62	28.34	15.87
DAN	80.00	26.83	6.25
DANN	84.73	15.14	13.96
Proposed	98.12	3.08	1.56

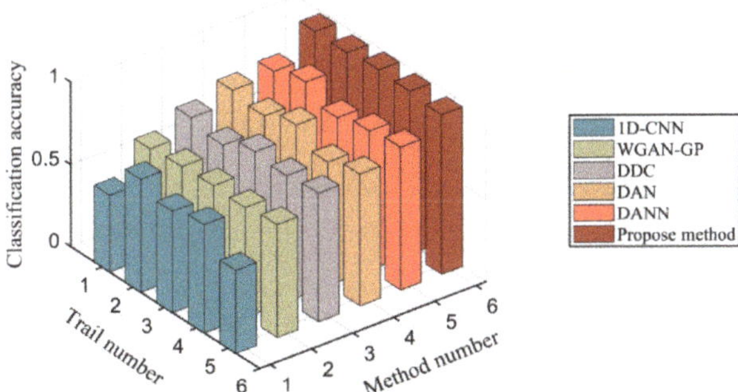

Figure 13. The classification accuracy comparison of multiple methods.

From the results shown in Table 4, due to the lack of domain adaptation procedure, the average classification accuracy of 1D-CNN model is just 58.53%, which is the smallest among all the methods. The average false alarm rate and missing alarm rate of 1D-CNN model are 30.69% and 33.33%, which are the highest among six methods. The detection results of 1D-CNN model further demonstrate that the mismatched distributions of the source and target domain data cause reduced performance in the target domain.

As shown in Table 4, the average classification accuracy of WGAN-GP model is 66.93%, which is the smallest in five deep domain adaptation methods but only higher than 1D-CNN. The average false alarm rate and missing alarm rate of WGAN-GP model are 23.88% and 30.28%, which is the highest among five deep domain adaptation methods. Due to the dispersions of specimens, sensor performance, and the installation process, the experimental signals of different specimens still have an obvious distinction (as shown in Figure 10), which deteriorates the synthetic data quality. Therefore, the large distribution discrepancy between the synthetic data and the target data leads the classifier trained on the synthetic data to show a poor classification performance on the target data. It can be inferred that the complexity of the target domain data has an important effect on the adversarial generation results of WGAN-GP model. Further, by comparing the detection results of 1D-CNN with WGAN-GP model, the classifier of WGAN-GP model presents a better classification performance because of the smaller distribution discrepancy between the synthetic and target data compared to that between the simulated and experimental data.

According to the detection results in Table 4, the average classification accuracy of DDC model and DAN model are 75.62% and 80.00%, which identifies the better domain adaptation performance of multiple adaptation layers and MK-MMD for the DAN model than that of the single adaptation layer and MMD for the DDC model. Based on the comparison results in Table 4, the average classification accuracy of DAN model and DANN model are 80.00% and 84.73%, which are close to each other, meaning that for the fatigue crack detection scenarios, the domain adaptation performance of minimizing MMD is equal to the adversarial training. The average classification accuracy of DAN model and DANN model are higher than WGAN-GP model, because the distribution discrepancy between the high-dimensional features vector for DAN and DANN model are much smaller than that between the synthetic and target data. Specially, the average classification accuracy of DAN and DANN model are smaller than MMD-DANN model, because DAN model reduces the distribution discrepancy just by minimizing MMD and DANN model reduces the distribution discrepancy just by adversarial training, but MMD-DANN model reduces the distribution discrepancy by combing MMD and adversarial training. The comparison results further verify the effectiveness of domain adaptation and adversarial training of the proposed MMD-DANN model.

A confusion matrix for the first training trail of the proposed model on the experimental dataset is shown in Figure 14. For 280 undamaged and 320 damaged experimental data, 3.57% of undamaged samples were misclassified as damaged conditions. All damaged samples are all classified accurately. It can be inferred that the proposed model can accurately classify undamaged and damaged conditions.

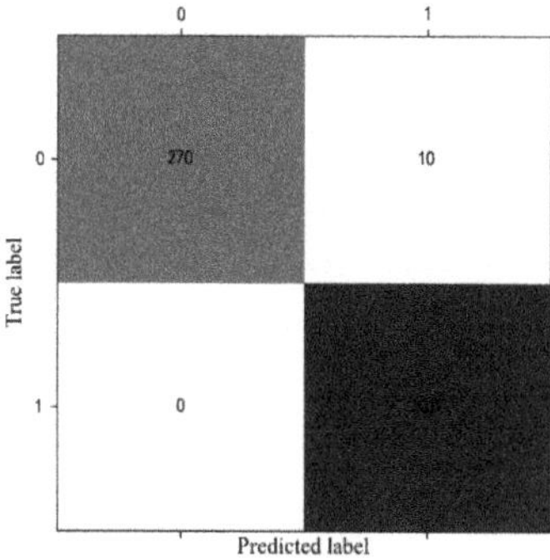

Figure 14. The confusion matrix of the proposed method on the experimental dataset.

In order to visually validate the transfer learning effectiveness of the proposed method, the t-distributed stochastic neighbor embedding (t-SNE) [43] technique was used to map the high-dimensional features vector into a two-dimensional space. The mapped transferrable features for six models are shown in Figure 15. From the result shown in Figure 15, the transferable learning features of the undamaged source data have an obvious gathering cluster because the undamaged source data are obtained with one undamaged model, and the transferable learning features of the damaged source data have multiple gathering clusters because of the damaged source data corresponding to different crack lengths. For the experimental target data, the undamaged and damaged data are scatter-distributed because of the dispersions among specimens.

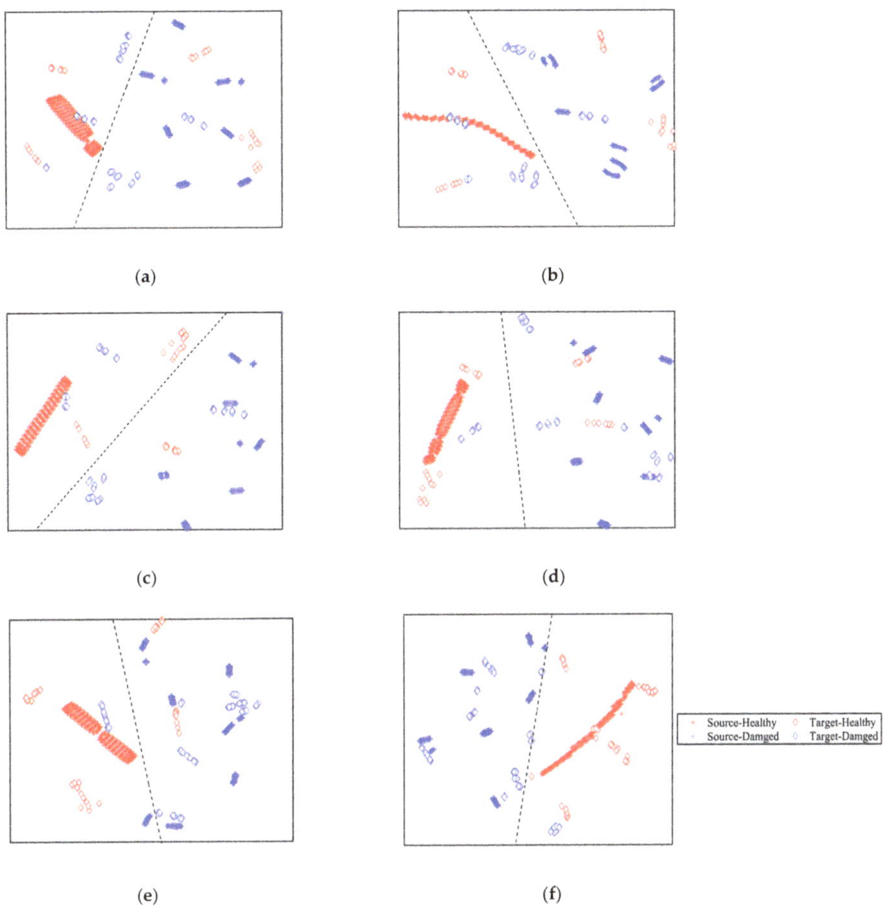

Figure 15. The visualization of the mapped features on the source and target domain: (**a**) 1D-CNN, (**b**) WGAN-GP, (**c**) DDC, (**d**) DAN, (**e**) DANN, and (**f**) MMD-DANN.

From Figure 15a, the transferable features extracted by 1D-CNN model suffer from poor distribution discrepancy. With the dash line plotted in Figure 15a, the source data can be accurately classified, while many undamaged and damaged target domain data are misclassified. Thus, 1D-CNN model trained on the source data cannot accurately classify the unlabeled target data. From Figure 15b, the transferable features of the target domain extracted by WGAN-GP model have a small among-class distance. The transferable features cannot be effectively obtained using the WGAN-GP model. In addition, compared with 1D-CNN model, with the dash line plotted in Figure 15b, less damaged source data are misclassified. Consequently, the classification accuracy of WGAN-GP model is higher than the 1D-CNN model. From Figure 15c–e, the transferable features learned by DAC,

DAN, and DANN model have a smaller cross-domain discrepancy. As a result, DAC, DAN, and DANN models showed higher classification accuracy on the unlabeled target data than 1D-CNN and WGAN-GP models. However, with the dash line plotted in Figure 15c–e, some undamaged samples are falsely alerted, and some damaged experimental samples are missing alerted. As shown in Figure 15f, the transferable features of two domains under the same class are projected into the same region, and the different conditions data are separated well. The result shows that the proposed MMD-DANN model not only effectively reduces the inter-domain distance but also enlarges the among-class distance. With the dash line plotted in Figure 15f, only a few damaged target data are misclassified. The corresponding crack length of the misclassified damaged target samples are all 3 mm, because the response signals with 3 mm crack length have a small variation compared to the undamaged signals. Consequently, the classification accuracy of the proposed model is the highest. The visualization results prove that the proposed MMD-DANN model outperforms the other methods and presents the best transfer performance. The proposed MMD-DANN model can be used as an effective domain adaptation tool to accurately detect fatigue crack for the unlabeled experimental samples.

4. Conclusions

In this paper, an automated fatigue crack detection method based on MMD-DANN model was proposed to accurately detect structural conditions for metal structures. To overcome the difficulty of time-consuming acquisitions of the labeled data in practice, FEM simulations were adopted to obtain simulated response Lamb wave signals with different healthy conditions, which are assigned as the labeled source domain data. Due to the distribution discrepancy between the simulated and experimental signals, the classifier model trained on the simulated data cannot be directly applicable to the experimental data. A novel unsupervised domain adaptation method based on MMD-DANN model was developed. By integrating MMD with the adversarial training of DANN model, the discriminative and domain-invariant features of the simulated source domain and the experimental target domain can be extracted. Following that, the classification knowledge of the labeled source domain can be generalized to the unlabeled target domain. Fatigue tests on center-hole metal specimens are implemented to validate the proposed method. Comparing with the data-driven intelligent method and other deep domain adaptation methods, the detection results on the experimental data show that the proposed MMD-DANN model presents higher classification accuracy and better transfer performance. The average classification accuracy on the experimental data for MMD-DANN model is 98.12%, and the false alarm rate and missing alarm rate are 3.08% and 1.56%, showing the domain adaptation effectiveness of the proposed MMD-DANN model.

The proposed method is a regular binary classification method which only classifies the undamaged and damaged conditions, and is unable to further classify damaged cases with different crack length. Therefore, in our future research, an automated damage quantification detection method for fatigue crack based on unsupervised deep domain adaptation will be studied. Moreover, unsupervised domain adaptation detection methods for complex multi-bolt joint specimens and lap specimens also need to be established.

Author Contributions: Conceptualization, L.W. and Y.Y.; methodology, L.W.; software, L.W., C.Z. and G.L.; validation, L.W. and C.Z.; review, C.Z. and J.Q.; supervision, Y.Y.; funding acquisition, J.Q. All authors have read and agreed to the published version of the manuscript.

Funding: This work was supported by National Natural Science Foundation of China (52175141 & 52235003 & 51921003), Natural Science Foundation of Jiangsu Province (BK20220133).

Institutional Review Board Statement: Not applicable.

Informed Consent Statement: Not applicable.

Data Availability Statement: Not applicable.

Conflicts of Interest: The authors declare no conflict of interest.

Appendix A

The pseudo code of the proposed MMD-DANN model is presented in the Algorithm A1.

Algorithm A1: MMD-DANN

Initialization: feature extractor G_f, label predictor G_y, and domain classifier G_y with parameters of $\theta_f, \theta_y, \theta_d$; mini - batch size M = 64; training step $N = 20,000$; training progress p; learning rate; $\mu = \eta_0/(1+10p)^{0.75}$, $\eta_0 = 1 \times 10^{-3}$; trade − off parameter $\alpha = 5$; trade − off parameter $\lambda = 2/\left(1+\mathrm{e}^{-10p}\right) - 1$

for $l = 1, \ldots, N$ **do**

Input a batch from the labeled source data $\mathcal{D}_s = \left\{x_s^i, y_s^i\right\}_{i=1}^m$, unlabeled target data $\mathcal{D}_t = \left\{x_t^j\right\}_{j=1}^n$

Compute training progress : $p = l/N$

Compute learning rate μ, trade-off parameters α, λ

Calculate label prediction loss : $L_y(\mathcal{D}_s) = -\mathrm{E}_{(x_s,y_s)\sim\mathcal{D}_s} \sum_{k=1}^{K} 1_{[k=y_s]} \log\left(G_y\left(G_f(x_s)\right)\right)$

Calculate domain classification loss : $L_d(\mathcal{D}_s, \mathcal{D}_t) = -\mathrm{E}_{x_s\sim\mathcal{D}_s}\left[\log\left(G_d\left(G_f(x_s)\right)\right)\right] - \mathrm{E}_{x_t\sim\mathcal{D}_t}\left[\log\left(1-G_d\left(G_f(x_t)\right)\right)\right]$

Calculate distribution discrepancy metric : $L_{MMD}(\mathcal{D}_s, \mathcal{D}_t) = \sqrt{\frac{1}{m^2}\sum_{i,j=1}^{m} k\left(x_s^i, x_s^j\right) - \frac{2}{mn}\sum_{i,j=1}^{m,n} k\left(x_s^i, x_t^j\right) + \frac{1}{n^2}\sum_{i,j=1}^{n} k\left(x_t^i, x_t^j\right)}$

Calculate loss function of feature extractor : $L_f(\mathcal{D}_s, \mathcal{D}_t) = L_y(\mathcal{D}_s) - \lambda L_d(\mathcal{D}_s, \mathcal{D}_t) + \alpha L_{MMD}(\mathcal{D}_s, \mathcal{D}_t)$

Update parameters: $\theta_f \leftarrow \theta_f - \mu\frac{\partial L_f}{\partial \theta_f}$, $\theta_y \leftarrow \theta_y - \mu\frac{\partial L_y}{\partial \theta_y}$, $\theta_d \leftarrow \theta_d - \mu\frac{\partial L_d}{\partial \theta_d}$

End for

Output: Predicted class label for target domain data $G_y\left(G_f(x_t)\right)$

Appendix B

The architectures of the 1D-CNN, WGAN-GP, DAC, DAN, and DANN model used in this paper are shown in Figure A1.

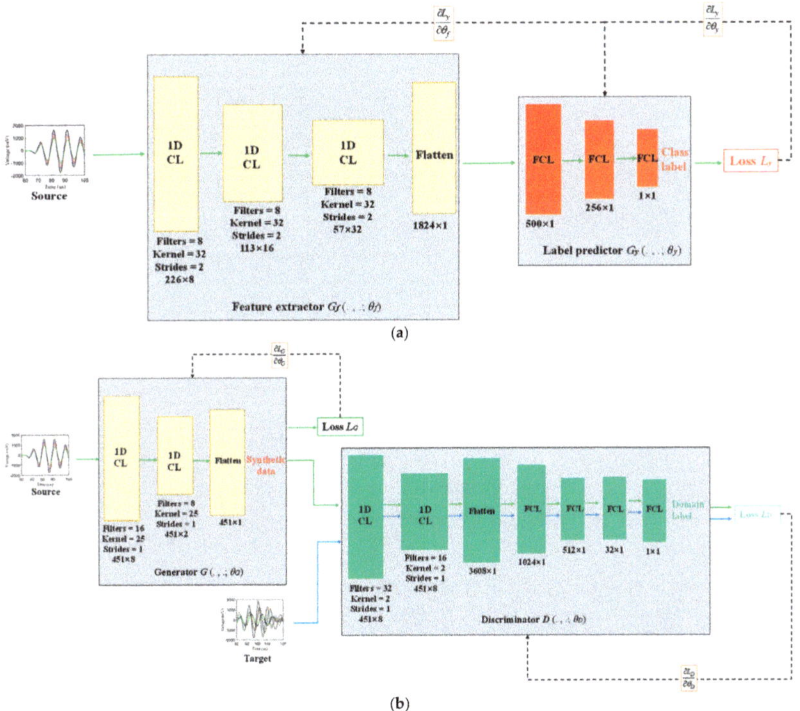

Figure A1. *Cont.*

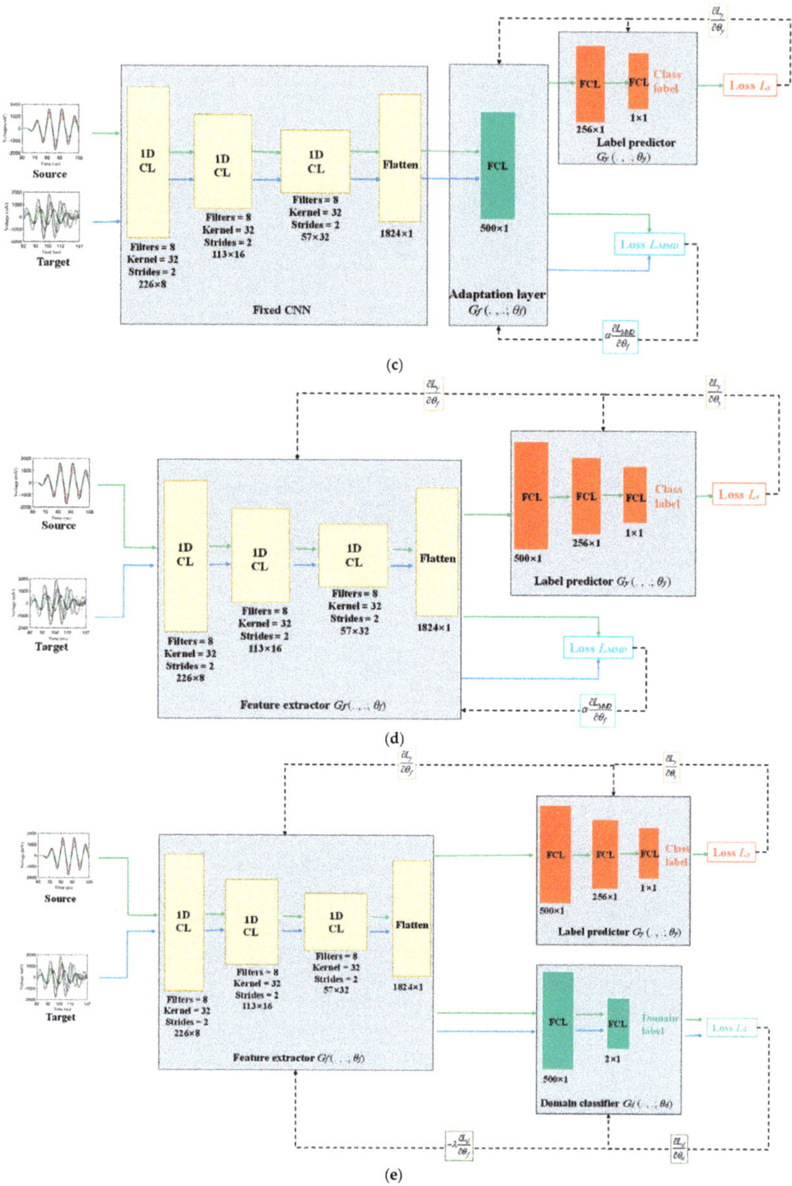

Figure A1. The architectures of (**a**) 1D-CNN, (**b**) WGAN-GP, (**c**) DAC, (**d**) DAN, and (**e**) DANN model used in the experiments.

References

1. Staszewski, W.; Boller, C.; Tomlinson, G. *Health Monitoring of Aerospace Structures: Smart Sensor Technologies and Signal Processing*, 1st ed.; John Wiley & Sons: Chichester, UK, 2003.
2. Qing, X.; Li, W.; Wang, Y.; Sun, H. Piezoelectric transducer-based structural health monitoring for aircraft applications. *Sensors* **2022**, *19*, 545. [CrossRef] [PubMed]
3. Berger, U. Onboard–SHM for life time prediction and damage detection on aircraft structure using fibre optical sensor and Lamb wave technology. In Proceedings of the 6th European Workshop on Structural Health Monitoring-Tu.2.A.1, Dresden, Germany, 3–6 July 2012.

4. Park, S.; Lee, J.; Yun, C.; Inman, D. Electro-mechanical impedance-based wireless structural health monitoring using PCA-data compression and k-means clustering algorithms. *J. Intell. Mater. Syst. Struct.* **2008**, *19*, 509–520.
5. Huang, C.; Zhou, X.; Rong, K.; Cao, J.; Zhang, J.; Wang, K. Smart coating based on frequency-selective spoof surface plasmon polaritons for crack monitoring. In Proceedings of the IEEE 3rd International Conference on Electronic Information and Communication Technology, Shenzhen, China, 13–15 November 2020; pp. 758–760.
6. Giurgiutiu, V.; Zagrai, A.; Bao, J. Piezoelectric wafer embedded active sensors for aging aircraft structural health monitoring. *Struct. Health Monit.* **2002**, *1*, 41–61.
7. Pandey, P.; Rai, A.; Mitra, M. Explainable 1-D conventional neural network for damage detection using Lamb wave. *Mech. Syst. Signal Process.* **2022**, *164*, 108220. [CrossRef]
8. Chen, H.; Huang, L.; Yang, L.; Chen, Y. Model-based method with nonlinear ultrasonic system identification for mechanical structural health assessment. *Trans. Emerg. Tel. Tech.* **2020**, *12*, e3955. [CrossRef]
9. Chen, H.; Chen, Y.; Yang, L. Intelligent early structural health prognosis with nonlinear system identification for RFID signal analysis. *Comput. Commun.* **2020**, *157*, 150–161. [CrossRef]
10. Monaco, E.; Memmolo, V.; Ricci, F.; Boffa, N. Guided waves based SHM systems for composites structural elements: Statistical analyses finalized at probability of detection definition and assessment. In Proceedings of the Health Monitoring of Structural and Biological Systems, San Diego, CA, USA, 9–12 March 2015.
11. Qiu, L.; Fang, F.; Yuan, S. Improved density peak clustering-based adaptive Gaussian mixture model for damage monitoring in aircraft structures under time-varying conditions. *Mech. Syst. Signal Process.* **2019**, *126*, 281–304. [CrossRef]
12. Atashipour, S.; Mirdamadi, H.; Hemasian, M.; Amirfattahi, R. An effective damage identification approach in thick steel beams based on guided ultrasonic waves for structural health monitoring applications. *J. Intell. Mater. Syst. Struct.* **2012**, *24*, 584–597. [CrossRef]
13. Li, R.; Gu, H.; Hu, B.; She, Z. Multi-feature fusion and damage identification of large generator stator insulation based on lamb wave detection and SVM method. *Sensors* **2019**, *19*, 3733.
14. Yang, Y.; Zhou, Y.; Wang, L. Integrated method of multiple machine-learning models for damage recognition of composite structures. *J. Data Acquis. Process.* **2022**, *35*, 278–287.
15. Lee, H.; Lim, H.; Skinner, T.; Chattopadhyay, A. Automated fatigue damage detection and classification technique for composite structures using Lamb waves and deep autoencoder. *Mech. Syst. Signal Process.* **2022**, *163*, 108148. [CrossRef]
16. Chen, J.; Wu, W.; Ren, Y.; Yuan, S. Fatigue crack evaluation with the guided wave–convolutional neural network ensemble and differential wavelet spectrogram. *Sensors* **2022**, *22*, 307. [CrossRef] [PubMed]
17. Wu, J.; Xu, X.; Liu, C.; Deng, C. Lamb wave-based damage detection of composite structures using deep convolutional neural network and continuous wavelet transform. *Compos. Struct.* **2021**, *276*, 114590. [CrossRef]
18. Sampath, S.; Jang, J.; Sonh, H. Ultrasonic Lamb wave mixing based fatigue crack detection using a deep learning model and higher-order spectral analysis. *Int. J. Fatigue* **2022**, *163*, 107028.
19. Yang, Y.; Wang, B.; Lyu, S.; Liu, Y. A deep-learning-based method for damage identification of composite laminates. *Aeronaut. Sci. Technol.* **2020**, *31*, 102–108.
20. Pan, S.; Qing, Y. A survey on transfer learning. *IEEE Trans. Knowl. Data Eng.* **2010**, *22*, 1345–1359.
21. Tan, C.; Sun, F.; Kong, T.; Zhang, W. A survey on deep transfer learning. *arXiv* **2018**, arXiv:1808.01974.
22. Gretton, A.; Borgwardt, K.; Rasch, M.; Scholkopf, B. A kernel two-sample test. *J. Mach. Learn. Res.* **2012**, *13*, 723–773.
23. Tzeng, E.; Hoffman, J.; Zhang, N.; Saenko, K. Deep domain confusion: Maximizing for domain invariance. *arXiv* **2014**, arXiv:1412.3474.
24. Long, M.; Cao, Y.; Wang, J.; Jordan, M. Learning transferable features with deep adaptation networks. *arXiv* **2015**, arXiv:1502.02791.
25. Nguyen, A.; Tran, T. KL guided domain adaptation. *arXiv* **2021**, arXiv:2106.07780.
26. Dong, Y.; Yao, K.; Gang, L.; Seide, F. KL-divergence regularized deep neural network adaptation for improved large vocabulary speech recognition. In Proceedings of the 2013 IEEE International Conference on Acoustics, Speech and Signal Processing (ICASSP), Vancouver, BC, Canada, 26–30 March 2013.
27. Minka, T. *Divergence Measures and Message Passing*; Microsoft Research Ltd.: Redmond, WA, USA, 2005.
28. Xu, P.; Gurram, P.; Whipps, G.; Chellappa, R. Wasserstein distance based domain adaptation for object detection. *arXiv* **2019**, arXiv:1909.08675.
29. Balaji, Y.; Chellappa, R.; Feizi, S. Normalized Wasserstein distance for mixture distributions with applications in adversarial learning and domain adaptation. *arXiv* **2019**, arXiv:1902.00415.
30. Ganin, Y.; Lempitsky, V. Unsupervised Domain Adaptation by Backpropagation. *arXiv* **2015**, arXiv:1409.7495.
31. Shao, J.; Huang, Z.; Zhu, J. Transfer learning method based on adversarial domain adaptation for bearing fault diagnosis. *IEEE Access* **2020**, *8*, 119421–119430. [CrossRef]
32. Goodfellow, I.; Pouget-Abadie, J.; Mirza, M.; Xu, B. Generative adversarial nets. *Adv. Neural Inf. Process. Syst.* **2017**, *1704*, 00028.
33. Arjovsky, M.; Chintala, S.; Bottou, L. Wasserstein GAN. *arXiv* **2017**, arXiv:1701.07875.
34. Gulrajani, I.; Ahmed, F.; Arjovsky, M.; Dumoulin, V. Improved training of Wasserstein GANs. *arXiv* **2018**, arXiv:1805.00778.
35. Alguri, K.; Harley, B. Transfer learning of ultrasonic guided waves using autoencoders: A preliminary study. *AIP Conf. Proc.* **2019**, *2102*, 050013.

36. Zhang, B.; Hong, X.; Liu, Y. Distribution adaptation deep transfer learning method for cross-structure health monitoring using guided waves. *Struct. Health Monit.* **2022**, *21*, 853–871.
37. Zhang, B.; Hong, X.; Liu, Y. Multi-task deep transfer learning method for guided wave based integrated health monitoring using piezoelectric transducers. *IEEE Sens. J.* **2020**, *20*, 1558–1748. [CrossRef]
38. Wang, Y.; Lyu, S.; Yang, Y.; Li, J. Damage recognition of composite structures based on domain adaptive model. *Acta Aeronaut. Astronaut. Sin.* **2022**, *71*, 65223.
39. Nair, V.; Hinton, G. Rectified linear units improve restricted Boltzmann machines. In Proceedings of the 27th International Conference on Machine Learning, Haifa, Israel, 21–24 June 2010.
40. Ioffe, S.; Szegedy, C. Batch normalization: Accelerating deep network training by reducing internal covariate shift. *arXiv* **2015**, arXiv:1502.03167.
41. Liu, B.; Fang, D. Domain-switching embedded nonlinear electromechanical finite element method for ferroelectric ceramics. *Sci. China Phys. Mech. Astron.* **2011**, *54*, 606–617.
42. Liu, Q.; Xiao, Y.; Zhang, H.; Ren, G. Baseline signal reconstruction for temperature compensation in Lamb wave-based damage detection. *Sensors* **2016**, *16*, 1273. [CrossRef] [PubMed]
43. Maaten, L.; Hinton, G. Visualizing data using t-SNE. *J. Mach. Res.* **2008**, *8*, 2579–2605.

Article

Wrinkle Detection in Carbon Fiber-Reinforced Polymers Using Linear Phase FIR-Filtered Ultrasonic Array Data

Tengfei Ma [1], Yang Li [2,*], Zhenggan Zhou [2,3] and Jia Meng [4]

1 School of Energy and Power Engineering, Beihang University, Beijing 100191, China
2 Ningbo Institute of Technology, Beihang University, Ningbo 315800, China
3 School of Mechanical Engineering and Automation, Beihang University, Beijing 100191, China
4 COMAC Shanghai Aircraft Manufacturing Co., Ltd., Shanghai 201324, China
* Correspondence: liyang19890327@buaa.edu.cn; Tel.: +86-10-82338668

Abstract: Carbon fiber-reinforced polymers (CFRP) are extensively used in aerospace applications. Out-of-plane wrinkles frequently occur in aerospace CFRP parts that are commonly large and complex. Wrinkles acting as failure initiators severely damage the mechanical performance of CFRP parts. Wrinkles have no significant acoustic impedance mismatch, reflecting weak echoes. The total focusing method (TFM) using weak reflection signals is vulnerable to noise, so our primary work is to design discrete-time filters to relieve the noise interference. Wrinkles in CFRP composites are geometric defects, and their direct detection requires high spatial precision. The TFM method is a time-domain delay-and-sum algorithm, and it requires that the time information of filtered signals has no change or can be corrected. A linear phase filter can avoid phase distortion, and its filtered signal can be corrected by shifting a constant time. We first propose a wrinkle detection method using linear phase FIR-filtered ultrasonic array data. Linear phase filters almost do not affect the wrinkle geometry of detection results and can relieve noise-induced dislocation. Four filters with different bandwidths have been designed and applied for wrinkle detection. The 2 MHz bandwidth filter is recommended as an optimum choice.

Keywords: ultrasonic array nondestructive testing; CFRP; out-of-plane wrinkle; full matrix; total focusing method; instantaneous phase; linear phase FIR filter

Citation: Ma, T.; Li, Y.; Zhou, Z.; Meng, J. Wrinkle Detection in Carbon Fiber-Reinforced Polymers Using Linear Phase FIR-Filtered Ultrasonic Array Data. *Aerospace* **2023**, *10*, 181. https://doi.org/10.3390/aerospace10020181

Academic Editors: Phong B. Dao, Lei Qiu, Liang Yu, Tadeusz Uhl and Minh-Quy Le

Received: 2 December 2022
Revised: 4 February 2023
Accepted: 11 February 2023
Published: 15 February 2023

1. Introduction

Carbon fiber-reinforced polymers (CFRP) are extensively used lightweight materials in aerospace applications as they exhibit rather high specific strength and good corrosion resistance [1]. CFRP parts used in the aerospace field are always complex curved surface products, such as aircraft engine fan blades, aircraft wings, and satellites' motor cases [2]. Out-of-plane wrinkles are common defects in these complex parts, with the carbon fiber plies deviating from the anticipated direction. Wrinkles act as failure initiators in CFRP parts, severely damaging the mechanical performance (like tensile [3], compressive [4], flexural, and fatigue strength [5]). Wrinkles frequently occur inside the CFRP laminated parts, almost impossibly inspected from the surface. A wide variety of non-destructive testing (NDT) methods is established for detecting and evaluating out-of-plane wrinkles. These methods include visual inspection [6], eddy current method [7], infrared thermography [8], X-ray micro-computed tomography [9], and ultrasonic testing (UT) [10,11]. Visual inspection can only examine the surface or cross-section wrinkles by optical microscopy or the naked eye. The eddy current method is mainly limited by its detection depth, and only near-surface waviness can be detected [12]. Infrared thermography detects out-of-plane wrinkles by stress analysis with cyclic loading. This method is cumbersome to evaluate large complex parts, and its result is non-intuitive and susceptible to other stress concentration points [8]. X-ray micro-computed tomography can provide an accurate 3D geometrical image of a detected part. However, the low X-ray absorption contrast between the carbon fibers and epoxy polymers limits the sample size. X-ray micro-computed tomography is unsuitable for aerospace CFRP parts which are large complex parts. The ultrasonic technique is the most powerful method to detect wrinkles for aerospace CFRP parts.

The ultrasonic technique is already used in the aerospace industry to detect delamination [13], barely visible impact damage (BVID) [14,15], and debonding [16]. Lamb waves are commonly used for the structural health monitoring (SHM) [17,18] of large structures because they can propagate long distances along the component curvature to achieve a quick inspection. However, the signal of lamb waves is complicated to analyze and interpret, and the result is non-intuitive for geometrical defects, namely wrinkles. Bulk waves, especially longitudinal waves, are usually used to detect and size flaws in CFRP parts. Our work only considers using longitudinal waves to detect out-of-plane wrinkles in this paper.

Unlike delamination and transverse cracks, wrinkles do not break material continuity, so they do not generate significant acoustic impedance mismatch. It is challenging to detect wrinkles because there is no strong reflection response when the ultrasound wave encounters wrinkles [19]. Researchers have tried to exploit the slight impedance mismatch between the carbon fiber plies and resin-rich interplies. The slight impedance mismatch generates weak reflection echoes that contain faint spatial information of resin-rich interplies. The most classical method is the instantaneous phase method using pulse-echo ultrasonic inspection, extracting the phase information from the multi-layered structure echoes [20,21]. The pulse-echo ultrasonic inspection has a convenient, inexpensive advantage, and its signals are easy to analyze and interpret. However, this pulse-echo method utilizes only one-dimensional information, namely the normal incidence reflection response of the CFRP parts. Other pulse-echo methods, such as the instantaneous amplitude method with low-pass (LP) filtering and the Wiener deconvolution method with spectral extrapolation [22], require a higher center frequency. High-frequency ultrasounds suffer severe attenuation in CFRP materials limiting the detection depth. Theoretically, the normal interplies can be correctly located by these one-dimensional ultrasonic signals, but the inter-ply incline caused by wrinkles induces estimation error or even failure. The lateral resolution of a spherically focused immersion transducer is limited by the ultrasonic beam width. The ultrasonic beam width cannot be fully optimal throughout a thick CFRP part.

Linear array probes with multiple elements at distinct positions in the space can provide multi-dimensional and multi-angle signals that can theoretically correctly estimate the inclined interplies' location. Phase array technology with the ultrasonic beam electronically steering at different angles is employed to detect and characterize wrinkles [23]. One steered angle can only inspect one side of wrinkles. A thorough wrinkle inspection using phase array technology requires repeat scans at different angles, which is time-consuming. The analysis and interpretation of scan data acquired by phased array technology with different inspection angles are also cumbersome for wrinkle characterization. The scattering matrix contains all the far-field scattered amplitudes of discontinuity defects from the given combination of incident and scattered direction. It is considered to carry information about the shape, orientation, and size of discontinuity defects (such as cracks or voids). The scattering matrix has been extracted to detect and characterize wrinkles [24], but this is an indirect detection method that allows statistically distinguishing wrinkles with different severity. A convolutional neural network (CNN) is the most common image recognition and analysis tool. It can achieve accurate classification without feature analysis and extraction manually. A CNN model with a short-time Fourier transform has been applied to analyzing ultrasonic data acquired by a linear array probe under full matrix capture (FMC) mode, and it achieves superior accuracy for wrinkle detection [19]. However, the CNN method is also indirect, and its results are statistical indicators without clear physical meaning. The scattering matrix and CNN methods cannot directly measure the maximum wrinkle angle. The maximum wrinkle angle is the crucial indicator of mechanical performance damage and the primary parameter to be estimated necessarily [25]. The total focusing method (TFM) is a post-processing imaging algorithm using data acquired by linear array probes under FMC mode [26]. The TFM method can achieve optimal focusing and provide a high lateral resolution. The instantaneous phase method can eliminate the amplitude information, highlighting the phase information related to the spatial information of the resin-rich interplies. Therefore, researchers [27] have proposed a TFM instantaneous phase method which is a direct method for wrinkle detection. A Fourier domain approach, the wavenumber algorithm, is also used to detect wrinkles [28]. Its primary advantage is superior computational performance. However, the advantage of the TFM method is its flexibility and versatility for arbitrary imaging geometries [29]. The TFM method is more suitable for complex geometrical parts in the aerospace industry. This study adopts the TFM instantaneous phase method to detect wrinkles in CFRP composites.

The wrinkle detection using weak reflection echoes is vulnerable to noise. However, the signals acquired by the ultrasonic phased array system on CFRP composites are contaminated by various noises, for example, electrical noise from the ultrasonic inspection system and structural noise from the heterogeneity in CFRP composites (such as voids, the fiber–matrix heterogeneity). In this study, discrete-time filters are designed to relieve noise interference. These designed filters filter the noise in the A-scan data acquired under FMC mode. The wrinkle detection using the TFM instantaneous phase method requires high spatial accuracy, so the filters must preserve the phase information of the A-scan data. The TFM method is a time-domain delay-and-sum algorithm, and its result highly depends on the time information of A-scan signals. A finite impulse response (FIR) filter can easily be designed as a linear phase filter that can avoid phase distortion [30]. The phase component of the frequency response is linear to frequency for a linear phase filter, and all frequency components of the filtered signal are shifted for a constant time. The linear phase filter can preserve the phase information, and the time information change in filtered signals can also be corrected by shifting a constant time. Bandpass filters are designed to allow through frequency components in a specified band, and their parameters are determined by the time-frequency analysis of the A-scans in FMC datasets. We first apply linear phase FIR filters to the TFM instantaneous phase method for wrinkle detection in this paper.

2. Testing Sample and Experimental Setup

2.1. Testing Samples

A CFRP sample with an induced wrinkle is prepared. Its size is 195 mm × 100 mm × 5.92 mm. This CFRP sample consists of 32 unidirectional plies in the non-wrinkle section. The average thickness of each ply is 0.165 mm. The fiber volume fraction (FVF) of this sample is approximatively estimated to be 60%. The longitudinal velocity is assumed to be 3000 m/s, so the resonant frequency of the ply is about 9 MHz. The stacking sequence is $[0/45/90/-45]4s$. The wrinkle defect is induced by inserting three additional narrow strips between the 30th and 31st ply. The narrow strips' angle is 90 degrees. The cure condition is one hour at 130 °C. The cross-section micrograph of the wrinkle and partially enlarged images of voids are shown in Figure 1. Many voids are randomly distributed in the resin-rich interplies, and a few are in the carbon fiber plies. The voids in the interplies seem a little larger than those in the carbon fiber plies. These voids induce material heterogeneity in CFRP composites, causing predominant structural noise.

2.2. Experimental Setup

The experimental system (Figure 2a) consists of a computer workstation, an ultrasonic array controller (AOS OEM-PA 128/128), a 10 MHz linear array probe (Doppler 10L64-0.3×5-D77 EJA557), and a flat wedge. The main features of the flat wedge are listed in Table 1. The flat wedge is used to avoid the near-surface dead zones of the probes. The 10 MHz linear array probe is chosen because its center frequency, close to the resonant frequency (9 MHz), can enhance the inter-ply reflection [19]. The whole 64 elements of the linear array probe are used to acquire FMC datasets with the same sampling rate of 100 MS/s. The pulser pulse width values are set at 50 ns for the 10 MHz probe. The pulser pulse voltage is set at 100 V, and the A-scan resolution is chosen 8 bits during the FMC datasets acquirement. The dataset acquirement process is shown in Figure 2b.

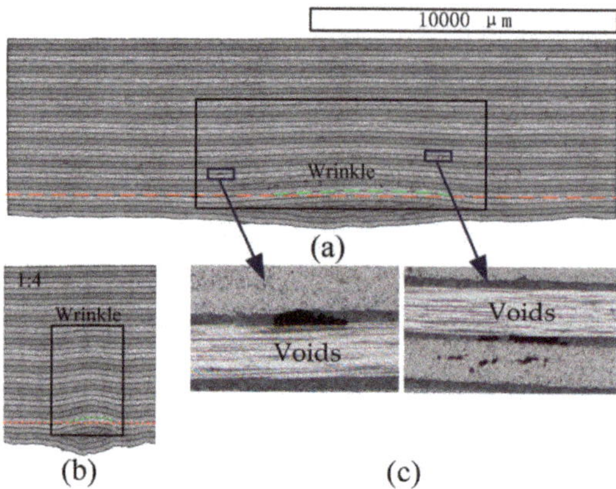

Figure 1. (**a**) A cross-section micrograph of the wrinkle sample; (**b**) a shortened length image to highlight the wrinkle defect; (**c**) partially enlarged images of voids.

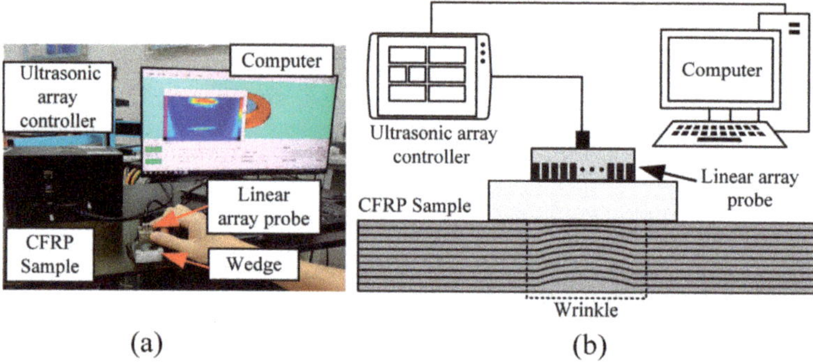

Figure 2. (**a**) Experimental system and (**b**) the system sketch used to acquire datasets under FMC mode.

Table 1. The main features of the used flat wedge.

Length	Width	Height	Ultrasonic Velocity
65 mm	40 mm	20 mm	2337 m/s

3. Methods

The TFM instantaneous phase method using weak reflection signals is vulnerable to noise, especially structural noise from voids. The noise contaminates weak inter-ply reflection signals and induces dislocation in the TFM instantaneous phase results. Our primary work is to design bandpass filters to relieve the noise interference in this paper. The center of the passband frequencies is determined by the frequency components of the inter-ply echoes. The energy of the front-face and back-wall echoes dominates the acquired A-scan signals, so a time-frequency analysis is used to analyze the frequency components of the inter-ply echoes. The designed filters should have no influence on the TFM method and avoid inducing phase distortion, so linear phase filters are designed to filter A-scan signals in FMC datasets.

3.1. Total Focusing Method

The total focusing method applies heuristic delay-and-sum beamforming on the FMC dataset, achieving focus in the target region and generating a detection image. The intensity of the detection image $I(x, z)$ at every point is calculated by [31]

$$I(x,z) = \sum_{t=1}^{N}\sum_{r=1}^{N} S_{t,r}\left(\frac{\sqrt{(x_t - x)^2 + z^2} + \sqrt{(x_r - x)^2 + z^2}}{c}\right) \quad (1)$$

where t and r are the serial numbers of the transmission and reception element, respectively; N is the number of elements; c is the longitudinal velocity here; x_t and x_r are the positions of the transmission and reception elements, respectively; (x, z) is the position of every point in the target region; $S_{t,r}$ is the A-scan signal in FMC datasets, received by element r (the transmission element t is firing). It should be noted that $S_{t,r}$ is not the Hilbert transform, or instantaneous amplitude, of the A-scan signal in this paper. The phase information in A-scan signals is crucial for wrinkle detection.

The delay-and-sum beamforming is implemented in the time domain, permitting flexibility and versatility, which makes the TFM method suitable for aerospace CFRP parts with complex curved surfaces [32]. The TFM method can focus and steer the ultrasonic beam at every point in the target region, achieving a high imaging resolution, which is vital for the continuously varying geometric defects, namely wrinkles. The TFM method makes good use of the multi-dimensional and multi-angle information in FMC datasets, making it capable of handling the inclined interplies.

3.2. Phase Extraction

The TFM result using A-scan signals contains the intensity and spatial information. The wrinkle detection only needs spatial information to map the wrinkle geometry. The lateral spatial information along the upper surface of the sample does not require additional process, but in the depth direction, the intensity information should be removed. The intensity fluctuation in the depth direction can be treated as a modulated signal [20].

$$S_{\text{depth}}(t) = a(t)\cos\phi(t) \quad (2)$$

where the amplitude $a(t)$ is mainly the intensity information related to the material attenuation and the reflection intensity; the phase $\phi(t)$ mainly contains the spatial information of the resin-rich interplies.

This section aims to extract the phase information from the intensity fluctuation in the depth direction. The Hilbert transform is a standard tool for extracting the phase information, but it requires that the amplitude $a(t)$ variation is sufficiently slow to ensure spectral disjointness [33]. The reflection intensity of the interplies is similar everywhere in the sample. The intensity variation due to the material attenuation is smooth. As a result, the amplitude $a(t)$ variation is mainly slow, except for the front-face and back-wall echoes. The Hilbert transform can be used as an effective tool to extract spatial information, just like the instantaneous phase method using pulse-echo ultrasonic inspection.

The real signal $S_{\text{depth}}(t)$ can be transformed into an analytic signal by

$$Z_{\text{depth}}(t) = S_{\text{depth}}(t) + i\mathcal{H}\left\{S_{\text{depth}}(t)\right\} \quad (3)$$

where $\mathcal{H}$ is the Hilbert transform [33],

$$\mathcal{H}\left\{S_{\text{depth}}(t)\right\} = \frac{1}{\pi}\text{p.v.}\left\{\int_{-\infty}^{+\infty} \frac{S_{\text{depth}}(\tau)}{t-\tau}\mathrm{d}\tau\right\}$$

The analytic associate of the real signal $S_{\text{depth}}(t)$ can be represented as

$$Z_{\text{depth}}(t) = a(t)e^{i\phi(t)} \tag{4}$$

where $\phi(t)$ is also called as the instantaneous phase. The cosine of the instantaneous phase, like the wrapped phase, contains the main spatial information. The cosine of the instantaneous phase is used to map the wrinkle geometry in this paper, and the cosine results are still referred to as the instantaneous phase images.

3.3. Time-Frequency Analysis

The filter should be designed according to the frequency components of the inter-ply echoes. However, the front-face and back-wall echoes are the main energy components in the acquired A-scan signals. The Fourier transform is not suitable because its result is the frequency components of a full A-scan signal. The time-frequency analysis, analyzing a signal in both the time and frequency domains, is appropriate in this case.

The short-time Fourier transform (STFT) is a classical method for time-frequency signal analysis [34]. It can provide a proper resolution in the time or frequency domains depending on the parameter setting. The "Spectrogram" function in MATLAB R2021b is used to analyze A-scan signals.

The uncertainty principle, also called the Gabor limit, states that the time-frequency analysis of signals cannot achieve a high resolution in both the time and frequency domains. A too short time window (high time resolution) will lead to a poor frequency resolution, while a too long time window (poor time resolution) results in a good frequency resolution. The frequency components of the inter-ply echoes are concerned, so a relatively long window (ten times the duration corresponding to the center frequency) is chosen for the STFT.

3.4. FIR Filter Design

The TFM method is a delay-and-sum algorithm in the time domain, and the time information of filtered signals should not change or can be corrected. Wrinkles in CFRP composites are geometric defects with the carbon fiber plies deviating from the anticipated direction. The deviating plies are continuously and gradually changed. The essence of direct wrinkle detection is locating carbon fiber plies and disclosing the deviating plies. Direct detection requires high spatial precision. The phase information is related to spatial information, so filters must avoid phase distortion. A linear phase filter can avoid phase distortion, and the filtered signal's frequency components are shifted for a constant time. The time information change induced by a linear phase filter can be corrected by shifting the resulting signal. An FIR filter can become a linear phase filter by making coefficients symmetric. The Parks–McClellan algorithm [35] is used to implement linear phase FIR bandpass filters whose normalized stopband frequencies and normalized passband frequencies are determined by the frequency components of the inter-ply echoes. In this paper, the passband frequencies are set as the values within which the desired frequencies are allowed to pass, and the stopband frequencies are set to be 1.11 times wider than the passband frequencies.

4. Results and Discussion

4.1. Frequency Components of the Inter-Ply Echoes

Figure 3a is the acquired A-scan signals with the 32nd element firing and all 64 elements receiving, and the partial signals related to the CFRP sample are shown in Figure 3b. The back-wall echoes' amplitude reduces when the reception element drifts away from the transmission element. The A-scan signals are almost symmetric about the transmission element (the 32nd element). The A-scan signals with the 1st element firing and all 64 elements receiving are shown in Figure 3c, and the partial signals related to the sample are shown in Figure 3c. The inspected CFRP sample can be assumed to be the same everywhere for analyzing the frequency components of the inter-ply echoes. Some A-scan signals with the 1st element firing are analyzed using the "Spectrogram" function with a relatively long window. Figure 4 shows the spectrograms of the sample-related signals with the 1st element firing and the 1st, 8th, 16th, 24th, 32nd, 40th, 48th, 56th, and 64th elements receiving. The signal with the ith element firing and the jth element receiving is expressed as the (i, j) signal for short from here on. The black lines in Figure 4 are the Time-frequency ridges representing the maximum energy frequency component at each time. It can be seen that the main frequency components of inter-ply signals are distributed around 9 MHz (red rectangular boxes in Figure 4).

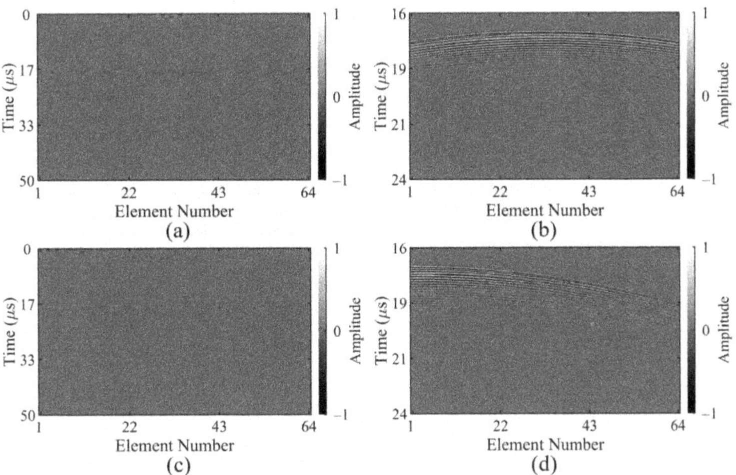

Figure 3. (**a**) Signals with the 32nd element firing and all 64 elements receiving, and (**b**) the partial signals related to the CFRP sample; (**c**) signals with the 1st element firing and all 64 elements receiving, and (**d**) the partial signals related to the CFRP sample.

4.2. The TFM Instantaneous Phase Method with Linear Phase FIR Filter

Using the Parks–McClellan algorithm, a linear phase FIR bandpass filter with a coefficient sequence symmetric has been designed. Its passband frequencies are set between 7 MHz~11 MHz, and its stopband frequencies are set between 6.78 MHz~11.22 MHz. The stopband frequencies are 1.11 times wider than the passband frequencies. The center of the passband frequencies is set at 9 MHz. The magnitude and phase responses of the designed filter are shown in Figure 5. The phase response is a linear function of frequency within the passband frequencies. The time information change caused by the designed linear phase filter can be corrected by shifting the filtered signals (Figure 6). The comparison between the raw and the corrected A-scans of the (1,1) and (1,16) signals is shown in Figure 6. As expected, although the amplitude changes, the phase and time information are almost unchanged. The spectrograms of the (1,1) and (1,16) filtered signals related to the sample show that the frequency components of the inter-ply echoes are relatively more concentrated around 9 MHz (Figure 7). The TFM and its related instantaneous phase images of the unfiltered signals are shown in Figure 8a,b, respectively, and The TFM and its related instantaneous phase results of filtered signals are shown in Figure 8c,d, respectively. The back-wall information almost disappears in the filtered result. This phenomenon is mainly attributed to the amplitude reduction of the back-wall echoes after filtering. The high-frequency components have a high inter-ply reflection coefficient and high CFRP attenuation [36], so the frequency components of back-wall echoes contain substantial low-frequency components. The low-frequency components are filtered out by the designed filter. As a result, the back-wall echoes' amplitude decreases, and the back-wall information becomes invisible in Figure 8c,d. In fact, the back-wall information is inessential, but the maximum wrinkle angle mainly affecting the mechanical performance is the primary parameter to be measured necessarily [25]. The instantaneous phase images can be used to map the wrinkle geometry. The dislocation in the instantaneous phase images is mainly induced by noise, so the TFM instantaneous phase image of the filtered signals has less dislocation than that of the unfiltered. The peaks ($\phi = \pi/2$) in the filtered TFM instantaneous phase image are extracted and indicated by red points, and those in the unfiltered are green points (Figure 9). Most peaks in the filtered and unfiltered are coincident, and these coincident points appear with yellow ones. As shown in Figure 9, the designed linear phase filter almost does not affect the wrinkle geometry of the detection result and relieves some noise-induced dislocation.

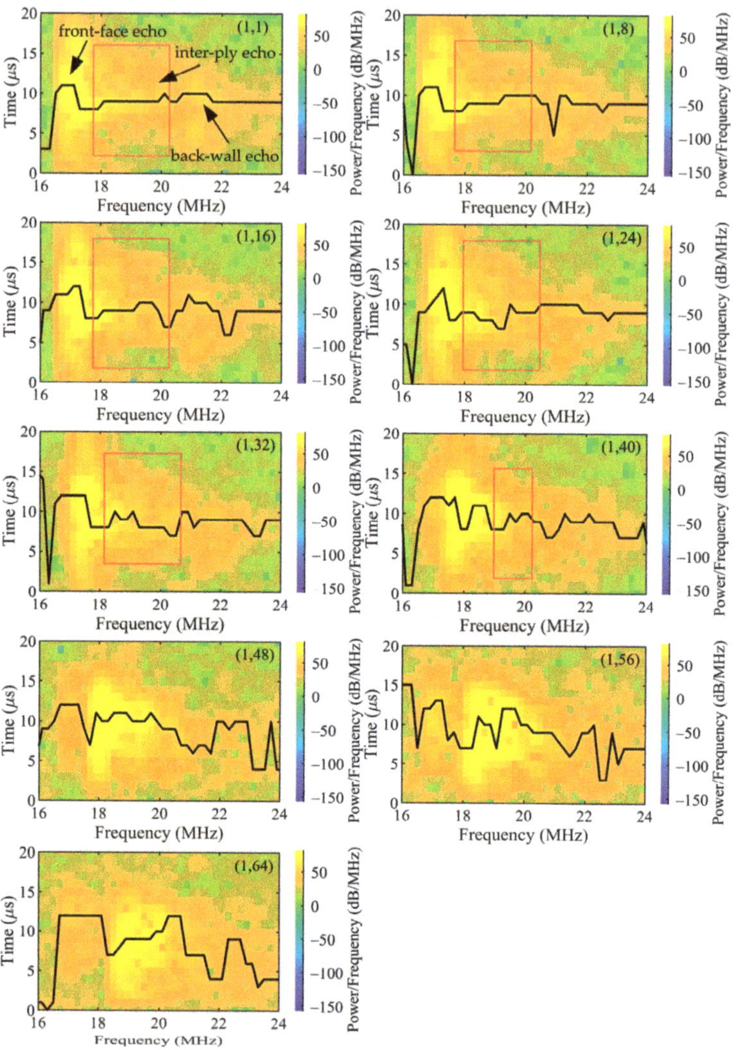

Figure 4. Spectrograms of the sample-related signals. Note: the signal with the *i*th element firing and the *j*th element receiving is expressed as the (*i*, *j*) signal for short, and the symbol (*i*, *j*) is labeled in the top right-hand corner of each spectrogram.

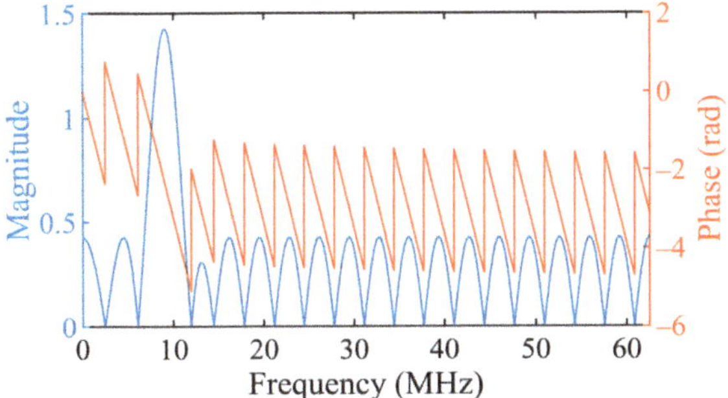

Figure 5. The magnitude and phase responses of the designed filter.

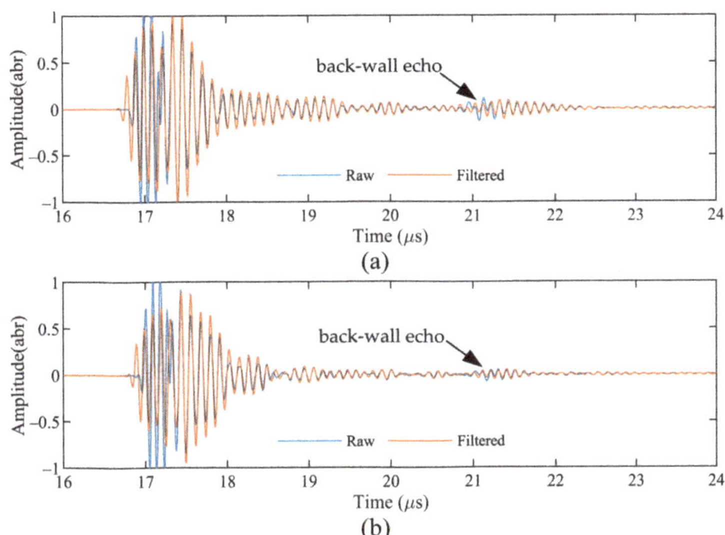

Figure 6. The comparison between the raw and the corrected filtered A-scans of the (**a**) (1,1) and (**b**) (1,16) signals.

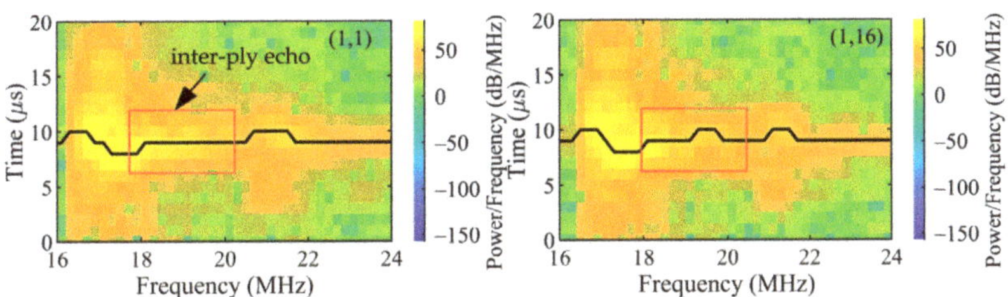

Figure 7. Spectrograms of the (1,1) and (1,16) filtered signals.

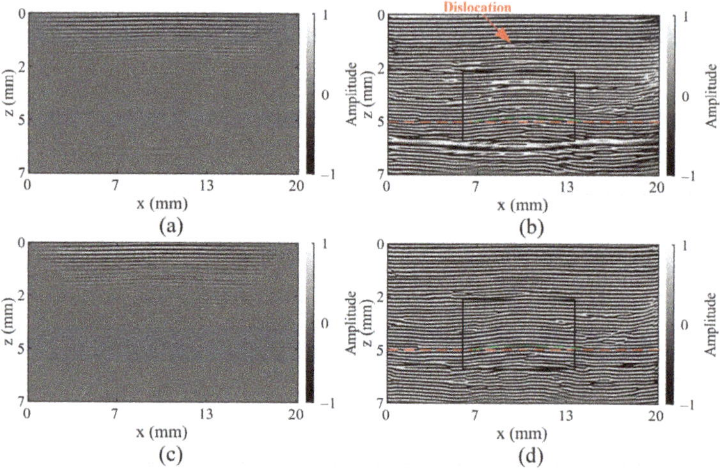

Figure 8. (**a**) The TFM and (**b**) its related instantaneous phase images of the unfiltered signals; (**c**) the TFM and (**d**) its related instantaneous phase images of the filtered signals.

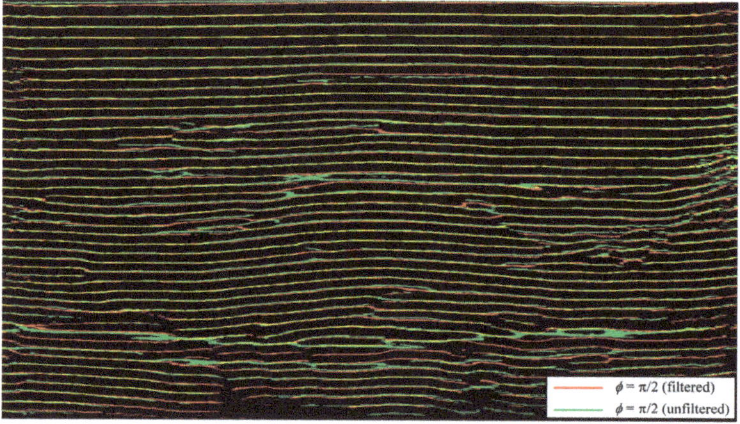

Figure 9. The comparison between the unfiltered and the filtered results.

4.3. The Filter Bandwidth

The filter bandwidth denotes the passband frequencies' range in this paper. In Section 3.2, the filter passband frequencies have been set between 7 MHz~11 MHz; the filter bandwidth is 4 MHz. In this section, linear phase FIR filters with different bandwidths have been designed, and their parameters are mainly listed in Table 2. The bandwidths of designed filters are 8 MHz, 4MHz, 2MHz, and 1 MHz, and their corresponding TFM instantaneous phase results are shown in Figure 10a–d, respectively. It can be seen that the result corresponding to a narrower bandwidth filter has less dislocation in Figure 10. A narrower bandwidth filter can filter out more noise. The comparison between the unfiltered and the 1MHz bandwidth-filtered results is shown in Figure 11. Although the wrinkle geometry of the detection result remains nearly unchanged, the back-wall information suffers further loss for the 1 MHz bandwidth filter. The result of the 1 MHz bandwidth filter is not better than that of the 2 MHz bandwidth filter. The 2 MHz bandwidth filter is recommended as an optimum choice for the TFM instantaneous phase method. The peaks ($\phi = \pi/2$) of the detection result from the 2 MHz bandwidth filter are overlaid on the micrograph of the wrinkle sample (Figure 12), and these peaks can excellently track the deviating plies (wrinkle defects).

Table 2. The parameters of FIR filters with different bandwidths.

Filter Name	Passband Frequencies	Stopband Frequencies	Filter Bandwidth
a	5 MHz~13 MHz	4.56 MHz~13.44 MHz	8 MHz
b	7 MHz~11 MHz	6.78 MHz~11.22 MHz	4 MHz
c	8 MHz~10 MHz	7.89 MHz~10.11 MHz	2 MHz
d	8.5 MHz~9.5MHz	8.45 MHz~9.55 MHz	1 MHz

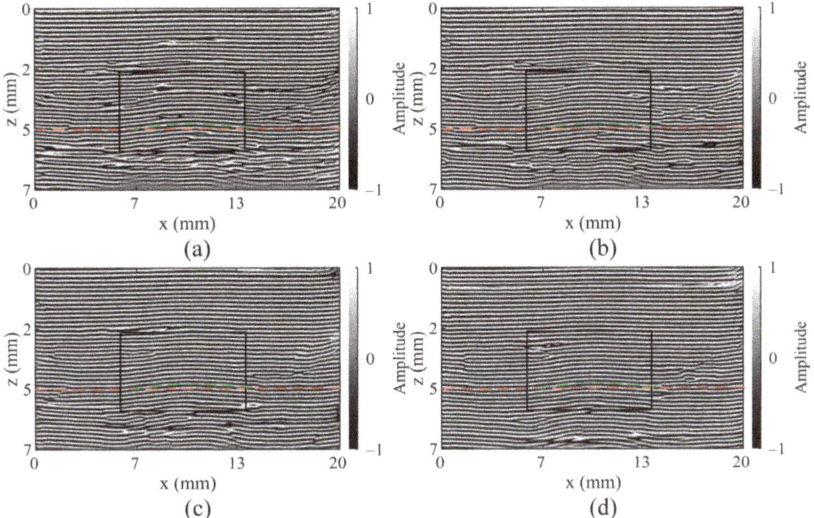

Figure 10. The TFM instantaneous phase results filtered by the (**a**) 8 MHz, (**b**) 4 MHz, (**c**) 2 MHz, and (**d**) 1 MHz bandwidth linear phase filters.

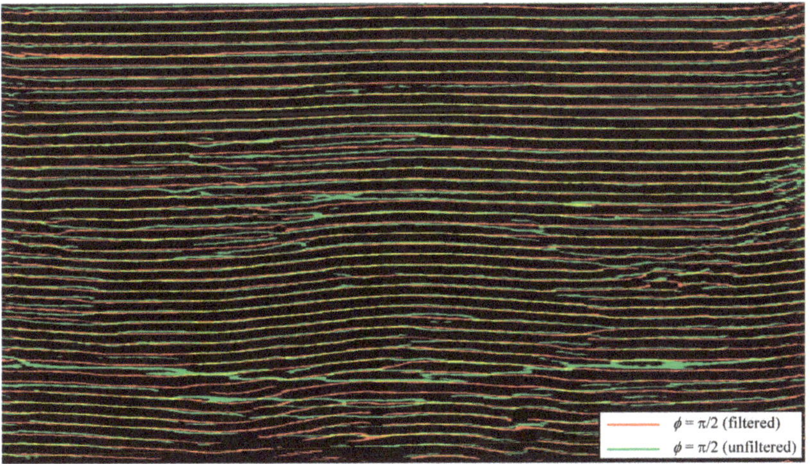

Figure 11. The comparison between the unfiltered and the filtered results.

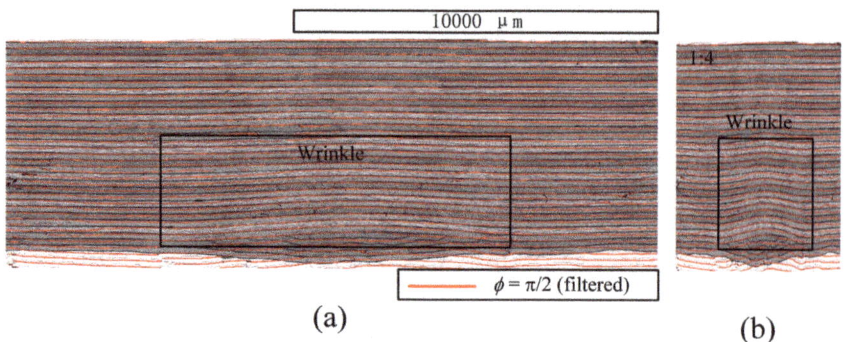

Figure 12. (**a**) The overlaid view of peaks ($\phi = \pi/2$) from the 2 MHz bandwidth linear phase filter, and (**b**) its shortened length image.

5. Conclusions

The TFM instantaneous phase method using weak reflection signals is vulnerable to noise, especially structural noise from voids. The noise contaminates weak inter-ply re-flection signals and induces dislocation in the TFM instantaneous phase results. The aim of this paper is to use suitable filters to relieve noise-induced dislocation for the TFM instantaneous phase method. Due to the character of the total focusing method (TFM) and the high spatial precision requirements of direct wrinkle detection, we have chosen to design linear phase filters. Linear phase filters can avoid phase distortion and preserve spatial information.

The linear phase filter almost has no effect on the wrinkle geometry of the detection results and relieves some noise-induced dislocation. The 2 MHz bandwidth filter is an optimum choice for the TFM instantaneous phase method. The filtered detection result can excellently track the deviating plies (namely wrinkle defects).

Author Contributions: T.M.: conceptualization, investigation, and writing—original draft. Y.L.: supervision and writing—review and editing, project administration, and funding acquisition. Z.Z.: supervision and writing—review and editing, project administration. J.M.: resource and supervision. All authors have read and agreed to the published version of the manuscript.

Funding: This research was funded by the Commercial Aircraft Corporation of China (COMAC), grant number COMAC-SFGS-2021-666.

Data Availability Statement: Not applicable.

Conflicts of Interest: The authors declare no conflict of interest.

References

1. Nartey, M.; Zhang, T.; Gong, B.; Wang, J.; Peng, S.; Wang, H.; Peng, H.-X. Understanding the impact of fibre wrinkle architectures on composite laminates through tailored gaps and overlaps. *Compos. Part B Eng.* **2020**, *196*, 108097. [CrossRef]
2. Matsui, J. Polymer matrix composites (PMC) in aerospace. *Adv. Compos. Mater.* **1995**, *4*, 197–208. [CrossRef]
3. Mukhopadhyay, S.; Jones, M.I.; Hallett, S.R. Tensile failure of laminates containing an embedded wrinkle; numerical and experimental study. *Compos. Part A Appl. Sci. Manuf.* **2015**, *77*, 219–228. [CrossRef]
4. Wilhelmsson, D.; Gutkin, R.; Edgren, F.; Asp, L. An experimental study of fibre waviness and its effects on compressive properties of unidirectional NCF composites. *Compos. Part A Appl. Sci. Manuf.* **2018**, *107*, 665–674. [CrossRef]
5. Kulkarni, P.; Mali, K.D.; Singh, S. An overview of the formation of fibre waviness and its effect on the mechanical performance of fibre reinforced polymer composites. *Compos. Part A Appl. Sci. Manuf.* **2020**, *137*, 106013. [CrossRef]
6. Elhajjar, R.F.; Shams, S.S.; Kemeny, G.J.; Stuessy, G. A hybrid numerical and imaging approach for characterizing defects in composite structures. *Compos. Part A Appl. Sci. Manuf.* **2016**, *81*, 98–104. [CrossRef]
7. Liao, Y.; Wang, J.; Zeng, Z.; Lin, J.; Dai, Y. Detection of fiber waviness in carbon fiber prepreg using Eddy current method. *Compos. Commun.* **2021**, *28*, 100981. [CrossRef]
8. Elhajjar, R.; Haj-Ali, R.; Wei, B.-S. An infrared thermoelastic stress analysis investigation for detecting fiber waviness in composite structures. *Polym.-Plast. Technol. Eng.* **2014**, *53*, 1251–1258. [CrossRef]

9. Hallander, P.; Akermo, M.; Mattei, C.; Petersson, M.; Nyman, T. An experimental study of mechanisms behind wrinkle development during forming of composite laminates. *Compos. Part A Appl. Sci. Manuf.* **2013**, *50*, 54–64. [CrossRef]
10. Smith, R.A.; Nelson, L.J.; Mienczakowski, M.J.; Challis, R.E. Automated analysis and advanced defect characterisation from ultrasonic scans of composites. *Insight-Non-Destr. Test. Cond. Monit.* **2009**, *51*, 82–87. [CrossRef]
11. Chakrapani, S.K.; Dayal, V.; Hsu, D.K.; Barnard, D.J.; Gross, A. Characterization of waviness in wind turbine blades using air coupled ultrasonics. In *AIP Conference Proceedings*; American Institute of Physics: College Park, ML, USA, 2011; pp. 956–962.
12. Kosukegawa, H.; Kiso, Y.; Hashimoto, M.; Uchimoto, T.; Takagi, T. Evaluation of detectability of differential type probe using directional eddy current for fibre waviness in CFRP. *Philos. Trans. A Math. Phys. Eng. Sci.* **2020**, *378*, 20190587. [CrossRef]
13. Zhang, Z.; Guo, S.; Li, Q.; Cui, F.; Malcolm, A.A.; Su, Z.; Liu, M. Ultrasonic detection and characterization of delamination and rich resin in thick composites with waviness. *Compos. Sci. Technol.* **2020**, *189*, 108016. [CrossRef]
14. Philibert, M.; Soutis, C.; Gresil, M.; Yao, K. Damage Detection in a Composite T-Joint Using Guided Lamb Waves. *Aerospace* **2018**, *5*, 40. [CrossRef]
15. Memmolo, V.; Boffa, N.D.; Maio, L.; Monaco, E.; Ricci, F. Damage Localization in Composite Structures Using a Guided Waves Based Multi-Parameter Approach. *Aerospace* **2018**, *5*, 111. [CrossRef]
16. Malinowski, P.H.; Tserpes, K.I.; Ecault, R.; Ostachowicz, W.M. Mechanical and Non-Destructive Study of CFRP Adhesive Bonds Subjected to Pre-Bond Thermal Treatment and De-Icing Fluid Contamination. *Aerospace* **2018**, *5*, 36. [CrossRef]
17. Díaz Valdés, S.H.; Soutis, C. A structural health monitoring system for laminated composites. In Proceedings of the International Design Engineering Technical Conferences and Computers and Information in Engineering Conference, Pittsburgh, PN, USA, 9–12 September 2001; pp. 2013–2021.
18. Su, Z.; Ye, L.; Lu, Y. Guided Lamb waves for identification of damage in composite structures: A review. *J. Sound Vib.* **2006**, *295*, 753–780. [CrossRef]
19. Zhang, H.; Peng, L.; Zhang, H.; Zhang, T.; Zhu, Q. Phased array ultrasonic inspection and automated identification of wrinkles in laminated composites. *Compos. Struct.* **2022**, *300*, 116170. [CrossRef]
20. Smith, R.A.; Nelson, L.J.; Mienczakowski, M.J.; Wilcox, P.D. Ultrasonic tracking of ply drops in composite laminates. In *AIP Conference Proceedings*; AIP Publishing LLC.: Melville, NY, USA, 2016; p. 050006.
21. Smith, R.A.; Nelson, L.J.; Mienczakowski, M.J.; Wilcox, P.D. Ultrasonic Analytic-Signal Responses From Polymer-Matrix Composite Laminates. *IEEE Trans. Ultrason. Ferroelectr. Freq. Control.* **2018**, *65*, 231–243. [CrossRef]
22. Yang, X.; Verboven, E.; Ju, B.-f.; Kersemans, M. Comparative study of ultrasonic techniques for reconstructing the multilayer structure of composites. *NDT E Int.* **2021**, *121*, 102460. [CrossRef]
23. Fernández-López, A.; Larrañaga-Valsero, B.; Guemes, A. Wrinkle detection with ultrasonic phased array technology. In Proceedings of the 6th International Symposium on NDT in Aerospace, Madrid, Spain, 12–14 November 2014; Department of Aeronautics, Polytechnic University of Madrid (UPM): Madrid, Spain, 2014.
24. Pain, D.; Drinkwater, B.W. Detection of Fibre Waviness Using Ultrasonic Array Scattering Data. *J. Nondestruct. Eval.* **2013**, *32*, 215–227. [CrossRef]
25. Xie, N.; Smith, R.A.; Mukhopadhyay, S.; Hallett, S.R. A numerical study on the influence of composite wrinkle defect geometry on compressive strength. *Mater. Des.* **2018**, *140*, 7–20. [CrossRef]
26. Lin, L.; Cao, H.; Luo, Z. Total focusing method imaging of multidirectional CFRP laminate with model-based time delay correction. *NDT E Int.* **2018**, *97*, 51–58. [CrossRef]
27. Larrañaga-Valsero, B.; Smith, R.A.; Boumda, R.T.; Fernández-López, A.; Güemes, A. Wrinkle characterisation from ultrasonic scans of composites. In Proceedings of the 55th Annual Conference of the British Institute of Non-Destructive Testing, NDT 2016, Nottingham, UK, 12–14 September 2016; pp. 508–521.
28. Zhang, H.; Ren, Y.; Song, J.; Zhu, Q.; Ma, X. The wavenumber imaging of fiber waviness in hybrid glass–carbon fiber reinforced polymer composite plates. *J. Compos. Mater.* **2021**, *55*, 4633–4643. [CrossRef]
29. Hunter, A.J.; Drinkwater, B.W.; Wilcox, P.D. The wavenumber algorithm for full-matrix imaging using an ultrasonic array. *IEEE Trans. Ultrason. Ferroelectr. Freq. Control* **2008**, *55*, 2450–2462. [CrossRef] [PubMed]
30. Oppenheim, A.V.; Schafer, R.W. *Discrete-Time Signal Processing*; Pearson: London, UK, 2010.
31. Holmes, C.; Drinkwater, B.W.; Wilcox, P.D. Post-processing of the full matrix of ultrasonic transmit–receive array data for non-destructive evaluation. *NDT E Int.* **2005**, *38*, 701–711. [CrossRef]
32. Luo, Z.; Zhang, S.; Jin, S.; Liu, Z.; Lin, L. Heterogeneous ultrasonic time-of-flight distribution in multidirectional CFRP corner and its implementation into total focusing method imaging. *Compos. Struct.* **2022**, *294*, 115789. [CrossRef]
33. Boashash, B. *Time-Frequency Signal Analysis and Processing: A Comprehensive Reference*; Academic Press: Cambridge, MA, USA, 2015.
34. Durak, L.; Arikan, O. Short-time Fourier transform: Two fundamental properties and an optimal implementation. *IEEE Trans. Signal Process.* **2003**, *51*, 1231–1242. [CrossRef]

35. Filip, S.-I. A robust and scalable implementation of the Parks-McClellan algorithm for designing FIR filters. *ACM Trans. Math. Softw. (TOMS)* **2016**, *43*, 1–24. [CrossRef]
36. Smith, R.A. Use of 3D Ultrasound Data Sets to Map the Localised Properties of Fibre-Reinforced Composites. Ph.D. Thesis, University of Nottingham, Nottingham, UK, 2010.

Article

On Cointegration Analysis for Condition Monitoring and Fault Detection of Wind Turbines Using SCADA Data

Phong B. Dao

Department of Robotics and Mechatronics, AGH University of Science and Technology, Al. Mickiewicza 30, 30-059 Krakow, Poland; phongdao@agh.edu.pl

Abstract: Cointegration theory has been recently proposed for condition monitoring and fault detection of wind turbines. However, the existing cointegration-based methods and results presented in the literature are limited and not encouraging enough for the broader deployment of the technique. To close this research gap, this paper presents a new investigation on cointegration for wind turbine monitoring using a four-year SCADA data set acquired from a commercial wind turbine. A gearbox fault is used as a testing case to validate the analysis. A cointegration-based wind turbine monitoring model is established using five process parameters, including the wind speed, generator speed, generator temperature, gearbox temperature, and generated power. Two different sets of SCADA data were used to train the cointegration-based model and calculate the normalized cointegrating vectors. The first training data set involves 12,000 samples recorded before the occurrence of the gearbox fault, whereas the second one includes 6000 samples acquired after the fault occurrence. Cointegration residuals—obtained from projecting the testing data (2000 samples including the gearbox fault event) on the normalized cointegrating vectors—are used in control charts for operational state monitoring and automated fault detection. The results demonstrate that regardless of which training data set was used, the cointegration residuals can effectively monitor the wind turbine and reliably detect the fault at the early stage. Interestingly, despite using different training data sets, the cointegration analysis creates two residuals which are almost identical in their shapes and trends. In addition, the gearbox fault can be detected by these two residuals at the same moment. These interesting findings have never been reported in the literature.

Keywords: wind turbine; condition monitoring; fault detection; cointegration; SCADA data

Citation: Dao, P.B. On Cointegration Analysis for Condition Monitoring and Fault Detection of Wind Turbines Using SCADA Data. *Energies* **2023**, *16*, 2352. https://doi.org/10.3390/en16052352

Academic Editor: Davide Astolfi

Received: 31 January 2023
Revised: 24 February 2023
Accepted: 27 February 2023
Published: 1 March 2023

1. Introduction

Due to the high demand of global energy consumption and the aggravation of environmental problems, wind energy has kept a progressively important role among other renewable energy sources and accordingly contributed an indispensable solution to solving world energy problems. The total installed capacity of the wind power sector in the world was reported to reach 837 GW by the end of 2021 [1]. In Poland, the total capacity of onshore wind power installations was up to 6.35 GW by the end of 2020, and it is expected to continue growing and reach between 8 GW to 10 GW by 2030 [2]. However, because wind turbines are typically situated in remote locations, operate under severe environments, and have load conditions varying over time, their failure rate and downtime are relatively high [3]. Hence, the sector faces many challenges related to high operations and maintenance (O&M) costs and downtime losses. These circumstances bring huge economic loss to the asset owners and also cause a negative influence on the sustainable development of the wind energy industry [4]. Therefore, it is important to develop condition monitoring solutions for wind turbines that can predict or detect incipient failures at the early stage [5].

Condition-based maintenance has been extensively deployed as an effective strategy to reduce O&M costs and improve the availability and efficiency of wind farms [3]. Vibration analysis and oil monitoring are two commonly used techniques which use large volumes of high-frequency data, including vibration signals and oil debris measurements collected from main turbine components [6–8]. Nevertheless, both techniques are sophisticated and expensive, since they require additional sensors and data acquisition systems being installed on the operating wind turbines [4]. Alternatively, wind turbine monitoring using data collected by the supervisory control and data acquisition (SCADA) systems has been considered as a cost-effective approach, as these systems are widely pre-installed in the majority of commercial wind turbines [4,5]. SCADA systems record the operation state information and environmental conditions from wind turbines on a regular basis. Compared with the vibration analysis and oil monitoring methods, the SCADA-based monitoring solutions offer users large amounts of data readily available for analysis without additional cost. As a result, much research has made use of SCADA data to develop reliable and cost-effective monitoring systems in recent years, as reported in [4,5,9–11]. However, because each wind farm often consists of a great number of wind turbines which are required to be monitored concomitantly, the operator has to deal with large volumes and diversity of SCADA data. To cope with this difficulty, most recently developed solutions have been based on the competences of artificial intelligence (AI) and machine learning (ML) techniques such as learning, classification, and adaptation [12]. Many advanced AI/ML methods, such as self-supervised health representation learning [13], anomaly decomposition based on multi-variable correlation extraction [14], and hierarchical hyper-parameter searching algorithm [15], have been recently developed. However, AI-based and ML-based algorithms are known to be sophisticated, require a lot of data to train algorithms, need extensive training time, and incur heavy computational cost [12,16,17]. Hence, more simple and computationally efficient solutions have been explored in recent years. Amongst these, the statistical approaches have been effectively exploited for wind turbine health monitoring and fault assessment, such as multivariate statistical hypothesis testing [9], nonparametric regression analysis [18], and the cointegration-based approach [19–26]. Recently, change-point detection methods [27,28], cumulative sum (CUSUM)-based methods [29,30], and the Wilcoxon rank sum test based method [31] have been proposed for SCADA-based wind turbine condition monitoring.

Cointegration, a technique originally developed in the field of econometrics [32,33], has been adopted for structural health monitoring (SHM) as a potential data-driven method to remove or compensate for common long-term trends instigated by effects of environmental and operational variability (EOV) in the measured data. Some selected examples of cointegration-based methods developed for SHM applications can be found in [34–41]. The main idea in applying cointegration for SHM is based on the analysis of nonstationary time series. When nonstationary data collected from a structure or process are cointegrated, it is possible to obtain one or several stationary cointegration residuals, which represent the undamaged (or normal operating) condition. Then, during the monitoring or testing process, if the residuals become nonstationary then one can infer that the current data are no longer representing the normal condition [34–36]. In addition, cointegration can effectively remove the common trends, induced by EOV effects, from the original data, leaving the residuals independent of EOV that still maintain their sensitivity to damage. To understand how common trends, induced by EOV effects, can be purged from the measured data by cointegration procedure and how a fault or damage can be detected using cointegration residuals, potential readers are referred to the work [42].

Recently, the cointegration technique has been proposed for the purpose of condition monitoring and fault detection of wind turbines, as reported in [19–26]. A cointegration-based method was developed in [19–21] to analyse a benchmark SCADA data set recorded from a 2 MW wind turbine drivetrain during 30 days under environmental and operational variations. A human-made gearbox fault was progressively created during the experimental and data acquisition process. The results proved that the proposed method can effectively analyse nonlinear data trends, continuously monitor the wind turbine and reliably detect abnormal problems. In [22], a cointegration-based method was reported to effectively monitor the abnormal state of generator and gearbox such that early warning of faults was possible. In [23,26], SCADA data acquired from a 1.5 MW wind turbine under varying environmental and operational conditions were used to establish a cointegration model for identifying a set of known gearbox fault data. The cointegration analysis was applied for vibration-based damage detection of a wind turbine blade under the influence of EOV [24]. The results demonstrated that cointegration could be used to detect the presence of damages under conditions not allowing for direct discrimination between damage and EOV. In [25], a Bayesian multivariate cointegration method was developed for vibration-based damage detection of wind turbine blades. The results showed that the method could effectively eliminate the influence of EOV and detect the progressive damage of the wind turbine blade. A common point of these previous works is that the operating condition of a given wind turbine can be monitored by means of observing the cointegration residuals, obtained from the cointegration process of SCADA data, in control charts. However, the existing cointegration-based methods and results presented in [19–26] are not sufficient and encouraging enough for the broader deployment of the technique in practical applications. This work aims to close this research gap through performing a new investigation on cointegration for wind turbine monitoring using a four-year SCADA data set acquired from a commercial wind turbine. A gearbox fault is used as a testing case to validate the analysis. A cointegration-based computation procedure, consisting of three stages, was developed for this purpose. In the first stage, a cointegration model of the monitored wind turbine is established using a set of process parameters. This model has the role of a wind turbine monitoring model. In the second stage, the Johansen's cointegration procedure [33] is deployed to train the cointegration-based monitoring model and calculate the normalized cointegrating vectors. In the third stage, SCADA data—acquired from the monitored wind turbine during the regular operating period for producing electricity—are projected on the normalized cointegrating vectors found in the second stage to form cointegration residuals used for on-line monitoring of the wind turbine. The monitoring scheme is based on the residual-based control chart technique, which is one of the most popular tools used for statistical process control.

Using this computation procedure, a cointegration-based wind turbine monitoring model has been established using five operational parameters, i.e., the wind speed, generator speed, generator temperature, gearbox temperature, and generated power. Two different sets of SCADA data, recorded before and after the occurrence of the gearbox fault, were used to train the cointegration-based model and calculate the normalized cointegrating vectors. The results demonstrate that regardless of which training data set was used, the cointegration residuals monitored the wind turbine accurately and detected the fault reliably at the early stage. Interestingly, despite using different training data sets, the cointegration analysis created two residuals which are almost identical in their shapes and trends. In addition, the gearbox fault was detected by these two residuals at the same moment. These interesting findings have never been reported in the literature.

The remaining parts of this paper are planned as follows. Section 2 gives a brief introduction of the cointegration theory. Section 3 presents a three-stage cointegration-based computation procedure for on-line wind turbine monitoring and fault detection. SCADA data used for validating the proposed cointegration-based monitoring method are described in Section 4. Section 5 presents the validation results and discussions. Finally, the paper is closed with conclusions and future work suggestions in Section 6.

2. A Brief Introduction of Cointegration Theory

In the previous studies [34,35], the basic theory of cointegration analysis and other relevant topics, such as stationarity of time series, cointegration, and common stochastic trends, were described and explained in detail. Hence, these concepts are not presented in depth in this paper. Potential readers are referred to those materials for detailed descriptions of the cointegration theory. Furthermore, to know and be familiar with how cointegration was previously applied for condition monitoring and fault detection of wind turbines, the readers are referred to some previous works [19–21]. In the following, only a brief introduction of nonstationarity and cointegration is presented and explained.

A nonstationary time series has its mean, variance, and covariance parameters generally change over time [43]. For example, a time series exhibiting a shift in its mean is a nonstationary process because it is a variable with a heteroscedastic variance over time. It is well known that a common way to transform a nonstationary time series into a stationary time series is by means of differencing. The number of differences required to make a given nonstationary time series become stationary is called the order of integration. A time series of order d is denoted as $I(d)$. Therefore, a nonstationary $I(1)$ time series becomes a stationary $I(0)$ time series by first-order differencing. In the case of a nonstationary $I(2)$ time series, a second-order differencing would be required to make it stationary. Generally, cointegration is characterized by two or more nonstationary $I(1)$ variables sharing a common long-run development, i.e., they do not drift away from each other except for transitory fluctuations. In other words, if a group of nonstationary time series variables have the propensity to establish and maintain a long-run equilibrium relationship, the cointegration analysis can be used to find this relationship.

Let $Y_t = (y_{1t}, y_{2t}, \ldots, y_{nt})^T$ denote an $(n \times 1)$ vector of $I(1)$ time series. This n-dimension vector is said to be linearly cointegrated if there exists a vector $\beta = (\beta_1, \beta_2, \ldots, \beta_n)^T$ such that

$$\beta^T Y_t = \beta_1 y_{1t} + \beta_2 y_{2t} + \cdots + \beta_n y_{nt} \sim I(0) \tag{1}$$

Equation (1) infers that the nonstationary time series in Y_t are cointegrated if there is (at least) a linear combination of those series that is stationary or has the $I(0)$ status. This linear combination, denoted as $u_t = \beta^T Y_t + c$, where c is a constant value, is referred to as a cointegration residual that represents a long-run equilibrium relationship between the cointegrated time series [43]. The vector β is referred to as a cointegrating vector. However, the cointegrating vector β is not unique, since for any scalar k, we have

$$k \cdot \beta^T Y_t = (\beta^*)^T Y_t \sim I(0) \tag{2}$$

A normalization assumption can be used to uniquely identify β. A typical normalization is [43]

$$\beta = (1, -\beta_2, \ldots, -\beta_n)^T \tag{3}$$

Using this normalization, the cointegrating relationship in Equation (1) can be rewritten as

$$\beta^T Y_t = y_{1t} - \beta_2 y_{2t} - \cdots - \beta_n y_{nt} \sim I(0) \tag{4}$$

or

$$y_{1t} = \beta_2 y_{2t} + \beta_3 y_{3t} + \cdots + \beta_n y_{nt} + \beta^T Y_t \tag{5}$$

The cointegration residual ($u_t = \beta^T Y_t + c$) is formed by projecting n vectors of a time series in Y_t on the normalized cointegrating vector β. This projection is equivalent to multiplying Y_t by β^T. The single cointegration relationship in Equation (1) can be extended to multiple cointegrations. In this case, Y_t is said to be cointegrated with r linearly independent cointegrating vectors (where $0 < r < n$) if there is an $(n \times r)$ matrix B such that

$$B^T Y_t = \begin{pmatrix} \beta_1^T Y_t \\ \vdots \\ \beta_r^T Y_t \end{pmatrix} = \begin{pmatrix} u_{1t} \\ \vdots \\ u_{rt} \end{pmatrix} \sim I(0) \tag{6}$$

The stationary linear combinations $u_{rt} = B^T Y_t + c_r$, where c_r is a constant vector, are known as the r cointegration residuals, which are formed by projecting n vectors of time series in Y_t on the cointegrating matrix B, or equivalently, by multiplying Y_t by B^T. When using the cointegration method, one of the most important points is to estimate (or calculate) suitable normalized cointegrating vectors so as to create stationary cointegration residuals together with common trends removed. The Johansen's cointegration method [33]—a sequential procedure based on the maximum likelihood estimation (MLE)—has been generally used for this purpose. The theory behind this method is sophisticated and thus not presented here. For more theoretical details of the Johansen's cointegration method, potential readers are referred to the original work [33]; a simpler description version can be found in [35]. The Johansen's cointegration procedure has been employed in this work, through applying the Econometrics Toolbox [44], to estimate the normalized cointegrating vectors.

3. On Cointegration for Condition Monitoring and Fault Detection of Wind Turbines

In the present work, the cointegration technique has been exploited for on-line condition monitoring and fault detection of wind turbines using SCADA data. The entire cointegration-based computation procedure, consisting of three stages, is shown in Figure 1. In the following, these stages are described and discussed.

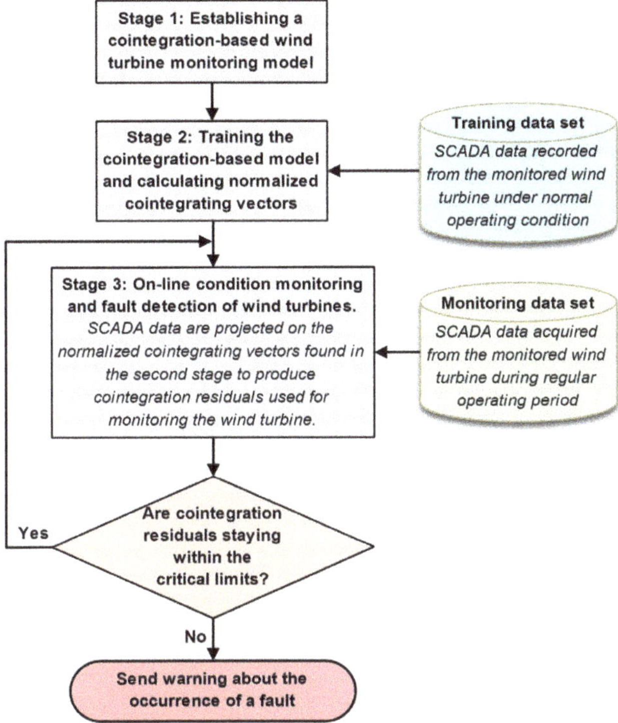

Figure 1. Flowchart of the cointegration-based computation procedure for wind turbine monitoring and fault detection.

3.1. Establishing a Cointegration-Based Wind Turbine Monitoring Model

The purpose of the first stage is to establish a cointegration model for a given wind turbine. Specifically, a number of key process parameters of the wind turbine are required to be selected to form the model. A cointegration model is described by Equation (4), where variables $y_{1t}, y_{2t}, \ldots, y_{nt}$ represent the wind turbine parameters. In general, important operational parameters, such as the wind speed, generator speed, generated power, generator temperature, generator voltage, generator current, gearbox temperature, gearbox oil sump temperature, rotor bearing temperature, and rotor speed, can be chosen for this purpose. In this study, a cointegration model of the monitored wind turbine, formed with a set of process parameters, has the role of a wind turbine monitoring model. It is noted that at least two parameters must be selected such that a cointegration-based wind turbine monitoring model can be established.

The cointegration-based wind turbine monitoring model does not require all important operational parameters, as named above, to be included in the model. However, it is suggested that the wind speed and generated power should be employed in the cointegration-based monitoring model. The reason is because the relationship between wind speed and turbine power output represents the wind turbine power curve, which is one of the most important characteristics commonly used for wind turbine selection, capacity factor estimation, wind energy assessment and forecasting, and turbine performance and health monitoring [45]. In addition, temperature parameters of the generator and gearbox should be included in the model because a fault or an abnormal event, associated with the generator or gearbox component, is substantially a progressive phenomenon, that is, the initial sign of a gearbox or generator fault could appear several days or weeks before the fault event occurred in reality and it might be manifested by the increase in the gearbox bearing and/or generator temperature [28,31].

3.2. *Training the Cointegration-Based Model and Calculating Normalized Cointegrating Vectors*

In the second stage, the Johansen's cointegration procedure [33] is deployed to train the cointegration-based monitoring model and calculate the normalized cointegrating vectors. The computation uses only SCADA data of several process parameters acquired from the monitored wind turbine under normal operating condition or a "healthy" state. In a simple description, the estimation of cointegrating vectors is executed in three steps. First is evaluating eigenvalues from the characteristic equation of a cointegration model. Next is sorting the eigenvalues from the largest to the smallest one. Then, the normalized cointegrating vectors are calculated from the sorted eigenvalues. Hence, the first and the last cointegrating vector are corresponding to the largest and the smallest eigenvalue, respectively. As reported in the previous works [19,34,35,41], the first cointegrating vector is said to create the most stationary cointegration residual. In other words, when projecting SCADA series stored in different process parameters on the first cointegrating vector, we obtain the first cointegration residual which is the most stationary combination of the cointegrated data. This cointegration residual has been considered as the best (or the most suitable) indicator used for fault and/or damage detection, as discussed in [19,34–36,40,41]. In this study, we also consider the first cointegration residual as the best feature and therefore use only this residual to monitor the health state of the wind turbine.

It is supposed that the training data set—selected for calculating the normalized cointegrating vectors—has a significant influence on the wind turbine health monitoring and fault detection results. As mentioned above, only the SCADA data recorded from a wind turbine operating in healthy condition should be used for this purpose. However, this requirement faces some challenges. First, model training and cointegrating vector calculation require sufficient amounts of normal operation data collected over a long period covering a representative range of wind turbine operating conditions. Certainly, when these data are scarce or when they are not representative for the turbine's current normal operation state, fault detection may not feasible because the cointegration-based monitoring model cannot be trained properly. This is the case for newly installed wind turbines at the initial stage of their operation life when the amount of normal operation data accumulated is small, which cannot provide sufficient information for training cointegration-based models. Moreover, due to many unavoidable reasons, such as wind turbine ageing, subsystem replacements, software updates, or sensor recalibration, the normal operation data collected months or years before might be outdated and so they are no longer representative of the turbine's current normal operation behaviour.

An alternative solution has been suggested by this work to deal with these challenges, that is, one may consider using several training data sets, which represent different normal operating modes of the wind turbine, to obtain different sets of normalized cointegrating vectors. Given that, more than one set of cointegration residuals can be employed to monitor the turbine and detect abnormal problems. This idea has been validated in this paper and the obtained results are presented in Section 5.

3.3. On-Line Condition Monitoring and Fault Detection of Wind Turbines

In the third stage, SCADA data—acquired from the monitored wind turbine during the regular operating period for producing electricity—are projected on the normalized cointegrating vectors found in the second stage to produce cointegration residuals used for monitoring the wind turbine. As explained in Section 2, this projection is simply equivalent to the multiplication of data vectors. Since SCADA data stored in each process parameter can be considered as a vector of time series, a cointegration residual (given by $u_t = \beta^T Y_t + c$) can be formed by multiplying vectors of SCADA series stored in different process parameters by one cointegrating vector. This implies that a cointegration residual also has the form of a sequence of time series. To obtain multiple cointegration residuals (denoted by $u_{rt} = B^T Y_t + c_r$), one can multiply vectors of SCADA series stored in different process parameters by r cointegrating vectors. This computation can be executed in a real-time manner on a computer-based monitoring system, which provides a simple on-line condition monitoring solution for wind turbines. As discussed in Section 3.2, only the first cointegration residual is used in this study to monitor the health condition of wind turbines. The creation of this residual is achieved by multiplying vectors of SCADA series, corresponding to the selected process parameters, by the first normalized cointegrating vector.

The possibility of using a cointegration-based monitoring model, in particular, the first cointegration residual, for on-line condition monitoring of wind turbines is explained here. When a new set of monitoring samples collected by the SCADA system are made available for analysis, these data are instantly projected on the first normalized cointegrating vector to create a new value of the first cointegration residual. This value is then compared with the critical limits, calculated as statistical confidence levels, of the control chart to determine whether the wind turbine is still operating under its normal condition. To present the monitoring process in an illustrative manner, the first cointegration residual is plotted against the critical region; if the residual crosses the upper or lower critical line, then it means that a fault would occur in the turbine.

4. Wind Turbine SCADA Data

The long-term monitoring campaign of the La Houte Bourne onshore wind farm in Villeneuve-d'Ascq, France, over eight years (from 1 January 2013 to 31 December 2020) has provided for public a plentiful open-access SCADA data source [46]. The wind farm has four wind turbines of the MM82 model, manufactured by Senvion. The technical details of the wind turbines are given in Table 1. There were 34 process parameters measured at an interval of 10 min for each wind turbine and in total 1,057,868 samples were recorded. The data acquired for the wind turbine (labelled as R80721) over four years (from 1 January 2013 to 31 December 2016) were selected for the analysis in this study. There were 210,095 data samples recorded for each parameter. Before analysing the data using the cointegration-based method, data pre-processing and outlier cleaning procedures were performed to remove all samples associated with unphysical, corrupted, or missing values. As a result, we attained 142,613 data samples for each parameter. This four-year data collection of the wind turbine R80721 was recently used to validate a new wind turbine health monitoring method which is based on the Wilcoxon rank sum test [31]. SCADA data of this wind turbine, including the wind speed, generator speed, generator temperature, gearbox bearing temperature, and generated power, are plotted in Figure 2. These five process parameters are used in this study to create a cointegration-based monitoring model for the selected wind turbine. The validation results of the developed model are presented in the following section.

Table 1. Technical information of wind turbines in the wind farm.

Technical Parameters	Value
Rated power	2050 kW
Cut-in wind speed	4 m/s
Cut-out wind speed	22 m/s
Rated wind speed	14.5 m/s
Operating temperature range	−20 °C to +35 °C
Rotor diameter	82 m
Rotor area	5281 m^2
Rotor blade length	40 m
Hub height	80 m

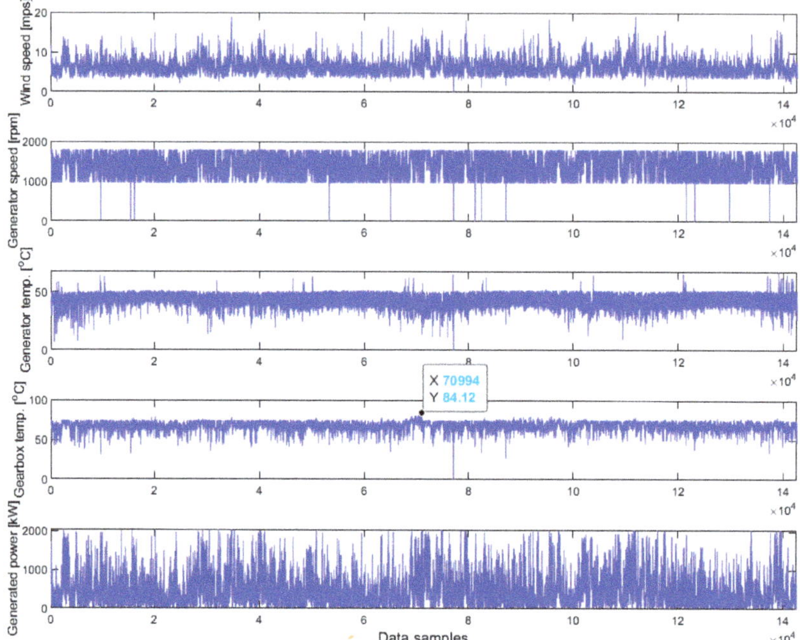

Figure 2. SCADA data plotted for the five wind turbine parameters used in this study to form a cointegration-based monitoring model.

It is important to mention that during the four-year monitoring period of interest, the gearbox bearing temperature of the wind turbine R80721 was raised up to a peak value of 84.12 °C at the data sample 70,994, as marked in Figure 2. It is assumed that a fault in the gearbox is substantially a progressive phenomenon and that the initial signs of the anomaly, mostly indicated by a sudden increase in the gearbox bearing temperature, could appear at least several hours before its actual occurrence. Hence, it is crucial that this gearbox fault can be accurately predicted or detected early before the temperature of the gearbox bearing goes up. The wind turbine power curve, formed by plotting the generated power against the wind speed measured at the hub height for all data, is shown in Figure 3. The power curve describes how much electrical power output is produced by a wind turbine at different wind speeds.

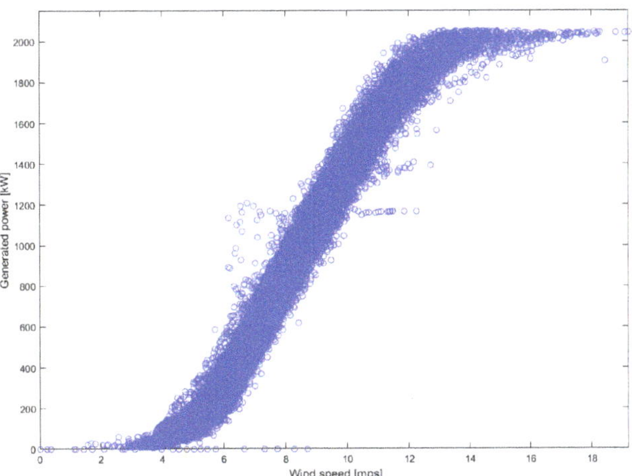

Figure 3. Wind turbine power curve.

5. Results and Discussion

This section presents the validation of the cointegration-based monitoring method, introduced in Section 3, using the wind turbine SCADA data, described in Section 4. A cointegration-based monitoring model for the wind turbine (R80721) was established using five process parameters, including the wind speed (y_{1t}), generator speed (y_{2t}), generator temperature (y_{3t}), gearbox temperature (y_{4t}), and generated power (y_{5t}), where $y_{1t}, y_{2t}, y_{3t}, y_{4t}, y_{5t}$ are variables representing the wind turbine parameters. Two different sets of SCADA data were used to train the cointegration-based model and calculate the normalized cointegrating vectors. The first training data set involves 12,000 samples recorded before the occurrence of the gearbox fault (case 1), whereas the second one includes 6000 samples acquired after the fault occurrence (case 2). It is noted that these two training data sets represent the periods when the given wind turbine was operating in normal condition or healthy state. Cointegration residuals—obtained from projecting the testing data (2000 samples including the gearbox fault event) on the normalized cointegrating vectors—are used in control charts for operational condition monitoring and automated fault/abnormal detection.

5.1. Results Obtained by Using the First Training Data Set (Case 1)

The three-stage cointegration-based computation procedure (presented in Section 3) was deployed for the case study. Following Equation (4), we first established a cointegration-based monitoring model for the wind turbine (R80721). This model has the form

$$\beta^T Y_t = y_{1t} - \beta_2 y_{2t} - \beta_3 y_{3t} - \beta_4 y_{4t} - \beta_5 y_{5t} \tag{7}$$

In the next stage, the cointegration-based wind turbine monitoring model was trained, and then the normalized cointegrating vectors were estimated using the Johansen's cointegration method [33]. SCADA data within the sample points [17,000–29,000], corresponding to 12,000 samples recorded before the gearbox fault occurrence, were used for this purpose. These data are plotted in Figure 4 for the five process parameters. The wind turbine power curve in this case is shown in Figure 5. The minimum and maximum values of each process parameter used for training the cointegrating vectors are provided in Table 2. As a result, we obtained four normalized cointegrating vectors (in the form of four column vectors), which are given as follows:

$$\begin{bmatrix} 1 & 1 & 1 & 1 \\ -2.5869 & -5.7813 & -2.9252 & -0.9384 \\ -0.0031 & -0.0034 & 0.0119 & 0.0065 \\ 0.8893 & -0.0300 & -0.0700 & 0.0433 \\ -0.5507 & 0.6266 & -0.4709 & 0.1699 \end{bmatrix}$$

where the first normalized cointegrating vector is specified by the first column, the second cointegrating vector is specified by the second column, and so on.

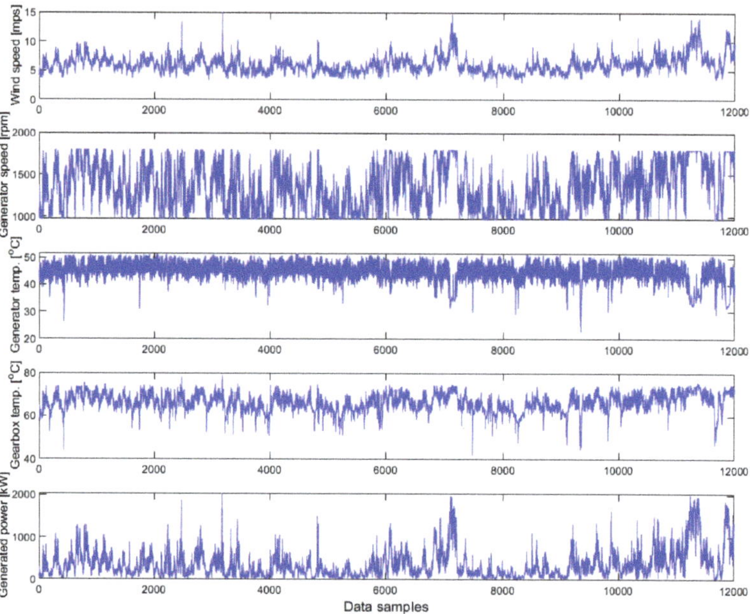

Figure 4. Wind turbine SCADA data corresponding to the first training data set (case 1).

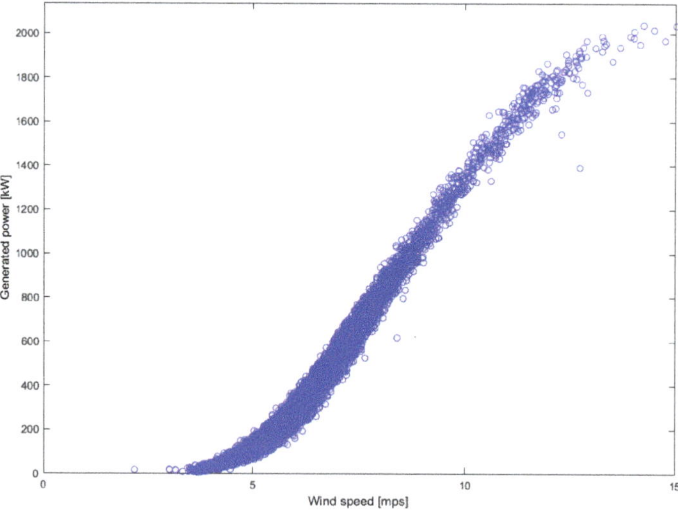

Figure 5. Wind turbine power curve corresponding to the first training data set (case 1).

Table 2. Parameters used for training the cointegrating vectors (case 1) and their limited values.

Parameters	Min Value	Max Value
Wind speed	2.18 mps	15.01 mps
Generator speed	969.83 rpm	1801.21 rpm
Generator temperature	22.76 °C	51.71 °C
Gearbox temperature	42.04 °C	78.85 °C
Generated power	6.82 kW	2038.92 kW

It is noted that in this case the constant vector c_r, where $r = 1, \ldots, 4$, is found as $\begin{bmatrix} 0.0229 & 0.0211 & 0.0090 & -0.0013 \end{bmatrix}$. As mentioned in Section 3.2, only the first cointegration residual is used in this study to monitor the health condition of wind turbines. This cointegration residual is created by multiplying five vectors of the SCADA series (i.e., testing data), corresponding to the five selected process parameters, by the first normalized cointegrating vector. Therefore, the first cointegration residual (u_{1t}) can be written as

$$\begin{aligned} u_{1t} = \beta^T Y_t + c_1 = y_{1t} - \beta_2 y_{2t} - \beta_3 y_{3t} - \beta_4 y_{4t} - \beta_5 y_{5t} + c_1 = \\ y_{1t} + 2.5869 y_{2t} + 0.0031 y_{3t} - 0.8893 y_{4t} + 0.5507 y_{5t} + 0.0229 \end{aligned} \tag{8}$$

As mentioned in Section 4, the abnormal temperature and fault in the gearbox bearing occurred at the data sample 70,994. Hence, we selected 2000 data samples consisting of the sample points from 70,000 to 72,000, i.e., covering the gearbox fault event, as the testing data for each process parameter. Next, 2000 data samples of five process parameters are inserted into variables $y_{1t}, y_{2t}, y_{3t}, y_{4t}, y_{5t}$ in Equation (8). This creates the first cointegration residual in the form of a time series with 2000 samples. The obtained cointegration residual is plotted together with the gearbox temperature in Figure 6 for the comparison. In addition, the residual is plotted against the 99.9% statistical confidence intervals. The confidence interval—with respect to the average of the residual—was calculated as $\nu \pm 3\sigma$, where ν and σ are the mean and standard deviation. The first 900 sample points of the residual were used for calculating the confidence interval. The two red dotted horizontal lines specify the critical limits of the confidence interval. During the monitoring, if the cointegration residual stays within these two lines, this means that the monitored wind turbine is operating in the healthy state. On the contrary, a fault would appear whenever the residual goes beyond the confidence levels. We can observe that the gearbox fault could be detected at data sample 922 in the residual timescale (or 70,922 in the SCADA data timescale). This implies that the anomaly was detected about 720 min (or 12 h) before its actual occurrence at the data sample 70,994.

5.2. Results Obtained by Using the Second Training Data Set (Case 2)

The computation procedure in Section 3 was applied for this case study. Regarding the first stage, we used the same cointegration-based monitoring model for the wind turbine (R80721), which was previously established for case 1 and given by Equation (7). However, in the second stage the cointegration-based wind turbine monitoring model was trained and then the normalized cointegrating vectors were estimated using SCADA data within the sample points [130,000–136,000], corresponding to 6000 samples recorded after the occurrence of the gearbox fault. These data are plotted in Figure 7 for the five process parameters. The wind turbine power curve in this case is shown in Figure 8. The minimum and maximum values of process parameters used for training the cointegrating vectors are given in Table 3. As a result, we obtained four normalized cointegrating vectors (in the form of four column vectors), which are listed below:

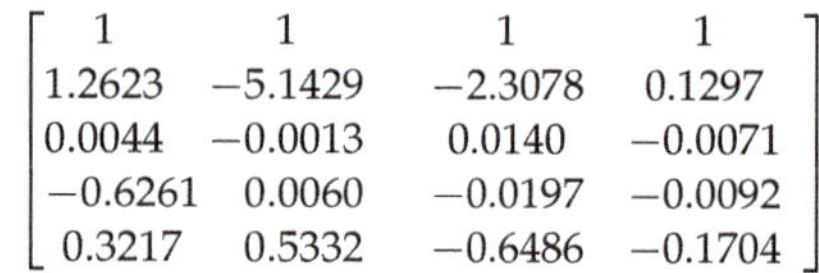

$$\begin{bmatrix} 1 & 1 & 1 & 1 \\ 1.2623 & -5.1429 & -2.3078 & 0.1297 \\ 0.0044 & -0.0013 & 0.0140 & -0.0071 \\ -0.6261 & 0.0060 & -0.0197 & -0.0092 \\ 0.3217 & 0.5332 & -0.6486 & -0.1704 \end{bmatrix}$$

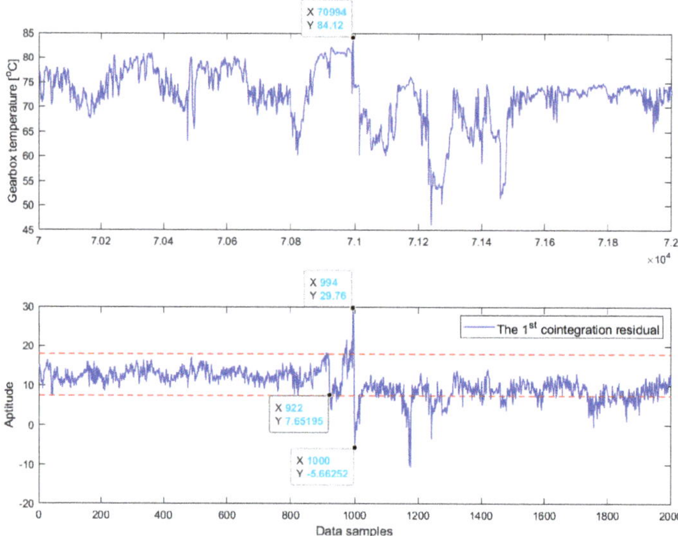

Figure 6. Early detection of the abnormal temperature in the gearbox bearing by monitoring the first cointegration residual.

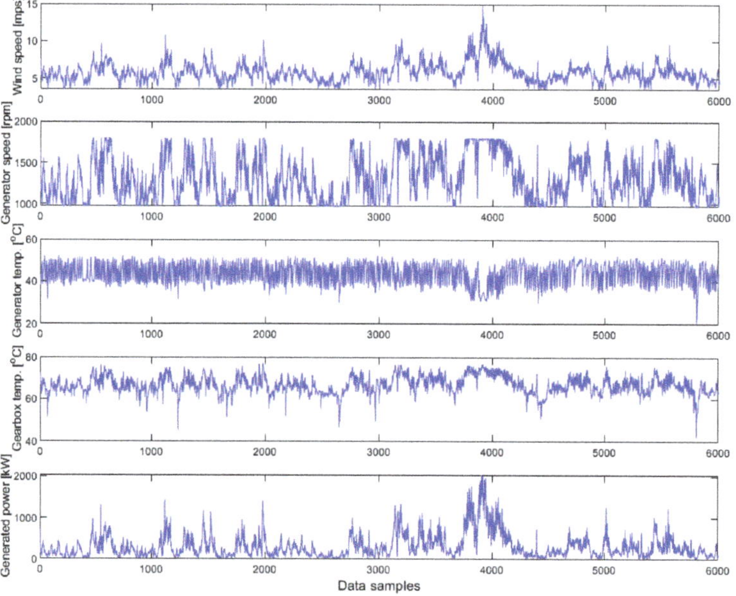

Figure 7. Wind turbine SCADA data corresponding to the second training data set (case 2).

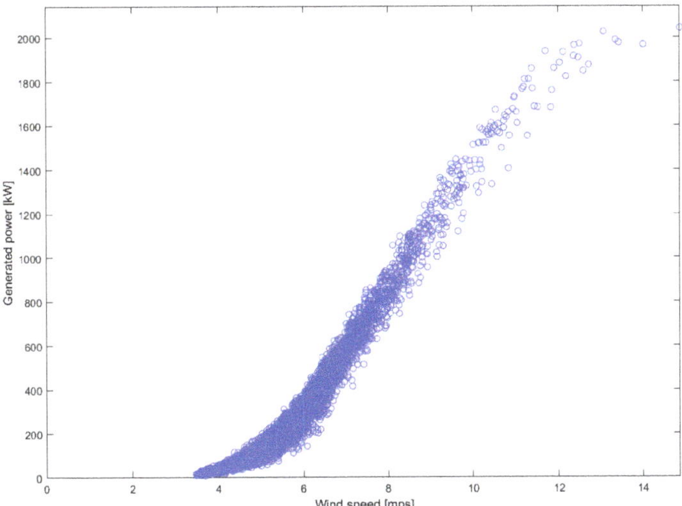

Figure 8. Wind turbine power curve corresponding to the second training data set (case 2).

Table 3. Parameters used for training the cointegrating vectors (case 2) and their limited values.

Parameters	Min Value	Max Value
Wind speed	3.47 mps	14.88 mps
Generator speed	969.83 rpm	1801.34 rpm
Generator temperature	20.89 °C	52.39 °C
Gearbox temperature	42.32 °C	77.09 °C
Generated power	10.41 kW	2042.91 kW

The constant vector c_r is equal to $\begin{bmatrix} -0.0148 & 0.0166 & 0.0063 & 0.0062 \end{bmatrix}$ in this case. Again, the first cointegration residual is created by multiplying five vectors of the SCADA series (i.e., testing data), corresponding to the five selected process parameters, by the first normalized cointegrating vector. Therefore, the first cointegration residual (u_{1t}) can be formed as

$$\begin{aligned} u_{1t} = \beta^T Y_t + c_1 = y_{1t} - \beta_2 y_{2t} - \beta_3 y_{3t} - \beta_4 y_{4t} - \beta_5 y_{5t} + c_1 = \\ y_{1t} - 1.2623 y_{2t} - 0.0044 y_{3t} + 0.6261 y_{4t} - 0.3217 y_{5t} - 0.0148 \end{aligned} \quad (9)$$

The same set of the testing data, i.e., 2000 data samples used for case 1 in Section 5.1, was also used in this case. After the testing data of five process parameters were inserted into variables $y_{1t}, y_{2t}, y_{3t}, y_{4t}, y_{5t}$ in Equation (9), we obtained the first cointegration residual in the form of a time series with 2000 samples. The residual is also plotted together with the gearbox temperature in Figure 9 to illustrate the fault detection. The same confidence interval was applied in this case. Interestingly, the abnormal temperature in the gearbox bearing was detected at the same moment as reported in case 1, that is, at the data sample 922 in the residual timescale (or 70,922 in the SCADA data timescale). Therefore, the anomaly was detected about 720 min (or 12 h) before its actual occurrence at the data sample 70,994.

5.3. Discussion

The first important point to be discussed here is that for both cases investigated where different sets of SCADA data were used to estimate the cointegrating vectors, the first cointegration residuals obtained in Figure 6 (case 1) and Figure 9 (case 2) exhibit the same behaviour. To ease the observation and comparison, these two residuals are plotted together in Figure 10. We can observe that their amplitudes, shapes, and trends are almost identical. In particular, the moment of fault detection (at the data sample 922 in the residual timescale) and the peak (at the data sample 994) are the same in both cases. Interestingly, the peak is at the same moment as the fault occurrence.

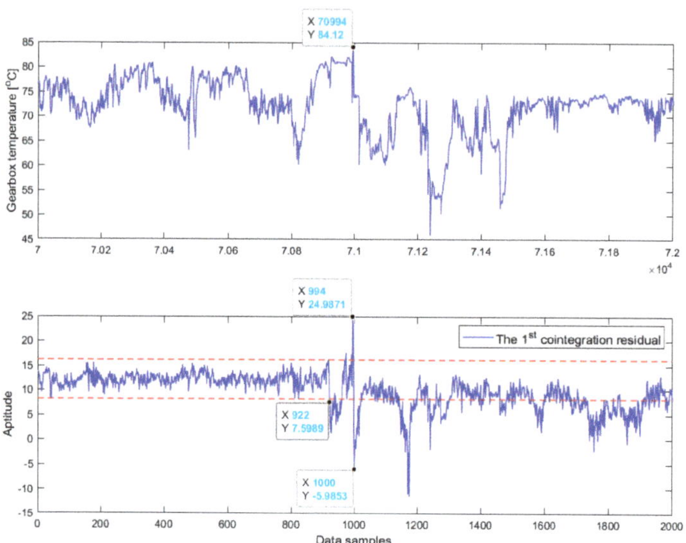

Figure 9. Early detection of the abnormal temperature in the gearbox bearing by monitoring the first cointegration residual.

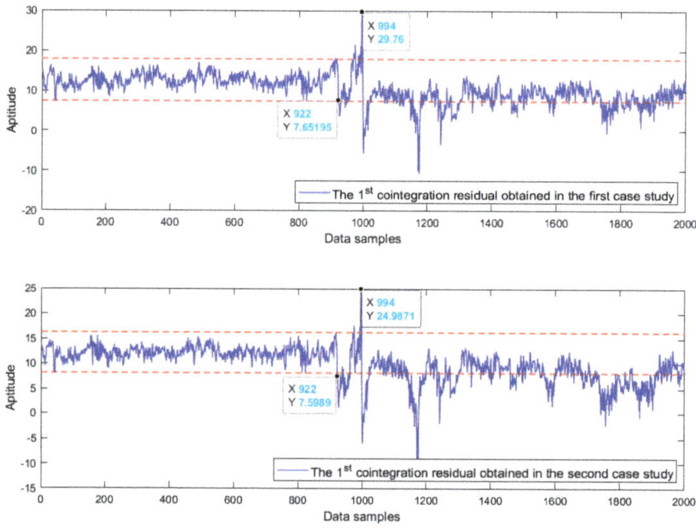

Figure 10. Comparison of the first cointegration residuals obtained from two cases: (**top**) case 1; (**bottom**) case 2.

Moreover, it is well known that a certain cointegration residual usually represents a long-run equilibrium relationship between the cointegrated time series [43]. In Figure 10, we can observe that for both cases the wind turbine exhibited a long-run equilibrium relationship until the moment (at the data sample 922) when the fault was detected. After passing the fault-related period, we can observe in both cases that the wind turbine established a new long-run equilibrium relationship approximately after data sample 1200.

6. Conclusions

This study has reported a new investigation on cointegration for wind turbine monitoring using a four-year SCADA data set acquired from a commercial wind turbine. We investigated for the first time what can be expected if two different sets of SCADA data, representing different normal operating modes of the given wind turbine, are used to train the cointegration-based monitoring model and calculate the normalized cointegrating vectors. The experimental results demonstrated that although different training data sets were used, the cointegration analysis created two residuals, having identical shapes and trends, which could detect the gearbox fault at the same moment. These findings have never been reported in the literature and would be helpful for the potential users of the method in the future.

In comparison with well-trained ML-based methods, the cointegration-based wind turbine monitoring solution may not provide very early warning signs about the fault occurrence. However, the simplicity of the proposed method is an essential factor in practical condition monitoring applications. Instead of analysing and interpreting many wind turbine parameters at the same time, by using this method, the wind turbine monitoring and fault detection process is as simple as observing the stability of a single cointegration residual in a control chart. This constitutes a simple and effective way to monitor the operating state and detect incipient failures of wind turbines in a wind farm. In addition, the use of multiple data sets to train the cointegration-based wind turbine monitoring model and calculate the normalized cointegrating vectors could improve the reliability of the condition monitoring and fault detection process.

In this study, the gearbox anomaly was detected about 12 h before its actual occurrence. However, it is expected in practice that the early fault detection should be at least some days or even weeks in advance for preventing wind turbine damages. Therefore, future study on adapting the cointegration-based monitoring method to make it possible for early fault prognosis in wind turbines has been planned. In addition, the training data sets were analysed without cleaning so that the wind turbine power curves contained a lot of outliers. The early fault detection would have improved if we had performed the power curve cleaning.

This study presents some promising results. However, some works can be suggested for the further development and validation of the method. First, the cointegration-based monitoring method should be validated using other SCADA data sets which involve different fault types associated with main turbine components. Second, it would be interesting to investigate if the normalized cointegrating vectors calculated for a wind turbine with sufficient training data can be reused for other wind turbines with scarce or limited operation data, especially for newly installed wind turbines. In other words, this future work will involve the transfer learning of cointegration-based normal behaviour models between wind turbines.

Funding: This research received no external funding.

Data Availability Statement: No new data were created in this study. The wind turbine SCADA data sets used in this study are mentioned in the Acknowledgments; the access to these data sets is given in Ref. [46].

Acknowledgments: The author would like to thank the ENGIE company for opening up and sharing SCADA data of the La Houte Borne wind farm.

Conflicts of Interest: The author declares no conflict of interest.

References

1. Global Wind Energy Council. Global Wind Report: Annual Market Update 2022. Published in April 2022. Available online: https://gwec.net/global-wind-report-2022/ (accessed on 10 September 2022).
2. Polish Wind Energy Association (PSEW); DWF Group; TPA Poland/Baker Tilly TPA. *Onshore Wind Energy in Poland Annual Report*; PSEW: Serock, Poland; DWF Group: Manchester, UK; TPA Poland/Baker Tilly TPA: Warsaw, Poland, 2021.
3. Kusiak, A.; Li, W. The prediction and diagnosis of wind turbine faults. *Renew. Energy* **2011**, *36*, 16–23. [CrossRef]
4. Tchakoua, P.; Wamkeue, R.; Ouhrouche, M.; Slaoui-Hasnaoui, F.; Tameghe, T.A.; Ekemb, G. Wind turbine condition monitoring: State-of-the-art review, new trends, and future challenges. *Energies* **2014**, *7*, 2595–2630. [CrossRef]
5. Tautz-Weinert, J.; Watson, S.J. Using SCADA data for wind turbine condition monitoring—A review. *IET Renew. Power Gener.* **2017**, *11*, 382–394. [CrossRef]
6. Salameh, J.P.; Cauet, S.; Etien, E.; Sakout, A.; Rambault, L. Gearbox condition monitoring in wind turbines: A review. *Mech. Syst. Signal Process.* **2018**, *111*, 251–264. [CrossRef]
7. Wang, T.; Han, Q.; Chu, F.; Feng, Z. Vibration based condition monitoring and fault diagnosis of wind turbine planetary gearbox: A review. *Mech. Syst. Signal Process.* **2019**, *126*, 662–685. [CrossRef]
8. Zhu, J.; Yoon, J.M.; He, D.; Bechhoefer, E. Online particle-contaminated lubrication oil condition monitoring and remaining useful life prediction for wind turbines. *Wind Energy* **2015**, *18*, 1131–1149. [CrossRef]
9. Pozo, F.; Vidal, Y.; Salgado, Ó. Wind Turbine Condition Monitoring Strategy through Multiway PCA and Multivariate Inference. *Energies* **2018**, *11*, 749. [CrossRef]
10. Zhang, S.; Lang, Z.Q. SCADA-data-based wind turbine fault detection: A dynamic model sensor method. *Control Eng. Pract.* **2020**, *102*, 104546. [CrossRef]
11. Jin, X.; Xu, Z.; Qiao, W. Condition monitoring of wind turbine generators using SCADA data analysis. *IEEE Trans. Sustain. Energy* **2021**, *12*, 202–210. [CrossRef]
12. Stetco, A.; Dinmohammadi, F.; Zhao, X.; Robu, V.; Flynn, D.; Barnes, M.; Keane, J.; Nenadic, G. Machine learning methods for wind turbine condition monitoring: A review. *Renew. Energy* **2019**, *133*, 620–635. [CrossRef]
13. Sun, S.; Wang, T.; Yang, H.; Chu, F. Condition monitoring of wind turbine blades based on self-supervised health representation learning: A conducive technique to effective and reliable utilization of wind energy. *Appl. Energy* **2022**, *313*, 118882. [CrossRef]
14. Wang, A.; Pei, Y.; Qian, Z.; Zareipour, H.; Jing, B.; An, J. A two-stage anomaly decomposition scheme based on multi-variable correlation extraction for wind turbine fault detection and identification. *Appl. Energy* **2022**, *321*, 119373. [CrossRef]
15. Zhang, Y.; Liu, W.; Wang, X.; Shaheer, M.A. A novel hierarchical hyper-parameter search algorithm based on greedy strategy for wind turbine fault diagnosis. *Expert Syst. Appl.* **2022**, *202*, 117473. [CrossRef]
16. Xiang, L.; Yang, X.; Hu, A.; Su, H.; Wang, P. Condition monitoring and anomaly detection of wind turbine based on cascaded and bidirectional deep learning networks. *Appl. Energy* **2022**, *305*, 117925. [CrossRef]
17. Schlechtingen, M.; Santos, I.F.; Achiche, S. Wind turbine condition monitoring based on SCADA data using normal behavior models. Part 1: System description. *Appl. Soft Comput.* **2013**, *13*, 259–270. [CrossRef]
18. Yampikulsakul, N.; Byon, E.; Huang, S.; Sheng, S.; You, M. Condition monitoring of wind power system with nonparametric regression analysis. *IEEE Trans. Energy Convers.* **2014**, *29*, 288–299.
19. Dao, P.B.; Staszewski, W.J.; Barszcz, T.; Uhl, T. Condition monitoring and fault detection in wind turbines based on cointegration analysis of SCADA data. *Renew. Energy* **2018**, *116*, 107–122. [CrossRef]
20. Dao, P.B.; Staszewski, W.J.; Uhl, T. Operational condition monitoring of wind turbines using cointegration method. In *Advances in Condition Monitoring of Machinery in Non-Stationary Operations, Applied Condition Monitoring*; Timofiejczuk, A., Chaari, F., Zimroz, R., Bartelmus, W., Haddar, M., Eds.; Springer: Cham, Switzerland, 2018; Volume 9, Chapter 21; pp. 223–233.
21. Dao, P.B. Condition monitoring of wind turbines based on cointegration analysis of gearbox and generator temperature data. *Diagnostyka* **2018**, *19*, 63–71. [CrossRef]
22. Sun, X.; Xue, D.; Li, R.; Li, X.; Cui, L.; Zhang, X.; Wu, W. Research on condition monitoring of key components in wind turbine based on cointegration analysis. *IOP Conf. Ser. Mater. Sci. Eng.* **2019**, *575*, 012015. [CrossRef]
23. Zhang, B.; Zhang, C.; Duan, H.; Ma, Y.; Li, J.; Cui, L. Realization of condition monitoring of gear box of wind turbine based on cointegration analysis. In *Advances in Asset Management and Condition Monitoring. Smart Innovation, Systems and Technologies*; Ball, A., Gelman, L., Rao, B., Eds.; Springer: Cham, Switzerland, 2020; Volume 166, pp. 281–291.
24. Qadri, B.A.; Ulriksen, M.D.; Damkilde, L.; Tcherniak, D. Cointegration for detecting structural blade damage in an operating wind turbine: An experimental study. In *Dynamics of Civil Structures, Conference Proceedings of the Society for Experimental Mechanics Series*; Pakzad, S., Ed.; Springer: Cham, Switzerland, 2020; Volume 2, pp. 173–180.
25. Xu, M.; Li, J.; Wang, S.; Yang, N.; Hao, H. Damage detection of wind turbine blades by Bayesian multivariate cointegration. *Ocean Eng.* **2022**, *258*, 111603. [CrossRef]
26. Zhang, C.; Zhao, G.; Wu, Y. Wind Turbine Condition Monitoring Based on SCADA Data Co-integration Analysis. In *Proceedings of IncoME-VI and TEPEN 2021. Mechanisms and Machine Science*; Zhang, H., Feng, G., Wang, H., Gu, F., Sinha, J., Eds.; Springer: Cham, Switzerland, 2023; Volume 117, pp. 97–103.
27. Letzgus, S. Change-point detection in wind turbine SCADA data for robust condition monitoring with normal behaviour models. *Wind Energy Sci.* **2020**, *5*, 1375–1397. [CrossRef]

28. Dao, P.B. Condition monitoring and fault diagnosis of wind turbines based on structural break detection in SCADA data. *Renew. Energy* **2022**, *185*, 641–654. [CrossRef]
29. Dao, P.B. A CUSUM-based approach for condition monitoring and fault diagnosis of wind turbines. *Energies* **2021**, *14*, 3236. [CrossRef]
30. Latiffianti, E.; Sheng, S.; Ding, Y. Wind turbine gearbox failure detection through cumulative sum of multivariate time series data. *Front. Energy Res.* **2022**, *10*, 904622. [CrossRef]
31. Dao, P.B. On Wilcoxon rank sum test for condition monitoring and fault detection of wind turbines. *Appl. Energy* **2022**, *318*, 119209. [CrossRef]
32. Engle, R.F.; Granger, C.W.J. Cointegration and error-correction: Representation, estimation and testing. *Econometrica* **1987**, *55*, 251–276. [CrossRef]
33. Johansen, S. Statistical analysis of cointegration vectors. *J. Econ. Dyn. Control* **1988**, *12*, 231–254. [CrossRef]
34. Cross, E.J.; Worden, K.; Chen, Q. Cointegration: A novel approach for the removal of environmental trends in structural health monitoring data. *Proc. R. Soc. A* **2011**, *467*, 2712–2732. [CrossRef]
35. Dao, P.B.; Staszewski, W.J. Cointegration approach for temperature effect compensation in Lamb wave based damage detection. *Smart Mater. Struct.* **2013**, *22*, 095002. [CrossRef]
36. Dao, P.B.; Staszewski, W.J.; Klepka, A. Stationarity-based approach for the selection of lag length in cointegration analysis used for structural damage detection. *Comput. Aided Civ. Infrastruct. Eng.* **2017**, *32*, 138–153. [CrossRef]
37. Coletta, G.; Miraglia, G.; Pecorelli, M.; Ceravolo, R.; Cross, E.J.; Surace, C.; Worden, K. Use of the cointegration strategies to remove environmental effects from data acquired on historical buildings. *Eng. Struct.* **2019**, *183*, 1014–1026. [CrossRef]
38. Salvetti, M.; Sbarufatti, C.; Cross, E.J.; Corbetta, M.; Worden, K.; Giglio, M. On the performance of a cointegration-based approach for novelty detection in realistic fatigue crack growth scenarios. *Mech. Syst. Signal Process.* **2019**, *123*, 84–101. [CrossRef]
39. He, H.; Wang, W.; Zhang, X. Frequency modification of continuous beam bridge based on co-integration analysis considering the effect of temperature and humidity. *Struct. Health Monit.* **2019**, *18*, 376–389. [CrossRef]
40. Tomé, E.S.; Pimentel, M.; Figueiras, J. Damage detection under environmental and operational effects using cointegration analysis—Application to experimental data from a cable-stayed bridge. *Mech. Syst. Signal Process.* **2020**, *135*, 106386. [CrossRef]
41. Turrisi, S.; Cigada, A.; Zappa, E. A cointegration-based approach for automatic anomalies detection in large-scale structures. *Mech. Syst. Signal Process.* **2022**, *166*, 108483. [CrossRef]
42. Dao, P.B.; Staszewski, W.J. Cointegration and how it works for structural health monitoring. *Measurement* **2023**, *209*, 112503. [CrossRef]
43. Zivot, E.; Wang, J. *Modeling Financial Time Series with S-PLUS*, 2nd ed.; Springer: New York, NY, USA, 2006.
44. LeSage, J.P. Econometrics Toolbox. Available online: www.spatial-econometrics.com (accessed on 20 November 2022).
45. Bilendo, F.; Meyer, A.; Badihi, H.; Lu, N.; Cambron, P.; Jiang, B. Applications and modeling techniques of wind turbine power curve for wind farms—A review. *Energies* **2023**, *16*, 180. [CrossRef]
46. ENGIE OpenData, SCADA Datasets of La Houte Bourne Wind Farm. Available online: https://opendata-renewables.engie.com/explore/index (accessed on 16 August 2022).

 *energies*

Editorial

Moving towards Preventive Maintenance in Wind Turbine Structural Control and Health Monitoring

Jersson X. Leon-Medina [1] and Francesc Pozo [1,2,*]

[1] Control, Data and Artificial Intelligence (CoDAlab), Department of Mathematics, Escola d'Enginyeria de Barcelona Est (EEBE), Campus Diagonal-Besòs (CDB), Universitat Politècnica de Catalunya (UPC), Eduard Maristany 16, 08019 Barcelona, Spain; jersson.xavier.leon@upc.edu

[2] Institute of Mathematics (IMTech), Universitat Politècnica de Catalunya (UPC), Pau Gargallo 14, 08028 Barcelona, Spain

* Correspondence: francesc.pozo@upc.edu

1. Introduction

In recent years, the scope of structural health monitoring in wind turbines has broadened due to the development of innovative data-driven methodologies. These methodologies enable the execution of condition monitoring and fault detection, which can provide support for predictive maintenance across the various components of a wind turbine.

In this Editorial, we present a selection of 10 papers published in *Energies* that contribute to the ongoing research on wind turbine condition monitoring. The papers were divided into three categories: wind fault detection and diagnosis, wind turbine condition monitoring, and wind turbine maintenance. The first category includes papers that focus on the detection and diagnosis of faults in wind turbines, such as those related to bearings, pitch actuator systems, and gearboxes. The second category comprises papers that discuss various aspects of wind turbine condition monitoring, including literature reviews, comparative studies on anomaly detection techniques, and methodologies that predict the remaining useful life of wind turbine components. The third category includes papers that aim to reduce the cost of wind turbine maintenance by developing new maintenance strategies. These papers contribute to the ongoing effort to develop reliable, efficient, and cost-effective techniques for monitoring the condition of wind turbines, with the ultimate goal of increasing their reliability, lifetime, and performance.

Citation: Leon-Medina, J.X.; Pozo, F. Moving towards Preventive Maintenance in Wind Turbine Structural Control and Health Monitoring. *Energies* **2023**, *16*, 2730. https://doi.org/10.3390/en16062730

Received: 22 February 2023
Accepted: 9 March 2023
Published: 15 March 2023

2. Wind Turbine Fault Detection and Diagnosis

The study conducted by Castellani et al. [1] presents an analysis of the failures in the bearings of a wind turbine. The novelty of this research lies in its focus on monitoring the vibrations in the wind turbine tower without the need for human interaction, while also ensuring that the developed procedure does not interfere with the normal operation of the wind turbine. The validation of this procedure was carried out on a set of five wind turbines, comprising three healthy structures, one damaged structure, and a recently repaired specimen. The methodology employed in the study involved processing accelerometer signals acquired in the tower of the wind turbines, which underwent several stages, including univariate data cleaning, feature extraction, multivariate data cleaning, statistical analysis, and damage detection based on the Mahalanobis distance. The results of the study indicate that the detectability of damage was successful, as demonstrated by the low number of false alarms in the predictions.

An investigation on the fault detection of a wind turbine pitch actuator system was conducted in [2]. The study employed an algorithm based on the interval observer theory and included three fault stages to authenticate the developed fault detection approach. These fault stages assessed the impact of hydraulic leakage, high air oil content, and pump wear. The study resulted in the development of a fault detection algorithm and a fault classification algorithm. To authenticate these algorithms, 500 experiments were performed, each with a random input. The accuracy values of the detection algorithm exceeded 62%, whereas the classification algorithm had an accuracy value higher than 83%.

In the study conducted by Santolamazza et al. [3], artificial neural networks (ANN) were utilized to detect faults in wind turbines, and SCADA data were used for modeling. The developed methodology consisted of several stages. Firstly, a preprocessing stage was carried out for the removal of abnormal samples, followed by the removal of outliers through the use of a clustering method which evaluated the Mahalanobis distance. Secondly, a model processing stage was carried out, in which the data were split into training, validation, and test sets. Feature selection and tuning of the hyper-parameters of the feed-forward neural network (FFNN) model was also conducted. Finally, a post-processing stage involving an alarm system was implemented to identify any anomalies in the deviations from the real-time baseline value of the output variable. The developed methodology was applied to two critical components of the wind turbine, the generator and the gearbox, with the generator showing the best fault-detection results. The paper also includes Table A1, which summarizes relevant fault detection studies in wind turbines, obtained from the literature. These studies are classified based on their components, methods, real case studies, and approach data type.

In [4], a fault diagnosis methodology of a wind turbine gearbox was developed using both time and frequency domain analysis. The methodology used data collected from accelerometers attached to the gearbox housing. An adaptive Variational Mode Decomposition (VMD) algorithm was optimized with the fast gray wolf optimizer (GWO) to decompose the signal obtained by the accelerometers. Next, a principal component analysis was performed in each domain to obtain a new feature vector. An ELM classifier was trained and tested with classification accuracies greater than 90%. The developed methodology was compared with other classifiers such as SVM, deep convolutional neural networks, a genetic algorithm back propagation neural network, and the ensemble empirical mode decomposition (EEMD) method for signal decomposition. Results indicated that the developed methodology outperformed all compared methods.

In the study by Liu et al. [5], fault indicators were investigated in three components of wind turbines: the converter, the generator, and the pitch system. The study utilized SCADA data obtained from 24 wind turbines of 1.5 MW. The developed methodology involved creating radar charts that varied depending on whether the system was healthy or unhealthy. Three different machine learning classification methods were compared, including the support vector machine (SVM), support vector regression (SVR), and a convolutional neural network (CNN) using ResNet50 structure as the backbone network. The results showed that the CNN achieved the best classification results in all cases, with accuracy values of 97.87% for the generator, 98.03% for the converter, and 98.41% for the pitch system.

3. Wind Turbine Condition Monitoring

In the publication by Maldonado et al. [6], the authors conducted a systematic literature review regarding the use of Supervisory Control and Data Acquisition (SCADA) data for wind turbine condition monitoring (CM). The authors emphasize the economic benefits of utilizing data obtained from the SCADA system, which eliminates the need to install additional sensors or equipment on the wind turbines. The literature review revealed that, among the 102 articles analyzed, 26% were focused on the gearbox failure of wind turbines. The gearbox is considered one of the most critical components, as it contributes to 20% of total downtime in wind turbines. Additionally, the review highlighted other components of wind turbines that have been the subject of different condition monitoring methodologies, including blades, tower, drive train and bearing, yaw system, and various electrical and electronic components. The authors also noted some of the challenges associated with wind turbine condition monitoring, such as the development of non-intrusive and sensorless CM systems, online and real-time CM, standardization of SCADA data, public availability of SCADA data, and the development of an assessment method for monitoring the overall condition of a wind turbine. Furthermore, the SCADA variables utilized for wind turbine condition monitoring include, but are not limited to, environment temperature, wind speed, active power, pitch angle, gearbox temperature, and rotor speed.

In the study by McKinnon et al. [7], a number of anomaly detection techniques were developed based on gearbox SCADA data for wind turbine condition monitoring. The data utilized in the study were obtained from 21 wind turbines, with each turbine providing two months of data. The first month was considered to represent the healthy state, occurring one year prior to the failure, while the second month represented the unhealthy state, occurring one month prior to the failure. The models were trained using various variables, including the temperature and pressure of gearbox components, as well as the generator, rotor, and ambient wind speed. Three different models were compared: the Isolation Forest (IF), One-Class Support Vector Machine (OCSVM), and Elliptical Envelope (EE). OCSVM and IF exhibited an average accuracy of 82%, whereas EE presented a lower accuracy of 77%.

In the study by Velandia et al. [8], the authors investigated the detection of faults in wind turbines. They employed three data preprocessing techniques to improve the performance of a classification task on a highly imbalanced dataset obtained from SCADA data over a period of seven months in a wind turbine. Principal component analysis was used as a dimensionality reduction method, data reshaping served as a data augmentation technique, and random oversampling was applied to deal with the imbalanced characteristic of the dataset. Three different classifiers were compared: RUSBoost, Support Vector Machines (SVM), and k-Nearest Neighbors (k-NN). The results of the research show that the F_1 scores of at least 95% were achieved.

4. Wind Turbine Maintenance

In [9] a data-centric methodology is described. This methodology is designed to enhance the prediction accuracy of the remaining useful life (RUL) of different parts of a wind turbine. This approach contrasts with the conventional model-centric approach, in which the hyperparameters of a machine learning model are adjusted by assessing various feature selection and data preprocessing techniques within a machine learning pipeline. The methodology was validated and trained with data collected from a wind farm between 2016 and 2017. The study focused on diagnosing and analyzing five wind turbine components: the generator, hydraulic group, generator bearing, transformer, and gearbox. Given the skewed nature of the wind turbine dataset, with a significant amount of healthy data and limited fault data, precision and recall metrics were used to assess the classification task. The data-centric methodology outperformed the model-centric approach.

The research described in [10] successfully accomplished the objective of reducing repair costs and the number of repairs in four distinct components of a wind turbine using a support vector machine (SVM) algorithm for condition-based maintenance. The components studied were the rotor, pitch system, gearbox, and generator. The research showed that the generator component incurred the highest maintenance costs. The rotor, on the other hand, presents the best opportunity for cost savings, as its single maintenance cost is relatively higher. One notable contribution of this research is that it includes a comparative scenario analysis versus separate periodic maintenance, demonstrating an improvement in cost savings of 32.5%.

5. Concluding Remarks and Perspectives

The 10 papers described in this Editorial provide a comprehensive overview of current research trends in wind turbine condition monitoring, fault diagnosis, and maintenance. Future research in these areas should focus on the development of more advanced methodologies and the integration of multiple data sources for more accurate and reliable predictions. The integration of machine learning and artificial intelligence techniques can help identify patterns and hidden correlations within the data, leading to more accurate fault detection and condition monitoring of wind turbines. In addition, future research will focus on the development of techniques that address the challenges associated with wind turbine condition monitoring, such as the development of non-intrusive and sensorless condition monitoring systems, the standardization of SCADA data, and the assessment of the overall condition of a wind turbine. With the increasing demand for renewable energy, the need for reliable and efficient wind turbines is paramount, and research in these areas can help further the development of more effective wind turbine systems, improving their performance, extending their lifespan, and reducing downtime and maintenance costs.

Author Contributions: Conceptualization, J.X.L.-M. and F.P.; methodology, J.X.L.-M. and F.P.; resources, F.P.; writing—original draft preparation, J.X.L.-M.; writing—review and editing, F.P.; supervision, F.P.; project administration, F.P.; funding acquisition, F.P. All authors have read and agreed to the published version of the manuscript.

Funding: This work has been partially funded by the Spanish Agencia Estatal de Investigación (AEI)—Ministerio de Economía, Industria y Competitividad (MINECO), and the Fondo Europeo de Desarrollo Regional (FEDER) through the research projects PID2021-122132OB-C21 and TED2021-129512B-I00; and by the Generalitat de Catalunya through the research projects 2021-SGR-01044.

Conflicts of Interest: The authors declare no conflict of interest. The funders had no role in the design of the study; in the collection, analyses, or interpretation of data; in the writing of the manuscript; or in the decision to publish the results.

References

1. Castellani, F.; Garibaldi, L.; Daga, A.P.; Astolfi, D.; Natili, F. Diagnosis of faulty wind turbine bearings using tower vibration measurements. *Energies* **2020**, *13*, 1474. [CrossRef]
2. Pujol-Vazquez, G.; Acho, L.; Gibergans-Báguena, J. Fault detection algorithm for wind turbines' pitch actuator systems. *Energies* **2020**, *13*, 2861. [CrossRef]
3. Santolamazza, A.; Dadi, D.; Introna, V. A data-mining approach for wind turbine fault detection based on SCADA data analysis using artificial neural networks. *Energies* **2021**, *14*, 1845. [CrossRef]
4. Li, H.; Fan, B.; Jia, R.; Zhai, F.; Bai, L.; Luo, X. Research on multi-domain fault diagnosis of gearbox of wind turbine based on adaptive variational mode decomposition and extreme learning machine algorithms. *Energies* **2020**, *13*, 1375. [CrossRef]
5. Liu, Z.; Xiao, C.; Zhang, T.; Zhang, X. Research on fault detection for three types of wind turbine subsystems using machine learning. *Energies* **2020**, *13*, 460. [CrossRef]
6. Maldonado-Correa, J.; Martín-Martínez, S.; Artigao, E.; Gómez-Lázaro, E. Using SCADA data for wind turbine condition monitoring: A systematic literature review. *Energies* **2020**, *13*, 3132. [CrossRef]
7. McKinnon, C.; Carroll, J.; McDonald, A.; Koukoura, S.; Infield, D.; Soraghan, C. Comparison of new anomaly detection technique for wind turbine condition monitoring using gearbox SCADA data. *Energies* **2020**, *13*, 5152. [CrossRef]
8. Velandia-Cardenas, C.; Vidal, Y.; Pozo, F. Wind turbine fault detection using highly imbalanced real SCADA data. *Energies* **2021**, *14*, 1728. [CrossRef]
9. Garan, M.; Tidriri, K.; Kovalenko, I. A Data-Centric Machine Learning Methodology: Application on Predictive Maintenance of Wind Turbines. *Energies* **2022**, *15*, 826. [CrossRef]
10. Kang, J.; Wang, Z.; Guedes Soares, C. Condition-based maintenance for offshore wind turbines based on support vector machine. *Energies* **2020**, *13*, 3518. [CrossRef]

 materials

Article

Ultrasonic Nonlinearity Experiment due to Plastic Deformation of Aluminum Plate Due to Bending Damage

Junpil Park [1], Mohammed Aslam [1] and Jaesun Lee [2,*]

[1] Extreme Environment Design and Manufacturing Engineering, Changwon National University, Changwon 51140, Republic of Korea; junpil@changwon.ac.kr (J.P.); aslam@changwon.ac.kr (M.A.)
[2] School of Mechanical Engineering, Changwon National University, Changwon 51140, Republic of Korea
* Correspondence: jaesun@changwon.ac.kr; Tel.: +82-55-213-3621

Abstract: The nonlinear ultrasonic evaluation technique is useful for assessing micro-defects and microstructure changes caused by fatigue or bending damage. In particular, the guided wave is advantageous for long-distance testing such as piping and plate. Despite these advantages, the study of nonlinear guided wave propagation has received relatively less attention compared to bulk wave techniques. Furthermore, there is a lack of research on the correlation between nonlinear parameters and material properties. In this study, the relationship between nonlinear parameters and plastic deformation resulting from bending damage was experimentally investigated using Lamb waves. The findings indicated an increase in the nonlinear parameter for the specimen, which was loaded within the elastic limit. Inversely, regions of maximum deflection in specimens with plastic deformation exhibited a decrease in the nonlinear parameter. This research is expected to be helpful for maintenance technology in the nuclear power plant and aerospace fields that require high reliability and accuracy.

Keywords: non-destructive testing; ultrasonic; nonlinearity; lamb wave; bending damage

Citation: Park, J.; Aslam, M.; Lee, J. Ultrasonic Nonlinearity Experiment due to Plastic Deformation of Aluminum Plate Due to Bending Damage. *Materials* **2023**, *16*, 4241. https://doi.org/10.3390/ma16124241

Academic Editors: Phong B. Dao, Tadeusz Uhl, Liang Yu, Lei Qiu and Minh-Quy Le

Received: 4 May 2023
Revised: 3 June 2023
Accepted: 6 June 2023
Published: 8 June 2023

1. Introduction

In technology-intensive industries such as aerospace, nuclear power, and petrochemical plants, the precision and reliability of equipment and mechanical systems are of utmost importance. Micro-defects in these structures grow rapidly, often surpassing the stage where they can be easily visualized. Current ultrasonic non-destructive testing methods are unable to diagnose defects prior to the formation of macro-cracks [1]. Recent research shows nonlinear ultrasonic techniques have higher sensitivity in detecting corrosion, microdefects, and changes in microstructure [2,3]. Conventional local inspection methods using bulk waves face limitations when it comes to inspecting large structures such as ships, aircraft, containment liner plates, and bridges. Recognizing this challenge, extensive research is being conducted to develop wide-range testing techniques utilizing induced ultrasonic waves [4–6]. In particular, nonlinear Lamb waves are advantageous for long-distance testing due to their unique characteristics and the ability to achieve higher sensitivity through various mode selections.

High-tech industries require materials with high reliability and stability. Nonlinear ultrasonic inspection techniques offer the potential to detect microstructural defects resulting from various factors such as micro-defects, corrosion, and plastic deformation due to fatigue or external loads. However, research on the impact of nonlinearity caused by changes in material properties remains insufficient. To bridge this gap, an experimental investigation utilizing Lamb wave-based inspection methods was conducted. The focus of the study was to explore the relationship between bending damage and the manifestation of nonlinearity in aluminum alloy materials. This research contributes to the understanding of material behavior and holds implications for improving the reliability and performance of materials in high-tech industries.

There has been continuous research and development in the field of nonlinear ultrasonic testing techniques, with a particular focus on studying and comparing the fatigue behavior and nonlinearity trends in various materials [7,8]. In particular, Jacobs L.J. and Qu J. studied the nonlinearity of nickel alloy steels using Rayleigh surface waves [9], and the nonlinearity of aluminum alloys was also investigated using a Lamb wave [10]. Deng et al. employed the second method of perturbation theory to investigate the second harmonics of the shear-horizontal (SH) mode in an isotropic plate featuring two free boundaries [11,12]. In addition, they studied the effects of nonlinearity on the cumulative effect, mode analysis, and phase matching, which are important experimental bases for nonlinear ultrasonic techniques. [13,14]. Li Weibin conducted experiments to characterize the thermal fatigue damage of composite laminates using second harmonic Lamb waves [15] and investigated the effect of material microstructure evolution on the acoustic nonlinear response of ultrasound in rolled copper and brass [16]. In addition, Li W. and Park J. conducted a study to detect micro-cracks in materials using non-contact electromagnetic ultrasonic transducers [17,18]. Lissenden presented a method for detecting local fatigue damage on an aluminum plate by utilizing a PVDF (polyvinylidene difluorine) sensor to simultaneously receive shear-horizontal waves and the secondary Lamb waves they generate. The experimental findings were further validated through finite element simulation [19,20]. Furthermore, there is an active research effort among many scholars to diagnose micro-damage through the utilization of second and third harmonics [21–23]. Additionally, researchers are exploring methods to enhance sensitivity by mitigating system nonlinearity [24], as well as investigating the effects of thermal aging on sensitivity [24]. However, it is important to note that these studies primarily focus on micro-damage assessment during the material's initial manufacturing process and intrinsic properties [25–27]. Currently, there is a lack of experimental verification regarding material deformation under conditions such as tension or compression due to bending.

Guided ultrasound has traditionally been employed for non-destructive testing of large areas, primarily focusing on diagnosing defects larger than the macro stage. However, limited research has been conducted on the application of nonlinear ultrasound beyond micro-scale defect detection, fatigue analysis, and nonlinearity studies under uniaxial tensile and compressive loads [28,29]. Plates are extensively utilized in various structures, and while some curved plates are manufactured using molds, many products are formed through bending to achieve the desired curvature. Unfortunately, there is a dearth of research exploring material damage resulting from such bending processes. To address this gap, this study experimentally investigates the relationship between bending damage and nonlinearity in aluminum alloy materials using Lamb waves. The study also employs simulations and theoretical analyses to establish a correlation between yield strength and bending, providing comprehensive verification of these findings. To enhance the reliability of the experiment, various experimental conditions were employed, and the key findings can be summarized as follows:

1. It was experimentally verified that the superimposition effect of nonlinear Lamb waves applied to an aluminum plate;
2. A comparative assessment of induced nonlinearities was conducted when subjecting the aluminum plate to bending loads within both the elastic and plastic regions;
3. The experimental investigation further examined nonlinearity tendencies specific to each mode of the Lamb wave on the aluminum plate subjected to bending loads in the plastic region;
4. A comparative analysis was performed to assess and analyze the induced nonlinearity due to compression plasticity and tensile plasticity in the aluminum plate subjected to bending.

2. Nonlinear Ultrasonic Theory

Nonlinear ultrasonic waves have the characteristics of generating high harmonics due to the deterioration of material and the change in plasticity due to fatigue and external forces when an acoustic wave passes through the material. The reason for this harmonic generation is that as the high-order elastic constants increase due to material defects, the stress–strain relationship changes into a nonlinear rather than a linear relationship. In other words, it is difficult to explain the plasticity of the stress–strain curve or materials that do not follow the elastic behavior with the linear ultrasonic theory, and a more accurate behavior can be obtained by applying the nonlinear ultrasonic theory.

Nonlinearity can be categorized into two main types: material nonlinearity and geometric nonlinearity. Material nonlinearity arises from the variations in the states of the constituent crystals within the material, deviating from ideal uniformity. This introduces nonlinearity into the material's behavior. On the other hand, geometric nonlinearity refers to the nonlinearity that arises due to significant deformations or displacements of the material, where the strain-displacement relationship becomes nonlinear.

The nonlinear ultrasonic equation can be expressed as Equation (1) [30].

$$\begin{aligned} u_{tt} - c_l^2 u_{aa} &= \left(3c_l^2 + C_{111}/\rho\right) u_a u_{aa} + \left(c_l^2 + C_{166}/\rho\right)(v_a v_{aa} + w_a w_{aa}) \\ v_{tt} - c_s^2 v_{aa} &= \left(c_l^2 + C_{166}/\rho\right)(u_a v_{aa} + v_a u_{aa}) \\ w_{tt} - c_s^2 w_{aa} &= \left(c_l^2 + C_{166}/\rho\right)(u_a w_{aa} + w_a u_{aa}) \end{aligned} \tag{1}$$

In contrast to the linear wave equation, the cubic elastic modulus is introduced as C_{111}, C_{166} and the solution of the nonlinear ultrasonic equation can be derived by the perturbation method. If this solution is expanded by the sum of the solution of the first fundamental frequency and the solution of the second harmonic frequency, it can be expressed as Equation (2).

$$\begin{aligned} u_{tt}^{(1)} - c_l^2 u_{aa}^{(1)} &= 0 \\ v_{tt}^{(1)} - c_s^2 v_{aa}^{(1)} &= 0 \\ w_{tt}^{(1)} - c_s^2 w_{aa}^{(1)} &= 0 \end{aligned} \tag{2}$$

Considering only the longitudinal component,

$$\begin{aligned} u_{tt}^{(1)} - c_l^2 u_{aa}^{(1)} &= 0 \\ u_{tt}^{(2)} - c_l^2 u_{aa}^{(2)} &= \left(3c_l^2 + C_{111}/\rho\right) u_a^{(1)} u_{aa}^{(1)} \end{aligned} \tag{3}$$

Equations (4) and (5) are derived by using the perturbation method to obtain the solution of the first fundamental frequency and the second harmonic frequency, respectively. Here, $u^{(1)}$ represents the primary wave solution, and $u^{(2)}$ represents the second-order wave solution.

$$u^{(1)} = A_1 \exp j(k_l a - wt) \tag{4}$$

$$u^{(2)} = A_2 \exp j(2k_l a - 2wt) \tag{5}$$

In the above equation, $A_2 = \frac{\beta_l A_1^2 k_l^2 a}{8C_l^2}$, $\beta_l = \frac{A_2}{A_1^2}\frac{8C_l^2}{K_l^2 x}$, where β_l is the longitudinal second harmonic nonlinear parameter. A_1 and A_2 are the fundamental and second harmonic amplitudes, respectively. The nonlinear parameter is expressed as the ratio of the second harmonic amplitude divided by the square of the fundamental amplitude.

3. Experimental Setup

In order to investigate the nonlinearity of Lamb waves in response to bending stress, an experimental study was conducted on an aluminum plate (Al5052). The test specimens consisted of three specimens: one with no load applied to the same specimen, one with stresses in the elastic region below yield strength, and one with stresses in the plastic region above yield strength.

Figure 1 illustrates the fixation of the specimen to the equipment, and the relationship between maximum deflection and corresponding stress at the mid-span of the specimen was determined using a strain gauge. In order to fabricate the specimen, the aluminum material properties are referenced as shown in Table 1. The schematic representation of the application of load and corresponding deflected shape is depicted in Figure 2.

Figure 1. Bending device.

Table 1. Aluminum properties (Al5052).

Yield Stress (σ_Y)	**81.0 MPa (0.2% Offset)**
Modulus of Elasticity (E)	70.3 GPa
Density (ρ)	2.68 g/cc
Poission's ratio (ν)	0.33
Shear Modulus (G)	25.9 GPa

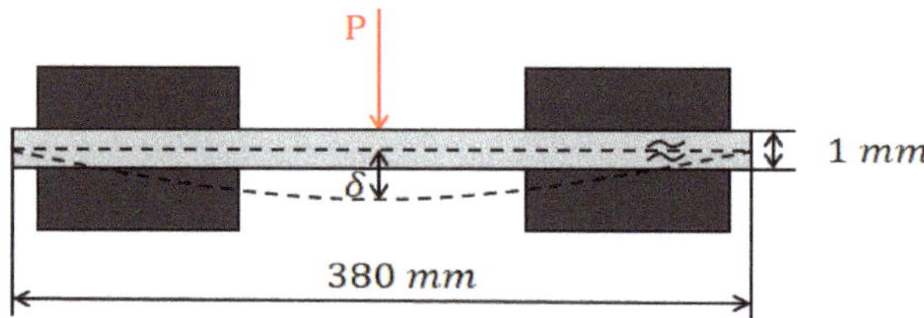

Figure 2. Application of load and corresponding deflected shape.

Prior to conducting the experiment as shown in Figure 3, the deflection of the specimen was determined using the Structural Analysis Tool (ABAQUS 2021 ver., Dassault Systèmes, Vélizy-Villacoublay, France). This analysis aimed to ascertain the deflection value at the mid-span for both loading conditions: within the elastic limit and beyond the yield strength (plastic region). The structural analysis revealed that the plate remained in the elastic range until the maximum stress reached 40.5 MPa, with a corresponding maximum deflection of 9.45 mm. Based on these findings, experimental tests were conducted on the specimens. For the plate loaded within the elastic limit, a stress equal to half of the yield strength was applied. In contrast, for the plate loaded beyond the yield strength, a stress of 0.2% offset beyond the yield strength was applied. The stresses and their corresponding deflections in the elastic and plastic regions are presented in Table 2. Figure 4 visually represents the specimens before and after the bending test, capturing the bending damage incurred. Following the bending test, the specimens were utilized for nonlinearity measurements.

A schematic representation of the nonlinear measurement system can be observed in Figure 5. The excitation signal is generated at a distance of 100 mm from the end of the specimen, and the signal was received at 10 mm intervals within a distance of 60 mm. The tone burst equipment (RAM-5000, RITEC Inc., Warwick, RI, USA) was employed for this purpose. The objective of this investigation was to explore the nonlinearity associated with bending stress. Therefore, particular attention was given to the region near the maximum deflection, where the signal reception was focused. Figure 6a,b show the phase matching of the antisymmetric mode, which was employed for frequency selection. Based on the velocity matching criteria of antisymmetric modes, the selected frequencies and their corresponding wave velocities are presented in Tables 3 and 4.

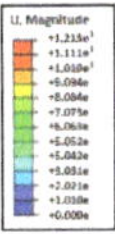

Figure 3. Displacement contour of the plate loaded beyond the elastic limit.

Table 2. Specimens information.

	Stress	Maximum Deflection	Thickness
Unstressed	-	-	1 mm
Elastic Region	40.5 MPa	9.45 mm	1 mm
Plastic region	89.6 MPa	12.13 mm	1 mm

Figure 4. Aluminum specimens (Al5052).

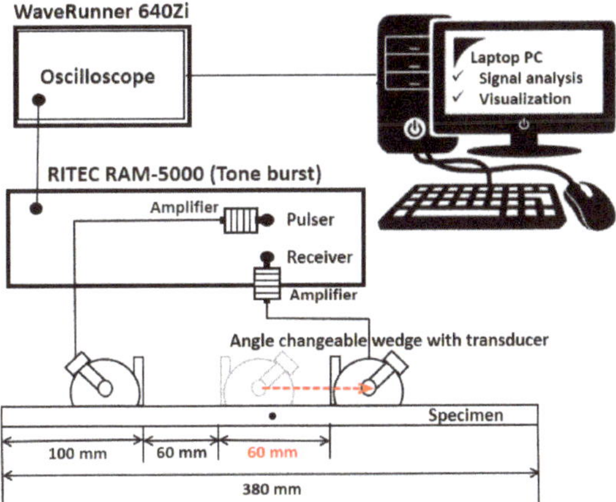

Figure 5. Schematic of experimental set up.

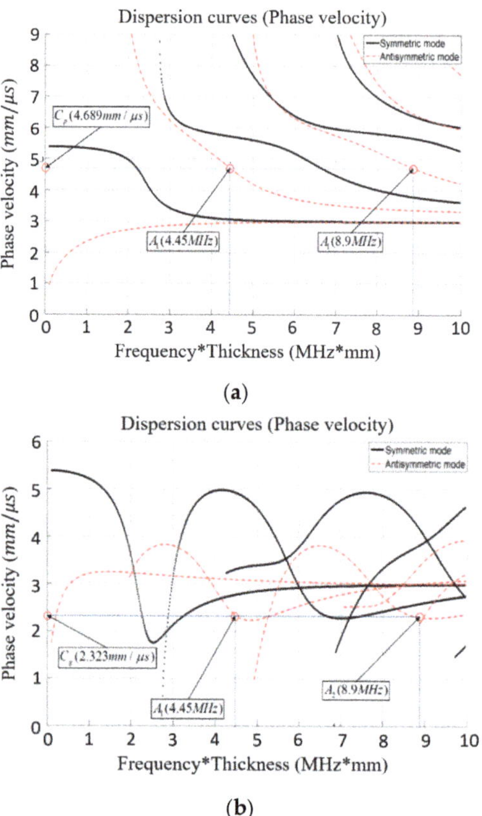

Figure 6. Mode selection: (**a**) Phase matching of antisymmetric mode (A_1); (**b**) Group velocity of antisymmetric mode (A_1).

Table 3. Frequency and velocity parameter in antisymmetric mode.

Mode	Antisymmetric
A_1 (Fundamental frequency)	4.45 MHz
A_2 (Second harmonic)	8.9 MHz
C_P (Phase velocity)	4.689 mm/μs
C_S (Group velocity)	2.323 mm/μs

Table 4. Frequency and velocity parameter in symmetric mode.

Mode	Antisymmetric
S_1 (Fundamental frequency)	4.45 MHz
S_1 (Second harmonic)	8.9 MHz
C_P (Phase velocity)	6.035 mm/μs
C_S (Group velocity)	4.510 mm/μs

When stress is applied to the center of the plate, as illustrated in Figure 7, the upper surface of the plate experiences compressive forces while the lower surface undergoes tensile forces. It is well known that the tensile strength of materials is typically lower than their compressive strength. Consequently, it is expected that the lower surface, subjected to tensile forces, may exhibit a relatively higher occurrence of micro-defects or microstructure changes.

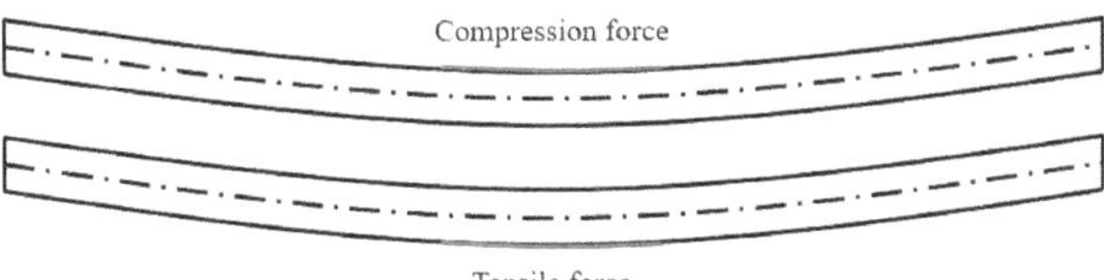

Figure 7. Compression and tensile force in specimen.

4. Experimental Result

In order to evaluate the presence of intrinsic material nonlinearity, initial nonlinear measurements were performed on an unstressed plate. To accomplish this, the nonlinearity of the specimen was measured in the antisymmetric mode, employing the same distance as described in the previous section (Figure 5).

Figure 8 illustrates the received signal of the A1 mode Lamb wave, and the comparison of nonlinearity was conducted through the application of a fast Fourier transform to the signal. Figure 9 presents the variation of nonlinear parameters with respect to distance. The nonlinear parameter obtained for this stress-free state was set as the control, and the cumulative effect was also observed. Although considerable variations in the degree of nonlinearity were observed across different locations, these differences were not deemed statistically significant. Furthermore, it should be noted that the values depicted in the graph represent the relative size of the parameter. Thus, it is important to examine the trend within each interval rather than focusing solely on the absolute values.

Figure 10 presents the variation of the nonlinear parameter of the Lamb wave after elastic recovery, where a stress of 40.5 MPa was applied to the center of the specimen, as well as after plastic deformation at a stress of 89.6 MPa. The nonlinear characteristics exhibited distinct behaviors. Significant variations in the degree of nonlinearity were observed at different locations for the unstressed plate (Figure 9). In contrast, after elastic recovery, the variation in nonlinearity was comparatively reduced. However, nonlinear tendencies are dominantly influenced by the overlap effect, and it is difficult to find a tendency different from the previous one. For the specimen with plastic deformation, the nonlinear parameter exhibits a slight increase up to 70 mm, followed by a subsequent decrease. Beyond 90 mm, the parameter remains relatively constant.

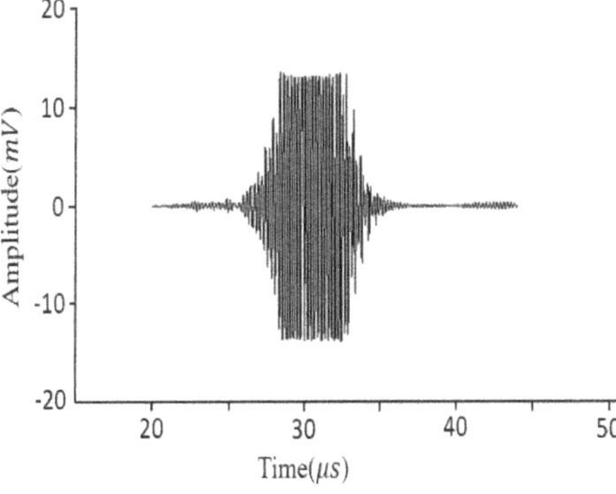

Figure 8. Source signal of A_1 mode (4.45 MHz) Lamb wave.

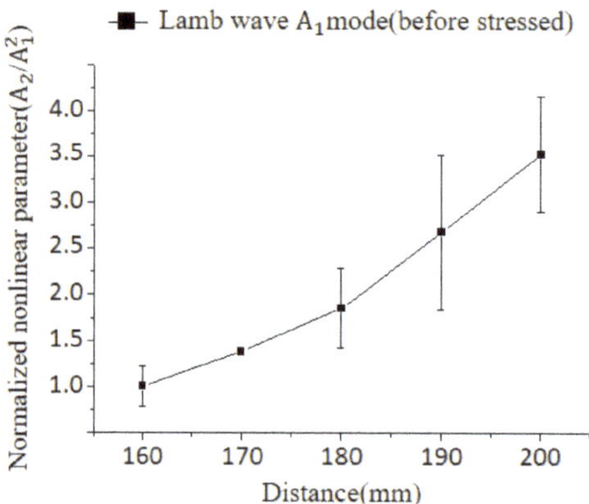

Figure 9. Nonlinear Lamb wave A_1 mode (4.45 MHz) applied to cumulative effect in Al5052 before stressed.

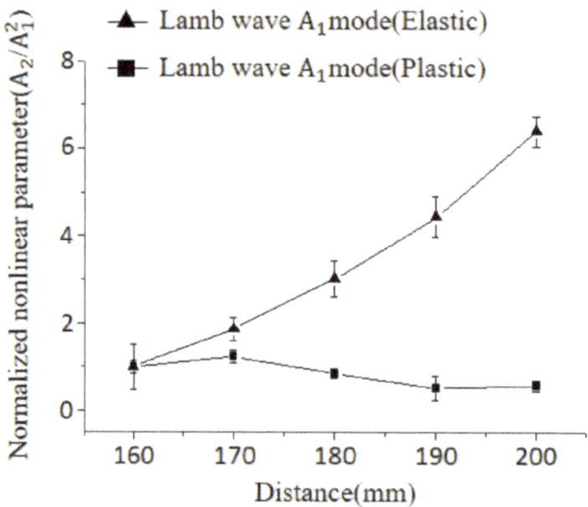

Figure 10. Comparison of nonlinear parameters for specimens after elastic recovery and plastic deformation (4.45 MHz).

The plate underwent bending, resulting in compressive stress on the upper surface and tensile stress on the bottom surface. To examine the influence of these stresses on nonlinearity, measurements were conducted at both the top and bottom surfaces. Figure 11 illustrates a graph comparing the nonlinear parameters attributed to compression and tensile stress.

Next, we investigated the dependences of nonlinearity for different Lamb wave modes on the specimen with plastic deformation. The upper surface (compression region) of the plate was scanned using the generated antisymmetric A1 mode and symmetric S1 mode.

Figure 12 depicts a comparison between the antisymmetric mode and the symmetric mode. In both cases, the nonlinearities show an initial increase followed by a decrease beyond 70 mm, with a subsequent decrease up to 90 mm. The trend in the variation of the nonlinear parameter is similar for both modes, indicating no significant difference in in-plane or out-of-plane displacements. Figure 13 illustrates the comparison between the symmetric and antisymmetric modes for the bottom surface (tension region) of the plate. The trend in variation is similar to that shown in Figure 12. However, it can be observed that the nonlinear amplitude is relatively higher for the antisymmetric mode compared to the symmetric mode.

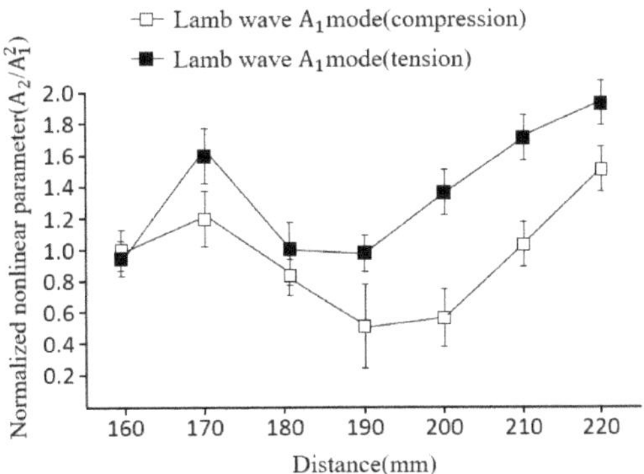

Figure 11. Comparison of nonlinear parameters due to compression and tensile stress (4.45 MHz).

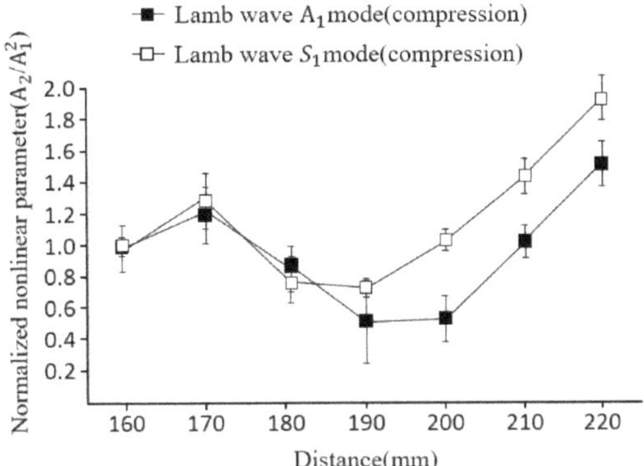

Figure 12. Comparison of nonlinearity between S1 and A1 modes for the upper surface of the plate (3.4 MHz).

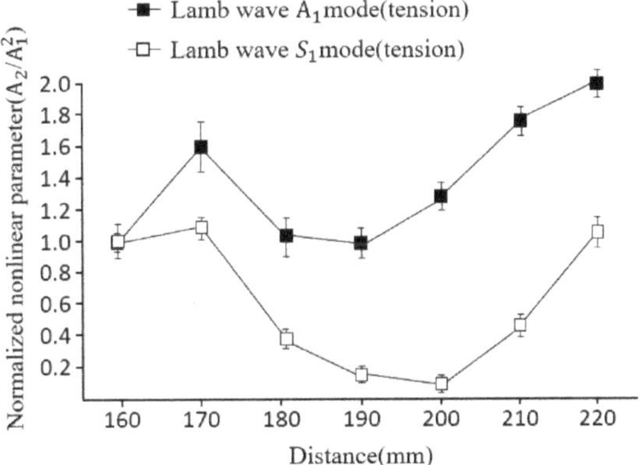

Figure 13. Comparison of nonlinearity between S1 and A1 modes for the bottom surface of the plate (3.4 MHz).

To precisely identify the location of plastic deformation, measurements were conducted at 2 mm intervals within the distance range of 70 mm to 80 mm. The measurements were performed using the antisymmetric Lamb wave mode. This approach aimed to capture the exact position of the plastic deformation, which exhibited a decreasing trend in nonlinearity between the 70 mm and 80 mm sections in the previous experiments.

Figure 14 reveals that the nonlinearity exhibits an increasing trend up to 76 mm, followed by a subsequent decrease. This suggests that plastic deformation likely occurred in the 76 mm section. The decrease in nonlinearity can be attributed to signal scattering caused by the presence of plastic deformation.

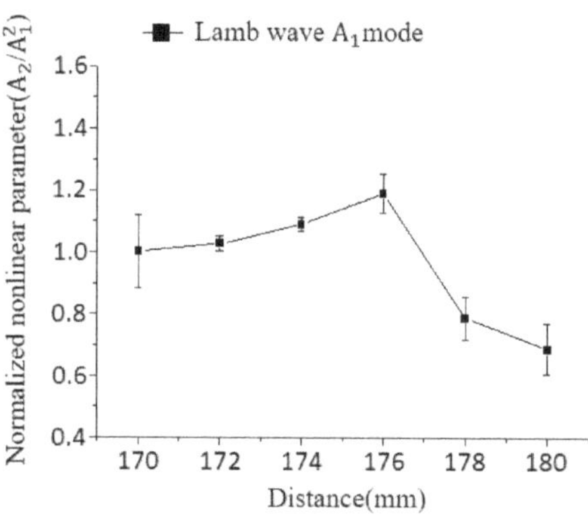

Figure 14. Variation of nonlinearity with distance, indicating maximum plastic deformation at 76 mm.

5. Conclusions

In previous studies, nonlinear ultrasonic techniques have been employed to investigate the detection of micro-defects in materials and the correlation between nonlinearity and tensile/compressive loads. However, limited research has been conducted on the behavior of materials subjected to bending stress during the manufacturing stage, particularly for products with curvature created through bending rather than molds. This study aimed to examine the response of nonlinear Lamb waves to bending stress in aluminum plates.

Experimental investigations were conducted on specimens with both compressive and tensile stresses, representing the plastic deformation state. The nonlinear tendencies of Lamb waves were analyzed and compared, considering the dominant influence of the symmetric mode on in-plane displacement as a control group against the antisymmetric mode, which predominantly affects out-of-plane displacement. Through these experimental verifications, the following findings were obtained:

1. The nonlinearity of the specimens showed distinct behavior depending on the stress applied. Specimens subjected to stress within the elastic region exhibited increased nonlinearity, while specimens under stress within the plastic region displayed a decrease in nonlinearity. This indicates that the material's response to bending stress varies depending on its elastic or plastic state;
2. When analyzing the specimens with plastic deformation, it was observed that the nonlinearity tended to decrease in sections experiencing both tensile and compressive forces. This suggests that the plastic deformation resulted in a reduction of the nonlinearity of the material, possibly due to signal scattering caused by the plastic deformation itself;
3. Both the symmetric and antisymmetric modes of Lamb waves exhibited similar nonlinear characteristics when subjected to compressive and tensile forces. This similarity suggests that the influence of stress on the nonlinearity of the material is consistent based on the chosen Lamb wave modes.

6. Discussion

Based on the aforementioned findings, the following deductions can be made. The nonlinearity of a material generally increases with the propagation distance of the ultrasonic waves. However, in the plastic deformation section, there is a decrease in nonlinearity followed by an increase in nonlinearity after the deformation. This decrease in nonlinearity can be attributed to amplitude attenuation or scattering caused by the deformation of the specimen's shape. Additionally, microscopic changes in the structure of the specimen resulting from plastic deformation are likely to have an impact. The nonlinearity of specimens with plastic deformation, regardless of whether they experienced tensile or compressive stresses, showed a decrease. This consistent behavior across different modes indicates a high level of reliability in the test results.

Based on the aforementioned observations, it can be concluded that the utilization of the wave mixing technique, which offers superior sensitivity and scanning electron microscopy (SEM), can yield more accurate and meaningful results. Further experimental investigations are required to explore the changes in nonlinearity due to varying physical properties. This study provides valuable insights into the behavior of Lamb waves in response to bending stress.

Author Contributions: J.P. and M.A. designed and performed the experiments. J.L. and M.A. conceived the original idea. J.P. developed the theory and performed the computations with support from J.L., and M.A. verified the analytical methods. M.A. wrote the draft study, J.P. completed the final study, and J.L. made the final review. All authors have read and agreed to the published version of the manuscript.

Funding: This work was supported by the National Research Foundation of Korea (NRF) grant funded by the Korea government (MSIT) (No. 2019R1A5A8083201) and This work was supported by Korea Institute of Energy Technology Evaluation and Planning (KETEP) grant funded by the Korea government (MOTIE) (No. 20217410100100).

Conflicts of Interest: The authors declare no conflict of interest.

References

1. Dace, G.E.; Thompson, R.B.; Brash, L.J.H. Nonlinear acoustics, a technique to determine microstructural changes in materials. In *Review of Progress in Quantitative Nondestructive Evaluation*; Thompson, D.O., Chimenti, D.E., Eds.; Plenum Press: New York, NY, USA, 1991; Volume 10B.
2. Zaitsev, V.Y.; Sutin, A.M.; Belyaeva, I.Y.; Nazarov, V.E. Nonlinear Interaction of Acoustical Waves Due to Cracks and Its Possible Usage for Cracks Detection. *J. Vib. Control.* **1995**, *1*, 335. [CrossRef]
3. Ekimov, A.E.; Didenkulov, I.N.; Kazakov, V.V. Modulation of torsional waves in a rod with a crack. *J. Acoust. Soc. Am.* **1999**, *106*, 1289. [CrossRef]
4. Park, J.; Lee, J.; Jeong, S.; Cho, Y. A study on guided wave propagation in a long distance curved pipe. *J. Mech. Sci. Technol.* **2019**, *33*, 4111–4117. [CrossRef]
5. Park, J.; Cho, Y. A study on guided wave tomographic imaging for defects on a curved structure. *J. Vis.* **2019**, *22*, 1081–1092. [CrossRef]
6. Yang, Z.; Yang, L.; Zhang, J.; Ma, S.; Tian, T.; Deng, D.; Wu, Z. Damage shape recognition algorithm of composite woven fabric plate based on guided waves. *Compos. Struct.* **2022**, *303*, 116351.
7. Bermes, C.J.; Kim, J.Y.J.; Qu, J.; Jacobs, L.J. Experimental characterization of material nonlinearity using Lamb waves. *Appl. Phys. Lett.* **2007**, *90*, 021901. [CrossRef]
8. Nagy, P.B. Fatigue damage assessment by nonlinear ultrasonic material characterization. *Ultrasonics* **1998**, *36*, 275–381. [CrossRef]
9. Kim, J.Y.; Qu, J.; Jacobs, J.L.; Littles, J.W.; Savage, M.F. Acoustic nonlinearity parameter due to micro plasticity. *J. Nondestruct. Eval.* **2006**, *25*, 29–37. [CrossRef]
10. Shui, G.; Kim, J.Y.; Qu, J.L.; Jacobs, J. A new technique for measuring the acoustic nonlinearity of material using Rayleigh waves. *NDT E Int.* **2008**, *41*, 326–329. [CrossRef]
11. Deng, M. Second-harmonic properties of horizontally polarized shear modes. *Jpn. J. Appl. Phys.* **1996**, *35*, 4004. [CrossRef]
12. Deng, M. Cumulative second-harmonic generation accompanying nonlinear shear horizontal mode propagation in a solid plate. *J. Appl. Phys.* **1998**, *84*, 3500.
13. Deng, M. Cumulative second-harmonic generation of Lamb-mode propagation in a solid plate. *J. Appl. Phys.* **1999**, *85*, 3051. [CrossRef]
14. Deng, M. Analysis of second-harmonic generation of Lamb modes using a modal analysis approach. *J. Appl. Phys.* **2003**, *94*, 4152. [CrossRef]
15. Li, W.; Cho, Y.; Achenbach, J.D. Detection of thermal fatigue in composites by second harmonic Lamb waves. *Smart Mater. Struct.* **2012**, *21*, 085019. [CrossRef]
16. Li, W.; Chen, B.; Qing, X.; Cho, Y. Characterization of microstructural evolution by ultrasonic nonlinear parameters adjusted by attenuation factor. *Metals* **2019**, *9*, 271. [CrossRef]

17. Li, W.; Jiang, C.; Deng, M. Thermal damage assessment of metallic plates using a nonlinear electromagnetic acoustic resonance technique. *NDT E Int.* **2019**, *108*, 102172. [CrossRef]
18. Park, J.; Lee, J.; Min, J.; Cho, Y. Defects inspection in wires by nonlinear ultrasonic-guided wave generated by electromagnetic sensors. *Appl. Sci.* **2020**, *10*, 4479. [CrossRef]
19. Cho, H.J.; Hasanian, M.; Shana, S.; Lissenden, C.J. Nonlinear guided wave technique for localized damage detection in plates with surface-bonded sensors to receive Lamb waves generated by shear-horizontal wave mixing. *NDT E Int.* **2019**, *102*, 35–46. [CrossRef]
20. Shan, S.; Hasanian, M.; Cho, H.; Lissenden, C.J.; Cheng, L. New nonlinear ultrasonic method for material characterization: Codirectional shear horizontal guided wave mixing in plate. *Ultrasonics* **2019**, *96*, 64–74. [CrossRef]
21. Li, W.; Xu, Y.; Hu, N.; Deng, M. Impact damage detection in composites using a guided wave mixing technique. *Meas. Sci. Technol.* **2019**, *31*, 014001. [CrossRef]
22. Li, W.; Shi, T.; Qin, X.; Deng, M. Detection and Location of Surface Damage Using Third-Order Combined Harmonic Waves Generated by Non-Collinear Ultrasonic Waves Mixing. *Sensors* **2021**, *21*, 6027. [CrossRef] [PubMed]
23. Aslam, M.; Park, J.; Lee, J. Micro inspection in a functionally graded plate structure using monlinear guided wave. *Structures* **2023**, *49*, 666–677. [CrossRef]
24. Ju, T.; Achenbach, J.D.; Jacobs, L.J.; Qu, J. Nondestructive evaluation of thermal aging of adhesive joints by using a nonlinear wave mixing technique. *NDT E Int.* **2019**, *103*, 62–67. [CrossRef]
25. Aslam, M.; Nagarajan, P.; Remanan, M. Nonlinear ultrasonic evaluation of damaged concrete based on mixed harmonic generation. *Struct. Control. Health Monit.* **2022**, *29*, e3110. [CrossRef]
26. Park, J.; Choi, J.; Lee, J. A feasibility study for a nonlinear guided wave mixing technique. *Appl. Sci.* **2021**, *11*, 6569. [CrossRef]
27. Hankai, Z.; Chung, T.N.; Andrei, K. Low-frequency Lamb wave mixing for fatigue damage evaluation using phase-reversal approach. *Ultrasonics* **2022**, *124*, 106768.
28. Radaj, S.; Zhang, S. Geometrically nonlinear behavior or spot welded joint in tensile and compressive shear loading. *Eng. Fract. Mech.* **1995**, *51*, 281–294. [CrossRef]
29. Lee, S.E.; Lim, H.J.; Jin, S.; Sohn, H.; Hong, J.H. Micro-Crack detection with nonlinear wave modulation technique and its application to loaded cracks. *NDT E Int.* **2019**, *107*, 102132. [CrossRef]
30. Norris, A.N. Finite-Amplitude Waves in Solids. In *Nolinear Acoustics*; Academic Press: New York, NY, USA, 1998; pp. 267–269.

applied sciences

Article

Enhancing Fire Detection Technology: A UV-Based System Utilizing Fourier Spectrum Analysis for Reliable and Accurate Fire Detection

Cong Tuan Truong, Thanh Hung Nguyen, Van Quang Vu *, Viet Hoang Do and Duc Toan Nguyen *

School of Mechanical Engineering, Hanoi University of Science and Technology, 1st Dai Co Viet Road, Hai Ba Trung District, Hanoi 100000, Vietnam; tuan.truongcong@hust.edu.vn (C.T.T.); hung.nguyenthanh@hust.edu.vn (T.H.N.); viethoang1340@gmail.com (V.H.D.)

* Correspondence: quang.vuvan1@hust.edu.vn (V.Q.V.); toan.nguyenduc@hust.edu.vn (D.T.N.)

Abstract: This study proposes a low-cost and reliable smart fire alarm system that utilizes ultraviolet (UV) detection technology with an aspherical lens to detect fires emitting photons in the 185–260 nm range. The system integrates the aspherical lens with an accelerator and a digital compass to determine the fire source's direction, allowing for safe evacuation and effective firefighting. Artificial intelligence is employed to reduce false alarms and achieve a low false alarm rate. The system's wide detection range and direction verification make it an effective fire detection solution. Upon detecting a fire, the system sends a warning signal via Wi-Fi or smartphone to the user. The proposed system's advantages include early warning, a low false alarm rate, and detection of a wide range of fires. Experimental results validate the system's design and demonstrate high accuracy, reliability, and practicality, making it a valuable addition to fire management and prevention. The proposed system utilizes a parabolic mirror to collect UV radiation into the detector and a simple classification model that uses Fourier transform algorithm to reduce false alarms. The results showed accuracies of approximately 95.45% and 93.65% for the flame and UVB lamp, respectively. The system demonstrated its effectiveness in detecting flames in the range of up to 50 m, making it suitable for various applications, including small and medium-sized buildings, homes, and vehicles.

Keywords: smart fire alarm system; ultraviolet detection technology; direction detection; low false alarm rate; fourier transform algorithm

Citation: Truong, C.T.; Nguyen, T.H.; Vu, V.Q.; Do, V.H.; Nguyen, D.T. Enhancing Fire Detection Technology: A UV-Based System Utilizing Fourier Spectrum Analysis for Reliable and Accurate Fire Detection. *Appl. Sci.* **2023**, *13*, 7845. https://doi.org/10.3390/app13137845

Academic Editors: Liang Yu, Phong B. Dao, Lei Qiu, Tadeusz Uhl and Minh-Quy Le

Received: 29 May 2023
Revised: 28 June 2023
Accepted: 2 July 2023
Published: 4 July 2023

1. Introduction

Fire alarm systems are essential safety requirements in various settings, such as factories, companies, seaports, and houses. The need for early fire warning has spurred the development of smart and sensitive alert systems. Currently, commercial flame detectors in use operate based on either heat measurement or smoke detection principles. Smoke sensors available in the market operate within a thermal band of −10 °C to 50 °C and have a slow response time of 10–30 s [1–4]. These sensors are triggered when the smoke's dimming falls within the range of 5%–20%. Heat sensors use heat resistance to obtain accurate heat measurements, but their response speed is low, ranging from 15 to 90 s. Additionally, they require contact with the heat source, and their operating temperature is between −10 °C and 50 °C. Although smoke and thermal sensors are widely used and low-cost, their response speed, sensitivity, and working range are limited. Smoke sensors only detect smoke and not the fire source itself, which can be dangerous when the fire grows on a large scale. These sensors work best in closed spaces [1–4]. However, in open spaces, they may fail to detect fires accurately due to air currents and other factors. As a result, new fire detection systems that utilize novel techniques and technologies, such as ultraviolet detection and Fourier spectrum analysis, have been developed to address these limitations and enhance fire safety.

Fire alarm systems are essential in various settings, such as factories, warehouses, seaports, and homes. The development of smart and sensitive alert systems that can detect fires at an early stage is crucial for the safety of people and property. While conventional fire detection systems use heat or smoke sensors, recent advancements have led to the development of more sophisticated systems that employ different techniques, such as image processing and infrared/ultraviolet sensors. Infrared (IR) sensors are commonly used in fire detection systems and can detect heat sources with temperatures of 200 °C to 300 °C above the base temperature. However, the limitation of IR and IR/UV sensors is that they may not be able to detect fires with low temperatures, and their sensitivity is affected by noise from heat sources in the ambient environment and solar radiation [5–8]. Despite these limitations, these sensors are widely used in many fire alarm systems due to their cost-effectiveness and relative reliability. Recently, fire detection systems based on image processing techniques have gained traction [9–12]. These systems use cameras and sophisticated algorithms to analyze the captured images and identify fires. For instance, an integrated long-wave IR and mid-wave IR bolometer camera with a dual-band filter has been reported to detect flames from 30 to 200 feet outdoors, depending on the fire source. However, this detector cannot detect flames in all directions [13–16]. Another fire monitoring system uses both IR sensing and color videos and can detect the burning of smokeless liquids, gas fires, and fires of materials that contain carbon and produce a lot of smoke. The cameras used to localize the fire can be integrated into the security system, but this device is difficult to install in hidden areas and is rather short-lived [17–23]. In summary, while traditional fire detection systems using heat and smoke sensors have their limitations, recent advancements in image processing and infrared/ultraviolet sensors have shown promising results in early fire detection. However, each system has its advantages and disadvantages, and the selection of a fire alarm system depends on various factors such as the size of the building, the type of materials stored, and the location of the fire detection system.

Uncooled IR detectors have become a popular choice for industrial applications due to their ability to detect gases of interest, whose strongest bands are found in the 1–5 micron range. Compared to other commercial IR sensors, uncooled IR detectors offer better response time, making them highly sensitive and able to provide both spatial and temporal information. However, their area of detection is often narrow, which can limit their effectiveness [24,25]. In recent years, Internet of Things (IoT) sensors have played a critical role in primary flame detection. A camera-based system that utilizes multiband IR imagery and thermal analysis has been developed for wildfire detection. This system incorporates LoRa-based wireless communication for long-distance monitoring. Researchers have used the Dempster-Shafer theory and Linear Discriminant Analysis for data analysis and detection of false positives. Despite its effectiveness, the mathematical prediction model for this system is quite complicated and can be difficult to compute [26].

In recent years, fire detection technology has advanced rapidly, with the development of new systems that offer improved accuracy and speed in fire detection. One such system is a Low-Earth Orbit system that utilizes long- and shortwave IR sensors to detect and measure actively burning fires on the Earth's surface [27]. This system is particularly useful for monitoring large areas, such as forests, where traditional ground-based fire detection systems may not be sufficient. In addition to satellite-based systems, smart fire warning and alarm systems are also becoming more sophisticated. These systems typically use multiple sensors, such as flame detectors, humidity, heat, and smoke sensors, among others, to detect fires. To process the data from these sensors and generate a fire alert, Adaptive Neuro-Fuzzy Inference System is commonly employed [28,29]. Despite the benefits of multi-sensor and IoT-based fire detection systems, these technologies also have some drawbacks. For example, they may suffer from slow response times, low sensitivity, high cost, and high system complexity, which can make them difficult to install and maintain.

This paper introduces a new approach to fire detection and warning systems that offers a smart, compact, and cost-effective solution. The system incorporates a UV detector that operates in the range of 185–260 nm to detect the signal from the fire source, making it suitable for identifying the burning of different materials at various rates. To eliminate background noise, the signal processing circuits have been designed to cancel out any sporadic noise that may arise due to the background. The system also determines the direction of the fire source, allowing for more precise and efficient fire extinguishing. The alarm signal can be generated in a high-frequency pulse shape (100 Hz) and transmitted to a computer via Wi-Fi or sent to the user's smartphone. Experimental results demonstrate that the system can detect a fire source within a range of 50 m. Furthermore, an artificial intelligence (AI) system has been developed to prevent false alarms. The proposed system is well-suited for use in warehouses, seaports, and factories, and can be integrated into a monitoring network. The innovative approach presented in this paper offers a new, effective solution for fire detection and warning systems that can enhance safety and reduce property damage.

2. System Design

The fire detection system proposed in Figure 1 comprises several essential components. A UV detector, specifically the Hamamatsu UVtron R12257 (Hamamatsu Photonics, Shizuoka, Japan), is utilized in the system. This detector demonstrates high sensitivity and boasts a rapid response time of just a few milliseconds, enabling it to effectively detect UV radiation emitted by flames. The sensor operates within a selective spectral range of 185–280 nm, eliminating the need for optical filters to block out visible light. Its output consists of a series of pulses that persist for a duration sufficient to identify genuine flames. While this pulse-based output helps filter out background noise within the sensor's field of view, it is important to consider potential interference sources. Heat-emitting devices, such as automobiles or processing machines, may emit minimal amounts of UV radiation, which can serve as a possible source of interference. The UV radiation is focused onto the sensor using a parabolic mirror, the Sigmakoki Inc. (Tokyo, Japan) TCPA-105C-15-SH18, mounted on a precise rotation stage, which scans the entire monitoring area. The scanning direction of the mirror is determined by a digital compass and an accelerometer that are integrated into the stage, and the output signal from the sensor and the position of the mirror are used to determine the fire's location. To eliminate sporadic noise that may occur due to the background, the signal processing circuits have been designed to cancel out any such noise. Additionally, the system determines the direction of the fire source, allowing for more accurate and effective fire extinguishing. The alarm signal generated by the system can be transferred to a computer via Wi-Fi or sent to the user's smartphone in a high-frequency pulse shape (100 Hz).

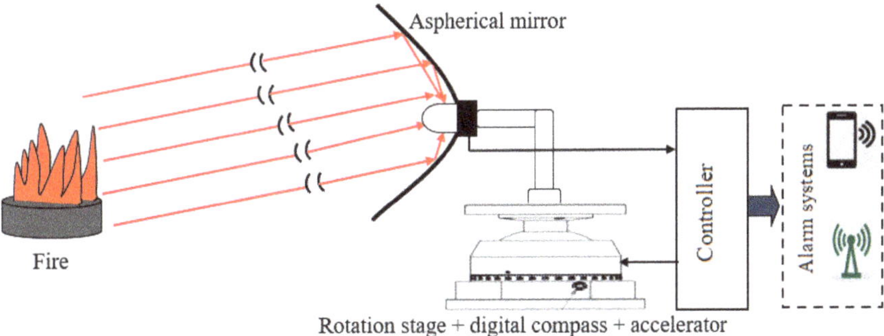

Figure 1. Fire detection system.

In the case of a fire, the proposed fire detection system immediately generates a warning signal. This signal can be transmitted to an alarm system, which can be accessed on a smartphone or via Wi-Fi. The warning signal can alert the user to the presence of a fire, allowing them to take immediate action. Additionally, the system's operation can be monitored in real-time from a monitoring center. This monitoring center can be set up to monitor the system's performance and status, ensuring that the system is functioning properly and that any problems or malfunctions are quickly detected and resolved. The monitoring center can also provide valuable data and information about the fire, such as its location, intensity, and spread, which can be used to help first responders and firefighters effectively combat the blaze. Furthermore, the real-time monitoring capability of the system can help prevent false alarms and provide a high level of accuracy in detecting fires, thus minimizing the risk of property damage and ensuring the safety of occupants.

The control circuit for the fire detection system is an essential component of the system, and it plays a significant role in ensuring the system's accuracy and effectiveness. As shown in Figure 2, the circuit consists of several key components that work together to provide a reliable and efficient solution for detecting fires. The UVtron control unit supplies high voltage to the sensor and produces short pulses warning signals when the sensor is exposed to UV radiation. The output pulses, which last for 10 ms, are sent to a microprocessor (Arduino Uno R3), which is responsible for processing the signals and controlling the system's operation.

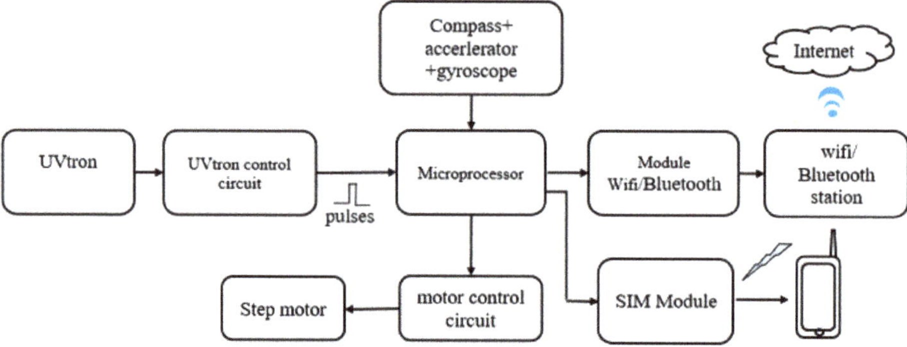

Figure 2. Control circuit for the fire detection system.

Since the effective area of the UV sensor is small, and the sensing field is limited to an angle of 70°–120°, photons radiate from various directions, making it challenging for them to hit the sensor's focal point. To overcome this issue, a stepper motor (MSK 4240-400) is attached to the microprocessor, which rotates the parabolic mirror, enhancing the system's sensitivity and effective working range. The direction of the mirror is determined using digital compass, gyroscope, and accelerometer signals, which allow for accurate detection and localization of the fire.

The microcontroller is linked to a SIM block (SIM900A Mlab, Hanoi, Vietnam) to send alarms to a mobile phone that is already installed. Additionally, the microcontroller has a Wi-Fi module (RF 433 MHz Transmitter/Receiver) that enables it to communicate with a smartphone through the internet. This allows for remote monitoring of the system's operation, providing real-time data on the fire's location and intensity. The integration of these components into a single circuit provides a compact and cost-effective solution for fire detection, making it suitable for use in various settings, such as warehouses, factories, and seaports.

3. Experiment

The fire detection system's effectiveness in detecting fires was evaluated through comprehensive testing under varying experimental conditions, considering both short-range and long-range capabilities. Table 1 provides an overview of the experimental conditions, while Figure 3 illustrates the system employed for the fire detection test. The experiments were conducted during daylight hours, acknowledging that environmental factors can influence the test results. Specifically, the presence of sunlight could emit UV radiation within the sensor's targeted range of 185–280 nm. To mitigate this, our fire detection system implementation ensures that the sensor is not directly exposed to sunlight. In practical applications, the device is installed overhead, facing downwards. For assessing the system's short-range capabilities, a candle was utilized as the test fire, positioned at distances ranging from 1 to 10 m. The distance between the fire sensor and the candle was accurately measured using a laser distance meter (Leica Disto D2, Leica Geosystems AG, St. Gallen, Switzerland), which provides a measurement range of up to 100 m. To identify the fire's location, the fire detection system was rotated around the test area, comparing the density of generated pulses. The outcomes of the experiment, specifically the results obtained at the fire's location, are presented in Figure 4.

To evaluate the system's long-range detection capability, a fire with a size of 300 mm × 300 mm was used as the test object. The experimental setup and results for the long-range test are shown in Figure 5, which includes the experiment setup in (a) and the stable warning signal in (b). The experiment demonstrated that the system can detect fires at distances of up to 50 m. It is worth noting that the warning signal remained stable throughout the long-range detection experiment, which demonstrates the system's ability to effectively detect fires even at extended distances.

The experimental results indicate that the proposed fire detection system is highly effective in detecting fires, both at short and long distances. The system's use of a UV detector with a rapid response time and the ability to cancel out background noise, coupled with the rotation of the parabolic mirror to enhance its sensitivity, enables it to accurately detect fires and provide timely warnings to prevent property damage and protect lives. The system's ability to send warning signals to a smartphone or via Wi-Fi and to be monitored in real-time from a monitoring center further enhances its capabilities and makes it suitable for use in warehouses, seaports, and factories.

Table 1. Condition of the fire detection experiment.

Measurement target size	Short range: 5 mm × 10 mm Long range: 300 mm × 300 mm
Testing distance	Short range: 1.5, 3.0, 5.0, 10.0 m Long range: 50 m
Detected wavelength range	185–280 nm
Power supply	12 V
Current supply	2 mA

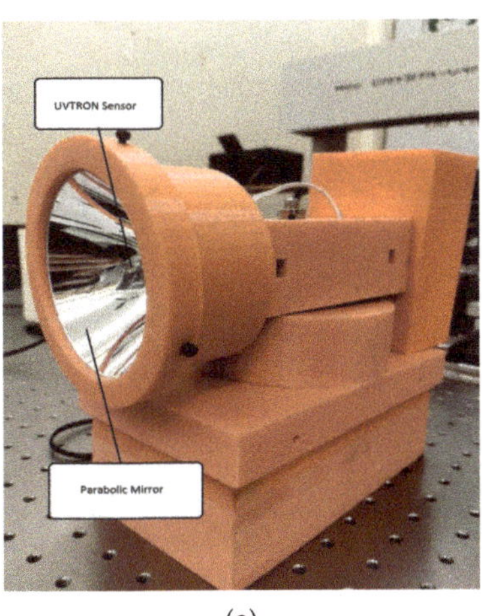

(a)

(b)

(c)

Figure 3. Experimental fire detection system and setup used to verify its measurement range and sensitivity: (**a**,**b**) experimental system in detail; (c) test of the fire detection experiment using a candle.

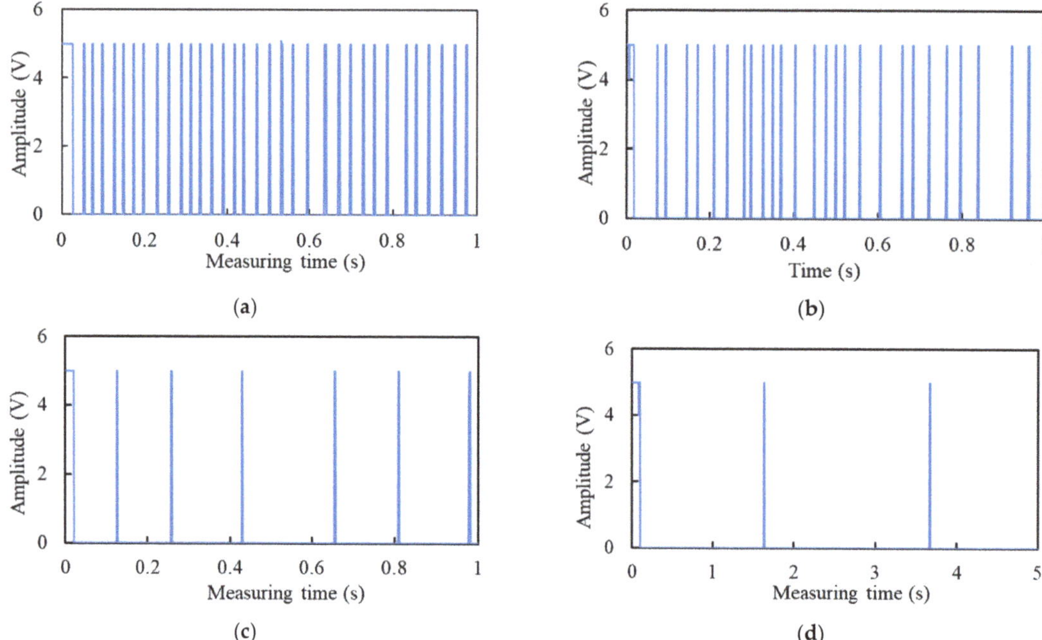

Figure 4. Experimental results for short-range detection: (**a**) 1.5 m; (**b**) 3 m; (**c**) 5 m; (**d**) 10 m.

The experimental results indicate that the fire detection system is capable of accurately detecting the presence of a fire source within a maximum distance of 50 m. The system's detection capability is optimized for specific dimensions of 300 mm × 300 mm, ensuring reliable and precise identification of fire sources within this defined area. The reliable and accurate detection of fires is critical for ensuring safety in various industries, including chemical, petroleum, and electrical power. False alarms can lead to costly disruptions, reduce efficiency, and cause unnecessary evacuations. Therefore, it is essential to develop techniques that can accurately distinguish between UV radiation emitted from flames and other sources.

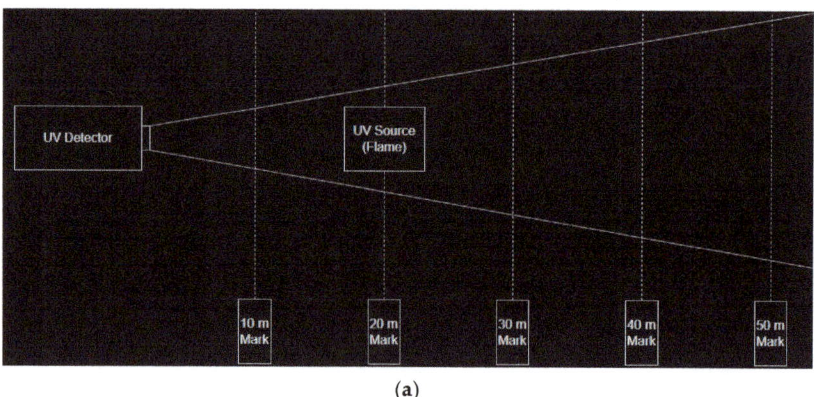

(**a**)

Figure 5. *Cont.*

(b)

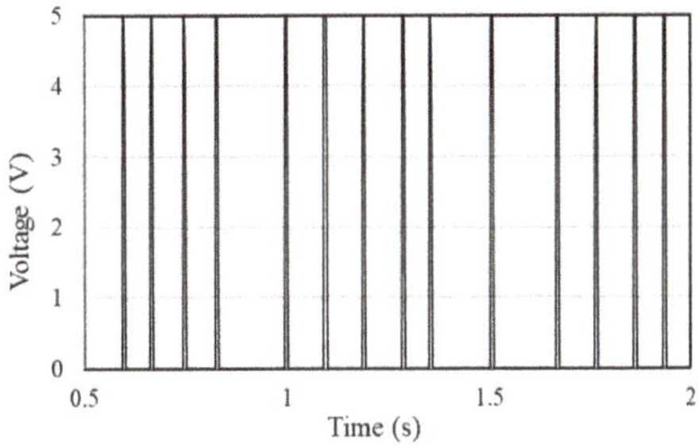

(c)

Figure 5. Diagram for experimental setup for fire detection at various distances (**a**); real look of experimental setup (**b**); illustration of output signal (**c**).

One approach to minimizing false alarms is to exploit the chaotic nature of radiation from an uncontrolled flame. Unlike other sources such as UV lamps, heaters, and sunlight, combustion involves nonlinear instabilities that result in chaotic emission. This characteristic can be used to distinguish flame signals from other signals that tend to be regular. This study used the Naïve Bayes classification algorithm with the signal spectral characteristics in the frequency domain to distinguish between signals from a flame and a controlled UV source (UVB lamp). The proposed approach was evaluated using a dataset obtained from a fire detection system equipped with a UV detector, and the results, presented in Figure 6, demonstrate the algorithm's robust accuracy and efficiency. These findings provide strong evidence for the potential of this technique in effectively reducing false alarms in fire detection systems.

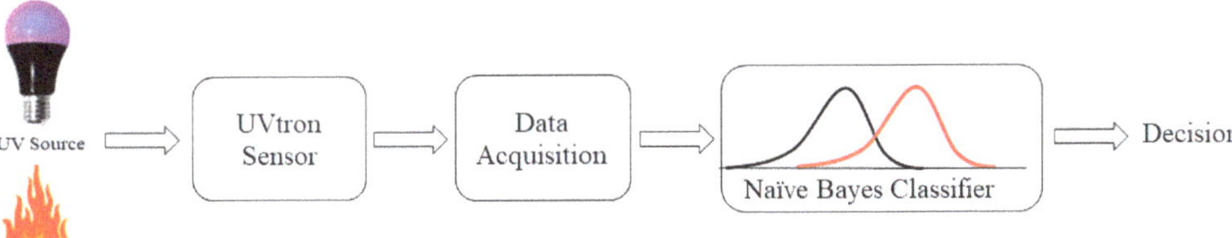

Figure 6. Naïve Bayes classifier model applied to the UVtron fire detection system. Distributions of two classes are represented by black and red line.

The study aimed to distinguish signals from a flame and a controlled UV source using Fourier spectrum analysis as the classification algorithm. The UVB lamp was modified by limiting the surface area of the radiation source with a pinhole of 10 mm × 10 mm in front of it. The resulting signal data obtained from the UVtron sensor appeared as a time series consisting of pulses lasting 10 ms, as shown in Figure 7. The sensor's output signal in response to the UV source consists of a series of fixed-width (10 ms) pulses, exhibiting varying amplitudes (between 4 VDC and 5 VDC) due to conditioning circuit operations. The pulse signals depicted in Figures 6 and 7 represent random selections from the overall output signal obtained during the measurements. The signal pulses from the flame were found to be more discrete than those from the UVB lamp, indicating that the sensor responds to the flame in a more chaotic manner. Fourier spectrum analysis was applied to obtain information on the signal's amplitude at different frequencies, where larger amplitude at any frequency indicated more signal pulses at that frequency and vice versa. Therefore, the Fourier spectrum analysis result for the signal output over a specific period of time was used as the data feature for classification in this study. It is important to note that in the case of the flame, the pulse duration steps for each distance were not uniform at different moments in time, and the time-step magnitude distribution was also non-uniform. Thus, Fourier spectrum analysis was chosen as a robust algorithm that could handle such non-uniformity in the pulse duration and magnitude distribution.

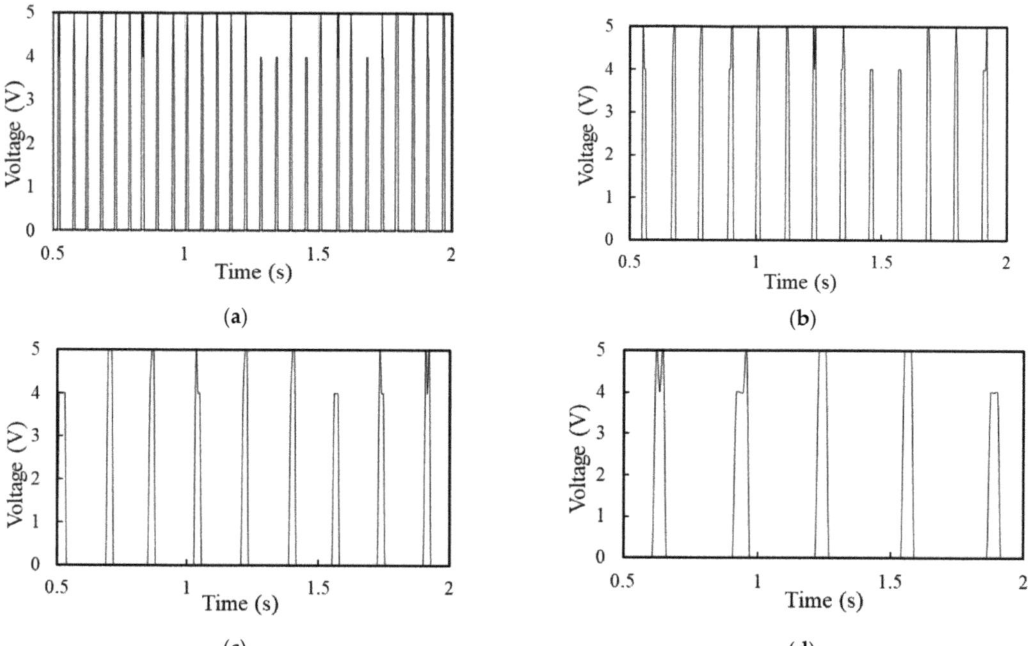

Figure 7. Output signals of the sensor responding to the UV lamp placed at different distances. (**a**) 1.5 m; (**b**) 3 m; (**c**) 5 m; (**d**) 10 m.

The signal pulses from the flame (Figure 7) were found to be more discrete than those from the UVB lamp, indicating that the sensor responds to the flame in a more chaotic manner. Fourier spectrum analysis was applied to obtain information on the signal's amplitude at different frequencies, where a larger the amplitude at any frequency indicating more signal pulses at that frequency, and vice versa. The data feature for classification was therefore the Fourier spectrum analysis result for the signal output over a specific period of time.

To extract the features, the time series of the output signal was divided into windows of duration 250 ms, corresponding to N = 100 samples each, and denoted as $\boldsymbol{x} = [x_1, x_2, \ldots, x_N]$. Fast Fourier Transform was then applied to each window, resulting in a vector of power of spectrum $\boldsymbol{X} = \left[X_1, X_2, \ldots, X_{\frac{N}{2}}\right]$ with a dimension of N/2 = 50. Here, each X_k represents the spectrum power at a frequency of $k \times F_1$, with F_1 = 4 Hz as the fundamental frequency. Figure 8 provides examples of the normalized Fourier spectrum analysis of the signals obtained from the flame and UVB lamp.

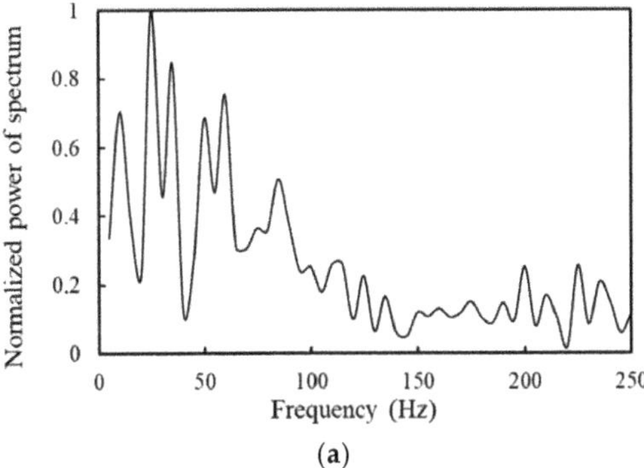

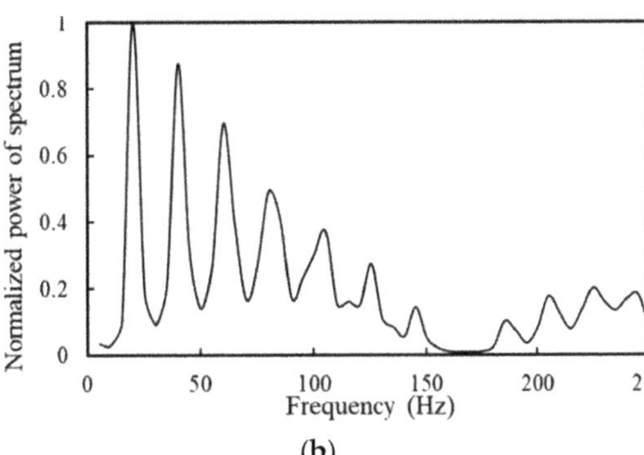

Figure 8. Normalized Fourier spectrum analysis results for the output signals of a flame (**a**) and UV lamp (**b**).

The Fourier spectrum analysis results of the flame, as observed in this study, were observed to exhibit significant variability over time, indicating that the signal feature space may be difficult to model using standard techniques. To address this challenge, constructing distributions for the vector of spectrum power X was considered as a viable solution to the classification problem. Specifically, the Fourier spectrum amplitude values for each frequency over time were treated as random variables, and the joint probability distribution for these values was constructed. In this case, the input features were the Fourier spectrum amplitude values for each frequency, and the classification models were used to classify new signals as either coming from the flame or the UVB lamp.

Due to the fact that the data that needs to be classified currently has exactly two classes, it is implied that the classification can be done with the Support Vector Machine (SVM) method, which is suitable for edge devices [12]. The advantages of this method can be listed as follows: it is effective in high-dimensional spaces, even in the cases where number of dimensions is greater than the number of samples; it is memory efficient due to the fact that it uses a subset of training points in the decision function; it offers flexibility in various kernel selection. In this work, the idea comes with using SVM for binary classification in which the data is labeled with $y = 1$ for the flame and $y = -1$ for the UBV lamp. The aim of this method is to find the function of a hyperplane over the feature vector X, which is the vector of spectrum power in this work:

$$f(X) = X^T\beta + b \tag{1}$$

Hence, it is necessary to specify the parameters β, b that must meet certain criteria. In practical, our data appears to not allow for a separating hyperplane. Therefore, the objective function to estimate β, b in this work presents the L^1-norm problem, which is constructed by adding slack variables ξ_j and a penalty parameter C, as follows:

$$\min_{\beta,b,\xi}\left(\frac{1}{2}\beta^T\beta + C\sum_j \xi_j^1\right) \tag{2}$$

subjected to the constraints:

$$y_j f(X_j) \geq 1 - \xi_j \text{ and } \xi_j \geq 0 \forall j \quad (3)$$

On the other hand, it was possible to capture the inherent variability of the flame signal over time and encode this information into a probabilistic model. This approach allowed for the development of a Naïve Bayes model, which is a type of probabilistic classifier that assumes the independence of the input features. Overall, this approach proved to be effective at accurately distinguishing between these two types of signals and may be useful in other applications where signal variability poses a challenge to traditional classification techniques.

The Naïve Bayes classifier is typically used when predictors are independent of each other within each class. However, it has been shown to perform well in practice even when the independence assumption is not valid. Specifically, the Naïve Bayes model estimated the probability density of predictor X (the vector of spectrum power) given class y-(flame or UVB lamp), denoted by $P(\mathbf{X} \mid y)$. Therefore, the Naïve Bayes model resulted in parameter vectors $\boldsymbol{\mu}$, $\boldsymbol{\sigma}$, $\boldsymbol{\mu}'$, $\boldsymbol{\sigma}'$ with dimensions of 50 for the following probability densities:

$$P(\mathbf{X}|y = flame) = \prod_{k=1}^{50} \frac{1}{\sigma_i \sqrt{2\pi}} \times \exp\left(-\frac{(x_i - \mu_i)^2}{2\sigma_i^2}\right) \quad (4)$$

$$P(\mathbf{X} \mid y = UV\ lamp) = \prod_{k=1}^{50} \frac{1}{\sigma'_i \sqrt{2\pi}} \times \exp\left(-\frac{(x_i - \mu'_i)^2}{2\sigma'^2_i}\right) \quad (5)$$

The present study utilized the parameter vectors obtained from assuming a Gaussian distribution for $P(\mathbf{X} \mid y)$ to classify signals into either flame or UVB lamp classes based on the probability of the spectrum power vector X belonging to each class, utilizing Bayes' theorem. The Naïve Bayes model was found to be an effective approach for classifying signals based on their Fourier spectrum analysis results.

The results of this study demonstrate the potential of using the SVM classifier and Naïve Bayes classifier based on Fourier spectrum analysis to distinguish between signals from a flame and a UVB lamp. The classifiers were trained using 1000 samples of both the flame and UVB lamp and were evaluated with 500 samples of each. By using the SVM classifier with the penalty parameter C = 10, the accuracies obtained in the testing phase were 95.15% and 91.25% for flame and UVB lamp classification, respectively. Meanwhile, the results of performing the Naïve Bayes classifier were 95.45% and 93.65% for flame and UVB lamp classification, respectively. These findings suggest that the SVM classifier and Naïve Bayes classifier are reliable methods for accurately differentiating between signals from a flame and a UVB lamp based on their Fourier spectrum analysis results.

The high accuracies achieved in this study offer a promising solution to the problem of false alarms in fire detection systems. By minimizing false alarms, the proposed system has the potential to save lives and minimize property damage. Additionally, the use of Fourier spectrum analysis to extract features of the flame and UVB lamp signals offers a robust and effective method for fire detection.

Future research could further investigate the effectiveness of this approach in more complex environments, such as those with multiple sources of UV radiation or in the presence of smoke. Additionally, exploring the potential for integrating this system with other fire detection methods could lead to even greater accuracy and reliability in fire detection.

4. Conclusions

In this study, we presented a low-cost and efficient fire detection system that utilized UV radiation detection and Fourier spectrum analysis. By utilizing a parabolic mirror and digital sensors, we were able to detect the position of the fire accurately. Furthermore, the proposed classification models, based on the Fourier transform algorithm, successfully extracted features of a fire and a UV source, which enabled us to minimize the occurrence of false alarms. The experimental results showed that the proposed system achieved high accuracies of 95.15% and 91.25% using the SVM classifier, and approximately 95.45% and 93.65% using the Naïve Bayes classifier for the flame and UVB lamp, respectively.

Our proposed system offers several benefits compared to existing fire detection systems. It is cost-effective, making it a viable option for small and medium-sized buildings, homes, and vehicles. Moreover, the system's simple design and high accuracy make it attractive for use in areas where expensive systems are not feasible. Our system has been shown to be effective in detecting flames up to 50 m away, making it suitable for a wide range of applications.

In conclusion, the proposed system has demonstrated the effectiveness of using UV radiation detection and Fourier spectrum analysis as an efficient method for fire detection. Future studies could explore ways to improve the system's performance in challenging environments and investigate its potential for integration with other fire detection systems. This study provides valuable insights into the development of low-cost and efficient fire detection systems that can contribute to the safety of individuals and properties.

Author Contributions: Conceptualization, C.T.T.; Methodology, V.Q.V.; Software, T.H.N., V.Q.V. and V.H.D.; Validation, V.Q.V. and D.T.N.; Formal analysis, C.T.T. and V.H.D.; Investigation, C.T.T., T.H.N. and D.T.N.; Resources, V.Q.V.; Data curation, C.T.T., T.H.N., V.Q.V. and V.H.D.; Writing—original draft, C.T.T.; Writing—review & editing, D.T.N.; Supervision, D.T.N.; Project administration, C.T.T.; Funding acquisition, C.T.T. All authors have read and agreed to the published version of the manuscript.

Funding: This work was funded by the Vietnam Ministry of Education and Training, Project number B2020-BKA-16.

Data Availability Statement: Not applicable.

Conflicts of Interest: The authors declare no conflict of interest.

References

1. Nolan, D.P. *Handbook of Fire and Explosion Protection Engineering Principles: For Oil, Gas, Chemical and Related Facilities*; William Andrew: Kidlington, UK, 2014.
2. Cheon, J.; Lee, J.; Lee, I.; Chae, Y.; Yoo, Y.; Han, G. A single-chip CMOS smoke and temperature sensor for an intelligent fire detector. *IEEE Sens. J.* **2009**, *9*, 914–921. [CrossRef]
3. Jee, S.W.; Lee, C.H.; Kim, S.K.; Lee, J.J.; Kim, P.Y. Development of a traceable fire alarm system based on the conventional fire alarm system. *Fire Technol.* **2014**, *50*, 805–822. [CrossRef]
4. Bakhoum, E.G. High-sensitivity miniature smoke detector. *IEEE Sens. J.* **2012**, *12*, 3031–3035. [CrossRef]
5. Wing, M.G.; Burnett, J.; Sessions, J. Remote sensing and unmanned aerial system technology for monitoring and quantifying forest fire impacts. *Int. J. Remote Sens.* **2014**, *4*, 18–35. [CrossRef]
6. Barmpoutis, P.; Papaioannou, P.; Dimitropoulos, K.; Grammalidis, N. A review on early forest fire detection systems using optical remote sensing. *Sensors* **2020**, *20*, 6442. [CrossRef]
7. Li, Y.; Yu, L.; Zheng, C.; Ma, Z.; Yang, S.; Song, F.; Zheng, K.; Ye, W.; Zhang, Y.; Wang, Y.; et al. Development and field deployment of a mid-infrared CO and CO_2 dual-gas sensor system for early fire detection and location. *Spectrochim. Acta A Mol. Biomol. Spectrosc.* **2022**, *270*, 120834. [CrossRef]
8. Prasojo, I.; Nguyen, P.T.; Shahu, N. Design of ultrasonic sensor and ultraviolet sensor implemented on a fire fighter robot using AT89S52. *JRC* **2020**, *1*, 55–58. [CrossRef]
9. Burnett, J.D.; Wing, M.G. A low-cost near-infrared digital camera for fire detection and monitoring. *Int. J. Remote Sens.* **2018**, *39*, 741–753. [CrossRef]
10. Cowlard, A.; Jahn, W.; Abecassis-Empis, C.; Rein, G.; Torero, J.L. Sensor assisted fire fighting. *Fire Technol.* **2020**, *46*, 719–741. [CrossRef]
11. Zarkasi, A.; Nurmaini, S.; Stiawan, D.; Amanda, C.D. Implementation of fire image processing for land fire detection using color filtering method. *J. Phys. Conf. Ser.* **2019**, *1196*, 012003. [CrossRef]
12. Ya'acob, N.; Najib, M.S.M.; Tajudin, N.; Yusof, A.L.; Kassim, M. Image processing based forest fire detection using infrared camera. *J. Phys. Conf. Ser.* **2021**, *1768*, 012014. [CrossRef]
13. Chowdary, V.; Gupta, M.K. Automatic forest fire detection and monitoring techniques: A survey. In *Intelligent Communication, Control and Devices, Proceedings of ICICCD 2017, Dehradun, India, 15–16 April 2017*; Springer: Singapore; pp. 1111–1117.

14. Sadi, M.; Zhang, Y.; Xie, W.F.; Hossain, F.A. Forest fire detection and localization using thermal and visual cameras. In Proceedings of the 2021 International Conference on Unmanned Aircraft Systems (ICUAS), Athens, Greece, 15–18 June 2021; pp. 744–749.
15. De Vivo, F.; Manuela, B.; Eric, J. Infra-red line camera data-driven edge detector in UAV forest fire monitoring. *Aerosp. Sci. Technol.* **2021**, *111*, 106574. [CrossRef]
16. Namburu, A.; Selvaraj, P.; Mohan, S.; Ragavanantham, S.; Eldin, E.T. Forest Fire Identification in UAV Imagery Using X-MobileNet. *Electronics* **2023**, *12*, 733. [CrossRef]
17. Wang, M.; Jiang, L.; Yue, P.; Yu, D.; Tuo, T. FASDD: An Open-access 100,000-level Flame and Smoke Detection Dataset for Deep Learning in Fire Detection. *Earth Syst. Sci. Data Discuss.* **2023**, 1–26. [CrossRef]
18. Almeida, J.S.; Jagatheesaperumal, S.K.; Nogueira, F.G.; de Albuquerque, V.H.C. EdgeFireSmoke++: A novel lightweight algorithm for real-time forest fire detection and visualization using internet of things-human machine interface. *Expert Syst. Appl.* **2023**, *221*, 119747. [CrossRef]
19. Singh, R.; Sharma, S.; Sharma, S.; Kaushik, S. Real-Time Fire Detection System Based on CNN Using Tensorflow and OpenCV. *J. Data Acquis. Process.* **2023**, *38*, 723.
20. Gaur, A.; Singh, A.; Kumar, A.; Kumar, A.; Kapoor, K. Video flame and smoke based fire detection algorithms: A literature review. *Fire Technol.* **2020**, *56*, 1943–1980. [CrossRef]
21. Huang, P.; Chen, M.; Chen, K.; Zhang, H.; Yu, L.; Liu, C. A combined real-time intelligent fire detection and forecasting approach through cameras based on computer vision method. *Process Saf. Environ. Prot.* **2022**, *164*, 629–638. [CrossRef]
22. Dang, J.; Yu, H.; Song, F.; Wang, Y.; Sun, Y.; Zheng, C. An early fire gas sensor based on 2.33 μm DFB laser. *Infrared Phys. Technol.* **2018**, *92*, 84–89. [CrossRef]
23. Hendel, I.G.; Ross, G.M. Efficacy of remote sensing in early forest fire detection: A thermal sensor comparison. *Can. J. Remote Sens.* **2020**, *46*, 414–428. [CrossRef]
24. Rizanov, S.; Stoynova, A.; Todorov, D. Single-pixel optoelectronic IR detectors in wireless wildfire detection systems. In Proceedings of the 2020 43rd International Spring Seminar on Electronics Technology (ISSE), Demanovska Valley, Slovakia, 14–15 May 2020; pp. 1–6.
25. Wooster, M.J.; Roberts, G.J.; Giglio, L.; Roy, D.P.; Freeborn, P.H.; Boschetti, L.; Justice, C.; Ichoku, C.; Schroeder, W.; Davies, D.; et al. Satellite remote sensing of active fires: History and current status, applications and future requirements. *Remote Sens. Environ.* **2021**, *267*, 112694. [CrossRef]
26. Kaur, H.; Sood, S.K. Adaptive neuro fuzzy inference system (ANFIS) based wildfire risk assessment. *J. Exp. Theor. Artif. Intell.* **2019**, *31*, 599–619. [CrossRef]
27. Rajan, M.S.; Dilip, G.; Kannan, N.; Namratha, M.; Majji, S.; Mohapatra, S.K.; Patnala, T.R.; Karanam, S.R. Diagnosis of fault node in wireless sensor networks using adaptive neuro-fuzzy inference system. *Appl. Nanosci.* **2023**, *13*, 1007–1015. [CrossRef]
28. Chatzopoulos-Vouzoglanis, K.; Reinke, K.J.; Soto-Berelov, M.; Jones, S.D. One year of near-continuous fire monitoring on a continental scale: Comparing fire radiative power from polar-orbiting and geostationary observations. *Int. J. Appl. Earth Obs. Geoinf.* **2023**, *117*, 103214. [CrossRef]
29. Xu, W.; Wooster, M.J. Sentinel-3 SLSTR active fire (AF) detection and FRP daytime product-Algorithm description and global intercomparison to MODIS, VIIRS and landsat AF data. *Sci. Remote Sens.* **2023**, *7*, 100087. [CrossRef]

Article

Modification and Evaluation of Attention-Based Deep Neural Network for Structural Crack Detection

Hangming Yuan [1], Tao Jin [2,*] and Xiaowei Ye [2]

[1] Polytechnic Institute, Zhejiang University, Hangzhou 310058, China
[2] Department of Civil Engineering, Zhejiang University, Hangzhou 310058, China
* Correspondence: cetaojin@zju.edu.cn

Abstract: Cracks are one of the safety-evaluation indicators for structures, providing a maintenance basis for the health and safety of structures in service. Most structural inspections rely on visual observation, while bridges rely on traditional methods such as bridge inspection vehicles, which are inefficient and pose safety risks. To alleviate the problem of low efficiency and the high cost of structural health monitoring, deep learning, as a new technology, is increasingly being applied to crack detection and recognition. Focusing on this, the current paper proposes an improved model based on the attention mechanism and the U-Net network for crack-identification research. First, the training results of the two original models, U-Net and lrassp, were compared in the experiment. The results showed that U-Net performed better than lrassp according to various indicators. Therefore, we improved the U-Net network with the attention mechanism. After experimenting with the improved network, we found that the proposed ECA-UNet network increased the Intersection over Union (IOU) and recall indicators compared to the original U-Net network by 0.016 and 0.131, respectively. In practical large-scale structural crack recognition, the proposed model had better recognition performance than the other two models, with almost no errors in identifying noise under the premise of accurately identifying cracks, demonstrating a stronger capacity for crack recognition.

Keywords: structural crack; deep learning; attention mechanism; structural health monitoring

Citation: Yuan, H.; Jin, T.; Ye, X. Modification and Evaluation of Attention-Based Deep Neural Network for Structural Crack Detection. *Sensors* **2023**, *23*, 6295. https://doi.org/10.3390/s23146295

Academic Editors: Phong B. Dao, Tadeusz Uhl, Liang Yu, Lei Qiu and Minh-Quy Le

Received: 5 June 2023
Revised: 6 July 2023
Accepted: 7 July 2023
Published: 11 July 2023

1. Introduction

With the development of China's economy and the continuously expanding investment in infrastructure, the number of large structures such as bridges and buildings has increased [1–3]. Some buildings are in a long-term state of overload, corrosion, etc., and are susceptible to functional barriers under the overlapping impact of natural disasters, resulting in serious accidents [4,5]. Cracks in the structure are some of the most important indicators of structural damage or destruction caused by aging and other reasons [6,7]. As time goes by, the width and number of cracks will gradually increase, affecting the safety, practicality, and durability of the structure [8]. If reliable inspections are conducted on cracks, this can effectively prevent serious damage to buildings and prolong the life of facilities through appropriate maintenance [9–13]. Traditional inspection methods mainly rely on visual inspection and bridge inspection vehicles. Among them, the efficiency and accuracy of manual visual inspection [14–16] are greatly affected by the experience of the inspectors, and the human eye has many limitations, which can easily cause omissions; bridge inspection vehicles have many safety hazards during operation, and they are prone to high costs, slow efficiency, and traffic congestion. Therefore, to speed up the inspection process and achieve reliable and consistent inspections, deep-learning networks have developed rapidly in the field of structural health, and the crack recognition ability for complex situations continues to improve [17,18].

Deep learning [19,20] has greatly improved the latest technological level in fields such as visual object recognition and object detection, providing precise analysis for crack detection in structures. Semantic segmentation networks such as DeepLab [21], SegNet [22], and FCN [23] have also been widely used in crack recognition and detection. To overcome the limitations of human resources for visual inspections and provide accurate detection of multiple types of crack damage, Cha et al. [24] introduced a detection method based on faster convolutional neural networks. Researchers developed a database containing 2366 images and used it for modification, training, validation, and testing to develop multiple types of damage detection. Due to its fast speed and high accuracy, a video-based near real-time damage-detection framework based on trained networks was proposed. Li et al. [25] established a database of 2750 images of concrete structure cracks, spalling, weathering, etc., which was manually annotated. They tested and compared the fully convolutional network (FCN) architecture using this database and used the SegNet-based method to demonstrate that this method can accurately detect multiple concrete damage areas at the pixel level. Cardellicchio et al. [26] collected an existing image database of defects in reinforced concrete bridges, and domain experts classified the most common types of defects. Several convolutional neural network (CNN) algorithms were applied to the dataset for automatic identification of all defects. Zhang et al. [27] developed a context-aware deep convolutional semantic-segmentation method, leveraging local cross-state and cross-space constraints for image block fusion. Yamane et al. [28] proposed a deep learning-based semantic segmentation method that accurately detects concrete crack regions and removes other artifacts in photographs of concrete structures under adverse conditions. Lee et al. [29] proposed a crack-detection network and crack image generation algorithm based on image-segmentation networks. The training and validation results demonstrate that this method possesses high robustness and accuracy. Li et al. [30] proposed a semi-supervised method for road-crack detection that uses unlabeled road images for training and employs adversarial learning and fully convolutional discriminators to improve accuracy.

U-Net [31], as the most classic representative network of the U-shaped network structure, can extract the input image features. In addition, U-Net's accuracy is often higher than that of other models, and its structure is simple, mainly divided into three parts: feature extraction, clipping, and upsampling. It is widely used in industrial defect detection and has achieved good results in image segmentation [32,33]. Although U-Net has achieved high segmentation accuracy and speed, traditional convolutional and pooling layers generally suffer from information loss during information transmission, which is affected by the background environment, resulting in blurred boundaries of the segmented target area and a lot of noise. Based on the above shortcomings, we are focusing on increasing the attention of the network on small target features, specifically for crack-detection problems. We propose to embed an attention mechanism into the existing model to improve the ability to recognize cracks [34,35].

Attention mechanisms in deep learning [36,37] are very similar to human visual attention mechanisms, which select more important information for the current target and remove redundant information. This allows the network to adaptively focus on the necessary information and can be achieved by using importance weight vectors to approximate the final target value through weighted vector summation. Attention mechanisms mainly include the SE (Squeeze-and-Excitation) attention mechanism [38,39], the CBAM (Convolutional Block Attention Module) attention mechanism [40], the CA (Channel Attention) attention mechanism [41], etc. The introduction of attention mechanisms can improve crack-image detection accuracy with a small increase in computational cost. This effectively extracts multi-scale features of cracks while capturing local features and the edge details of small cracks. Attention mechanisms can focus on key areas and reconstruct semantics, significantly improving the crack-segmentation ability of the U-Net model [42–45].

In this article, research on the improvement of model-recognition performance through the addition of attention mechanisms was conducted. A modification has been made to the U-Net using an ECA (Efficient Channel Attention) mechanism, and a performance comparison has been conducted with the original network. The improvements were supported by the indicators and image testing. Large image-recognition experiments were conducted on actual structural cracks, and the recognition results were compared and evaluated.

2. Method of Attention-Based Structural Crack Detection

The attention mechanism is similar to our eyes as we use them to focus on the data we want to pay attention to. Similarly, the attention mechanism acts like the eyes of a deep-learning network, which can inform the network about the specific image features that we want to focus on and thus, enable more accurate acquisition of image information. This article focuses on the scientific problem of extracting the semantic segmentation of structural cracks. It mainly compares different deep-learning network models and improves recognition performance by adding attention mechanisms. The ECA attention mechanism proposed by Wang et al. [46] can achieve significant accuracy with a small number of parameters. This module is an efficient attention-channeling module, which can avoid feature loss caused by dimensionality reduction in other attention mechanisms and efficiently capture information interaction between different channels. In terms of its structural characteristics, the ECA attention mechanism is more suitable for network models with simpler structures such as U-Net due to its lightweight structure. The structure of ECA is shown in Figure 1:

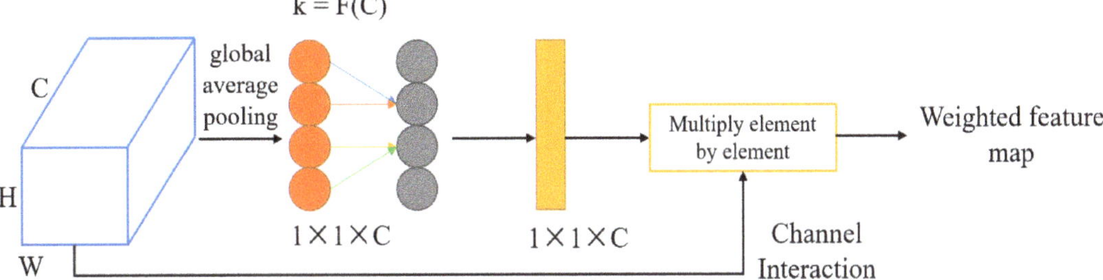

Figure 1. Diagram of the ECA structure.

The feature map is transformed from a matrix to a $1 \times 1 \times C$ vector through average pooling. The formula for the adaptive one-dimensional convolution kernel size k is shown in Formula (1). By adjusting the kernel size, the weight of each channel in the feature map is obtained. Then, the obtained weights are multiplied with each channel of the original input image to obtain the feature map with attention added.

$$k = |\log_2(C)/\gamma + b/\gamma|; \gamma = 2, b = 1 \quad (1)$$

Based on the research content of this article, the technical roadmap is shown in Figure 2 below. We trained three networks, lraspp, U-Net, and ECA-UNet, with existing public datasets. The performance of the three networks was analyzed based on the data obtained from the network training, and the effects before and after adding attention mechanisms were compared. Finally, actual cracks were used for image-segmentation and recognition–visualization comparison analysis in the real structure.

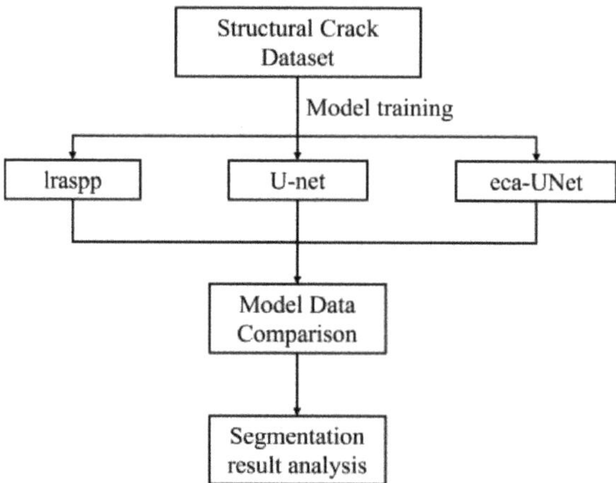

Figure 2. Technical roadmap.

3. Evaluation of Attention-Modified DNN

In this section, we conducted training and testing on three different deep-learning network models under the same conditions of batch size, learning rate, and iteration number. The study aimed to investigate the testing performance of different models based on their training results.

3.1. Evaluation Metrics

To verify the training effect of different models, we used precision, recall, and intersection over union (IOU) as evaluation metrics. Recall is the proportion of true positive samples in the model-predicted positive samples, usually indicating the model's recall performance, as shown in Formula (2); precision represents the proportion of true positive samples predicted by the model to be positive, as shown in Formula (3); IOU represents the degree of overlap between different class samples and labels, as shown in Formula (4).

$$\text{recall} = \text{TP}/(\text{TP} + \text{FP}) \tag{2}$$

$$\text{precision} = \text{FP}/(\text{TP} + \text{FP}) \tag{3}$$

$$\text{IOU} = \text{TP}/(\text{TP} + \text{FN} + \text{FP}) \tag{4}$$

In the formulas, TP denotes the number of true positive samples that were predicted correctly, and FP is the number of false positive samples that were predicted incorrectly. TN is the number of true negative samples that predicted correctly, and FN is the number of false negative samples that were predicted incorrectly.

3.2. Training and Analysis of the Original Model

This section first compares the performance of the U-Net network and the lraspp network. The training set used 5000 images with and without cracks from the bridge-crack library [47], with a crack-to-non-crack image ratio of 4:1. A validation set of 1000 crack images was used. The training dataset for the crack images included cracks in vertical, horizontal, and diagonal orientations. Part of the dataset is shown in Figure 3. Each crack image had a size of 256 × 256 pixels. After fine annotation with annotation tools, each crack image was paired with a PNG data label corresponding to the JPG format original image. All three models were iterated 200 times, and the training results are shown in Table 1. The lraspp model did not perform better overall than U-Net on the test set. In terms of precision, the value for lraspp was 0.810 and for U-Net was 0.921, indicating that lraspp was 0.111 less precise than U-Net. In terms of IOU, lraspp was 0.097 less than U-Net, and in terms of recall, lraspp was 0.051 less than U-Net, with lraspp having a value of 0.588 and U-Net having a value of 0.639.

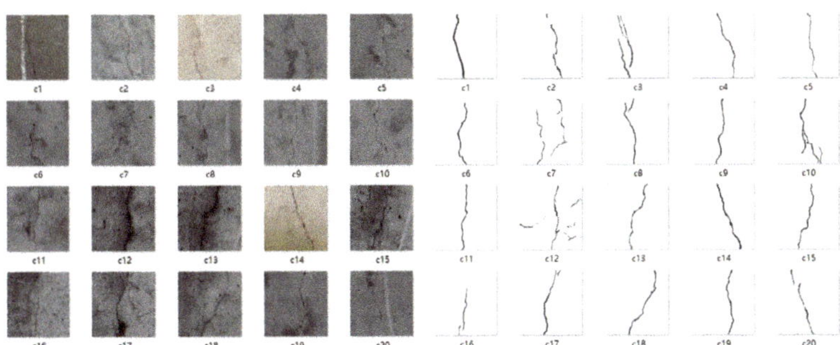

Figure 3. Example of the crack dataset.

Table 1. Model part of the training data.

Model Name	Precision	IOU	Recall
U-Net	0.921	0.676	0.639
lraspp	0.810	0.579	0.588

Based on the above data, U-Net performs better than lraspp in all three indicators. Therefore, we use the U-Net network with attention mechanism in the following text.

3.3. Improved U-Net Model Based on Attention Mechanism

The ECA-UNet after adding the attention mechanism is shown in Figure 4. On the original U-Net structure, we added the attention mechanism ECA to the sampling part of each layer. Because the downsampling part is the main feature-extraction network, adding the attention mechanism to the trunk-extraction part of the downsampling part will interfere with the weight of the original network on image–feature extraction; furthermore, it will cause the model to be unable to accurately judge and distribute the features. Therefore, we added the ECA attention mechanism to enhance the feature pick-up network; that is, by up-sampling each layer, a total of 4 points was added. The feature graph optimized by the attention mechanism was then fused with the five effective feature layers obtained by the backbone network, and finally the classification output was obtained through 1×1 convolution.

The training results are displayed in Figures 5–7 below, where Figures 5–7 show the precision, recall, and IOU curves of the three models after training. From the precision curve, it can be observed that lraspp oscillates around 65% without showing a significant upward trend. While ECA-UNet and U-Net overlap in the early stages, ECA-UNet has a tendency to oscillate downwards compared to U-Net after about 140 epochs, although both maintain around 80%. From the recall curve, it can be seen that lraspp performed well in terms of recall, but with more spikes in the curve than the other two models. A clear restrict relationship between recall and precision is noticeable, meaning that the model did not balance the relationship between these two indicators well. However, ECA-UNet performed better than U-Net after about 150 epochs, showing a gradual upward trend. From the IOU curve, it can be seen that lraspp performed poorly, not as well as the other two networks. The ECA-UNet curve overlaps with the U-Net curve to a high degree, and the upward trend is similar. Based on the above analysis, lraspp performed poorly in both metrics and did not balance the relative relationship between precision and recall well. On the other hand, ECA-UNet performed well in terms of recall and IOU, especially surpassing U-Net in terms of recall.

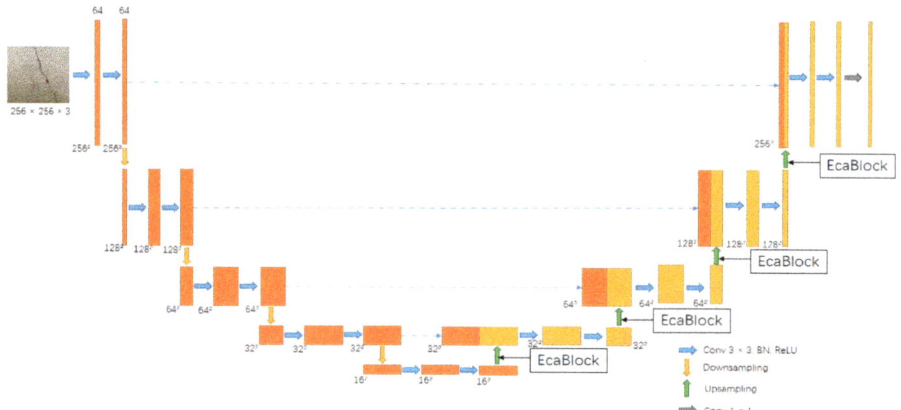

Figure 4. The schematic diagram of the improved ECA-UNet structure is shown as above.

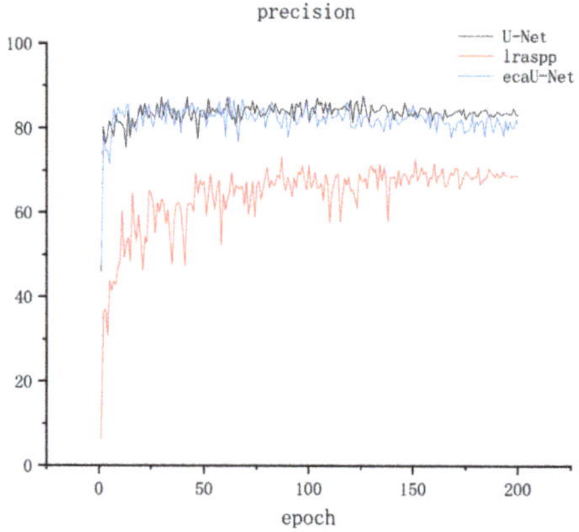

Figure 5. Precision metric result graph.

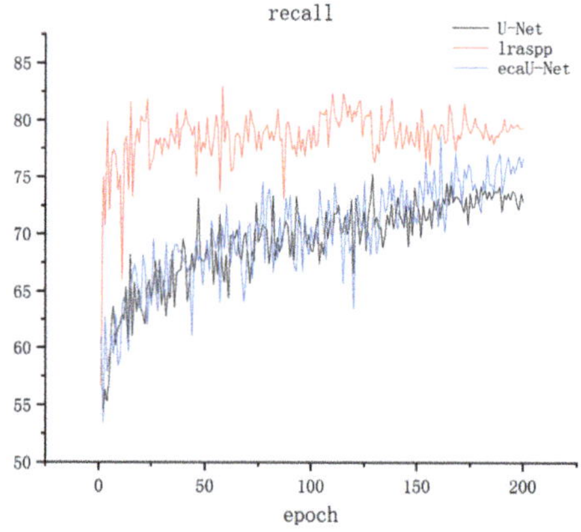

Figure 6. Recall metric result graph.

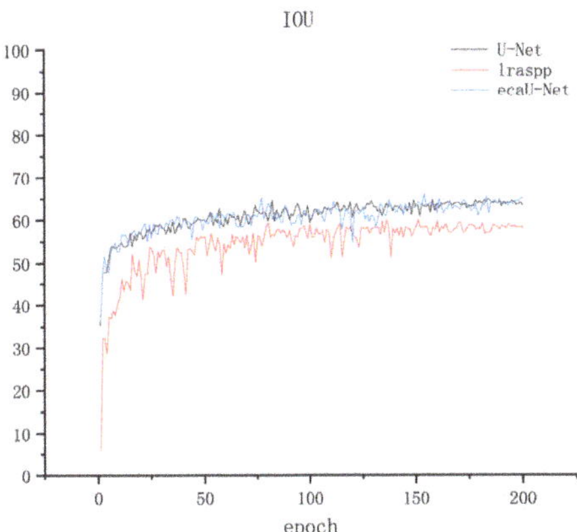

Figure 7. IOU metric result graph.

The improved ECA-UNet was trained with the same parameters as the above model, and the performance comparison with U-Net is shown in Table 2 as follows:

Table 2. U-Net and ECA-UNet result comparison.

Model Name	Precision	IOU	Recall
U-Net	0.921	0.676	0.639
ecaUNet	0.872	0.692	0.770

According to the table, the ECA-UNet scores 0.692 in the IOU metric, which is 0.016 higher than U-Net, and scores 0.770 in the recall metric, which is 0.131 higher than U-Net. The improved recall rate has been increased, improving the comprehensiveness of crack identification. The improved model shows a decrease in precision compared to the original model. This is because recall and precision have a constraining relationship, where an increase in one can lead to a decrease in the other. Therefore, we needed to balance these two indicators under existing conditions and ensure an improvement in the indicators while maintaining a relatively balanced state. Due to the large and diverse training dataset, as well as the unfamiliarity of the model with the features required by our needs during the process of learning-feature extraction, the model tended to have a higher error rate in recognition. On the other hand, the testing dataset had fewer images, and the model had already completed the learning process, gaining a better understanding of the desired features. Therefore, there could be cases where the metrics in the testing results are higher than the training output metrics. However, the training and testing data do not overlap, and the numerical relationship between the two does not have a significant correlation. When evaluating the performance of the model, it is insufficient to compare the testing results alone, as the training data do not affect the assessment of the model's recognition effectiveness. In terms of model running speed, under the same conditions, the testing time for a single image in both U-Net and ECA-U-Net is 0.058 s, while the testing time for lrassp is 0.062 s. This further highlights the advantages of ECA-U-Net, which has high accuracy and fast running speed.

4. Field Test of Raw Structural Crack Images

To more intuitively demonstrate the performance of each model in recognizing cracks, we used real-life structural cracks for crack recognition, which contained noise motifs other than cracks. The crack in Figure 8a is slightly inclined, with a physical width of 1.5 mm. The crack in Figure 8b is almost horizontal, with a physical width of 2.1 mm, and the crack in Figure 8c is vertical, with a physical width of 1.7 mm. Furthermore, the three cracks have disconnected points along the paths themselves, respectively. The raw structural images contain multiple noise motifs including water stains, spots, joint lines, concrete stripes, scratches, pits, etc., as marked with blue boxes. These images came from an on-site bridge structural inspection and were not contained in the bridge-crack library [47]. The results of recognition are shown in Figure 8.

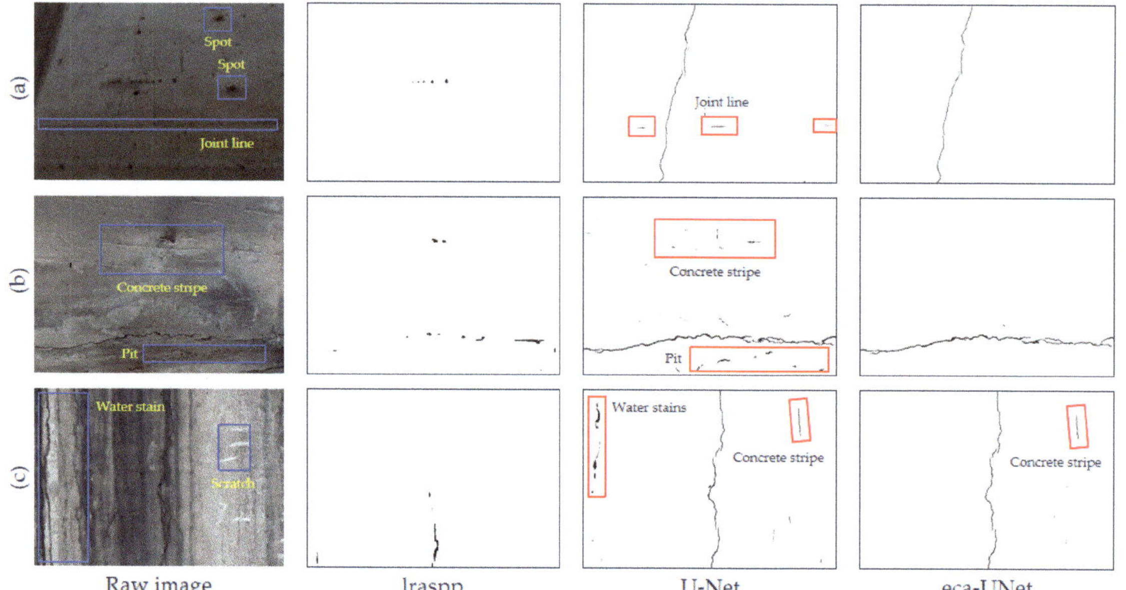

Figure 8. Recognition results of actual crack images by the model.

It can be seen that although the lraspp network was not cheated by all kinds of noise motifs in the raw images, it recognized very few parts of the crack regions. It missed the majority of the crack regions in Figure 8b,c, and it missed all the crack regions in Figure 8a. As for the U-Net, it almost successfully detected all the crack regions in all three of the testing images. Nonetheless, U-Net's recognition results contained a small number of non-crack noise motifs. Part of the joint line in Figure 8a was misidentified as cracks, and the concrete strips and the pitted area in Figure 8b were also misidentified as cracks. Moreover, part of the water stains and the concrete strip in Figure 8c were mistakenly recognized as cracks. This indicated that U-Net could achieve satisfactory performance in detecting the cracks but the robustness against noise motifs was not enough. When it came to the ECA-UNet, which was based on the U-Net and modified by adding the attention mechanism, a better performance was achieved. It accurately recognized the crack areas in the raw crack images. Yet, part of the concrete strip in Figure 8c still caused error detection.

Seen from the result, the recognition performance of lraspp on the feature of cracks is unsatisfactory, and the ability to capture features is weak. The U-Net network can successfully identify cracks, but it is still affected by some false crack noise interference with crack-like characteristics, such as the concrete strips and some of the water stains. The improved ECA-UNet based on the attention mechanism can accurately identify the crack area and is more robust against all kinds of noise motifs, although errors can still be detected.

5. Conclusions

This article investigated the performance of the ECA attention mechanism in improving the crack-detection capacity of the deep neural network. Three trained models were evaluated for their recognition performance on crack images of actual bridges. The following conclusions were drawn:

(i) Training of existing network public datasets: This article discusses training we conducted on two primary models, lraspp and U-Net, using a publicly available dataset of bridge cracks. The trained models were then tested for their generalization performance, and the results showed that the U-Net model performed better than the lraspp model in terms of data metrics. The precision, recall, and IOU values of the U-Net model were 0.111, 0.051, and 0.097 higher than those of the lraspp model, respectively.

(ii) Improvement of the U-Net network based on the ECA attention mechanism: U-Net performed well on the crack dataset, and based on this, an ECA attention mechanism was added to the upsampling part of the U-Net network to enhance the model's crack-detection performance. By keeping the original training parameters unchanged, the results of the training showed an increase of 0.131 in the recall rate and an improvement of 0.016 in the IOU compared to the original U-Net network, achieving improvements in both performance metrics.

(iii) Recognition of real structural cracks in raw images: In the recognition of actual structural crack images, it was observed that the lraspp network was almost insensitive to the crack feature and recognized hardly any cracks. Although the U-Net network was able to identify cracks, it also misjudged some false crack noise. The improved ECA-UNet network proposed in this paper showed better recognition performance than the other two networks and accurately identified cracks without minor mistakes.

(iv) This paper proposed a method to improve the crack-detection performance of the original U-Net model by integrating the ECA attention mechanism. Although the ECA-UNet achieved comparatively satisfactory results, more efforts are still required to improve the crack-detection performance. As for the improvement of detection performance, it can be seen from the testing results that the ECA-UNet can be cheated by noise motifs with linear geometry. Thus, the proposed network needs to be trained for robustness to exclude images with crack-like linear noise motifs. Furthermore, the network is quite large in terms of training parameters; therefore, how to reduce the size of the network and keep the crack detection performance is also an important aspect waiting for investigation. Moreover, attempts to embed the existing models into mobile devices for real-time crack identification are also pertinent to bringing this method into practical application.

Author Contributions: Conceptualization, T.J. and X.Y.; methodology, T.J. and H.Y.; validation, T.J., X.Y. and H.Y.; formal analysis, H.Y.; investigation, T.J. and X.Y.; resources, X.Y. and H.Y.; data curation, H.Y.; writing—original draft preparation, X.Y. and H.Y.; writing—review and editing, X.Y. and T.J.; visualization, X.Y.; supervision, T.J.; project administration, T.J. and X.Y.; funding acquisition, T.J. and X.Y. All authors have read and agreed to the published version of the manuscript.

Funding: The work described in this paper was jointly supported by the the China Postdoctoral Science Foundation (Grant No. 2022M712787) and the National Natural Science Foundation of China (Grant No. 52178306).

Institutional Review Board Statement: Not applicable.

Informed Consent Statement: Not applicable.

Data Availability Statement: The data that support the findings of this study are available from the corresponding author upon reasonable request.

Conflicts of Interest: The authors declare no conflict of interest.

References

1. Hamishebahar, Y.; Guan, H.; So, S.; Jo, J. A comprehensive review of deep learning-based crack detection approaches. *Appl. Sci.* **2022**, *12*, 1374. [CrossRef]
2. Munawar, H.S.; Hammad, A.W.A.; Haddad, A.; Soares, C.A.P.; Waller, S.T. Image-based crack detection methods: A review. *Infrastructures* **2021**, *6*, 115. [CrossRef]

3. Liu, Z.; Cao, Y.; Wang, Y.; Wang, W. Computer vision-based concrete crack detection using U-net fully convolutional networks. *Autom. Constr.* **2019**, *104*, 129–139. [CrossRef]
4. Cha, Y.J.; Choi, W.; Büyüköztürk, O. Deep learning-based crack damage detection using convolutional neural networks. *Comput.-Aided Civ. Infrastruct. Eng.* **2017**, *32*, 361–378. [CrossRef]
5. Azimi, M.; Eslamlou, A.D.; Pekcan, G. Data-driven structural health monitoring and damage detection through deep learning: State-of-the-art review. *Sensors* **2020**, *20*, 2778. [CrossRef]
6. Avci, O.; Abdeljaber, O.; Kiranyaz, S.; Hussein, M.; Gabbouj, M.; Inman, D.J. A review of vibration-based damage detection in civil structures: From traditional methods to Machine Learning and Deep Learning applications. *Mech. Syst. Signal Process.* **2021**, *147*, 107077. [CrossRef]
7. Malekloo, A.; Ozer, E.; AlHamaydeh, M.; Girolami, M. Machine learning and structural health monitoring overview with emerging technology and high-dimensional data source highlights. *Struct. Health Monit.* **2022**, *21*, 1906–1955. [CrossRef]
8. Su, C.; Wang, W. Concrete Cracks Detection Using Convolutional NeuralNetwork Based on Transfer Learning. *Math. Probl. Eng.* **2020**, *2020*, 7240129. [CrossRef]
9. Kim, J.J.; Kim, A.R.; Lee, S.W. Artificial neural network-based automated crack detection and analysis for the inspection of concrete structures. *Appl. Sci.* **2020**, *10*, 8105. [CrossRef]
10. Li, S.; Zhao, X. Image-based concrete crack detection using convolutional neural network and exhaustive search technique. *Adv. Civ. Eng.* **2019**, *2019*, 6520620. [CrossRef]
11. Chow, J.K.; Su, Z.; Wu, J.; Tan, P.; Mao, X.; Wang, Y. Anomaly detection of defects on concrete structures with the convolutional autoencoder. *Adv. Eng. Inform.* **2020**, *45*, 101105. [CrossRef]
12. Ye, X.W.; Jin, T.; Chen, P.Y. Structural crack detection using deep learning–based fully convolutional networks. *Adv. Struct. Eng.* **2019**, *22*, 3412–3419. [CrossRef]
13. Zhang, E.; Shao, L.; Wang, Y. Unifying transformer and convolution for dam crack detection. *Autom. Constr.* **2023**, *147*, 104712. [CrossRef]
14. Dorafshan, S.; Thomas, R.J.; Maguire, M. Comparison of deep convolutional neural networks and edge detectors for image-based crack detection in concrete. *Constr. Build. Mater.* **2018**, *186*, 1031–1045. [CrossRef]
15. Alipour, M.; Harris, D.K. Increasing the robustness of material-specific deep learning models for crack detection across different materials. *Eng. Struct.* **2020**, *206*, 110157. [CrossRef]
16. Fang, F.; Li, L.; Gu, Y.; Zhu, H.; Lim, J.-H. A novel hybrid approach for crack detection. *Pattern Recognit.* **2020**, *107*, 107474. [CrossRef]
17. Ai, D.; Jiang, G.; Kei, L.S.; Li, C. Automatic pixel-level pavement crack detection using information of multi-scale neighborhoods. *IEEE Access* **2018**, *6*, 24452–24463. [CrossRef]
18. Feng, C.; Zhang, H.; Wang, H.; Wang, S.; Li, Y. Automatic pixel-level crack detection on dam surface using deep convolutional network. *Sensors* **2020**, *20*, 2069. [CrossRef]
19. LeCun, Y.; Bengio, Y.; Hinton, G. Deep learning. *Nature* **2015**, *521*, 436–444. [CrossRef]
20. Ali, R.; Chuah, J.H.; Talip, M.S.A.; Mokhtar, N.; Shoaib, M.A. Structural crack detection using deep convolutional neural networks. *Autom. Constr.* **2022**, *133*, 103989. [CrossRef]
21. Chen, L.C.; Papandreou, G.; Kokkinos, I.; Murphy, K.; Yuille, A.L. Deeplab: Semantic image segmentation with deep convolutional nets, atrous convolution, and fully connected crfs. *IEEE Trans. Pattern Anal. Mach. Intell.* **2017**, *40*, 834–848. [CrossRef]
22. Badrinarayanan, V.; Kendall, A.; Cipolla, R. Segnet: A deep convolutional encoder-decoder architecture for image segmentation. *IEEE Trans. Pattern Anal. Mach. Intell.* **2017**, *39*, 2481–2495. [CrossRef] [PubMed]
23. Long, J.; Shelhamer, E.; Darrell, T. Fully convolutional networks for semantic segmentation. In Proceedings of the IEEE Conference on Computer Vision and Pattern Recognition, Boston, MA, USA, 7–12 June 2015; pp. 3431–3440.
24. Cha, Y.J.; Choi, W.; Suh, G.; Mahmoudkhani, S.; Büyüköztürk, O. Autonomous structural visual inspection using region-based deep learning for detecting multiple damage types. *Comput.-Aided Civ. Infrastruct. Eng.* **2018**, *33*, 731–747. [CrossRef]
25. Li, S.; Zhao, X.; Zhou, G. Automatic pixel-level multiple damage detection of concrete structure using fully convolutional network. *Comput.-Aided Civ. Infrastruct. Eng.* **2019**, *34*, 616–634. [CrossRef]
26. Cardellicchio, A.; Ruggieri, S.; Nettis, A.; Renò, V.; Uva, G. Physical interpretation of machine learning-based recognition of defects for the risk management of existing bridge heritage. *Eng. Fail. Anal.* **2023**, *149*, 107237. [CrossRef]
27. Zhang, X.; Rajan, D.; Story, B. Concrete crack detection using context-aware deep semantic segmentation network. *Comput.-Aided Civ. Infrastruct. Eng.* **2019**, *34*, 951–971. [CrossRef]
28. Yamane, T.; Chun, P. Crack detection from a concrete surface image based on semantic segmentation using deep learning. *J. Adv. Concr. Technol.* **2020**, *18*, 493–504. [CrossRef]
29. Lee, D.; Kim, J.; Lee, D. Robust concrete crack detection using deep learning-based semantic segmentation. *Int. J. Aeronaut. Space Sci.* **2019**, *20*, 287–299. [CrossRef]
30. Li, G.; Wan, J.; He, S.; Liu, Q.; Ma, B. Semi-supervised semantic segmentation using adversarial learning for pavement crack detection. *IEEE Access* **2020**, *8*, 51446–51459. [CrossRef]
31. Siddique, N.; Paheding, S.; Elkin, C.P.; Devabhaktuni, V. U-net and its variants for medical image segmentation: A review of theory and applications. *IEEE Access* **2021**, *9*, 82031–82057. [CrossRef]

32. Song, W.; Zheng, N.; Liu, X.; Qiu, L.; Zheng, R. An improved u-net convolutional networks for seabed mineral image segmentation. *IEEE Access* **2019**, *7*, 82744–82752. [CrossRef]
33. Zunair, H.; Hamza, A.B. Sharp U-Net: Depthwise convolutional network for biomedical image segmentation. *Comput. Biol. Med.* **2021**, *136*, 104699. [CrossRef] [PubMed]
34. Han, G.; Zhang, M.; Wu, W.; He, M.; Liu, K.; Qin, L.; Liu, X. Improved U-Net based insulator image segmentation method based on attention mechanism. *Energy Rep.* **2021**, *7*, 210–217. [CrossRef]
35. Wang, H.; Miao, F. Building extraction from remote sensing images using deep residual U-Net. *Eur. J. Remote Sens.* **2022**, *55*, 71–85. [CrossRef]
36. Guo, M.-H.; Xu, T.-X.; Liu, J.-J.; Liu, Z.-N.; Jiang, P.-T.; Mu, T.-J.; Zhang, S.-H.; Martin, R.R.; Cheng, M.-M.; Hu, S.-M. Attention mechanisms in computer vision: A survey. *Comput. Vis. Media* **2022**, *8*, 331–368. [CrossRef]
37. Li, C.; Fu, L.; Zhu, Q.; Zhu, J.; Fang, Z.; Xie, Y.; Guo, Y.; Gong, Y. Attention enhanced u-net for building extraction from farmland based on google and worldview-2 remote sensing images. *Remote Sens.* **2021**, *13*, 4411. [CrossRef]
38. Roy, A.G.; Navab, N.; Wachinger, C. Recalibrating fully convolutional networks with spatial and channel "squeeze and excitation" blocks. *IEEE Trans. Med. Imaging* **2018**, *38*, 540–549. [CrossRef]
39. Wang, L.; Peng, J.; Sun, W. Spatial–spectral squeeze-and-excitation residual network for hyperspectral image classification. *Remote Sens.* **2019**, *11*, 884. [CrossRef]
40. Chen, B.; Zhang, Z.; Liu, N.; Tan, Y.; Liu, X.; Chen, T. Spatiotemporal convolutional neural network with convolutional block attention module for micro-expression recognition. *Information* **2020**, *11*, 380. [CrossRef]
41. Li, H.; Qiu, K.; Chen, L.; Mei, X.; Hong, L.; Tao, C. SCAttNet: Semantic segmentation network with spatial and channel attention mechanism for high-resolution remote sensing images. *IEEE Geosci. Remote Sens. Lett.* **2020**, *18*, 905–909. [CrossRef]
42. Xu, G.; Han, X.; Zhang, Y.; Wu, C. Dam crack image detection model on feature enhancement and attention mechanism. *Water* **2022**, *15*, 64. [CrossRef]
43. Cui, X.; Wang, Q.; Dai, J.; Xue, Y.; Duan, Y. Intelligent crack detection based on attention mechanism in convolution neural network. *Adv. Struct. Eng.* **2021**, *24*, 1859–1868. [CrossRef]
44. Ren, J.; Zhao, G.; Ma, Y.; Zhao, D.; Liu, T.; Yan, J. Automatic Pavement Crack Detection Fusing Attention Mechanism. *Electronics* **2022**, *11*, 3622. [CrossRef]
45. Chu, H.; Wang, W.; Deng, L. Tiny-Crack-Net: A multiscale feature fusion network with attention mechanisms for segmentation of tiny cracks. *Comput.-Aided Civ. Infrastruct. Eng.* **2022**, *37*, 1914–1931. [CrossRef]
46. Liu, T.; Luo, R.; Xu, L.; Feng, D.; Cao, L.; Liu, S.; Guo, J. Spatial Channel Attention for Deep Convolutional Neural Networks. *Mathematics* **2022**, *10*, 1750. [CrossRef]
47. Ye, X.W.; Jin, T.; Li, Z.X.; Ma, S.Y.; Ding, Y.; Ou, Y.H. Structural crack detection from benchmark data sets using pruned fully convolutional networks. *J. Struct. Eng.* **2021**, *147*, 04721008. [CrossRef]

 applied sciences

MDPI

Article

Triangular Position Multi-Bolt Layout Structure Optimization

Xiaohan Lu [1], Min Zhu [1,*], Yilong Liu [1], Shengao Wang [1], Zijian Xu [1] and Shengnan Li [2]

[1] College of Nuclear Science and Technology, Naval University of Engineering, Wuhan 430030, China
[2] College of Power Engineering, Naval University of Engineering, Wuhan 430030, China
* Correspondence: min0zhu@163.com; Tel.: +86-13505358526

Abstract: Stress concentration often occurs around bolt holes in load-bearing joint structures of large complex equipment, ships, aerospace and other complex machinery fields, which is an important mechanical factor leading to the failure of joint structures. It is of great engineering significance to study the phenomenon of stress concentration on connected structures for the safety of large and complex equipment; meanwhile, the layout of bolts seriously affects the stress around holes. Many scholars have studied the layout optimization of multi-bolted structures through experiments and simulations, but few algorithms have been applied to the layout optimization of bolted structures. And most of the studied types of multi-bolt structures are symmetrical. Therefore, in this paper, the gray wolf algorithm is used to optimize the layout of nickel steel plate connectors with a bolt layout in triangular position, and the optimal objective function is found based on the hole circumferential stress of the nickel steel plate, maximum shear stress of the bolt and bending stress of the nickel steel plate. Comparing the optimal values obtained by the fruit fly optimization algorithm, particle swarm optimization algorithm, gray wolf optimization algorithm, multiverse optimization algorithm and wind driven optimization algorithm, the accuracy of selecting the gray wolf algorithm for optimization is verified. A multi-bolt connection structure model was established in ABAQUS, and the surface stress before and after optimization was compared to verify the correctness of the gray wolf algorithm applied to the structure layout optimization of the nickel steel flat bolt connection. The results show that under the force of 15 KN, compared with the original bolt structure layout, the optimized upper side nickel steel plate bore peripheral stress is reduced by 73.1 MPa, and the optimization rate is 24%; bolt stress is reduced by 47.7 MPa, and the optimization rate is 12.5%; when the load is less than 18 KN, the optimization effect of both the upper nickel steel plate and bolt group is more than 10%. When the load is greater than 18 KN, the optimization effect is reduced, and when the load is greater than 21 KN, the nickel steel plate has exceeded the yield limit. Due to the existence of fixed constraints, the optimization of the lower nickel steel plate is not obvious. The results of this study can provide data and theoretical support for the layout optimization of the nickel steel flat bolt connection structure, and help to improve reliability analysis and health monitoring in complex assembly fields such as large complex equipment and aerospace.

Keywords: bolted connection; layout optimization; algorithm; ABAQUS

Citation: Lu, X.; Zhu, M.; Liu, Y.; Wang, S.; Xu, Z.; Li, S. Triangular Position Multi-Bolt Layout Structure Optimization. *Appl. Sci.* **2023**, *13*, 8786. https://doi.org/10.3390/app13158786

Academic Editors: Phong B Dao, Lei Qiu, Liang Yu, Tadeusz Uhl and Minh-Quy Le

Received: 10 June 2023
Revised: 18 July 2023
Accepted: 24 July 2023
Published: 29 July 2023

1. Introduction

A bolted connection is an important connection method in large complex equipment, aerospace and other mechanical structures [1–3]. The service life of the bolt and normal operation of the mechanical structure are directly affected by the change in the preload force of the connection mode. The stress around the hole of the bolted connection is the key point that affects the construction life. In order to make the structure more stable and reduce the stress on the bearing structure, it is necessary to optimize the layout of the multi-bolt connection structure to achieve a better connection effect. Many scholars have carried out a lot of research on the optimization of bolt layout.

The stress distribution of the bolt-connected structure is affected by the thickness of the cover plate, aperture, nut diameter and other factors. When the value of the aperture and nut diameter is small, the stability of the bolt-connected structure can be improved [4]. Liu et al. [5] tested the mechanical properties of asymmetrical bolts and came to the conclusion that the performance of the copper gasket is stable and bearing capacity is higher than that of an aluminum gasket. Moreover, under the action of a copper gasket, the coefficient of friction imbalance is between 0.919 and 1.050, which can show good slip and energy dissipation under earthquake conditions.

It takes time and effort to study the effect of bolt layout on the mechanical properties of the connected structure by experiment, so many researchers optimize and analyze the performance of the bolted structure by the simulation method. Wu et al. [6] used a parametric model to adjust the layout of the bolt connection plate by changing the radius and angle of the bolt connection, and then analyzed its influence on the modal frequency. The results show that there are some order mode frequencies that affect the bolt layout, and these frequencies should be avoided. Pedersen [7] optimized the shape of the starting point and root of the thread, and also optimized the bolt rod and thread, which reduced the stress by 34%, reduced the stiffness of the bolt, increased the stiffness of the bolt connection and greatly extended the fatigue life.

In order to simulate the bolted structure more accurately, more and more scholars apply artificial intelligence to the simulation. Shen et al. [8] improved the genetic algorithm, applied the improved objective function to the connection structure, took the average acceleration at the corner of the component as the objective function of the algorithm, adopted the idea of parametric modeling, realized the interaction between MATLAB and NASTRAN, and reduced the stress value at the corner of the component. The layout of the bolt on the edge of the component is optimized and the safety of the component is improved. Han et al. [9] used the gray wolf algorithm to optimize the motion trajectory of each joint of the robotic arm, which takes time efficiency and smoothness as the objective function, and the results show that the gray wolf algorithm can effectively perform trajectory optimization. Liang et al. [10] improved the thresholds and weights of an Elman neural network by particle swarm algorithm and applied the method to clock difference forecasting, which can improve the stability and forecasting accuracy of a satellite clock. Wang et al. [11] carried out layout optimization experiments on bolt sets based on the firefly algorithm, and carried out simulation analysis on double-row bolts and three-row bolts, and the optimized bolt layout reduced the perimeter stress of the holes and improved the safety of the structure. However, the optimization of multi-bolt joint structures in the existing literature is limited to symmetric bolt structures [12–16], while there are few studies on asymmetric bolt structures, especially the optimization of a multi-bolt layout in a triangular position. In addition, researchers rarely apply the gray wolf optimization algorithm to the layout optimization of the connection structure of large complex equipment. Therefore, based on the gray wolf algorithm, this paper optimizes the bolt group of the triangular multi-bolt layout structure, seeks the optimal bolt margin and spacing, and reduces the surface stress of the connection structure and maximum stress on the surface of the bolt group; to prevent excessive stress around the bolt hole, resulting in fatigue damage of the bolt hole or plastic deformation of the connector, thus affecting the stability and reliability of the connection. Nickel steel has the advantages of strong compressive resistance, corrosion resistance, low temperature toughness and so on, and is widely used in aerospace and other large complex important equipment, and there is little research on nickel steel. Therefore, the connection structure of this paper chooses nickel steel.

Based on the above background, this paper analyzes the hole surrounding stress, maximum shear stress of the bolt and bending stress of the nickel steel plate; and adopts the gray wolf algorithm to optimize the bolt spacing and bolt diameter, which can accurately find the optimal bolt spacing and bolt radius. The results of the fruit fly optimization algorithm [17], particle swarm optimization algorithm [18], gray wolf optimization algorithm [19], multiverse optimization algorithm [20] and wind driven optimization algorithm [21] for triangular multi-bolt layout structure optimization were compared, and the accuracy of the gray wolf optimization algorithm was verified. The optimized bolts still met the requirements of the design specification, and the efficiency of the joint design is improved. ABAQUS-2022 software was used to establish the finite element model of the optimal solution, and the stress conditions of the nickel steel plate and bolts before and after optimization were compared to verify the correctness of the algorithm, objective function and constraint conditions. Based on the gray wolf algorithm, this paper selects the optimal triangular multi-bolt layout structure and corresponding bolt diameter, and reduces the hole stress and maximum stress on the surface of the nickel steel plate, which provides a new idea for the layout optimization of the asymmetric bolt-connected structure, improves the strength and stiffness of the structure, and extends the service life and reliability of the bolt-connected structure. When engineering design bolts and connecting structures, the optimal solution of the gray wolf algorithm can select a reasonable range of bolt diameters and bolt spacing, so that the stress around bolt holes is within a reasonable range, avoiding excessive use of materials and reducing material costs. It is helpful to improve the safety and reliability of load-bearing connection structures in complex machinery fields such as large complex equipment, ship sailing and aerospace.

2. Materials and Methods

2.1. Principles of the Grey Wolf Algorithm

2.1.1. Algorithm Introduction

The gray wolf algorithm is designed based on the hunting system and leadership level of the gray wolf [22]. The pyramid level of the algorithm is divided into four layers. The top gray wolf is responsible for decision-making events such as hunting and distribution of goods, called α. The second layer is the top gray wolf's think tank team, called β. It is mainly responsible for assisting in decision-making, and when the top position is vacant, the second layer of workers will take over the top position. The third layer of workers, called δ, listens to the transfer orders of the first two layers and is mainly responsible for tasks such as nursing and sentinel. The bottom worker, known as ω, mainly works to balance internal relationships.

The gray wolf algorithm has fast convergence speed, has high global search ability, can be applied in different problem domains, does not require complex computer models, and can automatically adapt to the characteristics of the problem and change of search space. To sum up, the gray wolf algorithm is selected in this paper to optimize the layout of the multi-bolt connection structure.

2.1.2. Round up Prey

The prey catching behavior of the gray wolf group is defined as [19]:

$$\vec{D} = \left|\vec{C}\vec{X}_p(t) - \vec{X}(t)\right| \tag{1}$$

$$\vec{X}(t+1) = \vec{X}_p(t) - \vec{A}\vec{D} \tag{2}$$

Formula (1) is the distance between the gray wolf and its prey, and Formula (2) is the formula for updating the position of the gray wolf. Where, t represents the current number of iterations, $\vec{A}$ and $\vec{C}$ are the position vector of the prey and the position vector of the gray wolf, respectively. $\vec{X}_p$ and $\vec{X}$ are the coefficient vector. The formula for calculating $\vec{A}$ and $\vec{C}$ are as follows [19]:

$$\vec{A} = 2\vec{a}\vec{r}_1 - \vec{a} \tag{3}$$

$$\vec{C} = 2\vec{r}_2 \tag{4}$$

where, $\vec{a}$ is the convergence factor, *r*ß1 and *r*ß2 are the random number between the interval [0–1].

2.1.3. Hunt

Grey wolves can identify the specific location of prey by themselves [23], and under the leadership of the gray wolves at the top of the pyramid, gradually surround the prey, so that each gray wolf can reach the best position. Therefore, in order to simulate the behavior of the gray wolf, we selected three optimal positions around the prey, used these three positions to judge the specific position of the prey, and forced other gray wolves to update their own positions according to the position of the optimal gray wolf. By closing in on the prey, they round it up. The principle of gray wolf individual location renewal is shown in Figure 1.

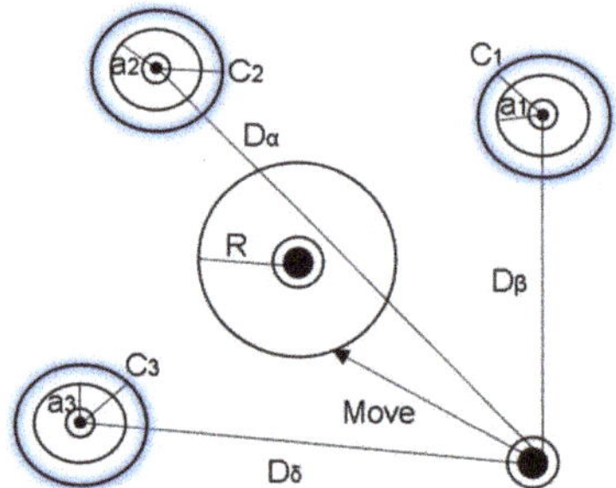

Figure 1. Schematic diagram of individual renewal position of a gray wolf.

Its mathematical model is described as follows [19]:

$$\begin{cases} \vec{D}_\alpha = \left|\vec{C}_1\vec{X}_\alpha - \vec{X}\right| \\ \vec{D}_\beta = \left|\vec{C}_2\vec{X}_\beta - \vec{X}\right| \\ \vec{D}_\delta = \left|\vec{C}_3\vec{X}_\delta - \vec{X}\right| \end{cases} \tag{5}$$

$\vec{D}_\alpha$,$\vec{D}_\beta$,$\vec{D}_\delta$ are the distance vector between α, β, δ and other gray wolves, $\vec{X}_\alpha$, $\vec{X}_\beta$, $\vec{X}_\delta$ are the vector of the former position of α, β, δ, $\vec{C}_1$, $\vec{C}_2$, $\vec{C}_3$ are the random vector, $\vec{X}$ represents the current position vector.

$$\begin{cases} \vec{X}_1 = \vec{X}_\alpha - A_1\vec{D}_\alpha \\ \vec{X}_2 = \vec{X}_\beta - A_2\vec{D}_\beta \\ \vec{X}_3 = \vec{X}_\delta - A_3\vec{D}_\delta \end{cases} \tag{6}$$

Formula (6) shows the direction and step length of an individual gray wolf ω moving toward α, β, δ respectively.

$$\vec{X}(t+1) = \frac{\vec{X}_1 + \vec{X}_2 + \vec{X}_3}{3} \tag{7}$$

Formula (7) is the optimal position of the bottom gray wolf ω.

2.1.4. Attack Prey

To simulate the movement of gray wolves as they approach prey, as $\vec{a}$ decreases, the range of $\vec{A}$ decreases, the value interval of $\vec{A}$ is $[-\alpha, \alpha]$. When $\vec{A}$ is located in this interval, the gray wolf can switch between its own position and the position of prey at will. When $|\vec{A}| < 1$, the gray wolf pack attacks and captures its prey.

2.2. Construct Objective Function

2.2.1. Bolt Mechanical Model

Apply a pull in the positive direction of X and a pull in the negative direction of Y at point K, both of which are 15 KN. Among them, the bolts are arranged in the position of an equilateral triangle, whose side length is a and centroid is O. The structural diagram is shown in Figure 2.

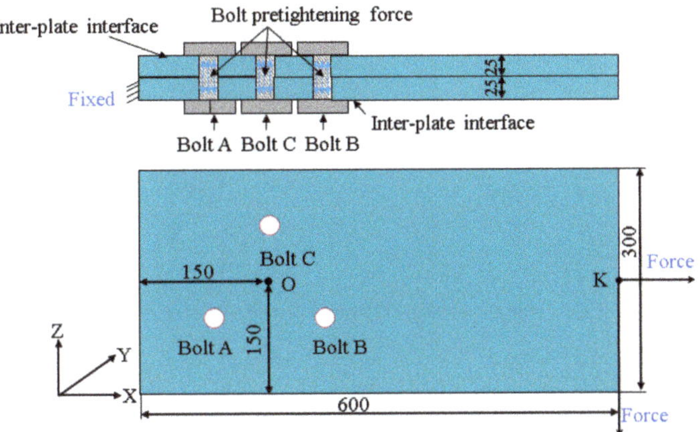

Figure 2. Schematic diagram of bolt structure.

By simplifying the force at point K to point O, the torque at point O is formed. Carrying out force analysis on bolts *A*, *B* and *C*:

$$F_A = \left(5 - \frac{2250\sqrt{3} \times \sin 30^{\circ}}{a}\right)\vec{i} + \left(-5 - \frac{2250\sqrt{3} \times \cos 30^{\circ}}{a}\right)\vec{j} \tag{8}$$

$$F_B = \left(5 - \frac{2250\sqrt{3} \times \cos 30^{\circ}}{a}\right)\vec{i} + \left(-5 + \frac{2250\sqrt{3} \times \sin 30^{\circ}}{a}\right)\vec{j} \tag{9}$$

$$F_C = \left(5 + \frac{2250\sqrt{3}}{a}\right)\vec{i} + (-5)\vec{j} \tag{10}$$

According to Equations (8)–(10), from F_A, F_B and F_C, it can be seen that the maximum force is bolt C:

$$F_{max} = F_C \tag{11}$$

At this time, the maximum shear stress on the bolt is:

$$\tau_{max} = \frac{F_{max}}{\pi d^2/4} \tag{12}$$

It can be seen from the analysis that bolt B receives the maximum torque. Bolt B is cut along the *Y*-axis to obtain the profile as shown in Figure 3a and the torque diagram of the plate as shown in Figure 3b.

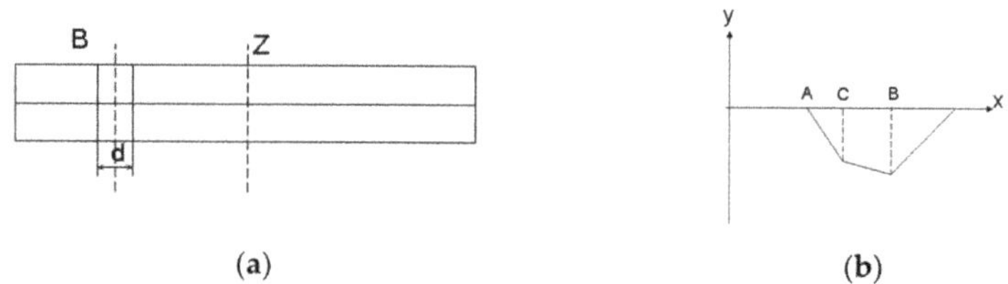

Figure 3. Bolt profile and torque diagram. (**a**) Profile of the nickel steel plate; (**b**) Torque diagram of nickel steel plate.

The torque and moment of inertia at bolt B are:

$$M = F_B\left[300 + (150 - \frac{a}{2})\right] \tag{13}$$

$$I = I_{max} - \left[I_{hole} + I_{offset}\right] = \frac{10 \times 300^3}{12} - \left[\frac{10d^3}{12} + \left(\frac{a}{2\sqrt{3}}\right)^2 \times 25\text{d}\right] \tag{14}$$

Then the critical bending stress of the nickel steel plate is:

$$\sigma_{max} = \frac{M}{I} \tag{15}$$

Formulas (11), (12) and (15) can be used to calculate the maximum stress on bolts, maximum shear stress and critical bending stress of the nickel steel plate, which lays a prerequisite for the selection of the objective function in the following algorithm.

2.2.2. Objective Function

In this paper, the simulation software ABAQUS was used to simulate the working condition of bolts [24], and the maximum stress, maximum shear stress and critical bending stress of bolts were considered at the same time [25]. In order to fit the three objective functions into one objective function, their superposition was processed:

$$Fitness = Fmax + \frac{4 \times Fmax}{\pi {x_2}^2} + \frac{6 \times F_B \times (450 - \frac{x_1}{2})}{1.35 \times 10^8 - (5{x_2}^3 + 45{x_1}^2 x_2)} + k(x_1, x_2) \tag{16}$$

where, x_1 is the distance variable a between bolts; x_2 is the diameter d of the bolt; $k(x_1, x_2)$ is the fitting factor added when fitting $F_{max}, \tau_{max}, \sigma_{max}$.

2.3. Selection Constraint

Before the algorithm is solved, the population number pop is set to 50, variable dimension dim is 2, maximum number of iterations maxIter is 500, upper boundary velocity is [2, 2] and lower boundary velocity is [−2, −2]. For M8-type bolts, the distance between the bolts and nickel steel plate, and distance between the bolts cannot be too small; according to the steel design code [26], the distance between its bolts is not less than 3d and the end distance of the bolts is not less than 1.5d. Also considering the diameter of small bolts, the variation range of the bolt diameter d is [1.4~36]. Therefore, the range of spacing a and bolt diameter d selected in this paper is:

2.8 mm < a < 142.5 mm

1.4 mm < d <3 6 mm

2.4. Finite Element Modeling and Boundary Conditions

The bolt specifications in this paper refer to GB/T191-197-2003 [27], and M8 bolts with a pitch of 1.25 mm and a screw length of 50 mm are selected. Nickel steel has important properties such as formability, weldability and ductility, and has good corrosion resistance. It is an important material, and widely used in large equipment, aerospace and other complex assemblies in the field of connection structures. At present, there is little research in the field of nickel steel plates, so the bolt role of the plate in this paper is selected from nickel steel material. The modulus of elasticity of both the bolt and nickel steel plate is 210,000 MPa, the Poisson's ratio is 0.283 and the density is 7999 $kg{\cdot}m^{-3}$; the maximum stress limit of the bolt is 640 MPa and maximum stress limit of the nickel steel plate is 340 MPa.

To verify the accuracy of the objective function as well as the algorithm, parametric modeling of the bolt and nickel steel flat plate was carried out based on ABAQUS with Python-3.90 software. Due to the difficulty of meshing the threads in ABAQUS and non-convergence of the calculations, the paper uses a simplified modeling of the bolts without threads [28] and makes the multi-bolt structure act on the bearing surface by applying a preload [29]. The load-bearing structure is two thin nickel steel plates, both of which are elastoplastic flat plates. The two fast nickel steel flat plates were superimposed to simulate the connecting action of complex equipment, as shown in Figure 4. The nickel steel plate is meshed using a hexahedral structured mesh with a cell type of C3D8R cells. In order to make the simulation easier to converge and more precise, the area around the bolt holes and nut of the nickel steel plate is divided to obtain a more precise mesh, as shown in Figure 5 below.

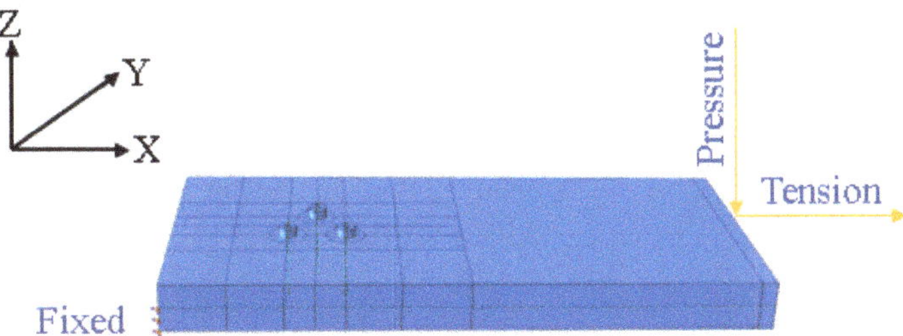

Figure 4. Bolted connection model.

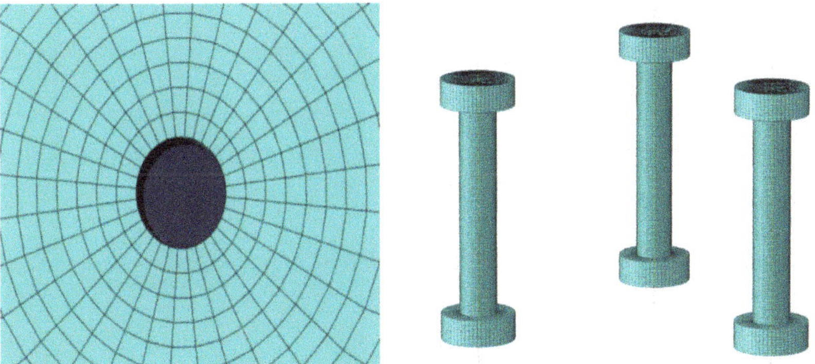

Figure 5. Stress of multiple bolts.

In this paper, a high quality mesh with 45,430 meshes is divided and partitioned at different locations to ensure the accuracy of the calculation results. At the same time, the finite element analysis method has been widely used in the simulation analysis of the model, and the reliability of its analysis results can be effectively guaranteed under the condition of ensuring high quality meshes.

In the solution stage, this paper sets four analysis steps, the first and second steps in the bolt's middle surface, gradually add the bolt load, that is, the preload. In the third and fourth steps, the current length is fixed in order to meet the actual engineering requirements, so that the bolt load changes with the deformation of the structure.

The contact properties are set to finite slip and the junction-surface discretization method is used. For contact, the harder nickel steel plate is used as the master surface and the bolt surface as the slave surface, defaulting to contact when the distance between the two is less than 0.03. The tangential contact is set with a friction coefficient of 0.3, and the normal contact is a hard contact to prevent penetration.

If a force of 15 KN is applied in the X and Y directions, the bolt will be severely deformed, thus affecting the simulation results, and the simulation mainly compares the stresses on the nickel steel plate and bolt before and after optimization, independent of the force direction of the plate. Therefore, in Step 3, the tension and pressure are applied to the upper nickel plate in the positive X-axis and negative Z-axis directions. The leftmost side of the lower nickel steel board applies a fixed restraint.

3. Results

3.1. Algorithm Solution

Based on the MATLAB(2020) software for the multi-bolt connection structure of the spacing a and diameter d of the bolt to find the best calculation, from Figure 6, it can be seen that the objective function and constraints are fully applicable to the gray wolf algorithm program, the iterative effect is good, and finally stable convergence to a fixed value is observed. Its convergence to a stable fitness value is not used as a criterion for judging bolts and nickel steel plates. As can be seen from Figure 7, through different iterations, the particles gradually converge to the best particle, whose best particle is the result sought. The algorithm results in a pitch a = 108.5629 mm and a diameter d = 9.0233 mm. The results optimized by firefly algorithm for the spacing of double-row bolts and triple-row bolts in previous studies [11] are not much different from the optimized bolt spacing in this paper, and are also consistent with the results optimized by Xiao et al. [30] for the layout optimization of bolt groups for flameproof boxes used in mines.

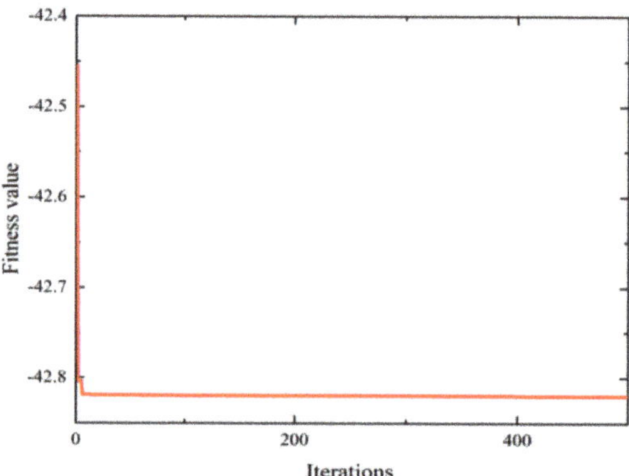

Figure 6. Results of algorithm operation.

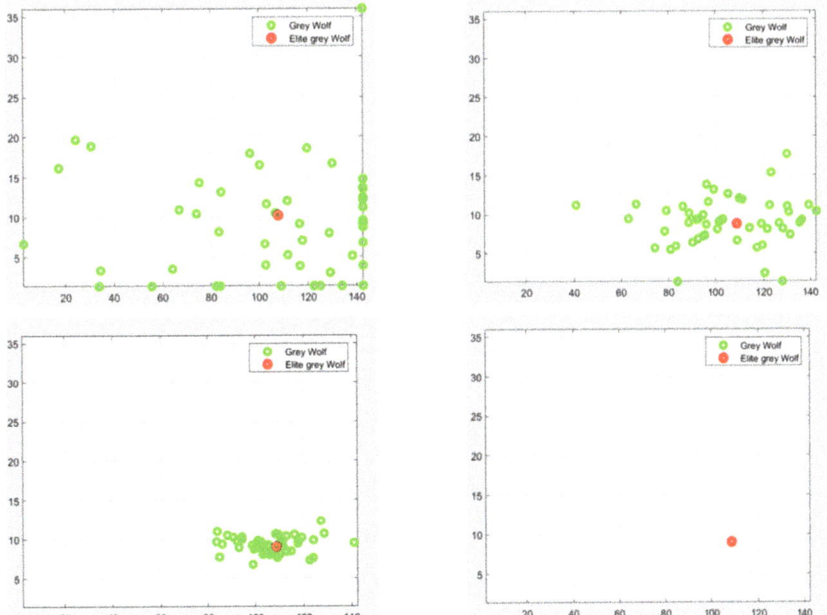

Figure 7. Different iteration times of the algorithm.

In this paper, the objective function as well as the boundary conditions are applied to the fruit fly optimization algorithm, particle swarm optimization algorithm, gray wolf optimization algorithm, multiverse optimization algorithm and wind driven optimization algorithm, the results of which are shown in Figure 8, with approximately the same algorithmic optimization curves. There is only a slight difference at the initial iteration, which is explained by the fact that the different algorithms contain different principles. As shown in Table 1, the final fitness values were all stable at −42.8195, with the gray wolf algorithm having the highest number of iterations, but none of them exceeded 500 iterations.

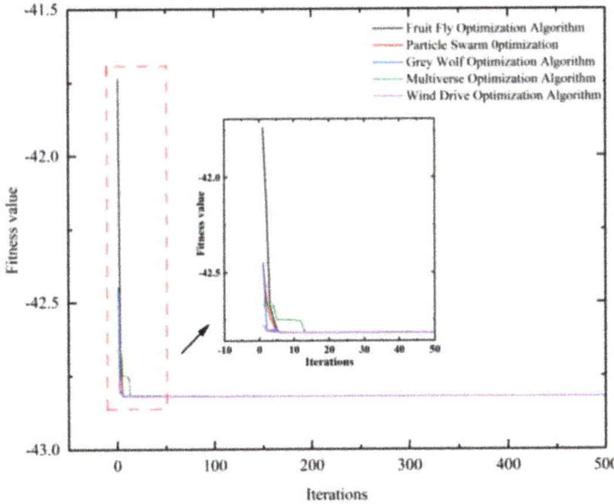

Figure 8. Optimization effects of different algorithms.

Table 1. The number of fastest iterations.

Algorithm	Fastest Iteration (Generation)	Stable Value
Fruit Fly Optimization Algorithm	70	−42.8195
Particle Swarm Optimization	77	−42.8195
Gray Wolf Optimization Algorithm	342	−42.8195
Multiverse Optimization Algorithm	315	−42.8195
Wind Driven Optimization Algorithm	150	−42.8195

Figure 9 shows the optimal result values calculated by various algorithms, and the results calculated by the particle swarm optimization algorithm, multiverse algorithm and gray wolf algorithm are roughly the same. The bolt spacing and bolt diameter calculated by the fruit fly optimization algorithm are both larger, while the bolt spacing and bolt diameter calculated by the wind driven optimization algorithm are larger and smaller. Because the core functions of various algorithms are different, there is a certain deviation, which is also a limitation of artificial intelligence. In this simulation, the maximum deviation of the bolt diameter is 0.59% and maximum deviation rate of bolt spacing is 0.12%, both of which have small deviations and can be ignored. The results show that the optimal solutions of the five algorithms selected in this paper are roughly the same, and the deviation rate is low, which will not cause certain impact on the subsequent simulation analysis. Therefore, this paper can choose the optimal solution calculated by the gray wolf optimization algorithm as the focus of the subsequent analysis and discussion.

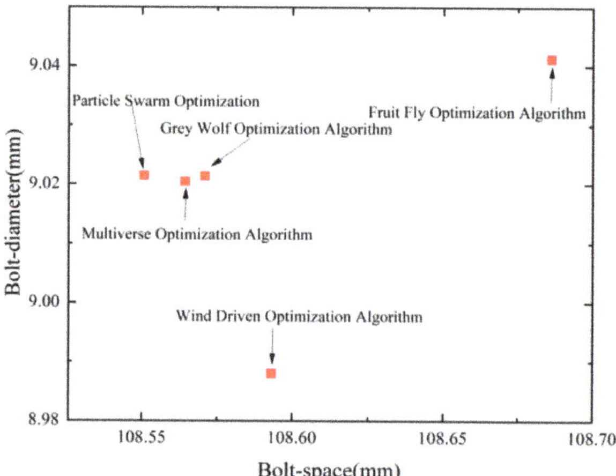

Figure 9. Algorithm optimization results.

3.2. Optimized Stress Comparison between Front and Rear Bolts

The ABAQUS-based simulation design can verify the correctness of the optimal solution selection of the gray wolf algorithm to reduce the stress concentration phenomenon in the structure and thus improve the safety performance of the aerostructure. The stress clouds of the upper nickel steel plate, lower nickel steel plate and bolts before and after optimization are shown in Figures 10–12.

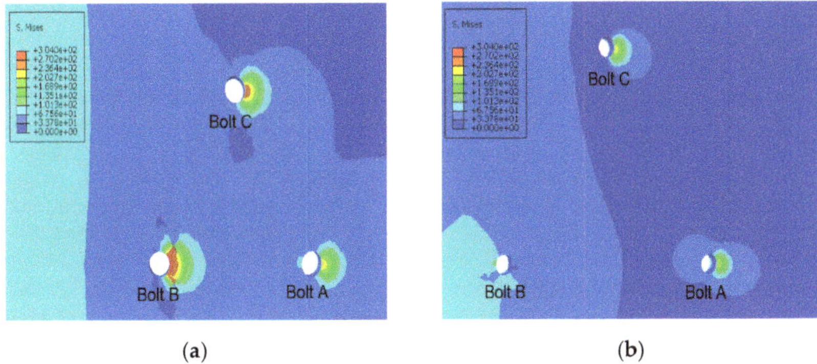

Figure 10. Upper nickel steel plate stress cloud. (**a**) Optimized pre-stress nephogram; (**b**) Optimized stress nephogram.

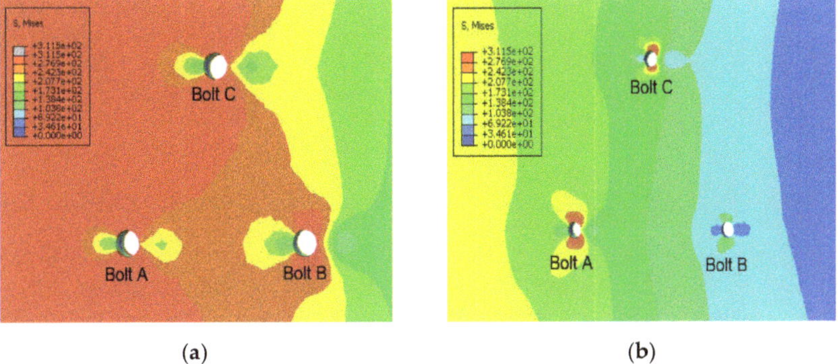

Figure 11. Lower nickel steel plate stress cloud. (**a**) Optimized pre-stress nephogram; (**b**) Optimized stress nephogram.

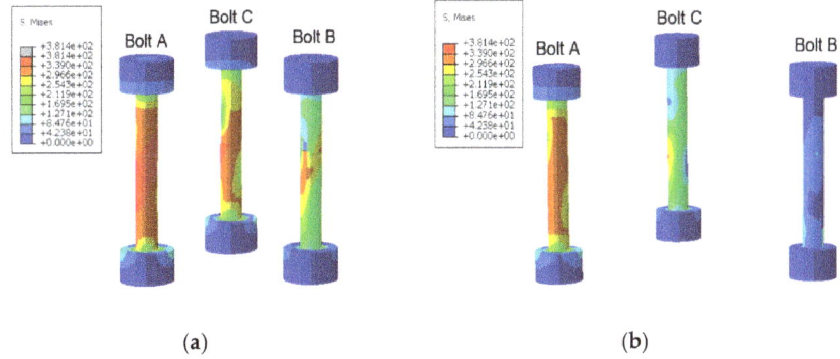

Figure 12. Stress clouds for bolt sets in triangular positions. (**a**) Optimized pre-stress nephogram; (**b**) Optimized stress nephogram.

The stress cloud diagram of the nickel steel plate above is shown in Figure 10. In order to make the stress nephogram visually compare the stress change and stress concentration through images, this paper takes the stress extreme value before optimization as the standard and changes the mechanism of the optimized nephogram. Before and after optimization, the maximum stress of the nickel steel plate is 304.0 MPa and 230.9 MPa, respectively, and the maximum stress is reduced by 73.1 MPa, with an optimization rate of 24%. A previous study [8] optimized the layout of the bolts on the connection structure based on the improved genetic algorithm, and the results showed that the optimized bolt layout reduces the maximum equivalent stress at corners constructed by 20.4%. In another study [11], the firefly algorithm was used to optimize bolts with double and triple rows, and the results showed that after optimization, the stress around the hole of the plate structure with triple rows of bolts was reduced by about 20%, and the stress reduction effect in this paper was better. Xiao et al. [31] adopted the improved genetic algorithm to optimize the equivalent stress of the aviation connection structure under the position of three rows of bolts under severe working conditions. The results showed that the maximum pore circumference stress was reduced by 31.03% under a static pressure load of 30 MPa, while the optimization efficiency only reached 23.36% under severe working conditions. Under the conditions of 15 KN tension and pressure studied in this paper, it is already a serious condition, and the optimization effect of both is the same, which proves the reliability and accuracy of the analysis in this paper. Among them, the stress around bolt hole B decreased significantly after optimization, and the maximum stress value around bolt hole B decreased from 304.0 MPa to 112.1 MPa, with a decrease rate of 63.1%, which greatly reduced the stress around bolt hole B. The stress around bolt hole C before and after optimization is 309.2 MPa and 302.8 MPa, respectively. Bolt hole C was originally the most stressed bolt, and its optimization effect is not obvious. The stress around bolt hole A is increased by 27.9 MPa, mainly because after optimization, the position of bolt hole A is closer to the edge of the nickel steel plate, and the position of concentrated force is closer to the vicinity of bolt A.

Figure 11 shows the stress cloud map of the lower nickel steel plate. Before and after optimization, the maximum stress of the nickel steel plate is 311.5 MPa and 302.8 MPa, respectively, and the maximum stress is reduced by 8.7 MPa, with an optimization rate of 2.8%. The maximum stress at bolt hole B decreased from 307.1 MPa to 177.5 MPa, and the optimization rate was 42.2%, which significantly improved the stress concentration phenomenon at bolt hole B. The maximum stress at bolt holes A and C decreased less, 8.9 MPa and 6.4 MPa, respectively, but the equivalent stress around them decreased significantly. According to the stress nephogram, the stress concentration around the three bolt holes has been significantly improved after the optimization of the triangular multi-bolt connection structure, and the bolt hole B has been greatly improved.

Figure 12 shows that before and after the optimization, the stresses on bolt A were 381.4 MPa and 333.7 MPa, respectively, both of which were the maximum values of the bolt stresses before and after the optimization, and the maximum stress was reduced by 47.7 MPa, and its optimization rate was 12.5%. The literature [31] applied the improved genetic algorithm to the layout optimization of an actual aerospace bolt connection structure, and the results showed that the maximum equivalent stress of the bolt was reduced by 10.14%, which is not much different from the results of this paper. The main reason for the highest stress value in bolt A is the left-hand fixation of the nickel steel plate below, which affects the forces in the connection structure. Bolt B has the most obvious optimization effect, with its maximum equivalent force dropping by 177.5 MPa and optimization rate of 57.2%, effectively extending the service life and reliability of bolt B. The maximum stress value at bolt C was reduced from 345.5 MPa to 301.0 MPa, with an optimization rate of 12.9%. In summary, the bolts in the triangular position were improved by the optimal solution calculated by the gray wolf algorithm, which improved the overall safety and stability. In the literature [32], the layout of the beam-column extended end-plate connection structure was optimized based on the firefly algorithm, and the maximum hole perimeter equivalent force of the connection structure was reduced by 41.2 MPa after optimization, with an optimization rate of 10.9%; however, the maximum equivalent force on its bolt surface was reduced by only 3.8%. Compared with this paper, which uses the gray wolf algorithm to optimize the multi-bolt layout in triangular positions, this method is more efficient and can provide an effective reference for engineers when designing multi-bolt connection structures for nickel steel flat plates.

Before and after the optimization, the stress concentration around the hole of the nickel steel plate is improved, especially the maximum stress at the bolt hole B, which reaches 40–60% of the optimized value. The overall optimization rate for the upper nickel steel plate is 24% and the overall optimization rate for the lower nickel steel plate is 2.8%, mainly due to the fixed restraint applied on the left side. The overall optimization rate for the bolt set was 12.5%.

In order to study the optimization under different force conditions, this paper changed the pressure and tension and added comparative analysis, respectively, 3 KN, 9 KN, 15 KN and 21 KN, as shown in Figure 13. The results show that under the 18 KN load, the nickel steel plate and bolt set above can achieve better optimization efficiency, and their optimization rates are above 10%. At 15 KN, the optimization effect is better. However, when the load is greater than 18 KN, the optimization efficiency is lower. When the load is greater than 21 KN, the nickel steel plate has exceeded the yield limit and presents a large deformation condition, which is not suitable for calculating its optimization efficiency. The optimization efficiency of the lower nickel steel plate is always lower, mainly because of the fixed constraints on the lower nickel steel plate. Overall, the analysis of this paper proves that the gray wolf algorithm is suitable for structural optimization of multi-bolt layouts in triangular positions, which helps to reduce the stress concentration phenomenon and contributes to the effective structural optimization design of large and complex equipment and aeronautical structures.

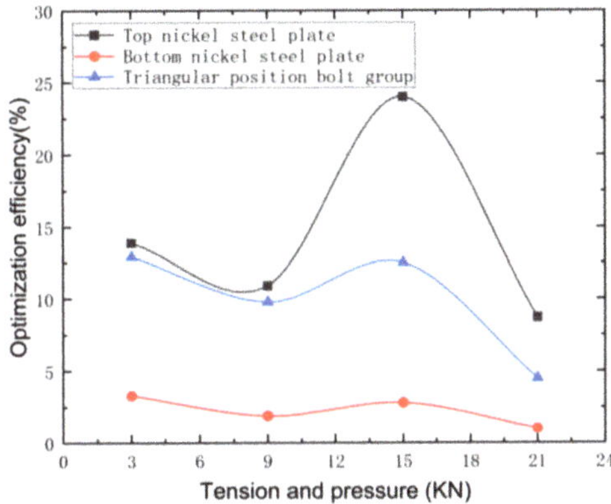

Figure 13. Comparison chart for optimisation of different force cases.

4. Conclusions

This paper applies the gray wolf algorithm to optimize the structural layout of multi-bolt positioning in triangular configurations. By analyzing the forces, the objective function is formulated based on the fitted functions of hole perimeter stress for nickel steel plates, maximum shear stress for bolts and bending stress for nickel steel plates. The value range for bolt diameter and spacing is determined according to design specifications. The optimal solutions calculated using the fruit fly optimization algorithm, particle swarm optimization algorithm, gray wolf optimization algorithm, multi-verse optimization algorithm and wind driven optimization algorithm are compared. ABAQUS simulation software is used to model and compare the stress on nickel steel plates and bolts before and after optimization, thereby verifying the correctness of the algorithm application. Through analysis, the following conclusions can be drawn:

(1) Before optimization, a traditional design method was used with a bolt spacing of 60 mm and a bolt diameter of 8 mm. After optimization, the bolt spacing is chosen as 108.5629 mm and bolt diameter is 9.0233 mm. The stress distribution of the newly

arranged nickel steel plate is more uniform, and the local stress values have also been reduced, resulting in a more stable overall stress state.

(2) Through this layout optimization, under pressures and tensions of 15 KN, the hole perimeter stress on the upper nickel steel plate decreased by 73.1 MPa, with an optimization rate of 24%. The hole perimeter stress on the lower nickel steel plate decreased by 8.7 MPa, with an optimization rate of 2.8%. The maximum equivalent stress on the bolts decreased by 47.7 MPa, with an optimization rate of 12.5%.

(3) When the load is less than 18 KN, the optimization effect of both the upper nickel steel plate and bolt assembly is above 10%. However, when the load exceeds 18 KN, the optimization effect decreases. When the load exceeds 21 KN, the nickel steel plate has exceeded its yield limit. The upper nickel steel plate shows significant optimization, but the optimization effect on the lower nickel steel plate is not obvious due to the existence of fixed constraints that restrict pressure.

(4) The chosen material for the connecting structure in this article is nickel steel plates, which are widely used in the load-bearing connection structures of large and complex equipment and aerospace and other complex mechanical fields. By optimizing the structural layout, fatigue damage of bolt holes or plastic deformation of connectors caused by excessive hole perimeter stress can be effectively prevented. This enhances the safety, reliability and stability of load-bearing connection structures.

(5) This article studied the non-symmetrical triangular structure of bolt connections, providing an effective method and ideas for the layout optimization of non-symmetrical bolt configurations, enriching the research achievements in related fields.

This paper proves the correctness of applying algorithms to optimize bolt layouts. However, it only focuses on a specific equilateral triangular multi-bolt structure. In future studies, the layout optimization of randomly positioned multi-bolt structures should be further analyzed. Moreover, this article only conducts simulations on nickel steel plates, and subsequent experiments and simulations should cover different metal materials to enhance the applicability of this article.

Author Contributions: X.L. conducted experiments, data collection and analysis, and study design. M.Z. designed and conceived the project. Y.L. wrote and edited the manuscript to verify the correctness of the method. S.W. reviewed and edited the manuscript. Z.X. and S.L. investigated the research background and conducted a literature search. All authors have read and agreed to the published version of the manuscript.

Funding: This research received no external funding.

Institutional Review Board Statement: Not applicable.

Informed Consent Statement: Not applicable.

Data Availability Statement: Data sharing is not applicable to this article.

Conflicts of Interest: The authors declare that they have no conflict of interest.

References

1. Zhao, W.Y.; Yang, J.P.; Dong, X.H.; Huang, B. Simulation Analysis and Research of Deep Loosening Machine Shovel Seat Based on ANSYS Workbench. *South China Agric. Mach.* **2023**, *54*, 10–14.
2. Zhao, L.; Cai, H.; Wang, Z.Q.; He, J.J.; Wang, P.; Wang, B.; Qin, C.P.; Wang, B.H. Phased Array Ultrasonic Detection Method for Bolt Holes of Two In-service Fans. *Therm. Power Gener.* **2023**, *52*, 67–72.
3. Wang, X.T. *Study on Stiffness and Strength of Single Lap Bolted Joint Plate under Shear Load*; Nanchang Hangkong University: Nanchang, China, 2021.
4. Huang, L. Numerical Study on Optimal Geometric Parameters for Multi-performance of Friction High-Strength Bolts. *Int. J. Steel Struct.* **2022**, *22*, 56–69. [CrossRef]
5. Liu, C.; Zhou, Z.G.; Fang, B.H.; Hu, S.J.; Zhang, B. Experimental study on Mechanical properties of unsymmetrical reaming bolted Connection. *Eng. Seism. Resist. Reinf.* **2022**, *44*, 37–44.
6. Wu, S.J.; Liu, W.G. Research on Optimal Anti-Resonance Layout of Bolted Plate. *Mach. Tool Hydraul.* **2020**, *48*, 164–168.
7. Pedersen, N.L. Overall bolt stress optimization. *J. Strain Anal. Eng. Des.* **2013**, *48*, 155–165. [CrossRef]

8. Shen, B.; Pan, Z.W.; Lin, H.; Li, W.Z. Bolt layout optimization design based on genetic algo-rithm. *IOP Conf. Ser. Mater. Sci. Eng.* **2019**, *576*, 012005. [CrossRef]
9. Han, J.G.; Wu, Z.; Sun, Y.L.; Zhang, W.Q. Multi-joint integrated Motion trajectory Optimization of Robot based on Gray Wolf Algorithm. *J. Nav. Univ. Eng.* **2020**, *32*, 90–95.
10. Liang, Y.F.; Xu, J.G.; Wu, M. GPS Fast clock difference prediction Method Optimized by Elman Neural Network based on Particle swarm Optimization Algorithm. *J. Nav. Univ. Eng.* **2022**, *34*, 41–47.
11. Wang, C.F.; Zhou, H.L. Optimization of Bolt Connection Structure Layout Based on Firefly Algorithm. *Mach. Tool Hydraul.* **2022**, *50*, 144–147.
12. Kwon, Y.D.; Lee, D.S. A study on design optimization of conical bolt in the tf coil structure of the kstar tokamak. *Int. J. Mod. Phys. B* **2008**, *20*, 4475–4480. [CrossRef]
13. Lalaina, R.; Jeet, D.; Patrick, O.; Grégoire, A. Coupled topology optimization of structure and connections for bolted mechanical systems. *Eur. J. Mech. A/Solids* **2022**, *93*, 104499.
14. Kim, M.; Kim, Y.; Kim, P.; Park, J. Design Optimization of Double-array Bolted Joints in Cylindrical Composite Structures. *Int. J. Aeronaut. Space Sci.* **2016**, *17*, 332–340. [CrossRef]
15. Zhao, S.; Li, D.; Xiang, J. Multi-objective optimum of composite bolted joints by using the multi-layer convex hull method. *Struct. Multidiscip. Optim.* **2018**, *58*, 1233–1242. [CrossRef]
16. Elik, L.D.; Ciftci, M.; Ozturk, Y. Effects of the external tendons on the system capacity and stud behavior in composite beams. *Structures* **2020**, *24*, 851–863.
17. Pan, W.C. Application of Fruit Fly optimization Algorithm to optimize generalized regression neural network for enterprise performance evaluation. *J. Taiyuan Univ. Technol.* **2011**, *29*, 1–5.
18. Poli, R.; Kennedy, J.; Blackwell, T. Particle swarm optimization. *Swarm Intell.* **2007**, *1*, 33–57. [CrossRef]
19. Mirjalili, S.; Mirjalili, M.S.; Lewis, A. Grey Wolf Optimizer. *Adv. Eng. Softw.* **2014**, *69*, 46–61. [CrossRef]
20. Mirjalili, S.; Mirjalili, M.S.; Hatamlou, A. Multi-Verse Optimizer: A nature-inspired algorithm for global optimization. *Neural Comput. Appl.* **2016**, *27*, 495–513. [CrossRef]
21. Zikri, B.; Muge, K.; Jeremy, A.B.; Douglas, H. The Wind Driven Optimization Technique and its Application in Electromagnetics. *IEEE Trans. Antennas Propag.* **2013**, *61*, 2745–2757.
22. Pan, C.S.; Si, Z.H.; Xi, L.; Lv, Y.N. A four-step decision-making grey wolf optimization algorithm. *Soft Comput.* **2021**, *25*, 14375–14391. [CrossRef]
23. Yu, X.W.; Huang, L.P.; Liu, Y.; Zhang, K.; Li, P.; Li, Y. WSN node location based on beetle antennae search to improve the Gray wolf algorithm. *Wirel. Netw.* **2022**, *28*, 539–549. [CrossRef]
24. Chen, B.C.; Lv, Z.M. Finite Element Analysis of Sealing Contact Based on Abaqus Gasket. *China Water Transp. (Second Half)* **2023**, *23*, 62–64.
25. He, H.B. Optimization Design of Bolted Steel Member Based on Genetic Algorithm. *Comput. Appl. Softw.* **2011**, *28*, 236–238.
26. Beijing Iron and Steel Design Research Institute. *Steel Structure Design Code GB 50017-2017 [S]*; China Building and Construction Press: Beijing, China, 2017.
27. *GB/T197-2003*; The General Administration of Quality Supervision, Inspection and Quarantine of the People's Republic of China and the Standardization Administration of China. Standards Press of China: Beijing, China, 2003.
28. Liao, R.; Sun, Y.; Liu, J.; Zhang, W. Applicability of damage models for failure analysis of threaded bolts. *Eng. Fract. Mech.* **2011**, *78*, 514–524. [CrossRef]
29. Yang, G.; Yang, L.; Chen, J.; Xiao, S.; Jiang, S. Competitive Failure of Bolt Loosening and Fatigue under Different Preloads. *Chin. J. Mech. Eng.* **2021**, *34*, 1–11. [CrossRef]
30. Xiao, L.J.; Gong, X.B.; Zhu, X.L.; Chi, Y.R. Optimal design of mining explosion-proof box bolt set layout. *Coal Eng.* **2016**, *7*, 129–131.
31. Xiao, W.Y.; He, E.M.; Hu, Y.Q. A new method for optimizing the bolt layout of connectors based on improved genetic algorithm. *J. Northwest. Polytech. Univ.* **2017**, *35*, 414–421.
32. Wang, C.F. *Mechanical Analysis and Layout Optimization Design of Bolted Connection Structure*; Hefei University of Technology: Hefei, China, 2021.

MDPI
St. Alban-Anlage 66
4052 Basel
Switzerland
www.mdpi.com

MDPI Books Editorial Office
E-mail: books@mdpi.com
www.mdpi.com/books

www.ingramcontent.com/pod-product-compliance
Lightning Source LLC
LaVergne TN
LVHW071618170726
843515LV00009B/2423

* 9 7 8 3 0 3 6 5 9 9 3 5 9 *